| Unless Recalled Earlier | |
|---|---|
| APR 3 0 2003 | |
| | |
| | |
| | |
| | |
| | |
| | |
| | |
| | |
| | |
| | |
| | |
| | |
| | |
| | |
| | |
| | |

**S**pringer Series in
**C**omputational
**M**athematics

**31**

Editorial Board

R. Bank, La Jolla (CA)
R.L. Graham, La Jolla (CA)
J. Stoer, Würzburg
R. Varga, Kent (Ohio)
H. Yserentant, Tübingen

Springer
*Berlin*
*Heidelberg*
*New York*
*Barcelona*
*Hong Kong*
*London*
*Milan*
*Paris*
*Tokyo*

Ernst Hairer
Christian Lubich
Gerhard Wanner

# Geometric Numerical Integration

## Structure-Preserving Algorithms for Ordinary Differential Equations

With 119 Figures

Springer

Ernst Hairer
Gerhard Wanner

Université de Genève
Section de Mathématiques, C.P. 240
2–4 rue du Lièvre
CH-1211 Genève 24, Switzerland
e-mail: Ernst.Hairer@math.unige.ch
        Gerhard.Wanner@math.unige.ch

QA
299.3
.G45
2002

Christian Lubich
Universität Tübingen
Mathematisches Institut
Auf der Morgenstelle 10
72076 Tübingen, Germany
e-mail: Lubich@na.uni-tuebingen.de


Cataloging-in-Publication Data applied for

Die Deutsche Bibliothek - CIP-Einheitsaufnahme
Hairer, Ernst:
Geometric numerical integration: structure preserving algorithms
for ordinary differential equations / Ernst Hairer ; Christian Lubich ;
Gerhard Wanner. - Berlin ; Heidelberg ; New York ; Barcelona ;
Hong Kong ; London ; Milan ; Paris ; Tokyo : Springer, 2002
  (Springer series in computational mathematics ; 31)
  ISBN 3-540-43003-2


Mathematics Subject Classification (2000): 65Lxx, 65P10, 70Fxx, 34Cxx


ISSN 0179-3632
ISBN 3-540-43003-2 Springer-Verlag Berlin Heidelberg New York

This work is subject to copyright. All rights are reserved, whether the whole or part of the material is concerned, specifically the rights of translation, reprinting, reuse of illustrations, recitation, broadcasting, reproduction on microfilm or in any other way, and storage in data banks. Duplication of this publication or parts thereof is permitted only under the provisions of the German Copyright Law of September 9, 1965, in its current version, and permission for use must always be obtained from Springer-Verlag. Violations are liable for prosecution under the German Copyright Law.

Springer-Verlag Berlin Heidelberg New York
a member of BertelsmannSpringer Science+Business Media GmbH
http://www.springer.de
© Springer-Verlag Berlin Heidelberg 2002
Printed in Germany

The use of general descriptive names, registered names, trademarks etc. in this publication does not imply, even in the absence of a specific statement, that such names are exempt from the relevant protective laws and regulations and therefore free for general use.

Cover design: *design&production*, Heidelberg
Typeset by the authors
Printed on acid-free paper           SPIN 10759237           46/3142LK-5 4 3 2 1 0

# Preface

> They throw geometry out the door, and it comes back through the window.
> (H.G.Forder, Auckland 1973, reading new mathematics at the age of 84)

The subject of this book is numerical methods that preserve geometric properties of the flow of a differential equation: symplectic integrators for Hamiltonian systems, symmetric integrators for reversible systems, methods preserving first integrals and numerical methods on manifolds, including Lie group methods and integrators for constrained mechanical systems, and methods for problems with highly oscillatory solutions. Structure preservation – with its questions as to where, how, and what for – is the unifying theme.

In the last few decades, the theory of numerical methods for general (non-stiff and stiff) ordinary differential equations has reached a certain maturity, and excellent general-purpose codes, mainly based on Runge-Kutta methods or linear multistep methods, have become available. The motivation for developing structure-preserving algorithms for special classes of problems came independently from such different areas of research as astronomy, molecular dynamics, mechanics, theoretical physics, and numerical analysis as well as from other areas of both applied and pure mathematics. It turned out that the preservation of geometric properties of the flow not only produces an improved qualitative behaviour, but also allows for a more accurate long-time integration than with general-purpose methods.

An important shift of view-point came about by ceasing to concentrate on the numerical approximation of a single solution trajectory and instead to consider a numerical method as a *discrete dynamical system* which approximates the flow of the differential equation – and so the geometry of phase space comes back again through the window. This view allows a clear understanding of the preservation of invariants and of methods on manifolds, of symmetry and reversibility of methods, and of the symplecticity of methods and various generalizations. These subjects are presented in Chapters IV through VII of this book. Chapters I through III are of an introductory nature and present examples and numerical integrators together with important parts of the classical order theories and their recent extensions. Chapter VIII deals with questions of numerical implementations and numerical merits of the various methods.

It remains to explain the relationship between geometric properties of the numerical method and the favourable error propagation in long-time integrations. This

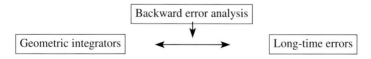

is done using the idea of *backward error analysis*, where the numerical one-step map is interpreted as (almost) the flow of a modified differential equation, which is constructed as an asymptotic series (Chapter IX). In this way, geometric properties of the numerical integrator translate into structure preservation on the level of the modified equations. Much insight and rigorous error estimates over long time intervals can then be obtained by combining this backward error analysis with KAM theory and related perturbation theories. This is explained in Chapters X through XII for Hamiltonian and reversible systems. The final Chapters XIII and XIV treat the numerical solution of differential equations with high-frequency oscillations and the long-time dynamics of multistep methods, respectively.

This book grew out of the lecture notes of a course given by Ernst Hairer at the University of Geneva during the academic year 1998/99. These lectures were directed at students in the third and fourth year. The reactions of students as well as of many colleagues, who obtained the notes from the Web, encouraged us to elaborate our ideas to produce the present monograph.

We want to thank all those who have helped and encouraged us to prepare this book. In particular, Martin Hairer for his valuable help in installing computers and his expertise in Latex and Postscript, Jeff Cash and Robert Chan for reading the whole text and correcting countless scientific obscurities and linguistic errors, Haruo Yoshida for making many valuable suggestions, Stéphane Cirilli for preparing the files for all the photographs, and Bernard Dudez, the irreplaceable director of the mathematics library in Geneva. We are also grateful to many friends and colleagues for reading parts of the manuscript and for valuable remarks and discussions, in particular to Assyr Abdulle, Melanie Beck, Sergio Blanes, John Butcher, Mari Paz Calvo, Begoña Cano, Philippe Chartier, David Cohen, Peter Deuflhard, Stig Faltinsen, Francesco Fassò, Martin Gander, Marlis Hochbruck, Bulent Karasözen, Wilhelm Kaup, Ben Leimkuhler, Pierre Leone, Frank Loose, Katina Lorenz, Robert McLachlan, Ander Murua, Alexander Ostermann, Truong Linh Pham, Sebastian Reich, Chus Sanz-Serna, Zaijiu Shang, Yifa Tang, Matt West, Will Wright.

We are especially grateful to Thanh-Ha Le Thi and Dr. Martin Peters from Springer-Verlag Heidelberg for assistance, in particular for their help in getting most of the original photographs from the Oberwolfach Archive and from Springer New York, and for clarifying doubts concerning the copyright.

Geneva and Tübingen, November 2001  The Authors

# Table of Contents

**I. Examples and Numerical Experiments** .................. 1
    I.1    Two-Dimensional Problems .......................... 1
          I.1.1    The Lotka-Volterra Model ..................... 1
          I.1.2    Hamiltonian Systems — the Pendulum ........... 4
    I.2    The Kepler Problem and the Outer Solar System ........ 7
          I.2.1    Exact Integration of the Kepler Problem ........ 7
          I.2.2    Numerical Integration of the Kepler Problem ... 9
          I.2.3    The Outer Solar System ....................... 10
    I.3    Molecular Dynamics .................................. 12
          I.3.1    The Störmer/Verlet Scheme ................... 13
          I.3.2    Numerical Experiments ....................... 15
    I.4    Highly Oscillatory Problems .......................... 16
          I.4.1    A Fermi-Pasta-Ulam Problem ................. 17
          I.4.2    Application of Classical Integrators ........... 18
    I.5    Exercises ............................................ 20

**II. Numerical Integrators** ................................... 23
    II.1    Runge-Kutta and Collocation Methods ................. 23
          II.1.1    Runge-Kutta Methods ........................ 24
          II.1.2    Collocation Methods ......................... 26
          II.1.3    Gauss and Lobatto Collocation ................ 30
          II.1.4    Discontinuous Collocation Methods ........... 31
    II.2    Partitioned Runge-Kutta Methods .................... 34
          II.2.1    Definition and First Examples ................. 34
          II.2.2    Lobatto IIIA - IIIB Pairs ....................... 36
          II.2.3    Nyström Methods ............................ 37
    II.3    The Adjoint of a Method ............................. 38
    II.4    Composition Methods ............................... 39
    II.5    Splitting Methods .................................... 41
    II.6    Exercises ............................................ 46

## III. Order Conditions, Trees and B-Series ... 47
- III.1 Runge-Kutta Order Conditions and B-Series ... 47
  - III.1.1 Derivation of the Order Conditions ... 47
  - III.1.2 B-Series ... 52
  - III.1.3 Composition of Methods ... 55
  - III.1.4 Composition of B-Series ... 57
  - III.1.5 The Butcher Group ... 60
- III.2 Order Conditions for Partitioned Runge-Kutta Methods ... 62
  - III.2.1 Bi-Coloured Trees and P-Series ... 62
  - III.2.2 Order Conditions for Partitioned Runge-Kutta Methods ... 64
  - III.2.3 Order Conditions for Nyström Methods ... 65
- III.3 Order Conditions for Composition Methods ... 67
  - III.3.1 Introduction ... 67
  - III.3.2 The General Case ... 69
  - III.3.3 Reduction of the Order Conditions ... 71
  - III.3.4 Order Conditions for Splitting Methods ... 76
- III.4 The Baker-Campbell-Hausdorff Formula ... 78
  - III.4.1 Derivative of the Exponential and Its Inverse ... 78
  - III.4.2 The BCH Formula ... 80
- III.5 Order Conditions via the BCH Formula ... 83
  - III.5.1 Calculus of Lie Derivatives ... 83
  - III.5.2 Lie Brackets and Commutativity ... 85
  - III.5.3 Splitting Methods ... 86
  - III.5.4 Composition Methods ... 88
- III.6 Exercises ... 90

## IV. Conservation of First Integrals and Methods on Manifolds ... 93
- IV.1 Examples of First Integrals ... 93
- IV.2 Quadratic Invariants ... 97
  - IV.2.1 Runge-Kutta Methods ... 97
  - IV.2.2 Partitioned Runge-Kutta Methods ... 98
  - IV.2.3 Nyström Methods ... 100
- IV.3 Polynomial Invariants ... 101
  - IV.3.1 The Determinant as a First Integral ... 101
  - IV.3.2 Isospectral Flows ... 103
- IV.4 Projection Methods ... 105
- IV.5 Numerical Methods Based on Local Coordinates ... 110
  - IV.5.1 Manifolds and the Tangent Space ... 110
  - IV.5.2 Differential Equations on Manifolds ... 112
  - IV.5.3 Numerical Integrators on Manifolds ... 112
- IV.6 Differential Equations on Lie Groups ... 115
- IV.7 Methods Based on the Magnus Series Expansion ... 118
- IV.8 Lie Group Methods ... 121
  - IV.8.1 Crouch-Grossman Methods ... 121
  - IV.8.2 Munthe-Kaas Methods ... 123

|       |        | IV.8.3 Further Coordinate Mappings .................... 125 |
|-------|--------|---|

      IV.9   Exercises ................................................. 128

## V. Symmetric Integration and Reversibility ..................... 131
    V.1   Reversible Differential Equations and Maps .................. 131
    V.2   Symmetric Runge-Kutta Methods .......................... 134
          V.2.1   Collocation and Runge-Kutta Methods ............... 134
          V.2.2   Partitioned Runge-Kutta Methods .................... 136
    V.3   Symmetric Composition Methods .......................... 137
          V.3.1   Symmetric Composition of First Order Methods ........ 138
          V.3.2   Symmetric Composition of Symmetric Methods ........ 142
          V.3.3   Effective Order and Processing Methods .............. 146
    V.4   Symmetric Methods on Manifolds .......................... 149
          V.4.1   Symmetric Projection .............................. 149
          V.4.2   Symmetric Methods Based on Local Coordinates ....... 154
    V.5   Energy – Momentum Methods and Discrete Gradients .......... 159
    V.6   Exercises ................................................. 164

## VI. Symplectic Integration of Hamiltonian Systems ................. 167
    VI.1   Hamiltonian Systems .................................... 168
          VI.1.1   Lagrange's Equations ............................. 168
          VI.1.2   Hamilton's Canonical Equations ..................... 169
    VI.2   Symplectic Transformations ............................... 170
    VI.3   First Examples of Symplectic Integrators ..................... 175
    VI.4   Symplectic Runge-Kutta Methods .......................... 178
          VI.4.1   Criterion of Symplecticity .......................... 178
          VI.4.2   Connection Between Symplectic and Symmetric Methods 181
    VI.5   Generating Functions .................................... 182
          VI.5.1   Existence of Generating Functions ................... 182
          VI.5.2   Generating Function for Symplectic Runge-Kutta
                  Methods ........................................ 184
          VI.5.3   The Hamilton-Jacobi Partial Differential Equation ...... 186
          VI.5.4   Methods Based on Generating Functions .............. 189
    VI.6   Variational Integrators ................................... 191
          VI.6.1   Hamilton's Principle .............................. 191
          VI.6.2   Discretization of Hamilton's Principle ................ 192
          VI.6.3   Symplectic Partitioned Runge-Kutta Methods Revisited . 195
          VI.6.4   Noether's Theorem ............................... 197
    VI.7   Characterization of Symplectic Methods ..................... 199
          VI.7.1   Symplectic P-Series (and B-Series) ................... 199
          VI.7.2   Irreducible Runge-Kutta Methods ................... 202
          VI.7.3   Characterization of Irreducible Symplectic Methods .... 203
          VI.7.4   Conjugate Symplecticity ........................... 204
    VI.8   Exercises ................................................. 206

## VII. Further Topics in Structure Preservation ... 209
- VII.1 Constrained Mechanical Systems ... 209
  - VII.1.1 Introduction and Examples ... 209
  - VII.1.2 Hamiltonian Formulation ... 211
  - VII.1.3 A Symplectic First Order Method ... 213
  - VII.1.4 SHAKE and RATTLE ... 216
  - VII.1.5 The Lobatto IIIA - IIIB Pair ... 218
  - VII.1.6 Splitting Methods ... 224
- VII.2 Poisson Systems ... 226
  - VII.2.1 Canonical Poisson Structure ... 226
  - VII.2.2 General Poisson Structures ... 228
  - VII.2.3 Simultaneous Linear Partial Differential Equations ... 231
  - VII.2.4 Coordinate Changes and the Darboux-Lie Theorem ... 234
  - VII.2.5 Poisson Integrators ... 237
  - VII.2.6 Lie-Poisson Systems ... 242
- VII.3 Volume Preservation ... 248
- VII.4 Exercises ... 253

## VIII. Structure-Preserving Implementation ... 255
- VIII.1 Dangers of Using Standard Step Size Control ... 255
- VIII.2 Reversible Adaptive Step Size Selection ... 258
- VIII.3 Time Transformations ... 261
  - VIII.3.1 Symplectic Integration ... 261
  - VIII.3.2 Reversible Integration ... 264
- VIII.4 Multiple Time Stepping ... 266
  - VIII.4.1 Fast-Slow Splitting: the Impulse Method ... 266
  - VIII.4.2 Averaged Forces ... 269
- VIII.5 Reducing Rounding Errors ... 272
- VIII.6 Implementation of Implicit Methods ... 275
  - VIII.6.1 Starting Approximations ... 275
  - VIII.6.2 Fixed-Point Versus Newton Iteration ... 279
- VIII.7 Exercises ... 284

## IX. Backward Error Analysis and Structure Preservation ... 287
- IX.1 Modified Differential Equation – Examples ... 287
- IX.2 Modified Equations of Symmetric Methods ... 292
- IX.3 Modified Equations of Symplectic Methods ... 293
  - IX.3.1 Existence of a Local Modified Hamiltonian ... 293
  - IX.3.2 Existence of a Global Modified Hamiltonian ... 294
  - IX.3.3 Poisson Integrators ... 297
- IX.4 Modified Equations of Splitting Methods ... 298
- IX.5 Modified Equations of Methods on Manifolds ... 300
- IX.6 Modified Equations for Variable Step Sizes ... 303
- IX.7 Rigorous Estimates – Local Error ... 304
  - IX.7.1 Estimation of the Derivatives of the Numerical Solution ... 306

|  | IX.7.2 | Estimation of the Coefficients of the Modified Equation . 307 |
|---|---|---|
|  | IX.7.3 | Choice of $N$ and the Estimation of the Local Error...... 310 |
| IX.8 | Long-Time Energy Conservation ............................ 312 | |
| IX.9 | Modified Equation in Terms of Trees......................... 314 | |
|  | IX.9.1 | B-Series of the Modified Equation .................... 315 |
|  | IX.9.2 | Extension to Partitioned Systems ..................... 317 |
| IX.10 | Modified Hamiltonian ..................................... 319 | |
|  | IX.10.1 | Elementary Hamiltonians ............................ 321 |
|  | IX.10.2 | Characterization of Symplectic P-Series............... 324 |
| IX.11 | Exercises ................................................. 325 | |

## X. Hamiltonian Perturbation Theory and Symplectic Integrators ..... 327

| X.1 | Completely Integrable Hamiltonian Systems .................. 328 | |
|---|---|---|
|  | X.1.1 | Local Integration by Quadrature ..................... 328 |
|  | X.1.2 | Completely Integrable Systems ...................... 331 |
|  | X.1.3 | Action-Angle Variables ............................. 335 |
|  | X.1.4 | Conditionally Periodic Flows........................ 337 |
|  | X.1.5 | The Toda Lattice – an Integrable System .............. 340 |
| X.2 | Transformations in the Perturbation Theory for Integrable Systems 342 | |
|  | X.2.1 | The Basic Scheme of Classical Perturbation Theory..... 343 |
|  | X.2.2 | Lindstedt-Poincaré Series........................... 344 |
|  | X.2.3 | Kolmogorov's Iteration............................. 348 |
|  | X.2.4 | Birkhoff Normalization Near an Invariant Torus ........ 350 |
| X.3 | Linear Error Growth and Near-Preservation of First Integrals .... 351 | |
| X.4 | Near-Invariant Tori on Exponentially Long Times .............. 355 | |
|  | X.4.1 | Estimates of Perturbation Series ..................... 355 |
|  | X.4.2 | Near-Invariant Tori of Perturbed Integrable Systems .... 359 |
|  | X.4.3 | Near-Invariant Tori of Symplectic Integrators .......... 360 |
| X.5 | Kolmogorov's Theorem on Invariant Tori .................... 361 | |
|  | X.5.1 | Kolmogorov's Theorem ............................ 361 |
|  | X.5.2 | KAM Tori under Symplectic Discretization ........... 366 |
| X.6 | Invariant Tori of Symplectic Maps .......................... 368 | |
|  | X.6.1 | A KAM Theorem for Symplectic Near-Identity Maps ... 369 |
|  | X.6.2 | Invariant Tori of Symplectic Integrators ............... 371 |
|  | X.6.3 | Strongly Non-Resonant Step Sizes ................... 371 |
| X.7 | Exercises ................................................. 372 | |

## XI. Reversible Perturbation Theory and Symmetric Integrators ....... 375

| XI.1 | Integrable Reversible Systems ............................. 375 | |
|---|---|---|
| XI.2 | Transformations in Reversible Perturbation Theory............. 379 | |
|  | XI.2.1 | The Basic Scheme of Reversible Perturbation Theory ... 379 |
|  | XI.2.2 | Reversible Perturbation Series ....................... 380 |
|  | XI.2.3 | Reversible KAM Theory ........................... 382 |
|  | XI.2.4 | Reversible Birkhoff-Type Normalization .............. 384 |
| XI.3 | Linear Error Growth and Near-Preservation of First Integrals .... 384 | |

XI.4 Invariant Tori under Reversible Discretization ................. 386
    XI.4.1 Near-Invariant Tori over Exponentially Long Times ..... 386
    XI.4.2 A KAM Theorem for Reversible Near-Identity Maps .... 387
XI.5 Exercises ............................................... 389

## XII. Dissipatively Perturbed Hamiltonian and Reversible Systems ...... 391
XII.1 Numerical Experiments with Van der Pol's Equation ........... 391
XII.2 Averaging Transformations ................................. 394
    XII.2.1 The Basic Scheme of Averaging ..................... 394
    XII.2.2 Perturbation Series ................................ 395
XII.3 Attractive Invariant Manifolds .............................. 396
XII.4 Weakly Attractive Invariant Tori of Perturbed Integrable Systems . 400
XII.5 Weakly Attractive Invariant Tori of Numerical Integrators ....... 401
    XII.5.1 Modified Equations of Perturbed Differential Equations . 402
    XII.5.2 Symplectic Methods ................................ 403
    XII.5.3 Symmetric Methods ................................ 405
XII.6 Exercises ............................................... 405

## XIII. Highly Oscillatory Differential Equations ....................... 407
XIII.1 Towards Longer Time Steps in Solving Oscillatory Differential Equations ............................................... 407
    XIII.1.1 The Störmer/Verlet Method vs. Multiple Time Scales ... 408
    XIII.1.2 Gautschi's and Deuflhard's Trigonometric Methods ..... 409
    XIII.1.3 The Impulse Method ............................... 411
    XIII.1.4 The Mollified Impulse Method ...................... 412
    XIII.1.5 Gautschi's Method Revisited ....................... 413
    XIII.1.6 Two-Force Methods ............................... 414
XIII.2 A Nonlinear Model Problem and Numerical Phenomena ........ 414
    XIII.2.1 Time Scales in the Fermi-Pasta-Ulam Problem ......... 415
    XIII.2.2 Numerical Methods ................................ 416
    XIII.2.3 Accuracy Comparisons ............................. 418
    XIII.2.4 Energy Exchange between Stiff Components .......... 419
    XIII.2.5 Near-Conservation of Total and Oscillatory Energy ..... 420
XIII.3 Principal Terms of the Modulated Fourier Expansion .......... 422
    XIII.3.1 Decomposition of the Exact Solution ................. 422
    XIII.3.2 Decomposition of the Numerical Solution ............. 424
XIII.4 Accuracy and Slow Exchange ................................ 426
    XIII.4.1 Convergence Properties on Bounded Time Intervals ..... 426
    XIII.4.2 Intra-Oscillatory and Oscillatory-Smooth Exchanges .... 431
XIII.5 Modulated Fourier Expansions .............................. 432
    XIII.5.1 Expansion of the Exact Solution .................... 433
    XIII.5.2 Expansion of the Numerical Solution ................. 435
    XIII.5.3 Expansion of the Velocity Approximation ............. 438
XIII.6 Almost-Invariants of the Modulated Fourier Expansions ........ 439
    XIII.6.1 The Hamiltonian of the Modulated Fourier Expansion ... 440

XIII.6.2 A Formal Invariant Close to the Oscillatory Energy ..... 441
XIII.6.3 Almost-Invariants of the Numerical Method ........... 443
XIII.7 Long-Time Near-Conservation of Total and Oscillatory Energy... 447
XIII.8 Energy Behaviour of the Störmer/Verlet Method ............... 449
XIII.9 Exercises ................................................. 452

# XIV. Dynamics of Multistep Methods ............................. 455
XIV.1 Numerical Methods and Experiments ....................... 455
    XIV.1.1 Linear Multistep Methods ......................... 455
    XIV.1.2 Multistep Methods for Second Order Equations ........ 457
    XIV.1.3 Partitioned Multistep Methods...................... 459
    XIV.1.4 Multi-Value or General Linear Methods .............. 460
XIV.2 Related One-Step Methods ................................ 461
    XIV.2.1 The Underlying One-Step Method ................... 461
    XIV.2.2 Formal Analysis for Weakly Stable Methods .......... 463
    XIV.2.3 Backward Error Analysis for Multistep Methods ....... 464
    XIV.2.4 Dynamics of Weakly Stable Methods ................ 467
    XIV.2.5 Invariant Manifold of the Augmented System ......... 468
XIV.3 Can Multistep Methods be Symplectic? ..................... 470
    XIV.3.1 Non-Symplecticity of the Underlying One-Step Method . 470
    XIV.3.2 Symplecticity in the Higher-Dimensional Phase Space .. 471
XIV.4 Symmetric Multi-Value Methods ........................... 474
    XIV.4.1 Definition of Symmetry ........................... 474
    XIV.4.2 A Useful Criterion for Symmetry ................... 476
XIV.5 Stability of the Invariant Manifold ......................... 477
    XIV.5.1 Partitioned General Linear Methods ................. 477
    XIV.5.2 The Linearized Augmented System ................. 479
    XIV.5.3 Dissipatively Perturbed Hamiltonian Systems ......... 481
    XIV.5.4 Numerical Instabilities and Resonances .............. 483
    XIV.5.5 Extension to Variable Step Sizes .................... 486
XIV.6 Exercises ................................................ 490

**Bibliography** ................................................. 493

**Index** ....................................................... 509

# Chapter I.
# Examples and Numerical Experiments

This chapter introduces some interesting examples of differential equations and illustrates different types of qualitative behaviour of numerical methods. We deliberately consider only very simple numerical methods of orders 1 and 2 to emphasize the qualitative aspects of the experiments. The same effects (on a different scale) occur with more sophisticated higher-order integration schemes. The experiments presented here should serve as a motivation for the theoretical and practical investigations of later chapters. The reader is encouraged to repeat the experiments or to invent similar ones.

## I.1 Two-Dimensional Problems

> Numerical applications of the case of two dependent variables are not easily obtained.
> (A.J. Lotka 1925, p. 79)

Differential equations in two dimensions already show many geometric properties and can be studied easily.

### I.1.1 The Lotka-Volterra Model

We start with an equation from mathematical biology which models the growth of animal species. If a real variable $u(t)$ is to represent the number of individuals of a certain species at time $t$, the simplest assumption about its evolution is $du/dt = u \cdot \alpha$, where $\alpha$ is the reproduction rate. A constant $\alpha$ leads to exponential growth. In the case of more species living together, the reproduction rates will also depend on the population numbers of the *other* species. For example, for two species with $u(t)$ denoting the number of predators and $v(t)$ the number of prey, a plausible assumption is made by the *Lotka-Volterra model*

$$\begin{aligned} \dot{u} &= u(v-2) \\ \dot{v} &= v(1-u), \end{aligned} \tag{1.1}$$

where the dots on $u$ and $v$ stand for differentiation with respect to time. (We have chosen the constants 2 and 1 in (1.1) arbitrarily.) A.J. Lotka (1925, Chap. VIII) used this model to study parasitic invasion of insect species, and, with its help, V. Volterra (1927) explained curious fishing data from the upper Adriatic Sea following World War I.

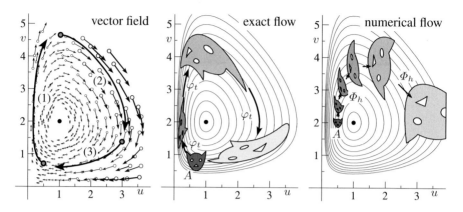

**Fig. 1.1.** Vector field, exact flow, and numerical flow for the Lotka-Volterra model (1.1)

Equations (1.1) constitute an autonomous system of differential equations. In general, we write such a system in the form

$$\dot{y} = f(y) . \tag{1.2}$$

Every $y$ represents a point in the *phase space*, in equation (1.1) above $y = (u, v)$ is in the phase plane $\mathbb{R}^2$. The vector-valued function $f(y)$ represents a *vector field* which, at any point of the phase space, prescribes the velocity (direction and speed) of the solution $y(t)$ that passes through that point (see the first picture of Fig. 1.1).

For the Lotka-Volterra model, we observe that the system cycles through three stages: (1) the prey population increases; (2) the predator population increases by feeding on the prey; (3) the predator population diminishes due to lack of food.

**Flow of the System.** A fundamental concept is the *flow* over time $t$. This is the mapping which, to any point $y_0$ in the phase space, associates the value $y(t)$ of the solution with initial value $y(0) = y_0$. This map, denoted by $\varphi_t$, is thus defined by

$$\varphi_t(y_0) = y(t) \quad \text{if} \quad y(0) = y_0. \tag{1.3}$$

The second picture of Fig. 1.1 shows the results of three iterations of $\varphi_t$ (with $t = 1.3$) for the Lotka-Volterra problem, for a set of initial values $y_0 = (u_0, v_0)$ forming an animal-shaped set $A$.[1]

**Invariants.** If we divide the two equations of (1.1) by each other, we obtain a single equation between the variables $u$ and $v$. After separation of variables we get

$$0 = \frac{1-u}{u}\dot{u} - \frac{v-2}{v}\dot{v} = \frac{d}{dt}I(u,v)$$

where

$$I(u,v) = \ln u - u + 2\ln v - v, \tag{1.4}$$

---

[1] This cat came to fame through Arnold (1963).

so that $I(u(t), v(t)) = \mathit{Const}$ for all $t$. We call the function $I$ an *invariant* of the system (1.1). Every solution of (1.1) thus lies on a level curve of (1.4). Some of these curves are drawn in the pictures of Fig. 1.1. Since the level curves are closed, all solutions of (1.1) are periodic.

**Explicit Euler Method.** The simplest of all numerical methods for the system (1.2) is the method formulated by Euler (1768),

$$y_{n+1} = y_n + h f(y_n). \tag{1.5}$$

It uses a constant step size $h$ to compute, one after the other, approximations $y_1, y_2, y_3, \ldots$ to the values $y(h), y(2h), y(3h), \ldots$ of the solution starting from a given initial value $y(0) = y_0$. The method is called the *explicit Euler method*, because the approximation $y_{n+1}$ is computed using an explicit evaluation of $f$ at the already known value $y_n$. Such a formula represents a mapping

$$\Phi_h : y_n \mapsto y_{n+1},$$

which we call the *discrete* or *numerical flow*. Some iterations of the discrete flow for the Lotka-Volterra problem (1.1) (with $h = 0.5$) are represented in the third picture of Fig. 1.1.

**Other Numerical Methods.** The *implicit Euler method*

$$y_{n+1} = y_n + h f(y_{n+1}), \tag{1.6}$$

is known for its all-damping stability properties. In contrast to (1.5), the approximation $y_{n+1}$ is defined implicitly by (1.6), and the implementation requires the numerical solution of a nonlinear system of equations.

Taking the mean of $y_n$ and $y_{n+1}$ in the argument of $f$, we get the *implicit midpoint rule*

$$y_{n+1} = y_n + h f\left(\frac{y_n + y_{n+1}}{2}\right). \tag{1.7}$$

It is a *symmetric* method, which means that the formula is left unaltered after exchanging $y_n \leftrightarrow y_{n+1}$ and $h \leftrightarrow -h$ (more on symmetric methods in Chap. V).

For *partitioned* systems

$$\begin{aligned} \dot{u} &= a(u, v) \\ \dot{v} &= b(u, v), \end{aligned} \tag{1.8}$$

such as the problem (1.1), we consider also a *partitioned* Euler method

$$\begin{aligned} u_{n+1} &= u_n + h a(u_{n+1}, v_n) \\ v_{n+1} &= v_n + h b(u_{n+1}, v_n), \end{aligned} \tag{1.9}$$

which treats the $u$-variable by the implicit and the $v$-variable by the explicit Euler method. In view of an important property of this method to be discussed in Chap. VI, we call it the *symplectic Euler method*.

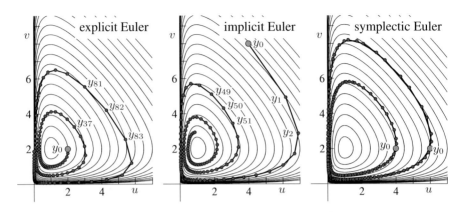

**Fig. 1.2.** Solutions of the Lotka-Volterra equations (1.1) (step sizes $h = 0.12$; initial values $(2, 2)$ for the explicit Euler method, $(4, 8)$ for the implicit Euler method, $(4, 2)$ and $(6, 2)$ for the symplectic Euler method.)

**Numerical Experiment.** Our first numerical experiment shows the behaviour of the various numerical methods applied to the Lotka-Volterra problem. In particular, we are interested in the preservation of the invariant $I$ over long times. Fig. 1.2 plots the numerical approximations of the first 125 steps with the above numerical methods applied to (1.1), all with constant step sizes. We observe that the explicit and implicit Euler methods show wrong qualitative behaviour. The numerical solution either spirals outwards or inwards. The symplectic Euler method, however, gives a numerical solution that lies apparently on a closed curve as does the exact solution. Note that the curves of the numerical and exact solutions do not coincide.

## I.1.2 Hamiltonian Systems — the Pendulum

A great deal of attention in this book will be addressed to Hamiltonian problems, and our next examples will be of this type. These problems are of the form

$$\dot{p} = -H_q(p, q), \qquad \dot{q} = H_p(p, q), \tag{1.10}$$

where the *Hamiltonian* $H(p_1, \ldots, p_d, q_1, \ldots q_d)$ represents the total energy; $q_i$ are the position coordinates and $p_i$ the momenta for $i = 1, \ldots, d$, with $d$ the number of degrees of freedom; $H_p$ and $H_q$ are the vectors of partial derivatives. One verifies easily by differentiation (see Sect. IV.1) that, along the solution curves of (1.10),

$$H\big(p(t), q(t)\big) = Const, \tag{1.11}$$

i.e., the Hamiltonian is an invariant or a *first integral*. More details about Hamiltonian systems and their derivation from Lagrangian mechanics will be given in Sect. VI.1.

**Pendulum.** The mathematical pendulum (mass $m = 1$, massless rod of length $\ell = 1$, gravitational acceleration $g = 1$) is a system with one degree of freedom having the Hamiltonian

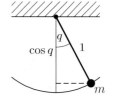

$$H(p,q) = \frac{1}{2}p^2 - \cos q, \qquad (1.12)$$

so that the equations of motion (1.10) become

$$\dot{p} = -\sin q, \qquad \dot{q} = p. \qquad (1.13)$$

Since the vector field (1.13) is $2\pi$-periodic in $q$, it is natural to consider $q$ as a variable on the circle $S^1$. Hence, the phase space of points $(p,q)$ becomes the cylinder $\mathbb{R} \times S^1$. Fig. 1.3 shows some level curves of $H(p,q)$. By (1.11), the solution curves of the problem (1.13) lie on such level curves.

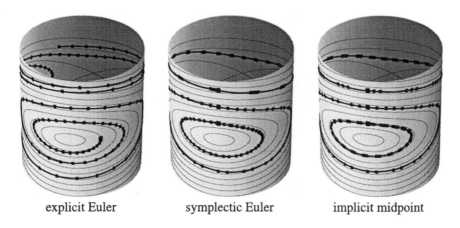

explicit Euler   symplectic Euler   implicit midpoint

**Fig. 1.3.** Solutions of the pendulum problem (1.13); explicit Euler with step size $h = 0.2$, initial value $(p_0, q_0) = (0, 0.5)$; symplectic Euler and implicit midpoint rule with $h = 0.3$ and initial values $q_0 = 0$, $p_0 = 0.7, 1.4, 2.1$.

**Numerical Experiment.** We apply the above numerical methods to the pendulum equations (see Fig. 1.3). Similar to the computations for the Lotka-Volterra equations, we observe that the numerical solutions of the explicit Euler and of the implicit Euler method (not drawn in Fig. 1.3) spiral either outwards or inwards. Only the symplectic Euler method and also the implicit midpoint rule show the correct qualitative behaviour. It can be observed that the numerical solution of the midpoint rule is closer to the exact solution than the other numerical solutions. The reason is that this is a method of order 2, whereas the Euler methods are only of order 1.

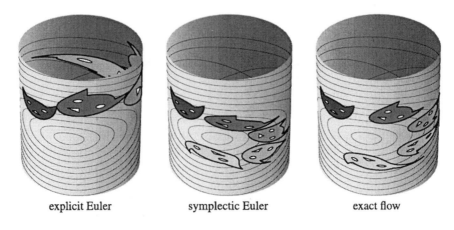

**Fig. 1.4.** Numerical and exact flow for the pendulum problem (1.13); step sizes $h = t = 1$.

**Area Preservation.** Figure 1.4 (right picture) illustrates that the exact flow of a Hamiltonian system (1.10) is area preserving. This can be explained as follows: the derivative of the flow $\varphi_t$ with respect to initial values $(p, q)$,

$$\varphi'_t(p, q) = \frac{\partial(p(t), q(t))}{\partial(p, q)},$$

satisfies the variational equation [2]

$$\dot{\varphi}'_t(p, q) = \begin{pmatrix} -H_{pq} & -H_{qq} \\ H_{pp} & H_{qp} \end{pmatrix} \varphi'_t(p, q),$$

where the second partial derivatives of $H$ are evaluated at $\varphi_t(p, q)$. In the case of one degree of freedom ($d = 1$), a simple computation shows that

$$\frac{d}{dt} \det \varphi'_t(p, q) = \frac{d}{dt} \left( \frac{\partial p(t)}{\partial p} \frac{\partial q(t)}{\partial q} - \frac{\partial p(t)}{\partial q} \frac{\partial q(t)}{\partial p} \right) = \ldots = 0.$$

Since $\varphi_0$ is the identity, this implies $\det \varphi'_t(p, q) = 1$ for all $t$, which means that the flow $\varphi_t(p, q)$ is an *area-preserving* mapping.

The first two pictures of Fig. 1.4 show numerical flows. The explicit Euler method is clearly seen not to preserve area but the symplectic Euler method is (this will be proved in Sect. VI.3). One of the aims of 'geometric integration' is the study of numerical integrators that preserve such types of qualitative behaviour of the exact flow.

---

[2] As is common in the study of mechanical problems, we use *dots* for denoting time-derivatives, and we use *primes* for denoting derivatives with respect to other variables.

## I.2 The Kepler Problem and the Outer Solar System

> I awoke as if from sleep, a new light broke on me. (J. Kepler; quoted from J.L.E. Dreyer, *A history of astronomy*, 1906, Dover 1953, p. 391)

One of the great achievements in the history of science was the discovery of J. Kepler (1609), based on many precise measurements of the positions of Mars by Tycho Brahe and himself, that the planets move in *elliptic orbits* with the sun at one of the foci (Kepler's first law)

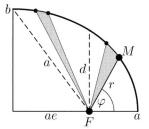

$$r = \frac{d}{1 + e \cos \varphi} = a + ae \cos E, \qquad (2.1)$$

(where $a$ = the great axis, $e$ = the eccentricity, $b = a\sqrt{1 - e^2}$, $d = b\sqrt{1 - e^2} = a(1 - e^2)$, $E$ = the eccentric anomaly) and that $r^2 \dot{\varphi} = Const$ (Kepler's second law).

Newton (*Principia* 1687) then *explained* this motion by his general law of gravitational attraction (proportional to $1/r^2$) and the relation between forces and acceleration (the "Lex II" of the *Principia*). This then opened the way for treating arbitrary celestial motions by solving differential equations.

**Two-Body Problem.** For computing the motion of two bodies which attract each other, we choose one of the bodies as the centre of our coordinate system; the motion will then stay in a plane (Exercise 3) and we can use two-dimensional coordinates $q = (q_1, q_2)$ for the position of the second body. Newton's laws, with a suitable normalization, then yield the following differential equations

$$\ddot{q}_1 = -\frac{q_1}{(q_1^2 + q_2^2)^{3/2}}, \qquad \ddot{q}_2 = -\frac{q_2}{(q_1^2 + q_2^2)^{3/2}}. \qquad (2.2)$$

This is equivalent to a Hamiltonian system with the Hamiltonian

$$H(p_1, p_2, q_1, q_2) = \frac{1}{2}\left(p_1^2 + p_2^2\right) - \frac{1}{\sqrt{q_1^2 + q_2^2}}, \qquad p_i = \dot{q}_i. \qquad (2.3)$$

### I.2.1 Exact Integration of the Kepler Problem

> Pour voir présentement que cette courbe $ABC$ ... est toûjours une Section Conique, ainsi que Mr. Newton l'a supposé, *pag. 55. Coroll.I.* sans le démontrer; il y faut bien plus d'adresse: (Joh. Bernoulli 1710, p. 475)

It is now interesting, inversely, to prove that *any* solution of (2.2) follows either an elliptic, parabolic or hyperbolic arc and to describe the solutions analytically. This was first done by Joh. Bernoulli (1710, full of sarcasm against Newton), and by Newton (1713, second edition of the *Principia*, without mentioning a word about Bernoulli).

The system has not only the total energy $H(p, q)$ as a first integral, but also the angular momentum
$$L(p_1, p_2, q_1, q_2) = q_1 p_2 - q_2 p_1. \tag{2.4}$$
This can be checked by differentiation and is nothing other than Kepler's second law. Hence, every solution of (2.2) satisfies the two relations
$$\frac{1}{2}\left(\dot{q}_1^2 + \dot{q}_2^2\right) - \frac{1}{\sqrt{q_1^2 + q_2^2}} = H_0, \qquad q_1 \dot{q}_2 - q_2 \dot{q}_1 = L_0, \tag{2.5}$$
where the constants $H_0$ and $L_0$ are determined by the initial values. Using polar coordinates $q_1 = r\cos\varphi$, $q_2 = r\sin\varphi$, this system becomes
$$\frac{1}{2}\left(\dot{r}^2 + r^2\dot{\varphi}^2\right) - \frac{1}{r} = H_0, \qquad r^2\dot{\varphi} = L_0. \tag{2.6}$$
For its solution we consider $r$ as a function of $\varphi$ and write $\dot{r} = \frac{dr}{d\varphi} \cdot \dot{\varphi}$. The elimination of $\dot{\varphi}$ in (2.6) then yields
$$\frac{1}{2}\left(\left(\frac{dr}{d\varphi}\right)^2 + r^2\right)\frac{L_0^2}{r^4} - \frac{1}{r} = H_0.$$
In this equation we use the substitution
$$\frac{1}{r} = u, \qquad dr = -\frac{du}{u^2}$$
which gives (with $' = d/d\varphi$)
$$\frac{1}{2}(u'^2 + u^2) - \frac{u}{L_0^2} - \frac{H_0}{L_0^2} = 0. \tag{2.7}$$
This is a "Hamiltonian" for the system
$$u'' + u = \frac{1}{d} \quad \text{i.e.,} \quad u = \frac{1}{d} + c_1 \cos\varphi + c_2 \sin\varphi = \frac{1 + e\cos(\varphi - \varphi^*)}{d} \tag{2.8}$$
where $d = L_0^2$ and the constant $e$ becomes, from (2.7),
$$e^2 = 1 + 2H_0 L_0^2 \tag{2.9}$$
(by Exercise 7, the expression $1 + 2H_0 L_0^2$ is non-negative). This is precisely formula (2.1). The angle $\varphi^*$ is determined by the initial values $r_0$ and $\varphi_0$. Equation (2.1) represents an elliptic orbit with eccentricity $e$ for $H_0 < 0$ (see Fig. 2.1), a parabola for $H_0 = 0$, and a hyperbola for $H_0 > 0$.

Finally, we must determine the variables $r$ and $\varphi$ as functions of $t$. With the relation (2.8) and $r = 1/u$, the second equation of (2.6) gives
$$\frac{d^2}{\left(1 + e\cos(\varphi - \varphi^*)\right)^2} d\varphi = L_0\, dt \tag{2.10}$$
which, after integration, represents an implicit equation for $\varphi(t)$.

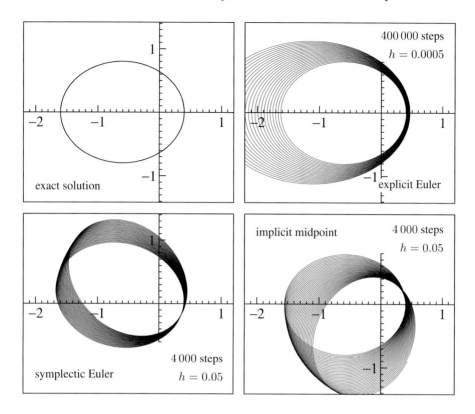

**Fig. 2.1.** Exact and numerical solutions of the Kepler problem (eccentricity $e = 0.6$)

## I.2.2 Numerical Integration of the Kepler Problem

For the problem (2.2) we choose, with $0 \leq e < 1$, the initial values

$$q_1(0) = 1 - e, \quad q_2(0) = 0, \quad \dot{q}_1(0) = 0, \quad \dot{q}_2(0) = \sqrt{\frac{1+e}{1-e}}. \quad (2.11)$$

This implies that $H_0 = -1/2$, $L_0 = \sqrt{1-e^2}$, $d = 1 - e^2$ and $\varphi^* = 0$. The period of the solution is $2\pi$ (Exercise 5). Fig. 2.1 shows the exact solution with eccentricity $e = 0.6$ and some numerical solutions. After our previous experience, it is no longer a surprise that the explicit Euler method spirals outwards and gives a completely wrong answer. For the symplectic Euler method and the implicit midpoint rule we take a step size 100 times larger in order to "see something". We see that the numerical solution does not distort the ellipse, but there is a *precession* effect, clockwise for the symplectic Euler method and anti-clockwise for the implicit midpoint rule. The same behaviour occurs for the exact solution of *perturbed* Kepler problems (Exercise 12) and has occupied astronomers for centuries.

Our next experiment (Fig. 2.2 and Table 2.1) studies the conservation of invariants and the global error. The main observation is that the error in the energy grows

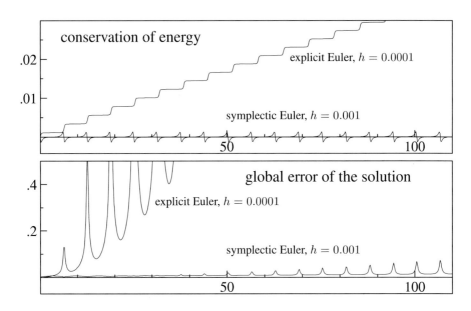

**Fig. 2.2.** Energy conservation and global error for the Kepler problem

**Table 2.1.** Qualitative long-time behaviour for the Kepler problem; $t$ is time, $h$ the step size.

| method | error in $H$ | error in $L$ | global error |
|---|---|---|---|
| explicit Euler | $\mathcal{O}(th)$ | $\mathcal{O}(th)$ | $\mathcal{O}(t^2 h)$ |
| symplectic Euler | $\mathcal{O}(h)$ | $0$ | $\mathcal{O}(th)$ |
| implicit midpoint | $\mathcal{O}(h^2)$ | $0$ | $\mathcal{O}(th^2)$ |

linearly for the explicit Euler method, and it remains bounded and small (no secular terms) for the symplectic Euler method. The global error, measured in the Euclidean norm, shows a quadratic growth (explicit Euler) compared to a linear growth (symplectic Euler and implicit midpoint rule). We remark that the angular momentum $L(p,q)$ is exactly conserved by the symplectic Euler and the implicit midpoint rule.

### I.2.3 The Outer Solar System

> The evolution of the entire planetary system has been numerically integrated for a time span of nearly 100 million years[3]. This calculation confirms that the evolution of the solar system as a whole is chaotic, ...
> (G.J. Sussman & J. Wisdom 1992)

---

[3] 100 million years is not much in astronomical time scales; it just goes back to "Jurassic Park".

We next apply our methods to the system which describes the motion of the five outer planets relative to the sun. This system has been studied extensively by astronomers. The problem is a Hamiltonian system (1.10) ($N$-body problem) with

$$H(p,q) = \frac{1}{2}\sum_{i=0}^{5} \frac{1}{m_i} p_i^T p_i - G\sum_{i=1}^{5}\sum_{j=0}^{i-1} \frac{m_i m_j}{\|q_i - q_j\|}. \quad (2.12)$$

Here $p$ and $q$ are the supervectors composed by the vectors $p_i, q_i \in \mathbb{R}^3$ (momenta and positions), respectively. The chosen units are: masses relative to the sun, so that the sun has mass 1. We have taken

$$m_0 = 1.00000597682$$

to take account of the inner planets. Distances are in astronomical units (1 [A.U.] = 149 597 870 [km]), times in earth days, and the gravitational constant is

$$G = 2.95912208286 \cdot 10^{-4}.$$

The initial values for the sun are taken as $q_0(0) = (0,0,0)^T$ and $\dot{q}_0(0) = (0,0,0)^T$. All other data (masses of the planets and the initial positions and initial velocities) are given in Table 2.2. The initial data is taken from "Ahnerts Kalender für Sternfreunde 1994", Johann Ambrosius Barth Verlag 1993, and they correspond to September 5, 1994 at 0h00.[4]

**Table 2.2.** Data for the outer solar system

| planet | mass | initial position | initial velocity |
|---|---|---|---|
| Jupiter | $m_1 = 0.000954786104043$ | −3.5023653<br>−3.8169847<br>−1.5507963 | 0.00565429<br>−0.00412490<br>−0.00190589 |
| Saturn | $m_2 = 0.000285583733151$ | 9.0755314<br>−3.0458353<br>−1.6483708 | 0.00168318<br>0.00483525<br>0.00192462 |
| Uranus | $m_3 = 0.0000437273164546$ | 8.3101420<br>−16.2901086<br>−7.2521278 | 0.00354178<br>0.00137102<br>0.00055029 |
| Neptune | $m_4 = 0.0000517759138449$ | 11.4707666<br>−25.7294829<br>−10.8169456 | 0.00288930<br>0.00114527<br>0.00039677 |
| Pluto | $m_5 = 1/(1.3 \cdot 10^8)$ | −15.5387357<br>−25.2225594<br>−3.1902382 | 0.00276725<br>−0.00170702<br>−0.00136504 |

---

[4] We thank Alexander Ostermann, who provided us with this data.

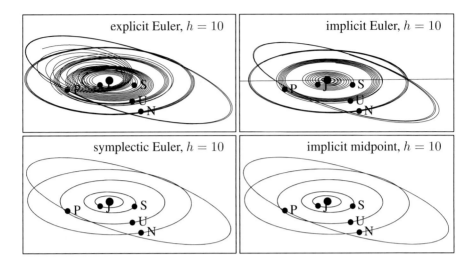

**Fig. 2.3.** Solutions of the outer solar system

To this system we applied our four methods, all with step size $h = 10$ (days) and over a time period of $200\,000$ days. The numerical solution (see Fig. 2.3) behaves similarly to that for the Kepler problem. With the explicit Euler method the planets have increasing energy, they spiral outwards, Jupiter approaches Saturn which leaves the plane of the two-body motion. With the implicit Euler method the planets (first Jupiter and then Saturn) fall into the sun and are thrown far away. Both the symplectic Euler method and the implicit midpoint rule show the correct behaviour. An integration over a much longer time of say several million years does not deteriorate this behaviour. Let us remark that Sussman & Wisdom (1992) have integrated the outer solar system with special geometric integrators.

## I.3 Molecular Dynamics

> We do not need exact classical trajectories to do this, but must lay great emphasis on energy conservation as being of primary importance for this reason. (M.P. Allen & D.J. Tildesley 1987)

Molecular dynamics requires the solution of Hamiltonian systems (1.10), where the total energy is given by

$$H(p,q) = \frac{1}{2} \sum_{i=1}^{N} \frac{1}{m_i} p_i^T p_i + \sum_{i=2}^{N} \sum_{j=1}^{i-1} V_{ij}\Big(\|q_i - q_j\|\Big), \qquad (3.1)$$

and $V_{ij}(r)$ are given potential functions. Here, $q_i$ and $p_i$ denote the positions and momenta of atoms and $m_i$ is the atomic mass of the $i$th atom. We remark that the outer solar system (2.12) is such an $N$-body system with $V_{ij}(r) = -Gm_i m_j / r$. In molecular dynamics the Lennard-Jones potential

$$V_{ij}(r) = 4\varepsilon_{ij}\left(\left(\frac{\sigma_{ij}}{r}\right)^{12} - \left(\frac{\sigma_{ij}}{r}\right)^{6}\right) \qquad (3.2)$$

is very popular ($\varepsilon_{ij}$ and $\sigma_{ij}$ are suitable constants depending on the atoms). This potential has an absolute minimum at distance $r = \sigma_{ij}\sqrt[6]{2}$. The force due to this potential strongly repels the atoms when they are closer than this value, and they attract each other when they are farther away.

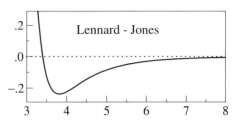

## I.3.1 The Störmer/Verlet Scheme

The Hamiltonian (3.1) is of the form $H(p,q) = T(p) + U(q)$, where $T(p)$ is a quadratic function. The Hamiltonian system becomes

$$\dot{q} = M^{-1}p, \qquad \dot{p} = -\nabla U(q),$$

where $M = \text{diag}(m_1 I, \ldots, m_N I)$ and $I$ is the 3-dimensional identity matrix, and $\nabla U = (\partial U/\partial q)^T$ is the gradient of $U$. This system is equivalent to the special second order differential equation

$$\ddot{q} = f(q), \qquad (3.3)$$

where the right-hand side $f(q) = -M^{-1}\nabla U(q)$ does not depend on $\dot{q}$. The most natural discretization of (3.3) is[5]

$$q_{n+1} - 2q_n + q_{n-1} = h^2 f(q_n). \qquad (3.4)$$

This basic method, or its equivalent formulation given below, is called the *Störmer method* in astronomy, the *Verlet method*[6] in molecular dynamics, the *leap-frog method* in the context of partial differential equations, and it may well have further names in other areas. C. Störmer (1907) used higher-order variants for numerical computations concerning the aurora borealis. L. Verlet (1967) proposed this method for computations in molecular dynamics, where it has become by far the most widely used integration scheme.

An approximation to the derivative $v = \dot{q}$ is simply obtained by

---

[5] *Attention.* In (3.4) and in the subsequent formulas $q_n$ denotes an approximation to $q(nh)$, whereas $q_i$ in (3.1) denotes the $i$th subvector of $q$.

[6] Irony of fate: Professor Loup Verlet, who later became interested in the history of science, discovered precisely "his" method in Newton's *Principia* (Book I, figure for Theorem I). Private communication.

**Fig. 3.1.** Carl Störmer (left picture), born: 3 September 1874 in Skien (Norway), died: 13 August 1957.
Loup Verlet (right picture), born: 24 May 1931 in Paris.

$$v_n = \frac{q_{n+1} - q_{n-1}}{2h}. \tag{3.5}$$

For the second order problem (3.3) one usually has given initial values $q(0) = q_0$ and $\dot{q}(0) = v_0$. However, one also needs $q_1$ in order to be able to start the integration with the 3-term recursion (3.4). Putting $n = 0$ in (3.4) and (3.5), an elimination of $q_{-1}$ gives

$$q_1 = q_0 + hv_0 + \frac{h^2}{2} f(q_0)$$

for the missing starting value.

The Störmer/Verlet method admits a *one-step formulation* which is useful for actual computations. Introducing the velocity approximation at the midpoint $v_{n+1/2} := v_n + \frac{h}{2} f(q_n)$, an elimination of $q_{n-1}$ (as above) yields

$$\begin{aligned} v_{n+1/2} &= v_n + \frac{h}{2} f(q_n) \\ q_{n+1} &= q_n + hv_{n+1/2} \\ v_{n+1} &= v_{n+1/2} + \frac{h}{2} f(q_{n+1}) \end{aligned} \tag{3.6}$$

which is an explicit one-step method $\Phi_h^V : (q_n, v_n) \mapsto (q_{n+1}, v_{n+1})$ for the first order system $\dot{q} = v, \dot{v} = f(q)$. If one is not interested in the values $v_n$ of the derivative, the first and third equations in (3.6) can be replaced by

$$v_{n+1/2} = v_{n-1/2} + h f(q_n).$$

## I.3.2 Numerical Experiments

As in Biesiadecki & Skeel (1993) we consider the interaction of seven argon atoms in a plane, where six of them are arranged symmetrically around a centre atom (frozen argon crystal). As a mathematical model we take the Hamiltonian (3.1) with $N = 7$, $m_i = m = 66.34 \cdot 10^{-27}$ [kg],

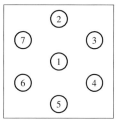

$$\varepsilon_{ij} = \varepsilon = 119.8\, k_B\ [\text{J}], \qquad \sigma_{ij} = \sigma = 0.341\ [\text{nm}],$$

where $k_B = 1.380658 \cdot 10^{-23}$ [J/K] is Boltzmann's constant (see Allen & Tildesley (1987), page 21). As units for our calculations we take masses in [kg], distances in nanometers (1 [nm] $= 10^{-9}$ [m]), and times in nanoseconds (1 [nsec] $= 10^{-9}$ [sec]). Initial positions (in [nm]) and initial velocities (in [nm/nsec]) are given in Table 3.1. They are chosen such that neighbouring atoms have a distance that is close to the one with lowest potential energy, and such that the total momentum is zero and therefore the centre of gravity does not move. The energy at the initial position is $H(p_0, q_0) \approx -1260.2\, k_B$ [J].

**Table 3.1.** Initial values for the simulation of a frozen argon crystal

| atom | 1 | 2 | 3 | 4 | 5 | 6 | 7 |
|---|---|---|---|---|---|---|---|
| position | 0.00 | 0.02 | 0.34 | 0.36 | −0.02 | −0.35 | −0.31 |
|  | 0.00 | 0.39 | 0.17 | −0.21 | −0.40 | −0.16 | 0.21 |
| velocity | −30 | 50 | −70 | 90 | 80 | −40 | −80 |
|  | −20 | −90 | −60 | 40 | 90 | 100 | −60 |

For computations in molecular dynamics one is usually not interested in the trajectories of the atoms, but one aims at macroscopic quantities such as temperature, pressure, internal energy, etc. Here we consider the total energy, given by the Hamiltonian, and the temperature which can be calculated from the formula (see Allen & Tildesley (1987), page 46)

$$T = \frac{1}{2Nk_B} \sum_{i=1}^{N} m_i \|\dot{q}_i\|^2. \tag{3.7}$$

We apply the explicit and symplectic Euler methods and also the Verlet method to this problem. Observe that for a Hamiltonian such as (3.1) all three methods are explicit, and all of them need only one force evaluation per integration step. In Fig. 3.2 we present the numerical results of our experiments. The integrations are done over an interval of length 0.2 [nsec]. The step sizes are indicated in femtoseconds (1 [fsec] $= 10^{-6}$ [nsec]).

The two upper pictures show the values $\big(H(p_n, q_n) - H(p_0, q_0)\big)/k_B$ as a function of time $t_n = nh$. For the exact solution, this value is precisely zero for all times.

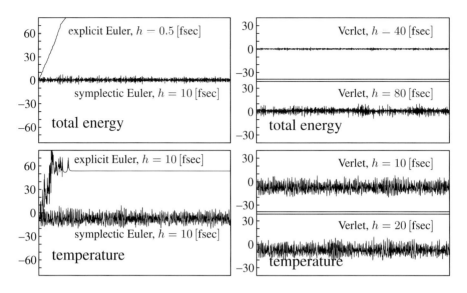

**Fig. 3.2.** Computed total energy and temperature of the argon crystal

Similar to earlier experiments we see that the symplectic Euler method is qualitatively correct, whereas the numerical solution of the explicit Euler method, although computed with a much smaller step size, is completely useless (see the citation at the beginning of this section). The Verlet method is qualitatively correct and gives much more accurate results than the symplectic Euler method (we shall see later that the Verlet method is of order 2). The two computations with the Verlet method show that the energy error decreases by a factor of 4 if the step size is reduced by a factor of 2 (second order convergence).

The two lower pictures of Fig. 3.2 show the numerical values of the temperature difference $T - T_0$ with $T$ given by (3.7) and $T_0 \approx 22.72$ [K] (initial temperature). In contrast to the total energy, this is not an exact invariant, but for our problem it fluctuates around a constant value. The explicit Euler method gives wrong results, but the symplectic Euler and the Verlet methods show the desired behaviour. This time a reduction of the step size does not reduce the amplitude of the oscillations, which indicates that the fluctuation of the exact temperature is of the same size.

## I.4 Highly Oscillatory Problems

In this section we discuss a system with almost-harmonic high-frequency oscillations. We show numerical phenomena of methods applied with step sizes that are not small compared to the period of the fastest oscillations.

## I.4.1 A Fermi-Pasta-Ulam Problem

> ... dealing with the behavior of certain nonlinear physical systems where the non-linearity is introduced as a perturbation to a primarily linear problem. The behavior of the systems is to be studied for times which are long compared to the characteristic periods of the corresponding linear problems.
> (E. Fermi, J. Pasta, S. Ulam 1955)

> In the early 1950s MANIAC-I had just been completed and sat poised for an attack on significant problems. ... Fermi suggested that it would be highly instructive to integrate the equations of motion numerically for a judiciously chosen, one-dimensional, harmonic chain of mass points weakly perturbed by nonlinear forces.
> (J. Ford 1992)

The problem of Fermi, Pasta & Ulam (1955) is a simple model for simulations in statistical mechanics which revealed highly unexpected dynamical behaviour. We consider a modification consisting of a chain of $m$ mass points, connected with alternating soft nonlinear and stiff linear springs, and fixed at the end points (see Galgani, Giorgilli, Martinoli & Vanzini (1992) and Fig. 4.1). The variables $q_1, \ldots, q_{2m}$

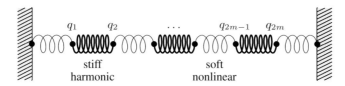

**Fig. 4.1.** Chain with alternating soft nonlinear and stiff linear springs

($q_0 = q_{2m+1} = 0$) stand for the displacements of the mass points, and $p_i = \dot q_i$ for their velocities. The motion is described by a Hamiltonian system with total energy

$$H(p,q) = \frac{1}{2}\sum_{i=1}^{m}(p_{2i-1}^2 + p_{2i}^2) + \frac{\omega^2}{4}\sum_{i=1}^{m}(q_{2i} - q_{2i-1})^2 + \sum_{i=0}^{m}(q_{2i+1} - q_{2i})^4,$$

where $\omega$ is assumed to be large. It is quite natural to introduce the new variables

$$\begin{aligned} x_i &= (q_{2i} + q_{2i-1})/\sqrt{2}, & x_{m+i} &= (q_{2i} - q_{2i-1})/\sqrt{2}, \\ y_i &= (p_{2i} + p_{2i-1})/\sqrt{2}, & y_{m+i} &= (p_{2i} - p_{2i-1})/\sqrt{2}, \end{aligned} \qquad (4.1)$$

where $x_i$ ($i=1,\ldots,m$) represents a scaled displacement of the $i$th stiff spring, $x_{m+i}$ a scaled expansion (or compression) of the $i$th stiff spring, and $y_i, y_{m+i}$ their velocities (or momenta). With this change of coordinates, the motion in the new variables is again described by a Hamiltonian system, with

$$H(y,x) = \frac{1}{2}\sum_{i=1}^{2m} y_i^2 + \frac{\omega^2}{2}\sum_{i=1}^{m} x_{m+i}^2 + \frac{1}{4}\Big((x_1 - x_{m+1})^4 + \sum_{i=1}^{m-1}(x_{i+1} - x_{m+i+1} - x_i - x_{m+i})^4 + (x_m + x_{2m})^4\Big). \qquad (4.2)$$

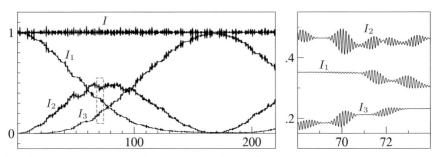

**Fig. 4.2.** Exchange of energy in the exact solution of the Fermi-Pasta-Ulam model. The picture to the right is an enlargement of the narrow rectangle in the left-hand picture.

Besides the fact that the equations of motion are Hamiltonian, so that the total energy is exactly conserved, they have a further interesting feature. Let

$$I_j(x_{m+j}, y_{m+j}) = \frac{1}{2}\left(y_{m+j}^2 + \omega^2 x_{m+j}^2\right) \tag{4.3}$$

denote the energy of the $j$th stiff spring. It turns out that there is an exchange of energy between the stiff springs, but the total oscillatory energy $I = I_1 + \ldots + I_m$ remains close to a constant value, in fact, $I\bigl((x(t), y(t))\bigr) = I\bigl((x(0), y(0))\bigr) + \mathcal{O}(\omega^{-1})$. We call $I(x, y)$ an *adiabatic invariant* of the Hamiltonian system. For an illustration of this property, we choose $m = 3$ (as in Fig. 4.1), $\omega = 50$,

$$x_1(0) = 1, \quad y_1(0) = 1, \quad x_4(0) = \omega^{-1}, \quad y_4(0) = 1,$$

and zero for the remaining initial values. Fig. 4.2 displays the energies $I_1, I_2, I_3$ of the stiff springs together with the total oscillatory energy $I = I_1 + I_2 + I_3$ as a function of time. The solution has been computed very carefully with high precision, so that the displayed oscillations can be considered as exact.

### I.4.2 Application of Classical Integrators

Which of the methods of the foregoing sections produce qualitatively correct approximations when the product of the step size $h$ with the high frequency $\omega$ is relatively large?

**Linear Stability Analysis.** To get an idea of the maximum admissible step size, we neglect the quartic term in the Hamiltonian (4.2), so that the differential equation splits into the two-dimensional problems $\dot{y}_i = 0$, $\dot{x}_i = y_i$ and

$$\dot{y}_{m+i} = -\omega^2 x_{m+i}, \quad \dot{x}_{m+i} = y_{m+i}. \tag{4.4}$$

Omitting the subscript, the solution of (4.4) is

$$\begin{pmatrix} y(t) \\ \omega x(t) \end{pmatrix} = \begin{pmatrix} \cos \omega t & -\sin \omega t \\ \sin \omega t & \cos \omega t \end{pmatrix} \begin{pmatrix} y(0) \\ \omega x(0) \end{pmatrix}.$$

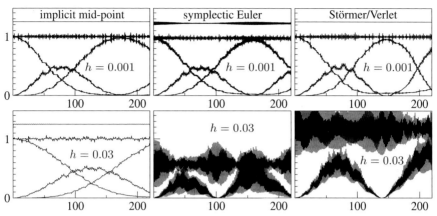

**Fig. 4.3.** Numerical solution for the FPU problem (4.2) with data as in Sect. I.4.1, obtained with the implicit midpoint rule (left), symplectic Euler (middle), and Störmer/Verlet scheme (right); the upper pictures use $h = 0.001$, the lower pictures $h = 0.03$; the first four pictures show the Hamiltonian $H - 0.8$ and the oscillatory energies $I_1, I_2, I_3, I$; the last two pictures only show $I_2$ and $I$.

The numerical solution of a one-step method applied to (4.4) yields

$$\begin{pmatrix} y_{n+1} \\ \omega x_{n+1} \end{pmatrix} = M(h\omega) \begin{pmatrix} y_n \\ \omega x_n \end{pmatrix}, \tag{4.5}$$

and the eigenvalues $\lambda_i$ of $M(h\omega)$ determine the long-time behaviour of the numerical solution. Stability (i.e., boundedness of the solution of (4.5)) requires the eigenvalues to be less than or equal to one in modulus. For the explicit Euler method we have $\lambda_{1,2} = 1 \pm ih\omega$, so that the energy $I_n = (y_n^2 + \omega^2 x_n^2)/2$ increases as $(1 + h^2\omega^2)^{n/2}$. For the implicit Euler method we have $\lambda_{1,2} = (1 \pm ih\omega)^{-1}$, and the energy decreases as $(1 + h^2\omega^2)^{-n/2}$. For the implicit midpoint rule, the matrix $M(h\omega)$ is orthogonal and therefore $I_n$ is exactly preserved for all $h$ and for all times. Finally, for the symplectic Euler method and for the Störmer/Verlet scheme we have

$$M(h\omega) = \begin{pmatrix} 1 & -h\omega \\ h\omega & 1 - h^2\omega^2 \end{pmatrix}, \quad M(h\omega) = \begin{pmatrix} 1 - \frac{h^2\omega^2}{2} & -\frac{h\omega}{2}\left(1 - \frac{h^2\omega^2}{4}\right) \\ \frac{h\omega}{2} & 1 - \frac{h^2\omega^2}{2} \end{pmatrix},$$

respectively. For both matrices, the characteristic polynomial is $\lambda^2 - (2 - h^2\omega^2)\lambda + 1$, so that the eigenvalues are of modulus one if and only if $|h\omega| \leq 2$.

**Numerical Experiments.** We apply several methods to the Fermi-Pasta-Ulam (FPU) problem, with $\omega = 50$ and initial data as given in Sect. I.4.1. The explicit and implicit Euler methods give completely wrong solutions even for very small step sizes. Fig. 4.3 presents the numerical results for $H, I, I_1, I_2, I_3$ obtained with the implicit midpoint rule, the symplectic Euler, and the Störmer/Verlet scheme. For the small step size $h = 0.001$ all methods give satisfactory results, although the energy exchange is not reproduced accurately over long times. The Hamiltonian $H$ and the total oscillatory energy $I$ are well conserved over much longer time intervals. The larger step size $h = 0.03$ has been chosen such that $h\omega = 1.5$ is close to the stability limit of the symplectic Euler and the Störmer/Verlet methods. The values of $H$ and $I$ are still bounded over very long time intervals, but the oscillations do not represent the true behaviour. Moreover, the average value of $I$ is no longer close to 1, as it is for the exact solution. These phenomena call for an explanation, and for numerical methods with an improved behaviour (see Chap. XIII).

## I.5 Exercises

1. Show that the Lotka-Volterra problem (1.1) in logarithmic scale, i.e., by putting $p = \log u$ and $q = \log v$, becomes a Hamiltonian system with the function (1.4) as Hamiltonian (see Fig. 5.1).

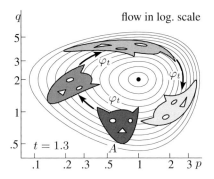

**Fig. 5.1.** Area preservation in logarithmic scale of the Lotka-Volterra flow

2. Apply the symplectic Euler method (or the implicit midpoint rule) to problems such as
$$\begin{pmatrix} \dot{u} \\ \dot{v} \end{pmatrix} = \begin{pmatrix} (v-2)/v \\ (1-u)/u \end{pmatrix}, \qquad \begin{pmatrix} \dot{u} \\ \dot{v} \end{pmatrix} = \begin{pmatrix} u^2 v(v-2) \\ v^2 u(1-u) \end{pmatrix}$$
with various initial conditions. Both problems have the same first integral (1.4) as the Lotka-Volterra problem and therefore their solutions are also periodic. Do the numerical solutions also show this behaviour?

3. A general two-body problem (sun and planet) is given by the Hamiltonian
$$H(p, p_S, q, q_S) = \frac{1}{2M} p_S^T p_S + \frac{1}{2m} p^T p - \frac{GmM}{\|q - q_S\|},$$
where $q_S, q \in \mathbb{R}^3$ are the positions of the sun (mass $M$) and the planet (mass $m$), $p_S, p \in \mathbb{R}^3$ are their momenta, and $G$ is the gravitational constant.

a) Prove that, in heliocentric coordinates $Q := q - q_S$, the equations of motion are
$$\ddot{Q} = -G(M+m)\frac{Q}{\|Q\|^3}.$$

b) Prove that $\frac{d}{dt}(Q(t) \times \dot{Q}(t)) = 0$, so that $Q(t)$ stays for all times $t$ in the plane $E = \{q\,;\, d^T q = 0\}$, where $d = Q(0) \times \dot{Q}(0)$.

*Conclusion.* The coordinates corresponding to a basis in $E$ satisfy the two-dimensional equations (2.2).

4. In polar coordinates, the two-body problem (2.2) becomes
$$\ddot{r} = -V'(r) \quad \text{with} \quad V(r) = \frac{L_0^2}{2r^2} - \frac{1}{r}$$
which is independent of $\varphi$. The angle $\varphi(t)$ can be obtained by simple integration from $\dot{\varphi}(t) = L_0/r^2(t)$.

5. Compute the period of the solution of the Kepler problem (2.2) and deduce from the result Kepler's "third law".
*Hint.* Comparing Kepler's second law (2.6) with the area of the ellipse gives $\frac{1}{2}L_0 T = ab\pi$. Then apply (2.7). The result is $T = 2\pi(2|H_0|)^{-3/2} = 2\pi a^{3/2}$.

6. Deduce Kepler's first law from equations (2.2) by the elegant method of Laplace (1799).
*Hint.* Multiplying (2.2) with (2.5) gives
$$L_0 \ddot{q}_1 = \frac{d}{dt}\left(\frac{q_2}{r}\right), \qquad L_0 \ddot{q}_2 = \frac{d}{dt}\left(-\frac{q_1}{r}\right),$$
and after integration
$$L_0 \dot{q}_1 = \frac{q_2}{r} + B, \qquad L_0 \dot{q}_2 = -\frac{q_1}{r} + A,$$
where $A$ and $B$ are integration constants. Then eliminate $\dot{q}_1$ and $\dot{q}_2$ by multiplying these equations by $q_2$ and $-q_1$ respectively and by subtracting them. The result is a quadratic equation in $q_1$ and $q_2$.

7. Whatever the initial values for the Kepler problem are, $1 + 2H_0 L_0^2 \geq 0$ holds. Hence, the value $e$ is well defined by (2.9).
*Hint.* $L_0$ is the area of the parallelogram spanned by the vectors $q(0)$ and $\dot{q}(0)$.

8. Show that not only does the symplectic Euler and the implicit midpoint rule preserve exactly the angular momentum for the Kepler problem (see Table 2.1), but the Störmer/Verlet scheme does as well.

9. *Implementation of the Störmer/Verlet scheme.* Explain why the use of the one-step formulation (3.6) is numerically more stable than that of the two-term recursion (3.4).

10. *Runge-Lenz-Pauli vector.* Prove that the function
$$A(p,q) = \begin{pmatrix} p_1 \\ p_2 \\ 0 \end{pmatrix} \times \begin{pmatrix} 0 \\ 0 \\ q_1 p_2 - q_2 p_1 \end{pmatrix} - \frac{1}{\sqrt{q_1^2 + q_2^2}} \begin{pmatrix} q_1 \\ q_2 \\ 0 \end{pmatrix}$$

22    I. Examples and Numerical Experiments

is a first integral of the Kepler problem, i.e., $A\big(p(t),q(t)\big) = Const$ along solutions of the problem. However, it is not a first integral of the perturbed Kepler problem of Exercise 12.

11. Add a column to Table 2.1 which shows the long-time behaviour of the error in the Runge-Lenz-Pauli vector (see Exercise 10) for the various numerical integrators.

12. Study numerically the solution of the perturbed Kepler problem with Hamiltonian

$$H(p_1,p_2,q_1,q_2) = \frac{1}{2}\left(p_1^2 + p_2^2\right) - \frac{1}{\sqrt{q_1^2 + q_2^2}} - \frac{\mu}{3\sqrt{(q_1^2 + q_2^2)^3}},$$

where $\mu$ is a positive or negative small number. Among others, this problem describes the motion of a planet in the Schwarzschild potential for Einstein's general relativity theory[7]. You will observe a precession of the perihelion, which, applied to the orbit of Mercury, represented the historically first verification of Einstein's theory (see e.g., Birkhoff 1923, p. 261-264).

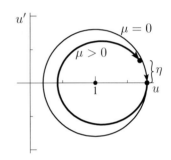

The precession can also be expressed analytically: the equation for $u = 1/r$ as a function of $\varphi$, corresponding to (2.8), here becomes

$$u'' + u = \frac{1}{d} + \mu u^2, \qquad (5.1)$$

where $d = L_0^2$. Now compute the derivative of this solution with respect to $\mu$, at $\mu = 0$ and $u = (1 + e\cos(\varphi - \varphi^*))/d$ after one period $t = 2\pi$. This leads to $\eta = \mu(e/d^2) \cdot 2\pi \sin\varphi$ (see the small picture). Then, for small $\mu$, the precession after one period is

$$\Delta\varphi = \frac{2\pi\mu}{d}. \qquad (5.2)$$

---

[7] We are grateful to Prof. Ruth Durrer for helpful hints about this subject.

# Chapter II.
# Numerical Integrators

After having seen in Chap. I some simple numerical methods and a variety of numerical phenomena that they exhibited, we now present more elaborate classes of numerical methods. We start with Runge-Kutta and collocation methods, and we introduce discontinuous collocation methods, which cover essentially all high-order implicit Runge-Kutta methods of interest. We then treat partitioned Runge-Kutta methods and Nyström methods, which can be applied to partitioned problems such as Hamiltonian systems. Finally we present composition and splitting methods.

## II.1 Runge-Kutta and Collocation Methods

**Fig. 1.1.** Carl David Tolmé Runge (left picture), born: 30 August 1856 in Bremen (Germany), died: 3 January 1927 in Göttingen (Germany).
Wilhelm Martin Kutta (right picture), born: 3 November 1867 in Pitschen, Upper Silesia (now Byczyna, Poland), died: 25 December 1944 in Fürstenfeldbruck (Germany).

Runge-Kutta methods form an important class of methods for the integration of differential equations. A special subclass, the collocation methods, allows for a particularly elegant access to order, symplecticity and continuous output.

## II.1.1 Runge-Kutta Methods

In this section, we treat non-autonomous systems of first-order ordinary differential equations

$$\dot{y} = f(t, y), \qquad y(t_0) = y_0. \tag{1.1}$$

The integration of this equation gives $y(t_1) = y_0 + \int_{t_0}^{t_1} f(t, y(t))\, dt$, and replacing the integral by the trapezoidal rule, we obtain

$$y_1 = y_0 + \frac{h}{2}\bigl(f(t_0, y_0) + f(t_1, y_1)\bigr). \tag{1.2}$$

This is the *implicit trapezoidal rule*, which, in addition to its historical importance for computations in partial differential equations (Crank-Nicolson) and in A-stability theory (Dahlquist), played a crucial role even earlier in the discovery of Runge-Kutta methods. It was the starting point of Runge (1895), who "predicted" the unknown $y_1$-value to the right by an Euler step, and obtained the first of the following formulas (the second being the analogous formula for the midpoint rule)

$$
\begin{aligned}
k_1 &= f(t_0, y_0) & k_1 &= f(t_0, y_0) \\
k_2 &= f(t_0 + h, y_0 + hk_1) & k_2 &= f(t_0 + \tfrac{h}{2}, y_0 + \tfrac{h}{2}k_1) \\
y_1 &= y_0 + \tfrac{h}{2}(k_1 + k_2) & y_1 &= y_0 + hk_2.
\end{aligned}
\tag{1.3}
$$

These methods have a nice geometric interpretation (which is illustrated in the first two pictures of Fig. 1.2 for a famous problem, the Riccati equation): they consist of polygonal lines, which assume the slopes prescribed by the differential equation evaluated at previous points.

*Idea of Heun (1900) and Kutta (1901)*: compute *several* polygonal lines, each starting at $y_0$ and assuming the various slopes $k_j$ on portions of the integration interval, which are proportional to some given constants $a_{ij}$; at the final point of each polygon evaluate a new slope $k_i$. The last of these polygons, with constants $b_i$, determines the numerical solution $y_1$ (see the third picture of Fig. 1.2). This idea leads to the class of *explicit* Runge-Kutta methods, i.e., formula (1.4) below with $a_{ij} = 0$ for $i \leq j$.

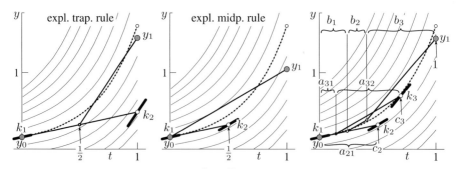

**Fig. 1.2.** Runge-Kutta methods for $\dot{y} = t^2 + y^2$, $y_0 = 0.46$, $h = 1$; dotted: exact solution.

Much more important for our purpose are *implicit* Runge-Kutta methods, introduced mainly in the work of Butcher (1963).

**Definition 1.1.** Let $b_i, a_{ij}$ ($i, j = 1, \ldots, s$) be real numbers and let $c_i = \sum_{j=1}^{s} a_{ij}$. An *s-stage Runge-Kutta method* is given by

$$k_i = f\left(t_0 + c_i h, y_0 + h \sum_{j=1}^{s} a_{ij} k_j\right), \quad i = 1, \ldots, s$$
$$y_1 = y_0 + h \sum_{i=1}^{s} b_i k_i. \tag{1.4}$$

Here we allow a full matrix $(a_{ij})$ of non-zero coefficients. In this case, the slopes $k_i$ can no longer be computed explicitly, and even do not necessarily exist. For example, for the problem set-up of Fig. 1.2 the implicit trapezoidal rule has no solution. However, the implicit function theorem assures that, for sufficiently small $h$, the nonlinear system (1.4) for the values $k_1, \ldots, k_s$ has a locally unique solution close to $k_i \approx f(t_0, y_0)$.

Since Butcher's work, the coefficients are usually displayed as follows:

$$\begin{array}{c|ccc} c_1 & a_{11} & \cdots & a_{1s} \\ \vdots & \vdots & & \vdots \\ c_s & a_{s1} & \cdots & a_{ss} \\ \hline & b_1 & \cdots & b_s \end{array} \tag{1.5}$$

**Definition 1.2.** A Runge-Kutta method (or a general one-step method) has *order p*, if for all sufficiently regular problems (1.1) the *local error* $y_1 - y(t_0 + h)$ satisfies

$$y_1 - y(t_0 + h) = \mathcal{O}(h^{p+1}) \quad \text{as} \quad h \to 0.$$

To check the order of a Runge Kutta method, one has to compute the Taylor series expansions of $y(t_0 + h)$ and $y_1$ around to $h = 0$. This leads to the following algebraic conditions for the coefficients for orders 1, 2, and 3:

$$\begin{array}{rll} & \sum_i b_i = 1 & \text{for order 1;} \\ \text{in addition} & \sum_i b_i c_i = 1/2 & \text{for order 2;} \\ \text{in addition} & \sum_i b_i c_i^2 = 1/3 & \\ \text{and} & \sum_{i,j} b_i a_{ij} c_j = 1/6 & \text{for order 3.} \end{array} \tag{1.6}$$

For higher orders, however, this problem represented a great challenge in the first half of the 20th century. We shall present an elegant theory in Sect. III.1 which allows order conditions to be derived.

Among the methods seen up to now, the explicit and implicit Euler methods

$$\begin{array}{c|c} 0 & \\ \hline & 1 \end{array} \qquad \begin{array}{c|c} 1 & 1 \\ \hline & 1 \end{array} \tag{1.7}$$

are of order 1, the implicit trapezoidal and midpoint rules as well as both methods
of Runge

$$
\begin{array}{c|cc}
0 & & \\
1 & 1/2 & 1/2 \\
\hline
 & 1/2 & 1/2
\end{array}
\qquad
\begin{array}{c|c}
0 & \\
1/2 & 1/2 \\
\hline
 & 1
\end{array}
\qquad
\begin{array}{c|cc}
0 & & \\
1 & 1 & \\
\hline
 & 1/2 & 1/2
\end{array}
\qquad
\begin{array}{c|cc}
0 & & \\
1/2 & 1/2 & \\
\hline
 & 0 & 1
\end{array}
$$

are of order 2. The most successful methods during more than half a century were
the 4th order methods of Kutta:

$$
\begin{array}{c|cccc}
0 & & & & \\
1/2 & 1/2 & & & \\
1/2 & 0 & 1/2 & & \\
1 & 0 & 0 & 1 & \\
\hline
 & 1/6 & 2/6 & 2/6 & 1/6
\end{array}
\qquad
\begin{array}{c|cccc}
0 & & & & \\
1/3 & 1/3 & & & \\
2/3 & -1/3 & 1 & & \\
1 & 1 & -1 & 1 & \\
\hline
 & 1/8 & 3/8 & 3/8 & 1/8
\end{array}
\qquad (1.8)
$$

### II.1.2 Collocation Methods

> The high speed computing machines make it possible to enjoy the advantages of intricate methods.   (P.C. Hammer & J.W. Hollingsworth 1955)

Collocation methods for ordinary differential equations have their origin, once again, in the implicit trapezoidal rule (1.2): Hammer & Hollingsworth (1955) discovered that this method can be interpreted as being generated by a *quadratic function* "which agrees in direction with that indicated by the differential equation at two points" $t_0$ and $t_1$ (see the picture to the right). This idea allows one to "see much-used methods in a new light" and allows various generalizations (Guillou & Soulé (1969), Wright (1970)). An interesting feature of collocation methods is that we not only get a discrete set of approximations, but also a *continuous approximation* to the solution.

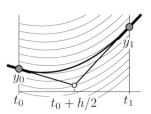

**Definition 1.3.** Let $c_1, \ldots, c_s$ be distinct real numbers (usually $0 \le c_i \le 1$). The *collocation polynomial* $u(t)$ is a polynomial of degree $s$ satisfying

$$
\begin{aligned}
u(t_0) &= y_0 \\
\dot{u}(t_0 + c_i h) &= f\big(t_0 + c_i h, u(t_0 + c_i h)\big), \quad i = 1, \ldots, s,
\end{aligned}
\qquad (1.9)
$$

and the numerical solution of the *collocation method* is defined by $y_1 = u(t_0 + h)$.

For $s = 1$, the polynomial has to be of the form $u(t) = y_0 + (t - t_0)k$ with

$$ k = f(t_0 + c_1 h, y_0 + h c_1 k). $$

We see that the explicit and implicit Euler methods and the midpoint rule are collocation methods with $c_1 = 0$, $c_1 = 1$ and $c_1 = 1/2$, respectively.

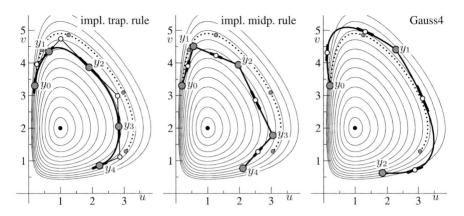

**Fig. 1.3.** Collocation solutions for the Lotka-Volterra problem (I.1.1); $u_0 = 0.2$, $v_0 = 3.3$; methods of order 2: four steps with $h = 0.4$; method of order 4: two steps with $h = 0.8$; dotted: exact solution.

For $s = 2$ and $c_1 = 0, c_2 = 1$ we find, of course, the implicit trapezoidal rule. The choice of Hammer & Hollingsworth for the collocation points is $c_{1,2} = 1/2 \pm \sqrt{3}/6$, the *Gaussian quadrature nodes* (see the picture to the right). We will see that the corresponding method is of order 4.

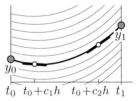

In Fig. 1.3 we illustrate the collocation idea with these methods for the Lotka-Volterra problem (I.1.1). One can observe that, in spite of the extremely large step sizes, the methods are quite satisfactory.

**Theorem 1.4 (Guillou & Soulé 1969, Wright 1970).** *The collocation method of Definition 1.3 is equivalent to the $s$-stage Runge-Kutta method (1.4) with coefficients*

$$a_{ij} = \int_0^{c_i} \ell_j(\tau)\,d\tau, \qquad b_i = \int_0^1 \ell_i(\tau)\,d\tau, \tag{1.10}$$

*where $\ell_i(\tau)$ is the Lagrange polynomial $\ell_i(\tau) = \prod_{l \neq i}(\tau - c_l)/(c_i - c_l)$.*

*Proof.* Let $u(t)$ be the collocation polynomial and define

$$k_i := \dot{u}(t_0 + c_i h).$$

By the Lagrange interpolation formula we have $\dot{u}(t_0 + \tau h) = \sum_{j=1}^{s} k_j \cdot \ell_j(\tau)$, and by integration we get

$$u(t_0 + c_i h) = y_0 + h \sum_{j=1}^{s} k_j \int_0^{c_i} \ell_j(\tau)\,d\tau.$$

Inserted into (1.9) this gives the first formula of the Runge-Kutta equation (1.4). Integration from 0 to 1 yields the second one. □

The above proof can also be read in reverse order. This shows that a Runge-Kutta method with coefficients given by (1.10) can be interpreted as a collocation method. Since $\tau^{k-1} = \sum_{j=1}^{s} c_j^{k-1} \ell_j(\tau)$ for $k = 1, \ldots, s$, the relations (1.10) are equivalent to the linear systems

$$
\begin{aligned}
C(q) : & \sum_{j=1}^{s} a_{ij} c_j^{k-1} = \frac{c_i^k}{k}, \quad k = 1, \ldots, q, \text{ all } i \\
B(p) : & \sum_{i=1}^{s} b_i c_i^{k-1} = \frac{1}{k}, \quad k = 1, \ldots, p,
\end{aligned}
\tag{1.11}
$$

with $q = s$ and $p = s$. What is the order of a Runge-Kutta method whose coefficients $b_i, a_{ij}$ are determined in this way?

Compared to the enormous difficulties that the first explorers had in constructing Runge-Kutta methods of orders 5 and 6, and also compared to the difficult algebraic proofs of the first papers of Butcher, the following general theorem and its proof, discovered in this form by Guillou & Soulé (1969), are surprisingly simple.

**Theorem 1.5 (Superconvergence).** *If the condition $B(p)$ holds for some $p \geq s$, then the collocation method (Definition 1.3) has order $p$. This means that the collocation method has the same order as the underlying quadrature formula.*

*Proof.* We consider the collocation polynomial $u(t)$ as the solution of a perturbed differential equation

$$\dot{u} = f(t, u) + \delta(t) \tag{1.12}$$

with defect $\delta(t) := \dot{u}(t) - f(t, u(t))$. Subtracting (1.1) from (1.12) we get after linearization that

$$\dot{u}(t) - \dot{y}(t) = \frac{\partial f}{\partial y}(t, y(t))\Big(u(t) - y(t)\Big) + \delta(t) + r(t), \tag{1.13}$$

where, for $t_0 \leq t \leq t_0 + h$, the remainder $r(t)$ is of size $\mathcal{O}(\|u(t) - y(t)\|^2) = \mathcal{O}(h^{2s+2})$ by Lemma 1.6 below. The variation of constants formula (see e.g., Hairer, Nørsett & Wanner (1993), p. 66) then yields

$$y_1 - y(t_0+h) = u(t_0+h) - y(t_0+h) = \int_{t_0}^{t_0+h} R(t_0+h, s)\Big(\delta(s) + r(s)\Big) ds, \tag{1.14}$$

where $R(t, s)$ is the resolvent of the homogeneous part of the differential equation (1.13), i.e., the solution of the matrix differential equation $\partial R(t,s)/\partial t = A(t) R(t, s)$, $R(s, s) = I$, with $A(t) = \partial f/\partial y(t, y(t))$. The integral over $R(t_0 + h, s) r(s)$ gives a $\mathcal{O}(h^{2s+3})$ contribution. The main idea now is to apply the quadrature formula $(b_i, c_i)_{i=1}^{s}$ to the integral over $g(s) = R(t_0 + h, s)\delta(s)$; because the defect $\delta(s)$ vanishes at the collocation points $t_0 + c_i h$ for $i = 1, \ldots, s$, this gives zero as the numerical result. Thus, the integral is equal to the quadrature error, which is bounded by $h^{p+1}$ times a bound of the $p$th derivative of the function $g(s)$. This derivative is bounded independently of $h$, because by Lemma 1.6 all derivatives of the collocation polynomial are bounded uniformly as $h \to 0$. Since, anyway, $p \leq 2s$, we get $y_1 - y(t_0 + h) = \mathcal{O}(h^{p+1})$ from (1.14). □

**Lemma 1.6.** *The collocation polynomial $u(t)$ is an approximation of order $s$ to the exact solution of (1.1) on the whole interval, i.e.,*

$$\|u(t) - y(t)\| \leq C \cdot h^{s+1} \qquad \text{for} \quad t \in [t_0, t_0 + h] \tag{1.15}$$

*and for sufficiently small $h$.*
  *Moreover, the derivatives of $u(t)$ satisfy for $t \in [t_0, t_0 + h]$*

$$\|u^{(k)}(t) - y^{(k)}(t)\| \leq C \cdot h^{s+1-k} \qquad \text{for} \quad k = 0, \ldots, s.$$

*Proof.* The collocation polynomial satisfies

$$\dot{u}(t_0 + \tau h) = \sum_{i=1}^{s} f\big(t_0 + c_i h, u(t_0 + c_i h)\big) \ell_i(\tau),$$

while the exact solution of (1.1) satisfies

$$\dot{y}(t_0 + \tau h) = \sum_{i=1}^{s} f\big(t_0 + c_i h, y(t_0 + c_i h)\big) \ell_i(\tau) + h^s E(\tau, h),$$

where the interpolation error $E(\tau, h)$ is bounded by $\max_{t \in [t_0, t_0+h]} \|y^{(s+1)}(t)\|/s!$ and its derivatives satisfy

$$\|E^{(k-1)}(\tau, h)\| \leq \max_{t \in [t_0, t_0+h]} \frac{\|y^{(s+1)}(t)\|}{(s-k+1)!}.$$

This follows from the fact that, by Rolle's theorem, the differentiated polynomial $\sum_{i=1}^{s} f(t_0 + c_i h, y(t_0 + c_i h)) \ell_i^{(k-1)}(\tau)$ can be interpreted as the interpolation polynomial of $h^{k-1} y^{(k)}(t_0 + \tau h)$ at $s-k+1$ points lying in $[t_0, t_0 + h]$. Integrating the difference of the above two equations gives

$$y(t_0 + \tau h) - u(t_0 + \tau h) = h \sum_{i=1}^{s} \Delta f_i \int_0^{\tau} \ell_i(\sigma)\, d\sigma + h^{s+1} \int_0^{\tau} E(\sigma, h)\, d\sigma \tag{1.16}$$

with $\Delta f_i = f\big(t_0 + c_i h, y(t_0 + c_i h)\big) - f\big(t_0 + c_i h, u(t_0 + c_i h)\big)$. Using a Lipschitz condition for $f(t, y)$, this relation yields

$$\max_{t \in [t_0, t_0+h]} \|y(t) - u(t)\| \leq h C L \max_{t \in [t_0, t_0+h]} \|y(t) - u(t)\| + \text{Const} \cdot h^{s+1},$$

implying the statement (1.15) for sufficiently small $h > 0$.
  The proof of the second statement follows from

$$h^k \Big( y^{(k)}(t_0 + \tau h) - u^{(k)}(t_0 + \tau h) \Big) = h \sum_{i=1}^{s} \Delta f_i\, \ell_i^{(k-1)}(\tau) + h^{s+1} E^{(k-1)}(\tau, h)$$

by using a Lipschitz condition for $f(t, y)$ and the estimate (1.15). □

## II.1.3 Gauss and Lobatto Collocation

**Gauss Methods.** If we take $c_1, \ldots, c_s$ as the zeros of the $s$th shifted Legendre polynomial

$$\frac{d^s}{dx^s}\left(x^s(x-1)^s\right),$$

the interpolatory quadrature formula has order $p = 2s$, and by Theorem 1.5, the Runge-Kutta (or collocation) method based on these nodes has the same order $2s$. For $s = 1$ we obtain the implicit midpoint rule. The Runge-Kutta coefficients for $s = 2$ (the method of Hammer & Hollingsworth 1955) and $s = 3$ are given in Table 1.1. The proof of the order properties for general $s$ was a sensational result of Butcher (1964a). At that time these methods were considered, at least by the editors of *Math. of Comput.*, to be purely academic without any practical value; 5 years later their $A$-stability was discovered, 12 years later their $B$-stability, and 25 years later their symplecticity. Thus, of all the papers in issue No. 85 of *Math. of Comput.*, the one most important to us is the one for which publication was the most difficult.

**Table 1.1.** Gauss methods of order 4 and 6

| | | |
|---|---|---|
| $\frac{1}{2} - \frac{\sqrt{3}}{6}$ | $\frac{1}{4}$ | $\frac{1}{4} - \frac{\sqrt{3}}{6}$ |
| $\frac{1}{2} + \frac{\sqrt{3}}{6}$ | $\frac{1}{4} + \frac{\sqrt{3}}{6}$ | $\frac{1}{4}$ |
| | $\frac{1}{2}$ | $\frac{1}{2}$ |

| | | | |
|---|---|---|---|
| $\frac{1}{2} - \frac{\sqrt{15}}{10}$ | $\frac{5}{36}$ | $\frac{2}{9} - \frac{\sqrt{15}}{15}$ | $\frac{5}{36} - \frac{\sqrt{15}}{30}$ |
| $\frac{1}{2}$ | $\frac{5}{36} + \frac{\sqrt{15}}{24}$ | $\frac{2}{9}$ | $\frac{5}{36} - \frac{\sqrt{15}}{24}$ |
| $\frac{1}{2} + \frac{\sqrt{15}}{10}$ | $\frac{5}{36} + \frac{\sqrt{15}}{30}$ | $\frac{2}{9} + \frac{\sqrt{15}}{15}$ | $\frac{5}{36}$ |
| | $\frac{5}{18}$ | $\frac{4}{9}$ | $\frac{5}{18}$ |

**Radau Methods.** Radau quadrature formulas have the highest possible order, $2s - 1$, among quadrature formulas with either $c_1 = 0$ or $c_s = 1$. The corresponding collocation methods for $c_s = 1$ are called Radau IIA methods. They play an important role in the integration of stiff differential equations (see Hairer & Wanner (1996), Sect. IV.8). However, they lack both *symmetry* and *symplecticity*, properties that will be the subjects of later chapters in this book.

**Lobatto IIIA Methods.** Lobatto quadrature formulas have the highest possible order with $c_1 = 0$ and $c_s = 1$. Under these conditions, the nodes must be the zeros of

$$\frac{d^{s-2}}{dx^{s-2}}\left(x^{s-1}(x-1)^{s-1}\right) \qquad (1.17)$$

and the quadrature order is $p = 2s - 2$. The corresponding collocation methods are called, for historical reasons, Lobatto IIIA methods. For $s = 2$ we have the implicit trapezoidal rule. The coefficients for $s = 3$ and $s = 4$ are given in Table 1.2.

**Table 1.2.** Lobatto IIIA methods of order 4 and 6

| 0 | 0 | 0 | 0 |
|---|---|---|---|
| $\frac{1}{2}$ | $\frac{5}{24}$ | $\frac{1}{3}$ | $-\frac{1}{24}$ |
| 1 | $\frac{1}{6}$ | $\frac{2}{3}$ | $\frac{1}{6}$ |
| | $\frac{1}{6}$ | $\frac{2}{3}$ | $\frac{1}{6}$ |

| 0 | 0 | 0 | 0 | 0 |
|---|---|---|---|---|
| $\frac{5-\sqrt{5}}{10}$ | $\frac{11+\sqrt{5}}{120}$ | $\frac{25-\sqrt{5}}{120}$ | $\frac{25-13\sqrt{5}}{120}$ | $\frac{-1+\sqrt{5}}{120}$ |
| $\frac{5+\sqrt{5}}{10}$ | $\frac{11-\sqrt{5}}{120}$ | $\frac{25+13\sqrt{5}}{120}$ | $\frac{25+\sqrt{5}}{120}$ | $\frac{-1-\sqrt{5}}{120}$ |
| 1 | $\frac{1}{12}$ | $\frac{5}{12}$ | $\frac{5}{12}$ | $\frac{1}{12}$ |
| | $\frac{1}{12}$ | $\frac{5}{12}$ | $\frac{5}{12}$ | $\frac{1}{12}$ |

### II.1.4 Discontinuous Collocation Methods

Collocation methods allow, as we have seen above, a very elegant proof of their order properties. By similar ideas, they also admit strikingly simple proofs for their $A$- and $B$-stability as well as for symplecticity, our subject in Chap. VI. However, not all method classes are of collocation type. It is therefore interesting to define a modification of the collocation idea, which allows us to extend all the above proofs to much wider classes of methods. This definition will also lead, later, to important classes of *partitioned* methods.

**Definition 1.7.** Let $c_2, \ldots, c_{s-1}$ be distinct real numbers (usually $0 \leq c_i \leq 1$), and let $b_1, b_s$ be two arbitrary real numbers. The corresponding *discontinuous collocation method* is then defined via a polynomial of degree $s - 2$ satisfying

$$u(t_0) = y_0 - hb_1\big(\dot{u}(t_0) - f(t_0, u(t_0))\big)$$
$$\dot{u}(t_0 + c_ih) = f\big(t_0 + c_ih, u(t_0 + c_ih)\big), \quad i = 2, \ldots, s-1, \qquad (1.18)$$
$$y_1 = u(t_1) - hb_s\big(\dot{u}(t_1) - f(t_1, u(t_1))\big).$$

The figure gives a geometric interpretation of the correction term in the first and third formulas of (1.18). The motivation for this definition will become clear in the proof of Theorem 1.9 below. Our first result shows that discontinuous collocation methods are equivalent to implicit Runge-Kutta methods.

**Theorem 1.8.** *The discontinuous collocation method of Definition 1.7 is equivalent to an s-stage Runge-Kutta method (1.4) with coefficients determined by $c_1 = 0$, $c_s = 1$, and*

$$a_{i1} = b_1, \qquad a_{is} = 0 \qquad \text{for } i = 1, \ldots, s, \tag{1.19}$$
$$C(s-2) \qquad \text{and} \qquad B(s-2),$$

*with the conditions $C(q)$ and $B(p)$ of (1.11).*

*Proof.* As in the proof of Theorem 1.4 we put $k_i := \dot{u}(t_0 + c_i h)$ (this time for $i = 2, \ldots, s-1$), so that $\dot{u}(t_0 + \tau h) = \sum_{j=2}^{s-1} k_j \cdot \ell_j(\tau)$ by the Lagrange interpolation formula. Here, $\ell_j(\tau)$ corresponds to $c_2, \ldots, c_{s-1}$ and is a polynomial of degree $s-3$. By integration and using the definition of $u(t_0)$ we get

$$\begin{aligned} u(t_0 + c_i h) &= u(t_0) + h \sum_{j=2}^{s-1} k_j \int_0^{c_i} \ell_j(\tau) \, d\tau \\ &= y_0 + h b_1 k_1 + h \sum_{j=2}^{s-1} k_j \left( \int_0^{c_i} \ell_j(\tau) \, d\tau - b_1 \ell_j(0) \right) \end{aligned}$$

with $k_1 = f(y_0)$. Inserted into (1.18) this gives the first formula of the Runge-Kutta equation (1.4) with $a_{ij} = \int_0^{c_i} \ell_j(\tau) \, d\tau - b_1 \ell_j(0)$. As for collocation methods, one checks that the $a_{ij}$ are uniquely determined by the condition $C(s-2)$. The formula for $y_1$ is obtained similarly. □

**Table 1.3.** Survey of discontinuous collocation methods

| type | characteristics | prominent examples |
|---|---|---|
| $b_1 = 0$, $b_s = 0$ | $(s-2)$-stage collocation | Gauss, Radau IIA, Lobatto IIIA |
| $b_1 = 0$, $b_s \neq 0$ | $(s-1)$-stage with $a_{is} = 0$ | methods of Butcher (1964b) |
| $b_1 \neq 0$, $b_s = 0$ | $(s-1)$-stage with $a_{i1} = b_1$ | Radau IA, Lobatto IIIC |
| $b_1 \neq 0$, $b_s \neq 0$ | $s$-stage with $a_{i1} = b_1$, $a_{is} = 0$ | Lobatto IIIB |

If $b_1 = 0$ in Definition 1.7, the entire first column in the Runge-Kutta tableau vanishes, so that the first stage can be removed, which leads to an equivalent method with $s-1$ stages. Similarly, if $b_s = 0$, we can remove the last stage. Therefore, we have all classes of methods, which are "continuous" either to the left, or to the right, or on both sides, as special cases in our definition.

In the case where $b_1 = b_s = 0$, the discontinuous collocation method (1.18) is equivalent to the $(s-2)$-stage collocation method based on $c_2, \ldots, c_{s-1}$ (see Table 1.3). The methods with $b_s = 0$ but $b_1 \neq 0$, which include the Radau IA and

**Table 1.4.** Lobatto IIIB methods of order 4 and 6

| 0 | $\frac{1}{6}$ | $-\frac{1}{6}$ | 0 |
|---|---|---|---|
| $\frac{1}{2}$ | $\frac{1}{6}$ | $\frac{1}{3}$ | 0 |
| 1 | $\frac{1}{6}$ | $\frac{5}{6}$ | 0 |
|  | $\frac{1}{6}$ | $\frac{2}{3}$ | $\frac{1}{6}$ |

| 0 | $\frac{1}{12}$ | $\frac{-1-\sqrt{5}}{24}$ | $\frac{-1+\sqrt{5}}{24}$ | 0 |
|---|---|---|---|---|
| $\frac{5-\sqrt{5}}{10}$ | $\frac{1}{12}$ | $\frac{25+\sqrt{5}}{120}$ | $\frac{25-13\sqrt{5}}{120}$ | 0 |
| $\frac{5+\sqrt{5}}{10}$ | $\frac{1}{12}$ | $\frac{25+13\sqrt{5}}{120}$ | $\frac{25-\sqrt{5}}{120}$ | 0 |
| 1 | $\frac{1}{12}$ | $\frac{11-\sqrt{5}}{24}$ | $\frac{11+\sqrt{5}}{24}$ | 0 |
|  | $\frac{1}{12}$ | $\frac{5}{12}$ | $\frac{5}{12}$ | $\frac{1}{12}$ |

Lobatto IIIC methods, are of interest for the solution of stiff differential equations (Hairer & Wanner 1996). The methods with $b_1 = 0$ but $b_s \neq 0$, introduced by Butcher (1964a, 1964b), are of historical interest. They were thought to be computationally attractive, because their last stage is explicit. In the context of geometric integration, much more important are methods for which both $b_1 \neq 0$ and $b_s \neq 0$.

**Lobatto IIIB Methods** (Table 1.4). We consider the quadrature formulas whose nodes are the zeros of (1.17). We have $c_1 = 0$ and $c_s = 1$. Based on $c_2, \ldots, c_{s-1}$ and $b_1, b_s$ we consider the discontinuous collocation method. This class of methods is called Lobatto IIIB (Ehle 1969), and it plays an important role in geometric integration in conjunction with the Lobatto IIIA methods of Sect. II.1.3 (see Theorem IV.2.3 and Theorem VI.4.5). These methods are of order $2s-2$, as the following result shows.

**Theorem 1.9 (Superconvergence).** *The discontinuous collocation method of Definition 1.7 has the same order as the underlying quadrature formula.*

*Proof.* We follow the lines of the proof of Theorem 1.5. With the polynomial $u(t)$ of Definition 1.7, and with the defect

$$\delta(t) := \dot{u}(t) - f(t, u(t))$$

we get (1.13) after linearization. The variation of constants formula then yields

$$u(t_0 + h) - y(t_0 + h) = R(t_0 + h, t_0)(u(t_0) - y_0) + \int_{t_0}^{t_0+h} R(t_0 + h, s)\big(\delta(s) + r(s)\big) ds,$$

which corresponds to (1.14) if $u(t_0) = y_0$. As a consequence of Lemma 1.10 below (with $k = 0$), the integral over $R(t_0 + h, s)r(s)$ gives a $\mathcal{O}(h^{2s-1})$ contribution. Since the defect $\delta(t_0 + c_i h)$ vanishes only for $i = 2, \ldots, s-1$, an application of the quadrature formula to $R(t_0 + h, s)\delta(s)$ yields $hb_1 R(t_0 + h, t_0)\delta(t_0) + hb_s \delta(t_0 + h)$ in addition to the quadrature error, which is $\mathcal{O}(h^{p+1})$. Collecting terms suitably, we obtain

$$u(t_1) - hb_s\delta(t_1) - y(t_1) = R(t_1, t_0)\big(u(t_0) + hb_1\delta(t_0) - y_0\big)$$
$$+ \mathcal{O}(h^{p+1}) + \mathcal{O}(h^{2s-1}),$$

which, after using the definitions of $u(t_0)$ and $u(t_1)$, proves $y_1 - y(t_1) = \mathcal{O}(h^{p+1}) + \mathcal{O}(h^{2s-1})$. □

**Lemma 1.10.** *The polynomial $u(t)$ of the discontinuous collocation method (1.18) satisfies for $t \in [t_0, t_0 + h]$ and for sufficiently small $h$*
$$\|u^{(k)}(t) - y^{(k)}(t)\| \leq C \cdot h^{s-1-k} \quad \text{for} \quad k = 0, \ldots, s-2.$$

*Proof.* The proof is essentially the same as that for Lemma 1.6. In the formulas for $\dot{u}(t_0 + \tau h)$ and $\dot{y}(t_0 + \tau h)$, the sum has to be taken from $i = 2$ to $i = s - 1$. Moreover, all $h^s$ become $h^{s-2}$. In (1.16) one has an additional term
$$y_0 - u(t_0) = hb_1\big(\dot{u}(t_0) - f(t_0, u(t_0))\big),$$
which, however, is just an interpolation error of size $\mathcal{O}(h^{s-1})$ and can be included in $Const \cdot h^{s-1}$. □

## II.2 Partitioned Runge-Kutta Methods

Some interesting numerical methods introduced in Chap. I (symplectic Euler and the Störmer/Verlet method) do not belong to the class of Runge-Kutta methods. They are important examples of so-called partitioned Runge-Kutta methods. In this section we consider differential equations in the partitioned form

$$\dot{y} = f(y, z), \qquad \dot{z} = g(y, z), \tag{2.1}$$

where $y$ and $z$ may be vectors of different dimensions.

### II.2.1 Definition and First Examples

The idea is to take two different Runge-Kutta methods, and to treat the $y$-variables with the first method $(a_{ij}, b_i)$, and the $z$-variables with the second method $(\widehat{a}_{ij}, \widehat{b}_i)$.

**Definition 2.1.** Let $b_i, a_{ij}$ and $\widehat{b}_i, \widehat{a}_{ij}$ be the coefficients of two Runge-Kutta methods. A *partitioned Runge-Kutta method* for the solution of (2.1) is given by

$$\begin{aligned}
k_i &= f\Big(y_0 + h\sum_{j=1}^s a_{ij}k_j,\ z_0 + h\sum_{j=1}^s \widehat{a}_{ij}\ell_j\Big), \\
\ell_i &= g\Big(y_0 + h\sum_{j=1}^s a_{ij}k_j,\ z_0 + h\sum_{j=1}^s \widehat{a}_{ij}\ell_j\Big), \\
y_1 &= y_0 + h\sum_{i=1}^s b_i k_i, \qquad z_1 = z_0 + h\sum_{i=1}^s \widehat{b}_i \ell_i.
\end{aligned} \tag{2.2}$$

Methods of this type were originally proposed by Hofer in 1976 and by Griepentrog in 1978 for problems with stiff and nonstiff parts (see Hairer, Nørsett & Wanner (1993), Sect. II.15). Their importance for Hamiltonian systems (see the examples of Chap. I) has been discovered only in the last decade.

An interesting example is the symplectic Euler method (I.1.9), where the implicit Euler method $b_1 = 1, a_{11} = 1$ is combined with the explicit Euler method $\widehat{b}_1 = 1, \widehat{a}_{11} = 0$. The Störmer/Verlet method (I.3.6) is of the form (2.2) with coefficients given in Table 2.1.

**Table 2.1.** Störmer/Verlet as a partitioned Runge-Kutta method

| 0 | 0 | 0 | | 1/2 | 1/2 | 0 |
|---|---|---|---|---|---|---|
| 1 | 1/2 | 1/2 | | 1/2 | 1/2 | 0 |
|   | 1/2 | 1/2 | |     | 1/2 | 1/2 |

The theory of Runge-Kutta methods can be extended in a straightforward manner to partitioned methods. Since (2.2) is a one-step method $(y_1, z_1) = \Phi_h(y_0, z_0)$, the Definition 1.2 of the order applies directly. Considering problems $\dot{y} = f(y)$, $\dot{z} = g(z)$ without any coupling terms, we see that the order of (2.2) cannot exceed $\min(p, \widehat{p})$, where $p$ and $\widehat{p}$ are the orders of the two methods.

**Conditions for Order Two.** Expanding the exact solution of (2.1) and the numerical solution (2.2) into Taylor series, we see that the method is of order 2 if the coupling conditions

$$\sum_{ij} b_i \widehat{a}_{ij} = 1/2, \qquad \sum_{ij} \widehat{b}_i a_{ij} = 1/2 \qquad (2.3)$$

are satisfied in addition to the usual Runge-Kutta order conditions for order 2. The method of Table 2.1 satisfies these conditions, and it is therefore of order 2. We also remark that (2.3) is automatically satisfied by partitioned methods that are based on the same quadrature nodes, i.e.,

$$c_i = \widehat{c}_i \qquad \text{for all } i \qquad (2.4)$$

where, as usual, $c_i = \sum_j a_{ij}$ and $\widehat{c}_i = \sum_j \widehat{a}_{ij}$.

**Conditions for Order Three.** The conditions for order three already become quite complicated, unless (2.4) is satisfied. In this case, we obtain the additional conditions

$$\sum_{ij} b_i \widehat{a}_{ij} c_j = 1/6, \qquad \sum_{ij} \widehat{b}_i a_{ij} c_j = 1/6. \qquad (2.5)$$

The order conditions for higher order will be discussed in Sect. III.2.2. It turns out that the number of coupling conditions increases very fast with order, and the proofs for high order are often very cumbersome. There is, however, a very elegant proof of the order for the partitioned method which is the most important one in connection with 'geometric integration', as we shall see now.

## II.2.2 Lobatto IIIA - IIIB Pairs

These methods generalize the Störmer/Verlet method to arbitrary order. Indeed, the left method of Table 2.1 is the trapezoidal rule, which is the Lobatto IIIA method with $s = 2$, and the method to the right is equivalent to the midpoint rule and, apart from the values of the $c_i$, is the Lobatto IIIB method with $s = 2$. Sun (1993b) and Jay (1996) discovered that for general $s$ the combination of the Lobatto IIIA and IIIB methods are suitable for Hamiltonian systems. The coefficients of the methods for $s = 3$ are given in Table 2.2. Using the idea of discontinuous collocation, we give a direct proof of the order for this pair of methods.

**Table 2.2.** Coefficients of the 3-stage Lobatto IIIA - IIIB pair

| | | | | | | | | |
|---|---|---|---|---|---|---|---|---|
| 0   | 0    | 0   | 0     | | 0   | 1/6 | −1/6 | 0 |
| 1/2 | 5/24 | 1/3 | −1/24 | | 1/2 | 1/6 | 1/3  | 0 |
| 1   | 1/6  | 2/3 | 1/6   | | 1   | 1/6 | 5/6  | 0 |
|     | 1/6  | 2/3 | 1/6   | |     | 1/6 | 2/3  | 1/6 |

**Theorem 2.2.** *The partitioned Runge-Kutta method composed of the $s$-stage Lobatto IIIA and the $s$-stage Lobatto IIIB method, is of order $2s - 2$.*

*Proof.* Let $c_1 = 0, c_2, \ldots, c_{s-1}, c_s = 1$ and $b_1, \ldots, b_s$ be the nodes and weights of the Lobatto quadrature. The partitioned Runge-Kutta method based on the Lobatto IIIA - IIIB pair can be interpreted as the discontinuous collocation method

$$\begin{aligned}
u(t_0) &= y_0 \\
v(t_0) &= z_0 - hb_1\big(\dot{v}(t_0) - g(u(t_0), v(t_0))\big) \\
\dot{u}(t_0 + c_i h) &= f\big(u(t_0 + c_i h), v(t_0 + c_i h)\big), & i &= 1, \ldots, s \\
\dot{v}(t_0 + c_i h) &= g\big(u(t_0 + c_i h), v(t_0 + c_i h)\big), & i &= 2, \ldots, s-1 \\
y_1 &= u(t_1) \\
z_1 &= v(t_1) - hb_s\big(\dot{v}(t_1) - g(u(t_1), v(t_1))\big),
\end{aligned} \quad (2.6)$$

where $u(t)$ and $v(t)$ are polynomials of degree $s$ and $s-2$, respectively. This is seen as in the proofs of Theorem 1.4 and Theorem 1.8. The superconvergence (order $2s - 2$) is obtained with exactly the same proof as for Theorem 1.9, where the functions $u(t)$ and $y(t)$ have to be replaced with $(u(t), v(t))^T$ and $(y(t), z(t))^T$, etc. Instead of Lemma 1.10 we use the estimates (for $t \in [t_0, t_0 + h]$)

$$\begin{aligned}
\|u^{(k)}(t) - y^{(k)}(t)\| &\le c \cdot h^{s-k} & \text{for} \quad k &= 0, \ldots, s, \\
\|v^{(k)}(t) - z^{(k)}(t)\| &\le c \cdot h^{s-1-k} & \text{for} \quad k &= 0, \ldots, s-2,
\end{aligned}$$

which can be proved by following the lines of the proofs of Lemma 1.6 and Lemma 1.10. □

## II.2.3 Nyström Methods

> Da bis jetzt die *direkte* Anwendung der Rungeschen Methode auf den wichtigen Fall von Differentialgleichungen zweiter Ordnung nicht behandelt war ...
> (E.J. Nyström 1925)

Second-order differential equations

$$\ddot{y} = g(t, y, \dot{y}) \qquad (2.7)$$

form an important class of problems. Most of the differential equations in Chap. I are of this form (e.g., the Kepler problem, the outer solar system, problems in molecular dynamics). This is mainly due to Newton's law that forces are proportional to second derivatives (acceleration). Introducing a new variable $z = \dot{y}$ for the first derivative, the problem (2.7) becomes equivalent to the partitioned system

$$\dot{y} = z, \qquad \dot{z} = g(t, y, z). \qquad (2.8)$$

A partitioned Runge-Kutta method (2.2) applied to this system yields

$$\begin{aligned} k_i &= z_0 + h \sum_{j=1}^{s} \widehat{a}_{ij} \ell_j, \\ \ell_i &= g\Big(t_0 + c_i h, y_0 + h \sum_{j=1}^{s} a_{ij} k_j, z_0 + h \sum_{j=1}^{s} \widehat{a}_{ij} \ell_j \Big), \\ y_1 &= y_0 + h \sum_{i=1}^{s} b_i k_i, \qquad z_1 = z_0 + h \sum_{i=1}^{s} \widehat{b}_i \ell_i. \end{aligned} \qquad (2.9)$$

If we insert the formula for $k_i$ into the others, we obtain Definition 2.3 with

$$\overline{a}_{ij} = \sum_{k=1}^{s} a_{ik} \widehat{a}_{kj}, \qquad \overline{b}_i = \sum_{k=1}^{s} b_k \widehat{a}_{ki}. \qquad (2.10)$$

**Definition 2.3.** Let $c_i, \overline{b}_i, \overline{a}_{ij}$ and $\widehat{b}_i, \widehat{a}_{ij}$ be real coefficients. A *Nyström method* for the solution of (2.7) is given by

$$\begin{aligned} \ell_i &= g\Big(t_0 + c_i h, y_0 + c_i h \dot{y}_0 + h^2 \sum_{j=1}^{s} \overline{a}_{ij} \ell_j, \dot{y}_0 + h \sum_{j=1}^{s} \widehat{a}_{ij} \ell_j \Big), \\ y_1 &= y_0 + h \dot{y}_0 + h^2 \sum_{i=1}^{s} \overline{b}_i \ell_i, \qquad \dot{y}_1 = \dot{y}_0 + h \sum_{i=1}^{s} \widehat{b}_i \ell_i. \end{aligned} \qquad (2.11)$$

For the important special case $\ddot{y} = g(t, y)$, where the vector field does not depend on the velocity, the coefficients $\widehat{a}_{ij}$ need not be specified. A Nyström method is of order $p$ if $y_1 - y(t_0 + h) = \mathcal{O}(h^{p+1})$ and $\dot{y}_1 - \dot{y}(t_0 + h) = \mathcal{O}(h^{p+1})$. It is not sufficient to consider $y_1$ alone. The order conditions will be discussed in Sect. III.2.3.

Notice that the Störmer/Verlet scheme (I.3.6) is a Nyström method for problems of the special form $\ddot{y} = g(t, y)$. We have $s = 2$, and the coefficients are $c_1 = 0, c_2 = 1, \overline{a}_{11} = \overline{a}_{12} = \overline{a}_{22} = 0, \overline{a}_{21} = 1/2, \overline{b}_1 = 1/2, \overline{b}_2 = 0$, and $\widehat{b}_1 = \widehat{b}_2 = 1/2$. With $q_{n+1/2} = q_n + \frac{h}{2} v_{n+1/2}$ the step $(q_{n-1/2}, v_{n-1/2}) \mapsto (q_{n+1/2}, v_{n+1/2})$ of (I.3.6) becomes a one-stage Nyström method with $c_1 = 1/2, \overline{a}_{11} = 0, \overline{b}_1 = \widehat{b}_1 = 1$.

## II.3 The Adjoint of a Method

We shall see in Chap. V that *symmetric* numerical methods have many important properties. The key for understanding symmetry is the concept of the *adjoint* method.

The flow $\varphi_t$ of an autonomous differential equation

$$\dot{y} = f(y), \qquad y(t_0) = y_0 \qquad (3.1)$$

satisfies $\varphi_{-t}^{-1} = \varphi_t$. This property is *not*, in general, shared by the one-step map $\Phi_h$ of a numerical method. An illustration is presented in the upper picture of Fig. 3.1 (a), where we see that the one-step map $\Phi_h$ for the explicit Euler method is different from the inverse of $\Phi_{-h}$, which is the implicit Euler method.

**Definition 3.1.** The *adjoint method* $\Phi_h^*$ of a method $\Phi_h$ is the inverse map of the original method with reversed time step $-h$, i.e.,

$$\Phi_h^* := \Phi_{-h}^{-1} \qquad (3.2)$$

(see Fig. 3.1 (b)). In other words, $y_1 = \Phi_h^*(y_0)$ is implicitly defined by $\Phi_{-h}(y_1) = y_0$. A method for which $\Phi_h^* = \Phi_h$ is called *symmetric*.

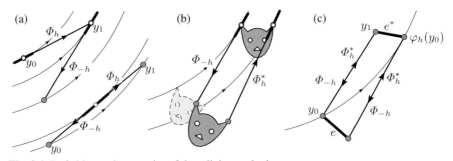

**Fig. 3.1.** Definition and properties of the adjoint method

The consideration of adjoint methods evolved independently from the study of symmetric integrators (Stetter (1973), p. 125, Wanner (1973)) and from the aim of constructing and analyzing stiff integrators from explicit ones (Cash (1975) calls them "the backward version" which were the first example of mono-implicit methods and Scherer (1977) calls them "reflected methods").

The adjoint method satisfies the usual properties such as $(\Phi_h^*)^* = \Phi_h$ and $(\Phi_h \circ \Psi_h)^* = \Psi_h^* \circ \Phi_h^*$ for any two one-step methods $\Phi_h$ and $\Psi_h$. The implicit Euler method is the adjoint of the explicit Euler method. The implicit midpoint rule is symmetric (see the lower picture of Fig. 3.1 (a)), and the trapezoidal rule and the Störmer/Verlet method are also symmetric.

The following theorem shows that the adjoint method has the same order as the original method, and, with a possible sign change, also the same leading error term.

**Theorem 3.2.** *Let $\varphi_t$ be the exact flow of (3.1) and let $\Phi_h$ be a one-step method of order $p$ satisfying*

$$\Phi_h(y_0) = \varphi_h(y_0) + C(y_0)h^{p+1} + \mathcal{O}(h^{p+2}). \tag{3.3}$$

The adjoint method $\Phi_h^*$ then has the same order $p$ and we have

$$\Phi_h^*(y_0) = \varphi_h(y_0) + (-1)^p C(y_0)h^{p+1} + \mathcal{O}(h^{p+2}). \tag{3.4}$$

*If the method is symmetric, its (maximal) order is even.*

*Proof.* The idea of the proof is exhibited in drawing (c) of Fig. 3.1. From a given initial value $y_0$ we compute $\varphi_h(y_0)$ and $y_1 = \Phi_h^*(y_0)$, whose difference $e^*$ is the local error of $\Phi_h^*$. This error is then 'projected back' by $\Phi_{-h}$ to become $e$. We see that $-e$ is the local error of $\Phi_{-h}$, i.e., by hypothesis (3.3),

$$e = (-1)^p C(\varphi_h(y_0))h^{p+1} + \mathcal{O}(h^{p+2}). \tag{3.5}$$

Since $\varphi_h(y_0) = y_0 + \mathcal{O}(h)$ and $e = (I + \mathcal{O}(h))e^*$, it follows that

$$e^* = (-1)^p C(y_0)h^{p+1} + \mathcal{O}(h^{p+2})$$

which proves (3.4). The statement for symmetric methods is an immediate consequence of this result, because $\Phi_h = \Phi_h^*$ implies $C(y_0) = (-1)^p C(y_0)$, and therefore $C(y_0)$ can be different from zero only for even $p$. □

## II.4 Composition Methods

The idea of composing methods has some tradition in several variants: composition of different Runge-Kutta methods with the same step size leading to the Butcher group, which is treated in Sect. III.1.3; cyclic composition of multistep methods for breaking the 'Dahlquist barrier' (see Stetter (1973), p. 216); composition of low order Runge-Kutta methods for increasing stability for stiff problems (Gentzsch & Schlüter (1978), Iserles (1984)). In the following, we consider the composition of a given basic one-step method (and, eventually, its adjoint method) with *different* step sizes. The aim is to increase the order while preserving some desirable properties of the basic method. This idea has mainly been developed in the papers of Suzuki (1990), Yoshida (1990), and McLachlan (1995).

Let $\Phi_h$ be a basic method and $\gamma_1, \ldots, \gamma_s$ real numbers. Then we call its composition with step sizes $\gamma_1 h, \gamma_2 h, \ldots, \gamma_s h$, i.e.,

$$\Psi_h = \Phi_{\gamma_s h} \circ \ldots \circ \Phi_{\gamma_1 h}, \tag{4.1}$$

the corresponding *composition method* (see Fig. 4.1 (a)).

**Theorem 4.1.** *Let $\Phi_h$ be a one-step method of order $p$. If*

$$\begin{aligned} \gamma_1 + \ldots + \gamma_s &= 1 \\ \gamma_1^{p+1} + \ldots + \gamma_s^{p+1} &= 0, \end{aligned} \tag{4.2}$$

*then the composition method (4.1) is at least of order $p+1$.*

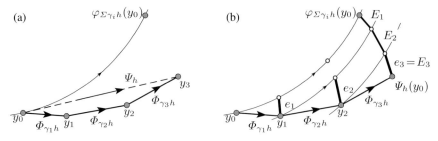

**Fig. 4.1.** Composition of method $\Phi_h$ with three step sizes

*Proof.* The proof is presented in Fig. 4.1(b) for $s = 3$. It is very similar to the proof of Theorem 3.2. By hypothesis

$$\begin{aligned} e_1 &= C(y_0) \cdot \gamma_1^{p+1} h^{p+1} + \mathcal{O}(h^{p+2}) \\ e_2 &= C(y_1) \cdot \gamma_2^{p+1} h^{p+1} + \mathcal{O}(h^{p+2}) \\ e_3 &= C(y_2) \cdot \gamma_3^{p+1} h^{p+1} + \mathcal{O}(h^{p+2}). \end{aligned} \quad (4.3)$$

We have, as before, $y_i = y_0 + \mathcal{O}(h)$ and $E_i = (I + \mathcal{O}(h))e_i$ for all $i$ and obtain, for $\sum \gamma_i = 1$,

$$\varphi_h(y_0) - \Psi_h(y_0) = E_1 + E_2 + E_3 = C(y_0)(\gamma_1^{p+1} + \gamma_2^{p+1} + \gamma_3^{p+1})h^{p+1} + \mathcal{O}(h^{p+2})$$

which shows that under conditions (4.2) the $\mathcal{O}(h^{p+1})$-term vanishes. $\square$

**Examples.** Equations (4.2) have no real solution for odd $p$. Therefore, the order increase is only possible for even $p$. In this case, the smallest $s$ which allows a solution is $s = 3$. We then have some freedom for solving the two equations. If we impose symmetry $\gamma_1 = \gamma_3$, then we obtain (Creutz & Gocksch 1989, Forest 1989, Suzuki 1990, Yoshida 1990)

$$\gamma_1 = \gamma_3 = \frac{1}{2 - 2^{1/(p+1)}}, \qquad \gamma_2 = -\frac{2^{1/(p+1)}}{2 - 2^{1/(p+1)}}. \quad (4.4)$$

This procedure can be repeated: we start with a symmetric method of order 2, apply (4.4) with $p = 2$ to obtain order 3; due to the symmetry of the $\gamma$'s this new method is in fact of order 4 (see Theorem 3.2). With this new method we repeat (4.4) with $p = 4$ and obtain a symmetric 9-stage composition method of order 6, then with $p = 6$ a 27-stage symmetric composition method of order 8, and so on. One obtains in this way *any* order, however, at the price of a terrible zig-zag of the step points (see Fig. 4.2).

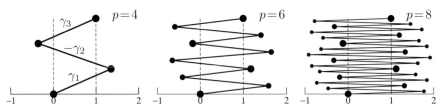

**Fig. 4.2.** Cascades of symmetric composition methods

If one desires methods with smaller values of $\gamma_i$, one has to increase $s$ even more. For example, for $s = 5$ the best solution of (4.2) has the sign structure $+ + - + +$ with $\gamma_1 = \gamma_2$. This leads to (Suzuki 1990)

$$\gamma_1 = \gamma_2 = \gamma_4 = \gamma_5 = \frac{1}{4 - 4^{1/(p+1)}}, \qquad \gamma_3 = -\frac{4^{1/(p+1)}}{4 - 4^{1/(p+1)}}. \tag{4.5}$$

The repetition of this algorithm for $p = 2, 4, 6, \ldots$ leads to a fractal structure of the step points (see Fig. 4.3).

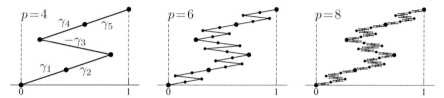

**Fig. 4.3.** Suzuki's 'fractal' composition methods

**Composition with the Adjoint Method.** If we replace the composition (4.1) by the more general formula

$$\Psi_h = \Phi_{\alpha_s h} \circ \Phi^*_{\beta_s h} \circ \ldots \circ \Phi^*_{\beta_2 h} \circ \Phi_{\alpha_1 h} \circ \Phi^*_{\beta_1 h}, \tag{4.6}$$

the condition for order $p + 1$ becomes, by using the result (3.4) and a similar proof as above,

$$\begin{aligned}\beta_1 + \alpha_1 + \beta_2 + \ldots + \beta_s + \alpha_s &= 1 \\ (-1)^p \beta_1^{p+1} + \alpha_1^{p+1} + (-1)^p \beta_2^{p+1} + \ldots + (-1)^p \beta_s^{p+1} + \alpha_s^{p+1} &= 0.\end{aligned} \tag{4.7}$$

This allows an order increase for odd $p$ as well. In particular, we see at once the solution $\alpha_1 = \beta_1 = 1/2$ for $p = s = 1$, which turns every consistent one-step method of order 1 into a second-order symmetric method

$$\Psi_h = \Phi_{h/2} \circ \Phi^*_{h/2}. \tag{4.8}$$

**Example 4.2.** If $\Phi_h$ is the explicit (resp. implicit) Euler method, then $\Psi_h$ in (4.8) becomes the implicit midpoint (resp. trapezoidal) rule.

**Example 4.3.** In a second-order problem $\dot{q} = v$, $\dot{v} = g(q)$, if $\Phi_h$ is the symplectic Euler method, which discretizes $q$ by the implicit Euler and $v$ by the explicit Euler method, then the composed method $\Psi_h$ in (4.8) is the Störmer/Verlet method.

## II.5 Splitting Methods

The splitting idea yields an approach that is completely different from Runge-Kutta methods. One decomposes the vector field into integrable pieces and treats them separately.

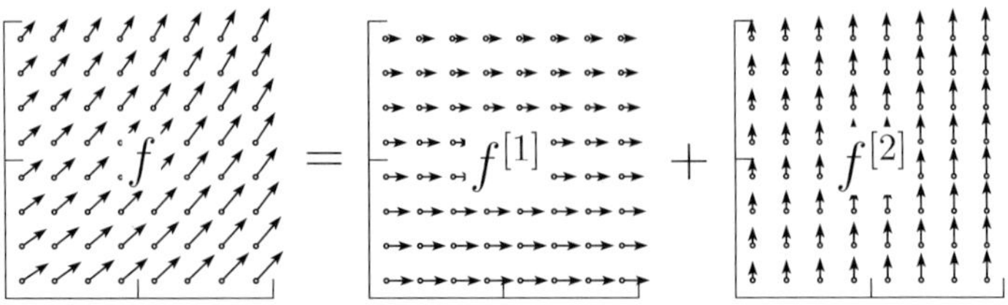

## II. Numerical Integrators

**Fig. 5.1.** A splitting of a vector field.

We consider an arbitrary system $\dot{y} = f(y)$ in $\mathbb{R}^n$, and suppose that the vector field is "split" as (see Fig. 5.1)

$$\dot{y} = f^{[1]}(y) + f^{[2]}(y). \tag{5.1}$$

If then, by chance, the exact flows $\varphi_t^{[1]}$ and $\varphi_t^{[2]}$ of the systems $\dot{y} = f^{[1]}(y)$ and $\dot{y} = f^{[2]}(y)$ can be calculated explicitly, we can, from a given initial value $y_0$, first solve the first system to obtain a value $y_{1/2}$, and from this value integrate the second system to obtain $y_1$. In this way we have introduced the numerical methods

$$\Phi_h^* = \varphi_h^{[2]} \circ \varphi_h^{[1]}$$
$$\Phi_h = \varphi_h^{[1]} \circ \varphi_h^{[2]} \tag{5.2}$$

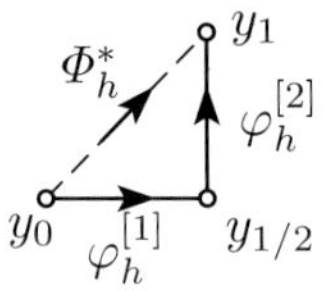

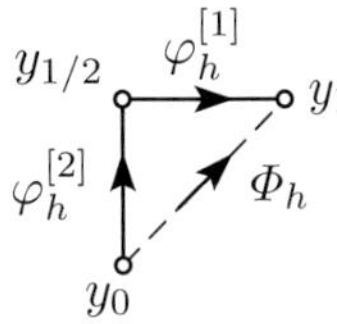

where one is the adjoint of the other. These formulas are often called the *Lie-Trotter splitting* (Trotter 1959). By Taylor expansion we find that $(\varphi_h^{[1]} \circ \varphi_h^{[2]})(y_0) = \varphi_h(y_0) + \mathcal{O}(h^2)$, so that both methods give approximations of order 1 to the solution of (5.1). Another idea is to use a symmetric version and put

$$\Phi_h^{[S]} = \varphi_{h/2}^{[1]} \circ \varphi_h^{[2]} \circ \varphi_{h/2}^{[1]}, \tag{5.3}$$

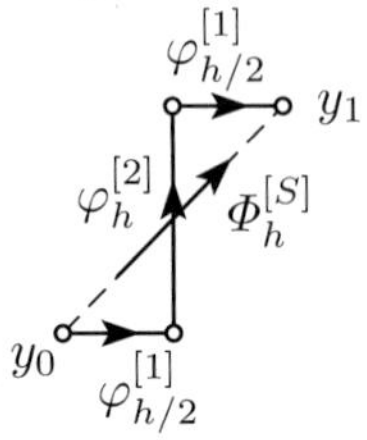

which is known as the *Strang splitting*[1] (Strang 1968), and sometimes as the *Marchuk splitting* (Marchuk 1968). By breaking up in (5.3) $\varphi_h^{[2]} = \varphi_{h/2}^{[2]} \circ \varphi_{h/2}^{[2]}$, we see that the Strang splitting $\Phi_h^{[S]} = \Phi_{h/2} \circ \Phi_{h/2}^*$ is the composition of the Lie-Trotter method and its adjoint with halved step sizes. The Strang splitting formula is therefore symmetric and of order 2 (see (4.8)).

**Example 5.1 (The Symplectic Euler and the Störmer/Verlet Schemes).** Suppose we have a Hamiltonian system with separable Hamiltonian $H(p,q) = T(p) + U(q)$.

---

[1] The article Strang (1968) deals with spatial discretizations of partial differential equations such as $u_t = Au_x + Bu_y$. There, the functions $f^{[i]}$ typically contain differences in only one spatial direction.

We consider this as the sum of *two* Hamiltonians, the first one depending only on $p$, the second one only on $q$. The corresponding Hamiltonian systems

$$\begin{aligned} \dot{p} &= 0 \\ \dot{q} &= T_p(p) \end{aligned} \quad \text{and} \quad \begin{aligned} \dot{p} &= -U_q(q) \\ \dot{q} &= 0 \end{aligned} \tag{5.4}$$

can be solved without problem to yield

$$\begin{aligned} p(t) &= p_0 \\ q(t) &= q_0 + t\, T_p(p_0) \end{aligned} \quad \text{and} \quad \begin{aligned} p(t) &= p_0 - t\, U_q(q_0) \\ q(t) &= q_0. \end{aligned} \tag{5.5}$$

Denoting the flows of these two systems by $\varphi_t^T$ and $\varphi_t^U$, we see that the symplectic Euler method (I.1.9) is just the composition $\varphi_h^T \circ \varphi_h^U$. Furthermore, the adjoint of the symplectic Euler method is $\varphi_h^U \circ \varphi_h^T$, and by Example 4.3 the Verlet scheme is $\varphi_{h/2}^U \circ \varphi_h^T \circ \varphi_{h/2}^U$, the Strang splitting (5.3). Anticipating the results of Chap. VI, the flows $\varphi_h^T$ and $\varphi_h^U$ are both symplectic transformations, and, since the composition of symplectic maps is again symplectic, this gives an elegant proof of the symplecticity of the 'symplectic' Euler method and the Verlet scheme.

**General Splitting Procedure.** In a similar way to the general idea of composition methods (4.6), we can form with arbitrary coefficients $a_1, b_1, a_2, \ldots, a_m, b_m$ (where, eventually, $a_1$ or $b_m$, or both, are zero)

$$\Psi_h = \varphi_{b_m h}^{[2]} \circ \varphi_{a_m h}^{[1]} \circ \varphi_{b_{m-1} h}^{[2]} \circ \ldots \circ \varphi_{a_2 h}^{[1]} \circ \varphi_{b_1 h}^{[2]} \circ \varphi_{a_1 h}^{[1]} \tag{5.6}$$

and try to increase the order of the scheme by suitably determining the free coefficients. An early contribution to this subject is the article of Ruth (1983), where, for the special case (5.4), a method (5.6) of order 3 with $m = 3$ is constructed. Forest & Ruth (1990) and Candy & Rozmus (1991) extend Ruth's technique and construct methods of order 4. One of their methods is just (4.1) with $\gamma_1, \gamma_2, \gamma_3$ given by (4.4) ($p = 2$) and $\Phi_h$ from (5.3). A systematic study of such methods started with the articles of Suzuki (1990, 1992) and Yoshida (1990).

A close connection between the theories of splitting methods (5.6) and of composition methods (4.6) was discovered by McLachlan (1995).

$$\begin{aligned} a_1 &= \beta_1 \\ b_1 &= \beta_1 + \alpha_1 \\ a_2 &= \alpha_1 + \beta_2 \\ b_2 &= \beta_2 + \alpha_2 \\ a_3 &= \alpha_2 + \beta_3 \\ b_3 &= \beta_3 \end{aligned} \tag{5.7}$$

Indeed, if we put $\beta_1 = a_1$ and break up $\varphi_{b_1 h}^{[2]} = \varphi_{\alpha_1 h}^{[2]} \circ \varphi_{\beta_1 h}^{[2]}$ (group property of the exact flow) where $\alpha_1$ is given in (5.7), further $\varphi_{a_2 h}^{[1]} = \varphi_{\beta_2 h}^{[1]} \circ \varphi_{\alpha_1 h}^{[1]}$ and so on (see (5.7)), we see, using (5.2), that $\Psi_h$ of (5.6) is identical with $\Psi_h$ of (4.6), where

$$\Phi_h = \varphi_h^{[1]} \circ \varphi_h^{[2]} \quad \text{so that} \quad \Phi_h^* = \varphi_h^{[2]} \circ \varphi_h^{[1]}. \tag{5.8}$$

A necessary and sufficient condition for the existence of $\alpha_i$ and $\beta_i$ satisfying (5.7) is that $\sum a_i = \sum b_i$, which is the consistency condition anyway for method (5.6).

**Combining Exact and Numerical Flows.** It may happen that the differential equation $\dot{y} = f(y)$ can be split according to (5.1), such that only the flow of, say, $\dot{y} = f^{[1]}(y)$ can be computed exactly. If $f^{[1]}(y)$ constitutes the dominant part of the vector field, it is natural to search for integrators that exploit this information. The above interpretation of splitting methods as composition methods allows us to construct such integrators. We just consider

$$\Phi_h = \varphi_h^{[1]} \circ \Phi_h^{[2]}, \qquad \Phi_h^* = \Phi_h^{[2]*} \circ \varphi_h^{[1]} \tag{5.9}$$

as the basis of the composition method (4.6). Here $\varphi_t^{[1]}$ is the exact flow of $\dot{y} = f^{[1]}(y)$, and $\Phi_h^{[2]}$ is some first-order integrator applied to $\dot{y} = f^{[2]}(y)$. Since $\Phi_h$ of (5.9) is consistent with (5.1), the resulting method (4.6) has the desired high order. It is given by

$$\Psi_h = \varphi_{\alpha_s h}^{[1]} \circ \Phi_{\alpha_s h}^{[2]} \circ \Phi_{\beta_s h}^{[2]*} \circ \varphi_{(\beta_s + \alpha_{s-1}) h}^{[1]} \circ \Phi_{\alpha_{s-1} h}^{[2]} \circ \ldots \circ \Phi_{\beta_1 h}^{[2]*} \circ \varphi_{\beta_1 h}^{[1]}. \tag{5.10}$$

Notice that replacing $\varphi_t^{[2]}$ with a low-order approximation $\Phi_t^{[2]}$ in (5.6) would not retain the high order of the composition, because $\Phi_t^{[2]}$ does not satisfy the group property.

**Splitting into More than Two Vector Fields.** Consider a differential equation

$$\dot{y} = f^{[1]}(y) + f^{[2]}(y) + \ldots + f^{[N]}(y), \tag{5.11}$$

where we assume that the flows $\varphi_t^{[j]}$ of the individual problems $\dot{y} = f^{[j]}(y)$ can be computed exactly. In this case there are many possibilities for extending (5.6) and for writing the method as a composition of $\varphi_{a_j h}^{[1]}, \varphi_{b_j h}^{[2]}, \varphi_{c_j h}^{[3]}, \ldots$ . This makes it difficult to find optimal compositions of high order. A simple and efficient way is to consider the first-order method

$$\Phi_h = \varphi_h^{[1]} \circ \varphi_h^{[2]} \circ \ldots \circ \varphi_h^{[N]}$$

together with its adjoint as the basis of the composition (4.6). Without any additional effort this yields splitting methods for (5.11) of arbitrary high order.

**A Numerical Example.** To demonstrate the numerical performance of the above methods, we choose the Kepler problem (I.2.2) with $e = 0.6$ and the initial values

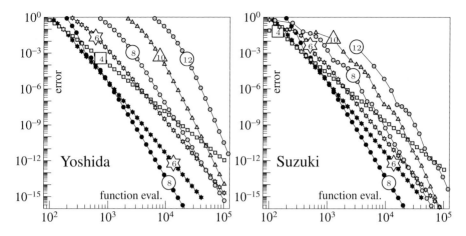

**Fig. 5.2.** Numerical results of Yoshida and Suzuki step sequences (grey symbols) compared to optimal methods (black symbols)

from (I.2.11). As integration interval we choose $[0, 7.5]$, a bit more than one revolution. The exact solution is obtained by carefully evaluating the integral (I.2.10), which gives

$$\varphi = 8.67002632314281495159108828552, \quad (5.12)$$

with the help of which we compute $r, \dot{\varphi}, \dot{r}$ from (I.2.8) and (I.2.6). This gives

$$\begin{aligned} q_1 &= -0.82816440269077081820 4757585370 \\ q_2 &= \phantom{-}0.77889809565863544708 1654480796 \\ p_1 &= -0.85638471534339535152 4486215030 \\ p_2 &= -0.16055215079983843525 4419104102 \,. \end{aligned} \quad (5.13)$$

As the basic method we use the Strang splitting (5.3) (i.e., the Verlet scheme) and compare in Fig. 5.2 the performances of the composition sequences of Yoshida (4.4) and those of Suzuki (4.5) for a large number of different equidistant basic step sizes and for orders $p = 4, 6, 8, 10, 12$. Each basic step is then divided into $3, 9, 27, 81, 243$ respectively $5, 25, 125, 625, 3125$ composition steps and the maximal final error is compared with the total number of function evaluations in double logarithmic scales. For each method and order, all the points lie asymptotically on a straight line with slope $-p$. Therefore, theoretically, a higher order method will become superior when the precision requirements become sufficiently high. But we see that for orders 10 and 12 these 'break even points' are far beyond any precision of practical interest, after some 40 or 50 digits. We also observe that the wild zig-zag of the Yoshida sequence (4.4) is a more serious handicap than the enormous number of small steps of the Suzuki sequence (4.5).

For later reference we have also included, in black symbols, the results obtained by the two methods (V.3.11) and (V.3.13) of orders 6 and 8, respectively, which will be the outcome of a more elaborate order theory in Chap. III.

## II.6 Exercises

1. Compute all collocation methods with $s = 2$ as a function of $c_1$ and $c_2$. Which of them are of order 3, which of order 4?
2. Prove that the collocation solution plotted in the right picture of Fig. 1.3 is composed of arcs of parabolas.
3. Let $b_1 = b_4 = 1/8$, $c_2 = 1/3$, $c_3 = 2/3$, and consider the corresponding discontinuous collocation method. Determine its order and find the coefficients of the equivalent Runge-Kutta method.
4. Compute the adjoint of the symplectic Euler method (I.1.9).
5. (Additive Runge-Kutta methods). Let $b_i, a_{ij}$ and $\widehat{b}_i, \widehat{a}_{ij}$ be the coefficients of two Runge-Kutta methods. An additive Runge-Kutta method for the solution of $\dot{y} = f^{[1]}(y) + f^{[2]}(y)$ is given by

$$k_i = f^{[1]}\left(y_0 + h\sum_{j=1}^{s} a_{ij}k_j\right) + f^{[2]}\left(y_0 + h\sum_{j=1}^{s} \widehat{a}_{ij}k_j\right)$$

$$y_1 = y_0 + h\sum_{i=1}^{s} b_i k_i.$$

Show that this can be interpreted as a partitioned Runge-Kutta method (2.2) applied to

$$\dot{y} = f^{[1]}(y) + f^{[2]}(z), \qquad \dot{z} = f^{[1]}(y) + f^{[2]}(z)$$

with $y(0) = z(0) = y_0$. Notice that $y(t) = z(t)$.

6. Let $\Phi_h$ denote the Störmer/Verlet scheme, and consider the composition

$$\Phi_{\gamma_{2k+1}h} \circ \Phi_{\gamma_{2k}h} \circ \ldots \circ \Phi_{\gamma_2 h} \circ \Phi_{\gamma_1 h}$$

with $\gamma_1 = \ldots = \gamma_k = \gamma_{k+2} = \ldots = \gamma_{2k+1}$. Compute $\gamma_1$ and $\gamma_{k+1}$ such that the composition gives a method of order 4. For several differential equations (pendulum, Kepler problem) study the global error of a constant step size implementation as a function of $k$.

7. Consider the composition method (4.1) with $s = 5$, $\gamma_5 = \gamma_1$, and $\gamma_4 = \gamma_2$. Among the solutions of

$$2\gamma_1 + 2\gamma_2 + \gamma_3 = 1, \qquad 2\gamma_1^3 + 2\gamma_2^3 + \gamma_3^3 = 0$$

find the one that minimizes $|2\gamma_1^5 + 2\gamma_2^5 + \gamma_3^5|$.

*Remark.* This property motivates the choice of the $\gamma_i$ in (4.5).

8. If the splitting method (5.6) is of order $p \geq 3$ for general $f^{[1]}$ and $f^{[2]}$, then at least one of the $\alpha_i$ is strictly negative, and also one of the $\beta_i$ is strictly negative.
*Remark.* This is a non-trivial result and has been proved independently by Sheng (1989) and Suzuki (1991); see also Goldman & Kaper (1996).

# Chapter III.
# Order Conditions, Trees and B-Series

In this chapter we present a compact theory of the order conditions of the methods presented in Chap. II, in particular Runge-Kutta methods, partitioned Runge-Kutta methods, and composition methods by using the notion of rooted trees and B-series. These ideas lead to algebraic structures which have recently found interesting applications in quantum field theory. The chapter terminates with the Baker-Campbell-Hausdorff formula, which allows another access to the order properties of composition and splitting methods.

Some parts of this chapter are rather short, but nevertheless self-contained. For more detailed presentations we refer to the monographs of Butcher (1987), of Hairer, Nørsett & Wanner (1993), and of Hairer & Wanner (1996). Readers mainly interested in geometric properties of numerical integrators may continue with Chapters IV, V or VI before returning to the technically more difficult jungle of trees.

## III.1 Runge-Kutta Order Conditions and B-Series

> Even the standard notation has been found to be too heavy in dealing with fourth and higher order processes, ...
> (R.H. Merson 1957)

In this section we derive the order conditions of Runge-Kutta methods by comparing the Taylor series of the exact solution of (1.1) with that of the numerical solution. The computation is much simplified, first by considering an *autonomous* system of equations (Gill 1951), and second, by the use of rooted trees (connected graphs without cycles and a distinguished vertex; Merson 1957). The theory has been developed by Butcher in the years 1963-72 (see Butcher (1987), Sect. 30) and by Hairer & Wanner in 1973-74 (see Hairer, Nørsett & Wanner (1993), Sections II.2 and II.12). Here we give new simplified proofs.

### III.1.1 Derivation of the Order Conditions

We consider an autonomous problem

$$\dot{y} = f(y), \qquad y(t_0) = y_0, \qquad (1.1)$$

where $f : \mathbb{R}^n \to \mathbb{R}^n$ is sufficiently differentiable. A problem $\dot{y} = f(t, y)$ can be brought into this form by appending the equation $\dot{t} = 1$. We develop the subsequent theory in four steps.

48    III. Order Conditions, Trees and B-Series

> Er sagte es klar und angenehm,
> was erstens, zweitens und drittens käm'.    (W. Busch, *Jobsiade* 1872)

**First Step.** We compute the higher derivatives of the solution $y$ at the initial point $t_0$. For this, we have from (1.1)

$$y^{(q)} = \bigl(f(y)\bigr)^{(q-1)} \tag{1.2}$$

and compute the latter derivatives by using the chain rule, the product rule, the symmetry of partial derivatives, and the notation $f'(y)$ for the derivative as a linear map (the Jacobian), $f''(y)$ the second derivative as a bilinear map and similarly for higher derivatives. This gives

$$\begin{aligned}
\dot{y} &= f(y) \\
\ddot{y} &= f'(y)\,\dot{y} \\
y^{(3)} &= f''(y)(\dot{y},\dot{y}) + f'(y)\,\ddot{y} \\
y^{(4)} &= f'''(y)(\dot{y},\dot{y},\dot{y}) + 3f''(y)(\ddot{y},\dot{y}) + f'(y)\,y^{(3)} \\
y^{(5)} &= f^{(4)}(y)(\dot{y},\dot{y},\dot{y},\dot{y}) + 6f'''(y)(\ddot{y},\dot{y},\dot{y}) + 4f''(y)(y^{(3)},\dot{y}) \\
&\quad + 3f''(y)(\ddot{y},\ddot{y}) + f'(y)\,y^{(4)},
\end{aligned} \tag{1.3}$$

and so on. The coefficients $3, 6, 4, 3, \ldots$ appearing in these expressions have a certain combinatorial meaning (number of partitions of a set of $q-1$ elements), but for the moment we need not know their values.

**Second Step.** We insert in (1.3) recursively the computed derivatives $\dot{y}, \ddot{y}, \ldots$ into the right side of the subsequent formulas. This gives for the first few

$$\begin{aligned}
\dot{y} &= f \\
\ddot{y} &= f'f \\
y^{(3)} &= f''(f,f) + f'f'f \\
y^{(4)} &= f'''(f,f,f) + 3f''(f'f,f) + f'f''(f,f) + f'f'f'f,
\end{aligned} \tag{1.4}$$

where the arguments $(y)$ have been suppressed. The expressions which appear in these formulas, denoted by $F(\tau)$, will be called the *elementary differentials*. We represent each of them by a suitable graph $\tau$ (a rooted tree) as follows:

Each $f$ becomes a vertex, a first derivative $f'$ becomes a vertex with one branch, and a $k$th derivative $f^{(k)}$ becomes a vertex with $k$ branches pointing upwards. The arguments of the $k$-linear mapping $f^{(k)}(y)$ correspond to trees that are attached on the upper ends of these branches. The tree to the right corresponds to $f''(f'f, f)$. Other trees are plotted in Table 1.1. In the above process, each insertion of an already known derivative consists of grafting the corresponding trees upon a new root as in Definition 1.1 below, and inserting the corresponding elementary differentials as arguments of $f^{(m)}(y)$ as in Definition 1.2.

## III.1 Runge-Kutta Order Conditions and B-Series

**Table 1.1.** Trees, elementary differentials, and coefficients

| $\|\tau\|$ | $\tau$ | graph | $\alpha(\tau)$ | $F(\tau)$ | $\gamma(\tau)$ | $\phi(\tau)$ | $\sigma(\tau)$ |
|---|---|---|---|---|---|---|---|
| 1 | $\bullet$ | $\bullet$ | 1 | $f$ | 1 | $\sum_i b_i$ | 1 |
| 2 | $[\bullet]$ |  | 1 | $f'f$ | 2 | $\sum_{ij} b_i a_{ij}$ | 1 |
| 3 | $[\bullet,\bullet]$ |  | 1 | $f''(f,f)$ | 3 | $\sum_{ijk} b_i a_{ij} a_{ik}$ | 2 |
| 3 | $[[\bullet]]$ |  | 1 | $f'f'f$ | 6 | $\sum_{ijk} b_i a_{ij} a_{jk}$ | 1 |
| 4 | $[\bullet,\bullet,\bullet]$ |  | 1 | $f'''(f,f,f)$ | 4 | $\sum_{ijkl} b_i a_{ij} a_{ik} a_{il}$ | 6 |
| 4 | $[[\bullet],\bullet]$ |  | 3 | $f''(f'f,f)$ | 8 | $\sum_{ijkl} b_i a_{ij} a_{ik} a_{jl}$ | 1 |
| 4 | $[[\bullet,\bullet]]$ |  | 1 | $f'f''(f,f)$ | 12 | $\sum_{ijkl} b_i a_{ij} a_{jk} a_{jl}$ | 2 |
| 4 | $[[[\bullet]]]$ |  | 1 | $f'f'f'f$ | 24 | $\sum_{ijkl} b_i a_{ij} a_{jk} a_{kl}$ | 1 |

**Definition 1.1 (Trees).** The set of (rooted) *trees* $T$ is recursively defined as follows:
a) the graph $\bullet$ with only one vertex (called the root) belongs to $T$;
b) if $\tau_1, \ldots, \tau_m \in T$, then the graph obtained by grafting the roots of $\tau_1, \ldots, \tau_m$ to a new vertex also belongs to $T$. It is denoted by

$$\tau = [\tau_1, \ldots, \tau_m],$$

and the new vertex is the root of $\tau$.

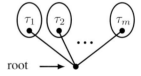

We further denote by $|\tau|$ the *order* of $\tau$ (the number of vertices), and by $\alpha(\tau)$ the coefficients appearing in the formulas (1.4). We remark that some of the trees among $\tau_1, \ldots, \tau_m$ may be equal and that $\tau$ does not depend on the ordering of $\tau_1, \ldots, \tau_m$. For example, we do not distinguish between $[[\bullet],\bullet]$ and $[\bullet,[\bullet]]$.

**Definition 1.2 (Elementary Differentials).** For a tree $\tau \in T$ the *elementary differential* is a mapping $F(\tau) : \mathbb{R}^n \to \mathbb{R}^n$, defined recursively by $F(\bullet)(y) = f(y)$ and

$$F(\tau)(y) = f^{(m)}(y)\Big(F(\tau_1)(y), \ldots, F(\tau_m)(y)\Big) \qquad \text{for} \quad \tau = [\tau_1, \ldots, \tau_m].$$

Examples of these constructions and the corresponding coefficients are seen in Table 1.1. With these definitions, we obtain from (1.4):

**Theorem 1.3.** *The $q$th derivative of the exact solution is given by*

$$y^{(q)}(t_0) = \sum_{|\tau|=q} \alpha(\tau) F(\tau)(y_0), \tag{1.5}$$

*where $\alpha(\tau)$ are positive integer coefficients.* □

**Third Step.** We now turn to the numerical solution of the Runge-Kutta method (II.1.4), which, by putting $hk_i = g_i$, we write as

$$g_i = hf(u_i) \tag{1.6}$$

and

$$u_i = y_0 + \sum_j a_{ij} g_j, \qquad y_1 = y_0 + \sum_i b_i g_i, \tag{1.7}$$

where $u_i$, $g_i$ and $y_1$ are functions of $h$. We develop the derivatives of (1.6), by Leibniz' rule, and obtain $g_i^{(q)} = h(f(u_i))^{(q)} + q \cdot (f(u_i))^{(q-1)}$. This gives, for $h = 0$,

$$g_i^{(q)} = q \cdot (f(u_i))^{(q-1)}, \tag{1.8}$$

the same expression as in (1.2), with $y$ just replaced by $u_i$ and with an extra factor $q$. Consequently, exactly as in (1.3),

$$\begin{aligned}
\dot{g}_i &= 1 \cdot f(y_0) \\
\ddot{g}_i &= 2 \cdot f'(y_0) \, \dot{u}_i \\
g_i^{(3)} &= 3 \cdot \left(f''(y_0)(\dot{u}_i, \dot{u}_i) + f'(y_0) \, \ddot{u}_i\right) \\
g_i^{(4)} &= 4 \cdot \left(f'''(y_0)(\dot{u}_i, \dot{u}_i, \dot{u}_i) + 3f''(y_0)(\ddot{u}_i, \dot{u}_i) + f'(y_0) \, u_i^{(3)}\right) \\
g_i^{(5)} &= 5 \cdot \left(f^{(4)}(y_0)(\dot{u}_i, \dot{u}_i, \dot{u}_i, \dot{u}_i) + 6f'''(y_0)(\ddot{u}_i, \dot{u}_i, \dot{u}_i) + 4f''(y_0)(u_i^{(3)}, \dot{u}_i)\right. \\
&\qquad \left. + 3f''(y_0)(\ddot{u}_i, \ddot{u}_i) + f'(y_0) \, u_i^{(4)}\right),
\end{aligned} \tag{1.9}$$

and so on. Here, the derivatives of $g_i$ and $u_i$ are evaluated at $h = 0$.

**Fourth Step.** We now insert recursively the derivatives $\dot{u}_i, \ddot{u}_i, \ldots$ into (1.9). This will give the next higher derivative of $g_i$, and, using

$$u_i^{(q)} = \sum_j a_{ij} \cdot g_j^{(q)}, \tag{1.10}$$

which follows from (1.7), also the next higher derivative of $u_i$. This process begins as

$$\begin{aligned}
\dot{g}_i &= 1 \cdot f & \dot{u}_i &= 1 \cdot \left(\textstyle\sum_j a_{ij}\right) \cdot f \\
\ddot{g}_i &= (1 \cdot 2) \left(\textstyle\sum_j a_{ij}\right) f'f & \ddot{u}_i &= (1 \cdot 2) \left(\textstyle\sum_{jk} a_{ij} a_{jk}\right) f'f
\end{aligned} \tag{1.11}$$

and so on. If we compare these formulas with the first lines of (1.4), we see that the results are precisely the same, apart from the extra factors. We denote the *integer factors* $1, 1 \cdot 2, \ldots$ by $\gamma(\tau)$ and the factors containing the $a_{ij}$'s by $\mathbf{g}_i(\tau)$ and $\mathbf{u}_i(\tau)$, respectively. We obtain by induction that the same happens in general, i.e. that, in contrast to (1.5),

$$g_i^{(q)}\big|_{h=0} = \sum_{|\tau|=q} \gamma(\tau) \cdot \mathbf{g}_i(\tau) \cdot \alpha(\tau) \, F(\tau)(y_0)$$
$$u_i^{(q)}\big|_{h=0} = \sum_{|\tau|=q} \gamma(\tau) \cdot \mathbf{u}_i(\tau) \cdot \alpha(\tau) \, F(\tau)(y_0), \tag{1.12}$$

where $\alpha(\tau)$ and $F(\tau)$ are *the same* quantities as before. This is seen by continuing the insertion process of the derivatives $u_i^{(q)}$ into the right-hand side of (1.9). For example, if $\dot u_i$ and $\ddot u_i$ are inserted into $3f''(\ddot u_i, \dot u_i)$, we will obtain the corresponding expression as in (1.4), multiplied by the two extra factors $\mathbf{u}_i(\mathord{\mathbf{\mathsf{f}}})$, brought in by $\ddot u_i$, and $\mathbf{u}_i(\bullet)$ from $\dot u_i$. For a general tree $\tau = [\tau_1, \ldots, \tau_m]$ this will be

$$\mathbf{g}_i(\tau) = \mathbf{u}_i(\tau_1) \cdot \ldots \cdot \mathbf{u}_i(\tau_m). \tag{1.13}$$

Second, the factors $\gamma(\mathord{\mathbf{\mathsf{f}}})$ and $\gamma(\bullet)$ will receive the additional factor $q = |\tau|$ from (1.9), i.e., we will have in general

$$\gamma(\tau) = |\tau|\, \gamma(\tau_1) \cdot \ldots \cdot \gamma(\tau_m). \tag{1.14}$$

Then, by (1.10),

$$\mathbf{u}_i(\tau) = \sum_j a_{ij}\, \mathbf{g}_j(\tau) = \sum_j a_{ij} \cdot \mathbf{u}_j(\tau_1) \cdot \ldots \cdot \mathbf{u}_j(\tau_m). \tag{1.15}$$

This formula can be re-used repeatedly, as long as some of the trees $\tau_1, \ldots, \tau_m$ are of order $> 1$. Finally, we have from the last formula of (1.7), that the coefficients for the numerical solution, which we denote by $\phi(\tau)$ and call the *elementary weights*, satisfy

$$\phi(\tau) = \sum_i b_i\, \mathbf{g}_i(\tau). \tag{1.16}$$

We summarize the result as follows:

**Theorem 1.4.** *The derivatives of the numerical solution of a Runge-Kutta method (II.1.4), for $h = 0$, are given by*

$$y_1^{(q)}\big|_{h=0} = \sum_{|\tau|=q} \gamma(\tau) \cdot \phi(\tau) \cdot \alpha(\tau) \, F(\tau)(y_0), \tag{1.17}$$

*where $\alpha(\tau)$ and $F(\tau)$ are the same as in Theorem 1.3, the coefficients $\gamma(\tau)$ satisfy $\gamma(\bullet) = 1$ and (1.14). The elementary weights $\phi(\tau)$ are obtained from the tree $\tau$ as follows: attach to every vertex a summation letter ('i' to the root), then $\phi(\tau)$ is the sum, over all summation indices, of a product composed of $b_i$, and factors $a_{jk}$ for each vertex 'j' directly connected with 'k' by an upwards directed branch.*

*Proof.* Repeated application of (1.15) followed by (1.16) shows that the elementary weight $\phi(\tau)$ is the collection of $\sum_i b_i$ from (1.16) and all $\sum_j a_{ij}$ of (1.15). □

**Theorem 1.5.** *The Runge-Kutta method has order $p$ if and only if*

$$\phi(\tau) = \frac{1}{\gamma(\tau)} \quad \text{for} \quad |\tau| \le p. \tag{1.18}$$

*Proof.* The comparison of Theorem 1.3 with Theorem 1.4 proves the sufficiency of condition (1.18). The necessity of (1.18) follows from the independence of the elementary differentials (see e.g., Hairer, Nørsett & Wanner (1993), Exercise 4 of Sect. II.2). □

**Example 1.6.** For the following tree of order 9 we have

$$\sum_{i,j,k,l,m,n,p,q,r} b_i a_{ij} a_{jm} a_{in} a_{ik} a_{kl} a_{lq} a_{lr} a_{kp} = \frac{1}{9 \cdot 2 \cdot 5 \cdot 3}$$

or, by using $\sum_j a_{ij} = c_i$,

$$\sum_{i,j,k,l} b_i c_i a_{ij} c_j a_{ik} c_k a_{kl} c_l^2 = \frac{1}{270}.$$

The quantities $\phi(\tau)$ and $\gamma(\tau)$ for all trees up to order 4 are given in Table 1.1. This also verifies the formulas (II.1.6) stated previously.

## III.1.2 B-Series

We now introduce the concept of B-series, which gives further insight into the behaviour of numerical methods and allows extensions to more general classes of methods.

Motivated by formulas (1.12) and (1.17) above, we consider the corresponding *series* as the objects of our study. This means, we study power series in $h^{|\tau|}$ containing elementary differentials $F(\tau)$ and arbitrary coefficients which are now written in the form $a(\tau)$. Such series will be called B-series. To move from (1.6) to (1.13) we need to prove a result stating that *a B-series inserted into $hf(\cdot)$ is again a B-series.* We start with

$$B(a,y) = y + a(\bullet)hf(y) + a(\mathbf{\mathit{l}})h^2(f'f)(y) + \ldots = y + \delta, \tag{1.19}$$

and get by Taylor expansion

$$hf(B(a,y)) = hf(y+\delta) = hf(y) + hf'(y)\delta + \frac{h}{2!}f''(y)(\delta,\delta) + \ldots. \tag{1.20}$$

Inserting $\delta$ from (1.19) and multiplying out, we obtain the expression

$$\begin{aligned} hf(B(a,y)) = hf &+ h^2 a(\bullet)f'f + h^3 a(\mathbf{\mathit{l}})f'f'f + \frac{h^3}{2!}a(\bullet)^2 f''(f,f) \\ &+ h^4 a(\bullet)a(\mathbf{\mathit{l}})f''(f'f,f) + \ldots. \end{aligned} \tag{1.21}$$

This beautiful formula is not yet perfect for two reasons. First, there is a denominator 2! in the fourth term. The origin of this lies in the *symmetry* of the tree $\vee$. We thus introduce the symmetry coefficients of Definition 1.7 (following Butcher 1987, Theorem 144A). Second, there is no first term $y$. We therefore allow the factor $a(\emptyset)$ in Definition 1.8.

**Definition 1.7 (Symmetry coefficients).** The symmetry coefficients $\sigma(\tau)$ are defined by $\sigma(\bullet) = 1$ and, for $\tau = [\tau_1, \ldots, \tau_m]$,

$$\sigma(\tau) = \sigma(\tau_1) \cdot \ldots \cdot \sigma(\tau_m) \cdot \mu_1! \mu_2! \cdot \ldots, \quad (1.22)$$

where the integers $\mu_1, \mu_2, \ldots$ count equal trees among $\tau_1, \ldots, \tau_m$.

**Definition 1.8 (B-Series).** For a mapping $a : T \cup \{\emptyset\} \to \mathbb{R}$ a formal series of the form

$$B(a, y) = a(\emptyset)y + \sum_{\tau \in T} \frac{h^{|\tau|}}{\sigma(\tau)} a(\tau) F(\tau)(y) \quad (1.23)$$

is called a *B-series*.[1]

The main results of the theory of B-series have their origin in the paper of Butcher (1972), although series expansions were not used there. B-series were then introduced by Hairer & Wanner (1974). The normalization used in Definition 1.8 is due to Butcher & Sanz-Serna (1996). The following fundamental lemma gives a second way of finding the order conditions.

**Lemma 1.9.** *Let* $a : T \cup \{\emptyset\} \to \mathbb{R}$ *be a mapping satisfying* $a(\emptyset) = 1$. *Then the corresponding B-series inserted into* $hf(\cdot)$ *is again a B-series. That is*

$$hf\big(B(a, y)\big) = B(a', y), \quad (1.24)$$

*where* $a'(\emptyset) = 0$, $a'(\bullet) = 1$, *and*

$$a'(\tau) = a(\tau_1) \cdot \ldots \cdot a(\tau_m) \quad \text{for} \quad \tau = [\tau_1, \ldots, \tau_m]. \quad (1.25)$$

*Proof.* Since $a(\emptyset) = 1$ we have $B(a, y) = y + \mathcal{O}(h)$, so that $hf\big(B(a, y)\big)$ can be expanded into a Taylor series around $y$. As in formulas (1.20) and (1.21), we get

---

[1] In this section we are not concerned about the convergence of the series. We shall see later in Chap. IX that the series converges for sufficiently small $h$, if $a(\tau)$ satisfies an inequality $|a(\tau)| \leq \gamma(\tau) cd^{|\tau|}$ and if $f(y)$ is an analytic function. If $f(y)$ is only $k$-times differentiable, then all formulas of this section remain valid for the truncated B-series $\sum_{\tau \in T, |\tau| \leq k} \cdot / \cdot$ with a suitable remainder term of size $\mathcal{O}(h^{k+1})$ added.

III. Order Conditions, Trees and B-Series

$$
\begin{aligned}
hf(B(a,y)) &= h \sum_{m \geq 0} \frac{1}{m!} f^{(m)}(y) \Big(B(a,y) - y\Big)^m \\
&= h \sum_{m \geq 0} \frac{1}{m!} \sum_{\tau_1 \in T} \cdots \sum_{\tau_m \in T} \frac{h^{|\tau_1|+\ldots+|\tau_m|}}{\sigma(\tau_1) \cdot \ldots \cdot \sigma(\tau_m)} \cdot a(\tau_1) \cdot \ldots \cdot a(\tau_m) \\
&\qquad\qquad\qquad\qquad\qquad \cdot f^{(m)}(y) \Big(F(\tau_1)(y), \ldots, F(\tau_m)(y)\Big) \\
&= \sum_{m \geq 0} \sum_{\tau_1 \in T} \cdots \sum_{\tau_m \in T} \frac{h^{|\tau|}}{\sigma(\tau)} \frac{\mu_1! \mu_2! \cdot \ldots}{m!} \cdot a'(\tau) F(\tau)(y) \\
&\qquad\qquad\qquad\qquad\qquad \text{with } \tau = [\tau_1, \ldots, \tau_m] \\
&= \sum_{\tau \in T} \frac{h^{|\tau|}}{\sigma(\tau)} a'(\tau) F(\tau)(y) = B(a', y).
\end{aligned}
$$

The last equality follows from the fact that there are $\binom{m}{\mu_1, \mu_2, \ldots}$ possibilities for writing the tree $\tau$ in the form $\tau = [\tau_1, \ldots, \tau_m]$. For example, the trees $[\bullet, \bullet, [\bullet]]$, $[\bullet, [\bullet], \bullet]$ and $[[\bullet], \bullet, \bullet]$ appear as different terms in the upper sum, but only as one term in the lower sum. □

**Back to the Order Conditions.** We present now a new derivation of the order conditions that is solely based on B-series and on Lemma 1.9. Let a Runge-Kutta method, say formulas (1.6) and (1.7), be given. All quantities in the defining formulas (eventually multiplied by $h$) are set up as B-series, $g_i = B(\mathbf{g}_i, y_0)$, $u_i = B(\mathbf{u}_i, y_0)$, $y_1 = B(\phi, y_0)$. Then, either the linearity and/or Lemma 1.9, translate the formulas of the method into corresponding formulas for the coefficients (1.13), (1.15), and (1.16). This justifies the ansatz as B-series, and proves Theorem 1.4 again.

Assuming the *exact* solution to be a B-series $B(e, y_0)$, a term-by-term derivation of this series and an application of Lemma 1.9 to (1.1) yields

$$\mathbf{e}(\tau) = \frac{1}{|\tau|} \mathbf{e}(\tau_1) \cdot \ldots \cdot \mathbf{e}(\tau_m).$$

Compared to definition (1.14) of $\gamma(\tau)$, we obtain

$$\mathbf{e}(\tau) = \frac{1}{\gamma(\tau)}. \tag{1.26}$$

This proves Theorem 1.3 again, together with the formula

$$\alpha(\tau) = \frac{|\tau|!}{\sigma(\tau) \cdot \gamma(\tau)}. \tag{1.27}$$

If the available tools are enriched by the more general composition law of Theorem 1.10 below, this procedure can be applied to yet larger classes of methods.

## III.1.3 Composition of Methods

The order theory for the composition of methods goes back to 1969, when Butcher used it to circumvent the order barrier for explicit 5th order 5 stage methods. It led to the seminal publication of Butcher (1972), where the general composition formula in (1.34) was expressed recursively.

**Composition of Runge-Kutta Methods.** Suppose that, starting from an initial value $y_0$, we compute a numerical solution $y_1$ using a Runge-Kutta method with coefficients $a_{ij}, b_i$ and step size $h$. Then, continuing from $y_1$, we compute a value $y_2$ using another method with coefficients $a_{ij}^*, b_i^*$ and the same step size. This composition of two methods is now considered as a *single* method (with coefficients $\widehat{a}_{ij}, \widehat{b}_i$). The problem is to derive the order properties of this new method, in particular to express the elementary weights $\widehat{\phi}(\tau)$ in terms of those of the original two methods.

If the value $y_1$ from the first method is inserted into the starting value for the second method, one sees that the coefficients of the combined method are given by (here written for two-stage methods)

$$\begin{array}{|cccc|}
\widehat{a}_{11} & \widehat{a}_{12} & & \\
\widehat{a}_{21} & \widehat{a}_{22} & & \\
\widehat{a}_{31} & \widehat{a}_{32} & \widehat{a}_{33} & \widehat{a}_{34} \\
\widehat{a}_{41} & \widehat{a}_{42} & \widehat{a}_{43} & \widehat{a}_{44} \\
\hline
\widehat{b}_1 & \widehat{b}_2 & \widehat{b}_3 & \widehat{b}_4
\end{array}
=
\begin{array}{|cccc|}
a_{11} & a_{12} & & \\
a_{21} & a_{22} & & \\
b_1 & b_2 & a_{11}^* & a_{12}^* \\
b_1 & b_2 & a_{21}^* & a_{22}^* \\
\hline
b_1 & b_2 & b_1^* & b_2^*
\end{array}
\quad (1.28)$$

and our problem is to compute the elementary weights of this scheme.

**Derivation.** The idea is to write the sum for $\widehat{\phi}(\tau)$, say for the tree $\curlyvee$, in full detail

$$\widehat{\phi}(\curlyvee) = \sum_{i=1}^{4}\sum_{j=1}^{4}\sum_{k=1}^{4}\sum_{l=1}^{4} \widehat{b}_i\, \widehat{a}_{ij}\, \widehat{a}_{ik}\, \widehat{a}_{kl} = \ldots \quad (1.29)$$

and to split each sum into the two different index sets. This leads to $2^{|\tau|}$ different expressions $\sum_{i=1}^{2}\sum_{j=1}^{2}\sum_{k=1}^{2}\sum_{l=1}^{2} ./. + \sum_{i=3}^{4}\sum_{j=1}^{2}\sum_{k=1}^{2}\sum_{l=1}^{2} ./. + \sum_{i=1}^{2}\sum_{j=3}^{4}\sum_{k=1}^{2}\sum_{l=1}^{2} ./. + \ldots$. We symbolize each expression by drawing the corresponding vertex of $\tau$ as a *bullet* for the first index set and as a *star* for the second. However, due to the zero pattern in the matrix in (1.28) (the upper right corner is missing), each term with "star above bullet" can be omitted, since the corresponding $\widehat{a}_{ij}$'s are zero. So the only combinations to be considered are those of Fig. 1.1. We finally insert the quantities from the right tableau in (1.28),

$$\widehat{\phi}(\curlyvee) = \sum b_i\, a_{ij}\, a_{ik}\, a_{kl} + \sum b_i^*\, b_j\, b_k\, a_{kl} + \sum b_i^*\, a_{ij}^*\, b_k\, a_{kl} + \sum b_i^*\, b_j\, a_{ik}^*\, b_l$$
$$+ \sum b_i^*\, a_{ij}^*\, a_{ik}^*\, b_l + \sum b_i^*\, b_j\, a_{ik}^*\, a_{kl}^* + \sum b_i^*\, a_{ij}^*\, a_{ik}^*\, a_{kl}^*,$$

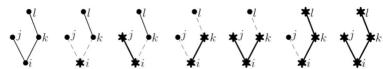
**Fig. 1.1.** Combinations with nonzero product.

and we observe that each factor of the type $b_j$ interrupts the summation, so that the terms decompose into factors of elementary weights of the individual methods as follows:

$$\widehat{\phi}(\vee\!\!\!\!\vee) = \phi(\vee\!\!\!\!\vee) + \phi^*(\bullet)\cdot\phi(\bullet)\phi(\mathbf{I}) + \phi^*(\mathbf{I})\cdot\phi(\mathbf{I}) + \phi^*(\mathbf{I})\cdot\phi(\bullet)\phi(\bullet)$$
$$+ \phi^*(\vee)\cdot\phi(\bullet) + \phi^*(\mathbf{V})\cdot\phi(\bullet) + \phi^*(\vee\!\!\!\!\vee).$$

The trees composed of the "star" nodes of $\tau$ in Fig. 1.1 constitute all possible "subtrees" $\theta$ (from the empty tree to $\tau$ itself) having the same root as $\tau$. This is the key for understanding the general result.

**Ordered Trees.** In order to formalize the procedure of Fig. 1.1, we introduce the set $OT$ of *ordered trees* recursively as follows: $\bullet \in OT$, and

if $\omega_1, \ldots, \omega_m \in OT$, then also the ordered $m$-tuple $(\omega_1, \ldots, \omega_m) \in OT$. (1.30)

As the name suggests, in the graphical representation of an ordered tree the order of the branches leaving cannot be permuted. Neglecting the ordering, a tree $\tau \in T$ can be considered as an equivalence class of ordered trees, denoted $\tau = \overline{\omega}$.

For example, the tree of Fig. 1.1 has two orderings, namely $\vee$ and $\vee$. We denote by $\nu(\tau)$ the number of possible orderings of the tree $\tau$. It is given by $\nu(\bullet) = 1$ and

$$\nu(\tau) = \frac{m!}{\mu_1!\mu_2!\cdots}\nu(\tau_1)\cdot\ldots\cdot\nu(\tau_m) \tag{1.31}$$

for $\tau = [\tau_1, \ldots, \tau_m]$, where the integers $\mu_1, \mu_2, \ldots$ are the numbers of equal trees among $\tau_1, \ldots, \tau_m$. This number is closely related to the symmetry coefficient $\sigma(\tau)$, because the product $\kappa(\tau) = \sigma(\tau)\nu(\tau)$ satisfies the recurrence relation

$$\kappa(\tau) = m!\,\kappa(\tau_1)\cdot\ldots\cdot\kappa(\tau_m). \tag{1.32}$$

We introduce the set $OST(\omega)$ of *ordered subtrees* of an ordered tree $\omega \in OT$ by

$$OST(\bullet) = \{\emptyset, \bullet\} \tag{1.33}$$
$$OST(\omega) = \{\emptyset\} \cup \{(\theta_1, \ldots, \theta_m)\,;\,\theta_i \in OST(\omega_i)\} \quad \text{for } \omega = (\omega_1, \ldots, \omega_m).$$

Each ordered subtree $\theta \in OST(\omega)$ is naturally associated with a tree $\overline{\theta} \in T$ obtained by neglecting the ordering and the $\emptyset$-components of $\theta$. For every tree $\tau \in T$ we choose, once and for all, an ordering. We denote this ordered tree by $\omega(\tau)$, and we put $OST(\tau) = OST(\omega(\tau))$.

For the tree of Fig. 1.1, considered as an ordered tree, the ordered subtrees correspond to the trees composed of the "star" nodes.

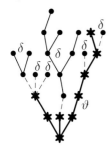

**The General Rule.** The general composition rule now becomes visible: for $\theta \in OST(\omega)$ we denote by $\omega \setminus \theta$ the "forest" collecting the trees left over when $\theta$ has been removed from the ordered tree $\omega$. For brevity we set $\tau \setminus \theta := \omega(\tau) \setminus \theta$. With the conventions $\phi^*(\theta) = \phi^*(\overline{\theta})$ and $\phi^*(\emptyset) = 1$ we then have

$$\widehat{\phi}(\tau) = \sum_{\theta \in OST(\tau)} \left( \phi^*(\theta) \cdot \prod_{\delta \in \tau \setminus \theta} \phi(\delta) \right). \tag{1.34}$$

This composition formula for the trees up to order 3 reads:

$$\widehat{\phi}(\bullet) = \phi^*(\emptyset) \cdot \phi(\bullet) + \phi^*(\bullet)$$
$$\widehat{\phi}(\mathord{\mathsf{J}}) = \phi^*(\emptyset) \cdot \phi(\mathord{\mathsf{J}}) + \phi^*(\bullet) \cdot \phi(\bullet) + \phi^*(\mathord{\mathsf{J}})$$
$$\widehat{\phi}(\mathord{\mathsf{V}}) = \phi^*(\emptyset) \cdot \phi(\mathord{\mathsf{V}}) + \phi^*(\bullet) \cdot \phi(\bullet)^2 + 2\phi^*(\mathord{\mathsf{J}}) \cdot \phi(\bullet) + \phi^*(\mathord{\mathsf{V}})$$
$$\widehat{\phi}(\mathord{\mathsf{\}}}) = \phi^*(\emptyset) \cdot \phi(\mathord{\mathsf{\}}}) + \phi^*(\bullet) \cdot \phi(\mathord{\mathsf{J}}) + \phi^*(\mathord{\mathsf{J}}) \cdot \phi(\bullet) + \phi^*(\mathord{\mathsf{\}}})$$

The tree $\tau = \mathord{\mathsf{V}}$ has the subtrees displayed in Fig. 1.2. It contains symmetries in that the third and fourth subtrees are topologically equivalent. This explains the factor 2 in the expression for the elementary weight.

**Fig. 1.2.** A tree with symmetry.

### III.1.4 Composition of B-Series

We now extend the above composition law to general B-series, i.e., we insert the B-series themselves into each other, as sketched in Fig. 1.3. This allows us to generalize Lemma 1.9 (because $hf(y)$ is a special B-series).

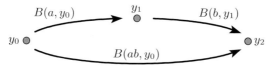

**Fig. 1.3.** Composition of B-series.

We start with an observation of Murua (see, e.g., Murua & Sanz-Serna (1999), p. 1083), namely that the proof of Lemma 1.9 remains the same if the function $hf(y)$ is replaced with any other function $hg(y)$; in this case (1.21) is replaced with

$$hg\bigl(B(a,y)\bigr) = hg + h^2 a(\bullet)g'f + h^3 a(\,\pounds\,)g'f'f + \frac{h^3}{2!}a(\bullet)^2 g''(f,f) \qquad (1.35)$$
$$+ h^4 a(\bullet)a(\,\pounds\,)g''(f'f,f) + \ldots .$$

Such series will reappear in Sect. III.3.1 below. Extending this idea further to, say, $f''(y)(v_1, v_2)$, where $v_1, v_2$ are two fixed vectors, we obtain

$$\begin{aligned} hf''\bigl(B(a,y)\bigr)(v_1,v_2) &= hf''(v_1,v_2) + h^2 a(\bullet)f'''(v_1,v_2,f) \qquad (1.36)\\ &+ h^3 a(\,\pounds\,)f'''(v_1,v_2,f'f) + \frac{1}{2!}h^3 a(\bullet)^2 f''''(v_1,v_2,f,f) \\ &+ h^4 a(\bullet)a(\,\pounds\,)f''''(v_1,v_2,f'f,f) + \ldots . \end{aligned}$$

This idea will lead to a direct proof of the following theorem of Hairer & Wanner (1974).

**Theorem 1.10.** *Let* $a : T \cup \{\emptyset\} \to \mathbb{R}$ *be a mapping satisfying* $a(\emptyset) = 1$ *and let* $b : T \cup \{\emptyset\} \to \mathbb{R}$ *be arbitrary. Then the B-series* $B(a,y)$ *inserted into* $B(b,\cdot)$ *is again a B-series*

$$B\Bigl(b, B(a,y)\Bigr) = B(ab, y), \qquad (1.37)$$

*where the group operation* $ab(\tau)$ *is as in (1.34), i.e.,*

$$ab(\tau) = \sum_{\theta \in OST(\tau)} b(\theta) \cdot a(\tau \setminus \theta) \quad \text{with} \quad a(\tau \setminus \theta) = \prod_{\delta \in \tau \setminus \theta} a(\delta). \qquad (1.38)$$

*Proof.* (a) In part (c) below we prove by induction on $|\vartheta|$, $\vartheta \in T$ that

$$\frac{h^{|\vartheta|}}{\sigma(\vartheta)} F(\vartheta)\bigl(B(a,y)\bigr) = \sum_{(\tau,\theta) \in A(\vartheta)} \frac{h^{|\tau|}}{\sigma(\tau)} a(\tau \setminus \theta) F(\tau)(y), \qquad (1.39)$$

where

$$A(\vartheta) = \bigl\{(\tau,\theta) \,;\, \tau \in T, \theta \in OST(\tau), \bar{\theta} = \vartheta\bigr\}.$$

Multiplying (1.39) by $b(\vartheta)$ and summing over all $\vartheta \in T$ yields the statement (1.37)-(1.38), because

$$\sum_{\vartheta \in T} \sum_{(\tau,\theta) \in A(\vartheta)} \cdot/\cdot = \sum_{\tau \in T} \sum_{\theta \in OST(\tau)} \cdot/\cdot .$$

(b) Choosing a different ordering of $\tau$ in the definition of $OST(\tau)$ yields the same sum in (1.39). Therefore (1.39) is equivalent to

$$\frac{h^{|\vartheta|}}{\sigma(\vartheta)} F(\vartheta)\bigl(B(a,y)\bigr) = \sum_{(\omega,\theta) \in \Omega(\vartheta)} \frac{h^{|\omega|}}{\sigma(\omega)\nu(\omega)} a(\omega \setminus \theta) F(\omega)(y), \qquad (1.40)$$

where

$$\Omega(\vartheta) = \bigl\{(\omega,\theta) \,;\, \omega \in OT, \theta \in OST(\omega), \bar{\theta} = \vartheta\bigr\},$$

and $\nu(\tau)$ is the number of orderings of the tree $\tau$, see (1.31). Functions defined on trees are naturally extended to ordered trees. In (1.40) we use $|\omega| = |\tau|$, $\sigma(\omega) = \sigma(\tau)$, $\nu(\omega) = \nu(\tau)$, $a(\omega \setminus \theta) = a(\tau \setminus \theta)$, and $F(\omega)(y) = F(\tau)(y)$ for $\bar{\omega} = \tau$.

(c) For $\vartheta = \bullet$ and $\omega = (\omega_1, \ldots, \omega_m)$ we have $a(\omega \setminus \theta) = a(\omega_1) \cdot \ldots \cdot a(\omega_m)$ if $\theta = \bullet$. Since we have a one-to-one correspondence $(\omega, \theta) \leftrightarrow \omega$ between $\Omega(\bullet)$ and $OT$, and since the expression in the sum of (1.40) is independent of the ordering of $\omega$, formula (1.40) is precisely Lemma 1.9.

To prove (1.40) for a general tree $\vartheta = [\vartheta_1, \ldots, \vartheta_l]$, we apply the idea put forward in (1.36) to $hf^{(l)}(B(a, y))(v_1, \ldots, v_l)$ with fixed $v_1, \ldots, v_l$, and obtain as in the proof of Lemma 1.9

$$hf^{(l)}(B(a,y))(v_1, \ldots, v_l) = \sum_{m \geq 0} \frac{1}{m!} \sum_{\tau_{l+1} \in T} \cdots \sum_{\tau_{l+m} \in T} \frac{h^{|\tau_{l+1}|+\ldots+|\tau_{l+m}|+1}}{\sigma(\tau_{l+1}) \cdot \ldots \cdot \sigma(\tau_{l+m})}$$

$$\cdot a(\tau_{l+1}) \cdot \ldots \cdot a(\tau_{l+m}) \cdot f^{(l+m)}(y)\Big(v_1, \ldots, v_l, F(\tau_{l+1})(y), \ldots, F(\tau_{l+m})(y)\Big).$$

Changing the sums over trees to sums over ordered trees we obtain

$$hf^{(l)}(B(a,y))(v_1, \ldots, v_l) = \sum_{m \geq 0} \frac{1}{m!} \sum_{\omega_{l+1} \in OT} \cdots \sum_{\omega_{l+m} \in OT} \frac{h^{|\omega_{l+1}|+\ldots+|\omega_{l+m}|+1}}{\kappa(\omega_{l+1}) \cdot \ldots \cdot \kappa(\omega_{l+m})}$$

$$\cdot a(\omega_{l+1}) \cdot \ldots \cdot a(\omega_{l+m}) \cdot f^{(l+m)}(y)\Big(v_1, \ldots, v_l, F(\omega_{l+1})(y), \ldots, F(\omega_{l+m})(y)\Big).$$

We insert $v_j = \frac{h^{|\vartheta_j|}}{\sigma(\vartheta_j)} F(\vartheta_j)(B(a, y))$ into this relation, and we apply our induction hypothesis

$$v_j = \frac{h^{|\vartheta_j|}}{\sigma(\vartheta_j)} F(\vartheta_j)\Big(B(a,y)\Big) = \sum_{(\omega_j, \theta_j) \in \Omega(\vartheta_j)} \frac{h^{|\omega_j|}}{\kappa(\omega_j)} a(\omega_j \setminus \theta_j) F(\omega_j)(y).$$

We then use the recursive definitions of $\sigma(\vartheta)$ and $F(\vartheta)(y)$ on the left-hand side. On the right-hand side we use the multilinearity of $f^{(l+m)}$, the recursive definitions of $|\omega|$, $\kappa(\omega)$, $F(\omega)(y)$ for $\omega = (\omega_1, \ldots, \omega_{l+m})$, and the facts that

$$a(\omega \setminus \theta) = a(\omega_1 \setminus \theta_1) \cdot \ldots \cdot a(\omega_l \setminus \theta_l) \cdot a(\omega_{l+1}) \cdot \ldots \cdot a(\omega_{l+m})$$

and

$$\sum_{(\omega_1, \theta_1) \in \Omega(\vartheta_1)} \cdots \sum_{(\omega_l, \theta_l) \in \Omega(\vartheta_l)} \sum_{\omega_{l+1} \in OT} \cdots \sum_{\omega_{l+m} \in OT} \cdot/\cdot = \frac{m! \mu_1! \mu_2! \cdot \ldots}{(l+m)!} \sum_{(\omega, \theta) \in \Omega_{l+m}(\vartheta)} \cdot/\cdot$$

where $\mu_1, \mu_2, \ldots$ count equal trees among $\vartheta_1, \ldots, \vartheta_l$, and $\Omega_{l+m}(\vartheta)$ consists of those pairs $(\omega, \theta) \in \Omega(\vartheta)$ for which $\omega$ is of the form $\omega = (\omega_1, \ldots, \omega_{l+m})$. The factorials appear, because to every $(l+m)$-tuple of the left-hand sum correspond $\binom{l+m}{m, \mu_1, \mu_2, \ldots}$ elements in $\Omega_{l+m}(\vartheta)$, obtained by permuting the order. This yields formula (1.40) and hence (1.39). □

**Example 1.11.** The composition laws for the trees of order $\leq 4$ are

$$ab(\bullet) = b(\emptyset) \cdot a(\bullet) + b(\bullet)$$
$$ab(\mathbf{\mathit{l}}) = b(\emptyset) \cdot a(\mathbf{\mathit{l}}) + b(\bullet) \cdot a(\bullet) + b(\mathbf{\mathit{l}})$$
$$ab(\mathbf{V}) = b(\emptyset) \cdot a(\mathbf{V}) + b(\bullet) \cdot a(\bullet)^2 + 2b(\mathbf{\mathit{l}}) \cdot a(\bullet) + b(\mathbf{V})$$
$$ab(\mathbf{\mathit{j}}) = b(\emptyset) \cdot a(\mathbf{\mathit{j}}) + b(\bullet) \cdot a(\mathbf{\mathit{l}}) + b(\mathbf{\mathit{l}}) \cdot a(\bullet) + b(\mathbf{\mathit{j}})$$
$$ab(\mathbf{W}) = b(\emptyset) \cdot a(\mathbf{W}) + b(\bullet) \cdot a(\bullet)^3 + 3b(\mathbf{\mathit{l}}) \cdot a(\bullet)^2 + 3b(\mathbf{V}) \cdot a(\bullet)$$
$$\qquad + b(\mathbf{W})$$
$$ab(\mathbf{\mathit{\Psi}}) = b(\emptyset) \cdot a(\mathbf{\mathit{\Psi}}) + b(\bullet) \cdot a(\bullet)a(\mathbf{\mathit{l}}) + b(\mathbf{\mathit{l}}) \cdot a(\mathbf{\mathit{l}}) + b(\mathbf{\mathit{l}}) \cdot a(\bullet)^2$$
$$\qquad + b(\mathbf{V}) \cdot a(\bullet) + b(\mathbf{\mathit{j}}) \cdot a(\bullet) + b(\mathbf{\mathit{\Psi}})$$
$$ab(\mathbf{Y}) = b(\emptyset) \cdot a(\mathbf{Y}) + b(\bullet) \cdot a(\mathbf{V}) + b(\mathbf{\mathit{l}}) \cdot a(\bullet)^2 + 2b(\mathbf{\mathit{j}}) \cdot a(\bullet)$$
$$\qquad + b(\mathbf{Y})$$
$$ab(\mathbf{\mathit{j}}) = b(\emptyset) \cdot a(\mathbf{\mathit{j}}) + b(\bullet) \cdot a(\mathbf{\mathit{j}}) + b(\mathbf{\mathit{l}}) \cdot a(\mathbf{\mathit{l}}) + b(\mathbf{\mathit{j}}) \cdot a(\bullet) + b(\mathbf{\mathit{j}})$$

**Remark 1.12.** The composition law (1.38) can alternatively be obtained from the corresponding formula (1.34) for Runge-Kutta methods by using the fact that B-series which represent Runge-Kutta methods are "dense" in the space of all B-series (see Theorem 306A of Butcher 1987).

### III.1.5 The Butcher Group

John C. Butcher,
born: 31 March 1933 in Auckland
(New Zealand)

The composition law (1.38) can be turned into a *group operation*, by introducing a *unit element*

$$e(\emptyset) = 1, \quad e(\tau) = 0 \quad \text{for } \tau \in T, \quad (1.41)$$

and by computing the *inverse element* of a given $a$. This is obtained recursively from the table of Example 1.11, by requiring $aa^{-1}(\tau) = 0$ and by inserting the previously known values of $a^{-1}(\vartheta)$. This gives for the first orders

$$a^{-1}(\bullet) = -a(\bullet)$$
$$a^{-1}(\mathbf{\mathit{l}}) = -a(\mathbf{\mathit{l}}) + a(\bullet)^2$$
$$a^{-1}(\mathbf{V}) = -a(\mathbf{V}) + 2a(\mathbf{\mathit{l}})a(\bullet) - a(\bullet)^3$$
$$a^{-1}(\mathbf{\mathit{j}}) = -a(\mathbf{\mathit{j}}) + 2a(\mathbf{\mathit{l}})a(\bullet) - a(\bullet)^3 \quad (1.42)$$

We can distinguish several realizations of this group:

$G_{RK}$ the set of Runge-Kutta schemes with composition (1.28);
$G_{EW}$ the set of elementary weights of Runge-Kutta schemes with the composition law (1.34);
$G_{TM}$ the set of tree mappings $a : T \cup \{\emptyset\} \to \mathbb{R}$ satisfying $a(\emptyset) = 1$ with composition (1.38);
$G_{BS}$ the set of B-series (1.23) satisfying $a(\emptyset) = 1$ with composition (1.37).

A technical difficulty concerns the group $G_{RK}$, where "reducible" schemes must be identified (by deleting unnecessary stages or by combining stages that give identical results) to the same "irreducible" method (see Butcher (1972), or Butcher & Wanner (1996), p. 140). The definition of $\phi(\tau)$ in Theorem 1.4 describes a group isomorphism from $G_{RK}$ to $G_{EW}$, further, $G_{EW}$ is a subgroup of $G_{TM}$ and Theorem 1.10 shows that formula (1.23) constitutes a group homomorphism from $G_{TM}$ to $G_{BS}$. Because the elementary differentials are independent (see, e.g., Hairer, Nørsett & Wanner (1993), Exercise 4 of Sect. II.2), the last two groups are isomorphic. The group $G_{RK}$ can also be extended by allowing "continuous" Runge-Kutta schemes with "infinitely many stages" (see Butcher (1972), or Butcher & Wanner (1996), p. 141). The term "Butcher group" was introduced by Hairer & Wanner (1974).

**Connection with Hopf Algebras and Quantum Field Theory.** A surprising connection between Runge-Kutta theory and renormalization in quantum field theory has been discovered recently by Brouder (2000). One denotes by a *Hopf algebra* a graded algebra which, besides the usual product, also possesses a *coproduct*, a tool used by H. Hopf (1941) [2] in his topological classification of certain manifolds. Hopf algebras generated by families of rooted trees proved to be extremely useful for simplifying the intricate combinatorics of renormalization (Kreimer 1998). Kreimer's Hopf algebra $\mathcal{H}$ is the space generated by linear combinations of families of rooted trees and the coproduct is a mapping $\triangle : \mathcal{H} \to \mathcal{H} \otimes \mathcal{H}$ which is, for the first trees, given by

$$\triangle(\bullet) = 1 \otimes \bullet + \bullet \otimes 1$$
$$\triangle(\mathord{\mathbf{\mathsf{l}}}) = 1 \otimes \mathord{\mathbf{\mathsf{l}}} + \bullet \otimes \bullet + \mathord{\mathbf{\mathsf{l}}} \otimes 1$$
$$\triangle(\mathsf{V}) = 1 \otimes \mathsf{V} + \bullet \otimes \bullet \bullet + 2 \mathord{\mathbf{\mathsf{l}}} \otimes \bullet + \mathsf{V} \otimes 1 \qquad (1.43)$$
$$\triangle(\mathord{\mathbf{\mathsf{\}}}}) = 1 \otimes \mathord{\mathbf{\mathsf{\}}}} + \bullet \otimes \mathord{\mathbf{\mathsf{l}}} + \mathord{\mathbf{\mathsf{l}}} \otimes \bullet + \mathord{\mathbf{\mathsf{\}}}} \otimes 1$$

It can be clearly seen, that this algebraic structure is precisely the one underlying the composition law of Example 1.11, so that the Butcher group $G_{TM}$ becomes the corresponding *character group*. The so-called *antipodes* of trees $\tau \in \mathcal{H}$, denoted by $S(\tau)$, are for the first trees

---
[2] Not to be confused with E. Hopf, the discoverer of the "Hopf bifurcation".

$$S(\bullet) = -\bullet$$
$$S(\mathord{\vcenter{\hbox{$\cdot$}}}) = -\mathord{\vcenter{\hbox{$\cdot$}}} + \bullet\bullet$$
$$S(\mathord{V}) = -\mathord{V} + 2\mathord{\cdot}\bullet - \bullet\bullet\bullet \qquad (1.44)$$
$$S(\mathord{\}}) = -\mathord{\}} + 2\mathord{\cdot}\bullet - \bullet\bullet\bullet$$

and, apparently, describes the *inverse element* (1.42) in the Butcher group.

## III.2 Order Conditions for Partitioned Runge-Kutta Methods

We now apply the ideas of the previous section to the creation of the order conditions for partitioned Runge-Kutta methods (II.2.2) of Sect. II.2. These results can then also be applied to Nyström methods.

### III.2.1 Bi-Coloured Trees and P-Series

Let us consider a partitioned system

$$\dot{y} = f(y, z), \qquad \dot{z} = g(y, z) \qquad (2.1)$$

(non-autonomous problems can be brought into this form by appending $\dot{t} = 1$). We start by computing the derivatives of its exact solution, which are to be inserted into the Taylor series expansion. By analogy with (1.4) we obtain in this case the derivatives of $y$ at $t_0$ as follows:

$$\dot{y} = f$$
$$\ddot{y} = f_y f + f_z g \qquad (2.2)$$
$$y^{(3)} = f_{yy}(f, f) + 2 f_{yz}(f, g) + f_{zz}(g, g) + f_y f_y f + f_y f_z g + f_z g_y f + f_z g_z g.$$

Here, $f_y, f_z, f_{yz}, \ldots$ denote partial derivatives and all terms are to be evaluated at $(y_0, z_0)$. Similar expressions are obtained for the derivatives of $z(t)$.

The terms occurring in these expressions are again called the *elementary differentials* $F(\tau)(y, z)$. For their graphical representation as a tree $\tau$, we distinguish between 'black' vertices for representing an $f$ and 'white' vertices for a $g$. Upwards pointing branches represent partial derivatives, with respect to $y$ if the branch leads to a black vertex, and with respect to $z$ if it leads to a white vertex. With this convention, the graph to the right corresponds to the expression $f_{zy}\bigl(g_{yz}(f, g), f\bigr)$ (see Table 2.1 for more examples).

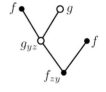

We denote by $TP$ the set of graphs obtained by the above procedure, and we call them (rooted) *bi-coloured trees*. The first graphs are $\bullet$ and $\circ$. By analogy with Definition 1.1, we denote by

III.2 Order Conditions for Partitioned Runge-Kutta Methods

**Table 2.1.** Bi-coloured trees, elementary differentials, and coefficients

| $|\tau|$ | $\tau$ | graph | $\alpha(\tau)$ | $F(\tau)$ | $\gamma(\tau)$ | $\phi(\tau)$ | $\sigma(\tau)$ |
|---|---|---|---|---|---|---|---|
| 1 | $\bullet$ | $\bullet$ | 1 | $f$ | 1 | $\sum_i b_i$ | 1 |
| 2 | $[\bullet]_y$ | | 1 | $f_y f$ | 2 | $\sum_{ij} b_i a_{ij}$ | 1 |
| 2 | $[\circ]_y$ | | 1 | $f_z g$ | 2 | $\sum_{ij} b_i \widehat{a}_{ij}$ | 1 |
| 3 | $[\bullet,\bullet]_y$ | | 1 | $f_{yy}(f,f)$ | 3 | $\sum_{ijk} b_i a_{ij} a_{ik}$ | 2 |
| 3 | $[\bullet,\circ]_y$ | | 2 | $f_{yz}(f,g)$ | 3 | $\sum_{ijk} b_i a_{ij} \widehat{a}_{ik}$ | 1 |
| 3 | $[\circ,\circ]_y$ | | 1 | $f_{zz}(g,g)$ | 3 | $\sum_{ijk} b_i \widehat{a}_{ij} \widehat{a}_{ik}$ | 2 |
| 3 | $[[\bullet]_y]_y$ | | 1 | $f_y f_y f$ | 6 | $\sum_{ijk} b_i a_{ij} a_{jk}$ | 1 |
| 3 | $[[\circ]_y]_y$ | | 1 | $f_y f_z g$ | 6 | $\sum_{ijk} b_i a_{ij} \widehat{a}_{jk}$ | 1 |
| 3 | $[[\bullet]_z]_y$ | | 1 | $f_z g_y f$ | 6 | $\sum_{ijk} b_i \widehat{a}_{ij} a_{jk}$ | 1 |
| 3 | $[[\circ]_z]_y$ | | 1 | $f_z g_z g$ | 6 | $\sum_{ijk} b_i \widehat{a}_{ij} \widehat{a}_{jk}$ | 1 |
| 1 | $\circ$ | $\circ$ | 1 | $g$ | 1 | $\sum_i \widehat{b}_i$ | 1 |
| 2 | $[\bullet]_z$ | | 1 | $g_y f$ | 2 | $\sum_{ij} \widehat{b}_i a_{ij}$ | 1 |
| | etc | etc | | etc | | etc | |

$$[\tau_1,\ldots,\tau_m]_y \quad \text{and} \quad [\tau_1,\ldots,\tau_m]_z, \quad \tau_1,\ldots,\tau_m \in TP$$

the bi-coloured trees obtained by connecting the roots of $\tau_1,\ldots,\tau_m$ to a new root, which is $\bullet$ in the first case, and $\circ$ in the second. Furthermore, we denote by $TP_y$ and $TP_z$ the subsets of $TP$ which are formed by trees with black and white roots, respectively. Hence, the trees of $TP_y$ correspond to derivatives of $y(t)$, whereas those of $TP_z$ correspond to derivatives of $z(t)$.

As in Definition 1.2 we denote the number of vertices of $\tau \in TP$ by $|\tau|$, the *order* of $\tau$. The symmetry coefficient $\sigma(\tau)$ is again defined by

$$\sigma(\bullet) = \sigma(\circ) = 1,$$

and, for $\tau = [\tau_1,\ldots,\tau_m]_y$ or $\tau = [\tau_1,\ldots,\tau_m]_z$, by

$$\sigma(\tau) = \sigma(\tau_1) \cdot \ldots \cdot \sigma(\tau_m) \cdot \mu_1! \mu_2! \ldots, \tag{2.3}$$

where the integers $\mu_1, \mu_2, \ldots$ count equal trees among $\tau_1,\ldots,\tau_m \in TP$. This is formally the same definition as in Sect. III.1. Observe, however, that $\sigma(\tau)$ depends on the colouring of the vertices. For example, we have $\sigma(\vee) = 2$, but $\sigma(\vee) = 1$. By analogy with Definition 1.8 we have:

**Definition 2.1 (P-Series).** For a mapping $a : TP \cup \{\emptyset_y, \emptyset_z\} \to \mathbb{R}$ a series of the form

$$P\Big(a, (y, z)\Big) = \begin{pmatrix} a(\emptyset_y)y + \displaystyle\sum_{\tau \in TP_y} \frac{h^{|\tau|}}{\sigma(\tau)} a(\tau) F(\tau)(y, z) \\ a(\emptyset_z)z + \displaystyle\sum_{\tau \in TP_z} \frac{h^{|\tau|}}{\sigma(\tau)} a(\tau) F(\tau)(y, z) \end{pmatrix}$$

is called a *P-series*.

The following results correspond to Lemma 1.9 and formula (1.26). They are obtained in exactly the same manner as the corresponding results for non-partitioned Runge-Kutta methods (Sect. III.1). We therefore omit their proofs.

**Lemma 2.2.** *Let* $a : TP \cup \{\emptyset_y, \emptyset_z\} \to \mathbb{R}$ *satisfy* $a(\emptyset_y) = a(\emptyset_z) = 1$. *Then*

$$h \begin{pmatrix} f\big(P(a, (y, z))\big) \\ g\big(P(a, (y, z))\big) \end{pmatrix} = P\Big(a', (y, z)\Big),$$

*where* $a'(\emptyset_y) = a'(\emptyset_z) = 0$, $a'(\bullet) = a'(\circ) = 1$, *and*

$$a'(\tau) = a(\tau_1) \cdot \ldots \cdot a(\tau_m), \tag{2.4}$$

*if either* $\tau = [\tau_1, \ldots, \tau_m]_y$ *or* $\tau = [\tau_1, \ldots, \tau_m]_z$. □

**Theorem 2.3 (P-Series of Exact Solution).** *The exact solution of (2.1) is a P-series* $\big(y(t_0 + h), z(t_0 + h)\big) = P\big(\mathbf{e}, (y_0, z_0)\big)$, *where* $\mathbf{e}(\emptyset_y) = \mathbf{e}(\emptyset_z) = 1$ *and*

$$\mathbf{e}(\tau) = \frac{1}{\gamma(\tau)} \quad \text{for all} \quad t \in TP \tag{2.5}$$

*where the* $\gamma(\tau)$ *have the same values as for mono-coloured trees.* □

### III.2.2 Order Conditions for Partitioned Runge-Kutta Methods

The next result corresponds to Theorem 1.4 and is a consequence of Lemma 2.2.

**Theorem 2.4 (P-Series of Numerical Solution).** *The numerical solution of a partitioned Runge-Kutta method (II.2.2) is a P-series* $(y_1, z_1) = P\big(\phi, (y_0, z_0)\big)$, *where* $\phi(\emptyset_y) = \phi(\emptyset_z) = 1$ *and*

$$\phi(\tau) = \begin{cases} \sum_{i=1}^{s} b_i \phi_i(\tau) & \text{for} \quad \tau \in TP_y \\ \sum_{i=1}^{s} \widehat{b}_i \phi_i(\tau) & \text{for} \quad \tau \in TP_z. \end{cases} \tag{2.6}$$

*The expression* $\phi_i(\tau)$ *is defined by* $\phi_i(\bullet) = \phi_i(\circ) = 1$ *and by*

$$\phi_i(\tau) = \psi_i(\tau_1) \cdot \ldots \cdot \psi_i(\tau_m) \quad \text{with} \quad \psi_i(\tau_k) = \begin{cases} \sum_{j_k=1}^{s} a_{ij_k} \phi_{j_k}(\tau_k) & \text{if } \tau_k \in TP_y \\ \sum_{j_k=1}^{s} \widehat{a}_{ij_k} \phi_{j_k}(\tau_k) & \text{if } \tau_k \in TP_z \end{cases} \tag{2.7}$$

*for* $\tau = [\tau_1, \ldots, \tau_m]_y$ *or* $\tau = [\tau_1, \ldots, \tau_m]_z$.

## III.2 Order Conditions for Partitioned Runge-Kutta Methods

*Proof.* These formulas result from Lemma 2.2 by writing $(hk_i, h\ell_i)$ from the formulas (II.2.2) as a P-series $(hk_i, h\ell_i) = P(\phi_i, (y_0, z_0))$ so that

$$\left(h \sum_j a_{ij} k_j, h \sum_j \widehat{a}_{ij} \ell_j\right) = P(\psi_i, (y_0, z_0))$$

is also a P-series. Observe that equation (2.6) corresponds to (1.16) (where $\mathbf{g}_i$ has to be replaced with $\phi_i$) and that formula (2.7) comprises (1.13) and (1.15), where we now write $\psi_i$ instead of $\mathbf{u}_i$. □

The expressions $\phi(\tau)$ are shown in Table 2.1 for all trees in $TP_y$ up to order $|\tau| \leq 3$. A similar table must be added for trees in $TP_z$, where all roots are white and all $b_i$ are replaced with $\widehat{b}_i$. The general rule is the following: attach to every vertex a summation index. Then, the expression $\phi(\tau)$ is a sum over all summation indices with the summand being a product of $b_i$ or $\widehat{b}_i$ (depending on whether the root '$i$' is black or white) and of $a_{jk}$ (if '$k$' is black) or $\widehat{a}_{jk}$ (if '$k$' is white), for each vertex '$k$' directly above '$j$'.

**Theorem 2.5 (Order Conditions).** *A partitioned Runge-Kutta method (II.2.2) has order $r$, i.e., $y_1 - y(t_0 + h) = \mathcal{O}(h^{r+1})$, $z_1 - z(t_0 + h) = \mathcal{O}(h^{r+1})$, if and only if*

$$\phi(\tau) = \frac{1}{\gamma(\tau)} \quad \text{for } \tau \in TP_y \cup TP_z \text{ with } |\tau| \leq r. \tag{2.8}$$

*Proof.* This corresponds to Theorem 1.5 and is seen by comparing the expansions of Theorems 2.4 and 2.3. □

**Example 2.6.** We see that not only does every individual Runge-Kutta method have to be of order $r$, but also the so-called *coupling conditions* between the coefficients of both methods must hold. The order conditions mentioned above (see formulas (II.2.3) and (II.2.5)) correspond to the trees ↑, ↓, ↑ and ↓. For the tree sketched below we obtain

$$\sum_{i,j,k,l,m,n,p,q,r} b_i \widehat{a}_{ij} \widehat{a}_{jm} \widehat{a}_{in} a_{ik} \widehat{a}_{kl} a_{lq} a_{lr} a_{kp} = \frac{1}{9 \cdot 2 \cdot 5 \cdot 3}$$

or, by using $\sum_j a_{ij} = c_i$ and $\sum_j \widehat{a}_{ij} = \widehat{c}_i$,

$$\sum_{i,j,k,l} b_i \widehat{c}_i \widehat{a}_{ij} \widehat{c}_j a_{ik} c_k \widehat{a}_{kl} c_l^2 = \frac{1}{270}.$$

### III.2.3 Order Conditions for Nyström Methods

A 'modern' order theory for Nyström methods (II.2.11) of Sect. II.2.3 was first given in 1976 by Hairer & Wanner (see Sect. II.14 of Hairer, Nørsett & Wanner 1993).

Later it turned out that these conditions are obtained easily by applying the theory of partitioned Runge-Kutta methods to the system

$$\dot{y} = z \qquad \dot{z} = g(y, z), \tag{2.9}$$

which is of the form (2.1). This function has the partial derivative $f_z = I$ and all other derivatives of $f$ are zero. As a consequence, many elementary differentials are zero and the corresponding order conditions can be omitted. The only trees remaining are those for which

"black vertices have at most one son and this son must be white". (2.10)

**Example 2.7.** The tree sketched below apparently satisfies condition (2.10) and the corresponding order condition becomes, by Theorem 2.4 and formula (2.8),

$$\sum_{i,j,k,\ldots,v} b_i \widehat{a}_{ij} \widehat{a}_{jk} a_{km} a_{kn} \widehat{a}_{kp} \widehat{a}_{jq} a_{qr} \widehat{a}_{rs} a_{j\ell} \widehat{a}_{\ell t} a_{tu} a_{tv} = \frac{1}{13 \cdot 12 \cdot 4 \cdot 3 \cdot 2 \cdot 4 \cdot 3}.$$

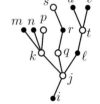

Due to property (2.10), each $a_{ik}$ inside the tree comes with a corresponding $\widehat{a}_{kj}$, and by (2.10), both factors contract to an $\overline{a}_{ij}$; similarly, the black root is only connected to one white vertex, the corresponding $b_i \widehat{a}_{ij}$ simplifies to $\overline{b}_j$. We thus get

$$\sum_{j,k,q,s,t} \overline{b}_j \widehat{a}_{jk} c_k^2 \widehat{c}_k \widehat{a}_{jq} \overline{a}_{qs} \overline{a}_{jt} c_t^2 = \frac{1}{13 \cdot 3456}.$$

Each of the above order conditions for a tree in $TP_y$ has a 'twin' in $TP_z$ of one order lower with the root cut off. For the above example this twin becomes

$$\sum_{j,k,q,s,t} b_j \widehat{a}_{jk} c_k^2 \widehat{c}_k \widehat{a}_{jq} \overline{a}_{qs} \overline{a}_{jt} c_t^2 = \frac{1}{3456}.$$

We need only consider the trees in $TP_z$ if

$$\overline{b}_i = b_i(1 - c_i)$$

is satisfied (see Lemma II.14.13 of Hairer, Nørsett & Wanner (1993), Sect. II.14).

**Remark 2.8.** Strictly speaking, the theory of partitioned methods is applicable to Nyström methods only if the matrix $(\widehat{a}_{ij})$ is invertible. However, since we arrive at expansions with a finite number of algebraic conditions, we can recover the singular case by a continuous perturbation of the coefficients.

**Equations without Friction.** Although condition (2.10) already eliminates many order conditions, Nyström methods for the general problem $\ddot{y} = g(y, \dot{y})$ cannot be much better than an excellent Runge-Kutta method applied pairwise to system (2.9).

There *is*, however, an important special case where much more progress is possible, namely equations of the type

$$\ddot{y} = g(y), \tag{2.11}$$

which corresponds to motion without friction. In this case, the function for $\dot{z}$ in (2.9) is *independent of* $z$, and in addition to (2.10) we have a second condition, namely

$$\text{"white vertices have only black sons"}. \tag{2.12}$$

Both conditions reduce the remaining trees drastically. Along each branch, there occur alternating black and white vertices. Ramifications only happen at white vertices. This case allows the construction of excellent numerical methods of high orders. For example, the following 13 trees

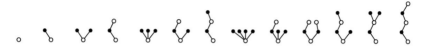

assure order 5, whereas ordinary Runge-Kutta theory requires 17 conditions for this order. See Hairer, Nørsett & Wanner (1993), pages 291f, for tables, examples and references.

## III.3 Order Conditions for Composition Methods

We have seen in the preceding chapter that composition methods of arbitrarily high order can be obtained with the use of Theorem II.4.1. However, as demonstrated in Fig. II.5.2, these methods are not attractive for high orders. This section is devoted to the derivation of order conditions, which then allow the construction of optimal high order composition methods.

The order conditions for these methods are often derived via the Baker-Campbell-Hausdorff formula. This will be the subject of Sect. III.5 below. Only very recently, Murua & Sanz-Serna (1999) have found an elegant theory based on the idea of B-series. This paper has largely inspired the subsequent presentation.

### III.3.1 Introduction

The principal tool in this section is the Taylor series expansion

$$\Phi_h(y) = y + hd_1(y) + h^2 d_2(y) + h^3 d_3(y) + \ldots \tag{3.1}$$

of the basic method. The only hypothesis which we require for this method is *consistency*, i.e., that

$$d_1(y) = f(y). \tag{3.2}$$

All other functions $d_i(y)$ are arbitrary.

The underlying idea for obtaining the expansions for composition methods is, in fact, very simple: we just insert the series (3.1), with varying values of $h$, into itself. All our experience from Sect. III.1.2 with the insertion of a B-series into a function will certainly be helpful. We demonstrate this for the case of the composition $\Psi_h = \Phi_{\alpha_2 h} \circ \Phi_{\alpha_1 h}$. Applied to an initial value $y_0$, this gives with (3.1)

$$y_1 = \Phi_{\alpha_1 h}(y_0) = y_0 + h\alpha_1 d_1(y_0) + h^2 \alpha_1^2 d_2(y_0) + \ldots$$
$$y_2 = \Phi_{\alpha_2 h}(y_1) = y_1 + h\alpha_2 d_1(y_1) + h^2 \alpha_2^2 d_2(y_1) + \ldots . \quad (3.3)$$

We now insert the first series into the second, in the same way as we did in (1.35). Then, for example, the term $h^2 \alpha_2^2 d_2(y_1)$ becomes

$$y_2 = \ldots + h^2 \alpha_2^2 d_2(y_0) + h^3 \alpha_2^2 \alpha_1 d_2'(y_0) d_1(y_0) \quad (3.4)$$
$$+ h^4 \alpha_2^2 \alpha_1^2 d_2'(y_0) d_2(y_0) + \frac{h^4}{2} \alpha_2^2 \alpha_1^2 d_2''(y_0)(d_1(y_0), d_1(y_0)) + \ldots$$

We see that we arrive at 'generalized' B-series, where the elementary differentials contain not only *one* function, but are composed of *infinitely many* functions and their derivatives. We symbolize the four terms written in (3.4) by the trees

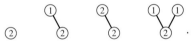

This leads us to the following definition.

**Definition 3.1 ($\infty$-Trees, $B_\infty$-series).** We extend Definitions 1.1 and 1.2 to $T_\infty$, the *set of all rooted trees where each vertex bears a positive integer* without any further restriction, and use the notation

$$①, ②, ③, \ldots = \text{ the trees with one vertex};$$
$$[\tau_1, \ldots, \tau_m]_i = \text{ the tree } \tau \text{ formed by a new root } ⓘ \text{ connected to } \tau_1, \ldots, \tau_m;$$
$$F(ⓘ)(y) = d_i(y);$$
$$F(\tau)(y) = d_i^{(m)}(y)(F(\tau_1)(y), \ldots, F(\tau_m)(y)) \text{ for } \tau \text{ as above};$$
$$|\tau| = 1 + |\tau_1| + \ldots + |\tau_m|, \quad \text{the number of vertices of } \tau;$$
$$||\tau|| = i + ||\tau_1|| + \ldots + ||\tau_m||, \quad \text{the sum of the labels of } \tau;$$
$$\sigma(\tau) = \mu_1! \mu_2! \cdot \ldots \cdot \sigma(\tau_1) \cdot \ldots \cdot \sigma(\tau_m),$$
$$\quad \text{where } \mu_1, \mu_2, \ldots \text{ count equal trees among } \tau_1, \ldots, \tau_m,$$
$$\quad \text{the *symmetry coefficient* respecting the labels};$$
$$i(\tau) = i, \quad \text{the *label of the root* of } \tau.$$

For a map $a : T_\infty \cup \{\emptyset\} \to \mathbb{R}$ we write

$$B_\infty(a, y) = a(\emptyset) y + \sum_{\tau \in T_\infty} \frac{h^{||\tau||}}{\sigma(\tau)} a(\tau) F(\tau)(y) \quad (3.5)$$

which extends the *notion of B-series* to the new situation.

## III.3 Order Conditions for Composition Methods

**Example 3.2.** For the tree

$$\tau = \vcenter{\hbox{(tree)}} \quad \Leftrightarrow \quad \tau = [\tau_1, \tau_2]_4 \quad \text{where} \quad \tau_1 = \text{①}, \quad \tau_2 = \vcenter{\hbox{(tree)}} \tag{3.6}$$

we have

$$F(\tau)(y) = d_4''(y)\big(d_1(y), d_7'''(y)\big(d_5(y), d_6(y), d_6(y)\big)\big)$$

$$\tau = [\text{①},[\text{⑤},\text{⑥},\text{⑥}]_7]_4, \quad |\tau| = 6, \quad \|\tau\| = 29, \quad \sigma(\tau) = 2, \quad i(\tau) = 4\,.$$

The above calculations for (3.4) are governed by the following lemma.

**Lemma 3.3.** *For a series $B_\infty(a, y)$ with $a(\emptyset) = 1$ we have*

$$h^i d_i \big(B_\infty(a, y)\big) = \sum_{\tau \in T_\infty, i(\tau) = i} \frac{h^{\|\tau\|}}{\sigma(\tau)} a'(\tau)\, F(\tau)(y), \tag{3.7}$$

*where $a'(\text{①}) = 1$ and*

$$a'(\tau) = a(\tau_1) \cdot \ldots \cdot a(\tau_m) \qquad \text{for } \tau = [\tau_1, \ldots, \tau_m]_i. \tag{3.8}$$

*Proof.* This is a straightforward extension of Lemma 1.9 with exactly the same proof. □

The preceding lemma leads directly to the order conditions for composition methods. However, if we continue with compositions of the type (II.4.1), we arrive at conditions without real solutions. We therefore turn to compositions including the adjoint method as well.

### III.3.2 The General Case

As in (II.4.6), we consider

$$\Psi_h = \Phi_{\alpha_s h} \circ \Phi^*_{\beta_s h} \circ \ldots \circ \Phi_{\alpha_2 h} \circ \Phi^*_{\beta_2 h} \circ \Phi_{\alpha_1 h} \circ \Phi^*_{\beta_1 h}, \tag{3.9}$$

and we obtain with the help of the above lemma the corresponding $B_\infty$-series.

**Lemma 3.4 (Recurrence Relations).** *The following compositions are $B_\infty$-series*

$$\begin{aligned}\big(\Phi^*_{\beta_k h} \circ \ldots \circ \Phi_{\alpha_1 h} \circ \Phi^*_{\beta_1 h}\big)(y) &= B_\infty(b_k, y) \\ \big(\Phi_{\alpha_k h} \circ \Phi^*_{\beta_k h} \circ \ldots \circ \Phi_{\alpha_1 h} \circ \Phi^*_{\beta_1 h}\big)(y) &= B_\infty(a_k, y).\end{aligned} \tag{3.10}$$

*Their coefficients are recursively given by $a_k(\emptyset) = 1$, $b_k(\emptyset) = 1$, $a_0(\tau) = 0$ for all $\tau \in T_\infty$, and*

$$\begin{aligned}b_k(\tau) &= a_{k-1}(\tau) - (-\beta_k)^{i(\tau)}\, b'_k(\tau), \\ a_k(\tau) &= b_k(\tau) + \alpha_k^{i(\tau)}\, b'_k(\tau).\end{aligned} \tag{3.11}$$

*Proof.* The coefficients $a_0(\tau)$ correspond to the identity map $B_\infty(a_0, y) = y$. The second formula of (3.11) follows from

$$B_\infty(a_k, y) = \Phi_{\alpha_k h}\big(B_\infty(b_k, y)\big) = B_\infty(b_k, y) + \sum_{i \geq 1} \alpha_k^i h^i d_i \big(B_\infty(b_k, y)\big),$$

and from an application of Lemma 3.3.

The relation $B_\infty(b_k, y) = \Phi^*_{\beta_k h}\big(B_\infty(a_{k-1}, y)\big)$, which involves the adjoint method, needs a little trick: we write it as $B_\infty(a_{k-1}, y) = \Phi_{-\beta_k h}\big(B_\infty(b_k, y)\big)$ (remember that $\Phi_h^* = \Phi_{-h}^{-1}$), apply Lemma 3.3 again, and reverse the formula. This gives the first equation of (3.11). □

Adding the equations of (3.11), we get

$$a_k(\tau) = a_{k-1}(\tau) + \big(\alpha_k^{i(\tau)} - (-\beta_k)^{i(\tau)}\big) b_k'(\tau). \tag{3.12}$$

Because of $b_k'(①) = 1$, we obtain

$$\begin{aligned}
a_k(①) &= \sum_{\ell=1}^{k} \big(\alpha_\ell^i - (-\beta_\ell)^i\big) \\
b_k(①) &= \sum_{\ell=1}^{k-1} \alpha_\ell^i - \sum_{\ell=1}^{k}(-\beta_\ell)^i = {\sum_{\ell=1}^{k}}' \big(\alpha_\ell^i - (-\beta_\ell)^i\big).
\end{aligned} \tag{3.13}$$

The fact that, for $b_k(①)$, the sum of $(-\beta_\ell)^i$ is from 1 to $k$, but the sum of $\alpha_\ell^i$ is only from 1 to $k-1$, has been *indicated by a prime* attached to the summation symbol. Continuing to apply the formulas (3.11) and (3.12) to more and more complicated trees, we quickly understand the general rule for the coefficients of an arbitrary tree.

**Example 3.5.** The tree $\tau$ in (3.6) gives

$$a_s(\tau) = \sum_{k=1}^{s} (\alpha_k^4 - \beta_k^4) {\sum_{\ell=1}^{k}}' (\alpha_\ell + \beta_\ell)$$
$$\cdot {\sum_{m=1}^{k}}' (\alpha_m^7 + \beta_m^7) {\sum_{n=1}^{m}}' (\alpha_n^5 + \beta_n^5) \bigg({\sum_{p=1}^{m}}' (\alpha_p^6 - \beta_p^6)\bigg)^2. \tag{3.14}$$

**The Order Conditions.** The exact solution of $\dot{y} = f(y)$ is a $B$-series $y(t_0 + h) = B(\mathbf{e}, y_0)$ (see (1.26)). Since $d_1(y) = f(y)$, every $B$-series is also a $B_\infty$-series with $\mathbf{e}(\tau) = 0$ for trees with at least one label different from 1. Therefore, we also have $y(t_0 + h) = B_\infty(\mathbf{e}, y_0)$, where the coefficients $\mathbf{e}(\tau)$ satisfy $\mathbf{e}(①) = 1$, $\mathbf{e}(\tau) = 0$ if $i(\tau) > 1$, and

$$\mathbf{e}(\tau) = \frac{1}{|\tau|} \mathbf{e}(\tau_1) \cdot \ldots \cdot \mathbf{e}(\tau_m) \qquad \text{for } \tau = [\tau_1, \ldots, \tau_m]_1. \tag{3.15}$$

**Theorem 3.6.** *The composition method $\Psi_h(y) = B_\infty(a_s, y)$ of (3.9) has order $p$ if*

$$a_s(\tau) = \mathbf{e}(\tau) \quad \text{for } \tau \in T_\infty \text{ with } ||\tau|| \leq p. \tag{3.16}$$

*Proof.* This follows from a comparison of the $B_\infty$-series for the numerical and the exact solution. For the necessity of (3.16), the independence of the elementary differentials has to be studied as in Exercise 3. □

## III.3.3 Reduction of the Order Conditions

The order conditions of the foregoing section are indeed beautiful, but for the moment they are not of much use, because of the enormous number of trees in $T_\infty$ of a certain order. For example, there are 166 trees in $T_\infty$ with $||\tau|| \leq 6$. Fortunately, the equations are not all independent, as we shall see now.

**Definition 3.7 (Butcher 1972, Murua & Sanz-Serna 1999).** For two trees in $T_\infty$, $u = [u_1, \ldots, u_m]_i$ and $v = [v_1, \ldots, v_l]_j$, we denote

$$u \circ v := [u_1, \ldots, u_m, v]_i, \qquad u \times v := [u_1, \ldots, u_m, v_1, \ldots, v_l]_{i+j} \tag{3.17}$$

and call them the *Butcher product* and *merging product*, respectively (see Fig. 3.1).

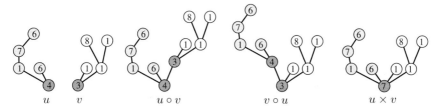

**Fig. 3.1.** The Butcher product and the merging product

The merging product is associative and commutative, the Butcher product is neither of the two. To simplify the notation, we write products of *several* factors without parentheses, when we mean evaluation from left to right:

$$u \circ v_1 \circ v_2 \circ \ldots \circ v_s = (((u \circ v_1) \circ v_2) \circ \ldots) \circ v_s. \tag{3.18}$$

Here the factors $v_1, \ldots, v_s$ can be freely permuted.

All subsequent results concern properties of $a_k(\tau)$ as well as $b_k(\tau)$, valid for all $k$. To avoid writing all formulas twice, we replace $a_k(\tau)$ and $b_k(\tau)$ everywhere by a neutral symbol $c(\tau)$.

**Lemma 3.8 (Switching Lemma).** *All $a_k, b_k$ of Lemma 3.4 satisfy, for all $u, v \in T_\infty$, the relation*

$$c(u \circ v) + c(v \circ u) = c(u) \cdot c(v) - c(u \times v). \tag{3.19}$$

*Proof.* The recursion formulas (3.11) are of the form

$$a(\tau) = b(\tau) + \alpha^{i(\tau)} b'(\tau). \tag{3.20}$$

We arrange this formula, for all five trees of Fig. 3.1, as follows:

$$\begin{aligned}
&\phantom{=}\ a(u \circ v) \quad + \quad a(v \circ u) \quad + \quad a(u \times v) \quad - \quad a(u)a(v) \\
&= \ b(u \circ v) \quad + \quad b(v \circ u) \quad + \quad b(u \times v) \quad - \quad b(u)b(v) \\
&\phantom{=}\ + \ \alpha^{i(u)} b'(u \circ v) + \alpha^{i(v)} b'(v \circ u) + \alpha^{i(u)+i(v)} b'(u \times v) \\
&\phantom{=}\ - \ \alpha^{i(u)} b'(u)b(v) - \alpha^{i(v)} b'(v)b(u) - \alpha^{i(u)}\alpha^{i(v)} b'(u)b'(v)\ .
\end{aligned}$$

Because of $b'(u \circ v) = b'(u)b(v)$ and $b'(u \times v) = b'(u)b'(v)$, the last two rows cancel, hence

$$a(\tau) \text{ satisfies (3.19)} \quad \Leftrightarrow \quad b(\tau) \text{ satisfies (3.19)}. \tag{3.21}$$

Thus, beginning with $a_0$, then $b_1$, then $a_1$, etc., all $a_k$ and $b_k$ must satisfy (3.19). □

The Switching Lemma 3.8 reduces considerably the number of order conditions. Since the right-hand expression involves only trees with $|\tau| < |u \circ v|$, and since relation (3.19) is also satisfied by $e(\tau)$, an induction argument shows that the order conditions (3.16) for the trees $u \circ v$ and $v \circ u$ are equivalent. The operation $u \circ v \mapsto v \circ u$ consists simply in switching the root from one vertex to the next. By repeating this argument, we see that we can freely move the root inside the graph, and of all these trees, only one needs to be retained. For order 6, for example, there remain 68 conditions out of the original 166.

Our next results show how relation (3.19) also generates a considerable amount of reductions of the order conditions. These ideas (for the special situation of symplectic methods) have already been exploited by Calvo & Hairer (1995b).

**Lemma 3.9.** *Assume that all $b_k$ of Lemma 3.4 satisfy a relation of the form*

$$\sum_{i=1}^{N} A_i \prod_{j=1}^{m_i} c(u_{ij}) = 0 \tag{3.22}$$

*with all $m_i > 0$. Then, for any tree $w$, all $a_k$ and $b_k$ satisfy the relation*

$$\sum_{i=1}^{N} A_i\, c(w \circ u_{i1} \circ u_{i2} \circ \ldots \circ u_{i,m_i}) = 0. \tag{3.23}$$

*Proof.* The relation (3.20), written for the tree $w \circ u_{i1} \circ u_{i2} \circ \ldots \circ u_{i,m_i}$, is

$$a(w \circ u_{i1} \circ \ldots \circ u_{i,m_i}) = b(w \circ u_{i1} \circ \ldots \circ u_{i,m_i})$$
$$+ \alpha^{i(w)} b'(w) b(u_{i1}) \cdot \ldots \cdot b(u_{i,m_i}).$$

Multiplying with $A_i$ and summing over $i$, this shows that, under the hypothesis (3.22) for $b$, the relation (3.23) holds for $b$ if and only if it holds for $a$. The coefficients $a_0(\tau) = 0$ for the identity map satisfy (3.22) and (3.23) because $m_i > 0$. Starting from this, we again conclude (3.23) recursively for all $a_k$ and $b_k$. □

The following lemma [3] extends formula (3.19) to the case of *several* factors.

**Lemma 3.10.** *For any three trees $u, v, w$ all $a_k, b_k$ of Lemma 3.4 satisfy a relation*

$$c(u \circ v \circ w) + c(v \circ u \circ w) + c(w \circ u \circ v) = c(u) \cdot c(v) \cdot c(w) + \ldots, \quad (3.24)$$

*where the dots indicate a linear combination of products $\prod_j c(v_j)$ with $|v_1| + |v_2| + \ldots < |u| + |v| + |w|$ and, for each term, at least one of the $v_j$ possesses a label larger than one. The general formula, for $m$ trees $u_1, \ldots, u_m$, is*

$$\sum_{i=1}^{m} c(u_i \circ u_1 \circ \ldots \circ u_{i-1} \circ u_{i+1} \circ \ldots \circ u_m) = \prod_{i=1}^{m} c(u_i) + \ldots. \quad (3.25)$$

*Proof.* We apply Lemma 3.9 to (3.19) and obtain

$$c(w \circ (u \circ v)) + c(w \circ (v \circ u)) = c(w \circ u \circ v) - c(w \circ (u \times v)). \quad (3.26)$$

Next, we apply the Switching Lemma 3.8 to the trees to the left and get

$$c(w \circ (u \circ v)) + c(u \circ v \circ w) = c(w) \cdot c(u \circ v) - c(w \times (u \circ v))$$
$$c(w \circ (v \circ u)) + c(v \circ u \circ w) = c(w) \cdot c(v \circ u) - c(w \times (v \circ u)).$$

Adding these formulas and subtracting (3.26) gives

$$c(u \circ v \circ w) + c(v \circ u \circ w) + c(w \circ u \circ v) = c(w)(c(u \circ v) + c(v \circ u)) + \ldots$$

which becomes (3.24) after another use of the Switching Lemma. Thereby, everything which goes into "$+ \ldots$" contains somewhere a merging product, whose roots introduce necessarily labels larger than one.

Continuing like this, we get recursively (3.25) for all $m$. □

In order that the further simplifications do not turn into chaos, we fix, once and for all, a *total order relation* (written $<$) on $T_\infty$, where we only require that the order respects the number of vertices, i.e., that

$$u < v \quad \text{whenever} \quad |u| < |v|. \quad (3.27)$$

Similar to the strategy introduced by Hall (1950) for simplifying bracket expressions in Lie algebras, we define the following subset of $T_\infty$.

---

[3] due to A. Murua, private communication, Feb. 2001

**Definition 3.11 (Hall Set).** The *Hall set* corresponding to an order relation (3.27) is a subset $\mathcal{H} \subset T_\infty$ defined by

$$\textcircled{i} \in \mathcal{H} \quad \text{for } i = 1, 2, 3, \ldots$$
$$\tau \in \mathcal{H} \quad \Leftrightarrow \quad \text{there exist } u, v \in \mathcal{H}, u > v, \text{ such that } \tau = u \circ v.$$

**Example 3.12.** The trees in the subsequent table are ordered from left to right with respect to $|\tau|$, and from top to bottom within fixed $|\tau|$. There remain finally 22 conditions for order 6.

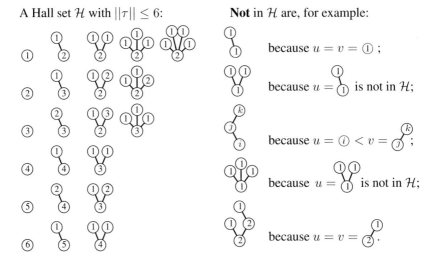

**Theorem 3.13 (Murua & Sanz-Serna 1999).** *For each $\tau \in T_\infty$ there are constants $A_i$, integers $m_i$ and trees $u_{ij} \in \mathcal{H}$ such that for all $a_k, b_k$ of Lemma 3.4 we have*

$$c(\tau) = \sum_{i=1}^{N} A_i \prod_{j=1}^{m_i} c(u_{ij}), \qquad u_{ij} \in \mathcal{H}, \ |u_{i1}| + \ldots + |u_{i,m_i}| \leq |\tau|. \tag{3.28}$$

*Proof.* We proceed by induction on $|\tau|$. For $\tau = \textcircled{i}$ the statement is trivial, because $\textcircled{i} \in \mathcal{H}$. We thus consider $\tau \in T_\infty$ with $|\tau| \geq 2$, write it as $\tau = u \circ v$, and conclude through the following two steps.

*First Step.* We apply the induction hypothesis (3.28) to $v$, i.e.,

$$c(v) = \sum_i B_i \prod_j c(v_{ij}), \qquad v_{ij} \in \mathcal{H}, \ \sum_j |v_{ij}| \leq |v|. \tag{3.29}$$

To this, we apply Lemma 3.9 followed by the Switching Lemma 3.8:

$$c(\tau) = c(u \circ v) = \sum_i B_i \, c(u \circ v_{i1} \circ v_{i2} \ldots \circ v_{i,n_i})$$
$$= -\sum_i B_i \, c\bigl(v_{in_i} \circ (u \circ v_{i1} \circ \ldots \circ v_{i,n_i-1})\bigr) + \ldots .$$

The "$+\ldots$" indicate terms containing trees to which we can apply our induction hypothesis. Inside the above expressions, we apply the induction hypothesis to the trees $u \circ v_{i1} \circ \ldots \circ v_{i,n_i-1}$, followed once again by Lemma 3.9. We arrive at a huge double sum which constitutes a linear combination of expressions of the form

$$c\bigl(u_1 \circ u_2 \circ \ldots \circ u_m\bigr) \tag{3.30}$$

and of terms "$+\ldots$" covered by the induction hypothesis. The point of the above dodges was *to make sure that all $u_1, u_2, \ldots, u_m$ are in $\mathcal{H}$*.

*Second Step.* It remains to reduce an expression (3.30) to the form required by (3.28). The trees $u_2, \ldots, u_m$ can be permuted arbitrarily; we arrange them in increasing order $u_2 \leq \ldots \leq u_m$.

*Case 1.* If $u_1 > u_2$, then by definition $u_1 \circ u_2 = w \in \mathcal{H}$ and we absorb the second factor into the first and obtain a product $w \circ u_3 \circ \ldots \circ u_m$ with *fewer* factors.

*Case 2.* If $u_1 < u_2 \leq \ldots$, we shuffle the factors with the help of Lemma 3.10 and obtain for (3.30) the expression

$$-\sum_{i=2}^{m} c\bigl(u_i \circ u_1 \circ \ldots\bigr) + \prod_{i=1}^{m} c(u_i) + \ldots.$$

With the first terms we return to Case 1, the second term is precisely as in (3.28), and the terms "$+\ldots$" are covered by the induction hypothesis.

*Case 3.* Now let $u_1 = u_2 < \ldots$. In this case, the formula (3.25) of Lemma 3.10 contains the term (3.30) twice. We group both together, so that (3.30) becomes

$$-\frac{1}{2}\sum_{i=3}^{m} c\bigl(u_i \circ u_1 \circ u_1 \circ \ldots\bigr) + \frac{1}{2}\prod_{i=1}^{m} c(u_i) + \ldots$$

and we go back to Case 1. If the first *three* trees are equal, we group three equal terms together and so on.

The whole reduction process is repeated until all Butcher products have disappeared. □

**Theorem 3.14 (Murua & Sanz-Serna 1999).** *The composition method $\Psi_h(y) = B_\infty(a_s, y)$ of (3.9) has order $p$ if and only if*

$$a_s(\tau) = \mathbf{e}(\tau) \qquad \text{for } \tau \in \mathcal{H} \text{ with } \|\tau\| \leq p.$$

*The coefficients $\mathbf{e}(\tau)$ are those of Theorem 3.6.*

*Proof.* We have seen in Sect. II.4 that composition methods of arbitrarily high order exist. Since the coefficients $A_i$ of (3.28) do not depend on the mapping $c(\tau)$, this together with Theorem 3.6 implies that the relation (3.28) is also satisfied by the mapping $\mathbf{e}$ for the exact solution. This proves the statement. □

**Example 3.15.** The order conditions for orders $p = 1, \ldots, 4$ become, with the trees of Example 3.12 and the rule of (3.14), as follows:

Order 1: ① $\quad \sum_{k=1}^{s} (\alpha_k + \beta_k) = 1$

Order 2: ② $\quad \sum_{k=1}^{s} (\alpha_k^2 - \beta_k^2) = 0$

Order 3: ③ $\quad \sum_{k=1}^{s} (\alpha_k^3 + \beta_k^3) = 0$

$\quad\quad\quad\quad$ ② on ① $\quad \sum_{k=1}^{s} (\alpha_k^2 - \beta_k^2) \sum_{\ell=1}^{k}{}' (\alpha_\ell + \beta_\ell) = 0 \quad\quad (3.31)$

Order 4: ④ $\quad \sum_{k=1}^{s} (\alpha_k^4 - \beta_k^4) = 0$

$\quad\quad\quad\quad$ ③ on ① $\quad \sum_{k=1}^{s} (\alpha_k^3 + \beta_k^3) \sum_{\ell=1}^{k}{}' (\alpha_\ell + \beta_\ell) = 0$

$\quad\quad\quad\quad$ ① ① on ② $\quad \sum_{k=1}^{s} (\alpha_k^2 - \beta_k^2) \Big(\sum_{\ell=1}^{k}{}' (\alpha_\ell + \beta_\ell)\Big)^2 = 0,$

where, as above, a *prime* attached to a summation symbol indicates that the sum of $\alpha_\ell^i$ is only from 1 to $k-1$, whereas the sum of $(-\beta_\ell)^i$ is from 1 to $k$. Similarly, the remaining trees of Example 3.12 with $||\tau|| = 5$ and $||\tau|| = 6$ give the additional conditions for order 5 and 6.

We shall see in Sect. V.3 how further reductions and numerical values are obtained under various assumptions of symmetry.

### III.3.4 Order Conditions for Splitting Methods

Splitting methods, introduced in Sect. II.5, are based on differential equations of the form
$$\dot{y} = f_1(y) + f_2(y), \quad\quad (3.32)$$
where the flows $\varphi_t^{[1]}$ and $\varphi_t^{[2]}$ of the systems $\dot{y} = f_1(y)$ and $\dot{y} = f_2(y)$ are assumed to be known exactly. In this situation, the method
$$\Phi_h = \varphi_h^{[1]} \circ \varphi_h^{[2]}$$
is of first order and, together with its adjoint $\Phi_h^* = \varphi_h^{[2]} \circ \varphi_h^{[1]}$, can be used as the basic method in the composition (3.9). This yields
$$\Psi_h = \varphi_{a_{s+1}h}^{[1]} \circ \varphi_{b_s h}^{[2]} \circ \varphi_{a_s h}^{[1]} \circ \ldots \circ \varphi_{b_2 h}^{[2]} \circ \varphi_{a_2 h}^{[1]} \circ \varphi_{b_1 h}^{[2]} \circ \varphi_{a_1 h}^{[1]} \quad\quad (3.33)$$
where

III.3 Order Conditions for Composition Methods

$$b_i = \alpha_i + \beta_i, \qquad a_i = \alpha_{i-1} + \beta_i \qquad (3.34)$$

with the conventions $\alpha_0 = 0$ and $\beta_{s+1} = 0$. Consequently, the splitting method (3.33) is a special case of (3.9) and we have the following obvious result.

**Theorem 3.16.** *Suppose that the composition method (3.9) is of order $p$ for all basic methods $\Phi_h$, then the splitting method (3.33) with $a_i, b_i$ given by (3.34) is of the same order $p$.* □

We now want to establish the reciprocal result. To every consistent splitting method (3.33), i.e., with coefficients satisfying $\sum_i a_i = \sum_i b_i = 1$, there exist unique $\alpha_i, \beta_i$ such that (3.34) holds. Does the corresponding composition method have the same order?

**Theorem 3.17.** *If a consistent splitting method (3.33) is of order $p$ at least for problems of the form (3.32) with the integrable splitting*

$$f_1(y) = \begin{pmatrix} g_1(y_2) \\ 0 \end{pmatrix}, \quad f_2(y) = \begin{pmatrix} 0 \\ g_2(y_1) \end{pmatrix} \quad \text{where} \quad y = \begin{pmatrix} y_1 \\ y_2 \end{pmatrix}, \qquad (3.35)$$

*then the corresponding composition method has the same order $p$ for an arbitrary basic method $\Phi_h$.*

*Proof.* McLachlan (1995) proves this result in the setting of Lie algebras. We give here a proof using the tools of this section.

a) The flows corresponding to the two vector fields $f_1$ and $f_2$ of (3.35) are $\varphi_t^{[1]}(y) = y + tf_1(y)$ and $\varphi_t^{[2]}(y) = y + tf_2(y)$, respectively. Consequently, the method $\Phi_h = \varphi_h^{[1]} \circ \varphi_h^{[2]}$ can be written in the form (3.1) with

$$d_1(y) = f_1(y) + f_2(y), \qquad d_{k+1}(y) = \frac{1}{k!} f_1^{(k)}(y)\big(f_2(y), \ldots, f_2(y)\big). \qquad (3.36)$$

The idea is to construct, for every tree $\tau \in \mathcal{H}$, functions $g_1(y_2)$ and $g_2(y_1)$ such that the first component of $F(\tau)(0)$ is non-zero whereas the first component of $F(\sigma)(0)$ vanishes for all $\sigma \in T_\infty$ different from $\tau$. This construction will be explained in part (b) below. Since the local error of the composition method is a $B_\infty$-series with coefficients $a_s(\tau) - e(\tau)$, this implies that the order conditions for $\tau \in \mathcal{H}$ with $\|\tau\| \le p$ are necessary already for this very special class of problems. Theorem 3.14 thus proves the statement.

b) For the construction of the functions $g_1(y_2)$ and $g_2(y_1)$ we have to understand the structure of $F(\tau)(y)$ with $d_k(y)$ given by (3.36). Consider for example the tree $\tau \in T_\infty$ of Fig. 3.2, for which we have $F(\tau)(y) = d_2''(y)\big(d_1(y), d_3(y)\big)$. Inserting $d_k(y)$ from (3.36), we get by Leibniz' rule a linear combination of eight expressions ($i \in \{1, 2\}$)

$$f_1'''(f_2, f_i, f_1''(f_2, f_2)), \qquad f_1''(f_2'f_i, f_1''(f_2, f_2)),$$
$$f_1''(f_i, f_2'f_1''(f_2, f_2)), \qquad f_1'f_2''(f_i, f_1''(f_2, f_2)),$$

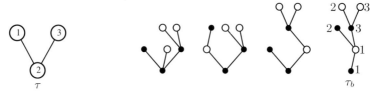

**Fig. 3.2.** Trees for illustrating the equivalence of the order conditions between composition and splitting methods.

each of which can be identified with a bi-coloured tree (see Sect. III.2.1, a vertex • corresponds to $f_1$ and ○ to $f_2$). The trees corresponding to these expressions with $i = 1$ are shown in Fig. 3.2. Due to the special form of $d_k(y)$ in (3.36) and due to the fact that in trees of the Hall set $\mathcal{H}$ the vertex ① can appear only at the end of a branch, there is always at least one bi-coloured tree where the vertices • are separated by those of ○ and vice versa. We now select such a tree, denoted by $\tau_b$, and we label the black and white vertices with $\{1, 2, \ldots\}$. We then let $y_1 = (y_1^1, \ldots, y_n^1)^T$ and $y_2 = (y_1^2, \ldots, y_m^2)^T$, where $n$ and $m$ are the numbers of vertices • and ○ in $\tau_b$, respectively. Inspired by "Exercise 4" of Hairer, Nørsett & Wanner (1993), page 155, we define the $i$th component of $g_1(y_2)$ as the product of all $y_j^2$ where $j$ runs through the labels of the vertices directly above the vertex • with label $i$. The function $g_2(y_1)$ is defined similarly. For the example of Fig. 3.2, the tree $\tau_b$ yields

$$g_1(y_2) = \begin{pmatrix} y_1^2 \\ y_2^2 y_3^2 \\ 1 \end{pmatrix}, \qquad g_2(y_1) = \begin{pmatrix} y_2^1 y_3^1 \\ 1 \\ 1 \end{pmatrix}.$$

One can check that with this construction the bi-coloured tree $\tau_b$ is the only one for which the first component of the elementary differential evaluated at $y = 0$ is different from zero. This in turn implies that among all trees of $T_\infty$ only the tree $\tau$ has a non-vanishing first component in its elementary differential. □

## III.4 The Baker-Campbell-Hausdorff Formula

This section treats the Baker-Campbell-Hausdorff (short BCH or CBH) formula on the composition of exponentials. It was proposed in 1898 by J.E. Campbell and proved independently by Baker (1905) and Hausdorff (1906). This formula will provide an alternative approach to the order conditions of composition (Sect. II.4) and splitting methods (Sect. II.5). For its derivation we shall use the inverse of the derivative of the exponential function.

### III.4.1 Derivative of the Exponential and Its Inverse

Elegant formulas for the derivative of $\exp$ and for its inverse can be obtained by the use of matrix commutators $[\Omega, A] = \Omega A - A\Omega$. If we suppose $\Omega$ fixed, this expression defines a linear operator $A \mapsto [\Omega, A]$

### III.4 The Baker-Campbell-Hausdorff Formula

$$\operatorname{ad}_\Omega(A) = [\Omega, A], \tag{4.1}$$

which is called the adjoint operator (see Varadarajan (1974), Sect. 2.13). Let us start by computing the derivatives of $\Omega^k$. The product rule for differentiation becomes

$$\left(\frac{d}{d\Omega}\Omega^k\right)H = H\Omega^{k-1} + \Omega H \Omega^{k-2} + \ldots + \Omega^{k-1}H, \tag{4.2}$$

and this equals $kH\Omega^{k-1}$ if $\Omega$ and $H$ commute. Therefore, it is natural to write (4.2) as $kH\Omega^{k-1}$ to which are added correction terms involving commutators and iterated commutators. In the cases $k=2$ and $k=3$ we have

$$H\Omega + \Omega H = 2H\Omega + \operatorname{ad}_\Omega(H)$$
$$H\Omega^2 + \Omega H \Omega + \Omega^2 H = 3H\Omega^2 + 3\bigl(\operatorname{ad}_\Omega(H)\bigr)\Omega + \operatorname{ad}_\Omega^2(H),$$

where $\operatorname{ad}_\Omega^i$ denotes the iterated application of the linear operator $\operatorname{ad}_\Omega$. With the convention $\operatorname{ad}_\Omega^0(H) = H$ we obtain by induction on $k$ that

$$\left(\frac{d}{d\Omega}\Omega^k\right)H = \sum_{i=0}^{k-1} \binom{k}{i+1}\bigl(\operatorname{ad}_\Omega^i(H)\bigr)\Omega^{k-i-1}. \tag{4.3}$$

This is seen by applying Leibniz' rule to $\Omega^{k+1} = \Omega \cdot \Omega^k$ and by using the identity $\Omega\bigl(\operatorname{ad}_\Omega^i(H)\bigr) = \bigl(\operatorname{ad}_\Omega^i(H)\bigr)\Omega + \operatorname{ad}_\Omega^{i+1}(H)$.

**Lemma 4.1.** *The derivative of* $\exp \Omega = \sum_{k \geq 0} \frac{1}{k!}\Omega^k$ *is given by*

$$\left(\frac{d}{d\Omega}\exp\Omega\right)H = \bigl(d\exp_\Omega(H)\bigr)\exp\Omega,$$

*where*

$$d\exp_\Omega(H) = \sum_{k \geq 0} \frac{1}{(k+1)!}\operatorname{ad}_\Omega^k(H). \tag{4.4}$$

*The series (4.4) converges for all matrices $\Omega$.*

*Proof.* Multiplying (4.3) by $(k!)^{-1}$ and summing, then exchanging the sums and putting $j = k - i - 1$ yields

$$\left(\frac{d}{d\Omega}\exp\Omega\right)H = \sum_{k \geq 0}\frac{1}{k!}\sum_{i=0}^{k-1}\binom{k}{i+1}\bigl(\operatorname{ad}_\Omega^i(H)\bigr)\Omega^{k-i-1}$$

$$= \sum_{i \geq 0}\sum_{j \geq 0}\frac{1}{(i+1)!\,j!}\bigl(\operatorname{ad}_\Omega^i(H)\bigr)\Omega^j.$$

The convergence of the series follows from the boundedness of the linear operator $\operatorname{ad}_\Omega$ (we have $\|\operatorname{ad}_\Omega\| \leq 2\|\Omega\|$). □

**Lemma 4.2 (Baker 1905).** *If the eigenvalues of the linear operator* $\operatorname{ad}_\Omega$ *are different from* $2\ell\pi i$ *with* $\ell \in \{\pm 1, \pm 2, \ldots\}$, *then* $d\exp_\Omega$ *is invertible. Furthermore, we have for* $\|\Omega\| < \pi$ *that*

$$d\exp_\Omega^{-1}(H) = \sum_{k \geq 0} \frac{B_k}{k!} \operatorname{ad}_\Omega^k(H), \qquad (4.5)$$

*where* $B_k$ *are the Bernoulli numbers, defined by* $\sum_{k \geq 0}(B_k/k!)x^k = x/(e^x - 1)$.

*Proof.* The eigenvalues of $d\exp_\Omega$ are $\mu = \sum_{k \geq 0} \lambda^k/(k+1)! = (e^\lambda - 1)/\lambda$, where $\lambda$ is an eigenvalue of $\operatorname{ad}_\Omega$. By our assumption, the values $\mu$ are non-zero, so that $d\exp_\Omega$ is invertible. By definition of the Bernoulli numbers, the composition of (4.5) with (4.4) gives the identity. Convergence for $\|\Omega\| < \pi$ follows from $\|\operatorname{ad}_\Omega\| \leq 2\|\Omega\|$ and from the fact that the radius of convergence of the series for $x/(e^x - 1)$ is $2\pi$. □

## III.4.2 The BCH Formula

Let $A$ and $B$ be two arbitrary (in general non-commuting) matrices. The problem is to find a matrix $C(t)$, such that

$$\exp(tA)\exp(tB) = \exp C(t). \qquad (4.6)$$

In order to get a first idea of the form of $C(t)$, we develop the expression to the left in a series: $\exp(tA)\exp(tB) = I + t(A+B) + \frac{t^2}{2}(A^2 + 2AB + B^2) + \mathcal{O}(t^3) =: I + X$. For sufficiently small $t$ (hence $\|X\|$ is small), the series expansion of the logarithm $\log(I + X) = X - X^2/2 + \ldots$ yields a matrix $C(t) = \log(I + X) = t(A + B) + \frac{t^2}{2}(A^2 + 2AB + B^2 - (A + B)^2) + \mathcal{O}(t^3)$, which satisfies (4.6). This series has a positive radius of convergence, because it is obtained by elementary operations of convergent series.

The main problem of the derivation of the BCH formula is to get explicit formulas for the coefficients of the series for $C(t)$, and to express the coefficients of $t^2, t^3, \ldots$ in terms of commutators. With the help of the following lemma, recurrence relations for these coefficients will be obtained, which allow for an easy computation of the first terms.

**Lemma 4.3.** *Let $A$ and $B$ be (non-commuting) matrices. Then, (4.6) holds, where $C(t)$ is the solution of the differential equation*

$$\dot{C} = A + B + \frac{1}{2}[A - B, C] + \sum_{k \geq 2} \frac{B_k}{k!} \operatorname{ad}_C^k(A + B) \qquad (4.7)$$

*with initial value $C(0) = 0$. Recall that $\operatorname{ad}_C A = [C, A] = CA - AC$, and that $B_k$ denote the Bernoulli numbers as in Lemma 4.2.*

### III.4 The Baker-Campbell-Hausdorff Formula

John Edward Campbell[4]     Henry Frederick Baker[5]     Felix Hausdorff[6]

*Proof.* We follow Varadarajan (1974), Sect. 2.15, and we consider for small $s$ and $t$ a smooth matrix function $Z(s,t)$ such that

$$\exp(sA)\exp(tB) = \exp Z(s,t). \tag{4.8}$$

Using Lemma 4.1, the derivative of (4.8) with respect to $s$ is

$$A\exp(sA)\exp(tB) = d\exp_{Z(s,t)}\Big(\frac{\partial Z}{\partial s}(s,t)\Big)\exp Z(s,t),$$

so that

$$\frac{\partial Z}{\partial s} = d\exp_Z^{-1}(A) = A - \frac{1}{2}[Z,A] + \sum_{k\geq 2}\frac{B_k}{k!}\operatorname{ad}_Z^k(A). \tag{4.9}$$

We next take the inverse of (4.8)

$$\exp(-tB)\exp(-sA) = \exp\bigl(-Z(s,t)\bigr),$$

and differentiate this relation with respect to $t$. As above we get

$$\frac{\partial Z}{\partial t} = d\exp_{-Z}^{-1}(B) = B + \frac{1}{2}[Z,B] + \sum_{k\geq 2}\frac{B_k}{k!}\operatorname{ad}_Z^k(B), \tag{4.10}$$

because $\operatorname{ad}_{-Z}^k(B) = (-1)^k\operatorname{ad}_Z^k(B)$ and the Bernoulli numbers satisfy $B_k = 0$ for odd $k > 2$. A comparison of (4.6) with (4.8) gives $C(t) = Z(t,t)$. The stated differential equation for $C(t)$ therefore follows from $\dot{C}(t) = \frac{\partial Z}{\partial s}(t,t) + \frac{\partial Z}{\partial t}(t,t)$, and from adding the relations (4.9) and (4.10). □

---

[4] John Edward Campbell, born: 27 May 1862 in Lisburn, Co Antrim (Ireland), died: 1 October 1924 in Oxford (England).

[5] Henry Frederick Baker, born: 3 July 1866 in Cambridge (England), died: 17 March 1956 in Cambridge.

[6] Felix Hausdorff, born: 8 November 1869 in Breslau, Silesia (now Wrocław, Poland), died: 26 January 1942 in Bonn (Germany).

Using Lemma 4.3 we can compute the first Taylor coefficients of $C(t)$,

$$\exp(tA)\exp(tB) = \exp\Big(tC_1 + t^2C_2 + t^3C_3 + t^4C_4 + t^5C_5 + \ldots\Big). \quad (4.11)$$

Inserting this expansion of $C(t)$ into (4.7) and comparing like powers of $t$ gives

$$
\begin{aligned}
C_1 &= A + B \\
C_2 &= \tfrac{1}{4}[A - B, A + B] = \tfrac{1}{2}[A, B] \\
C_3 &= \tfrac{1}{6}\Big[A - B, \tfrac{1}{2}[A, B]\Big] = \tfrac{1}{12}\Big[A, [A, B]\Big] + \tfrac{1}{12}\Big[B, [B, A]\Big] \\
C_4 &= \ldots = \tfrac{1}{24}\Big[A, [B, [B, A]]\Big] \\
C_5 &= \ldots = -\tfrac{1}{720}\Big[A, [A, [A, [A, B]]]\Big] - \tfrac{1}{720}\Big[B, [B, [B, [B, A]]]\Big] \\
&\quad + \tfrac{1}{360}\Big[A, [B, [B, [B, A]]]\Big] + \tfrac{1}{360}\Big[B, [A, [A, [A, B]]]\Big] \\
&\quad + \tfrac{1}{120}\Big[A, [A, [B, [B, A]]]\Big] + \tfrac{1}{120}\Big[B, [B, [A, [A, B]]]\Big].
\end{aligned} \quad (4.12)
$$

Here, the dots ... in the formulas for $C_4$ and $C_5$ indicate simplifications with the help of the Jacobi identity

$$[A, [B, C]] + [C, [A, B]] + [B, [C, A]] = 0, \quad (4.13)$$

which is verified by straightforward calculation. For higher order the expressions soon become very complicated.

**The Symmetric BCH Formula.** For the construction of symmetric splitting methods it is convenient to use a formula for the composition

$$\exp\Big(\tfrac{t}{2}A\Big)\exp(tB)\exp\Big(\tfrac{t}{2}A\Big) = \exp\Big(tS_1 + t^3S_3 + t^5S_5 + \ldots\Big). \quad (4.14)$$

Since the inverse of the left-hand side is obtained by changing the sign of $t$, the same must be true for the right-hand side. This explains why only odd powers of $t$ are present in (4.14). Applying the BCH formula (4.11) to $\exp(\tfrac{t}{2}A)\exp(\tfrac{t}{2}B) = \exp C(t)$ and a second time to $\exp(C(t))\exp(-C(-t))$ yields for the coefficients of (4.14) (Yoshida 1990)

$$
\begin{aligned}
S_1 &= A + B \\
S_3 &= -\tfrac{1}{24}\Big[A, [A, B]\Big] + \tfrac{1}{12}\Big[B, [B, A]\Big] \\
S_5 &= \tfrac{7}{5760}\Big[A, [A, [A, [A, B]]]\Big] - \tfrac{1}{720}\Big[B, [B, [B, [B, A]]]\Big] \\
&\quad + \tfrac{1}{360}\Big[A, [B, [B, [B, A]]]\Big] + \tfrac{1}{360}\Big[B, [A, [A, [A, B]]]\Big] \\
&\quad - \tfrac{1}{480}\Big[A, [A, [B, [B, A]]]\Big] + \tfrac{1}{120}\Big[B, [B, [A, [A, B]]]\Big].
\end{aligned} \quad (4.15)
$$

## III.5 Order Conditions via the BCH Formula

Using the BCH formula we present an alternative approach to the order conditions of splitting and composition methods. The main idea is to write the flow of a differential equation formally as the exponential of the Lie derivative.

### III.5.1 Calculus of Lie Derivatives

For a differential equation

$$\dot{y} = f^{[1]}(y) + f^{[2]}(y), \tag{5.1}$$

it is convenient to study the composition of the flows $\varphi_t^{[1]}$ and $\varphi_t^{[2]}$ of the systems

$$\dot{y} = f^{[1]}(y), \qquad \dot{y} = f^{[2]}(y), \tag{5.2}$$

respectively. We introduce the differential operators (*Lie derivative*)

$$D_i = \sum_j f_j^{[i]}(y) \frac{\partial}{\partial y_j} \tag{5.3}$$

which means that for differentiable functions $F : \mathbb{R}^n \to \mathbb{R}^m$ we have

$$D_i F(y) = F'(y) f^{[i]}(y). \tag{5.4}$$

It follows from the chain rule that, for the solutions $\varphi_t^{[i]}(y_0)$ of (5.2),

$$\frac{d}{dt} F\big(\varphi_t^{[i]}(y_0)\big) = (D_i F)\big(\varphi_t^{[i]}(y_0)\big), \tag{5.5}$$

and applying this operator iteratively we get

$$\frac{d^k}{dt^k} F\big(\varphi_t^{[i]}(y_0)\big) = (D_i^k F)\big(\varphi_t^{[i]}(y_0)\big). \tag{5.6}$$

Consequently, the Taylor series of $F\big(\varphi_t^{[i]}(y_0)\big)$, developed at $t = 0$, becomes

$$F\big(\varphi_t^{[i]}(y_0)\big) = \sum_{k \ge 0} \frac{t^k}{k!} (D_i^k F)(y_0) = \exp(tD_i) F(y_0). \tag{5.7}$$

Now, putting $F(y) = \mathrm{Id}(y) = y$, the identity map, this is the Taylor series of the solution itself

$$\varphi_t^{[i]}(y_0) = \sum_{k \ge 0} \frac{t^k}{k!} (D_i^k \mathrm{Id})(y_0) = \exp(tD_i)\mathrm{Id}(y_0). \tag{5.8}$$

If the functions $f^{[i]}(y)$ are not analytic, but only $N$-times continuously differentiable, the series (5.8) has to be truncated and a $\mathcal{O}(h^N)$ remainder term has to be included.

**Lemma 5.1 (Gröbner 1960).** *Let $\varphi_s^{[1]}$ and $\varphi_t^{[2]}$ be the flows of the differential equations $\dot y = f^{[1]}(y)$ and $\dot y = f^{[2]}(y)$, respectively. For their composition we then have*

$$\left(\varphi_t^{[2]} \circ \varphi_s^{[1]}\right)(y_0) = \exp(sD_1)\exp(tD_2)\operatorname{Id}(y_0).$$

*Proof.* This is precisely formula (5.7) with $i=1$, $t$ replaced with $s$, and with $F(y) = \varphi_t^{[2]}(y) = \exp(tD_2)\operatorname{Id}(y_0)$. □

**Remark 5.2.** Notice that the indices 1 and 2 as well as $s$ and $t$ to the left and right in the identity of Lemma 5.1 are permuted. Gröbner calls this phenomenon, which sometimes leads to some confusion in the literature, the 'Vertauschungssatz'.

**Remark 5.3.** The statement of Lemma 5.1 can be extended to more than two flows. If $\varphi_t^{[j]}$ is the flow of a differential equation $\dot y = f^{[j]}(y)$, then we have

$$\left(\varphi_u^{[m]} \circ \ldots \circ \varphi_t^{[2]} \circ \varphi_s^{[1]}\right)(y_0) = \exp(sD_1)\exp(tD_2)\cdot\ldots\cdot\exp(uD_m)\operatorname{Id}(y_0).$$

This follows by induction on $m$.

In general, the two operators $D_1$ and $D_2$ do not commute, so that the composition $\exp(tD_1)\exp(tD_2)\operatorname{Id}(y_0)$ is different from $\exp\bigl(t(D_1+D_2)\bigr)\operatorname{Id}(y_0)$, which represents the solution $\varphi_t(y_0)$ of $\dot y = f(y) = f^{[1]}(y) + f^{[2]}(y)$. The relation of Lemma 5.1 suggests the use of the BCH formula. However, $D_1$ and $D_2$ are unbounded differential operators so that the series expansions that appear cannot be expected to converge. A formal application of the BCH formula with $tA$ and $tB$ replaced with $sD_1$ and $tD_2$, respectively, yields

$$\exp(sD_1)\exp(tD_2) = \exp\bigl(D(s,t)\bigr), \tag{5.9}$$

where the differential operator $D(s,t)$ is obtained from (4.11) as

$$\begin{aligned} D(s,t) &= sD_1 + tD_2 + \frac{st}{2}[D_1,D_2] + \frac{s^2 t}{12}\Bigl[D_1,[D_1,D_2]\Bigr] \\ &\quad + \frac{st^2}{12}\Bigl[D_2,[D_2,D_1]\Bigr] + \frac{s^2 t^2}{24}\Bigl[D_1,\bigl[D_2,[D_2,D_1]\bigr]\Bigr] + \ldots. \end{aligned} \tag{5.10}$$

The *Lie bracket* for differential operators is calculated exactly as for matrices, namely, $[D_1, D_2] = D_1 D_2 - D_2 D_1$. But how can we interpret (5.9) rigorously? Expanding both sides in Taylor series we see that

$$\exp(sD_1)\exp(tD_2) = I + sD_1 + tD_2 + \frac{1}{2}\bigl(s^2 D_1^2 + 2st D_1 D_2 + t^2 D_2^2\bigr) + \ldots \tag{5.11}$$

and

$$\begin{aligned} \exp\bigl(D(s,t)\bigr) &= I + D(s,t) + \frac{1}{2}D(s,t)^2 + \ldots \\ &= I + sD_1 + tD_2 + \frac{1}{2}\Bigl((sD_1 + tD_2)^2 + st[D_1,D_2]\Bigr) + \ldots. \end{aligned}$$

By derivation of the BCH formula we have a formal identity, i.e., both series have exactly the same coefficients. Moreover, every finite truncation of the series can be applied without any difficulties to sufficiently differentiable functions $F(y)$. Consequently, for $N$-times differentiable functions the relation (5.9) holds true, if both sides are replaced by their truncated Taylor series and if a $\mathcal{O}(h^N)$ remainder is added ($h = \max(|s|, |t|)$).

## III.5.2 Lie Brackets and Commutativity

If we apply $D_2$ to a function $F$, followed by an application of $D_1$, we will obtain partial derivatives of $F$ of first and second orders. However, if we subtract from this the same expression with $D_1$ and $D_2$ reversed, the second derivatives will cancel (this was already remarked upon by Jacobi (1862), p. 39: "differentialia partialia secunda functionis $f$ non continere") and we see that the Lie bracket

$$[D_1, D_2] = D_1 D_2 - D_2 D_1 = \sum_i \left( \sum_j \left( \frac{\partial f_i^{[2]}}{\partial y_j} f_j^{[1]} - \frac{\partial f_i^{[1]}}{\partial y_j} f_j^{[2]} \right) \right) \frac{\partial}{\partial y_i} \quad (5.12)$$

is again a linear differential operator. So, from two vector fields $f^{[1]}$ and $f^{[2]}$ we obtain a *third* vector field $f^{[3]}$.

The *geometric meaning* of the new vector field can be deduced from Lemma 5.1. We see by subtracting (5.11) from itself, once as it stands and once with $sD_1$ and $tD_2$ permuted, that

$$\varphi_t^{[2]} \circ \varphi_s^{[1]}(y_0) - \varphi_s^{[1]} \circ \varphi_t^{[2]}(y_0) = st\, [D_1, D_2]\,\mathrm{Id}(y_0) + \ldots = st\, f^{[3]}(y_0) + \ldots \quad (5.13)$$

(see the picture), where "$+\ldots$" are terms of order $\geq 3$. This leads us to the following result.

**Lemma 5.4.** *Let $f^{[1]}(y)$ and $f^{[2]}(y)$ be defined on an open set. The corresponding flows $\varphi_s^{[1]}$ and $\varphi_t^{[2]}$ commute everywhere for all sufficiently small $s$ and $t$, if and only if*

$$[D_1, D_2] = 0. \quad (5.14)$$

*Proof.* The "only if" part is clear from (5.13). For proving the "if" part, we take $s$ and $t$ fixed, and subdivide, for a given $n$, the integration intervals into $n$ equidistant parts $\Delta s = s/n$ and $\Delta t = t/n$. This allows us to transform the solution $\varphi_t^{[2]} \circ \varphi_s^{[1]}(y_0)$ by a discrete homotopy in $n^2$ steps into the solution $\varphi_s^{[1]} \circ \varphi_t^{[2]}(y_0)$, each time appending a small rectangle of size $\mathcal{O}(n^{-2})$. If we denote such an intermediate stage by

$$\Gamma_k = \ldots \circ \varphi_{j_2 \Delta t}^{[2]} \circ \varphi_{i_2 \Delta s}^{[1]} \circ \varphi_{j_1 \Delta t}^{[2]} \circ \varphi_{i_1 \Delta s}^{[1]}(y_0)$$

then we have $\Gamma_0 = \varphi_t^{[2]} \circ \varphi_s^{[1]}(y_0)$ and $\Gamma_{n^2} = \varphi_s^{[1]} \circ \varphi_t^{[2]}(y_0)$ (see Fig. 5.1). Now, for $n \to \infty$, we have the estimate

$$|\Gamma_{k+1} - \Gamma_k| \leq \mathcal{O}(n^{-3}),$$

because the error terms in (5.13) are of order 3 at least, and because of the differentiability of the solutions with respect to initial values. Thus, by the triangle inequality $|\Gamma_{n^2} - \Gamma_0| \leq \mathcal{O}(n^{-1})$ and the result is proved. □

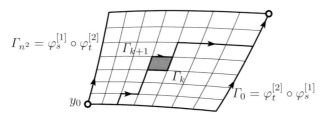

**Fig. 5.1.** Estimation of commuting solutions

### III.5.3 Splitting Methods

We follow the approach of Yoshida (1990) for obtaining the order conditions of splitting methods (II.5.6). The idea is the following: with the use of Lemma 5.1 we write the method as a product of exponentials, then we apply formally the Baker-Campbell-Hausdorff formula to get one exponential of a series in powers of $h$. Finally, we compare this series with $h(D_1 + D_2)$, which corresponds to the exact solution of (5.1).

The splitting method (II.5.6), viz.,

$$\Psi_h = \varphi^{[2]}_{b_m h} \circ \varphi^{[1]}_{a_m h} \circ \varphi^{[2]}_{b_{m-1} h} \circ \ldots \circ \varphi^{[1]}_{a_2 h} \circ \varphi^{[2]}_{b_1 h} \circ \varphi^{[1]}_{a_1 h}, \qquad (5.15)$$

is a composition of expressions $\varphi^{[2]}_{b_j h} \circ \varphi^{[1]}_{a_j h}$ which, by Lemma 5.1 and by (5.9), can be written as an exponential

$$\varphi^{[2]}_{b_j h} \circ \varphi^{[1]}_{a_j h} = \exp\Big(a_j h E_1^1 + b_j h E_2^1 + a_j b_j h^2 E_1^2 \\ + a_j^2 b_j h^3 E_1^3 + a_j b_j^2 h^3 E_2^3 + a_j^2 b_j^2 h^4 E_1^4 + \ldots \Big) \mathrm{Id}, \qquad (5.16)$$

where we use the abbreviations

$$E_1^1 = D_1, \quad E_2^1 = D_2, \quad E_1^2 = \frac{1}{2}[D_1, D_2], \quad E_1^3 = \frac{1}{12}[D_1, [D_1, D_2]],$$

$$E_2^3 = \frac{1}{12}[D_2, [D_2, D_1]], \quad E_1^4 = \frac{1}{24}[D_1[D_2, [D_2, D_1]]],$$

and the dots indicate $\mathcal{O}(h^5)$ expressions.

We next define $\Psi^{(j)}$ recursively by

$$\Psi^{(0)} = \mathrm{Id}, \qquad \Psi^{(j)} = \varphi^{[2]}_{b_j h} \circ \varphi^{[1]}_{a_j h} \circ \Psi^{(j-1)}, \qquad (5.17)$$

so that $\Psi^{(m)}$ is equal to our method (5.15). Aiming to write $\Psi^{(j)}$ also as an exponential of differential operators, we are confronted with computing commutators of the expressions $E_i^j$. We see that $[E_1^1, E_2^1] = 2E_1^2$, $[E_1^1, E_1^2] = 6E_1^3$, $[E_2^1, E_1^2] = -6E_2^3$, $[E_1^1, E_2^3] = 2E_1^4$, and $[E_2^1, a_1^3] = -2E_1^4$ as a consequence of the Jacobi identity (4.13). But the other commutators cannot be expressed in terms of $E_i^j$. We therefore introduce

$$E_2^4 = \frac{1}{24}[D_1, [D_1, [D_1, D_2]]], \quad E_3^4 = \frac{1}{24}[D_2, [D_2, [D_2, D_1]]].$$

This allows us to formulate the following result.

**Lemma 5.5.** *The method $\Psi^{(j)}$, defined by (5.17), can be formally written as*

$$\Psi^{(j)} = \exp\Big(c_{1,j}^1 h E_1^1 + c_{2,j}^1 h E_2^1 + c_{1,j}^2 h^2 E_1^2 + c_{1,j}^3 h^3 E_1^3$$
$$+ c_{2,j}^3 h^3 E_2^3 + c_{1,j}^4 h^4 E_1^4 + c_{2,j}^4 h^4 E_2^4 + c_{3,j}^4 h^4 E_3^4 + \ldots\Big)\mathrm{Id},$$

*where all coefficients are zero for $j = 0$, and where for $j \geq 1$*

$$c_{1,j}^1 = c_{1,j-1}^1 + a_j, \qquad c_{2,j}^1 = c_{2,j-1}^1 + b_j,$$
$$c_{1,j}^2 = c_{1,j-1}^2 + a_j b_j + c_{1,j-1}^1 b_j - c_{2,j-1}^1 a_j,$$
$$c_{1,j}^3 = c_{1,j-1}^3 + a_j^2 b_j + 2c_{1,j-1}^1 a_j b_j - 3c_{1,j-1}^2 a_j$$
$$+ (c_{1,j-1}^1)^2 b_j - c_{1,j-1}^1 c_{2,j-1}^1 a_j + c_{2,j-1}^1 a_j^2,$$
$$c_{2,j}^3 = c_{2,j-1}^3 + a_j b_j^2 - 4c_{2,j-1}^1 a_j b_j + 3c_{1,j-1}^2 b_j$$
$$+ (c_{2,j-1}^1)^2 a_j - c_{1,j-1}^1 c_{2,j-1}^1 b_j + c_{1,j-1}^1 b_j^2,$$

*and similar but more complicated formulas for $c_{i,j}^4$.*

*Proof.* Due to the reversed order in Lemma 5.1 we have to compute $\exp(A)\exp(B)$, where $A$ is the argument of the exponential for $\Psi^{(j-1)}$ and $B$ is that of (5.16). The rest is a tedious but straightforward application of the BCH formula. One has to use repeatedly the formulas for $[E_i^j, E_k^l]$, stated before Lemma 5.5. □

**Theorem 5.6.** *The splitting method (5.15) is of order $p$ if*

$$c_{1,m}^1 = c_{2,m}^1 = 1, \qquad c_{\ell,m}^k = 0 \quad \text{for } k = 2, \ldots, p \text{ and all } \ell. \tag{5.18}$$

*The coefficients $c_{\ell,m}^k$ are those defined in Lemma 5.5.*

*Proof.* This is an immediate consequence of Lemma 5.5, because the conditions of order $p$ imply that the Taylor series expansion of $\Psi^{(m)}(y_0)$ coincides with that of the solution $\varphi_h(y_0) = \exp(h(D_1 + D_2))y_0$ up to terms of size $\mathcal{O}(h^p)$. □

A simplification in the order conditions arises for symmetric methods (5.15), that is, for coefficients satisfying $a_{m+1-i} = a_i$ and $b_{m-i} = b_i$ for all $i$ (and $b_m = 0$). By Theorem II.3.2, it is sufficient to consider the order conditions (5.18) for odd $k$ only.

## III.5.4 Composition Methods

We now consider composition methods (II.4.6), viz.,

$$\Psi_h = \Phi_{\alpha_s h} \circ \Phi^*_{\beta_s h} \circ \ldots \circ \Phi^*_{\beta_2 h} \circ \Phi_{\alpha_1 h} \circ \Phi^*_{\beta_1 h}, \qquad (5.19)$$

where $\Phi_h$ is a first-order method for $\dot{y} = f(y)$ and $\Phi^*_h$ is its adjoint. We assume

$$\Phi_h = \exp\left(hC_1 + h^2 C_2 + h^3 C_3 + \ldots\right) \mathrm{Id} \qquad (5.20)$$

with differential operators $C_i$, and such that $C_1$ is the Lie derivative operator corresponding to $\dot{y} = f(y)$. For the splitting method $\Phi_h = \varphi^{[2]}_h \circ \varphi^{[1]}_h$ this follows from (5.16), and for general one-step methods this is a consequence of Sect. IX.1 on backward error analysis. The adjoint method then satisfies

$$\Phi^*_h = \exp\left(hC_1 - h^2 C_2 + h^3 C_3 - \ldots\right) \mathrm{Id}. \qquad (5.21)$$

From now on the procedure is similar to that of Sect. III.5.3. We define $\Psi^{(j)}$ recursively by

$$\Psi^{(0)} = \mathrm{Id}, \qquad \Psi^{(j)} = \Phi_{\alpha_j h} \circ \Phi^*_{\beta_j h} \circ \Psi^{(j-1)}, \qquad (5.22)$$

so that $\Psi^{(m)}$ becomes (5.19). We apply the BCH formula to obtain

$$\begin{aligned}\Phi_{\alpha_j h} \circ \Phi^*_{\beta_j h} &= \exp\left(\beta_j h C_1 - \beta_j^2 h^2 C_2 + \ldots\right) \exp\left(\alpha_j h C_1 + \alpha_j^2 h^2 C_2 + \ldots\right) \mathrm{Id} \\ &= \exp\Big((\alpha_j + \beta_j)h E_1^1 + (\alpha_j^2 - \beta_j^2)h^2 E_1^2 \\ &\quad (\alpha_j^3 + \beta_j^3)h^3 E_1^3 + \tfrac{1}{2}\alpha_j \beta_j(\alpha_j + \beta_j) h^3 E_2^3 + \ldots\Big) \mathrm{Id}\end{aligned}$$

where

$$E_1^k = C_k, \qquad E_2^3 = [C_1, C_2].$$

We then have the following result.

**Lemma 5.7.** *The method $\Psi^{(j)}$ of (5.22) can be formally written as*

$$\Psi^{(j)} = \exp\left(\gamma^1_{1,j} h E_1^1 + \gamma^2_{1,j} h^2 E_1^2 + \gamma^3_{1,j} h^3 E_1^3 + \gamma^3_{2,j} h^3 E_2^3 + \ldots\right) \mathrm{Id},$$

*where all coefficients are zero for $j = 0$, and where for $j = 1, \ldots, m$*

$$\begin{aligned}\gamma^1_{1,j} &= \gamma^1_{1,j-1} + \alpha_j + \beta_j \\ \gamma^2_{1,j} &= \gamma^2_{1,j-1} + \alpha_j^2 - \beta_j^2 \\ \gamma^3_{1,j} &= \gamma^3_{1,j-1} + \alpha_j^3 + \beta_j^3 \\ \gamma^3_{2,j} &= \gamma^3_{2,j-1} + \tfrac{1}{2}\alpha_j \beta_j(\alpha_j + \beta_j) + \tfrac{1}{2}\gamma^1_{1,j-1}(\alpha_j^2 - \beta_j^2) - \tfrac{1}{2}\gamma^2_{1,j-1}(\alpha_j + \beta_j).\end{aligned}$$

*Proof.* Similar to Lemma 5.5, the result follows using the BCH formula. □

**Theorem 5.8.** *The composition method (5.19) is of order $p$ if*

$$\gamma^1_{1,m} = 1, \qquad \gamma^k_{\ell,m} = 0 \quad \text{for } k = 2, \ldots, p \text{ and all } \ell. \tag{5.23}$$

*The coefficients $\gamma^k_{\ell,m}$ are those defined in Lemma 5.7.* □

It is interesting to see how these order conditions are related to those obtained with the use of trees. The conditions $\gamma^1_{1,m} = 1$ and $\gamma^2_{1,m} = \gamma^3_{1,m} = 0$ are identical to the first three order conditions of Example 3.15. The remaining condition for order 3, $\gamma^3_{2,m} = 0$, reads

$$\sum_{k=1}^{m} \alpha_k \beta_k (\alpha_k + \beta_k) + \sum_{k=1}^{m} (\alpha_k^2 - \beta_k^2) \sum_{i=1}^{k-1} (\alpha_i + \beta_i) - \sum_{k=1}^{m} (\alpha_k + \beta_k) \sum_{i=1}^{k-1} (\alpha_i^2 - \beta_i^2)$$

$$= \sum_{k=1}^{m} (\alpha_k^2 - \beta_k^2) {\sum_{i=1}^{k}}' (\alpha_i + \beta_i) - \sum_{k=1}^{m} (\alpha_k + \beta_k) {\sum_{i=1}^{k}}' (\alpha_i^2 - \beta_i^2) = 0.$$

This condition is just the difference of the order conditions for the trees ② ∘ ① and ① ∘ ②, whose sum is zero by the Switching Lemma 3.8. Therefore the condition $\gamma^3_{2,m} = 0$ is equivalent to (though more complicated than) the fourth condition of Example 3.15.

**Symmetric Composition of Symmetric Methods.** Consider now a composition

$$\Psi_h = \Phi_{\gamma_m h} \circ \ldots \circ \Phi_{\gamma_2 h} \circ \Phi_{\gamma_1 h} \circ \Phi_{\gamma_2 h} \circ \ldots \circ \Phi_{\gamma_m h}, \tag{5.24}$$

where $\Phi_h$ is a symmetric method that can be written as

$$\Phi_h = \exp\left(hS_1 + h^3 S_3 + h^5 S_5 + \ldots\right) \text{Id}$$

with $S_1$ the Lie derivative operator corresponding to $\dot{y} = f(y)$. For the Strang splitting $\Phi_h = \varphi^{[1]}_{h/2} \circ \varphi^{[2]}_h \circ \varphi^{[1]}_{h/2}$ such an expansion follows from the symmetric BCH formula (4.14), and for general symmetric one-step methods from Sect. IX.2. The derivation of the order conditions is similar to the above with $\Psi^{(j)}$ defined by

$$\Psi^{(1)} = \Phi_{\gamma_1 h}, \qquad \Psi^{(j)} = \Phi_{\gamma_j h} \circ \Psi^{(j-1)} \circ \Phi_{\gamma_j h},$$

so that $\Psi^{(m)}$ becomes (5.24).

**Lemma 5.9.** *The method $\Psi^{(j)}$ can be formally written as*

$$\Psi^{(j)} = \exp\left(\sigma^1_{1,j} h E^1_1 + \sigma^3_{1,j} h^3 E^3_1 + \sigma^5_{1,j} h^5 E^5_1 + \sigma^5_{2,j} h^5 E^5_2 + \ldots\right) \text{Id},$$

*where $E^k_1 = S_k$, $E^5_2 = [S_1[S_1, S_3]]$, and where $\sigma^k_{1,1} = \gamma^k_1$, $\sigma^5_{2,1} = 0$, and*

$$\sigma^k_{1,j} = \sigma^k_{1,j-1} + 2\gamma^k_j$$

$$\sigma^5_{2,j} = \sigma^5_{2,j-1} + \frac{1}{6}\left(\gamma^3_j (\sigma^1_{1,j-1})^2 - \gamma_j \sigma^1_{1,j-1} \sigma^3_{1,j-1} - \gamma^2_j \sigma^3_{1,j-1} + \gamma^4_j \sigma^1_{1,j-1}\right).$$

*Proof.* The result is a consequence of the symmetric BCH formula (4.14) with $\gamma_j h S_1 + \gamma_j^3 h^3 S_3 + \ldots$ and $\sigma_{1,j-1}^1 h E_1^1 + \sigma_{1,j-1}^3 h E_1^3 + \ldots$ in the roles of $\frac{t}{2} A$ and $tB$, respectively. □

**Theorem 5.10.** *The composition method (5.24) is of order $p$ if*

$$\sigma_{1,m}^1 = 1, \qquad \sigma_{\ell,m}^k = 0 \quad \text{for odd } k = 3, \ldots, p \text{ and all } \ell. \tag{5.25}$$

*The coefficients $\sigma_{\ell,m}^k$ are those defined in Lemma 5.9.* □

Symmetric composition methods up to order 10 will be constructed and discussed in Sect. V.3.

## III.6 Exercises

1. Find all trees of orders 5 and 6.
2. (A. Cayley 1857). Denote the number of trees of order $q$ by $a_q$. Prove that

$$a_1 + a_2 x + a_3 x^2 + a_4 x^3 + \ldots = (1-x)^{-a_1}(1-x^2)^{-a_2}(1-x^3)^{-a_3} \cdot \ldots .$$

| $q$   | 1 | 2 | 3 | 4 | 5 | 6  | 7  | 8   | 9   | 10  |
|-------|---|---|---|---|---|----|----|-----|-----|-----|
| $a_q$ | 1 | 1 | 2 | 4 | 9 | 20 | 48 | 115 | 286 | 719 |

3. Independency of the elementary differentials: show that for every $\tau \in T$ there is a system (1.1) such that the first component of $F(\tau)(0)$ equals 1, and the first component of $F(u)(0)$ is zero for all trees $u \neq \tau$.
   *Hint.* Consider a monotonic labelling of $\tau$, and define $y_i'$ as the product over all $y_j$, where $j$ runs through all labels of vertices that lie directly above the vertex '$i$'. For the first labelling of the tree of Exercise 4 this would be $\dot{y}_1 = y_2 y_3$, $\dot{y}_2 = 1$, $\dot{y}_3 = y_4$, and $\dot{y}_4 = 1$.
4. Prove that the coefficient $\alpha(\tau)$ of Definition 1.2 is equal to the number of possible monotonic labellings of the vertices of $\tau$, starting with the label 1 for the root. For example, the tree $[[\bullet],\bullet]$ has three different monotonic labellings.

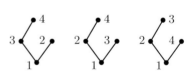

   In addition, deduce, from (1.22), the recursion formula

$$\alpha(\tau) = \binom{|\tau|-1}{|\tau_1|, \ldots, |\tau_m|} \alpha(\tau_1) \cdot \ldots \cdot \alpha(\tau_m) \frac{1}{\mu_1! \mu_2! \ldots}, \tag{6.1}$$

   where the integers $\mu_1, \mu_2, \ldots$ count equal trees among $\tau_1, \ldots, \tau_m$ and

$$\binom{|\tau|-1}{|\tau_1|, \ldots, |\tau_m|} = \frac{(|\tau|-1)!}{|\tau_1|! \cdot \ldots \cdot |\tau_m|!}.$$

denotes the multinomial coefficient.

*Remark.* In the theoretical physics literature, the coefficients $\alpha(\tau)$ are written $CM(\tau)$ and called "Connes-Moscovici weights".

5. If we denote by $N(\tau)$ the number of elements in $OST(\tau)$, then show that

$$N(\bullet) = 2$$
$$N([\tau_1, \ldots, \tau_m]) = 1 + N(\tau_1) \cdot \ldots \cdot N(\tau_m). \tag{6.2}$$

Use this result to compute the number of subtrees of the christmas tree decorating formula (1.34). *Answer:* 6865.

6. Prove that the elementary differentials for partitioned problems are independent. For a given tree $(\tau \in TP$, find a problem (2.1) such that a certain component of $F(\tau)(p,q)$ vanishes for all $u \in TP$ except for $\tau$.
*Hint.* Consider the construction of Exercise 3, and define the partitioning of $y$ into $(p,q)$ according to the colours of the vertices.

7. The number of order conditions for partitioned Runge-Kutta methods (II.2.2) is $2a_r$ for order $r$, where $a_r$ is given by (see Hairer, Nørsett & Wanner (1993), page 311)

| $r$   | 1 | 2 | 3 | 4  | 5   | 6   | 7    | 8    | 9     | 10     |
|-------|---|---|---|----|-----|-----|------|------|-------|--------|
| $a_r$ | 1 | 2 | 7 | 26 | 107 | 458 | 2058 | 9498 | 44987 | 216598 |

Find a formula similar to that of Exercise 2.

8. For the special second order differential equation $\ddot{y} = g(y)$, and for a Nyström method

$$\begin{aligned}
\ell_i &= g\Big(y_0 + c_i h \dot{y}_0 + h^2 \sum_{j=1}^{s} a_{ij} \ell_j\Big), \\
y_1 &= y_0 + h\dot{y}_0 + h^2 \sum_{i=1}^{s} \beta_i \ell_i, \qquad \dot{y}_1 = \dot{y}_0 + h \sum_{i=1}^{s} b_i \ell_i,
\end{aligned} \tag{6.3}$$

consider the simplifying assumption

$$CN(\eta): \quad \sum_{j=1}^{s} a_{ij} c_j^{k-2} = \frac{c_i^k}{k(k-1)}, \qquad k = 2, \ldots, \eta,$$

$$DN(\zeta): \quad \sum_{i=1}^{s} b_i c_i^{k-2} a_{ij} = b_j \Big(\frac{c_j^k}{k(k-1)} - \frac{c_j}{k-1} + \frac{1}{k}\Big), \qquad k = 2, \ldots, \zeta.$$

Prove that if the quadrature formula $(b_i, c_i)$ is of order $p$, if $\beta_i = b_i(1 - c_i)$ for all $i$, and if the simplifying assumptions $CN(\eta)$, $DN(\zeta)$ are satisfied with $2\eta + 2 \geq p$ and $\zeta + \eta \geq p$, then the Nyström method has order $p$.

9. *Nyström methods of maximal order $2s$.* Prove that there exists a one-parameter family of $s$-stage Nyström methods (6.3) for $\ddot{y} = g(y)$, which have order $2s$.

*Hint.* Consider the Gaussian quadrature formula and define the coefficients $a_{ij}$ by $CN(s)$ and by

$$\sum_{i=1}^{s} b_i c_i^{k-2} a_{is} = b_j \left( \frac{c_s^k}{k(k-1)} - \frac{c_s}{k-1} + \frac{1}{k} \right)$$

for $k = 2, \ldots, s$.

10. Prove that the coefficient $C_4$ in the series (4.11) of the Baker-Campbell-Hausdorff formula is given by $C_4 = [A, [B, [B, A]]]/24$.
11. Prove that the series (4.11) converges for $|t| < \ln 2/(\|A\| + \|B\|)$.
12. By Theorem 5.10 four order conditions have to be satisfied such that the symmetric composition method (5.24) is of order 6. Prove that these conditions are equivalent to the four conditions of Example V.3.15. (Care has to be taken due to the different meaning of the $\gamma_i$.)

# Chapter IV.
# Conservation of First Integrals and Methods on Manifolds

This chapter deals with the conservation of invariants (first integrals) by numerical methods, and with numerical methods for differential equations on manifolds. Our investigation will follow two directions. We first investigate which of the methods introduced in Chap. II conserve invariants automatically. We shall see that most of them conserve linear invariants, a few of them quadratic invariants, and none of them conserves cubic or general nonlinear invariants. We then construct new classes of methods, which are adapted to known invariants and which force the numerical solution to satisfy them. In particular, we study projection methods and methods based on local coordinates of the manifold defined by the invariants. We discuss in some detail the case where the manifold is a Lie group.

## IV.1 Examples of First Integrals

> Je nomme intégrale une équation $u = Const.$ telle que sa différentielle $du = 0$ soit vérifiée identiquement par le système des équations différentielles proposées ...     (C.G.J. Jacobi 1840, p. 350)

We consider differential equations

$$\dot{y} = f(y), \tag{1.1}$$

where $y$ is a vector or possibly a matrix.

**Definition 1.1.** A non-constant function $I(y)$ is called a *first integral* of (1.1) if

$$I'(y)f(y) = 0 \qquad \text{for all } y. \tag{1.2}$$

This implies that *every* solution $y(t)$ of (1.1) satisfies $I(y(t)) = I(y_0) = Const.$ Synonymously with "first integral", the terms *invariant* or *conserved quantity* or *constant of motion* are also used.

In Chap. I we have seen many examples of differential equations with invariants. For example, the Lotka-Volterra problem (I.1.1) has $I(u,v) = \ln u - u + 2\ln v - v$ as first integral. The pendulum equation (I.1.13) has $H(p,q) = p^2/2 - \cos q$, and the Kepler problem (I.2.2) has two first integrals, namely $H$ and $L$ of (I.2.3) and (I.2.4).

**Example 1.2 (Conservation of the Total Energy).** Hamiltonian systems are of the form
$$\dot{p} = -H_q(p,q), \qquad \dot{q} = H_p(p,q),$$
where $H_q = \nabla_q H = (\partial H/\partial q)^T$ and $H_p = \nabla_p H = (\partial H/\partial p)^T$ are the column vectors of partial derivatives. The Hamiltonian function $H(p,q)$ is a first integral. This follows at once from $H'(p,q) = (\partial H/\partial p, \partial H/\partial q)$ and
$$\frac{\partial H}{\partial p}\left(-\frac{\partial H}{\partial q}\right)^T + \frac{\partial H}{\partial q}\left(\frac{\partial H}{\partial p}\right)^T = 0.$$

**Example 1.3 (Conservation of the Total Linear and Angular Momentum of N-Body Systems).** We consider a system of $N$ particles interacting pairwise with potential forces which depend on the distances of the particles. This is formulated as a Hamiltonian system with total energy (I.3.1), viz.,
$$H(p,q) = \frac{1}{2}\sum_{i=1}^{N}\frac{1}{m_i}p_i^T p_i + \sum_{i=2}^{N}\sum_{j=1}^{i-1}V_{ij}\left(\|q_i - q_j\|\right).$$

Here $q_i, p_i \in \mathbb{R}^3$ represent the position and momentum of the $i$th particle of mass $m_i$, and $V_{ij}(r)$ ($i > j$) is the interaction potential between the $i$th and $j$th particle. The equations of motion read
$$\dot{q}_i = \frac{1}{m_i}p_i, \qquad \dot{p}_i = \sum_{j=1}^{N}\nu_{ij}(q_i - q_j)$$

where, for $i > j$, we have $\nu_{ij} = \nu_{ji} = -V'_{ij}(r_{ij})/r_{ij}$ with $r_{ij} = \|q_i - q_j\|$, and $\nu_{ii}$ is arbitrary, say $\nu_{ii} = 0$. The conservation of the total *linear momentum* $P = \sum_{i=1}^{N}p_i$ and the *angular momentum* $L = \sum_{i=1}^{N}q_i \times p_i$ is a consequence of the symmetry relation $\nu_{ij} = \nu_{ji}$:
$$\frac{d}{dt}\sum_{i=1}^{N}p_i = \sum_{i=1}^{N}\sum_{j=1}^{N}\nu_{ij}(q_i - q_j) = 0$$
$$\frac{d}{dt}\sum_{i=1}^{N}q_i \times p_i = \sum_{i=1}^{N}\frac{1}{m_i}p_i \times p_i + \sum_{i=1}^{N}\sum_{j=1}^{N}q_i \times \nu_{ij}(q_i - q_j) = 0.$$

**Example 1.4 (Conservation of Mass in Chemical Reactions).** Suppose that three substances $A, B, C$ undergo a chemical reaction such as[1]

$$A \xrightarrow{0.04} B \qquad \text{(slow)}$$
$$B + B \xrightarrow{3\cdot 10^7} C + B \qquad \text{(very fast)}$$
$$B + C \xrightarrow{10^4} A + C \qquad \text{(fast)}.$$

---

[1] This *Robertson problem* is very popular in testing codes for stiff differential equations.

We denote the masses (or concentrations) of the substances $A$, $B$, $C$ by $y_1$, $y_2$, $y_3$, respectively. By the mass action law this leads to the equations

$$
\begin{aligned}
A: &\quad \dot{y}_1 = -0.04\, y_1 + 10^4\, y_2 y_3 \\
B: &\quad \dot{y}_2 = \phantom{-}0.04\, y_1 - 10^4\, y_2 y_3 - 3\cdot 10^7\, y_2^2 \\
C: &\quad \dot{y}_3 = \phantom{-0.04\, y_1 - 10^4\, y_2 y_3 - } 3\cdot 10^7\, y_2^2
\end{aligned}
$$

We see that $\dot{y}_1 + \dot{y}_2 + \dot{y}_3 = 0$, hence the total mass $I(y) = y_1 + y_2 + y_3$ is an invariant of the system.

As was noted by Shampine (1986), such linear invariants are generally conserved by numerical integrators.

**Theorem 1.5 (Conservation of Linear Invariants).** *All explicit and implicit Runge-Kutta methods conserve linear invariants. Partitioned Runge-Kutta methods (II.2.2) conserve linear invariants if $b_i = \widehat{b}_i$ for all $i$, or if the invariant depends only on $p$ or only on $q$.*

*Proof.* Let $I(y) = d^T y$ with a constant vector $d$, so that $d^T f(y) = 0$ for all $y$. In the case of Runge-Kutta methods we thus have $d^T k_i = 0$, and consequently $d^T y_1 = d^T y_0 + h d^T (\sum_{i=1}^{s} b_i k_i) = d^T y_0$. The statement for partitioned methods is proved similarly. □

Next we consider differential equations of the form

$$\dot{Y} = A(Y)Y, \tag{1.3}$$

where $Y$ can be a vector or a matrix (not necessarily a square matrix). We then have the following result.

**Theorem 1.6.** *If $A(Y)$ is skew-symmetric for all $Y$ (i.e., $A^T = -A$), then the quadratic function $I(Y) = Y^T Y$ is an invariant. In particular, if the initial value $Y_0$ consists of orthonormal columns (i.e., $Y_0^T Y_0 = I$), then the columns of the solution $Y(t)$ of (1.3) remain orthonormal for all $t$.*

*Proof.* The derivative of $I(Y)$ is $I'(Y)H = Y^T H + H^T Y$. Thus, we have $I'(Y)f(Y) = I'(Y)(A(Y)Y) = Y^T A(Y)Y + Y^T A(Y)^T Y$ for all $Y$ which vanishes, because $A(Y)$ is skew-symmetric. This proves the statement. □

**Example 1.7 (Rigid Body).** The motion of a free rigid body, whose centre of mass is at the origin, is described by the Euler equations

$$
\begin{aligned}
\dot{y}_1 &= a_1 y_2 y_3, & a_1 &= (I_2 - I_3)/(I_2 I_3) \\
\dot{y}_2 &= a_2 y_3 y_1, & a_2 &= (I_3 - I_1)/(I_3 I_1) \\
\dot{y}_3 &= a_3 y_1 y_2, & a_3 &= (I_1 - I_2)/(I_1 I_2)
\end{aligned}
\tag{1.4}
$$

where the vector $y = (y_1, y_2, y_3)^T$ represents the angular momentum in the body frame, and $I_1, I_2, I_3$ are the principal moments of inertia (Euler (1758); see Marsden & Ratiu (1994, Chap. 15) for a detailed description). This problem can be written as

$$\begin{pmatrix} \dot{y}_1 \\ \dot{y}_2 \\ \dot{y}_3 \end{pmatrix} = \begin{pmatrix} 0 & y_3/I_3 & -y_2/I_2 \\ -y_3/I_3 & 0 & y_1/I_1 \\ y_2/I_2 & -y_1/I_1 & 0 \end{pmatrix} \begin{pmatrix} y_1 \\ y_2 \\ y_3 \end{pmatrix}, \tag{1.5}$$

which is of the form (1.3) with a skew-symmetric matrix $A(Y)$. By Theorem 1.6, $y_1^2 + y_2^2 + y_3^2$ is an invariant. A second quadratic invariant is

$$H(y_1, y_2, y_3) = \frac{1}{2} \left( \frac{y_1^2}{I_1} + \frac{y_2^2}{I_2} + \frac{y_3^2}{I_3} \right),$$

which represents the kinetic energy.

Inspired by the cover page of Marsden & Ratiu (1994), we present in Fig. 1.1 the sphere with some of the solutions of (1.4) corresponding to $I_1 = 2$, $I_2 = 1$ and $I_3 = 2/3$. They lie on the intersection of the sphere with the ellipsoid given by $H(y_1, y_2, y_3) = Const$. In the left picture we have included the numerical solution (30 steps) obtained by the implicit midpoint rule with step size $h = 0.3$ and initial value $y_0 = (\cos(1.1), 0, \sin(1.1))^T$. It stays exactly on a solution curve. This follows from the fact that the implicit midpoint rule preserves quadratic invariants exactly (Sect. IV.2).

For the explicit Euler method (right picture of Fig. 1.1, 320 steps with $h = 0.05$ and the same initial value) we see that the numerical solution shows a wrong qualitative behaviour (it should lie on a closed curve). The numerical solution even drifts away from the sphere.

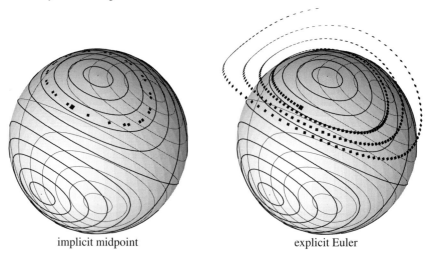

implicit midpoint            explicit Euler

**Fig. 1.1.** Solutions of the Euler equations (1.4) for the rigid body

## IV.2 Quadratic Invariants

Quadratic invariants appear often in applications. Examples are the conservation law of angular momentum in $N$-body systems (Example 1.3), the two invariants of the rigid body motion (Example 1.7), and the invariant $Y^T Y$ of Theorem 1.6. We therefore consider differential equations (1.1) and quadratic functions

$$Q(y) = y^T C y, \qquad (2.1)$$

where $C$ is a symmetric square matrix. It is an invariant of (1.1) if $y^T C f(y) = 0$ for all $y$.

### IV.2.1 Runge-Kutta Methods

We shall give a complete characterization of Runge-Kutta methods which automatically conserve all quadratic invariants. We first of all consider the Gauss collocation methods.

**Theorem 2.1.** *The Gauss methods of Sect. II.1.3 (collocation based on the shifted Legendre polynomials) conserve quadratic invariants.*

*Proof.* Let $u(t)$ be the collocation polynomial of the Gauss methods (Definition II.1.3). Since $\frac{d}{dt} Q(u(t)) = 2 u(t)^T C \dot{u}(t)$, it follows from $u(t_0) = y_0$ and $u(t_0 + h) = y_1$ that

$$y_1^T C y_1 - y_0^T C y_0 = 2 \int_{t_0}^{t_0 + h} u(t)^T C \dot{u}(t)\, dt. \qquad (2.2)$$

The integrand $u(t)^T C \dot{u}(t)$ is a polynomial of degree $2s - 1$, which is integrated without error by the $s$-stage Gaussian quadrature formula. It therefore follows from the collocation condition

$$u(t_0 + c_i h)^T C \dot{u}(t_0 + c_i h) = u(t_0 + c_i h)^T C f\big(u(t_0 + c_i h)\big) = 0$$

that the integral in (2.2) vanishes. □

Since the implicit midpoint rule is the special case $s = 1$ of the Gauss methods, the preceding theorem explains its good behaviour for the rigid body simulation in Fig 1.1.

**Theorem 2.2 (Cooper 1987).** *If the coefficients of a Runge-Kutta method satisfy*

$$b_i a_{ij} + b_j a_{ji} = b_i b_j \qquad \text{for all } i, j = 1, \ldots, s, \qquad (2.3)$$

*then it conserves quadratic invariants.*[2]

---

[2] For irreducible methods, the conditions of Theorem 2.2 and Theorem 2.4 are also necessary for the conservation of all quadratic invariants. This follows from the discussion in Sect. VI.7.2.

*Proof.* The proof is the same as that for B-stability, given independently by Burrage & Butcher and Crouzeix in 1979 (see Hairer & Wanner (1996), Sect. IV.12).

The relation $y_1 = y_0 + h \sum_{i=1}^{s} b_i k_i$ of Definition II.1.1 yields

$$y_1^T C y_1 = y_0^T C y_0 + h \sum_{i=1}^{s} b_i k_i^T C y_0 + h \sum_{j=1}^{s} b_j y_0^T C k_j + h^2 \sum_{i,j=1}^{s} b_i b_j k_i^T C k_j. \quad (2.4)$$

We then write $k_i = f(Y_i)$ with $Y_i = y_0 + h \sum_{j=1}^{s} a_{ij} k_j$. The main idea is to compute $y_0$ from this relation and to insert it into the central expressions of (2.4). This yields (using the symmetry of $C$)

$$y_1^T C y_1 = y_0^T C y_0 + 2h \sum_{i=1}^{s} b_i Y_i^T C f(Y_i) + h^2 \sum_{i,j=1}^{s} (b_i b_j - b_i a_{ij} - b_j a_{ji}) k_i^T C k_j.$$

The condition (2.3) together with the assumption $y^T C f(y) = 0$, which states that $y^T C y$ is an invariant of (1.1), imply $y_1^T C y_1 = y_0^T C y_0$. □

The criterion (2.3) is very restrictive. One finds that among all collocation and discontinuous collocation methods (Definition II.1.7) only the Gauss methods satisfy this criterion (Exercise 6). On the other hand, it is possible to construct other high-order Runge-Kutta methods satisfying (2.3). The key for such a construction is the $W$-transformation (see Hairer & Wanner (1996), Sect. IV.5), which is exploited in the articles of Sun (1993a) and Hairer & Leone (2000).

## IV.2.2 Partitioned Runge-Kutta Methods

We next consider partitioned Runge-Kutta methods for systems $\dot{y} = f(y, z)$, $\dot{z} = g(y, z)$. Usually such methods cannot conserve general quadratic invariants (Exercise 4). We therefore concentrate on quadratic invariants of the form

$$Q(y, z) = y^T D z, \quad (2.5)$$

where $D$ is a matrix of the appropriate dimensions. Observe that the angular momentum of $N$-body systems (Example 1.3) is of this form.

**Theorem 2.3 (Sun 1993b).** *The Lobatto IIIA - IIIB pair conserves all quadratic invariants of the form (2.5). In particular, this is true for the Störmer/Verlet scheme (see Sect. II.2.2).*

*Proof.* Let $u(t)$ and $v(t)$ be the (discontinuous) collocation polynomials of the Lobatto IIIA and Lobatto IIIB methods, respectively (see Sect. II.2.2). In analogy to the proof of Theorem 2.1 we have

$$Q\big(u(t_0 + h), v(t_0 + h)\big) - Q\big(u(t_0), v(t_0)\big)$$
$$= \int_{t_0}^{t_0+h} \Big( Q\big(\dot{u}(t), v(t)\big) + Q\big(u(t), \dot{v}(t)\big) \Big) dt. \quad (2.6)$$

Since $u(t)$ is of degree $s$ and $v(t)$ of degree $s-2$, the integrand of (2.6) is a polynomial of degree $2s-3$. Hence, an application of the Lobatto quadrature yields the exact result. Using the fact that $Q(y,z)$ is an invariant of the differential equation, i.e., $Q(f(y,z),z)+Q(y,g(y,z)) \equiv 0$, we thus obtain for the integral in (2.6)

$$hb_1 Q(u(t_0), \delta(t_0)) + hb_s Q(u(t_0+h), \delta(t_0+h)),$$

where $\delta(t) = \dot{v}(t) - g(u(t), v(t))$ denotes the defect. It now follows from $u(t_0) = y_0$, $u(t_0+h) = y_1$ (definition of Lobatto IIIA) and from $v(t_0) = z_0 - hb_1\delta(t_0)$, $v(t_0+h) = z_1 + hb_s\delta(t_0+h)$ (definition of Lobatto IIIB) that $Q(y_1, z_1) - Q(y_0, z_0) = 0$, which proves the theorem. □

Exchanging the role of the IIIA and IIIB methods also leads to an integrator that preserves quadratic invariants of the form (2.5). The following characterization extends Theorem 2.2 to partitioned Runge-Kutta methods.

**Theorem 2.4.** *If the coefficients of a partitioned Runge-Kutta method (II.2.2) satisfy*

$$b_i \widehat{a}_{ij} + \widehat{b}_j a_{ji} = b_i \widehat{b}_j \quad \text{for } i,j = 1,\ldots,s, \tag{2.7}$$

$$b_i = \widehat{b}_i \quad \text{for } i = 1,\ldots,s, \tag{2.8}$$

*then it conserves quadratic invariants of the form (2.5).*
*If the partitioned differential equation is of the special form $\dot{y} = f(z)$, $\dot{z} = g(y)$, then the condition (2.7) alone implies that invariants of the form (2.5) are conserved.*

*Proof.* The proof is nearly identical to that of Theorem 2.2. Instead of (2.4) we get

$$y_1^T Dz_1 = y_0^T Dz_0 + h \sum_{i=1}^s b_i k_i^T Dz_0 + h \sum_{j=1}^s \widehat{b}_j y_0^T D\ell_j + h^2 \sum_{i,j=1}^s b_i \widehat{b}_j k_i^T D\ell_j.$$

Denoting by $(Y_i, Z_i)$ the arguments of $k_i = f(Y_i, Z_i)$ and $\ell_i = g(Y_i, Z_i)$, the same trick as in the proof of Theorem 2.2 gives

$$y_1^T Dz_1 = y_0^T Dz_0 + h \sum_{i=1}^s b_i f(Y_i, Z_i)^T DZ_i + h \sum_{j=1}^s \widehat{b}_j Y_j^T Dg(Y_j, Z_j)$$

$$+ h^2 \sum_{i,j=1}^s (b_i \widehat{b}_j - b_i \widehat{a}_{ij} - \widehat{b}_j a_{ji}) k_i^T D\ell_j. \tag{2.9}$$

Since (2.5) is an invariant, we have $f(y,z)^T Dz + y^T Dg(y,z) = 0$ for all $y$ and $z$. Consequently, the two conditions (2.7) and (2.8) imply $y_1^T Dz_1 = y_0^T Dz_0$.

For the special case where $f$ depends only on $z$ and $g$ only on $y$, the assumption $f(z)^T Dz + y^T Dg(y) = 0$ (for all $y,z$) implies that $f(z)^T Dz = -y^T Dg(y) = \text{Const}$. Therefore, the condition (2.8) is no longer necessary for the proof of the statement. □

## IV.2.3 Nyström Methods

An important class of partitioned differential equations is $\dot y = z$, $\dot z = g(y)$ or, equivalently,
$$\ddot y = g(y). \tag{2.10}$$
Many examples of Chap. I are of this form, in particular the $N$-body problem of Example 1.3 for which the angular momentum is a quadratic first integral. Nyström methods (Definition II.2.3),

$$\begin{aligned}
\ell_i &= g\Big(y_0 + c_i h \dot y_0 + h^2 \sum_{j=1}^{s} a_{ij} \ell_j\Big), \\
y_1 &= y_0 + h \dot y_0 + h^2 \sum_{i=1}^{s} \beta_i \ell_i, \qquad \dot y_1 = \dot y_0 + h \sum_{i=1}^{s} b_i \ell_i,
\end{aligned} \tag{2.11}$$

are adapted to the numerical solution of (2.10) and it is interesting to investigate which methods within this class can conserve quadratic invariants.

**Theorem 2.5.** *If the coefficients of the Nyström method (2.11) satisfy*
$$\begin{aligned}
\beta_i &= b_i(1 - c_i) & \text{for } i &= 1, \ldots, s, \\
b_i(\beta_j - a_{ij}) &= b_j(\beta_i - a_{ji}) & \text{for } i, j &= 1, \ldots, s,
\end{aligned} \tag{2.12}$$
*then it conserves all quadratic invariants of the form $y^T D \dot y$.*

*Proof.* The quadratic form $Q(y, \dot y) = y^T D \dot y$ is a first integral of (2.10) if and only if
$$\dot y^T D \dot y + y^T D g(y) = 0 \qquad \text{for all } y, \dot y \in \mathbb{R}^n. \tag{2.13}$$
This implies that $D$ is skew-symmetric and that $y^T D g(y) = 0$.

In the same way as for the proofs of Theorems 2.2 and 2.4 we now compute $y_1^T D \dot y_1$ using the formulas of (2.11) and we substitute $y_0$ by $Y_i - c_i h \dot y_0 - h^2 \sum_j a_{ij} \ell_j$, where $Y_i$ denotes the argument of $g$ in (2.11). This yields

$$\begin{aligned}
y_1^T D \dot y_1 &= y_0^T D \dot y_0 + h \dot y_0^T D \dot y_0 + h \sum_{i=1}^{s} b_i Y_i^T D \ell_i \\
&\quad + h^2 \sum_{i=1}^{s} \beta_i \ell_i^T D \dot y_0 + h^2 \sum_{i=1}^{s} b_i(1 - c_i) \dot y_0^T D \ell_i \\
&\quad + h^3 \sum_{i,j=1}^{s} b_i(\beta_j - a_{ij}) \ell_j^T D \ell_i.
\end{aligned}$$

Using the skew-symmetry of $D$ and $Y_i^T D \ell_i = Y_i^T D g(Y_i) = 0$, the condition (2.12) implies the conservation property $y_1^T D \dot y_1 = y_0^T D \dot y_0$. □

**Remark 2.6 (Composition Methods).** If a method $\Phi_h$ conserves quadratic invariants (e.g., the mid-point rule by Theorem 2.1 or the Störmer/Verlet scheme by Theorem 2.3 or a Nyström method of Theorem 2.5), then so does the composition method

$$\Psi_h = \Phi_{\gamma_s h} \circ \ldots \circ \Phi_{\gamma_1 h}. \tag{2.14}$$

This obvious property is one of the most important motivations for considering composition methods.

## IV.3 Polynomial Invariants

We consider two classes of problems with polynomial invariants for degree higher than two. First, we treat linear problems for which the determinant of the resolvent is an invariant, and we show that (partitioned) Runge-Kutta methods cannot conserve them automatically. Second, we study isospectral flows.

### IV.3.1 The Determinant as a First Integral

We consider quasi-linear problems

$$\dot{Y} = A(Y)Y, \qquad Y(0) = Y_0 \tag{3.1}$$

where $Y$ and $A(Y)$ are $n \times n$ matrices. In the following we denote the trace of a matrix $A = (a_{ij})_{i,j=1}^n$ by trace $A = \sum_{i=1}^n a_{ii}$.

**Lemma 3.1.** *If* trace $A(Y) = 0$ *for all* $Y$, *then* $g(Y) := \det Y$ *is an invariant of the matrix differential equation (3.1).*

*Proof.* It follows from

$$\det(Y + \varepsilon A Y) = \det(I + \varepsilon A) \det Y = \big(1 + \varepsilon \operatorname{trace} A + \mathcal{O}(\varepsilon^2)\big) \det Y$$

that $g'(Y)(AY) = \operatorname{trace} A \cdot \det Y$ (this is the *Abel-Liouville-Jacobi-Ostrogradskii identity*). Hence, the determinant $g(Y) = \det Y$ is an invariant of the differential equation (3.1) if trace $A(Y) = 0$ for all $Y$. □

Since $\det Y$ represents the volume of the parallelepiped generated by the columns of the matrix $Y$, the conservation of the invariant $g(Y) = \det Y$ is related to volume preservation. This topic will be further discussed in Sect. VII.3. Here, we consider $\det Y$ as a polynomial invariant of degree $n$, and we investigate whether Runge-Kutta methods can automatically conserve this invariant for $n \geq 3$. The key lemma for this study is the following.

**Lemma 3.2 (Feng Kang & Shang Zai-jiu 1995).** *Let $R(z)$ be a differentiable function defined in a neighbourhood of $z = 0$, and assume that $R(0) = 1$ and $R'(0) = 1$. Then, we have for $n \geq 3$*

$$\det R(A) = 1 \quad \text{for all } n \times n \text{ matrices } A \text{ satisfying trace } A = 0, \tag{3.2}$$

*if and only if $R(z) = \exp(z)$.*

*Proof.* The "if" part follows from Lemma 3.1, because for constant $A$ the solution of $\dot Y = AY, Y(0) = I$ is given by $Y(t) = \exp(At)$.

For the proof of the "only if" part, we consider diagonal matrices of the form $A = \mathrm{diag}(\mu, \nu, -(\mu+\nu), 0, \ldots, 0)$, which have trace $A = 0$, and for which

$$R(A) = \mathrm{diag}\Big(R(\mu), R(\nu), R(-(\mu+\nu)), R(0), \ldots, R(0)\Big).$$

The assumptions $R(0) = 1$ and (3.2) imply

$$R(\mu)R(\nu)R(-(\mu+\nu)) = 1 \tag{3.3}$$

for all $\mu, \nu$ close to 0. Putting $\nu = 0$, this relation yields $R(\mu)R(-\mu) = 1$ for all $\mu$, and therefore (3.3) can be written as

$$R(\mu)R(\nu) = R(\mu+\nu) \qquad \text{for all } \mu, \nu \text{ close to 0.} \tag{3.4}$$

This functional equation can only be satisfied by the exponential function. This is seen as follows: from (3.4) we have

$$\frac{R(\mu+\varepsilon) - R(\mu)}{\varepsilon} = R(\mu)\frac{R(\varepsilon) - R(0)}{\varepsilon}.$$

Taking the limit $\varepsilon \to 0$ we obtain $R'(\mu) = R(\mu)$, because $R'(0) = 1$. This implies $R(\mu) = \exp(\mu)$. $\square$

**Theorem 3.3.** *For $n \geq 3$, no Runge-Kutta method can conserve all polynomial invariants of degree $n$.*

*Proof.* It is sufficient to consider linear problems $\dot Y = AY$ with constant matrix $A$ satisfying trace $A = 0$, so that $g(Y) = \det Y$ is a polynomial invariant of degree $n$. Applying a Runge-Kutta method to such a differential equation yields $Y_1 = R(hA)Y_0$, where

$$R(z) = 1 + zb^T(I - z\mathcal{A})^{-1}\mathbb{1}$$

($b^T = (b_1, \ldots, b_s)$, $\mathbb{1} = (1, \ldots, 1)^T$ and $\mathcal{A} = (a_{ij})$ is the matrix of Runge-Kutta coefficients) is the so-called stability function. It is seen to be rational. By Lemma 3.2 it is therefore not possible that $\det R(hA) = 1$ for all $A$ with trace $A = 0$. $\square$

This negative result motivates the search for new methods which can conserve polynomial invariants (see Sects. IV.4, IV.8 and VII.3). We consider here another interesting class of problems with polynomial invariants of degree higher than two.

## IV.3.2 Isospectral Flows

Such flows are created by a matrix differential equation

$$\dot{L} = [B(L), L], \qquad L(0) = L_0 \tag{3.5}$$

where $L_0$ is a given symmetric matrix, $B(L)$ is skew-symmetric for all $L$, and $[B, L] = BL - LB$ is the commutator of $B$ and $L$. Many interesting problems can be written in this form. We just mention the Toda system, the continuous realization of QR-type algorithms, projected gradient flows, and inverse eigenvalue problems (see Chu (1992) and Calvo, Iserles & Zanna (1997) for long lists of references).

**Lemma 3.4 (Lax 1968, Flaschka 1974).** *Let $L_0$ be symmetric and assume that $B(L)$ is skew-symmetric for all $L$. Then, the solution $L(t)$ of (3.5) is a symmetric matrix, and its eigenvalues are independent of t.*

*Proof.* The symmetry of $L(t)$ follows from the fact that the commutator of a skew-symmetric with a symmetric matrix gives a symmetric matrix.

To prove the isospectrality of the flow, we define $U(t)$ by

$$\dot{U} = B(L(t))U, \qquad U(0) = I. \tag{3.6}$$

Then, we have $(d/dt)(U^{-1}LU) = U^{-1}(\dot{L} - BL + LB)U = 0$, and hence $U(t)^{-1}L(t)U(t) = L_0$ for all $t$, so that $L(t) = U(t)L_0 U(t)^{-1}$ is the solution of (3.5). This proves the result. □

Note that, since $B(L)$ is skew-symmetric, the matrix $U(t)$ of (3.6) is orthogonal by Theorem 1.6.

Lemma 3.4 shows that the characteristic polynomial $\det(L - \lambda I) = \sum_{i=0}^{n} a_i \lambda^i$ and hence the coefficients $a_i$ also are independent of $t$. These coefficients are all polynomial invariants (e.g., $a_0 = \det L$, $a_{n-1} = \pm \text{trace } L$). Because of Theorem 3.3 there is no hope that Runge-Kutta methods applied to (3.5) can conserve these invariants automatically for $n \geq 3$.

**Isospectral Methods.** The proof of Lemma 3.4, however, suggests an interesting approach for the numerical solution of (3.5). For $n = 0, 1, \ldots$ we solve numerically

$$\dot{U} = B(UL_n U^T)U, \qquad U(0) = I \tag{3.7}$$

and we put $L_{n+1} = \widehat{U} L_n \widehat{U}^T$, where $\widehat{U}$ is the numerical approximation $\widehat{U} \approx U(h)$ after one step (cf. Calvo, Iserles & Zanna 1999). If $B(L)$ is skew-symmetric for all matrices $L$, then $U^T U$ is a quadratic invariant of (3.7) and the methods of Sect. IV.2 will produce an orthogonal $\widehat{U}$. Consequently, $L_{n+1}$ and $L_n$ have exactly the same eigenvalues, and they remain symmetric.

Diele, Lopez & Politi (1998) suggest the use of the Cayley transform $U = (I - Y)^{-1}(I + Y)$, which transforms (3.7) into

$$\dot{Y} = \frac{1}{2}(I - Y)B(UL_n U^T)(I + Y), \qquad Y(0) = 0,$$

and the orthogonality of $U$ into the skew-symmetry of $Y$ (see Lemma 8.8 below). Since all (also explicit) Runge-Kutta methods preserve the skew-symmetry of $Y$, which is a linear invariant, this yields an approach to explicit isospectral methods.

**Connection with the QR Algorithm.** In a diversion from the main theme of this section, we now show the relationship of the flow of (3.5) with the QR algorithm for the symmetric eigenvalue problem. Starting from a real symmetric matrix $A_0$, the basic *QR algorithm* (without shifts) computes a sequence of orthogonally similar matrices $A_1, A_2, A_3, \ldots$, expected to converge towards a diagonal matrix carrying the eigenvalues of $A_0$. Iteratively for $k = 0, 1, 2, \ldots$, one computes the QR decomposition of $A_k$:

$$A_k = Q_k R_k$$

with $Q_k$ orthogonal, $R_k$ upper triangular (the decomposition becomes unique if the diagonal elements of $R_k$ are taken positive). Then, $A_{k+1}$ is obtained by reversing the order of multiplication:

$$A_{k+1} = R_k Q_k.$$

It is an easy exercise to show that $Q(k) = Q_0 Q_1 \ldots Q_{k-1}$ is the matrix in the orthogonal similarity transformation between $A_0$ and $A_k$:

$$A_k = Q(k)^T A_0 Q(k) \tag{3.8}$$

and the same matrix $Q(k)$ is the orthogonal factor in the QR decomposition of $A_0^k$:

$$A_0^k = Q(k) R(k). \tag{3.9}$$

Consider now, for an arbitrary real function $f$ defined on the eigenvalues of a real symmetric matrix $L_0$, the QR decomposition

$$\exp(tf(L_0)) = Q(t) R(t) \tag{3.10}$$

and define

$$L(t) := Q(t)^T L_0 Q(t). \tag{3.11}$$

The relations (3.8) and (3.9) then show that for integer times $t = k$, the matrix $\exp\bigl(f(L(k))\bigr) = Q(k)^T \exp\bigl(f(L_0)\bigr) Q(k)$ coincides with the $k$th matrix in the QR algorithm starting from $A_0 = \exp(f(L_0))$:

$$\exp(f(L(k))) = A_k. \tag{3.12}$$

Now, how is all this related to the system (3.5)? Differentiating (3.11) as in the proof of Lemma 3.4 shows that $L(t)$ solves a differential equation of the form $\dot L = [B, L]$ with the skew-symmetric matrix $B = -Q^T \dot Q$. At first sight, however, $B$ is a function of $t$, not of $L$. On the other hand, differentiation of (3.10) yields (omitting the argument $t$ where it is clear from the context)

$$f(L_0) Q R = f(L_0) \exp(tf(L_0)) = \exp(tf(L_0)) f(L_0) = \dot Q R + Q \dot R,$$

and since $f(L) = Q^T f(L_0) Q$ by (3.11), this becomes

$$f(L) = Q^T \dot{Q} + \dot{R} R^{-1}.$$

Here the left-hand side is a symmetric matrix, and the right-hand side is the sum of a skew-symmetric and an upper triangular matrix. It follows that the skew-symmetric matrix $B = -Q^T \dot{Q}$ is given by

$$B(L) = f(L)_+ - f(L)_+^T, \tag{3.13}$$

where $f(L)_+$ denotes the part of $f(L)$ above the diagonal. Hence, $L(t)$ is the solution of an autonomous system (3.5) with a skew-symmetric $B(L)$.

For $f(x) = x$ and assuming $L_0$ symmetric and tridiagonal, the flow of (3.5) with (3.13) is known as the *Toda flow*. The QR iterates $A_0 = \exp(L_0), A_1, A_2, \ldots$ of the exponential of $L_0$ are seen to be equal to the exponentials of the solution $L(t)$ of the Toda equations at integer times: $A_k = \exp(L(k))$, a discovery of Symes (1982). An interesting connection of the Toda equations with a mechanical system will be discussed in Sect. X.1.5.

For $f(x) = \log x$, the above arguments show that the QR iteration itself, starting from a positive definite symmetric tridiagonal matrix, is the evaluation $A_k = L(k)$ at integer times of a solution $L(t)$ of the differential equation (3.5) with $B$ given by (3.13). This relationship was explored in a series of papers by Deift, Li, Nanda & Tomei (1983, 1989, 1993).

Notwithstanding the mathematical beauty of this relationship, it must be remarked that the practical QR algorithm (with shifts and deflation) follows a different path.

## IV.4 Projection Methods

> Und bist du nicht willig, so brauch ich Gewalt.
> (J.W. Goethe, *Der Erlkönig*)

Suppose we have an $(n-m)$-dimensional submanifold of $\mathbb{R}^n$,

$$\mathcal{M} = \{y \,;\, g(y) = 0\} \tag{4.1}$$

($g : \mathbb{R}^n \to \mathbb{R}^m$), and a differential equation $\dot{y} = f(y)$ with the property that

$$y_0 \in \mathcal{M} \quad \text{implies} \quad y(t) \in \mathcal{M} \text{ for all } t. \tag{4.2}$$

We want to emphasize that this assumption is weaker than the requirement that all components $g_i(y)$ of $g(y)$ are invariants in the sense of Definition 1.1. In fact, assumption (4.2) is equivalent to $g'(y) f(y) = 0$ for $y \in \mathcal{M}$, whereas Definition 1.1 requires $g'(y) f(y) = 0$ for all $y \in \mathbb{R}^n$. In the situation of (4.2) we call $g(y)$ a *weak invariant*, and we say that $\dot{y} = f(y)$ is a differential equation on the manifold $\mathcal{M}$.

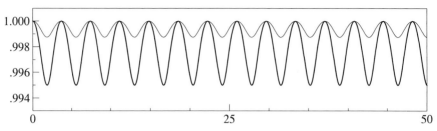

**Fig. 4.1.** The implicit midpoint rule applied to the differential equation (4.3). The picture shows the numerical values for $q_1^2 + q_2^2$ obtained with step size $h = 0.1$ (thick line) and $h = 0.05$ (thin line).

**Example 4.1.** Consider the pendulum equation written in Cartesian coordinates:

$$\begin{aligned} \dot{q}_1 &= p_1, & \dot{p}_1 &= -q_1 \lambda, \\ \dot{q}_2 &= p_2, & \dot{p}_2 &= -1 - q_2 \lambda, \end{aligned} \tag{4.3}$$

where $\lambda = (p_1^2 + p_2^2 - q_2)/(q_1^2 + q_2^2)$. One can check by differentiation that $q_1 p_1 + q_2 p_2$ (orthogonality of the position and velocity vectors) is an invariant in the sense of Definition 1.1. However, $q_1^2 + q_2^2$ (length of the pendulum) is only a weak invariant. The experiment of Fig. 4.1 shows that even methods which conserve quadratic first integrals (cf. Sect. IV.2) do not conserve the quadratic weak invariant $q_1^2 + q_2^2$. No numerical method that is allowed to evaluate the vector field $f(y)$ outside $\mathcal{M}$ can be expected to conserve weak invariants exactly. This is one of the motivations for considering the methods of this and the subsequent sections.

A natural approach to the numerical solution of differential equations on manifolds is by projection (see e.g., Hairer & Wanner (1996), Sect. VII.2, Eich-Soellner & Führer (1998), Sect. 5.3.3).

**Algorithm 4.2 (Standard Projection Method).** *Assume that $y_n \in \mathcal{M}$. One step $y_n \mapsto y_{n+1}$ is defined as follows (see Fig. 4.2):*

- *Compute $\widetilde{y}_{n+1} = \Phi_h(y_n)$, where $\Phi_h$ is an arbitrary one-step method applied to $\dot{y} = f(y)$;*
- *project the value $\widetilde{y}_{n+1}$ onto the manifold $\mathcal{M}$ to obtain $y_{n+1} \in \mathcal{M}$.*

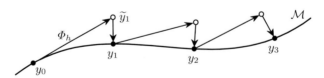

**Fig. 4.2.** Illustration of the standard projection method.

For $y_n \in \mathcal{M}$ the distance of $\widetilde{y}_{n+1}$ to the manifold $\mathcal{M}$ is of the size of the local error, i.e., $\mathcal{O}(h^{p+1})$. Therefore, the projection does not deteriorate the convergence order of the method.

For the computation of $y_{n+1}$ we have to solve the constrained minimization problem

$$\|y_{n+1} - \widetilde{y}_{n+1}\| \to \min \quad \text{subject to} \quad g(y_{n+1}) = 0. \tag{4.4}$$

In the case of the Euclidean norm, a standard approach is to introduce Lagrange multipliers $\lambda = (\lambda_1, \ldots, \lambda_m)^T$, and to consider the Lagrange function $\mathcal{L}(y_{n+1}, \lambda) = \|y_{n+1} - \widetilde{y}_{n+1}\|^2/2 - g(y_{n+1})^T \lambda$. The necessary condition $\partial \mathcal{L}/\partial y_{n+1} = 0$ then leads to the system

$$\begin{aligned} y_{n+1} &= \widetilde{y}_{n+1} + g'(\widetilde{y}_{n+1})^T \lambda \\ 0 &= g(y_{n+1}). \end{aligned} \tag{4.5}$$

We have replaced $y_{n+1}$ with $\widetilde{y}_{n+1}$ in the argument of $g'(y)$ in order to save some evaluations of $g'(y)$. Inserting the first relation of (4.5) into the second gives a nonlinear equation for $\lambda$, which can be efficiently solved by simplified Newton iterations:

$$\Delta\lambda_i = -\Big(g'(\widetilde{y}_{n+1})g'(\widetilde{y}_{n+1})^T\Big)^{-1} g\Big(\widetilde{y}_{n+1} + g'(\widetilde{y}_{n+1})^T \lambda_i\Big), \qquad \lambda_{i+1} = \lambda_i + \Delta\lambda_i.$$

For the choice $\lambda_0 = 0$ the first increment $\Delta\lambda_0$ is of size $\mathcal{O}(h^{p+1})$, so that the convergence is usually extremely fast. Often, one simplified Newton iteration is sufficient.

**Example 4.3.** As a first example we consider the perturbed Kepler problem (see Exercise I.12) with Hamiltonian function

$$H(p,q) = \frac{1}{2}(p_1^2 + p_2^2) - \frac{1}{\sqrt{q_1^2 + q_2^2}} - \frac{0.005}{2\sqrt{(q_1^2 + q_2^2)^3}},$$

and initial values $q_1(0) = 1 - e$, $q_2(0) = 0$, $p_1(0) = 0$, $p_2(0) = \sqrt{(1+e)/(1-e)}$ (eccentricity $e = 0.6$) on the interval $0 \le t \le 200$. The exact

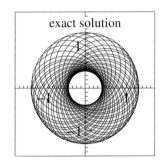
exact solution

solution (plotted to the right) is approximately an ellipse that rotates slowly around one of its foci. For this problem we know two first integrals: the Hamiltonian function $H(p,q)$ and the angular momentum $L(p,q) = q_1 p_2 - q_2 p_1$.

We apply the explicit Euler method and the symplectic Euler method (I.1.9), both with constant step size $h = 0.03$. The result is shown in Fig. 4.3. The numerical solution of the explicit Euler method (without projection) is completely wrong. The projection onto the manifold $\{H(p,q) = H(p_0, q_0)\}$ improves the numerical solution, but it still has a wrong qualitative behaviour. Only projection onto both invariants, $H(p,q) = Const$ and $L(p,q) = Const$ gives the correct behaviour. The symplectic Euler method already shows the correct behaviour without any projections (see Chap. IX for an explanation). Surprisingly, a projection onto $H(p,q) = Const$ destroys this behaviour, the numerical solution approaches the centre and the simplified Newton iterations fail to converge beyond $t = 25.23$. Projection onto both invariants re-establishes the correct behaviour.

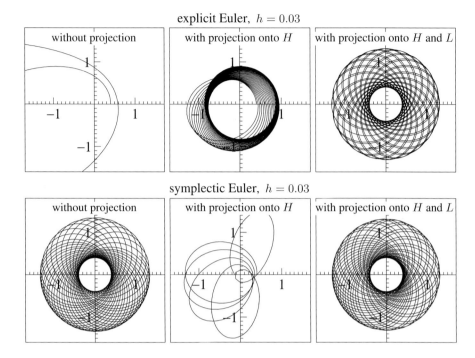

**Fig. 4.3.** Numerical solutions obtained with and without projections

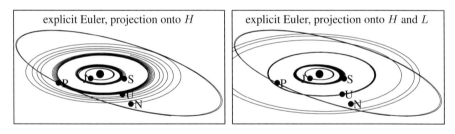

**Fig. 4.4.** Explicit Euler method with projections applied to the outer solar system, step size $h = 10$ (days), interval $0 \leq t \leq 200\,000$.

**Example 4.4 (Outer Solar System).** Having encountered excellent experience with projections onto $H$ and $L$ for the perturbed Kepler problem (Example 4.3), let us apply the same idea to a more realistic problem in celestial mechanics. We consider the outer solar system as described in Sect. I.2. The numerical solution of the explicit Euler method applied with constant step size $h = 10$, once with projection onto $H = Const$ and once with projection onto $H = Const$ and $L = Const$, is shown in Fig. 4.4 (observe that the conservation of the angular momentum $L(p,q) = \sum_{i=1}^{N} q_i \times p_i$ consists of three first integrals). We see a slight improvement in the orbits of Jupiter, Saturn and Uranus (compared to the explicit

Euler method without projections, see Fig. I.2.3), but the orbit of Neptune becomes even worse. There is no doubt that this problem contains a structure which cannot be correctly simulated by methods that only preserve the total energy $H$ and the angular momentum $L$.

**Example 4.5 (Volume Preservation).** Consider the matrix differential equation $\dot{Y} = A(Y)Y$, where trace $A(Y) = 0$ for all $Y$. We know from Lemma 3.1 that $g(Y) = \det Y$ is an invariant which cannot be automatically conserved by Runge-Kutta methods. Here, we show how we can enforce this invariant by projection. Let $\widetilde{Y}_{n+1}$ be the numerical approximation obtained with an arbitrary one-step method. We consider the Frobenius norm $\|Y\|_F = \sqrt{\sum_{i,j} |y_{ij}|^2}$ for measuring the distance to the manifold $\{Y\,;\,g(Y) = 0\}$. Using $g'(Y)(AY) = \text{trace}\,A \det Y$ (see the proof of Lemma 3.1) with $A$ chosen such that the product $AY$ contains only one non-zero element, the projection step (4.5) is seen to become (Exercise 9)

$$Y_{n+1} = \widetilde{Y}_{n+1} + \mu \widetilde{Y}_{n+1}^{-T} \tag{4.6}$$

with the scalar $\mu = \lambda \det \widetilde{Y}_{n+1}$. This leads to the scalar nonlinear equation $\det(\widetilde{Y}_{n+1} + \mu \widetilde{Y}_{n+1}^{-T}) = \det Y_n$, for which simplified Newton iterations become

$$\det(\widetilde{Y}_{n+1} + \mu_i \widetilde{Y}_{n+1}^{-T})\Big(1 + (\mu_{i+1} - \mu_i)\,\text{trace}\big((\widetilde{Y}_{n+1}^T \widetilde{Y}_{n+1})^{-1}\big)\Big) = \det Y_n.$$

If the $QR$-decomposition of $\widetilde{Y}_{n+1}$ is available from the computation of $\det \widetilde{Y}_{n+1}$, the value of $\text{trace}\big((\widetilde{Y}_{n+1}^T \widetilde{Y}_{n+1})^{-1}\big)$ can be computed efficiently with $\mathcal{O}(n^3/3)$ flops (see e.g., Golub & Van Loan (1989), Sect. 5.3.9).

The above projection is preferable to $Y_{n+1} = c\widetilde{Y}_{n+1}$, where $c \in \mathbb{R}$ is chosen such that $\det Y_{n+1} = \det Y_n$. This latter projection is already ill-conditioned for diagonal matrices with entries that differ by several magnitudes.

**Example 4.6 (Orthogonal Matrices).** As a final example let us consider $\dot{Y} = F(Y)$, where the solution $Y(t)$ is known to be an orthogonal matrix (e.g., the problems treated in Theorem 1.6 and in the proof of Lemma 3.4) or, more generally, an $n \times k$ matrix satisfying $Y^T Y = I$ (Stiefel manifold). The projection step (4.4) requires the solution of the problem

$$\|Y - \widetilde{Y}\|_F \to \min \qquad \text{subject to} \qquad Y^T Y = I, \tag{4.7}$$

where $\widetilde{Y}$ is a given matrix. This projection can be computed as follows: compute the singular value decomposition $\widetilde{Y} = U^T \Sigma V$, where $U^T$ and $V$ are $n \times k$ and $k \times k$ matrices with orthonormal columns, $\Sigma = \text{diag}(\sigma_1, \ldots, \sigma_k)$, and the singular values $\sigma_1 \geq \ldots \geq \sigma_k$ are all close to 1. Then the solution of (4.7) is given by $Y = U^T V$ (see Exercise 10 for some hints). This procedure is equivalent to the one proposed by D. Higham (1997): the orthogonal projection is the first factor of the *polar decomposition* $\widetilde{Y} = YR$ (where $Y$ has orthonormal columns and $R$ is symmetric positive definite). The equivalence is seen from the polar decomposition $\widetilde{Y} = (U^T V)(V^T \Sigma V)$.

A related procedure, where the first factor of the $QR$ decomposition of $\widetilde{Y}$ is used instead of that of the polar decomposition, is proposed in Dieci, Russell & van Vleck (1994).

As a conclusion to the above numerical experiments we see that a projection can give excellent results, but can also destroy the good long-time behaviour of the solution if applied inappropriately. If the original method already preserves some structure, then projection to a subset of invariants may destroy the good long-time behaviour. An important modification for reversible differential equations (symmetric projections) will be presented in Sect. V.4.1.

## IV.5 Numerical Methods Based on Local Coordinates

A second important class of methods for the numerical treatment of differential equations on manifolds uses local coordinates. Before explaining the ideas, we find it appropriate to discuss in more detail manifolds and differential equations on manifolds.

### IV.5.1 Manifolds and the Tangent Space

In Sect. IV.4 we assumed that locally (in a neighbourhood $U$ of $a \in \mathbb{R}^n$) a manifold is given by constraints, i.e.,

$$\mathcal{M} = \{y \in U \; ; \; g(y) = 0\}, \tag{5.1}$$

where $g : U \to \mathbb{R}^m$ is differentiable, $g(a) = 0$, and $g'(a)$ has full rank $m$.

Here, we use local parameters to characterize a manifold. Let $\psi : V \to \mathbb{R}^n$ be differentiable ($V \subset \mathbb{R}^{n-m}$ is a neighbourhood of 0), $\psi(0) = a$, and assume that $\psi'(0)$ has full rank $n - m$. Then, a manifold is locally given by

$$\mathcal{M} = \{y = \psi(z) \; ; \; z \in V\} \tag{5.2}$$

provided that $V$ is sufficiently small, so that $\psi : V \to \psi(V)$ is bijective with continuous inverse. The variables $z$ are called *parameters* or *local coordinates* of the manifold.

As an example, consider the unit sphere which, in the form (5.1), is given by the function $g(y_1, y_2, y_3) = y_1^2 + y_2^2 + y_3^2 - 1$. There are many possible choices of local coordinates. Away from the equator (i.e., $y_3 = 0$), we can take $z = (z_1, z_2)^T := (y_1, y_2)^T$ and $\psi(z) = \left(z_1, z_2, \pm\sqrt{1 - z_1^2 - z_2^2}\right)^T$. Alternatively, we can consider spherical coordinates $\psi(\alpha, \beta) = \left(\cos\alpha \sin\beta, \sin\alpha \sin\beta, \cos\beta\right)^T$ away from the north and south poles (i.e., $y_1 = y_2 = 0, y_3 = \pm 1$).

The tangent to a curve (or the tangent plane to a surface) is an affine space passing through the contact point $a \in \mathcal{M}$. It is convenient to place the origin at $a$, so that we obtain a vector space. More precisely, for a manifold $\mathcal{M}$ we define the *tangent space* at $a \in \mathcal{M}$ as

$$T_a\mathcal{M} = \left\{v \in \mathbb{R}^n \,\middle|\, \begin{array}{l}\text{there exists a differentiable path } \gamma : (-\varepsilon, \varepsilon) \to \mathbb{R}^n \\ \text{with } \gamma(t) \in \mathcal{M} \text{ for all } t, \ \gamma(0) = a, \ \dot{\gamma}(0) = v\end{array}\right\}. \quad (5.3)$$

**Lemma 5.1.** *If the manifold $\mathcal{M}$ is given by (5.1), where $g : U \to \mathbb{R}^m$ is differentiable, $g(a) = 0$, and $g'(a)$ has full rank $m$, then we have*

$$T_a\mathcal{M} = \ker g'(a) = \{v \in \mathbb{R}^n \,|\, g'(a)v = 0\}. \quad (5.4)$$

*If $\mathcal{M}$ is given by (5.2), where $\psi : V \to \mathbb{R}^n$ is differentiable, $\psi(0) = a$, and $\psi'(0)$ has full rank $n - m$, then we have*

$$T_a\mathcal{M} = \operatorname{Im} \psi'(0) = \{\psi'(0)w \,|\, w \in \mathbb{R}^{n-m}\}. \quad (5.5)$$

*Proof.* a) For a path $\gamma(t)$ satisfying $\gamma(0) = a$ and $g(\gamma(t)) = 0$ it follows by differentiation that $g'(a)\dot{\gamma}(0) = 0$. Consequently, we have $T_a\mathcal{M} \subset \ker g'(a)$.

Consider now the function $F(t, u) = g(a + tv + g'(a)^T u)$. We have $F(0, 0) = 0$ and an invertible $\partial F/\partial u(0, 0) = g'(a)g'(a)^T$, so that by the implicit function theorem the relation $F(t, u) = 0$ can be solved locally for $u = u(t)$. If $v \in \ker g'(a)$, it follows that $\dot{u}(0) = 0$, and the path $\gamma(t) = a + tv + g'(a)^T u(t)$ satisfies all requirements of (5.3), so that also $T_a\mathcal{M} \supset \ker g'(a)$.

b) Assume $\mathcal{M}$ to be given by (5.2). For an arbitrary $\eta : (-\varepsilon, \varepsilon) \to \mathbb{R}^m$ satisfying $\eta(0) = 0$, the path $\gamma(t) = \psi(\eta(t))$ lies in $\mathcal{M}$ and satisfies $\dot{\gamma}(0) = \psi'(0)\dot{\eta}(0)$. This proves $\operatorname{Im} \psi'(0) \subset T_a\mathcal{M}$.

The assumption on the rank of $\psi'(0)$ implies that, after a reordering of the components, we have $\psi(z) = (\psi_1(z), \psi_2(z))^T$, where $\psi_1(z)$ is a local diffeomorphism (by the inverse function theorem). We show that every smooth path $\gamma(t)$ in $\mathcal{M}$ can be written as $\gamma(t) = \psi(\eta(t))$ with some smooth $\eta(t)$. This then implies $T_a\mathcal{M} \subset \operatorname{Im} \psi'(0)$. To prove this we split $\gamma(t) = (\gamma_1(t), \gamma_2(t))^T$ according to the partitioning of $\psi$, and we define $\eta(t) = \psi_1^{-1}(\gamma_1(t))$. Since for $\gamma(t) \in \mathcal{M}$ the second part $\gamma_2(t)$ is uniquely determined by $\gamma_1(t)$, this proves $\gamma(t) = \psi(\eta(t))$. □

The proof of the preceding lemma shows the equivalence of the representations (5.1) and (5.2) of manifolds in $\mathbb{R}^n$. Let $\mathcal{M}$ be given by (5.1), and assume that the columns of $Q$ form an orthogonal basis of $T_a\mathcal{M}$. As in part (a) of the proof of Lemma 5.1 the condition $g(a + Qz + g'(a)^T u) = 0$ defines locally (close to $z = 0$) a function $u(z)$ which satisfies $u(0) = 0$ and $u'(0) = 0$. Hence, the manifold $\mathcal{M}$ is also given by (5.2) with the function $\psi(z) = a + Qz + g'(a)^T u(z)$.

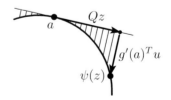

On the other hand, let $\mathcal{M}$ be given by (5.2). Part (b) of the proof of Lemma 5.1 shows that $y = \psi(z)$ can be partitioned into $y_1 = \psi_1(z)$ and $y_2 = \psi_2(z)$, where $\psi_1$ is a local diffeomorphism. Consequently, $\mathcal{M}$ is also given by (5.1) with $g(y) = y_2 - \psi_2(\psi_1^{-1}(y_1))$.

## IV.5.2 Differential Equations on Manifolds

In Sect. IV.4 we introduced differential equations on a manifold as problems satisfying (4.2). With the help of Lemma 5.1 we are now in a position to characterize such problems without knowledge of the solutions.

**Theorem 5.2.** *Let $\mathcal{M}$ be a submanifold of $\mathbb{R}^n$. The problem $\dot{y} = f(y)$ is a differential equation on the manifold $\mathcal{M}$ (i.e., it satisfies (4.2)) if and only if*

$$f(y) \in T_y\mathcal{M} \qquad \text{for all } y \in \mathcal{M}. \tag{5.6}$$

*Proof.* The necessity of (5.6) follows from the definition of $T_y\mathcal{M}$, because the exact solution of the differential equation lies in $\mathcal{M}$ and has $f(y)$ as derivative.

To prove the sufficiency, we assume (5.6) and let $\mathcal{M}$ be locally, near $y_0$, be given by a parametrization $y = \psi(z)$ as in (5.2). We try to write the solution of $\dot{y} = f(y)$, $y(0) = y_0 = \psi(z_0)$ as $y(t) = \psi(z(t))$. If this is at all possible, then $z(t)$ must satisfy

$$\psi'(z)\dot{z} = f(\psi(z))$$

which, by assumption (5.6) and the second part of Lemma 5.1, is equivalent to

$$\dot{z} = \psi'(z)^+ f(\psi(z)), \tag{5.7}$$

where $A^+ = (A^T A)^{-1} A^T$ denotes the pseudo-inverse of a matrix with full column rank. Conversely, define $z(t)$ as the solution of (5.7) with $z(0) = z_0$, which is known to exist locally in $t$ by the standard existence and uniqueness theory of ordinary differential equations on $\mathbb{R}^m$. Then $y(t) = \psi(z(t))$ is the solution of $\dot{y} = f(y)$ with $y(0) = y_0$. Hence, the solution $y(t)$ remains in $\mathcal{M}$. □

We remark that the sufficiency proof of Theorem 5.2 only requires the function $f(y)$ to be defined on $\mathcal{M}$. Due to the equivalence of $\dot{y} = f(y)$ with (5.7) the problem is transported to the space of local coordinates. The standard local theory for ordinary differential equations on an Euclidean space (existence and uniqueness of solutions, ...) can thus be extended in a straightforward way to differential equations on manifolds, i.e., $\dot{y} = f(y)$ with $f : \mathcal{M} \to \mathbb{R}^n$ satisfying (5.6).

## IV.5.3 Numerical Integrators on Manifolds

Whereas the projection methods of Sect. IV.4 require the function $f(y)$ of the differential equation to be defined in a neighbourhood of $\mathcal{M}$ (see Fig. 4.2), the numerical methods of this section evaluate $f(y)$ only on the manifold $\mathcal{M}$. The idea is to apply the numerical integrator in the parameter space rather than in the space where $\mathcal{M}$ is embedded.

**Algorithm 5.3 (Local Coordinates Approach).** *Assume that $y_n \in \mathcal{M}$ and that $\psi$ is a local parametrization of $\mathcal{M}$ satisfying $\psi(z_n) = y_n$. One step $y_n \mapsto y_{n+1}$ is defined as follows (see Fig. 5.1):*

- Compute $\tilde{z}_{n+1} = \Phi_h(z_n)$, the result of the method $\Phi_h$ applied to (5.7);
- define the numerical solution by $y_{n+1} = \psi(\tilde{z}_{n+1})$.

It is important to remark that the parametrization $y = \psi(z)$ can be changed at every step.

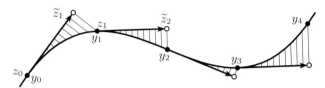

**Fig. 5.1.** The numerical solution of differential equations on manifolds via local coordinates.

As indicated at the beginning of Sect. IV.5.1, there are many possible choices of local coordinates. Consider the pendulum equation of Example 4.1, where $\mathcal{M} = \{(q_1, q_2, p_1, p_2) \mid q_1^2 + q_2^2 = 1, \ q_1 p_1 + q_2 p_2 = 0\}$. A standard parametrization here is $q_1 = \sin \alpha$, $q_2 = -\cos \alpha$, $p_1 = \omega \cos \alpha$, and $p_2 = \omega \sin \alpha$. In the new coordinates $(\alpha, \omega)$ the problem becomes simply $\dot{\alpha} = \omega$, $\dot{\omega} = -\sin \alpha$. Other typical choices are the exponential map $\psi(Z) = \exp(Z)$ for differential equations on Lie groups, and the Cayley transform $\psi(Z) = (I - Z)^{-1}(I + Z)$ for quadratic Lie groups. This will be studied in more detail in Sect. IV.8 below. Here we discuss two commonly used choices which do not use a special structure of the manifold.

**Generalized Coordinate Partitioning.** We assume that the manifold is given by (5.1). If $g : \mathbb{R}^n \to \mathbb{R}^m$ has a Jacobian with full rank $m$ at $y = a$, we can find a partitioning $y = (y_1, y_2)$, such that $\partial g / \partial y_2(a)$ is invertible. In this case we can choose the components of $y_1$ as local coordinates. The function $y = \psi(z)$ is then given by $y_1 = z$ and $y_2 = \psi_2(z)$, where $\psi_2(z)$ is implicitly defined by $g(z, \psi_2(z)) = 0$. This approach has been promoted by Wehage & Haug (1982) in the context of constrained mechanical systems, and the partitioning is found by Gaussian elimination with full pivoting applied to the matrix $g'(a)$. Another way of finding the partitioning is by the use of the QR decomposition with column change.

**Tangent Space Parametrization.** Let the manifold $\mathcal{M}$ be given by (5.1), and collect the vectors of an orthogonal basis of $T_a\mathcal{M}$ in the matrix $Q$. We then consider the parametrization
$$\psi_a(z) = a + Qz + g'(a)^T u(z), \tag{5.8}$$
where $u(z)$ is defined by $g(\psi_a(z)) = 0$, exactly as in the discussion after the proof of Lemma 5.1. Differentiating (5.8) yields
$$(Q + g'(a)^T u'(z))\dot{z} = \dot{y} = f(y) = f(\psi_a(z)).$$
Since $Q^T Q = I$ and $g'(a)Q = 0$, this relation is equivalent to the differential equation
$$\dot{z} = Q^T f(\psi_a(z)), \tag{5.9}$$

which corresponds to (5.7). If we apply a numerical method to (5.9), every function evaluation requires the projection of an element of the tangent space onto the manifold. This procedure is illustrated in Fig. 5.1, and was originally proposed by Potra & Rheinboldt (1991) for the solution of the Euler-Lagrange equations of constrained multibody systems (see also Hairer & Wanner (1996), p. 476).

**Example 5.4 (Stiefel manifold).** We consider problems $\dot Y = F(Y)$ on the Stiefel manifold

$$\mathcal{V}_{n,k} = \{Y \in \mathbb{R}^{n \times k} \mid Y^T Y = I\}. \tag{5.10}$$

Such problems arise naturally in the computation of Lyapunov exponents of differential equations (e.g., Dieci, Russell & van Vleck 1997). The use of projection methods has been explained in Example 4.6. Let us discuss here the application of the tangent space parametrization.

For the Stiefel manifold (5.10), hereafter denoted by $\mathcal{V}$, we have

$$T_A \mathcal{V} = \{H \in \mathbb{R}^{n \times k} \mid H^T A + A^T H = 0\}. \tag{5.11}$$

Using the standard Euclidean inner product on $\mathbb{R}^{n \times k}$, i.e., $\langle Y, Z \rangle = \text{trace}(Y^T Z)$, the orthonormal space to $T_A \mathcal{V}$ is given by

$$N_A \mathcal{V} = \{K \in \mathbb{R}^{n \times k} \mid K \perp T_A \mathcal{V}\} = \{AS \mid S \text{ symmetric } k \times k \text{ matrix}\}. \tag{5.12}$$

The orthogonality $AS \perp T_A \mathcal{V}$ follows from the fact that the trace of the product of a symmetric with a skew-symmetric matrix vanishes (Exercise 12). Moreover, an arbitrary matrix $X \in \mathbb{R}^{n \times k}$ admits the decomposition

$$X = P_A X + AS \quad \text{with} \quad P_A X = AT + (I - AA^T)X, \tag{5.13}$$

where $S = (A^T X + X^T A)/2$ and $T = (A^T X - X^T A)/2$ are the symmetric and the skew-symmetric part of $A^T X$ respectively. One can check in a straightforward way that $P_A X \in T_A \mathcal{V}$ is the orthogonal projection of $X$ onto the tangent space $T_A \mathcal{V}$.

Consequently, for $Z \in T_A \mathcal{V}$, the function

$$\psi_A(Z) = A + Z + AS(Z) \tag{5.14}$$

is a local parametrization of $\mathcal{V}$, if the symmetric matrix $S = S(Z)$ is such that $\psi_A(Z)^T \psi_A(Z) = I$, i.e., $S$ has to be a solution of the algebraic Riccati equation

$$S^2 + 2S + SA^T Z + Z^T AS + Z^T Z = 0. \tag{5.15}$$

Observe that for $k = 1$, where the Stiefel manifold reduces to the unit sphere in $\mathbb{R}^n$, the equation (5.15) is a scalar quadratic equation and can be easily solved. For $k > 1$, it can be solved iteratively using the scheme ($S_0 = 0$)

$$(I + Z^T A)S_n + S_n(I + A^T Z) = -Z^T Z - S_{n-1}^2.$$

Using a Schur decomposition $A^T Z = U^T R U$ (where $U$ is orthogonal and $R$ upper triangular), the elements of $U S_n U^T$ can be computed successively starting from the left upper corner. We refer to the monograph of Mehrmann (1991) for a detailed discussion of the solution of linear and algebraic Riccati equations.

With the parametrization $\psi_A(Z)$ of (5.14) the transformed differential equation becomes $\dot{Z} + A \frac{d}{dt} S(Z) = F(\psi_A(Z))$ which, after an orthogonal projection onto the tangent space, yields

$$\dot{Z} = P_A F(\psi_A(Z)), \tag{5.16}$$

in complete analogy to (5.9). The numerical solution of (5.16) requires, for every function evaluation, the solution of the Riccati equation (5.15) and the computation of a projection onto the tangent space, each needing $\mathcal{O}(nk^2)$ operations. Compared to the projection method of Example 4.6 the overhead (i.e., the computation apart from the evaluation of $F(Y)$) is more expensive, but the approach of this section has the advantage that all evaluations of $F$ are exactly on the manifold $\mathcal{V}$.

## IV.6 Differential Equations on Lie Groups

Theorem 1.6 and Lemma 3.1 are particular cases of a more general result which can be conveniently formulated with the concept of Lie groups and Lie algebras (see Olver (1986) and Varadarajan (1974) for an introduction to these subjects).

A *Lie group* is a group $G$ which is a differentiable manifold, and for which the product is a differentiable mapping $G \times G \to G$. We restrict our considerations to *matrix Lie groups*, that is, Lie groups which are subgroups of $GL(n)$, the group of invertible $n \times n$ matrices with the usual matrix product as the group operation.

Marius Sophus Lie[3]

**Example 6.1.** An important example of a Lie group is the group

$$O(n) = \{Y \in \text{GL}(n) \mid Y^T Y = I\}$$

of all orthogonal matrices. It is the zero set of $g(Y) = Y^T Y - I$, where we consider $g$ as a mapping from the set of all $n \times n$ matrices (i.e., $\mathbb{R}^{n \cdot n}$) to the set of all symmetric matrices (which can be identified with $\mathbb{R}^{n(n+1)/2}$). The derivative $g'(Y)$ is surjective for $Y \in O(n)$, because for any symmetric matrix $K$ the choice $H =$

---

[3] Marius Sophus Lie, born: 17 December 1842 in Nordfjordeid (Norway), died: 18 February 1899.

**Table 6.1.** Some matrix Lie groups and their corresponding Lie algebras

| | Lie group | | | Lie algebra | |
|---|---|---|---|---|---|
| GL($n$) | = | $\{Y \mid \det Y \neq 0\}$ <br> general linear group | $\mathfrak{gl}(n)$ | = | $\{A \mid \text{arbitrary matrix}\}$ <br> Lie algebra of $n \times n$ matrices |
| SL($n$) | = | $\{Y \mid \det Y = 1\}$ <br> special linear group | $\mathfrak{sl}(n)$ | = | $\{A \mid \text{trace}(A) = 0\}$ <br> special linear Lie algebra |
| O($n$) | = | $\{Y \mid Y^T Y = I\}$ <br> orthogonal group | $\mathfrak{so}(n)$ | = | $\{A \mid A^T + A = 0\}$ <br> skew-symmetric matrices |
| SO($n$) | = | $\{Y \in \text{O}(n) \mid \det Y = 1\}$ <br> special orthogonal group | $\mathfrak{so}(n)$ | = | $\{A \mid A^T + A = 0\}$ <br> skew-symmetric matrices |
| Sp($n$) | = | $\{Y \mid Y^T J Y = J\}$ <br> symplectic group | $\mathfrak{sp}(n)$ | = | $\{A \mid JA + A^T J = 0\}$ |

$YK/2$ solves the equation $g'(Y)H = K$. Therefore, the matrix $g'(Y)$ has full rank (cf. (5.1)) so that O($n$) defines a differentiable manifold of dimension $n^2 - n(n+1)/2 = n(n-1)/2$. The set O($n$) is also a group with unit element $I$ (the identity). Since the matrix multiplication is a differentiable mapping, O($n$) is a Lie group.

Table 6.1 lists further prominent examples. The matrix $J$ appearing in the definition of the symplectic group is the matrix determining the symplectic structure on $\mathbb{R}^n$ (see Sect. VI.2).

As the following lemma shows, the tangent space $\mathfrak{g} = T_I G$ at the identity $I$ of a matrix Lie group $G$ is closed under forming commutators of its elements. This makes $\mathfrak{g}$ an algebra, the *Lie algebra* of the Lie group $G$.

**Lemma 6.2 (Lie Bracket and Lie Algebra).** *Let $G$ be a matrix Lie group and let $\mathfrak{g} = T_I G$ be the tangent space at the identity. The Lie bracket (or commutator)*

$$[A, B] = AB - BA \tag{6.1}$$

*defines an operation $\mathfrak{g} \times \mathfrak{g} \to \mathfrak{g}$ which is bilinear, skew-symmetric ($[A, B] = -[B, A]$), and satisfies the Jacobi identity*

$$[A, [B, C]] + [C, [A, B]] + [B, [C, A]] = 0. \tag{6.2}$$

*Proof.* By definition of the tangent space, for $A, B \in \mathfrak{g}$, there exist differentiable paths $\alpha(t), \beta(t)$ ($|t| < \varepsilon$) in $G$ such that $\alpha(t) = I + tA(t)$ with a continuous function $A(t)$ with $A(0) = A$, and similarly $\beta(t) = I + tB(t)$ with $B(0) = B$. Now consider the path $\gamma(t)$ in $G$ defined by

$$\gamma(t) = \alpha(\sqrt{t})\beta(\sqrt{t})\alpha(\sqrt{t})^{-1}\beta(\sqrt{t})^{-1}, \quad t \geq 0.$$

An elementary computation then yields

$$\gamma(t) = I + t[A, B] + o(t).$$

With the extension $\gamma(t) = \gamma(-t)^{-1}$ for negative $t$, this is a differentiable path in $G$ satisfying $\gamma(0) = I$ and $\dot{\gamma}(0) = [A, B]$. Hence $[A, B] \in \mathfrak{g}$ by definition of the tangent space. The properties of the Lie bracket can be verified in a straightforward way. □

**Example 6.3.** Consider again the orthogonal group $\mathrm{O}(n)$. Since the derivative of $g(Y) = Y^T Y - I$ at the identity is $g'(I)H = I^T H + H^T I = H + H^T$, it follows from the first part of Lemma 5.1 that the Lie algebra corresponding to $\mathrm{O}(n)$ consists of all skew-symmetric matrices. The right column of Table 6.1 gives the Lie algebras of the other Lie groups listed there.

The following basic lemma shows that the exponential map yields a local parametrization of the Lie group near the identity, with the Lie algebra (a linear space) as the parameter space.

**Lemma 6.4 (Exponential Map).** *Consider a matrix Lie group $G$ and its Lie algebra $\mathfrak{g}$. The matrix exponential is a map*

$$\exp : \mathfrak{g} \to G,$$

*i.e., for $A \in \mathfrak{g}$ we have $\exp(A) \in G$. Moreover, $\exp$ is a local diffeomorphism in a neighbourhood of $A = 0$.*

*Proof.* For $A \in \mathfrak{g}$, it follows from the definition of the tangent space $\mathfrak{g} = T_I G$ that there exists a differentiable path $\alpha(t)$ in $G$ satisfying $\alpha(0) = I$ and $\dot{\alpha}(0) = A$. For a fixed $Y \in G$, the path $\gamma(t) := \alpha(t)Y$ is in $G$ and satisfies $\gamma(0) = Y$ and $\dot{\gamma}(0) = AY$. Consequently, $AY \in T_Y G$ and $\dot{Y} = AY$ defines a differential equation on the manifold $G$. The solution $Y(t) = \exp(tA)$ is therefore in $G$ for all $t$.

Since $\exp(H) - \exp(0) = H + \mathcal{O}(H^2)$, the derivative of the exponential map at $A = 0$ is the identity, and it follows from the inverse function theorem that $\exp$ is a local diffeomorphism close to $A = 0$. □

The proof of Lemma 6.4 shows that for a matrix Lie group $G$ the tangent space at $Y \in G$ has the form

$$T_Y G = \{AY \mid A \in \mathfrak{g}\}. \tag{6.3}$$

By Theorem 5.2, differential equations on a matrix Lie group (considered as a manifold) can therefore be written as

$$\dot{Y} = A(Y)Y \tag{6.4}$$

where $A(Y) \in \mathfrak{g}$ for all $Y \in G$. The following theorem summarizes this discussion, and extends the statements of Theorem 1.6 and Lemma 3.1 to more general matrix Lie groups.

**Theorem 6.5.** *Let $G$ be a matrix Lie group and $\mathfrak{g}$ its Lie algebra. If $A(Y) \in \mathfrak{g}$ for all $Y \in G$ and if $Y_0 \in G$, then the solution of (6.4) satisfies $Y(t) \in G$ for all $t$.* □

If in addition $A(Y) \in \mathfrak{g}$ for all matrices $Y$, and if

$$G = \{Y \mid g(Y) = Const\}$$

is one of the Lie groups of Table 6.1, then $g(Y)$ is an invariant of the differential equation (6.4) in the sense of Definition 1.1.

## IV.7 Methods Based on the Magnus Series Expansion

Wilhelm Magnus[4]

Before we discuss the numerical solution of differential equations (6.4) on Lie groups, let us give an explicit formula for the solution of linear matrix differential equations

$$\dot{Y} = A(t)Y. \qquad (7.1)$$

No assumption on the matrix $A(t)$ is made for the moment (apart from continuous dependence on $t$). For the scalar case, the solution of (7.1) with $Y(0) = Y_0$ is given by

$$Y(t) = \exp\left(\int_0^t A(\tau)\, d\tau\right) Y_0. \qquad (7.2)$$

Also in the case where the matrices $A(t)$ and $\int_0^t A(\tau)\, d\tau$ commute, (7.2) is the solution of (7.1). In the general non-commutative case we follow the approach of Magnus (1954) and we search for a matrix function $\Omega(t)$ such that

$$Y(t) = \exp\bigl(\Omega(t)\bigr) Y_0$$

solves (7.1). The main ingredient for the solution will be the inverse of the derivative of the matrix exponential. It has been studied in Sect. III.4, Lemma III.4.2, and is given by

$$d\exp_{\Omega}^{-1}(H) = \sum_{k \geq 0} \frac{B_k}{k!} \operatorname{ad}_{\Omega}^k(H), \qquad (7.3)$$

where $B_k$ are the Bernoulli numbers, and $\operatorname{ad}_{\Omega}(A) = [\Omega, A] = \Omega A - A\Omega$ is the adjoint operator introduced in (III.4.1).

---

[4] Wilhelm Magnus, born: 5 February 1907 in Berlin (Germany), died: 15 October 1990.

**Theorem 7.1 (Magnus 1954).** *The solution of the differential equation (7.1) can be written as* $Y(t) = \exp(\Omega(t))Y_0$ *with* $\Omega(t)$ *defined by*

$$\dot{\Omega} = d\exp_\Omega^{-1}(A(t)), \qquad \Omega(0) = 0. \tag{7.4}$$

*As long as* $\|\Omega(t)\| < \pi$, *the convergence of the* $d\exp_\Omega^{-1}$ *expansion (7.3) is assured.*

*Proof.* Comparing the derivative of $Y(t) = \exp(\Omega(t))Y_0$,

$$\dot{Y}(t) = \left(\frac{d}{d\Omega}\exp\Omega(t)\right)\dot{\Omega}(t)Y_0 = \left(d\exp_{\Omega(t)}(\dot{\Omega}(t))\right)\exp(\Omega(t))Y_0,$$

with (7.1) we obtain $A(t) = d\exp_{\Omega(t)}(\dot{\Omega}(t))$. Applying the inverse operator $d\exp_\Omega^{-1}$ to this relation yields the differential equation (7.4) for $\Omega(t)$. The statement on the convergence is a consequence of Lemma III.4.2. □

The first few Bernoulli numbers are $B_0 = 1$, $B_1 = -1/2$, $B_2 = 1/6$, $B_3 = 0$. The differential equation (7.4) therefore becomes

$$\dot{\Omega} = A(t) - \frac{1}{2}[\Omega, A(t)] + \frac{1}{12}[\Omega, [\Omega, A(t)]] + \ldots,$$

which is nonlinear in $\Omega$. Applying Picard fixed point iteration after integration yields

$$\begin{aligned}\Omega(t) &= \int_0^t A(\tau)\,d\tau - \frac{1}{2}\int_0^t \left[\int_0^\tau A(\sigma)\,d\sigma, A(\tau)\right] d\tau \\ &\quad + \frac{1}{4}\int_0^t \left[\int_0^\tau \left[\int_0^\sigma A(\mu)\,d\mu, A(\sigma)\right] d\sigma, A(\tau)\right] d\tau \\ &\quad + \frac{1}{12}\int_0^t \left[\int_0^\tau A(\sigma)\,d\sigma, \left[\int_0^\tau A(\mu)\,d\mu, A(\tau)\right]\right] d\tau + \ldots,\end{aligned} \tag{7.5}$$

which is the so-called *Magnus expansion*. For smooth matrices $A(t)$ the remainder in (7.5) is of size $\mathcal{O}(t^5)$ so that the truncated series inserted into $Y(t) = \exp(\Omega(t))Y_0$ gives an excellent approximation to the solution of (7.1) for small $t$.

**Numerical Methods Based on the Magnus Expansion.** Iserles & Nørsett (1999) study the general form of the Magnus expansion (7.5), and they relate the iterated integrals and the rational coefficients in (7.5) to binary trees. For a numerical integration of

$$\dot{Y} = A(t)Y, \qquad Y(t_0) = Y_0 \tag{7.6}$$

(where $Y$ is a matrix or a vector) they propose using $Y_{n+1} = \exp(h\Omega_n)Y_n$, where $h\Omega_n$ is a suitable approximation of $\Omega(h)$ given by (7.5) with $A(t_n + \tau)$ instead of $A(\tau)$. Of course, the Magnus expansion has to be truncated and the integrals have to be approximated by numerical quadrature.

We follow here the collocation approach suggested by Zanna (1999). The idea is to replace $A(t)$ locally by an interpolation polynomial

$$\widehat{A}(t) = \sum_{i=1}^{s} \ell_i(t)\, A(t_n + c_i h),$$

and to solve $\dot{Y} = \widehat{A}(t)Y$ on $[t_n, t_n + h]$ by the use of the truncated series (7.5).

**Theorem 7.2.** *Consider a quadrature formula* $(b_i, c_i)_{i=1}^{s}$ *of order* $p \geq s$, *and let* $Y(t)$ *and* $Z(t)$ *be solutions of* $\dot{Y} = A(t)Y$ *and* $\dot{Z} = \widehat{A}(t)Z$, *respectively, satisfying* $Y(t_n) = Z(t_n)$. *Then,* $Z(t_n + h) - Y(t_n + h) = \mathcal{O}(h^{p+1})$.

*Proof.* We write the differential equation for $Z$ as $\dot{Z} = A(t)Z + \bigl(\widehat{A}(t) - A(t)\bigr)Z$ and use the variation of constants formula to get

$$Z(t_n + h) - Y(t_n + h) = \int_{t_n}^{t_n + h} R(t_n + h, \tau)\bigl(\widehat{A}(\tau) - A(\tau)\bigr)Z(\tau)\, d\tau.$$

Applying our quadrature formula to this integral gives zero as result, and the remainder is of size $\mathcal{O}(h^{p+1})$. Details of the proof are as for Theorem II.1.5. □

**Example 7.3.** As a first example, we use the midpoint rule ($c_1 = 1/2$, $b_1 = 1$). In this case the interpolation polynomial is constant, and the method becomes

$$Y_{n+1} = \exp\bigl(hA(t_n + h/2)\bigr) Y_n, \tag{7.7}$$

which is of order 2.

**Example 7.4.** The two-stage Gauss quadrature is given by $c_{1,2} = 1/2 \pm \sqrt{3}/6$, $b_{1,2} = 1/2$. The interpolation polynomial is of degree one and we have to apply (7.5) in order to get an approximation $Y_{n+1}$. Since we are interested in a fourth order approximation, we can neglect the remainder term (indicated by ... in (7.5)). Computing analytically the iterated integrals over products of $\ell_i(t)$ we obtain

$$Y_{n+1} = \exp\left(\frac{h}{2}(A_1 + A_2) + \frac{\sqrt{3}\,h^2}{12}[A_2, A_1]\right) Y_n, \tag{7.8}$$

where $A_1 = A(t_n + c_1 h)$ and $A_2 = A(t_n + c_2 h)$. This is a method of order four. The terms of (7.5) with triple integrals give $\mathcal{O}(h^4)$ expressions, whose leading term vanishes by the symmetry of the method (Exercise V.7). Therefore, they need not be considered.

Theorem 7.2 allows us to obtain methods of arbitrarily high order. A straightforward use of the expansion (7.5) yields an expression with a large number of commutators. Munthe-Kaas & Owren (1999) and Blanes, Casas & Ros (2000a) construct higher order methods with a reduced number of commutators. For example, for order 6 the required number of commutators is reduced from 7 to 4.

Let us remark that all numerical methods of this section are of the form $Y_{n+1} = \exp(h\Omega_n)Y_n$, where $\Omega_n$ is a linear combination of $A(t_n + c_i h)$ and of their commutators. If $A(t) \in \mathfrak{g}$ for all $t$, then also $h\Omega_n$ lies in the Lie algebra $\mathfrak{g}$, so that the numerical solution stays in the Lie group $G$ if $Y_0 \in G$ (this is a consequence of Lemma 6.4).

# IV.8 Lie Group Methods

Consider a differential equation

$$\dot{Y} = A(Y)Y, \qquad Y(0) = Y_0 \tag{8.1}$$

on a matrix Lie group $G$. This means that $Y_0 \in G$ and that $A(Y) \in \mathfrak{g}$ for all $Y \in G$. Since this is a special case of differential equations on a manifold, projection methods (Sect. IV.4) as well as methods based on local coordinates (Sect. IV.5) are well suited for their numerical treatment. Here we present further approaches which also yield approximations that lie on the manifold.

All numerical methods of this section can be extended in a straightforward way to non-autonomous problems $\dot{Y} = A(t, Y)Y$ with $A(t, Y) \in \mathfrak{g}$ for all $t$ and all $Y \in G$. Just to simplify the notation we restrict ourselves to the formulation (8.1).

## IV.8.1 Crouch-Grossman Methods

> The discipline of Lie-group methods owes a great deal to the pioneering work of Peter Crouch and his co-workers ...
> (A. Iserles, H.Z. Munthe-Kaas, S.P. Nørsett & A. Zanna 2000)

The numerical approximation of explicit Runge-Kutta methods is obtained by a composition of the following two basic operations: (i) an evaluation of the vector field $f(Y) = A(Y)Y$ and (ii) a computation of an update of the form $Y + haf(Z)$. For example, the left method of (II.1.3) consists of the following steps: evaluate $K_1 = f(Y_0)$; compute $\widetilde{Y}_1 = Y_0 + hK_1$; evaluate $K_2 = f(\widetilde{Y}_1)$; compute $Y_{1/2} = Y_0 + \frac{h}{2}K_1$; compute $Y_1 = Y_{1/2} + \frac{h}{2}K_2$.

In the context of differential equations on Lie groups, these methods have the disadvantage that, even when $Y \in G$ and $Z \in G$, the update $Y + haA(Z)Z$ is in general not in the Lie group. The idea of Crouch & Grossman (1993) is to replace the "update" operation with $\exp(haA(Z))Y$.

**Definition 8.1.** Let $b_i, a_{ij}$ $(i, j = 1, \ldots, s)$ be real numbers. An explicit $s$-*stage Crouch-Grossman method* is given by

$$\begin{aligned} Y^{(i)} &= \exp(ha_{i,i-1}K_{i-1}) \cdot \ldots \cdot \exp(ha_{i1}K_1)Y_n, \qquad K_i = A(Y^{(i)}), \\ Y_{n+1} &= \exp(hb_s K_s) \cdot \ldots \cdot \exp(hb_1 K_1)Y_n. \end{aligned}$$

For example, the method of Runge described above ($s = 2$, $a_{21} = 1$, $b_1 = b_2 = 1/2$) leads to

$$Y_{n+1} = \exp\left(\frac{h}{2}K_2\right)\exp\left(\frac{h}{2}K_1\right)Y_n, \tag{8.2}$$

where $K_1 = A(Y_n)$ and $K_2 = A\big(\exp(hK_1)Y_n\big)$.

By construction, the methods of Crouch-Grossman give rise to approximations $Y_n$ which lie exactly on the manifold defined by the Lie group. But what can be said about their order of accuracy?

**Theorem 8.2.** *Let $c_i = \sum_j a_{ij}$. A Crouch-Grossman method has order $p$ ($p \leq 3$) if the following order conditions are satisfied:*

$$\text{order 1:} \quad \sum_i b_i = 1 \tag{8.3}$$

$$\text{order 2:} \quad \sum_i b_i c_i = 1/2 \tag{8.4}$$

$$\text{order 3:} \quad \sum_i b_i c_i^2 = 1/3 \tag{8.5}$$

$$\sum_{ij} b_i a_{ij} c_j = 1/6 \tag{8.6}$$

$$\sum_i b_i^2 c_i + 2 \sum_{i<j} b_i c_i b_j = 1/3. \tag{8.7}$$

*Proof.* As in the case of Runge-Kutta methods, the order conditions can be found by comparing the Taylor series expansions of the exact and the numerical solution. In addition to the conditions stated in the theorem, this leads to relations such as

$$\sum_i b_i^2 c_i + 2 \sum_{i<j} b_i b_j c_j = \frac{2}{3}. \tag{8.8}$$

Adding this equation to (8.7) we find $2 \sum_{ij} b_i c_i b_j = 1$, which is satisfied by (8.3) and (8.4). Hence, the relation (8.8) is already a consequence of the conditions stated in the theorem. □

**Table 8.1.** Crouch-Grossman methods of order 3

| 0 | | |
|---|---|---|
| $-1/24$ | $-1/24$ | |
| $17/24$ | $161/24$ | $-6$ |
| | $1$ | $-2/3$ | $2/3$ |

| 0 | | | |
|---|---|---|---|
| $3/4$ | $3/4$ | | |
| $17/24$ | $119/216$ | $17/108$ | |
| | $13/51$ | $-2/3$ | $24/17$ |

Crouch & Grossman (1993) present several solutions of the system (8.3)–(8.7), one of which is given in the left array of Table 8.1. The construction of higher order Crouch-Grossman methods is very complicated ("... any attempt to analyze algorithms of order greater than three will be very complex, ...", Crouch & Grossman, 1993).

The theory of order conditions for Runge-Kutta methods (Sect. III.1) has been extended to Crouch-Grossman methods by Owren & Marthinsen (1999). It turns out that the order conditions for classical Runge-Kutta methods form a subset of those for Crouch-Grossman methods. The first new condition is (8.7). For a method of order 4, thirteen conditions (including those of Theorem 8.2) have to be satisfied. Solving these equations, Owren & Marthinsen (1999) construct a 4th order method with $s = 5$ stages.

## IV.8.2 Munthe-Kaas Methods

These methods were developed in a series of papers by Munthe-Kaas (1995, 1998, 1999). The main motivation behind the first of these papers was to develop a theory of Runge-Kutta methods in a coordinate-free framework. After attempts that led to new order conditions (as for the Crouch-Grossman methods), Munthe-Kaas (1999) had the idea to write the solution as $Y(t) = \exp\big(\Omega(t)\big)Y_0$ and to solve numerically the differential equation for $\Omega(t)$. It sounds awkward to replace the differential equation (8.1) by a more complicated one. However, the nonlinear invariants $g(Y) = 0$ of (8.1) defining the Lie group are replaced with linear invariants $g'(I)(\Omega) = 0$ defining the Lie algebra, and we know from Sect. IV.1 that essentially all numerical methods automatically conserve linear invariants.

It follows from the proof of Theorem 7.1 that the solution of (8.1) can be written as $Y(t) = \exp\big(\Omega(t)\big)Y_0$, where $\Omega(t)$ is the solution of $\dot{\Omega} = d\exp_\Omega^{-1}\big(A\big(Y(t)\big)\big)$, $\Omega(0) = 0$. Since it is not practical to work with the operator $d\exp_\Omega^{-1}$, we truncate the series (7.3) suitably and consider the differential equation

$$\dot{\Omega} = A\big(\exp(\Omega)Y_0\big) + \sum_{k=1}^{q} \frac{B_k}{k!} \operatorname{ad}_\Omega^k\Big(A\big(\exp(\Omega)Y_0\big)\Big), \qquad \Omega(0) = 0. \qquad (8.9)$$

This leads to the following method.

**Algorithm 8.3 (Munthe-Kaas 1999).** *Consider the problem (8.1) with $A(Y) \in \mathfrak{g}$ for $Y \in G$. Assume that $Y_n$ lies in the Lie group $G$. Then, the step $Y_n \mapsto Y_{n+1}$ is defined as follows:*

- *consider the differential equation (8.9) with $Y_n$ instead of $Y_0$, and apply a Runge-Kutta method (explicit or implicit) to get an approximation $\Omega_1 \approx \Omega(h)$,*
- *then define the numerical solution by $Y_{n+1} = \exp(\Omega_1)Y_n$.*

Before analyzing this algorithm, we emphasize its close relationship with Algorithm 5.3. In fact, if we identify the Lie algebra $\mathfrak{g}$ with $\mathbb{R}^k$ (where $k$ is the dimension of the vector space $\mathfrak{g}$), the mapping $\psi(\Omega) = \exp(\Omega)Y_n$ is a local parametrization of the Lie group $G$ (see Lemma 6.4). Apart from the truncation of the series in (8.9), Algorithm 8.3 is a special case of Algorithm 5.3.

Important properties of the Munthe-Kaas methods are given in the next two theorems.

**Theorem 8.4.** *Let $G$ be a matrix Lie group and $\mathfrak{g}$ its Lie algebra. If $A(Y) \in \mathfrak{g}$ for $Y \in G$ and if $Y_0 \in G$, then the numerical solution of the Lie group method of Algorithm 8.3 lies in $G$, i.e., $Y_n \in G$ for all $n = 0, 1, 2, \ldots$ .*

*Proof.* It is sufficient to prove that for $Y_0 \in G$ the numerical solution $\Omega_1$ of the Runge-Kutta method applied to (8.9) lies in $\mathfrak{g}$. Since the Lie bracket $[\Omega, A]$ is an operation $\mathfrak{g} \times \mathfrak{g} \to \mathfrak{g}$, and since $\exp(\Omega)Y_0 \in G$ for $\Omega \in \mathfrak{g}$, the right-hand expression of (8.9) is in $\mathfrak{g}$ for $\Omega \in \mathfrak{g}$. Hence, (8.9) is a differential equation on the vector space $\mathfrak{g}$ with solution $\Omega(t) \in \mathfrak{g}$. All operations in a Runge-Kutta method give results in $\mathfrak{g}$, so that the numerical approximation $\Omega_1$ also lies in $\mathfrak{g}$. $\square$

**Theorem 8.5.** *If the Runge-Kutta method is of (classical) order $p$ and if the truncation index in (8.9) satisfies $q \geq p - 2$, then the method of Algorithm 8.3 is of order $p$.*

*Proof.* For sufficiently smooth $A(Y)$ we have $\Omega(t) = tA(Y_0) + \mathcal{O}(t^2)$, $Y(t) = Y_0 + \mathcal{O}(t)$ and $[\Omega(t), A(Y(t))] = \mathcal{O}(t^2)$. This implies that $\text{ad}^k_{\Omega(t)}(A(Y(t))) = \mathcal{O}(t^{k+1})$, so that the truncation of the series in (8.9) induces an error of size $\mathcal{O}(h^{q+2})$ for $|t| \leq h$. Hence, for $q + 2 \geq p$, this truncation does not affect the order of convergence. □

The most simple Lie group method is obtained if we take the explicit Euler method as basic discretization and $q = 0$ in (8.9). This leads to the so-called *Lie-Euler method*

$$Y_{n+1} = \exp(hA(Y_n))Y_n. \qquad (8.10)$$

This is also a special case of the Crouch-Grossman methods of Definition 8.1.

Taking the implicit midpoint rule as the basic discretization and again $q = 0$ in (8.9), we obtain the *Lie midpoint rule*

$$Y_{n+1} = \exp(\Omega)Y_n, \qquad \Omega = hA(\exp(\Omega/2)Y_n). \qquad (8.11)$$

This is an implicit equation in $\Omega$ and has to be solved by fixed point iteration or by Newton-type methods.

**Example 8.6.** We take the coefficients of the right array of Table 8.1. They give rise to 3rd order Munthe-Kaas and 3rd order Crouch-Grossman methods. We apply both methods with the large step size $h = 0.35$ to the system (1.5) which is already of the form (8.1). Observe that $Y_0$ is a vector in $\mathbb{R}^3$ and not a matrix, but all results of this section remain valid for this case. For the computation of the matrix exponential we use the Rodrigues formula (Exercise 19). The numerical results (first 1000 steps) are shown in Fig. 8.1. We see that the numerical solution stays on the manifold (sphere),

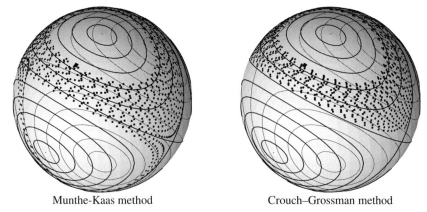

Munthe-Kaas method        Crouch–Grossman method

**Fig. 8.1.** Solutions of the Euler equations (1.4) for the rigid body

but on the sphere the qualitative behaviour is not correct. A similar behaviour could be observed for projection methods (the orthogonal projection consists simply in dividing the approximation $\widetilde{Y}_{n+1}$ by its norm) and by the methods based on local coordinates.

Crouch-Grossman methods and Munthe-Kaas methods are very similar. If they are based on the same set of Runge-Kutta coefficients, both methods use $s$ evaluations of the matrix $A(Y)$. The Crouch-Grossman methods require in general the computation of $s(s+1)/2$ matrix exponentials, whereas the Munthe-Kaas methods require only $s$ of them. On the other hand, Munthe-Kaas methods need also the computations of a certain number of commutators which increases with $q$ in (8.9). In such a comparison one has to take into account that every classical Runge-Kutta method defines a Munthe-Kaas method of the same order, but Crouch-Grossman methods of high order are very difficult to obtain, and need more stages for the same order (if $p \geq 4$).

## IV.8.3 Further Coordinate Mappings

The methods of Algorithm 8.3 are based on the local parametrization $\psi(\Omega) = \exp(\Omega)Y_n$. For all Lie groups, this is a diffeomorphism between the Lie group and the corresponding Lie algebra. Are there other, computationally more efficient parametrizations that can be used in special situations?

**The Cayley Transform.** Lie groups of the form

$$G = \{Y \mid Y^T P Y = P\}, \tag{8.12}$$

where $P$ is a given constant matrix, are called *quadratic Lie groups*. The corresponding Lie algebra is given by $\mathfrak{g} = \{\Omega \mid P\Omega + \Omega^T P = 0\}$. The orthogonal group $\mathrm{O}(n)$ and the symplectic group $\mathrm{Sp}(n)$ are prominent special cases (see Table 6.1). For such groups we have the following analogue of Lemma 6.4.

**Lemma 8.7.** *For a quadratic Lie group $G$, the Cayley transform*

$$\operatorname{cay} \Omega = (I - \Omega)^{-1}(I + \Omega)$$

*maps elements of $\mathfrak{g}$ into $G$. Moreover, it is a local diffeomorphism near $\Omega = 0$.*

*Proof.* For $\Omega \in \mathfrak{g}$ (i.e., $P\Omega + \Omega^T P = 0$) we have $P(I + \Omega) = (I - \Omega)^T P$ and also $P(I - \Omega)^{-1} = (I + \Omega)^{-T} P$. For $Y = (I - \Omega)^{-1}(I + \Omega)$ this immediately implies $Y^T P Y = P$. □

The use of the Cayley transform for the numerical integration of differential equations on Lie groups has been proposed by Lewis & Simo (1994) and Diele, Lopez & Peluso (1998) for the orthogonal group, and by Lopez & Politi (2001) for general quadratic groups. It is based on the following result, which is an adaptation of Lemma III.4.1 and Lemma III.4.2 to the Cayley transform.

**Lemma 8.8.** *The derivative of* cay $\Omega$ *is given by*

$$\left(\frac{d}{d\Omega}\,\mathrm{cay}\,\Omega\right)H = \Big(d\,\mathrm{cay}\,_\Omega(H)\Big)\,\mathrm{cay}\,\Omega,$$

*where*

$$d\,\mathrm{cay}\,_\Omega(H) = 2(I - \Omega)^{-1}H(I + \Omega)^{-1}. \tag{8.13}$$

*For the inverse of* $d\,\mathrm{cay}\,_\Omega$ *we have*

$$d\,\mathrm{cay}\,_\Omega^{-1}(H) = \frac{1}{2}(I - \Omega)H(I + \Omega). \tag{8.14}$$

*Proof.* By the usual rules of calculus we obtain

$$\left(\frac{d}{d\Omega}\,\mathrm{cay}\,\Omega\right)H = (I - \Omega)^{-1}H(I - \Omega)^{-1}(I + \Omega) + (I - \Omega)^{-1}H,$$

and a simple algebraic manipulation proves the statements. □

The numerical approach for solving (8.1) in the case of quadratic Lie groups is an adaptation of the Algorithm 8.3. We consider the local parametrization $Y = \psi(\Omega) = \mathrm{cay}\,(\Omega)Y_n$, and we apply one step of a numerical method to the differential equation $\dot{\Omega} = d\,\mathrm{cay}\,_\Omega^{-1}A\big(\mathrm{cay}\,(\Omega)Y_n\big)$ which, by (8.14), is equivalent to

$$\dot{\Omega} = \frac{1}{2}(I - \Omega)A\Big(\mathrm{cay}\,(\Omega)Y_n\Big)(I + \Omega).$$

This equation replaces (8.9) in the Algorithm 8.3. Since no truncation of an infinite series is necessary here, this approach is a special case of Algorithm 5.3.

**Canonical Coordinates of the Second Kind.** For a basis $\{C_1, C_2, \ldots, C_d\}$ of the Lie algebra $\mathfrak{g}$ the coordinates $z_1, \ldots, z_d$ of the local parametrization $\psi(z) = \exp\big(\sum_{i=1}^d z_i C_i\big)$ of the Lie group $G$ are called *canonical coordinates of the first kind*. Here we are interested in the parametrization

$$\psi(z) = \exp(z_1 C_1)\exp(z_2 C_2) \cdot \ldots \cdot \exp(z_d C_d), \tag{8.15}$$

and we call $z = (z_1, \ldots, z_d)^T$ *canonical coordinates of the second kind* (Varadarajan 1974). The use of these coordinates in connection with the numerical solution of differential equations on Lie groups has been promoted by Celledoni & Iserles (2001) and Owren & Marthinsen (2001). The idea behind this choice is that, due to a sparse structure of the $C_i$, the computation of $\exp(z_1 C_1), \ldots, \exp(z_d C_d)$ may be much cheaper than the computation of $\exp(\sum_i z_i C_i)$.

With the change of coordinates $y = \psi(z)$, the differential equation (8.1) becomes $\psi'(z)\dot{z} = A\big(\psi(z)\big)\psi(z)$, which is equivalent to

$$\begin{aligned}
A\big(\psi(z)\big) &= \sum_{i=1}^d \dot{z}_i\,\exp(z_1 C_1)\cdot\ldots\cdot\exp(z_{i-1}C_{i-1}) \\
&\qquad\qquad \cdot C_i \cdot \exp(-z_{i-1}C_{i-1})\cdot\ldots\cdot\exp(-z_1 C_1) \\
&= \sum_{i=1}^d \dot{z}_i\,\big(F_1 \circ \ldots \circ F_{i-1}\big)C_i,
\end{aligned} \tag{8.16}$$

where we use the notation $F_j C = \exp(z_j C_j)\, C \exp(-z_j C_j)$ for the linear operator $F_j : \mathfrak{g} \to \mathfrak{g}$; see Exercise 14. We need to compute $\dot z_1, \ldots, \dot z_d$ from (8.16), and this will usually be a computationally expensive task. However, for several Lie algebras and for well chosen bases this can be done very efficiently. The crucial idea is the following: we let $\widehat F_j$ be defined by

$$\widehat F_j C_i = \begin{cases} F_j C_i & \text{if } i > j \\ C_i & \text{if } i \le j, \end{cases} \tag{8.17}$$

and we assume that

$$\big(F_1 \circ \ldots \circ F_{i-1}\big) C_i = \big(\widehat F_1 \circ \ldots \circ \widehat F_{i-1}\big) C_i, \qquad i = 2, \ldots, d. \tag{8.18}$$

Under this assumption, we have $\big(F_1 \circ \ldots \circ F_{i-1}\big) C_i = \big(\widehat F_1 \circ \ldots \circ \widehat F_{i-1}\big) C_i = \big(\widehat F_1 \circ \ldots \circ \widehat F_{d-1}\big) C_i$, and the relation (8.16) becomes

$$\big(\widehat F_1 \circ \ldots \circ \widehat F_{d-1}\big)\Big(\sum_{i=1}^{d} \dot z_i C_i\Big) = A\big(\psi(z)\big). \tag{8.19}$$

In the situations which we have in mind, the operators $\widehat F_j$ can be efficiently inverted, and Algorithm 5.3 can be applied to the solution of (8.1).

The main difficulty of using this coordinate transform is to find a suitable ordering of a basis such that condition (8.18) is satisfied. The following lemma simplifies this task. We use the notation $\alpha_k(C)$ for the coefficient in the representation $C = \sum_{k=1}^{d} \alpha_k(C) C_k$.

**Lemma 8.9.** *Let $\{C_1, \ldots, C_d\}$ be a basis of the Lie algebra $\mathfrak{g}$. If for every pair $j < i$ and for $k < j$ we have*

$$\alpha_k(F_j C_i) \neq 0 \quad \Longrightarrow \quad F_\ell C_k = C_k \quad \text{for } \ell \text{ satisfying } k \le \ell < j, \tag{8.20}$$

*then the relation (8.18) holds for all $i = 2, \ldots, d$.*

*Proof.* We write $\widehat F_{i-1} C_i = F_{i-1} C_i = \sum_k \alpha_k(F_{i-1} C_i) C_k$. It follows from the definition of $\widehat F_j$ and from (8.20) that $(\widehat F_{i-2} \circ \widehat F_{i-1}) C_i = (F_{i-2} \circ F_{i-1}) C_i$. A repeated application of this argument proves the statement. □

Owren & Marthinsen (2001) have studied Lie algebras that admit a basis satisfying (8.18) for all $z$. We present here one of their examples.

**Example 8.10 (Special Linear Group).** Consider the differential equation (8.1) on the Lie group $\mathrm{SL}(n) = \{Y \mid \det Y = 1\}$, i.e., the matrix $A(Y)$ lies in $\mathfrak{sl}(n) = \{A \mid \operatorname{trace} A = 0\}$. As a basis of the Lie algebra $\mathfrak{sl}(n)$ we choose $E_{ij} = e_i e_j^T$ for $i \neq j$, and $D_i = e_i e_i^T - e_{i+1} e_{i+1}^T$ for $1 \le i < n$ (here, $e_i = (0, \ldots, 1, \ldots, 0)^T$ denotes the vector whose only non-zero element is in the $i$th position). Following Owren & Marthinsen (2001) we order the elements of this basis as

$$E_{12} < \ldots < E_{1n} < E_{23} < \ldots < E_{2n} < \ldots < E_{n-1,n}$$
$$< E_{21} < \ldots < E_{n1} < E_{32} < \ldots < E_{n2} < \ldots < E_{n,n-1}$$
$$< D_1 < \ldots < D_{n-1}.$$

With the use of Lemma 8.9 one can check in a straightforward way that the relation (8.18) is satisfied. In nearly all situations $\alpha_k(F_j C_i) = 0$ for $k < j < i$, so that (8.18) represents an empty condition. Consequently, the $\dot{z}_i$ can be computed from (8.19). Due to the sparsity of the matrices $E_{ij}$ and $D_i$, the computation of $\widehat{F}_i^{-1}$ can be done very efficiently.

## IV.9 Exercises

1. Prove that the symplectic Euler method (I.1.9) conserves quadratic invariants of the form (2.5). Explain the "0" entries of Table (I.2.1).
2. Prove that under the condition (2.3) a Runge-Kutta method preserves all invariants of the form $I(y) = y^T C y + d^T y + c$.
3. Prove that an $s$-stage diagonally implicit Runge-Kutta method (i.e., $a_{ij} = 0$ for $i < j$) satisfies the condition (2.3) if and only if it is equivalent to a composition $\Phi_{b_s h} \circ \ldots \circ \Phi_{b_1 h}$ based on the implicit midpoint rule.
4. Prove the following statements: a) If a partitioned Runge-Kutta method conserves general quadratic invariants $p^T C p + 2p^T D q + q^T E q$, then each of the two Runge-Kutta methods has to conserve quadratic invariants separately.
   b) If both methods, $\{b_i, a_{ij}\}$ and $\{\widehat{b}_i, \widehat{a}_{ij}\}$ are irreducible, satisfy (2.3) and if (2.7)-(2.8) hold, then we have $b_i = \widehat{b}_i$ and $a_{ij} = \widehat{a}_{ij}$ for all $i, j$.
5. Prove that the Gauss methods are the only collocation methods satisfying (2.3). *Hint.* Use the ideas of the proof of Lemma 13.9 in Hairer & Wanner (1996).
6. Discontinuous collocation methods with either $b_1 \neq 0$ or $b_s \neq 0$ (Definition II.1.7) cannot satisfy the criterion (2.3).
7. (Sanz-Serna & Abia 1991, Saito, Sugiura & Mitsui 1992). The condition (2.3) acts as simplifying assumption for the order conditions of Runge-Kutta methods. Assume that the order conditions are satisfied for the trees $u$ and $v$. Prove that it is satisfied for $u \circ v$ if and only if it is satisfied for $v \circ u$, and that it is automatically satisfied for trees of the form $u \circ u$.
   *Remark.* $u \circ v$ denotes the Butcher product introduced in Sect. VI.7.1.
8. If $L_0$ is a symmetric, tridiagonal matrix that is sufficiently close to $\Lambda = \text{diag}(\lambda_1, \ldots, \lambda_n)$, where $\lambda_1 > \lambda_2 > \ldots > \lambda_n$ are the eigenvalues of $L_0$, then the solution of (3.5) with $B(L) = L_+ - L_+^T$ converges exponentially fast to the diagonal matrix $\Lambda$. Hence, the numerical solution of (3.5) gives an algorithm for the computation of the eigenvalues of the matrix $L_0$.
   *Hint.* Let $\beta_1, \ldots, \beta_n$ be the entries in the diagonal of $L$, and $\alpha_1, \ldots, \alpha_{n-1}$ those in the subdiagonal. Assume that $|\beta_k(0) - \lambda_k| \leq R/3$ and $|\alpha_k(0)| \leq R$ with some sufficiently small $R$. Prove that $\beta_k(t) - \beta_{k+1}(t) \geq \mu - R$ and $|\alpha_k(t)| \leq R e^{-(\mu - R)t}$ for all $t \geq 0$, where $\mu = \min_k(\lambda_k - \lambda_{k+1}) > 0$.

9. Elaborate Example 4.5 for the special case where $Y$ is a matrix of dimension 2. In particular, show that (4.6) is the same as (4.5), and check the formulas for the simplified Newton iterations.

10. Show that for given $\widetilde{Y}$ the solution of the problem (4.7) is $Y = U^T V$, where $\widetilde{Y} = U^T \Sigma V$ is the singular value decomposition of $\widetilde{Y}$.
    *Hint.* Since $\|U^T S V\|_F = \|S\|_F$ holds for all orthogonal matrices $U$ and $V$, it is sufficient to consider the case $\widetilde{Y} = (\Sigma, 0)^T$ with $\Sigma = \operatorname{diag}(\sigma_1, \ldots, \sigma_k)$. Prove that $\|(\Sigma, 0)^T - Y\|_F^2 \geq \sum_{i=1}^k (\sigma_i - 1)^2$ for all matrices $Y$ satisfying $Y^T Y = I$.

11. (Brenan, Campbell & Petzold (1996), Sect. 2.5.3). Consider the differential equation $\dot{y} = f(y)$ with known invariants $g(y) = \mathit{Const}$, and assume that $g'(y)$ has full rank. Prove by differentiation of the constraints that, for initial values satisfying $g(y_0) = 0$, the solution of the differential-algebraic equation (DAE)
$$\begin{aligned} \dot{y} &= f(y) + g'(y)^T \mu \\ 0 &= g(y) \end{aligned}$$
also solves the differential equation $\dot{y} = f(y)$.
    *Remark.* Most methods for DAEs (e.g., stiffly accurate Runge-Kutta methods or BDF methods) lead to numerical integrators that preserve exactly the constraints $g(y) = 0$. The difference from the projection method of Sect. IV.4 is that here the internal stages also satisfy the constraint.

12. If $S$ is symmetric and $T$ is skew-symmetric, then we have $\operatorname{trace}(ST) = 0$.
    *Hint.* Transform $S$ by an orthogonal matrix to diagonal form and use the fact that the trace is invariant under similarity transformations.

13. Prove that $\operatorname{SL}(n)$ is a Lie group of dimension $n^2 - 1$, and that $\mathfrak{sl}(n)$ is its Lie algebra (see Table 6.1 for the definitions of $\operatorname{SL}(n)$ and $\mathfrak{sl}(n)$).

14. Let $G$ be a matrix Lie group and $\mathfrak{g}$ its Lie algebra. Prove that for $Y \in G$ and $A \in \mathfrak{g}$ we have $Y A Y^{-1} \in \mathfrak{g}$.
    *Hint.* Consider the path $\gamma(t) = Y \alpha(t) Y^{-1}$.

15. Consider a problem $\dot{Y} = A(Y)Y$, for which $A(Y) \in \mathfrak{so}(n)$ whenever $Y \in \operatorname{O}(n)$, but where $A(Y)$ is an arbitrary matrix for $Y \notin \operatorname{O}(n)$.
    a) Prove that $Y_0 \in \operatorname{O}(n)$ implies $Y(t) \in \operatorname{O}(n)$ for all $t$.
    b) Show by a counter-example that the numerical solution of the implicit midpoint rule does not necessarily stay in $\operatorname{O}(n)$.

16. (Feng Kang & Shang Zai-jiu 1995). Let $R(z) = (1 + z/2)/(1 - z/2)$ be the stability function of the implicit midpoint rule. Prove that for $A \in \mathfrak{sl}(3)$ we have
$$\det R(hA) = 1 \quad \Leftrightarrow \quad \det A = 0.$$

17. (Iserles & Nørsett 1999). Introducing $y_1 = y$ and $y_2 = \dot{y}$, write the problem
$$\ddot{y} + ty = 0, \qquad y(0) = 1, \qquad \dot{y}(0) = 0$$
in the form (7.6). Then apply the numerical method of Example 7.4 with different step sizes on the interval $0 \leq t \leq 100$. Compare the result with that obtained by fourth order classical (explicit or implicit) Runge-Kutta methods.

*Remark.* If $A(t)$ in (7.6) (or $A(t,y)$ in (8.1)) are much smoother than the solution $y(t)$, then Lie group methods are usually superior to standard integrators, because Lie group methods approximate $A(t)$, whereas standard methods approximate the solution $y(t)$ by polynomials.

18. Deduce the BCH formula from the Magnus expansion (IV.7.5).
    *Hint.* For constant matrices $A$ and $B$ consider the matrix function $A(t)$ defined by $A(t) = B$ for $0 \le t \le 1$ and $A(t) = A$ for $1 \le t \le 2$.

19. (Rodrigues formula, see Marsden & Ratiu (1994), page 261). Prove that

$$\exp(\Omega) = I + \frac{\sin\alpha}{\alpha}\Omega + \frac{1}{2}\left(\frac{\sin(\alpha/2)}{\alpha/2}\right)^2\Omega^2 \quad \text{for} \quad \Omega = \begin{pmatrix} 0 & -\omega_3 & \omega_2 \\ \omega_3 & 0 & -\omega_1 \\ -\omega_2 & \omega_1 & 0 \end{pmatrix}$$

where $\alpha = \sqrt{\omega_1^2 + \omega_2^2 + \omega_3^2}$. This formula allows for an efficient implementation of the Lie group methods in $O(3)$.

20. The solution of $\dot{Y} = A(Y)Y$, $Y(0) = Y_0$, is given by $Y(t) = \exp(\Omega(t))Y_0$, where $\Omega(t)$ solves the differential equation (8.9). Compute the first terms of the $t$-expansion of $\Omega(t)$.
    *Result.* $\Omega(t) = tA(Y_0) + \frac{t^2}{2}A'(Y_0)A(Y_0)Y_0 + \frac{t^3}{6}\big((A'(Y_0))^2 A(Y_0)Y_0^2 + A'(Y_0)A(Y_0)^2 Y_0 + A''(Y_0)(A(Y_0)Y_0, A(Y_0)Y_0) - \frac{1}{2}[A(Y_0), A'(Y_0)A(Y_0)Y_0]\big)$.

21. Consider the 2-stage Gauss method of order $p = 4$. In the corresponding Lie group method, eliminate the presence of $\Omega$ in $[\Omega, A]$ by iteration, and neglect higher order commutators. Show that this leads to

$$\Omega_1 = h\left(\frac{1}{4}A_1 + \left(\frac{1}{4} - \frac{\sqrt{3}}{6}\right)A_2\right) - \frac{h^2}{2}\left(-\frac{1}{12} + \frac{\sqrt{3}}{24}\right)[A_1, A_2]$$

$$\Omega_2 = h\left(\left(\frac{1}{4} + \frac{\sqrt{3}}{6}\right)A_1 + \frac{1}{4}A_2\right) - \frac{h^2}{2}\left(\frac{1}{12} + \frac{\sqrt{3}}{24}\right)[A_1, A_2]$$

$$y_1 = \exp\left(h\left(\frac{1}{2}A_1 + \frac{1}{2}A_2\right) - h^2\frac{\sqrt{3}}{12}[A_1, A_2]\right)y_0,$$

where $A_i = A(Y_i)$ and $Y_i = \exp(\Omega_i)y_0$. Prove that this is a Lie group method of order 4. Is it symmetric?

22. In Zanna (1999) a Lie group method similar to that of Exercise 21 is presented. The only difference is that the coefficients $(-1/12 + \sqrt{3}/24)$ and $(1/12 + \sqrt{3}/24)$ in the formulas for $\Omega_1$ and $\Omega_2$ are replaced with $(-5/72 + \sqrt{3}/24)$ and $(5/72 + \sqrt{3}/24)$, respectively. Is there an error somewhere? Are both methods of order 4?

# Chapter V.
# Symmetric Integration and Reversibility

Symmetric methods of this chapter and symplectic methods of the next chapter play a central role in the geometric integration of differential equations. We discuss reversible differential equations and reversible maps, and we explain how symmetric integrators are related to them. We study symmetric Runge-Kutta and composition methods, and we show how standard approaches for solving differential equations on manifolds can be symmetrized. A theoretical explanation of the excellent long-time behaviour of symmetric methods applied to reversible differential equations will be given in Chap. XI.

## V.1 Reversible Differential Equations and Maps

Conservative mechanical systems have the property that inverting the initial direction of the velocity vector and keeping the initial position does not change the solution trajectory, it only inverts the direction of motion. Such systems are 'reversible'. We extend this notion to more general situations.

**Definition 1.1.** Let $\rho$ be an invertible linear transformation in the phase space of $\dot{y} = f(y)$. This differential equation and the vector field $f(y)$ are called $\rho$-reversible if
$$\rho f(y) = -f(\rho y) \qquad \text{for all } y. \qquad (1.1)$$

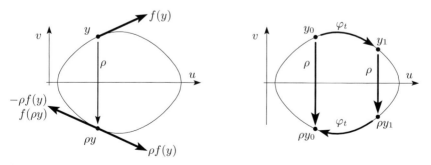

**Fig. 1.1.** Reversible vector field (left picture) and reversible map (right picture)

This property is illustrated in the left picture of Fig. 1.1. For $\rho$-reversible differential equations the exact flow $\varphi_t(y)$ satisfies

$$\rho \circ \varphi_t = \varphi_{-t} \circ \rho = \varphi_t^{-1} \circ \rho \tag{1.2}$$

(see the picture to the right in Fig. 1.1). The right identity is a consequence of the group property $\varphi_t \circ \varphi_s = \varphi_{t+s}$, and the left identity follows from

$$\frac{d}{dt}(\rho \circ \varphi_t)(y) = \rho f\big(\varphi_t(y)\big) = -f\big((\rho \circ \varphi_t)(y)\big)$$

$$\frac{d}{dt}(\varphi_{-t} \circ \rho)(y) = -f\big((\varphi_{-t} \circ \rho)(y)\big),$$

because all expressions of (1.2) satisfy the same differential equation with the same initial value $(\rho \circ \varphi_0)(y) = (\varphi_0 \circ \rho)(y) = \rho y$. Formula (1.2) motivates the following definition.

**Definition 1.2.** A map $\Phi(y)$ is called $\rho$-reversible if

$$\rho \circ \Phi = \Phi^{-1} \circ \rho.$$

**Example 1.3.** An important example is the partitioned system

$$\dot{u} = f(u, v), \qquad \dot{v} = g(u, v), \tag{1.3}$$

where $f(u, -v) = -f(u, v)$ and $g(u, -v) = g(u, v)$. Here, the transformation $\rho$ is given by $\rho(u, v) = (u, -v)$. If we call a vector field or a map *reversible* (without specifying the transformation $\rho$), we mean that it is $\rho$-reversible with this particular $\rho$. All second order differential equations $\ddot{u} = g(u)$ written as $\dot{u} = v$, $\dot{v} = g(u)$ are reversible. As a first implication of reversibility on the dynamics we mention the following fact: if $u$ and $v$ are scalar, and if (1.3) is reversible, then any solution that crosses the $u$-axis twice is periodic (Exercise 5, see also the solution of the pendulum problem in Fig. I.1.3).

It is natural to search for numerical methods that produce a reversible numerical flow when they are applied to a reversible differential equation. We then expect the numerical solution to have long-time behaviour similar to that of the exact solution; see Chap. XI for more precise statements. It turns out that the $\rho$-reversibility of a numerical one-step method is closely related to the concept of symmetry.

> Thus the method is theoretically *symmetrical* or *reversible*, a terminology we have never seen applied elsewhere.
> (P.C. Hammer & J.W. Hollingsworth 1955)

**Definition 1.4.** A numerical one-step method $\Phi_h$ is called *symmetric* or *time-reversible*,[1] if it satisfies

$$\Phi_h \circ \Phi_{-h} = \text{id} \qquad \text{or equivalently} \qquad \Phi_h = \Phi_{-h}^{-1}.$$

---

[1] The study of symmetric methods has its origin in the development of extrapolation methods (Gragg 1965, Stetter 1973), because the global error admits an asymptotic expansion in even powers of $h$. The notion of time-reversible methods is more common in the Computational Physics literature (Buneman 1967).

With the Definition II.3.1 of the adjoint method (i.e., $\Phi_h^* = \Phi_{-h}^{-1}$), the condition for symmetry reads $\Phi_h = \Phi_h^*$. A method $y_1 = \Phi_h(y_0)$ is symmetric if exchanging $y_0 \leftrightarrow y_1$ and $h \leftrightarrow -h$ leaves the method unaltered. In Chap. I we have already encountered the implicit midpoint rule (I.1.7) and the Störmer/Verlet scheme (I.3.6), both of which are symmetric. Many more symmetric methods will be given in the following sections.

**Theorem 1.5.** *If a numerical method, applied to a $\rho$-reversible differential equation, satisfies*

$$\rho \circ \Phi_h = \Phi_{-h} \circ \rho, \tag{1.4}$$

*then the numerical flow $\Phi_h$ is a $\rho$-reversible map if and only if $\Phi_h$ is a symmetric method.*

*Proof.* As a consequence of (1.4) the numerical flow $\Phi_h$ is $\rho$-reversible if and only if $\Phi_{-h} \circ \rho = \Phi_h^{-1} \circ \rho$. Since $\rho$ is an invertible transformation, this is equivalent to the symmetry of the method $\Phi_h$. □

Similarly, it is also true that a symmetric method is $\rho$-reversible if and only if the $\rho$-compatibility condition (1.4) holds.

Compared to the symmetry of the method, condition (1.4) is much less restrictive. It is automatically satisfied by most numerical methods. Let us briefly discuss the validity of (1.4) for different classes of methods.

- *Runge-Kutta methods* (explicit or implicit) satisfy (1.4) without any restriction other than (1.1) on the vector field (Stoffer 1988). Let us illustrate the proof with the explicit Euler method $\Phi_h(y_0) = y_0 + hf(y_0)$:

$$(\rho \circ \Phi_h)(y_0) = \rho y_0 + h\rho f(y_0) = \rho y_0 - hf(\rho y_0) = \Phi_{-h}(\rho y_0).$$

- *Partitioned Runge-Kutta methods* applied to a partitioned system (1.3) satisfy the condition (1.4) if $\rho(u,v) = \big(\rho_1(u), \rho_2(v)\big)$ with invertible $\rho_1$ and $\rho_2$. The proof is the same as for Runge-Kutta methods. Notice that the mapping $\rho(u,v) = (u,-v)$ of Example 1.3 is of this special form.
- *Composition methods.* If two methods $\Phi_h$ and $\Psi_h$ satisfy (1.4), then so does the adjoint $\Phi_h^*$ and the composition $\Phi_h \circ \Psi_h$. Consequently, the composition methods (3.1) and (3.2) below, which compose a basic method $\Phi_h$ and its adjoint with different step sizes, have the property (1.4) provided the basic method $\Phi_h$ has it.
- *Splitting methods* are based on a splitting $\dot{y} = f^{[1]}(y) + f^{[2]}(y)$ of the differential equation. If both vector fields, $f^{[1]}(y)$ and $f^{[2]}(y)$, satisfy (1.1), then their exact flows $\varphi_h^{[1]}$ and $\varphi_h^{[2]}$ satisfy (1.2). In this situation, the splitting method (II.5.6) has the property (1.4).
- For *differential equations on manifolds* we have to assume that $\rho$ maps $\mathcal{M}$ to $\mathcal{M}$. Otherwise, condition (1.1) does not make sense. For the projection method of Algorithm IV.4.2 with orthogonal projection onto the manifold we have: if the basic method satisfies (1.4) and if $\rho$ is an orthogonal matrix, then it satisfies (1.4) as well. This follows from the fact that the tangent and normal spaces satisfy

$T_{\rho y}\mathcal{M} = \rho T_y \mathcal{M}$ and $N_{\rho y}\mathcal{M} = \rho^{-T} N_y \mathcal{M}$, respectively. A similar result holds for methods based on local coordinates, if the local parametrization is well chosen. For example, this is the case if $\rho\psi(z)$ is the parametrization at $\rho y_0$ whenever $\psi(z)$ is the parametrization at $y_0$.

## V.2 Symmetric Runge-Kutta Methods

We give a characterization of symmetric methods of Runge-Kutta type and mention some important examples.

### V.2.1 Collocation and Runge-Kutta Methods

Symmetric collocation methods are characterized by the symmetry of the collocation points with respect to the midpoint of the integration step.

**Theorem 2.1.** *The adjoint method of a collocation method (Definition II.1.3) based on $c_1,\ldots,c_s$ is a collocation method based on $c_1^*,\ldots,c_s^*$, where*

$$c_i^* = 1 - c_{s+1-i}. \tag{2.1}$$

*In the case that $c_i = 1 - c_{s+1-i}$ for all $i$, the collocation method is symmetric.*

*The adjoint method of a discontinuous collocation method (Definition II.1.7) based on $b_1, b_s$ and $c_2,\ldots,c_{s-1}$ is a discontinuous collocation method based on $b_1^*, b_s^*$ and $c_2^*,\ldots,c_{s-1}^*$, where*

$$b_1^* = b_s, \quad b_s^* = b_1 \quad \text{and} \quad c_i^* = 1 - c_{s+1-i}. \tag{2.2}$$

*In the case that $b_1 = b_s$ and $c_i = 1 - c_{s+1-i}$ for all $i$, the discontinuous collocation method is symmetric.*

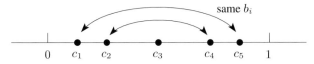

**Fig. 2.1.** Symmetry of collocation methods

*Proof.* Exchanging $(t_0, y_0) \leftrightarrow (t_1, y_1)$ and $h \leftrightarrow -h$ in the definition of a collocation method we get $u(t_1) = y_1$, $\dot{u}(t_1 - c_i h) = f\big(t_1 - c_i h, u(t_1 - c_i h)\big)$, and $y_0 = u(t_1 - h)$. Inserting $t_1 = t_0 + h$ this yields the collocation method based on $c_i^*$ of (2.1). Observe that the $c_i^*$ can be arbitrarily permuted. For discontinuous collocation methods the proof is similar. □

The preceding theorem immediately yields the following result.

**Corollary 2.2.** *The Gauss formulas (Table II.1.1), as well as the Lobatto IIIA (Table II.1.2) and Lobatto IIIB formulas (Table II.1.4) are symmetric integrators.* □

**Theorem 2.3 (Stetter 1973, Wanner 1973).** *The adjoint method of an s-stage Runge-Kutta method (II.1.4) is again an s-stage Runge-Kutta method. Its coefficients are given by*

$$a^*_{ij} = b_{s+1-j} - a_{s+1-i, s+1-j}, \qquad b^*_i = b_{s+1-i}. \qquad (2.3)$$

*If*

$$a_{s+1-i, s+1-j} + a_{ij} = b_j \qquad \text{for all } i, j, \qquad (2.4)$$

*then the Runge-Kutta method (II.1.4) is symmetric.*[2]

*Proof.* Exchanging $y_0 \leftrightarrow y_1$ and $h \leftrightarrow -h$ in the Runge-Kutta formulas yields

$$k_i = f\left(y_0 + h \sum_{j=1}^{s}(b_j - a_{ij})k_j\right), \qquad y_1 = y_0 + h \sum_{i=1}^{s} b_i k_i. \qquad (2.5)$$

Since the values $\sum_{j=1}^{s}(b_j - a_{ij}) = 1 - c_i$ appear in reverse order, we replace $k_i$ by $k_{s+1-i}$ in (2.5), and then we substitute all indices $i$ and $j$ by $s+1-i$ and $s+1-j$, respectively. This proves (2.3).

The assumption (2.4) implies $a^*_{ij} = a_{ij}$ and $b^*_i = b_i$, so that $\Phi^*_h = \Phi_h$. □

Explicit Runge-Kutta methods cannot fulfill condition (2.4) with $i = j$, and it is not difficult to see that no explicit Runge-Kutta can be symmetric (Exercise 2). Let us therefore turn our attention to *diagonally implicit Runge-Kutta methods* (DIRK), for which $a_{ij} = 0$ for $i < j$, but with diagonal elements that can be non-zero. In this case condition (2.4) becomes

$$a_{ij} = b_j = b_{s+1-j} \quad \text{for } i > j, \qquad a_{jj} + a_{s+1-j, s+1-j} = b_j. \qquad (2.6)$$

The Runge-Kutta tableau of such a method is thus of the form (e.g., for $s=5$)

$$
\begin{array}{c|ccccc}
c_1 & a_{11} & & & & \\
c_2 & b_1 & a_{22} & & & \\
c_3 & b_1 & b_2 & a_{33} & & \\
1-c_2 & b_1 & b_2 & b_3 & a_{44} & \\
1-c_1 & b_1 & b_2 & b_3 & b_2 & a_{55} \\
\hline
 & b_1 & b_2 & b_3 & b_2 & b_1
\end{array} \qquad (2.7)
$$

with $a_{33} = b_3/2$, $a_{44} = b_2 - a_{22}$, and $a_{55} = b_1 - a_{11}$. If one of the $b_i$ vanishes, then the corresponding stage does not influence the numerical result. This stage can therefore be suppressed, so that the method is equivalent to one with fewer stages. Our next result shows that methods (2.7) can be interpreted as the composition of $\theta$-methods, which are defined as

---

[2] For irreducible Runge-Kutta methods, the condition (2.4) is also necessary for symmetry (after a suitable permutation of the stages).

136    V. Symmetric Integration and Reversibility

$$\Phi_h^\theta(y_0) = y_1, \quad \text{where} \quad y_1 = y_0 + hf\big((1-\theta)y_0 + \theta y_1\big). \tag{2.8}$$

Observe that the adjoint of the $\theta$-method is $\Phi_h^{\theta *} = \Phi_h^{1-\theta}$.

**Theorem 2.4.** *A diagonally implicit Runge-Kutta method satisfying the symmetry condition (2.4) and $b_i \neq 0$ is equivalent to a composition of $\theta$-methods*

$$\Phi_{b_1 h}^{\alpha_1 *} \circ \Phi_{b_2 h}^{\alpha_2 *} \circ \ldots \circ \Phi_{b_2 h}^{\alpha_2} \circ \Phi_{b_1 h}^{\alpha_1}, \tag{2.9}$$

*where $\alpha_i = a_{ii}/b_i$.*

*Proof.* Since the $\theta$-method is a Runge-Kutta method with tableau

$$\begin{array}{c|c} \theta & \theta \\ \hline & 1 \end{array}$$

this follows from the discussion in Sect. III.1.3. We have used $\Phi_{b_{s+1-i}h}^{\alpha_{s+1-i}} = \Phi_{b_i h}^{\alpha_i *}$ which holds, because $b_{s+1-i} = b_i$ and $\alpha_{s+1-i} = 1 - \alpha_i$ by (2.6). □

A more detailed discussion of such methods is therefore postponed to Sect. V.3 on symmetric composition methods.

### V.2.2 Partitioned Runge-Kutta Methods

Applying partitioned Runge-Kutta methods (II.2.2) to general partitioned systems

$$\dot{y} = f(y, z), \qquad \dot{z} = g(y, z), \tag{2.10}$$

it is obvious that for their symmetry both Runge-Kutta methods have to be symmetric (because $\dot{y} = f(y)$ and $\dot{z} = g(z)$ are special cases of (2.10)). The proof of the following result is identical to that of Theorem 2.3 and therefore omitted.

**Theorem 2.5.** *If the coefficients of both Runge-Kutta methods $b_i, a_{ij}$ and $\widehat{b}_i, \widehat{a}_{ij}$ satisfy the condition (2.4), then the partitioned Runge-Kutta method (II.2.2) is symmetric.* □

As a consequence of this theorem we obtain that the Lobatto IIIA-IIIB pair (see Sect. II.2.2) and, in particular, the Störmer/Verlet scheme are symmetric integrators.

An interesting feature of partitioned Runge-Kutta methods is the possibility of having *explicit, symmetric* methods for problems of the form

$$\dot{y} = f(z), \qquad \dot{z} = g(y). \tag{2.11}$$

Second order differential equations $\ddot{y} = g(y)$, written in the form $\dot{y} = z, \dot{z} = g(y)$ have this structure, and also all Hamiltonian systems with separable Hamiltonian $H(p,q) = T(p) + V(q)$. It is not possible to get explicit symmetric integrators with non-partitioned Runge-Kutta methods (Exercise 2).

The Störmer/Verlet method (Table II.2.1) applied to (2.11) reads

$$z_{1/2} = z_0 + \frac{h}{2} g(y_0)$$
$$y_1 = y_0 + h f(z_{1/2})$$
$$z_1 = z_{1/2} + \frac{h}{2} g(y_1)$$

and is the composition $\Phi^*_{h/2} \circ \Phi_{h/2}$, where

$$\begin{pmatrix} y_1 \\ z_1 \end{pmatrix} = \Phi_h \begin{pmatrix} y_0 \\ z_0 \end{pmatrix}, \qquad \begin{matrix} y_1 = y_0 + hf(z_1) \\ z_1 = z_0 + hg(y_0) \end{matrix} \qquad (2.12)$$

is the symplectic Euler method and

$$\begin{pmatrix} y_1 \\ z_1 \end{pmatrix} = \Phi^*_h \begin{pmatrix} y_0 \\ z_0 \end{pmatrix}, \qquad \begin{matrix} y_1 = y_0 + hf(z_0) \\ z_1 = z_0 + hg(y_1) \end{matrix} \qquad (2.13)$$

its adjoint. All these methods are obviously explicit. How can they be extended to higher order? The idea is to consider partitioned Runge-Kutta methods based on diagonally implicit methods such as in (2.7). If $a_{ii} \cdot \widehat{a}_{ii} = 0$, then one component of the $i$th stage is given explicitly and, due to the special structure of (2.11), the other component is also obtained in a straightforward manner. In order to achieve $a_{ii} \cdot \widehat{a}_{ii} = 0$ with a symmetric partitioned method, we have to assume that $s$, the number of stages, is even.

**Theorem 2.6.** *A partitioned Runge-Kutta method, based on two diagonally implicit methods satisfying $a_{ii} \cdot \widehat{a}_{ii} = 0$ and (2.4) with $b_i \neq 0$ and $\widehat{b}_i \neq 0$, is equivalent to a composition of $\Phi_{b_i h}$ and $\Phi^*_{b_i h}$ with $\Phi_h$ and $\Phi^*_h$ given by (2.12) and (2.13), respectively.* □

For example, the partitioned method

$$\begin{array}{c|cccc} 0 \\ b_1 & b_2 \\ b_1 & b_2 & 0 \\ b_1 & b_2 & b_2 & b_1 \\ \hline & b_1 & b_2 & b_2 & b_1 \end{array} \qquad \begin{array}{c|cccc} \widehat{b}_1 \\ \widehat{b}_1 & 0 \\ \widehat{b}_1 & \widehat{b}_2 & \widehat{b}_2 \\ \widehat{b}_1 & \widehat{b}_2 & \widehat{b}_2 & 0 \\ \hline & \widehat{b}_1 & \widehat{b}_2 & \widehat{b}_2 & \widehat{b}_1 \end{array}$$

satisfies the assumptions of the preceding theorem. Since the methods have identical stages, the numerical result only depends on $\widehat{b}_1$, $b_1 + \widehat{b}_2$, $\widehat{b}_2 + b_3$, $b_3 + \widehat{b}_4$, and $\widehat{b}_4$. Therefore, we can assume that $\widehat{b}_i = b_i$ and the method is equivalent to the composition $\Phi^*_{b_1 h} \circ \Phi_{b_2 h} \circ \Phi^*_{b_2 h} \circ \Phi_{b_1 h}$.

# V.3 Symmetric Composition Methods

In Sect. II.4 the idea of composition methods is introduced, and a systematic way of obtaining high-order methods is outlined. These methods, based on (II.4.4) or on

(II.4.5), turn out to be symmetric, but they require too many stages. A theory of order conditions for general composition methods is developed in Sect. III.3. Here, we apply this theory to the construction of high-order symmetric methods. We mainly follow two lines.

- *Symmetric composition of first order methods.*

$$\Psi_h = \Phi_{\alpha_s h} \circ \Phi^*_{\beta_s h} \circ \ldots \circ \Phi^*_{\beta_2 h} \circ \Phi_{\alpha_1 h} \circ \Phi^*_{\beta_1 h}, \tag{3.1}$$

where $\Phi_h$ is an arbitrary first order method. In order to make this method symmetric, we assume $\alpha_s = \beta_1$, $\alpha_{s-1} = \beta_2$, etc.
- *Symmetric composition of symmetric methods.*

$$\Psi_h = \Phi_{\gamma_s h} \circ \Phi_{\gamma_{s-1} h} \circ \ldots \circ \Phi_{\gamma_2 h} \circ \Phi_{\gamma_1 h}, \tag{3.2}$$

where $\Phi_h$ is a symmetric second order method and $\gamma_s = \gamma_1$, $\gamma_{s-1} = \gamma_2$, etc.

## V.3.1 Symmetric Composition of First Order Methods

Because of Lemma 3.2 below, every method (3.2) is a special case of method (3.1). In this subsection we concentrate on methods that are of the form (3.1) but not of the form (3.2).

For constructing methods (3.1) of a certain order, one has to solve the system of nonlinear equations given in Theorem III.3.14 (see also Example III.3.15). The symmetry assumption on the coefficients considerably simplifies this system.

**Theorem 3.1.** *If the coefficients of method (3.1) satisfy $\alpha_{s+1-i} = \beta_i$ for all $i$, then it is sufficient to consider those trees with odd $\|\tau\|$.*

*Proof.* This is a consequence of Theorem II.3.2 (the maximal order of symmetric methods is even). In fact, if the condition for order 1 is satisfied, it is automatically of order 2. If, in addition, the conditions for order 3 are satisfied, it is automatically of order 4, etc. □

> It may come as a surprise that the popular leapfrog ... can be beaten, but only slightly.
> (R.I. McLachlan 1995)

**Methods of Order 2.** The only remaining condition for order two is $\sum_{k=1}^{s}(\alpha_k + \beta_k) = 1$, and, for $s = 1$, the symmetry requirement leads to $\Phi_{h/2} \circ \Phi^*_{h/2}$. Depending on the choice of $\Phi_h$, this method is equivalent to the midpoint rule, the trapezoidal rule, or the Störmer/Verlet scheme, all very famous and frequently used. However, McLachlan (1995) discovered that the case $s = 2$ can be slightly more advantageous. We obtain

$$\Phi_{\alpha h} \circ \Phi^*_{(1/2-\alpha)h} \circ \Phi_{(1/2-\alpha)h} \circ \Phi^*_{\alpha h}, \tag{3.3}$$

where $\alpha$ is a free parameter, which can serve for clever tuning.

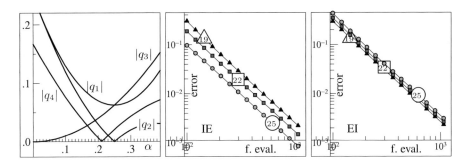

**Fig. 3.1.** The error functions $|q_i(\alpha)|$ defined in (3.5) (left picture). Work-precision diagrams for the Kepler problem (as in Fig. II.5.2) and for method (3.3) with $\alpha = 0.25$ (Störmer/Verlet), $\alpha = 0.1932$ (McLachlan), and $\alpha = 0.22$. 'IE': method $\Phi_h$ treats position by implicit Euler, velocity by explicit Euler; 'EI': method $\Phi_h$ treats position by explicit Euler, velocity by implicit Euler.

**Minimizing the Local Error of Composition Methods.** Subtracting the $B_\infty$-series of the numerical and the exact solutions (see Sect. III.3.2), we obtain

$$\Psi_h(y) - \varphi_h(y) = h^{p+1} \sum_{\|\tau\|=p+1} \frac{1}{\sigma(\tau)} \big(a(\tau) - e(\tau)\big) F(\tau)(y) + \mathcal{O}(h^{p+2}).$$

Assuming that the basic method has an expansion $\Phi_h(y) = y + hf(y) + h^2 d_2(y) + h^3 d_3(y) + \ldots$, we obtain for method (3.3), similar to (III.3.3), the local error

$$h^3 \Big(q_1(\alpha) d_3(y) + q_2(\alpha)(d_2' f)(y) + q_3(\alpha)(f' d_2)(y) \qquad (3.4)$$
$$+ \frac{1}{2} q_4(\alpha)(f''(f,f))(y) + q_5(\alpha)(f' f' f)(y)\Big) + \mathcal{O}(h^4),$$

which contains one term for each of the 5 trees $\tau \in T_\infty$ with $\|\tau\| = 3$. The $q_i(\alpha)$ are the polynomials

$$q_1(\alpha) = \frac{1}{4}(1 - 6\alpha + 12\alpha^2), \qquad q_2(\alpha) = \frac{1}{4}(-1 + 6\alpha - 8\alpha^2),$$
$$q_3(\alpha) = -\alpha^2, \qquad q_4(\alpha) = \frac{1}{6}(1 - 6\alpha + 6\alpha^2), \qquad q_5(\alpha) = \frac{1}{3} q_1(\alpha), \qquad (3.5)$$

which are plotted in the left picture of Fig. 3.1. If we allow arbitrary basic methods and arbitrary problems, all elementary differentials in the local error are independent, and there is no overall optimal value for $\alpha$. We see that the modulus of $q_1(\alpha)$ and $q_2(\alpha)$ are minimal for $\alpha = 1/4$, which is precisely the value corresponding to a double application of $\Phi_{h/2} \circ \Phi_{h/2}^*$ with halved step size. But the values $|q_3(\alpha)|$ and $|q_4(\alpha)|$ become smaller with decreasing $\alpha$ (close to $\alpha = 1/4$). McLachlan (1995) therefore minimizes some norm of the error (see Exercise 4) and arrives at the value $\alpha = 0.1932$.

In the numerical experiment of Fig. 3.1 we apply method (3.3) with three different values of $\alpha$ to the Kepler problem (with data as in Fig. II.5.2 and the symplectic

Euler method for $\Phi_h$). Once we treat the position variable by the implicit Euler method and the velocity variable by the explicit Euler method (central picture), and once the other way round (right picture). We notice that the method which is best in one case is worst in the other.

This simple experiment shows that choosing the free parameters of the method by minimizing some arbitrary measure of the error coefficients is problematic. For higher order methods there are many more expressions in the dominating term of the local error (for example: 29 terms for $||\tau|| = 5$). The corresponding functions $q_i$ give a lot of information on the local error, and they indicate the region of parameters that produce good methods. But, unless more information is known about the problem (second order differential equation, nearly integrable systems), one usually minimizes, for orders of 8 or 10, just the maximal values of the $\alpha_i$, $\beta_i$, or $\gamma_i$ (Kahan & Li 1997).

**Methods of Order 4.** Theorem 3.1 and Example III.3.15 give 3 conditions for order 4. Therefore, we put $s = 3$ in (3.1) and assume symmetry $\beta_1 = \alpha_3$, $\beta_2 = \alpha_2$, and $\beta_3 = \alpha_1$. This leads to the conditions

$$\alpha_1 + \alpha_2 + \alpha_3 = \frac{1}{2}, \qquad \alpha_1^3 + \alpha_2^3 + \alpha_3^3 = 0, \qquad (\alpha_3^2 - \alpha_1^2)(\alpha_1 + \alpha_2) = 0.$$

Since with $\alpha_1 + \alpha_2 = 0$ or with $\alpha_1 + \alpha_3 = 0$ the first two of these equations are not compatible, the unique solution of this system is

$$\alpha_1 = \alpha_3 = \frac{1}{2\left(2 - 2^{1/3}\right)}, \qquad \alpha_2 = -\frac{2^{1/3}}{2\left(2 - 2^{1/3}\right)}.$$

We observe that $\beta_i = \alpha_i$ for all $i$. Therefore, $\Phi_{\alpha_i h} \circ \Phi^*_{\beta_i h}$ can be grouped together in (3.1) and we have obtained a method of type (3.2), which is actually method (II.4.4) with $p = 2$.

Again, the solutions with the minimal number of stages do not give the best methods (remember the good performance of Suzuki's fourth order method (II.4.5) in Fig. II.5.2), so we look for 4th order methods with larger $s$. McLachlan (1995) has constructed a method for $s = 5$ with particularly small error terms and nice coefficients

$$\beta_1 = \alpha_5 = \frac{14 - \sqrt{19}}{108}, \qquad \alpha_1 = \beta_5 = \frac{146 + 5\sqrt{19}}{540}, \tag{3.6}$$

$$\beta_2 = \alpha_4 = \frac{-23 - 20\sqrt{19}}{270}, \qquad \alpha_2 = \beta_4 = \frac{-2 + 10\sqrt{19}}{135}, \qquad \beta_3 = \alpha_3 = \frac{1}{5},$$

which he recommends "for all uses".

In Fig. 3.2 we compare the numerical performances of all these methods on our already well-known example in both variants (implicit-explicit and vice-versa). We see that the best methods in *one* picture may be worse in the other. For comparison, the results are surrounded by 'ghosts in grey' representing good formulae from the next lower (order 2) and the next higher (order 6) class of methods.

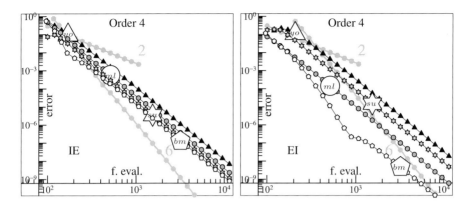

**Fig. 3.2.** Work-precision diagrams for methods of order 4 as in Fig. 3.1; 'yo': method (II.4.4) of Yoshida; 'su': method (II.4.5) of Suzuki; 'ml': McLachlan (3.6); 'bm': method (3.7); in grey: neighbouring order methods Störmer/Verlet (order 2) and $p6\ s9$ (order 6).

**Methods Tuned for Special Problems.** In the case where one is applying a *special* method to a *special* problem (e.g., to second order differential equations or to small perturbations of integrable systems), more spectacular gains of efficiency are possible. For example, Blanes & Moan (2002) have constructed the following fourth order method with $s = 6$

$$\begin{aligned}
\beta_1 &= \alpha_6 = 0.082984406417405, & \alpha_1 &= \beta_6 = 0.16231455076687, \\
\beta_2 &= \alpha_5 = 0.23399525073150, & \alpha_2 &= \beta_5 = 0.37087741497958, \\
\beta_3 &= \alpha_4 = -0.40993371990193, & \alpha_3 &= \beta_4 = 0.059762097006575,
\end{aligned} \quad (3.7)$$

which, when correctly applied to second order differential equations (right picture of Fig. 3.2) exhibits excellent performance.

Further methods, adapted to the integration of second order differential equations, have been constructed by Forest (1992), McLachlan & Atela (1992), Calvo & Sanz-Serna (1993), Okunbor & Skeel (1994), and McLachlan (1995). Another important situation, which allows a tuning of the parameters, are near-integrable systems such as the perturbed two-body motion (e.g., the outer solar system considered in Chap. I). If the differential equation can be split into $\dot{y} = f^{[1]}(y) + f^{[2]}(y)$, where $\dot{y} = f^{[1]}(y)$ is exactly integrable and $f^{[2]}(y)$ is small compared to $f^{[1]}(y)$, special integrators should be used. We refer to Kinoshita, Yoshida & Nakai (1991), Wisdom & Holman (1991), Saha & Tremaine (1992), and McLachlan (1995b) for more details and for the parameters of such integrators.

**Methods of Order 6.** By Theorem 3.1 and Example III.3.12 a method (3.1) has to satisfy 9 conditions for order 6. It turns out that these order conditions have already a solution with $s = 7$, but all known solutions with $s \leq 8$ are equivalent to methods of type (3.2). With order 6 we are apparently close to the point where the enormous simplifications of the order conditions due to Theorem 3.3 below start to outperform the freedom of choosing different values for $\alpha_i$ and $\beta_i$. We therefore continue our discussion by considering only the special case (3.2).

## V.3.2 Symmetric Composition of Symmetric Methods

The introduction of more symmetries into the method simplifies considerably the order conditions. These simplifications can be best understood with a sort of "Choleski decomposition" of symmetric methods (Murua & Sanz-Serna 1999).

**Lemma 3.2.** *For every symmetric method $\Phi_h(y)$ that admits an expansion in powers of $h$, there exists $\widehat{\Phi}_h(y)$ such that*
$$\Phi_h(y) = \big(\widehat{\Phi}_{h/2} \circ \widehat{\Phi}^*_{h/2}\big)(y).$$

*Proof.* Since $\Phi_h(y) = y + \mathcal{O}(h)$ is close to the identity, the existence of a unique method $\widehat{\Phi}_h(y) = y + h d_1(y) + h^2 d_2(y) + \ldots$ satisfying $\Phi_h = \widehat{\Phi}_{h/2} \circ \widehat{\Phi}_{h/2}$ follows from Taylor expansion and from a comparison of like powers of $h$.

If $\Phi_h(y)$ is symmetric, we have in addition
$$\Phi_h = \Phi^{-1}_{-h} = \widehat{\Phi}^{-1}_{-h/2} \circ \widehat{\Phi}^{-1}_{-h/2},$$
and $\widehat{\Phi}_{h/2} = \widehat{\Phi}^{-1}_{-h/2} = \widehat{\Phi}^*_{h/2}$ follows from the uniqueness of $\widehat{\Phi}_h$. □

We let $\Phi_h$ be a symmetric method, and we consider the composition
$$\Psi_h = \Phi_{\gamma_s h} \circ \ldots \circ \Phi_{\gamma_2 h} \circ \Phi_{\gamma_1 h}. \tag{3.8}$$

Using the method $\widehat{\Phi}_h$ of Lemma 3.2, this composition method is equivalent to (3.1) ($\Phi_h$ replaced with $\widehat{\Phi}_h$) with
$$\alpha_i = \beta_i = \frac{\gamma_i}{2}. \tag{3.9}$$

**Theorem 3.3.** *For composition methods (3.8) with symmetric $\Phi_h$ it is sufficient to consider the order conditions for $\tau \in \mathcal{H}$ where all vertices of $\tau$ have odd indices.*

*Proof.* If $i(\tau)$ is even, it follows from $\alpha_k = \beta_k$ and from (III.3.11) that
$$a_s(\tau) = a_{s-1}(\tau) = \ldots = a_1(\tau) = a_0(\tau) = 0.$$

Since $e(\tau) = 0$ for such trees, the corresponding order condition is automatically satisfied. Any other vertex with an even index can be brought to the root by applying the Switching Lemma III.3.8. □

After this reduction, only 7 conditions survive for order 6 from the trees displayed in Example III.3.12. A further reduction in the number of order conditions is achieved by assuming *symmetric coefficients* in method (3.8), i.e.,
$$\gamma_{s+1-j} = \gamma_j \quad \text{for all } j. \tag{3.10}$$

This implies that the overall method $\Psi_h$ is symmetric, so that the order conditions for trees with an even $\|\tau\|$ need not be considered. This proves the following result.

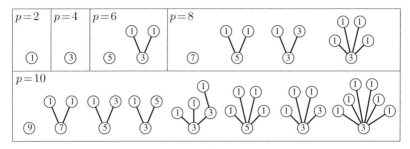

**Fig. 3.3.** Symmetric Composition of Symmetric Methods up to order 10

**Theorem 3.4.** *For composition methods (3.8) with symmetric $\Phi_h$, satisfying (3.10), it is sufficient to consider the order conditions for $\tau \in \mathcal{H}$ where all vertices of $\tau$ have odd indices and where $\|\tau\|$ is odd.* □

Figure 3.3 shows the remaining order conditions for methods up to order 10. We see that for order 6 there remain only 4 conditions, much less than the 166 that we started with (Theorem III.3.6).

**Example 3.5.** The rule of (III.3.14) leads to the following conditions for *symmetric* composition of *symmetric* methods:

| | | | | | |
|---|---|---|---|---|---|
| Order 2: | ① | $\sum_{k=1}^{s} \gamma_k = 1$ | | | |
| Order 4: | ③ | $\sum_{k=1}^{s} \gamma_k^3 = 0$ | | | |
| Order 6: | ⑤ | $\sum_{k=1}^{s} \gamma_k^5 = 0$ | | $\sum_{k=1}^{s} \gamma_k^3 \left(\sum_{\ell=1}^{k}{}' \gamma_\ell\right)^2 = 0$ | |
| Order 8: | ⑦ | $\sum_{k=1}^{s} \gamma_k^7 = 0$ | | $\sum_{k=1}^{s} \gamma_k^5 \left(\sum_{\ell=1}^{k}{}' \gamma_\ell\right)^2 = 0$ | |
| | | $\sum_{k=1}^{s} \gamma_k^3 \sum_{\ell=1}^{k}{}' \gamma_\ell \sum_{m=1}^{k}{}' \gamma_m^3 = 0$ | | $\sum_{k=1}^{s} \gamma_k^3 \left(\sum_{\ell=1}^{k}{}' \gamma_\ell\right)^4 = 0.$ | |

Here, similar to Example III.3.15, a *prime* attached to a summation symbol indicates that the last term $\gamma_\ell^i$ is taken as $\gamma_\ell^i/2$.

**Methods of Order 4.** The methods (II.4.4) and (II.4.5) are both of the form (3.8), and those with $p=2$ yield methods of order 4. We have seen in the experiment of Fig. II.5.2 that the method (II.4.5) yields more precise approximations; see also Fig. 3.2. We do not know of any 4th order method of type (3.2) that is significantly better than method (3.1) with coefficients (3.6).

**Methods of Order 6.** If we search for a minimal stage solution of the four equations for order 6, we apparently need four free parameters $\gamma_1, \gamma_2, \gamma_3, \gamma_4$; then $\gamma_5, \gamma_6, \gamma_7$ are determined by symmetry. Equation ① gives $\gamma_4 = 1 - 2(\gamma_1 + \gamma_2 + \gamma_3)$. So we end up with three equations for the three unknowns $\gamma_1, \gamma_2, \gamma_3$. A numerical search for this problem produces three solutions, the best of which has been discovered by many authors, in particular by Yoshida (1990), and is as follows:

$$\begin{aligned} \gamma_1 = \gamma_7 &= \phantom{-}0.78451361047755726381949763 \\ \gamma_2 = \gamma_6 &= \phantom{-}0.23557321335935813368479318 \\ \gamma_3 = \gamma_5 &= -1.17767998417887100694641568 \\ \gamma_4 &= \phantom{-}1.31518632068391121888424973 \end{aligned} \quad p6\ s7 \qquad (3.11)$$

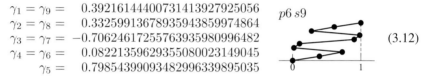

Using computer algebra, Koseloff (1996) proves that the nonlinear system for $\gamma_1, \gamma_2, \gamma_3$ has not more than three real solutions.

Similar to the situation for order 4, where relaxing the minimal number of stages allowed a significant increase of performance, we also might expect to obtain better methods of order 6 in this way. McLachlan (1995) increases $s$ by two and constructs good methods with small error coefficients. By minimizing $\max_i |\gamma_i|$, Kahan & Li (1997) obtain the following excellent method [3]

$$\begin{aligned} \gamma_1 = \gamma_9 &= \phantom{-}0.39216144400731413927925056 \\ \gamma_2 = \gamma_8 &= \phantom{-}0.33259913678935943859974864 \\ \gamma_3 = \gamma_7 &= -0.70624617255763935980996482 \\ \gamma_4 = \gamma_6 &= \phantom{-}0.08221359629355080023149045 \\ \gamma_5 &= \phantom{-}0.79854399093482996639895035 \end{aligned} \quad p6\ s9 \qquad (3.12)$$

This method produces, with a comparable number of total steps, errors which are typically smaller than those of method (3.11). Numerical results of these two methods are given in Fig. 3.4.

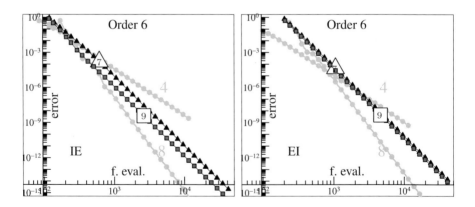

**Fig. 3.4.** Work-precision diagrams for methods of order 6 for the Kepler problem as in Fig. 3.1; '7': method $p6\ s7$ of (3.11); '9': method $p6\ s9$ of (3.12); in grey: neighbouring order methods (3.6) (order 4) and $p8\ s17$ (order 8)

---

[3] The authors are grateful to S. Blanes for this reference.

**Methods of Order 8.** For order 8, Fig. 3.3 represents 8 equations to solve. This indicates that the minimal value of $s$ is 15. A numerical search for solutions $\gamma_1, \ldots, \gamma_8$ of these equations produces hundreds of solutions. We choose among all these the solution with the smallest $\max(|\gamma_i|)$. The coefficients, which were originally given by Suzuki & Umeno (1993), Suzuki (1994), and later by McLachlan (1995), are as follows:

$$
\begin{aligned}
\gamma_1 = \gamma_{15} &= \phantom{-}0.74167036435061295344822780 \\
\gamma_2 = \gamma_{14} &= -0.40910082580003159399730010 \\
\gamma_3 = \gamma_{13} &= \phantom{-}0.19075471029623837995387626 \\
\gamma_4 = \gamma_{12} &= -0.57386247111608226665638773 \\
\gamma_5 = \gamma_{11} &= \phantom{-}0.29906418130365592384446354 \\
\gamma_6 = \gamma_{10} &= \phantom{-}0.33462491824529818378495798 \\
\gamma_7 = \gamma_9 &= \phantom{-}0.31529309239676659663205666 \\
\gamma_8 &= -0.79688793935291635401978884
\end{aligned}
\tag{3.13}
$$

By putting $s = 17$ we obtain one degree of freedom in solving the equations. This allows an improvement on the foregoing method. The best known solution, slightly better than a method of McLachlan (1995), has been found by Kahan & Li (1997) and is given by

$$
\begin{aligned}
\gamma_1 = \gamma_{17} &= \phantom{-}0.13020248308889008087881763 \\
\gamma_2 = \gamma_{16} &= \phantom{-}0.56116298177510838456196441 \\
\gamma_3 = \gamma_{15} &= -0.38947496264484728640807860 \\
\gamma_4 = \gamma_{14} &= \phantom{-}0.15884190655515560089621075 \\
\gamma_5 = \gamma_{13} &= -0.39590389413323757733623154 \\
\gamma_6 = \gamma_{12} &= \phantom{-}0.18453964097831570709183254 \\
\gamma_7 = \gamma_{11} &= \phantom{-}0.25837438768632204729397911 \\
\gamma_8 = \gamma_{10} &= \phantom{-}0.29501172360931029887096624 \\
\gamma_9 &= -0.60550853383003451169892108
\end{aligned}
\tag{3.14}
$$

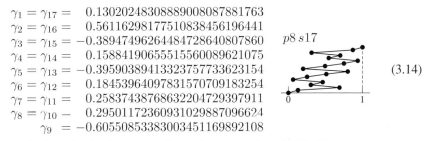

Numerical results, in the same style as above, are given in Fig. 3.5.

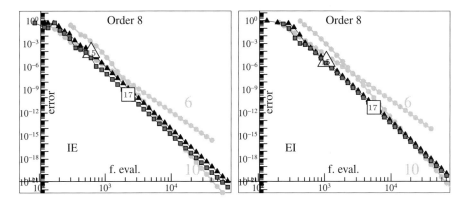

**Fig. 3.5.** Work-precision diagrams for methods of order 8 for the Kepler problem as in Fig. 3.1; '15': method $p8\,s15$ of (3.13); '17': method $p8\,s17$ of (3.14); in grey: neighbouring order methods $p6\,s9$ (order 6) and $p10\,s33$ (order 10)

**Methods of Order 10.** The first methods of order 10 were given by Kahan & Li (1997) with $s = 31$ and $s = 33$, which could be improved on after some nights of computer search. We present a method for $s = 33$ (see Fig. 3.5 for a comparison with eighth order methods):

$$
\begin{aligned}
\gamma_1 = \gamma_{33} &= \phantom{-}0.09040619368607278492161150 \\
\gamma_2 = \gamma_{32} &= \phantom{-}0.53591815953030120213784983 \\
\gamma_3 = \gamma_{31} &= \phantom{-}0.35123257547493978187517736 \\
\gamma_4 = \gamma_{30} &= -0.31116802097815835426086544 \\
\gamma_5 = \gamma_{29} &= -0.52556314194263510431065549 \\
\gamma_6 = \gamma_{28} &= \phantom{-}0.14447909410225247647345695 \\
\gamma_7 = \gamma_{27} &= \phantom{-}0.02983588609748235818064083 \\
\gamma_8 = \gamma_{26} &= \phantom{-}0.17786179923739805133592238 \\
\gamma_9 = \gamma_{25} &= \phantom{-}0.09826906939341637652532377 \\
\gamma_{10} = \gamma_{24} &= \phantom{-}0.46179986210411860873242126 \\
\gamma_{11} = \gamma_{23} &= -0.33377845599881851314531820 \\
\gamma_{12} = \gamma_{22} &= \phantom{-}0.07095684836524793621031152 \\
\gamma_{13} = \gamma_{21} &= \phantom{-}0.23666960070126868771909819 \\
\gamma_{14} = \gamma_{20} &= -0.49725977950660985445028388 \\
\gamma_{15} = \gamma_{19} &= -0.30399616617237257346546356 \\
\gamma_{16} = \gamma_{18} &= \phantom{-}0.05246957188100069574521612 \\
\gamma_{17} &= \phantom{-}0.44373380805019087955111365
\end{aligned}
\tag{3.15}
$$

### V.3.3 Effective Order and Processing Methods

> There has recently been a revival of interest in the concept of "effective order".
>
> (J.C. Butcher 1998)

The concept of effective order was introduced by Butcher (1969) with the aim of constructing 5th order explicit Runge-Kutta methods with 5 stages. The idea is to search for a computationally efficient method $K_h$ such that with a suitable $\chi_h$,

$$\Psi_h = \chi_h \circ K_h \circ \chi_h^{-1} \tag{3.16}$$

has an order higher than that of $K_h$. The method $K_h$ is called the *kernel*, and $\chi_h$ can be interpreted as a transformation in the phase space, close to the identity. Because of

$$\Psi_h^N = \chi_h \circ K_h^N \circ \chi_h^{-1},$$

an implementation of $\Psi_h$ over $N$ steps with constant step size $h$ has the same computational efficiency as $K_h$. The computation of $\chi_h^{-1}$ has only to be done once at the beginning of the integration, and $\chi_h$ has to be evaluated only at output points, which can be performed on another processor. In the article López-Marcos, Sanz-Serna & Skeel (1996) the notion of *preprocessing* for the step $\chi_h^{-1}$ and *postprocessing* for $\chi_h$ is introduced.

**Example 3.6 (Störmer/Verlet as Processed Symplectic Euler Method).** Consider a split differential equation, let $\Phi_h^{[LT]} = \varphi_h^{[2]} \circ \varphi_h^{[1]}$ be the Lie-Trotter formula or symplectic Euler method (see Sect. II.5), and $\Phi_h^{[S]} = \varphi_{h/2}^{[1]} \circ \varphi_h^{[2]} \circ \varphi_{h/2}^{[1]}$ the Strang splitting or Störmer/Verlet scheme. As a consequence of the group property of the exact flow, we have

$$\Phi_h^{[S]} = \varphi_{h/2}^{[1]} \circ \Phi_h^{[LT]} \circ \varphi_{-h/2}^{[1]} = \chi_h \circ \Phi_h^{[LT]} \circ \chi_h^{-1}$$

with $\chi_h = \varphi_{h/2}^{[1]}$. Hence, applying the Lie-Trotter formula with processing yields a second order approximation.

Since the use of geometric integrators requires constant step sizes, it is quite natural that Butcher's idea of effective order has been revived in this context. A systematic search for processed composition methods started with the works of Wisdom, Holman & Touma (1996), McLachlan (1996), and Blanes, Casas & Ros (1999, 2000b).

Let us explain the technique of processing in the situation where the kernel $K_h$ is a symmetric composition

$$K_h = \Phi_{\gamma_s h} \circ \ldots \circ \Phi_{\gamma_2 h} \circ \Phi_{\gamma_1 h} \qquad (\gamma_{s+1-i} = \gamma_i \text{ for all } i) \qquad (3.17)$$

of a symmetric method $\Phi_h$. We suppose that the processor is of the form

$$\chi_h = \Phi_{\delta_r h} \circ \ldots \circ \Phi_{\delta_2 h} \circ \Phi_{\delta_1 h}, \qquad (3.18)$$

such that its inverse is given by (use the symmetry $\Phi_h^{-1} = \Phi_{-h}$)

$$\chi_h^{-1} = \Phi_{-\delta_1 h} \circ \Phi_{-\delta_2 h} \circ \ldots \circ \Phi_{-\delta_r h}. \qquad (3.19)$$

**Order Conditions.** The composite method $\Psi_h = \chi_h \circ K_h \circ \chi_h^{-1}$ is of the form $\Psi_h = \Phi_{\varepsilon_{2r+s} h} \circ \ldots \circ \Phi_{\varepsilon_2 h} \circ \Phi_{\varepsilon_1 h}$ with

$$(\varepsilon_{2r+s}, \ldots, \varepsilon_2, \varepsilon_1) = (\delta_r, \ldots, \delta_1, \gamma_s, \ldots, \gamma_1, -\delta_1, \ldots, -\delta_r). \qquad (3.20)$$

Theorem 3.3 thus tells us that only the order conditions corresponding to $\tau \in \mathcal{H}$, whose vertices have odd indices, have to be considered. Unfortunately, the sequence $\{\varepsilon_i\}$ of (3.20) does not satisfy the symmetry relation (3.10), unless all $\delta_i$ vanish. However, if we require

$$\chi_{-h}(y) = \chi_h(y) + \mathcal{O}(h^{p+1}), \qquad (3.21)$$

we see that $\chi_h^{-1}(y) = \chi_h^*(y) + \mathcal{O}(h^{p+1})$, and the method $\Psi_h = \chi_h \circ K_h \circ \chi_h^{-1}$ is symmetric up to terms of order $\mathcal{O}(h^{p+1})$. Consequently, the reduction of Theorem 3.4 is valid, so that for order $p$ only the trees of Fig. 3.3 have to be considered.

For the first tree of Example 3.5 the order condition is

$$1 = \sum_{k=1}^{2r+s} \varepsilon_k = \sum_{k=1}^{s} \gamma_k,$$

and we see that this is a condition on the kernel $K_h$ only. Similarly, for odd $i$ we have

$$0 = \sum_{k=1}^{2r+s} \varepsilon_k^i = \sum_{k=1}^{s} \gamma_k^i, \qquad (3.22)$$

so that also the trees ③, ⑤, ⑦, ... give conditions on $K_h$ and cannot be influenced by the processor. We next consider the trees of Example 3.5 with three vertices, whose order condition is

$$0 = \sum_{k=1}^{2r+s} \varepsilon_k^i {\sum_{\ell=1}^{k}}' \varepsilon_\ell^j {\sum_{m=1}^{k}}' \varepsilon_m^q.$$

We split the sums according to the partitioning into $\delta_i, \gamma_i, -\delta_i$ in (3.20), and we denote the expressions appearing in Example 3.5 by $a(\tau)$ and those corresponding to $\chi_h$ and $\chi_h^{-1}$ by $b(\tau)$ and $b^{-1}(\tau)$, respectively. Using the abbreviations $\tau_i$ for the tree with one vertex labelled $i$, $\tau_{ij}$ for the tree with two vertices labelled $i$ (the root) and $j$, and by $\tau_{ijq}$ the trees with three vertices labelled $i$ (root), $j$ and $q$ (vertices that are directly connected to the root), this yields

$$\begin{aligned}
0 &= b^{-1}(\tau_{ijq}) + a(\tau_i)b^{-1}(\tau_j)b^{-1}(\tau_q) + a(\tau_{ij})b^{-1}(\tau_q) \\
&\quad + a(\tau_{iq})b^{-1}(\tau_j) + a(\tau_{ijq}) + b(\tau_i)b^{-1}(\tau_j)b^{-1}(\tau_q) \\
&\quad + b(\tau_i)b^{-1}(\tau_j)a(\tau_q) + b(\tau_i)a(\tau_j)b^{-1}(\tau_q) + b(\tau_i)a(\tau_j)a(\tau_q) \\
&\quad + b(\tau_{ij})b^{-1}(\tau_q) + b(\tau_{ij})a(\tau_q) + b(\tau_{iq})b^{-1}(\tau_j) + b(\tau_{iq})a(\tau_j) + b(\tau_{ijq}).
\end{aligned} \qquad (3.23)$$

How can we simplify this long expression? First of all, we imagine $K_h$ to be the identity (either $s = 0$ or all $\gamma_i = 0$), so that $\Psi_h = \chi_h \circ \chi_h^{-1}$ becomes the identity. In this situation, the terms involving $a(\tau)$ are not present in (3.23), and we obtain

$$0 = b^{-1}(\tau_{ijq}) + b(\tau_i)b^{-1}(\tau_j)b^{-1}(\tau_q) + b(\tau_{ij})b^{-1}(\tau_q) + b(\tau_{iq})b^{-1}(\tau_j) + b(\tau_{ijq}).$$

We can thus remove all terms in (3.23) that do not contain a factor $a(\tau)$. Now observe that by (3.21), $\chi_h(y)$ as well as $\chi_h^{-1}(y)$ have an expansion in even powers of $h$. Therefore, $b(\tau)$ and $b^{-1}(\tau)$ vanish for all $\tau$ with odd $\|\tau\|$. Formula (3.23) thus simplifies considerably and yields

$$0 = a(\tau_{311}) + 2b(\tau_{31})a(\tau_1), \qquad (3.24)$$

$$0 = a(\tau_{511}) + 2b(\tau_{51})a(\tau_1), \qquad (3.25)$$

$$0 = a(\tau_{313}) + b(\tau_{31})a(\tau_3) + b(\tau_{33})a(\tau_1). \qquad (3.26)$$

A similar computation for the last tree in Example 3.5 gives (in an obvious notation)

$$0 = a(\tau_{31111}) + 4b(\tau_{31})a(\tau_1)^3 + 4b(\tau_{3111})a(\tau_1). \qquad (3.27)$$

Since $a(\tau_1) = \sum_{i=1}^{s} \gamma_i = 1$, the conditions (3.24), (3.25) and (3.27) can be interpreted as conditions on the processor, namely on $b(\tau_{31})$, $b(\tau_{51})$ and $b(\tau_{3111})$. We already have $a(\tau_3) = 0$ from (3.22), and an application of the Switching Lemma III.3.8 gives $b(\tau_{33}) = \frac{1}{2}(b(\tau_3)^2 - b(\tau_6))$. The term $b(\tau_3)$ vanishes by (3.21) and $b(\tau_6) = 0$ is a consequence of the proof of Theorem 3.3. Therefore (3.26) is equivalent to $a(\tau_{313}) = 0$. We summarize our computation in the following theorem.

**Theorem 3.7.** *The processing method $\Psi_h = \chi_h \circ K_h \circ \chi_h^{-1}$ is of order $p$ ($p \leq 8$), if*
- *the coefficients $\gamma_i$ of the kernel satisfy the conditions of the left column in Example 3.5, i.e., 3 conditions for order 6, and 5 conditions for order 8;*
- *the coefficients $\delta_i$ of the processor are such that (3.21) holds (4 conditions for order 6, and 8 conditions for order 8), and in addition condition (3.24) for order 6, and (3.24), (3.25), (3.27) for order 8 are satisfied.* □

**Remark 3.8.** Although we have presented the computations only for $p \leq 8$, the result is general. All trees $\tau \in \mathcal{H}$, which are not of the form $\tau = u \circ \text{\textcircled{1}}$, give rise to conditions on the kernel $K_h$ (for a similar result in the context of Runge-Kutta methods see Butcher & Sanz-Serna (1996)). The remaining conditions have to be satisfied by the coefficients of the processor. Due to the reduced number of order conditions, it is relatively easy to construct high order kernels. However, the difficulty in constructing a suitable processor increases rapidly with the order.

The application of the processing technique is two-fold. A first possibility is to take one of the high-order composition methods of the form (3.2), e.g., one of those presented in Sect. V.3.2, and to exploit the freedom in the coefficients of the processor to make the error constants smaller.

Another possibility is to start from the beginning and to construct a method $K_h$ with coefficients satisfying only the conditions of Theorem 3.7. Methods of effective order 6 and 8 have been constructed in this way by Blanes (2001).

# V.4 Symmetric Methods on Manifolds

Numerical methods for differential equations on manifolds have been introduced in Sections IV.4 and IV.5. The presented algorithms are in general not symmetric. We discuss here suitable symmetric modifications which often have an improved long-time behaviour. We consider a differential equation

$$\dot{y} = f(y), \qquad f(y) \in T_y\mathcal{M} \tag{4.1}$$

on a manifold $\mathcal{M}$, and we assume that the manifold is either given as the zero set of a function $g(y)$ or by means of a suitable parametrization $y = \varphi(z)$.

## V.4.1 Symmetric Projection

Due to the projection at the end of an integration step, the standard projection method (Algorithm IV.4.2) is not symmetric (see Fig. IV.4.2). In order to make the overall algorithm symmetric, one has to apply a kind of "inverse projection" at the beginning of each integration step. This idea has first been used by Ascher & Reich (1999) to enforce conservation of energy, and it has been applied in more general contexts by Hairer (2000).

**Algorithm 4.1 (Symmetric Projection Method).** *Assume that $y_n \in \mathcal{M}$. One step $y_n \mapsto y_{n+1}$ is defined as follows (see Fig. 4.1, right picture):*

- $\widetilde{y}_n = y_n + G(y_n)^T \mu$ *where $g(y_n) = 0$ (perturbation step);*
- $\widetilde{y}_{n+1} = \Phi_h(\widetilde{y}_n)$ *(symmetric one-step method applied to $\dot{y} = f(y)$);*
- $y_{n+1} = \widetilde{y}_{n+1} + G(y_{n+1})^T \mu$ *with $\mu$ such that $g(y_{n+1}) = 0$ (projection step).*

Here, $G(y) = g'(y)$ denotes the Jacobian of $g(y)$. It is important to take a symmetric method in the second step, and the same vector $\mu$ in the perturbation and projection steps.

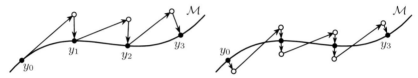

**Fig. 4.1.** Standard projection (left picture) compared to symmetric projection (right)

**Existence of the Numerical Solution.** The vector $\mu$ and the numerical approximation $y_{n+1}$ are implicitly defined by

$$F(h, y_{n+1}, \mu) = \begin{pmatrix} y_{n+1} - \Phi_h\big(y_n + G(y_n)^T \mu\big) - G(y_{n+1})^T \mu \\ g(y_{n+1}) \end{pmatrix} = 0. \quad (4.2)$$

Since $F(0, y_n, 0) = 0$ and since

$$\frac{\partial F}{\partial (y_{n+1}, \mu)}(0, y_n, 0) = \begin{pmatrix} I & -2G(y_n)^T \\ G(y_n) & 0 \end{pmatrix} \quad (4.3)$$

is invertible (provided that $G(y_n)$ has full rank), an application of the implicit function theorem proves the existence of the numerical solution for sufficiently small step size $h$. The simple structure of the matrix (4.3) can also be exploited for an efficient solution of the nonlinear system (4.2) using simplified Newton iterations. If the basic method $\Phi_h$ is itself implicit, the nonlinear system (4.2) should be solved in tandem with $\widetilde{y}_{n+1} = \Phi_h(\widetilde{y}_n)$.

**Order.** For a study of the local error we let $y_n := y(t_n)$ be a value on the exact solution $y(t)$ of (4.1). If the basic method $\Phi_h$ is of order $p$, i.e., if $y(t_n + h) - \Phi_h\big(y(t_n)\big) = \mathcal{O}(h^{p+1})$, we have $F\big(h, y(t_{n+1}), 0\big) = \mathcal{O}(h^{p+1})$. Compared to (4.2) the implicit function theorem yields

$$y_{n+1} - y(t_{n+1}) = \mathcal{O}(h^{p+1}) \quad \text{and} \quad \mu = \mathcal{O}(h^{p+1}).$$

This proves that the symmetric projection method of Algorithm 4.1 has the same order as the underlying one-step method $\Phi_h$.

**Symmetry of the Algorithm.** Exchanging $h \leftrightarrow -h$ and $y_n \leftrightarrow y_{n+1}$ in the Algorithm 4.1 yields

$$\begin{aligned}
\widetilde{y}_n &= y_{n+1} + G(y_{n+1})^T \mu, & g(y_{n+1}) &= 0, \\
\widetilde{y}_{n+1} &= \Phi_{-h}(\widetilde{y}_n), & & \\
y_n &= \widetilde{y}_{n+1} + G(y_n)^T \mu, & g(y_n) &= 0.
\end{aligned}$$

The auxiliary variables $\mu$, $\widetilde{y}_n$, and $\widetilde{y}_{n+1}$ can be arbitrarily renamed. If we replace them with $-\mu$, $\widetilde{y}_{n+1}$, and $\widetilde{y}_n$, respectively, we get the formulas of the original algorithm provided that the method $\Phi_h$ of the intermediate step is symmetric. This proves the symmetry of the algorithm.

Various modifications of the perturbation and projection steps are possible without destroying the symmetry. For example, one can replace the arguments $y_n$ and $y_{n+1}$ in $G(y)$ with $(y_n + y_{n+1})/2$. It might be advantageous to use a constant direction, i.e., $\widetilde{y}_n = y_n + A^T \mu$, $y_{n+1} = \widetilde{y}_{n+1} + A^T \mu$ with a constant matrix $A$. In this case the matrix $G(y)A^T$ has to be invertible along the solution in order to guarantee the existence of the numerical solution.

**Reversibility.** From Theorem 1.5 we know that symmetry alone does not imply the $\rho$-reversibility of the numerical flow. The method must also satisfy the compatibility condition (1.4). It is straightforward to check that this condition is satisfied if the integrator $\Phi_h$ of the intermediate step of Algorithm 4.1 satisfies (1.4) and, in addition,

$$\rho G(y)^T = G(\rho y)^T \sigma \tag{4.4}$$

holds with some constant invertible matrix $\sigma$. In many interesting situations we have $g(\rho y) = \sigma^{-T} g(y)$ with a suitable $\sigma$, so that (4.4) follows by differentiation if $\rho \rho^T = I$. Similarly, when a projection with constant direction $y = \widetilde{y} + A^T \mu$ is applied, the matrix $A$ has to satisfy $\rho A^T = A^T \sigma$ for a suitably chosen invertible matrix $\sigma$ (see the experiment of Example 4.4 below).

**Example 4.2.** Let us consider the equations of motion of a rigid body as described in Example IV.1.7. They constitute a differential equation on the manifold

$$\mathcal{M} = \{(y_1, y_2, y_3) \mid y_1^2 + y_2^2 + y_3^2 = 1\},$$

and it is $\rho$-reversible with respect to $\rho(y_1, y_2, y_3) = (-y_1, y_2, y_3)$, and also with respect to $\rho(y_1, y_2, y_3) = (y_1, -y_2, y_3)$ and $\rho(y_1, y_2, y_3) = (y_1, y_2, -y_3)$. For a numerical simulation we take $I_1 = 2$, $I_2 = 1$, $I_3 = 2/3$, and the initial value $y_0 = (\cos(0.9), 0, \sin(0.9))$. We apply the trapezoidal rule (II.1.2) with the large step size $h = 1$ in three different versions.

The upper picture of Fig. 4.2 shows the result of a direct application of the trapezoidal rule. The numerical solution lies apparently on a closed curve, but it does not

152   V. Symmetric Integration and Reversibility

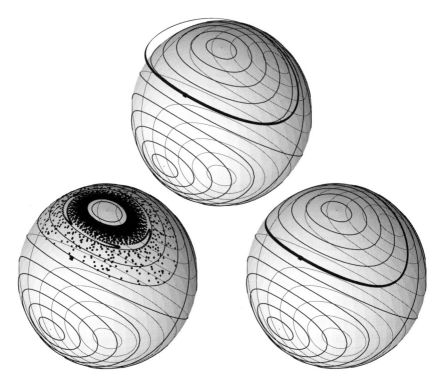

**Fig. 4.2.** Numerical simulation of the rigid body equations. The three pictures correspond to a direct application (upper), to the standard projection (lower left), and to the symmetric projection (lower right) of the trapezoidal rule; 5000 steps with step size $h = 1$.

lie exactly on the manifold $\mathcal{M}$. This can be seen as follows: the trapezoidal rule $\Phi_h^T$ is conjugate to the implicit midpoint rule $\Phi_h^M$ via a half-step of the explicit Euler method $\chi_{h/2}$. In fact the relations

$$\Phi_h^T = \chi_{h/2}^* \circ \chi_{h/2} \qquad \text{and} \qquad \Phi_h^M = \chi_{h/2} \circ \chi_{h/2}^*$$

hold, so that

$$\Phi_h^T = \chi_{h/2}^{-1} \circ \Phi_h^M \circ \chi_{h/2} \qquad \text{and} \qquad (\Phi_h^T)^N = \chi_{h/2}^{-1} \circ (\Phi_h^M)^N \circ \chi_{h/2}.$$

Consequently, the trajectory of the trapezoidal rule is obtained from the trajectory of the midpoint rule by a simple change of coordinates. On the other hand, the numerical solution of the midpoint rule lies exactly on a solution curve because it conserves quadratic invariants (Theorem IV.2.1).

Using standard orthogonal projection (Algorithm IV.4.2) we obviously obtain a numerical solution lying on the manifold $\mathcal{M}$. But as we can see from the lower left picture of Fig. 4.2, it does not remain near a closed curve and converges to a fixed point. The lower right picture shows that the use of the symmetric orthogonal projection (Algorithm 4.1) recovers the property of remaining near the closed solution curve.

**Example 4.3 (Numerical Experiment with Constant Direction of Projection).**
We consider the pendulum equation in Cartesian coordinates (see Example IV.4.1),

$$\dot{q}_1 = p_1, \qquad \dot{p}_1 = -q_1\lambda,$$
$$\dot{q}_2 = p_2, \qquad \dot{p}_2 = -1 - q_2\lambda, \qquad (4.5)$$

with $\lambda = (p_1^2 + p_2^2 - q_2)/(q_1^2 + q_2^2)$. This is a problem on the manifold

$$\mathcal{M} = \{(q_1, q_2, p_1, p_2) \mid q_1^2 + q_2^2 = 1, \; q_1 p_1 + q_2 p_2 = 0\}.$$

It is $\rho$-reversible with respect to $\rho(q_1, q_2, p_1, p_2) = (q_1, q_2, -p_1, -p_2)$ and also with respect to $\rho(q_1, q_2, p_1, p_2) = (-q_1, q_2, p_1, -p_2)$.

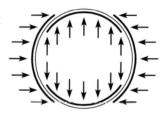

We apply two kinds of symmetric projection methods. First, we consider an orthogonal projection onto $\mathcal{M}$ as in Algorithm 4.1. Second, we project parallel to coordinate axes. More precisely, we fix the first components in position and velocity if the angle of the pendulum is close to $0$ or $\pi$ (vertical projection in the picture to the right), and we fix the second components if the angle is close to $\pm\pi/2$ (horizontal projection). The regions where the direction of projection changes, are overlapping.

We notice in Fig. 4.3 that for the orthogonal projection method the energy error remains bounded, and this is also true for integrations over much longer time intervals. This is in agreement with the observation of Chap. I, where symmetric methods showed an excellent long-time behaviour when applied to reversible differential equations.

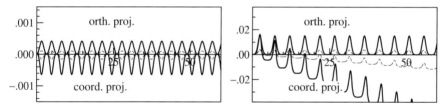

**Fig. 4.3.** Global error in the total energy for two different projection methods – orthogonal and coordinate projection – with the trapezoidal rule as basic integrator. Initial values for the position are $(\cos 0.8, -\sin 0.8)$ (left picture) and $(\cos 0.8, \sin 0.8)$ (right picture); zero initial values in the velocity; step sizes are $h = 0.1$ (solid) and $h = 0.05$ (thin dashed);.

For the coordinate projection, however, we observe a bounded energy error only for the initial value that is close to equilibrium (no change in the direction of the projection is necessary). As soon as the direction has to be changed (right picture of Fig. 4.3) a linear drift in the energy error becomes visible. Hence, care has to be taken with the choice of the projection. For an explanation of this phenomenon we refer to Chap. IX on backward error analysis and to Chap. XI on perturbation theory of reversible mappings.

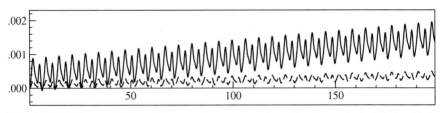

**Fig. 4.4.** Global error in the total energy for a symmetric projection method violating (1.4). Initial values for the position are $(\cos 0.8, -\sin 0.8)$ and $(0,0)$ for the velocity; step sizes are $h = 0.1$ (solid) and $h = 0.05$ (thin dashed).

**Example 4.4 (A Symmetric but Non-Reversible Projection Method).** We consider the pendulum equation as in Example 4.3. This time, however, we apply a projection $\widetilde{y}_n = y_n + A^T \mu$, $y_{n+1} = \widetilde{y}_{n+1} + A^T \mu$ with

$$A = \begin{pmatrix} \varepsilon & 1 & 0 & 0 \\ \varepsilon & 0 & 0 & 1 \end{pmatrix}, \qquad \varepsilon = 0.2.$$

For $\varepsilon = 0$ this corresponds to the vertical projection used in Example 4.3. For $\varepsilon \neq 0$ there is no matrix $\sigma$ such that $\rho A^T = A^T \sigma$ holds for one of the mappings $\rho$ that make the problem $\rho$-reversible. Hence condition (1.4) is violated, and the method is thus not $\rho$-reversible. The initial values are chosen such that $g'(y)A^T$ is invertible and well-conditioned along the solution. Although the projection direction need not be changed during the integration and the method is symmetric, the long-time behaviour is disappointing as shown in Fig. 4.4. This experiment illustrates that condition (1.4) is also important for a qualitatively correct simulation.

## V.4.2 Symmetric Methods Based on Local Coordinates

Numerical methods for differential equations on manifolds that are based on local coordinates (Algorithm IV.5.3) are in general not symmetric. For example, if we consider the parametrization (IV.5.8) with respect to the tangent space at $y_0$, the adjoint method would be parametrized by the tangent space at $y_1$. We can circumvent this difficulty by the following algorithm (Hairer 2001).

**Algorithm 4.5 (Symmetric Local Coordinates Approach).** *Assume that $y_n \in \mathcal{M}$ and that $\psi_a$ is a local parametrization of $\mathcal{M}$ satisfying $\psi_a(0) = a$ (close to $y_n$). One step $y_n \mapsto y_{n+1}$ is defined as follows (see Fig. 4.5):*

- *find $z_n$ (close to 0) such that $\psi_a(z_n) = y_n$;*
- *$\widetilde{z}_{n+1} = \Phi_h(z_n)$ (symmetric one-step method applied to (IV.5.7);*
- *$y_{n+1} = \psi_a(\widetilde{z}_{n+1})$;*
- *choose $a$ in the parametrization such that $z_n + \widetilde{z}_{n+1} = 0$.*

*It is important to remark that the parametrization $y = \psi_a(z)$ is in general changed in every step.*

**Fig. 4.5.** Symmetric use of local tangent space parametrization

This algorithm is illustrated in Fig. 4.5 for the tangent space parametrization (IV.5.8), given by

$$\psi_a(z) = a + Q(a)z + g'(a)^T u_a(z), \tag{4.6}$$

where the columns of $Q(a)$ form an orthogonal basis of $T_a\mathcal{M}$ and the function $u_a(z)$ is such that $\psi_a(z) \in \mathcal{M}$. It satisfies $u_a(0) = 0$ and $u'_a(0) = 0$.

**Existence of the Numerical Solution.** In Algorithm 4.5 the values $a \in \mathcal{M}$ and $z_n$ are implicitly determined by

$$F(h, z_n, a) = \begin{pmatrix} z_n + \Phi_h(z_n) \\ \psi_a(z_n) - y_n \end{pmatrix} = 0, \tag{4.7}$$

and the numerical solution is then explicitly given by $y_{n+1} = \psi_a\bigl(\Phi_h(z_n)\bigr)$. For more clarity we also use here the notation $\psi(z, a) = \psi_a(z)$. If the parametrization $\psi(z, a)$ is differentiable, we have

$$\frac{\partial F}{\partial (z_n, a)}(0, 0, y_n) = \begin{pmatrix} 2I & 0 \\ \frac{\partial \psi}{\partial z}(0, y_n) & \frac{\partial \psi}{\partial a}(0, y_n) \end{pmatrix}. \tag{4.8}$$

Since $\psi(z, a) \in \mathcal{M}$ for all $z$ and $a \in \mathcal{M}$, the derivative with respect to $a$ lies in the tangent space. Assume now that the parametrization $\psi(z, a)$ is such that the restriction of $\frac{\partial \psi}{\partial a}(0, y_n)$ onto the tangent space $T_{y_n}\mathcal{M}$ is bijective. Then, the matrix (4.8) is invertible on $\mathbb{R}^d \times T_{y_n}\mathcal{M}$ ($d$ denotes the dimension of the manifold). The implicit function theorem thus proves the existence of a numerical solution $(z_n, a)$ close to $(0, y_n)$. In the case where $\psi_a(z)$ is given by (4.6), the matrix

$$\frac{\partial \psi}{\partial a}(0, a) = I - g'(a)^T \bigl(g'(a)g'(a)^T\bigr)^{-1} g'(a)$$

is a projection onto the tangent space $T_a\mathcal{M}$ and satisfies the above assumptions provided that $g'(a)$ has full rank.

**Order.** We let $y_n := y(t_n)$ be a value on the exact solution $y(t)$ of (4.1). Then we fix $a \in \mathcal{M}$ as follows: we replace the upper part of the definition (4.7) of $F(h, z_n, a)$ with $z_n + \varphi_h^{(z)}(z_n)$, where $\varphi_t^{(z)}$ denotes the exact flow of the differential equation for $z(t)$ equivalent to (4.1). The above considerations show that such an $a$ exists; let us call it $a^*$. If $\Phi_h$ is of order $p$, we then have $F\bigl(h, z(t_n), a^*\bigr) = \mathcal{O}(h^{p+1})$. An application of the implicit function theorem thus gives $z_n - z(t_n) = \mathcal{O}(h^{p+1})$, implying $\widetilde{z}_{n+1} - z(t_{n+1}) = \mathcal{O}(h^{p+1})$, and finally also $y_{n+1} - y(t_{n+1}) = \mathcal{O}(h^{p+1})$. This proves order $p$ for the method defined by Algorithm 4.5.

**Symmetry.** Exchanging $h \leftrightarrow -h$ and $y_n \leftrightarrow y_{n+1}$ in Algorithm 4.5 yields

$$\psi_a(z_n) = y_{n+1}, \quad \widetilde{z}_{n+1} = \Phi_{-h}(z_n), \quad y_n = \psi_a(\widetilde{z}_{n+1}), \quad z_n + \widetilde{z}_{n+1} = 0.$$

If we also exchange the auxiliary variables $z_n$ and $\widetilde{z}_{n+1}$ and if we use the symmetry of the basic method $\Phi_h$, we regain the original formulas. This proves the symmetry of the algorithm. Again various kinds of modifications are possible. For example, the condition $z_n + \widetilde{z}_{n+1} = 0$ can be replaced with $z_n + \widetilde{z}_{n+1} = \chi(h, z_n, \widetilde{z}_{n+1})$. If $\chi(-h, v, u) = \chi(h, u, v)$, the symmetry of Algorithm 4.5 is not destroyed.

**Reversibility.** In general, we cannot expect the method of Algorithm 4.5 to satisfy the $\rho$-compatibility condition (1.4), which is needed for $\rho$-reversibility. However, if the parametrization is such that

$$\rho \psi_a(z) = \psi_{\rho a}(\sigma z) \quad \text{for some invertible } \sigma, \tag{4.9}$$

we shall show that the compatibility condition (1.4) holds. We first prove that for a $\rho$-reversible problem $\dot{y} = f(y)$ the differential equation (IV.5.7), written as $\dot{z} = F_a(z)$, is $\sigma$-reversible in the sense that $\sigma F_a(z) = -F_{\rho a}(\sigma z)$. This follows from $\rho \psi'_a(z) = \psi'_{\rho a}(\sigma z) \sigma$ (which is seen by differentiation of (4.9)) and from $f(\psi_{\rho a}(\sigma z)) = -\rho f(\psi_a(z))$, because

$$\psi'_a(z) F_a(z) = f(\psi_a(z)) \quad \Longrightarrow \quad \psi'_{\rho a}(\sigma z) \sigma F_a(z) = -f(\psi_{\rho a}(\sigma z)).$$

If the basic method $\Phi_h$ satisfies $\sigma \circ \Phi_h = \Phi_{-h} \circ \sigma$ when applied to $\dot{z} = F_a(z)$ (e.g., for all Runge-Kutta methods), the formulas of Algorithm 4.5 satisfy

$$\rho y_n = \rho \psi_a(z_n) = \psi_{\rho a}(\sigma z_n), \quad \sigma \widetilde{z}_{n+1} = \Phi_{-h}(\sigma z_n),$$
$$\psi_{\rho a}(\sigma \widetilde{z}_{n+1}) = \rho \psi_a(\widetilde{z}_{n+1}) = \rho y_{n+1}, \quad \sigma z_n + \sigma \widetilde{z}_{n+1} = 0.$$

This proves that, starting with $\rho y_n$ and a negative step size $-h$, the Algorithm 4.5 produces $\rho y_{n+1}$, where $y_{n+1}$ is just the result obtained with initial value $y_n$ and step size $h$. But this is nothing other than the $\rho$-compatibility condition (1.4) for Algorithm 4.5.

In order to verify condition (4.9) for the tangent space parametrization (4.6), we write it as $\psi_a(Z) = a + Z + N(Z)$, where $Z$ is an arbitrary element of the tangent space $T_a \mathcal{M}$ and $N(Z)$ is orthogonal to $T_a \mathcal{M}$ such that $\psi_a(Z) \in \mathcal{M}$. Since $\rho T_a \mathcal{M} = T_{\rho a} \mathcal{M}$ and since, for a $\rho$ satisfying $\rho \rho^T = I$, the vector $\rho N(Z)$ is orthogonal to $T_{\rho a} \mathcal{M}$, we have $\rho \psi_a(Z) = \psi_{\rho a}(\rho Z)$. This proves (4.9) for the tangent space parametrization of a manifold.

**Example 4.6.** We repeated the experiment of Example 4.2 with Algorithm IV.5.3, using tangent space parametrization and the trapezoidal rule as basic integrator, and compared it to the symmetrized version of Algorithm 4.5. We were surprised to see that both algorithms worked equally well and gave a numerical solution lying near a closed curve. An explanation is given in Exercise 11. There it is shown that for the

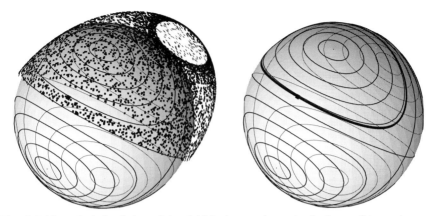

**Fig. 4.6.** Numerical simulation of the rigid body equations; standard use of tangent space parametrization with the trapezoidal rule as basic method (left picture) and its symmetrized version (right picture); $5000$ steps with step size $h = 0.4$.

special situation where $\mathcal{M}$ is a sphere, the standard algorithm is also symmetric for the trapezoidal rule. Let us therefore modify the problem slightly.

We consider the rigid body equations (IV.1.4) as a differential equation on the manifold

$$\mathcal{M} = \left\{ (y_1, y_2, y_3) \,\Big|\, \frac{y_1^2}{I_1} + \frac{y_2^2}{I_2} + \frac{y_3^2}{I_3} = Const \right\} \quad (4.10)$$

with parameters and initial data as in Example 4.2, and we apply the standard and the symmetrized method based on tangent space parametrization. The result is shown in Fig. 4.6. In both cases the numerical solution lies on the manifold (by definition of the method), but only the symmetric method has a correct long-time behaviour.

**Symmetric Lie Group Methods.** We turn our attention to particular problems

$$\dot{Y} = A(Y)Y, \qquad Y(0) = Y_0, \quad (4.11)$$

where $A(Y)$ is in the Lie algebra $\mathfrak{g}$ whenever $Y$ is in the corresponding Lie group $G$. The exact solution then evolves on the manifold $G$. Munthe-Kaas methods (Sect. IV.8.2) are in general not symmetric, even if the underlying Runge-Kutta method is symmetric. This is due to the unsymmetric use of the local coordinates $Y = \exp(\Omega)Y_0$. However, accidentally, the Lie group method based on the implicit midpoint rule

$$Y_{n+1} = \exp(\Omega)Y_n, \qquad \Omega = hA\big(\exp(\Omega/2)Y_n\big) \quad (4.12)$$

is symmetric. This can be seen as usual by exchanging $h \leftrightarrow -h$ and $Y_n \leftrightarrow Y_{n+1}$ (and also $\Omega \leftrightarrow -\Omega$ for the auxiliary variable). Numerical computations with the rigid body equations (considered as a problem on the sphere) shows an excellent long-time behaviour for this method similar to that of the right picture in Fig. 4.6. In contrast to the implicit midpoint rule (I.1.7), the numerical solution of (4.12) does not lie exactly on the ellipsoid (4.10); see Exercise 12.

For the construction of further symmetric Lie group methods we can apply the ideas of Algorithm 4.5. As local parametrization we choose

$$\psi_U(\Omega) = \exp(\Omega)U, \qquad (4.13)$$

where $U = \exp(\Theta)Y_n$ plays the role of the midpoint on the manifold. We put $Z_n = -\Theta$ so that $\psi_U(Z_n) = Y_n$. With this starting value $Z_n$ we apply any symmetric Runge-Kutta method to the differential equation

$$\dot{\Omega} = A(\psi_U(\Omega)) + \sum_{k=1}^{q} \frac{B_k}{k!} \operatorname{ad}_\Omega^k \bigl(A(\psi_U(\Omega))\bigr), \qquad \Omega(0) = -\Theta, \qquad (4.14)$$

(cf. (IV.8.9)) and thus obtain $\widetilde{Z}_{n+1}$. According to Algorithm 4.5, $\Theta$ is implicitly determined by the condition $Z_n + \widetilde{Z}_{n+1} = 0$, and the numerical approximation is obtained from

$$Y_{n+1} = \psi_U(\widetilde{Z}_{n+1}) = \exp(\widetilde{Z}_{n+1}) \exp(\Theta) Y_n = \exp(2\Theta) Y_n.$$

The method obtained in this way is identical to Algorithm 2 of Zanna, Engø & Munthe-Kaas (2001). With the coefficients of the 2-stage Gauss method (Table II.1.1) and with $q = 1$ in (4.14) we thus get

$$\Omega_1 = -h\frac{\sqrt{3}}{6}\Bigl(A_2 - \frac{1}{2}[\Omega_2, A_2]\Bigr), \qquad \Omega_2 = h\frac{\sqrt{3}}{6}\Bigl(A_1 - \frac{1}{2}[\Omega_1, A_1]\Bigr)$$

$$Y_{n+1} = \exp(2\Theta)Y_n = \exp\Bigl(\frac{h}{2}(A_1 + A_2) - \frac{h}{4}([\Omega_1, A_1] + [\Omega_2, A_2])\Bigr)Y_n,$$

where $A_i = A\bigl(\exp(\Omega_i)\exp(\Theta)Y_n\bigr)$. This is a symmetric Lie group method of order four. We can reduce the number of commutators by replacing $\Omega_i$ in the right-hand expression with its dominating term. This yields

$$\Omega_1 = -h\frac{\sqrt{3}}{6}A_2 + \frac{h^2}{24}[A_1, A_2], \qquad \Omega_2 = h\frac{\sqrt{3}}{6}A_1 - \frac{h^2}{24}[A_1, A_2]$$

$$Y_{n+1} = \exp\Bigl(\frac{h}{2}(A_1 + A_2) - h^2\frac{\sqrt{3}}{12}[A_1, A_2]\Bigr)Y_n \qquad (4.15)$$

(cf. Exercise IV.21). Although we have neglected terms of size $\mathcal{O}(h^4)$, the method remains of order four, because the order of symmetric methods is always even.

For any linear invertible transformation $\rho$, the parametrization (4.13) satisfies

$$\rho\psi_U(\Omega) = \rho\exp(\Omega)U = \exp(\rho\Omega\rho^{-1})\rho U = \psi_{\rho U}(\sigma U)$$

with $\sigma\Omega = \rho\Omega\rho^{-1}$. Hence (4.9) holds true. If the problem (4.11) is $\rho$-reversible, i.e., $\rho A(Y) = -A(\rho Y)\rho$, then the truncated differential equation (4.14) is $\sigma$-reversible for all choices of the truncation index $q$. Moreover, after the simplifications that lead to method (4.15), the $\rho$-compatibility condition (1.4) is also satisfied.

The following variant is also proposed in Zanna, Engø & Munthe-Kaas (2001). Instead of computing $\Theta$ from the relation $Z_n + \widetilde{Z}_{n+1} = 0$, $\Theta$ is determined by

$$Z_n + \widetilde{Z}_{n+1} = h \sum_{i=1}^{s} e_i \Big( A_i - \frac{1}{2}[\Omega_i, A_i] + \ldots \Big).$$

If the coefficients satisfy $e_{s+1-i} = -e_i$, this modification gives symmetric Lie group methods.

## V.5 Energy – Momentum Methods and Discrete Gradients

> Conventional numerical methods, when applied to the ordinary differential equations of motion of classical mechanics, conserve the total energy and angular momentum only to the order of the truncation error. Since these constants of motion play a central role in mechanics, it is a great advantage to be able to conserve them exactly.
> (R.A. LaBudde & D. Greenspan 1976)

This section is concerned with numerical integrators for the equations of motion of classical mechanics which conserve both the total energy and angular momentum. Their construction is related to the concept of discrete gradients. The methods considered are symmetric, which is incidental but useful: in our view their good long-time behaviour is a consequence of their symmetry (and reversibility) more than of their exact conservation properties; see the disappointing behaviour of the non-symmetric energy- and momentum-conserving projection method in Example IV.4.4.

**A Modified Midpoint Rule.** Consider first a single particle of mass $m$ in $\mathbb{R}^3$, with position coordinates $q(t) \in \mathbb{R}^3$, moving in a central force field with potential $U(q) = V(\|q\|)$ (e.g., $V(r) = -1/r$ in the Kepler problem). With the momenta $p(t) = m \dot q(t)$, the equations of motion read

$$\dot q = \frac{1}{m} p, \qquad \dot p = -\nabla U(q) = -V'(\|q\|) \frac{q}{\|q\|}.$$

Constants of motion are the total energy $H = T(p) + U(q)$, with $T(p) = \|p\|^2/(2m)$, and the angular momentum $L = q \times p$:

$$\frac{d}{dt}(q \times p) = \dot q \times p + q \times \dot p = \frac{1}{m} p \times p - V'(\|q\|) \frac{1}{\|q\|} q \times q = 0.$$

We know from Sect. IV.2 that the implicit midpoint rule conserves the quadratic invariant $L = q \times p$, and Theorem IV.2.4 (or a simple direct calculation) shows that $L$ remains actually conserved by any modification of the form

$$q_{n+1} = q_n + \frac{h}{m} p_{n+1/2} \qquad p_{n+1/2} = \tfrac{1}{2}(p_n + p_{n+1})$$
$$p_{n+1} = p_n - \kappa h \nabla U(q_{n+1/2}) \quad \text{with} \quad q_{n+1/2} = \tfrac{1}{2}(q_n + q_{n+1}) \qquad (5.1)$$

where $\kappa$ is an arbitrary real number. Simo, Tarnow & Wong (1992) introduce this additional parameter $\kappa$ and determine it so that the total energy is conserved: $H(p_{n+1}, q_{n+1}) = H(p_n, q_n)$. With the notation $F_{n+1/2} = -\nabla U(q_{n+1/2}) = -V'(\|q_{n+1/2}\|)/\|q_{n+1/2}\| \cdot q_{n+1/2}$ we have

$$T(p_{n+1}) = T(p_n + \kappa h F_{n+1/2}) = T(p_n) + \frac{\kappa h}{m} p_{n+1/2}^T F_{n+1/2} ,$$

and hence the condition for conservation of the total energy $H = T + U$ becomes

$$\kappa \frac{h}{m} p_{n+1/2}^T F_{n+1/2} = U(q_n) - U(q_{n+1}) .$$

This gives a reasonable method even if $p_{n+1/2}^T F_{n+1/2}$ is arbitrarily close to zero. This is seen as follows: let $\sigma = -\kappa V'(\|q_{n+1/2}\|)/\|q_{n+1/2}\|$ so that $\kappa F_{n+1/2} = \sigma q_{n+1/2}$. The above condition for energy conservation then reads

$$\sigma \frac{h}{m} p_{n+1/2}^T q_{n+1/2} = V(\|q_n\|) - V(\|q_{n+1}\|) ,$$

where we note further that

$$\frac{h}{m} p_{n+1/2}^T q_{n+1/2} = (q_{n+1} - q_n)^T \tfrac{1}{2}(q_{n+1} + q_n)$$
$$= \tfrac{1}{2}(\|q_{n+1}\|^2 - \|q_n\|^2) = (\|q_{n+1}\| - \|q_n\|) \tfrac{1}{2}(\|q_{n+1}\| + \|q_n\|) .$$

These formulas give

$$\sigma = -\frac{V(\|q_{n+1}\|) - V(\|q_n\|)}{\|q_{n+1}\| - \|q_n\|} \frac{1}{\tfrac{1}{2}(\|q_{n+1}\| + \|q_n\|)} , \qquad (5.2)$$

with which method (5.1) becomes

$$q_{n+1} = q_n + \frac{h}{m} p_{n+1/2}$$
$$p_{n+1} = p_n - h \frac{V(\|q_{n+1}\|) - V(\|q_n\|)}{\|q_{n+1}\| - \|q_n\|} \frac{q_{n+1/2}}{\tfrac{1}{2}(\|q_{n+1}\| + \|q_n\|)} . \qquad (5.3)$$

This is a second-order symmetric method which conserves the total energy and the angular momentum. It evaluates only the potential $U(q) = V(\|q\|)$. The force $-\nabla U(q) = -V'(\|q\|) \frac{q}{\|q\|}$ is approximated by finite differences.

The energy- and momentum-conserving method (5.3) first appeared in LaBudde & Greenspan (1974). The method (5.1) or (5.3) is the starting point for extensions in several directions to other problems of mechanics and other methods; see Simo,

Tarnow & Wong (1992), Simo & Tarnow (1992), Lewis & Simo (1994, 1996), Gonzalez & Simo (1996), Gonzalez (1996), and Reich (1996b). In the following we consider a direct generalization to systems of particles, also given in LaBudde & Greenspan (1974).

**An Energy-Momentum Method for N-Body Systems.** We consider a system of $N$ particles interacting pairwise with potential forces which depend on the distances between the particles. As in Example IV.1.3, this is formulated as a Hamiltonian system with total energy

$$H(p,q) = \frac{1}{2}\sum_{i=1}^{N}\frac{1}{m_i} p_i^T p_i + \sum_{i=2}^{N}\sum_{j=1}^{i-1} V_{ij}\big(\|q_i - q_j\|\big). \tag{5.4}$$

As an extension of method (5.3), we consider the following method (where we now write the time step number as a superscript for notational convenience)

$$\begin{aligned} q_i^{n+1} &= q_i^n + \frac{h}{m_i} p_i^{n+1/2} \\ p_i^{n+1} &= p_i^n + h \sum_{j=1}^{N} \sigma_{ij}\big(q_i^{n+1/2} - q_j^{n+1/2}\big) \end{aligned} \tag{5.5}$$

where $p_i^{n+1/2} = \frac{1}{2}(p_i^n + p_i^{n+1})$, $q_i^{n+1/2} = \frac{1}{2}(q_i^n + q_i^{n+1})$, and for $i > j$,

$$\sigma_{ij} = \sigma_{ji} = -\frac{V_{ij}(r_{ij}^{n+1}) - V_{ij}(r_{ij}^n)}{r_{ij}^{n+1} - r_{ij}^n} \frac{1}{\frac{1}{2}(r_{ij}^n + r_{ij}^{n+1})} \tag{5.6}$$

with $r_{ij}^n = \|q_i^n - q_j^n\|$, and $\sigma_{ii} = 0$. This method has the following properties.

**Theorem 5.1 (LaBudde & Greenspan 1974).** *The method (5.5) with (5.6) is a second-order symmetric implicit method which conserves the total linear momentum $P = \sum_{i=1}^{N} p_i$, the total angular momentum $L = \sum_{i=1}^{N} q_i \times p_i$, and the total energy $H$.*

*Proof.* A comparison of (5.6) with the equations of motion shows that the method is of order 2. Similar to the continuous case (Example IV.1.3), the conservation of linear and angular momentum is obtained as a consequence of the symmetry $\sigma_{ij} = \sigma_{ji}$ for all $i,j$. For the linear momentum we have

$$\sum_{i=1}^{N} p_i^{n+1} = \sum_{i=1}^{N} p_i^n + h\sum_{i=1}^{N}\sum_{j=1}^{N} \sigma_{ij}\big(q_i^{n+1/2} - q_j^{n+1/2}\big) = \sum_{i=1}^{N} p_i^n.$$

For the proof of the conservation of the angular momentum we observe that the first equation of (5.5) together with $p_i^{n+1/2} = \frac{1}{2}(p_i^{n+1} + p_i^n)$ yields

$$\big(q_i^{n+1} - q_i^n\big) \times \big(p_i^{n+1} + p_i^n\big) = 0 \tag{5.7}$$

for all $i$. The second equation of (5.5) together with $q_i^{n+1/2} = \frac{1}{2}(q_i^{n+1} + q_i^n)$ gives

$$\sum_{i=1}^{N}(q_i^{n+1} + q_i^n) \times (p_i^{n+1} - p_i^n) = 0, \tag{5.8}$$

because $\sigma_{ij} = \sigma_{ji}$ and therefore $\sum_{i,j=1}^{N} \sigma_{ij}\, q_i^{n+1/2} \times q_j^{n+1/2} = 0$. Adding the sum over $i$ of (5.7) to the equation (5.8) proves the statement $\sum_{i=1}^{N} q_i^{n+1} \times p_i^{n+1} = \sum_{i=1}^{N} q_i^n \times p_i^n$.

It remains to show the energy conservation. Now, the kinetic energy $T(p) = \frac{1}{2}\sum_{i=1}^{N} m_i^{-1} p_i^T p_i$ at step $n+1$ is

$$T(p^{n+1}) = T(p^n) + \sum_{i=1}^{N} \left(\frac{h}{m_i} p_i^{n+1/2}\right)^T \sum_{j=1}^{N} \sigma_{ij}\left(q_i^{n+1/2} - q_j^{n+1/2}\right)$$

$$= T(p^n) + \sum_{i=1}^{N}\sum_{j=1}^{N} \sigma_{ij}\left(q_i^{n+1} - q_i^n\right)^T \left(q_i^{n+1/2} - q_j^{n+1/2}\right).$$

Using once more the symmetry $\sigma_{ij} = \sigma_{ji}$, the double sum reduces to

$$\frac{1}{2}\sum_{i=1}^{N}\sum_{j=1}^{N} \sigma_{ij} \left(\left(q_i^{n+1} - q_j^{n+1}\right) - \left(q_i^n - q_j^n\right)\right)^T \frac{1}{2}\left(\left(q_i^{n+1} - q_j^{n+1}\right) + \left(q_i^n - q_j^n\right)\right)$$

$$= \sum_{i=2}^{N}\sum_{j=1}^{i-1} \sigma_{ij} \frac{1}{2}\left(\left(r_{ij}^{n+1}\right)^2 - \left(r_{ij}^n\right)^2\right).$$

On the other hand, the change in the potential energy is

$$U(q^{n+1}) - U(q^n) = \sum_{i=2}^{N}\sum_{j=1}^{i-1} \left(V_{ij}(r_{ij}^{n+1}) - V_{ij}(r_{ij}^n)\right),$$

and hence (5.6) yields the conservation of the total energy $H = T + U$. □

**Discrete-Gradient Methods.** The methods (5.3) and (5.5) are of the form

$$y_{n+1} = y_n + h\overline{B}(y_{n+1}, y_n)\,\overline{\nabla} H(y_{n+1}, y_n) \tag{5.9}$$

where $\overline{B}(\widehat{y}, y)$ is a skew-symmetric matrix for all $\widehat{y}, y$, and $\overline{\nabla} H(\widehat{y}, y)$ is a *discrete gradient* of $H$, that is, a continuous function of $(\widehat{y}, y)$ satisfying

$$\begin{aligned}\overline{\nabla} H(\widehat{y}, y)^T (\widehat{y} - y) &= H(\widehat{y}) - H(y) \\ \overline{\nabla} H(y, y) &= \nabla H(y)\,.\end{aligned} \tag{5.10}$$

The symmetry of the methods is seen from the properties $\overline{B}(\widehat{y}, y) = \overline{B}(y, \widehat{y})$ and $\overline{\nabla} H(\widehat{y}, y) = \overline{\nabla} H(y, \widehat{y})$. For example, for method (5.3) we have, with $y = (p, q)$ and $\widehat{y} = (\widehat{p}, \widehat{q})$,

$$\overline{B}(\widehat{y}, y) = \begin{pmatrix} 0 & -I_3 \\ I_3 & 0 \end{pmatrix} \quad \text{and} \quad \overline{\nabla} H(\widehat{y}, y) = \begin{pmatrix} \frac{1}{2}(\widehat{p} + p) \\ \sigma(\widehat{q}, q) \frac{1}{2}(\widehat{q} + q) \end{pmatrix}$$

where $\sigma(\widehat{q}, q)$ is given by the expression (5.2) with $(\widehat{q}, q)$ in place of $(q_{n+1}, q_n)$ or by the corresponding limit as $\|\widehat{q}\| \to \|q\|$.

The discrete-gradient method (5.9) is consistent with the differential equation

$$\dot{y} = B(y) \nabla H(y) \tag{5.11}$$

with the skew-symmetric matrix $B(y) = \overline{B}(y, y)$. This system conserves $H$, since

$$\frac{d}{dt} H(y) = \nabla H(y)^T \dot{y} = \nabla H(y)^T B(y) \nabla H(y) = 0 ,$$

and, as was noted by Gonzalez (1996) and McLachlan, Quispel & Robidoux (1999), $H$ is also conserved by method (5.9).

**Theorem 5.2.** *The discrete-gradient method (5.9) conserves the invariant $H$ of the system (5.11).*

*Proof.* The definitions (5.10) of a discrete gradient and of the method (5.9) give

$$\begin{aligned} H(y_{n+1}) - H(y_n) &= \overline{\nabla} H(y_{n+1}, y_n)^T (y_{n+1} - y_n) \\ &= h \, \overline{\nabla} H(y_{n+1}, y_n)^T \, \overline{B}(y_{n+1}, y_n) \, \overline{\nabla} H(y_{n+1}, y_n) = 0 , \end{aligned}$$

where the last equality follows from the skew-symmetry of $\overline{B}(y_{n+1}, y_n)$. □

**Example 5.3.** The Lotka-Volterra system (I.1.1) can be written as

$$\begin{pmatrix} \dot{u} \\ \dot{v} \end{pmatrix} = \begin{pmatrix} 0 & -uv \\ uv & 0 \end{pmatrix} \nabla H(u, v)$$

with the invariant $H(u, v) = \ln u - u + 2 \ln v - v$ of (I.1.4). Possible choices of a discrete gradient are the *coordinate increment discrete gradient* (Itoh & Abe 1988)

$$\overline{\nabla} H(\widehat{u}, \widehat{v}; u, v) = \begin{pmatrix} \dfrac{H(\widehat{u}, v) - H(u, v)}{\widehat{u} - u} \\ \dfrac{H(\widehat{u}, \widehat{v}) - H(\widehat{u}, v)}{\widehat{v} - v} \end{pmatrix} \tag{5.12}$$

and the *midpoint discrete gradient* (Gonzalez 1996)

$$\overline{\nabla} H(\widehat{y}, y) = \nabla H(\overline{y}) + \frac{H(\widehat{y}) - H(y) - \nabla H(\overline{y})^T \Delta y}{\|\Delta y\|^2} \Delta y \tag{5.13}$$

with $\overline{y} = \frac{1}{2}(\widehat{y} + y)$ and $\Delta y = \widehat{y} - y$. In contrast to (5.12), the discrete gradient (5.13) yields a symmetric discretization.

A systematic study of discrete-gradient methods is given in Gonzalez (1996) and McLachlan, Quispel & Robidoux (1999).

## V.6 Exercises

1. Prove that (after a suitable permutation of the stages) the condition $c_{s+1-i} = 1 - c_i$ (for all $i$) is also necessary for a collocation method to be symmetric.
2. Prove that explicit Runge-Kutta methods cannot be symmetric.
   *Hint.* If a one-step method applied to $\dot{y} = \lambda y$ yields $y_1 = R(h\lambda)y_0$ then, a necessary condition for the symmetry of the method is $R(z)R(-z) = 1$ for all complex $z$.
3. Consider an irreducible diagonally implicit Runge-Kutta method (irreducible in the sense of Sect. VI.7.2). Prove that the condition (2.4) is necessary for the symmetry of the method. No permutation of the stages has to be performed.
4. Let $\Phi_h = \varphi_h^{[1]} \circ \varphi_h^{[2]}$, where $\varphi_t^{[i]}$ represents the exact flow of $\dot{y} = f^{[i]}(y)$. In the situation of Theorem III.3.17 prove that the local error (3.4) of the composition method (3.3) has the form

$$h^3 \left( \frac{1}{24}(6\alpha - 1)[D_2, [D_2, D_1]] + \frac{1}{12}(1 - 6\alpha + 6\alpha^2)[D_1, [D_1, D_2]] \right) Id(y),$$

where, as usual, $D_i g(y) = g'(y) f^{[i]}(y)$. The value $\alpha = 0.1932$ is found by minimizing the expression $(6\alpha - 1)^2 + 4(1 - 6\alpha + 6\alpha^2)^2$ (McLachlan 1995).
5. For the linear transformation $\rho(p, q) = (-p, q)$, consider a $\rho$-reversible problem (1.3) with scalar $p$ and $q$. Prove that every solution which crosses the $q$-axis twice is periodic.
6. Prove that if a numerical method conserves quadratic invariants (IV.2.1), then so does its adjoint.
7. For the numerical solution of $\dot{y} = A(t)y$ consider the method $y_n \mapsto y_{n+1}$ defined by $y_{n+1} = z(t_n + h)$, where $z(t)$ is the solution of

$$\dot{z} = \widehat{A}(t)z, \qquad z(t_n) = y_n,$$

and $\widehat{A}(t)$ is the interpolation polynomial based on symmetric nodes $c_1, \ldots, c_s$, i.e., $c_{s+1-i} + c_i = 1$ for all $i$.
   a) Prove that this method is symmetric.
   b) Show that $y_{n+1} = \exp(\Omega(h))y_n$ holds, where $\Omega(h)$ has an expansion in odd powers of $h$. This justifies the omission of the terms involving triple integrals in Example IV.7.4.
8. If $\Phi_h$ stands for the implicit midpoint rule, what are the Runge-Kutta coefficients of the composition method (3.8)? The general theory of Sect. III.1 gives three order conditions for order 4 (those for the trees of order 2 and 4 are automatically satisfied by the symmetry of the method). Are they compatible with the two conditions of Example 3.5?
9. Make a numerical comparison of our favourite composition methods $p6\ s9$, $p8\ s17$, and $p10\ s33$ for the Lorenz problem

$$\begin{array}{lll} y_1' = -\sigma(y_1 - y_2) & y_1(0) = 10 & \sigma = 10 \\ y_2' = -y_1 y_3 + r y_1 - y_2 & y_2(0) = -20 & r = 28 \\ y_3' = -y_1 y_2 - b y_3 & y_3(0) = 20 & b = 8/3 \end{array} \qquad (6.1)$$

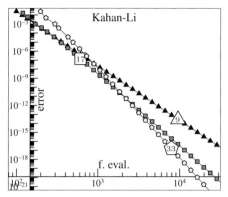

**Fig. 6.1.** Comparison of various composition methods applied to the Lorenz equations.

with exact solution

$$\begin{aligned} y_1(1) &= 8.63569270989250601793054871456 \\ y_2(1) &= 2.79866338792745705202308349547 \\ y_3(1) &= 33.36063508973142157789186299 82 \end{aligned} \quad (6.2)$$

by composing for $0 \leq t \leq 1$ the second order *symmetric splitting* scheme (see Kahan & Li 1997) which, for the time-stepping $y_i \mapsto Y_i$, is given by

$$\begin{aligned} Y_1 - y_1 &= \frac{h}{2}\bigl(-\sigma(y_1 + Y_1 - y_2 - Y_2)\bigr) \\ Y_2 - y_2 &= \frac{h}{2}\bigl(-y_1 Y_3 - Y_1 y_3 + r y_1 + r Y_1 - y_2 - Y_2\bigr) \\ Y_3 - y_3 &= \frac{h}{2}\bigl(y_1 Y_2 + Y_1 y_2 - b y_3 - b Y_3\bigr). \end{aligned} \quad (6.3)$$

This method requires, for each step, the solution of a *linear* system only. The results (see Fig. 6.1) show that the eighth order method is less brilliant here than it was in the Kepler problem.

10. *Symmetrized order conditions* (Suzuki 1992). Prove that for methods (3.8) of order four with $\gamma_i$ satisfying (3.10)

$$\sum_{k=1}^{s} \gamma_k^3 \Bigl(\sideset{}{'}\sum_{\ell=1}^{k} \gamma_\ell\Bigr)^2 = 0 \quad \Longleftrightarrow \quad \sum_{k=1}^{s} \gamma_k^3 \Bigl(\sideset{}{'}\sum_{\ell=1}^{k} \gamma_\ell\Bigr)\Bigl(\sideset{'}{}\sum_{\ell=k}^{s} \gamma_\ell\Bigr) = 0.$$

The prime after (before) a sum sign indicates that the term with highest (lowest) index is divided by 2. Prove also that the order conditions given in Suzuki (1992) for order $p \leq 8$ are equivalent to those of Example 3.5. Is this also true for order $p = 10$?

*Hint.* Use relations like $\sideset{}{'}\sum_{\ell=1}^{k} \gamma_\ell = 1 - \sideset{'}{}\sum_{\ell=k}^{s} \gamma_\ell$.

11. Let $\mathcal{M} = \{(y_1, y_2, y_3) \mid y_1^2 + y_2^2 + y_3^2 = 1\}$, and consider for $a \in \mathcal{M}$ the tangent space parametrization

$$\psi_a(z) = a + z + au_a(z),$$

where, for $z \in T_a \mathcal{M}$, the real value $u_a(z)$ is determined by the requirement $\psi_a(z) \in \mathcal{M}$. Prove that Algorithm IV.5.3, with the trapezoidal rule in the role of $\Phi_h$, is a symmetric method.

*Hint.* Since $z$ is a linear combination of $a$ and $\psi_a(z)$, it is uniquely determined by $a^T z$ (which is zero) and $\psi_a(z)^T z$.

12. (Zanna, Engø & Munthe-Kaas 2001). Verify numerically that the Lie group method (4.12) based on the implicit midpoint rule does not conserve general quadratic first integrals. One can consider the rigid body equations in the form (IV.1.5).

# Chapter VI.
# Symplectic Integration of Hamiltonian Systems

**Fig. 0.1.** Sir William Rowan Hamilton, born: 4 August 1805 in Dublin, died: 2 September 1865. Famous for research in optics, mechanics, and for the invention of quaternions.

Hamiltonian systems form the most important class of ordinary differential equations in the context of 'Geometric Numerical Integration'. An outstanding property of these systems is the symplecticity of the flow. As indicated in the following diagram,

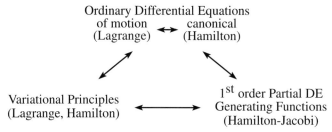

Hamiltonian theory operates in three different domains (equations of motion, partial differential equations and variational principles) which are all interconnected. Each of these viewpoints, which we will study one after the other, leads to the construction of methods preserving the symplecticity.

## VI.1 Hamiltonian Systems

Hamilton's equations appeared first, among thousands of other formulas, and inspired by previous research in optics, in Hamilton (1834). Their importance was immediately recognized by Jacobi, who stressed and extended the fundamental ideas, so that, a couple of years later, all the long history of research of Galilei, Newton, Euler and Lagrange, was, in the words of Jacobi (1842), "to be considered as an introduction". The next mile-stones in the exposition of the theory were the monumental three volumes of Poincaré (1892,1893,1899) on celestial mechanics, Siegel's "Lectures on Celestial Mechanics" (1956), English enlarged edition by Siegel & Moser (1971), and the influential book of V.I. Arnold (1989; first Russian edition 1974). Beyond that, Hamiltonian systems became fundamental in many branches of physics. One such area, the dynamics of particle accelerators, actually motivated the construction of the first symplectic integrators (Ruth 1983).

### VI.1.1 Lagrange's Equations

> Équations différentielles pour la solution de tous les problèmes de Dynamique.
> (J.-L. Lagrange 1788)

The problem of computing the dynamics of general mechanical systems began with Galilei (published 1638) and Newton's *Principia* (1687). The latter allowed one to reduce the movement of free mass points (the "mass points" being such planets as Mars or Jupiter) to the solution of differential equations (see Sect. I.2). But the movement of more complicated systems such as rigid bodies or bodies attached to each other by rods or springs, were the subject of long and difficult developments, until Lagrange (1760, 1788) found an elegant way of treating such problems in general.

Joseph-Louis Lagrange[1]

We suppose that the position of a mechanical system with $d$ degrees of freedom is described by $q = (q_1, \ldots, q_d)^T$ as *generalized coordinates* (this can be for example Cartesian coordinates, angles, arc lengths along a curve, etc.). The theory is then built upon two pillars, namely an expression

$$T = T(q, \dot q) \qquad (1.1)$$

which represents the *kinetic energy* (and which is often of the form $\frac{1}{2}\dot q^T M(q)\dot q$ where $M(q)$ is symmetric and positive definite), and by a function

---

[1] Joseph-Louis Lagrange, born: 25 January 1736 in Turin, Sardinia-Piedmont (now Italy), died: 10 April 1813 in Paris.

$$U = U(q) \tag{1.2}$$

representing the *potential energy*. Then, after denoting by

$$L = T - U \tag{1.3}$$

the corresponding *Lagrangian*, the coordinates $q_1(t), \ldots, q_d(t)$ obey the differential equations

$$\frac{d}{dt}\left(\frac{\partial L}{\partial \dot{q}}\right) = \frac{\partial L}{\partial q}, \tag{1.4}$$

which constitute the *Lagrange equations* of the system. A numerical (or analytical) integration of these equations allows one to predict the motion of any such system from given initial values ("Ce sont ces équations qui serviront à déterminer la courbe décrite par le corps $M$ et sa vitesse à chaque instant"; Lagrange 1760, p. 369).

**Example 1.1.** For a mass point of mass $m$ in $\mathbb{R}^3$ with Cartesian coordinates $x = (x_1, x_2, x_3)^T$ we have $T(\dot{x}) = m(\dot{x}_1^2 + \dot{x}_2^2 + \dot{x}_3^2)/2$. We suppose the point to move in a conservative force field $F(x) = -\nabla U(x)$. Then, the Lagrange equations (1.4) become $m\ddot{x} = F(x)$, which is Newton's second law. The equations (I.2.2) for the planetary motion are precisely of this form.

**Example 1.2 (Pendulum).** For the mathematical pendulum of Sect. I.1 we take the angle $\alpha$ as coordinate. The kinetic and potential energies are given by $T = m(\dot{x}^2 + \dot{y}^2)/2 = m\ell^2\dot{\alpha}^2/2$ and $U = mgy = -mg\ell\cos\alpha$, respectively, so that the Lagrange equations become $-mg\ell\sin\alpha - m\ell^2\ddot{\alpha} = 0$ or equivalently $\ddot{\alpha} + \frac{g}{\ell}\sin\alpha = 0$.

## VI.1.2 Hamilton's Canonical Equations

> An diese *Hamiltonsche* Form der Differentialgleichungen werden die ferneren Untersuchungen, welche den Kern dieser Vorlesung bilden, anknüpfen; das Bisherige ist als Einleitung dazu anzusehen.
> (C.G.J. Jacobi 1842, p. 143)

Hamilton (1834) simplified the structure of Lagrange's equations and turned them into a form that has remarkable symmetry, by

- introducing Poisson's variables, the conjugate *momenta*

$$p_k = \frac{\partial L}{\partial \dot{q}_k}(q, \dot{q}) \quad \text{for} \quad k = 1, \ldots, d, \tag{1.5}$$

- considering the *Hamiltonian*

$$H := p^T \dot{q} - L(q, \dot{q}) \tag{1.6}$$

as a function of $p$ and $q$, i.e., taking $H = H(p, q)$ obtained by expressing $\dot{q}$ as a function of $p$ and $q$ via (1.5).

Here it is, of course, required that (1.5) defines, for every $q$, a continuously differentiable bijection $\dot{q} \leftrightarrow p$. This map is called the *Legendre transform*.

**Theorem 1.3.** *Lagrange's equations (1.4) are equivalent to Hamilton's equations*

$$\dot p_k = -\frac{\partial H}{\partial q_k}(p,q), \qquad \dot q_k = \frac{\partial H}{\partial p_k}(p,q), \qquad k=1,\ldots,d. \tag{1.7}$$

*Proof.* The definitions (1.5) and (1.6) for the momenta $p$ and for the Hamiltonian $H$ imply that

$$\frac{\partial H}{\partial p} = \dot q^T + p^T\frac{\partial \dot q}{\partial p} - \frac{\partial L}{\partial \dot q}\frac{\partial \dot q}{\partial p} = \dot q^T,$$

$$\frac{\partial H}{\partial q} = p^T\frac{\partial \dot q}{\partial q} - \frac{\partial L}{\partial q} - \frac{\partial L}{\partial \dot q}\frac{\partial \dot q}{\partial q} = -\frac{\partial L}{\partial q}.$$

The Lagrange equations (1.4) are therefore equivalent to (1.7). □

**Case of Quadratic $T$.** In the case that $T = \frac{1}{2}\dot q^T M(q)\dot q$ is quadratic, where $M(q)$ is a symmetric and positive definite matrix, we have, for a fixed $q$, $p = M(q)\dot q$, so that the existence of the Legendre transform is established. Further, by replacing the variable $\dot q$ by $M(q)^{-1}p$ in the definition (1.6) of $H(p,q)$, we obtain

$$H(p,q) = p^T M(q)^{-1}p - L\big(q, M(q)^{-1}p\big)$$

$$= p^T M(q)^{-1}p - \frac{1}{2}p^T M(q)^{-1}p + U(q) = \frac{1}{2}p^T M(q)^{-1}p + U(q)$$

and the Hamiltonian is $H = T + U$, which is the *total energy* of the mechanical system.

In Chap. I we have seen several examples of Hamiltonian systems, e.g., the pendulum (I.1.13), the Kepler problem (I.2.2), the outer solar system (I.2.12), etc. In the following we consider Hamiltonian systems (1.7) where the Hamiltonian $H(p,q)$ is arbitrary, and so not necessarily related to a mechanical problem.

## VI.2 Symplectic Transformations

> The name "complex group" formerly advocated by me in allusion to line complexes, ... has become more and more embarrassing through collision with the word "complex" in the connotation of complex number. I therefore propose to replace it by the Greek adjective "symplectic."
> (H. Weyl (1939), p. 165)

A first property of Hamiltonian systems, already seen in Example 1.2 of Sect. IV.1, is that the Hamiltonian $H(p,q)$ is a *first integral* of the system (1.7). In this section we shall study another important property – the *symplecticity* of its flow. The basic objects to be studied are two-dimensional parallelograms lying in $\mathbb{R}^{2d}$. We suppose the parallelogram to be spanned by two vectors

$$\xi = \begin{pmatrix} \xi^p \\ \xi^q \end{pmatrix}, \qquad \eta = \begin{pmatrix} \eta^p \\ \eta^q \end{pmatrix}$$

in the $(p,q)$ space ($\xi^p, \xi^q, \eta^p, \eta^q$ are in $\mathbb{R}^d$) as

$$P = \{t\xi + s\eta \mid 0 \le t \le 1,\ 0 \le s \le 1\}.$$

In the case $d = 1$ we consider the *oriented area*

$$\text{or.area}(P) = \det\begin{pmatrix} \xi^p & \eta^p \\ \xi^q & \eta^q \end{pmatrix} = \xi^p \eta^q - \xi^q \eta^p \tag{2.1}$$

(see left picture of Fig. 2.1). In higher dimensions, we replace this by the *sum of the oriented areas of the projections of $P$ onto the coordinate planes* $(p_i, q_i)$, i.e., by

$$\omega(\xi, \eta) := \sum_{i=1}^{d} \det\begin{pmatrix} \xi_i^p & \eta_i^p \\ \xi_i^q & \eta_i^q \end{pmatrix} = \sum_{i=1}^{d}(\xi_i^p \eta_i^q - \xi_i^q \eta_i^p). \tag{2.2}$$

This defines a bilinear map acting on vectors of $\mathbb{R}^{2d}$, which will play a central role for Hamiltonian systems. In matrix notation, this map has the form

$$\omega(\xi, \eta) = \xi^T J \eta \quad \text{with} \quad J = \begin{pmatrix} 0 & I \\ -I & 0 \end{pmatrix} \tag{2.3}$$

where $I$ is the identity matrix of dimension $d$.

**Definition 2.1.** A linear mapping $A : \mathbb{R}^{2d} \to \mathbb{R}^{2d}$ is called *symplectic* if

$$A^T J A = J$$

or, equivalently, if $\omega(A\xi, A\eta) = \omega(\xi, \eta)$ for all $\xi, \eta \in \mathbb{R}^{2d}$.

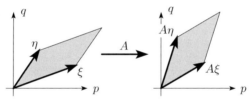

**Fig. 2.1.** Symplecticity (area preservation) of a linear mapping.

In the case $d = 1$, where the expression $\omega(\xi, \eta)$ represents the area of the parallelogram $P$, symplecticity of a linear mapping $A$ is therefore the *area preservation* of $A$ (see Fig. 2.1). In the general case $(d > 1)$, symplecticity means that the sum of the oriented areas of the projections of $P$ onto $(p_i, q_i)$ is the same as that for the transformed parallelograms $A(P)$.

We now turn our attention to nonlinear mappings. Differentiable functions can be locally approximated by linear mappings. This justifies the following definition.

**Definition 2.2.** A differentiable map $g : U \to \mathbb{R}^{2d}$ (where $U \subset \mathbb{R}^{2d}$ is an open set) is called *symplectic* if the Jacobian matrix $g'(p, q)$ is everywhere symplectic, i.e., if

$$g'(p,q)^T J g'(p,q) = J \quad \text{or} \quad \omega(g'(p,q)\xi, g'(p,q)\eta) = \omega(\xi, \eta).$$

Let us give a geometric interpretation of symplecticity for nonlinear mappings. Consider a 2-dimensional sub-manifold $M$ of the $2d$-dimensional set $U$, and suppose that it is given as the image $M = \psi(K)$ of a compact set $K \subset \mathbb{R}^2$, where

$\psi(s, t)$ is a continuously differentiable function. The manifold $M$ can then be considered as the limit of a union of small parallelograms spanned by the vectors

$$\frac{\partial \psi}{\partial s}(s, t)\, ds \quad \text{and} \quad \frac{\partial \psi}{\partial t}(s, t)\, dt.$$

For one such parallelogram we consider (as above) the sum over the oriented areas of its projections onto the $(p_i, q_i)$ plane. We then sum over all parallelograms of the manifold. In the limit this gives the expression

$$\Omega(M) = \iint_K \omega\left(\frac{\partial \psi}{\partial s}(s, t), \frac{\partial \psi}{\partial t}(s, t)\right) ds\, dt. \tag{2.4}$$

The transformation formula for double integrals implies that $\Omega(M)$ is independent of the parametrization $\psi$ of $M$.

**Lemma 2.3.** *If the mapping $g : U \to \mathbb{R}^{2d}$ is symplectic on $U$, then it preserves the expression $\Omega(M)$, i.e.,*

$$\Omega(g(M)) = \Omega(M)$$

*holds for all 2-dimensional manifolds $M$ that can be represented as the image of a continuously differentiable function $\psi$.*

*Proof.* The manifold $g(M)$ can be parametrized by $g \circ \psi$. We have

$$\Omega(g(M)) = \iint_K \omega\left(\frac{\partial(g \circ \psi)}{\partial s}(s, t), \frac{\partial(g \circ \psi)}{\partial t}(s, t)\right) ds\, dt = \Omega(M),$$

because $(g \circ \psi)'(s, t) = g'(\psi(s, t))\psi'(s, t)$ and $g$ is a symplectic transformation. □

For $d = 1$, $M$ is already a subset of $\mathbb{R}^2$ and we choose $K = M$ with $\psi$ the identity map. In this case, $\Omega(M) = \iint_M ds\, dt$ represents the area of $M$. Hence, Lemma 2.3 states that all symplectic mappings (also nonlinear ones) are *area preserving*.

We are now able to prove the main result of this section. We use the notation $y = (p, q)$, and we write the Hamiltonian system (1.7) in the form

$$\dot{y} = J^{-1} \nabla H(y), \tag{2.5}$$

where $J$ is the matrix of (2.3) and $\nabla H(y) = H'(y)^T$.

Recall that the flow $\varphi_t : U \to \mathbb{R}^{2d}$ of a Hamiltonian system is the mapping that advances the solution by time $t$, i.e., $\varphi_t(p_0, q_0) = (p(t, p_0, q_0), q(t, p_0, q_0))$, where $p(t, p_0, q_0), q(t, p_0, q_0)$ is the solution of the system corresponding to initial values $p(0) = p_0, q(0) = q_0$.

**Theorem 2.4 (Poincaré 1899).** *Let $H(p, q)$ be a twice continuously differentiable function on $U \subset \mathbb{R}^{2d}$. Then, for each fixed $t$, the flow $\varphi_t$ is a symplectic transformation wherever it is defined.*

*Proof.* The derivative $\partial \varphi_t / \partial y_0$ (with $y_0 = (p_0, q_0)$) is a solution of the variational equation which, for the Hamiltonian system (2.5), is of the form $\dot{\Psi} = J^{-1} \nabla^2 H(\varphi_t(y_0)) \Psi$, where $\nabla^2 H(p, q)$ is the Hessian matrix of $H(p, q)$ ($\nabla^2 H(p, q)$

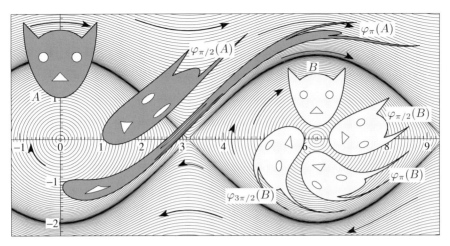

**Fig. 2.2.** Area preservation of the flow of Hamiltonian systems

is symmetric). We therefore obtain

$$\frac{d}{dt}\left(\left(\frac{\partial \varphi_t}{\partial y_0}\right)^T J\left(\frac{\partial \varphi_t}{\partial y_0}\right)\right) = \left(\frac{\partial \varphi_t}{\partial y_0}\right)'^T J\left(\frac{\partial \varphi_t}{\partial y_0}\right) + \left(\frac{\partial \varphi_t}{\partial y_0}\right)^T J\left(\frac{\partial \varphi_t}{\partial y_0}\right)'$$

$$= \left(\frac{\partial \varphi_t}{\partial y_0}\right)^T \nabla^2 H\big(\varphi_t(y_0)\big) J^{-T} J\left(\frac{\partial \varphi_t}{\partial y_0}\right) + \left(\frac{\partial \varphi_t}{\partial y_0}\right)^T \nabla^2 H\big(\varphi_t(y_0)\big) \left(\frac{\partial \varphi_t}{\partial y_0}\right) = 0,$$

because $J^T = -J$ and $J^{-T}J = -I$. Since the relation

$$\left(\frac{\partial \varphi_t}{\partial y_0}\right)^T J\left(\frac{\partial \varphi_t}{\partial y_0}\right) = J \qquad (2.6)$$

is satisfied for $t = 0$ ($\varphi_0$ is the identity map), it is satisfied for all $t$ and all $(p_0, q_0)$, as long as the solution remains in the domain of definition of $H$. □

**Example 2.5.** We illustrate this theorem with the pendulum problem (Example 1.2) using the normalization $m = \ell = g = 1$. We have $q = \alpha$, $p = \dot\alpha$, and the Hamiltonian is given by

$$H(p, q) = p^2/2 - \cos q.$$

Fig. 2.2 shows level curves of this function, and it also illustrates the area preservation of the flow $\varphi_t$. Indeed, by Theorem 2.4 and Lemma 2.3, the areas of $A$ and $\varphi_t(A)$ as well as those of $B$ and $\varphi_t(B)$ are the same, although their appearance is completely different.

We next show that symplecticity of the flow is a characteristic property for Hamiltonian systems. We call a differential equation $\dot y = f(y)$ *locally Hamiltonian*, if for every $y_0 \in U$ there exists a neighbourhood where $f(y) = J^{-1}\nabla H(y)$ for some function $H$.

**Theorem 2.6.** *Let $f : U \to \mathbb{R}^{2d}$ be continuously differentiable. Then, $\dot y = f(y)$ is locally Hamiltonian if and only if its flow $\varphi_t(y)$ is symplectic for all $y \in U$ and for all sufficiently small $t$.*

*Proof.* The necessity follows from Theorem 2.4. We therefore assume that the flow $\varphi_t$ is symplectic, and we have to prove the local existence of a function $H(y)$ such that $f(y) = J^{-1}\nabla H(y)$. Differentiating (2.6) and using the fact that $\partial\varphi_t/\partial y_0$ is a solution of the variational equation $\dot{\Psi} = f'(\varphi_t(y_0))\Psi$, we obtain

$$\frac{d}{dt}\left(\left(\frac{\partial\varphi_t}{\partial y_0}\right)^T J \left(\frac{\partial\varphi_t}{\partial y_0}\right)\right) = \left(\frac{\partial\varphi_t}{\partial y_0}\right)\left(f'(\varphi_t(y_0))^T J + Jf'(\varphi_t(y_0))\right)\left(\frac{\partial\varphi_t}{\partial y_0}\right) = 0.$$

Putting $t = 0$, it follows from $J = -J^T$ that $Jf'(y_0)$ is a symmetric matrix for all $y_0$. The Integrability Lemma 2.7 below shows that $Jf(y)$ can be written as the gradient of a function $H(y)$. □

The following integrability condition for the existence of a potential was already known to Euler and Lagrange (see e.g., Euler's *Opera Omnia*, vol. 19. p. 2-3, or Lagrange (1760), p. 375).

**Lemma 2.7 (Integrability Lemma).** *Let $D \subset \mathbb{R}^n$ be open and $f : D \to \mathbb{R}^n$ be continuously differentiable, and assume that the Jacobian $f'(y)$ is symmetric for all $y \in D$. Then, for every $y_0 \in D$ there exists a neighbourhood and a function $H(y)$ such that*

$$f(y) = \nabla H(y) \qquad (2.7)$$

*on this neighbourhood. In other words, the differential form $f_1(y)\,dy_1 + \ldots + f_n(y)\,dy_n = dH$ is a total differential.*

*Proof.* Assume $y_0 = 0$, and consider a ball around $y_0$ which is contained in $D$. On this ball we define

$$H(y) = \int_0^1 y^T f(ty)\,dt + \text{Const}.$$

Differentiation with respect to $y_k$, and using the symmetry assumption $\partial f_i/\partial y_k = \partial f_k/\partial y_i$ yields

$$\frac{\partial H}{\partial y_k}(y) = \int_0^1 \left(f_k(ty) + y^T \frac{\partial f}{\partial y_k}(ty)t\right)dt = \int_0^1 \frac{d}{dt}\left(tf_k(ty)\right)dt = f_k(y),$$

which proves the statement. □

For $D = \mathbb{R}^{2d}$ or for star-shaped regions $D$, the above proof shows that the function $H$ of Lemma 2.7 is globally defined. Hence the Hamiltonian of Theorem 2.6 is also globally defined in this case. This remains valid for simply connected sets $D$. A counter-example, which shows that the existence of a global Hamiltonian in Theorem 2.6 is not true for general $D$, is given in Exercise 7.

An important property of symplectic transformations, which goes back to Jacobi (1836, "Theorem X"), is that they preserve the Hamiltonian character of the differential equation. Such transformations have been termed *canonical* since the 19th century. The next theorem shows that canonical and symplectic transformations are the same.

**Theorem 2.8.** *Let $\psi : U \to V$ be a change of coordinates such that $\psi$ and $\psi^{-1}$ are continuously differentiable functions. If $\psi$ is symplectic, the Hamiltonian system $\dot{y} = J^{-1}\nabla H(y)$ becomes in the new variables $z = \psi(y)$*

$$\dot{z} = J^{-1}\nabla K(z) \quad \text{with} \quad K(z) = H(y). \tag{2.8}$$

*Conversely, if $\psi$ transforms every Hamiltonian system to another Hamiltonian system via (2.8), then $\psi$ is symplectic.*

*Proof.* Since $\dot{z} = \psi'(y)\dot{y}$ and $\psi'(y)^T \nabla K(z) = \nabla H(y)$, the Hamiltonian system $\dot{y} = J^{-1}\nabla H(y)$ becomes

$$\dot{z} = \psi'(y) J^{-1} \psi'(y)^T \nabla K(z) \tag{2.9}$$

in the new variables. It is equivalent to (2.8) if

$$\psi'(y) J^{-1} \psi'(y)^T = J^{-1}. \tag{2.10}$$

Multiplying this relation from the right by $\psi'(y)^{-T}$ and from the left by $\psi'(y)^{-1}$ and then taking its inverse yields $J = \psi'(y)^T J \psi'(y)$, which shows that (2.10) is equivalent to the symplecticity of $\psi$.

For the inverse relation we note that (2.9) is Hamiltonian for all $K(z)$ if and only if (2.10) holds. □

## VI.3 First Examples of Symplectic Integrators

Since symplecticity is a characteristic property of Hamiltonian systems (Theorem 2.6), it is natural to search for numerical methods that share this property. Pioneering work on symplectic integration is due to de Vogelaere (1956), Ruth (1983), and Feng Kang (1985).

**Definition 3.1.** A numerical one-step method is called *symplectic* if the one-step map $y_1 = \Phi_h(y_0)$ is symplectic whenever the method is applied to a smooth Hamiltonian system.

**Example 3.2.** We consider the pendulum problem of Example 2.5 with the same initial sets as in Fig. 2.2. We apply six different numerical methods to this problem: the explicit Euler method (I.1.5), the symplectic Euler method (I.1.9), and the implicit Euler

Feng Kang[2]

---

[2] Feng Kang, born: 9 September 1920 in Nanjing (China), died: 17 August 1993 in Beijing; picture obtained from Yuming Shi with the help of Yifa Tang.

176    VI. Symplectic Integration of Hamiltonian Systems

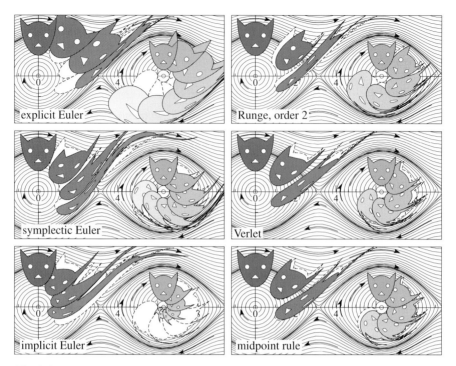

**Fig. 3.1.** Area preservation of numerical methods for the pendulum; same initial sets as in Fig. 2.2; first order methods (left column): $h = \pi/4$; second order methods (right column): $h = \pi/3$; dashed: exact flow.

method (I.1.6), as well as the second order method of Runge (II.1.3) (the right one), the Störmer/Verlet scheme (I.3.6), and the implicit midpoint rule (I.1.7). For two sets of initial values $(p_0, q_0)$ we compute several steps with step size $h = \pi/4$ for the first order methods, and $h = \pi/3$ for the second order methods. One clearly observes in Fig. 3.1 that the explicit Euler, the implicit Euler and the second order explicit method of Runge are not symplectic (not area preserving). We shall prove below that the other methods are symplectic. A different proof of their symplecticity (using generating functions) will be given in Sect. VI.5.

**Theorem 3.3.** *The so-called symplectic Euler method*

$$p_{n+1} = p_n - h\frac{\partial H}{\partial q}(p_{n+1}, q_n), \qquad q_{n+1} = q_n + h\frac{\partial H}{\partial p}(p_{n+1}, q_n) \qquad (3.1)$$

*is a symplectic method of order* 1.

*Proof.* Differentiation of (3.1) with respect to $(p_n, q_n)$ yields

$$\begin{pmatrix} I + hH_{qp}^T & 0 \\ -hH_{pp} & I \end{pmatrix} \begin{pmatrix} \partial(p_{n+1}, q_{n+1}) \\ \partial(p_n, q_n) \end{pmatrix} = \begin{pmatrix} I & -hH_{qq} \\ 0 & I + hH_{qp} \end{pmatrix},$$

where the matrices $H_{qp}, H_{pp}, \ldots$ of partial derivatives are all evaluated at $(p_{n+1}, q_n)$. This relation allows us to compute $\frac{\partial(p_{n+1}, q_{n+1})}{\partial(p_n, q_n)}$ and to check in a straightforward way the symplecticity condition $\left(\frac{\partial(p_{n+1}, q_{n+1})}{\partial(p_n, q_n)}\right)^T J \left(\frac{\partial(p_{n+1}, q_{n+1})}{\partial(p_n, q_n)}\right) = J$. □

The same proof shows that the adjoint method of (3.1),

$$p_{n+1} = p_n - h \frac{\partial H}{\partial q}(p_n, q_{n+1}), \qquad q_{n+1} = q_n + h \frac{\partial H}{\partial p}(p_n, q_{n+1}) \tag{3.2}$$

is also symplectic.

**Theorem 3.4.** *The implicit midpoint rule*

$$y_{n+1} = y_n + h J^{-1} \nabla H\big((y_{n+1} + y_n)/2\big) \tag{3.3}$$

*is a symplectic method of order 2.*

*Proof.* Differentiation of (3.3) yields

$$\left(I - \frac{h}{2} J^{-1} \nabla^2 H\right) \left(\frac{\partial y_{n+1}}{\partial y_n}\right) = \left(I + \frac{h}{2} J^{-1} \nabla^2 H\right).$$

Again it is straightforward to verify that $\left(\frac{\partial y_{n+1}}{\partial y_n}\right)^T J \left(\frac{\partial y_{n+1}}{\partial y_n}\right) = J$. Due to its symmetry, the midpoint rule is known to be of order 2 (see Theorem II.3.2). □

**Theorem 3.5.** *The Störmer/Verlet scheme (I.3.6) is a symplectic method of order 2.*

*Proof.* This is an immediate consequence of the fact that the Störmer/Verlet scheme is the composition of the symplectic Euler method with its adjoint (cf. Example II.4.3). Order 2 follows from its symmetry. □

The next two theorems are a consequence of the fact that the composition of symplectic transformations is again symplectic. They are also used to prove the existence of symplectic methods of arbitrarily high order, and to explain why the theory of composition methods of Chapters II and III is so important for geometric integration.

**Theorem 3.6.** *Let $\Phi_h$ denote the symplectic Euler method (3.1). Then, the composition method (II.4.6) is symplectic for every choice of the parameters $\alpha_i, \beta_i$.*
*If $\widehat{\Phi}_h$ is symplectic and symmetric (e.g., the implicit midpoint rule or the Störmer/Verlet scheme), then the composition method (V.3.8) is symplectic too.* □

**Theorem 3.7.** *Assume that the Hamiltonian is given by $H(y) = H_1(y) + H_2(y)$, and consider the splitting*

$$\dot{y} = J^{-1} \nabla H(y) = J^{-1} \nabla H_1(y) + J^{-1} \nabla H_2(y).$$

*The splitting method (II.5.6) is then symplectic.* □

## VI.4 Symplectic Runge-Kutta Methods

The systematic study of symplectic Runge-Kutta methods started around 1988, and a complete characterization has been found independently by Lasagni (1988) (using the approach of generating functions), and by Sanz-Serna (1988) and Suris (1988) (using the ideas of the classical papers of Burrage & Butcher (1979) and Crouzeix (1979) on algebraic stability).

### VI.4.1 Criterion of Symplecticity

We follow the approach of Bochev & Scovel (1994), which is based on the following important lemma.

**Lemma 4.1.** *For Runge-Kutta methods and for partitioned Runge-Kutta methods the following diagram commutes:*

$$
\begin{array}{ccc}
\dot{y} = f(y),\ y(0) = y_0 & \longrightarrow & \begin{array}{l}\dot{y} = f(y),\ y(0) = y_0 \\ \dot{\Psi} = f'(y)\Psi,\ \Psi(0) = I\end{array} \\
\Big\downarrow \text{method} & & \Big\downarrow \text{method} \\
\{y_n\} & \longrightarrow & \{y_n, \Psi_n\}
\end{array}
$$

*(horizontal arrows mean a differentiation with respect to $y_0$). Therefore, the numerical result $y_n, \Psi_n$, obtained from applying the method to the problem augmented by its variational equation, is equal to the numerical solution for $\dot{y} = f(y)$ augmented by its derivative $\Psi_n = \partial y_n/\partial y_0$.*

*Proof.* The result is proved by implicit differentiation. Let us illustrate this for the explicit Euler method

$$y_{n+1} = y_n + hf(y_n).$$

We consider $y_n$ and $y_{n+1}$ as functions of $y_0$, and we differentiate with respect to $y_0$ the equation defining the numerical method. For the Euler method this gives

$$\frac{\partial y_{n+1}}{\partial y_0} = \frac{\partial y_n}{\partial y_0} + hf'(y_n)\frac{\partial y_n}{\partial y_0},$$

which is exactly the relation that we get from applying the method to the variational equation. Since $\partial y_0/\partial y_0 = I$, we have $\partial y_n/\partial y_0 = \Psi_n$ for all $n$. □

The main observation now is that the symplecticity condition (2.6) is a quadratic first integral of the variational equation: we write the Hamiltonian system together with its variational equation as

$$\dot{y} = J^{-1}\nabla H(y), \qquad \dot{\Psi} = J^{-1}\nabla^2 H(y)\Psi. \tag{4.1}$$

It follows from

$$(J^{-1}\nabla^2 H(y)\Psi)^T J\Psi + \Psi^T J(J^{-1}\nabla^2 H(y)\Psi) = 0$$

(see also the proof of Theorem 2.4) that $\Psi^T J\Psi$ is a quadratic first integral of the augmented system (4.1).

Therefore, every Runge-Kutta method that preserves quadratic first integrals, is a symplectic method. From Theorem IV.2.1 and Theorem IV.2.2 we thus obtain the following results.

**Theorem 4.2.** *The Gauss collocation methods of Sect. II.1.3 are symplectic.* □

**Theorem 4.3.** *If the coefficients of a Runge-Kutta method satisfy*

$$b_i a_{ij} + b_j a_{ji} = b_i b_j \qquad \text{for all } i, j = 1, \ldots, s, \tag{4.2}$$

*then it is symplectic.* □

Similar to the situation in Theorem V.2.4, diagonally implicit, symplectic Runge-Kutta methods are composition methods.

**Theorem 4.4.** *A diagonally implicit Runge-Kutta method satisfying the symplecticity condition (4.2) and $b_i \neq 0$ is equivalent to the composition*

$$\Phi^M_{b_s h} \circ \ldots \circ \Phi^M_{b_2 h} \circ \Phi^M_{b_1 h},$$

*where $\Phi^M_h$ stands for the implicit midpoint rule.*

*Proof.* For $i = j$ condition (4.2) gives $a_{ii} = b_i/2$ and, together with $a_{ji} = 0$ (for $i > j$), implies $a_{ij} = b_j$. This proves the statement. □

The assumption "$b_i \neq 0$" is not restrictive in the sense that for diagonally implicit Runge-Kutta methods satisfying (4.2) the internal stages corresponding to "$b_i = 0$" do not influence the numerical result and can be removed.

To understand the symplecticity of partitioned Runge-Kutta methods, we write the solution $\Psi$ of the variational equation as

$$\Psi = \begin{pmatrix} \Psi^p \\ \Psi^q \end{pmatrix}.$$

Then, the Hamiltonian system together with its variational equation (4.1) is a partitioned system with variables $(p, \Psi^p)$ and $(q, \Psi^q)$. Every component of

$$\Psi^T J\Psi = (\Psi^p)^T \Psi^q - (\Psi^q)^T \Psi^p$$

is of the form (IV.2.5), so that Theorem IV.2.3 and Theorem IV.2.4 yield the following results.

**Theorem 4.5.** *The Lobatto IIIA - IIIB pair is a symplectic method.* □

**Theorem 4.6.** *If the coefficients of a partitioned Runge-Kutta method (II.2.2) satisfy*

$$b_i \widehat{a}_{ij} + \widehat{b}_j a_{ji} = b_i \widehat{b}_j \quad \text{for } i,j = 1,\ldots,s, \qquad (4.3)$$

$$b_i = \widehat{b}_i \quad \text{for } i = 1,\ldots,s, \qquad (4.4)$$

*then it is symplectic.*

*If the Hamiltonian is of the form $H(p,q) = T(p) + U(q)$, i.e., it is separable, then the condition (4.3) alone implies the symplecticity of the numerical flow.* □

We have seen in Sect. V.2.2 that within the class of partitioned Runge-Kutta methods it is possible to get explicit, symmetric methods for separable systems $\dot{y} = f(z)$, $\dot{z} = g(y)$. A similar result holds for symplectic methods. However, as in Theorem V.2.6, such methods are not more general than composition or splitting methods as considered in Sect. II.5. This has first been observed by Okunbor & Skeel (1992).

**Theorem 4.7.** *Consider a partitioned Runge-Kutta method based on two diagonally implicit methods (i.e., $a_{ji} = \widehat{a}_{ji} = 0$ for $i > j$), assume $a_{ii} \cdot \widehat{a}_{ii} = 0$ for all $i$, and apply it to a separable Hamiltonian system with $H(p,q) = T(p) + U(q)$. If (4.3) holds, then the numerical result is the same as that obtained from the splitting method (II.5.6).*

*By (II.5.7), such a method is equivalent to a composition of symplectic Euler steps.*

*Proof.* We first notice that the stage values $k_i = f(Z_i)$ (for $i$ with $b_i = 0$) and $\ell_i = g(Y_i)$ (for $i$ with $\widehat{b}_i = 0$) do not influence the numerical solution and can be removed. This yields a scheme with non-zero $b_i$ and $\widehat{b}_i$, but with possibly non-square matrices $(a_{ij})$ and $(\widehat{a}_{ij})$.

Since the method is explicit for separable problems, one of the reduced matrices $(a_{ij})$ or $(\widehat{a}_{ij})$ has a row consisting only of zeros. Assume that it is the first row of $(a_{ij})$, so that $a_{1j} = 0$ for all $j$. The symplecticity condition thus implies $\widehat{a}_{i1} = \widehat{b}_1 \neq 0$ for all $i \geq 1$, and $a_{i1} = b_1 \neq 0$ for $i \geq 2$. This then yields $\widehat{a}_{22} \neq 0$, because otherwise the first two stages of $(\widehat{a}_{ij})$ would be identical and one could be removed. By our assumption we get $a_{22} = 0$, $\widehat{a}_{i2} = \widehat{b}_2 \neq 0$ for $i \geq 2$, and $a_{i2} = b_2$ for $i \geq 3$. Continuing this procedure we see that the method becomes

$$\ldots \circ \varphi^{[2]}_{\widehat{b}_2 h} \circ \varphi^{[1]}_{b_2 h} \circ \varphi^{[2]}_{\widehat{b}_1 h} \circ \varphi^{[1]}_{b_1 h},$$

where $\varphi^{[1]}_t$ and $\varphi^{[2]}_t$ are the exact flows corresponding to the Hamiltonians $T(p)$ and $U(q)$, respectively. □

The necessity of the conditions of Theorem 4.3 and Theorem 4.6 for symplectic (partitioned) Runge-Kutta methods will be discussed at the end of this chapter in Sect. VI.7.3.

A second order differential equation $\ddot{y} = g(y)$, augmented by its variational equation, is again of this special form. Furthermore, the diagram of Lemma 4.1 commutes for Nyström methods, so that Theorem IV.2.5 yields the following result originally obtained by Suris (1988, 1989).

**Theorem 4.8.** *If the coefficients of a Nyström method (IV.2.11) satisfy*

$$\begin{aligned} B_i &= b_i(1-c_i) &\text{for } i=1,\ldots,s, \\ b_i(B_j - a_{ij}) &= b_j(B_i - a_{ji}) &\text{for } i,j=1,\ldots,s, \end{aligned} \quad (4.5)$$

*then it is symplectic.* □

## VI.4.2 Connection Between Symplectic and Symmetric Methods

There exist symmetric methods that are not symplectic, and there exist symplectic methods that are not symmetric. For example, the *trapezoidal rule*

$$y_1 = y_0 + \frac{h}{2}\Big(f(y_0) + f(y_1)\Big) \qquad (4.6)$$

is symmetric, but it does not satisfy the condition (4.2) for symplecticity. In fact, this is true of all Lobatto IIIA methods (see Example II.2.2). On the other hand, any composition $\Phi_{\gamma_1 h} \circ \Phi_{\gamma_2 h}$ ($\gamma_1 + \gamma_2 = 1$) of symplectic methods is symplectic but symmetric only if $\gamma_1 = \gamma_2$.

However, for (non-partitioned) Runge-Kutta methods and for quadratic Hamiltonians $H(y) = \frac{1}{2} y^T C y$ ($C$ is a symmetric real matrix), where the corresponding system (2.5) is linear,

$$\dot{y} = J^{-1} C y, \qquad (4.7)$$

we shall see that both concepts are equivalent.

A Runge-Kutta method, applied with step size $h$ to a linear system $\dot{y} = Ly$, is equivalent to

$$y_1 = R(hL) y_0, \qquad (4.8)$$

where the rational function $R(z)$ is given by

$$R(z) = 1 + z b^T (I - zA)^{-1} \mathbb{1}, \qquad (4.9)$$

$A = (a_{ij})$, $b^T = (b_1, \ldots, b_s)$, and $\mathbb{1}^T = (1, \ldots, 1)$. The function $R(z)$ is called the *stability function* of the method, and it is familiar to us from the study of stiff differential equations (see e.g., Hairer & Wanner (1996), Chap. IV.3).

For the explicit Euler method, the implicit Euler method and the implicit midpoint rule, the stability function $R(z)$ is given by

$$1 + z, \qquad \frac{1}{1-z}, \qquad \frac{1+z/2}{1-z/2}.$$

**Theorem 4.9.** *For Runge-Kutta methods the following statements are equivalent:*
- *the method is symmetric for linear problems $\dot{y} = Ly$;*
- *the method is symplectic for problems (4.7) with symmetric $C$;*
- *the stability function satisfies $R(-z)R(z) = 1$ for all complex $z$.*

*Proof.* The method $y_1 = R(hL)y_0$ is symmetric, if and only if $y_0 = R(-hL)y_1$ holds for all initial values $y_0$. But this is equivalent to $R(-hL)R(hL) = I$.

Since $\Phi'_h(y_0) = R(hL)$, symplecticity of the method for the problem (4.7) is defined by $R(hJ^{-1}C)^T JR(hJ^{-1}C) = J$. For $R(z) = P(z)/Q(z)$ this is equivalent to
$$P(hJ^{-1}C)^T JP(hJ^{-1}C) = Q(hJ^{-1}C)^T JQ(hJ^{-1}C). \tag{4.10}$$

By the symmetry of $C$, the matrix $L := J^{-1}C$ satisfies $L^T J = -JL$ and hence also $(L^k)^T J = J(-L)^k$ for $k = 0, 1, 2, \ldots$. Consequently, (4.10) is equivalent to
$$P(-hJ^{-1}C)P(hJ^{-1}C) = Q(-hJ^{-1}C)Q(hJ^{-1}C),$$
which is nothing other than $R(-hJ^{-1}C)R(hJ^{-1}C) = I$. □

## VI.5 Generating Functions

> ... by which the study of the motions of all free systems of attracting or repelling points is reduced to the search and differentiation of one central relation, or characteristic function. (W.R. Hamilton 1834)

> Professor Hamilton hat ... das merkwürdige Resultat gefunden, dass ... sich die Integralgleichungen der Bewegung ... sämmtlich durch die partiellen Differentialquotienten einer einzigen Function darstellen lassen. (C.G.J. Jacobi 1837)

We enter here the second heaven of Hamiltonian theory, the realm of partial differential equations and generating functions. The starting point of this theory was the discovery of Hamilton, that a certain "characteristic" function $S$ allows one to describe the movement of the problem entirely and that $S$ is the solution of a partial differential equation, now called the *Hamilton-Jacobi differential equation*.

It was noticed later, especially by Siegel (see Siegel & Moser 1971, §3), that such a function $S$ is directly connected to any symplectic map. It received the name *generating function*.

### VI.5.1 Existence of Generating Functions

We now consider a fixed Hamiltonian system and a fixed time interval and denote by the column vectors $p$ and $q$ the *initial values* $p_1, \ldots, p_d$ and $q_1, \ldots, q_d$ at $t_0$ of a trajectory. The *final values* at $t_1$ are written as $P$ and $Q$. We thus have a mapping $(p,q) \mapsto (P,Q)$ which, as we know, is symplectic on an open set $U$.

**Theorem 5.1.** *A mapping $\varphi : (p,q) \mapsto (P,Q)$ is symplectic if and only if there exists locally a function $S(p,q)$ such that*
$$P^T dQ - p^T dq = dS. \tag{5.1}$$

*This means that $P^T dQ - p^T dq$ is a total differential.*

*Proof.* We split the Jacobian of $\varphi$ into the natural $2 \times 2$ block matrix

$$\frac{\partial(P,Q)}{\partial(p,q)} = \begin{pmatrix} P_p & P_q \\ Q_p & Q_q \end{pmatrix}.$$

Inserting this into (2.6) and multiplying out shows that the three conditions

$$P_p^T Q_p = Q_p^T P_p, \qquad P_p^T Q_q - I = Q_p^T P_q, \qquad Q_q^T P_q = P_q^T Q_q \tag{5.2}$$

are equivalent to symplecticity. We now insert $dQ = Q_p\, dp + Q_q\, dq$ into the left-hand side of (5.1) and obtain

$$\left(P^T Q_p, \ P^T Q_q - p^T\right) \begin{pmatrix} dp \\ dq \end{pmatrix} = \begin{pmatrix} Q_p^T P \\ Q_q^T P - p \end{pmatrix}^T \begin{pmatrix} dp \\ dq \end{pmatrix}.$$

To apply the Integrability Lemma 2.7, we just have to verify the symmetry of the Jacobian of the coefficient vector,

$$\begin{pmatrix} Q_p^T P_p & Q_p^T P_q \\ Q_q^T P_p - I & Q_q^T P_q \end{pmatrix} + \sum_i P_i \frac{\partial^2 Q_i}{\partial(p,q)^2}. \tag{5.3}$$

Since the Hessians of $Q_i$ are symmetric anyway, it is immediately clear that the symmetry of the matrix (5.3) is equivalent to the symplecticity conditions (5.2). □

**Reconstruction of the Symplectic Map from $S$.** Up to now we have considered all functions as depending on $p$ and $q$. The essential idea now is to introduce new coordinates; namely (5.1) suggests using $z = (q, Q)$ instead of $y = (p, q)$. This is a well-defined local change of coordinates $y = \psi(z)$ if $p$ can be expressed in terms of the coordinates $(q, Q)$, which is possible by the implicit function theorem if $\frac{\partial Q}{\partial p}$ is invertible. Abusing our notation we again write $S(q, Q)$ for the transformed function $S(\psi(z))$. Then, by comparing the coefficients of $dS = \frac{\partial S(q,Q)}{\partial q} dq + \frac{\partial S(q,Q)}{\partial Q} dQ$ with (5.1), we arrive at

$$P = \frac{\partial S}{\partial Q}(q, Q), \qquad p = -\frac{\partial S}{\partial q}(q, Q). \tag{5.4}$$

If the transformation $(p, q) \mapsto (P, Q)$ is symplectic, it can be reconstructed from the scalar function $S(q, Q)$ by the relations (5.4). By Theorem 5.1 the converse is also true: *any sufficiently smooth and nondegenerate function $S(q, Q)$ "generates" via (5.4) a symplectic mapping $(p, q) \mapsto (P, Q)$.* This gives us a powerful tool for creating symplectic methods.

**Mappings Close to the Identity.** We are mainly interested in the situation where the mapping $(p, q) \mapsto (P, Q)$ is close to the identity. In this case, the choices $(p, Q)$ or $(P, q)$ or $((P+p)/2, (Q+q)/2)$ of independent variables are more convenient and lead to the following characterizations.

**Lemma 5.2.** *Let $(p, q) \mapsto (P, Q)$ be a smooth transformation, close to the identity. It is symplectic if and only if one of the following conditions holds locally:*
- $Q^T dP + p^T dq = d(P^T q + S^1)$ *for some function* $S^1(P, q)$;
- $P^T dQ + q^T dp = d(p^T Q - S^2)$ *for some function* $S^2(p, Q)$;
- $(Q - q)^T d(P + p) - (P - p)^T d(Q + q) = 2\, dS^3$
  *for some function* $S^3\big((P + p)/2, (Q + q)/2\big)$.

*Proof.* Using $d(P^T Q) = P^T dQ + Q^T dP$, the first characterization with $S^1 = P^T(Q - q) - S$ is equivalent to (5.1). For the second characterization we use $d(p^T q) = p^T dq + q^T dp$. The last one follows from the fact that (5.1) is equivalent to $(Q - q)^T d(P + p) - (P - p)^T d(Q + q) = d\big((P + p)^T(Q - q) - 2S\big)$. □

The generating functions $S^1$, $S^2$, and $S^3$ have been chosen such that we obtain the identity mapping when they are replaced with zero. Comparing the coefficient functions of $dq$ and $dP$ in the first characterization of Lemma 5.2, we obtain

$$p = P + \frac{\partial S^1}{\partial q}(P, q), \qquad Q = q + \frac{\partial S^1}{\partial P}(P, q). \tag{5.5}$$

Whatever the scalar function $S^1(P, q)$ is, the relation (5.5) defines a symplectic transformation $(p, q) \mapsto (P, Q)$. For $S^1(P, q) := hH(P, q)$ we recognize the symplectic Euler method (I.1.9). This is an elegant proof of the symplecticity of this method. The second characterization leads to the adjoint of the symplectic Euler method.

The third characterization of Lemma 5.2 can be written as

$$\begin{aligned} P &= p - \partial_2 S^3\big((P + p)/2, (Q + q)/2\big), \\ Q &= q + \partial_1 S^3\big((P + p)/2, (Q + q)/2\big), \end{aligned} \tag{5.6}$$

which, for $S^3 = hH$, is nothing other than the implicit midpoint rule (I.1.7) applied to a Hamiltonian system. We have used the notation $\partial_1$ and $\partial_2$ for the derivative with respect to the first and second argument, respectively. The system (5.6) can also be written in compact form as

$$Y = y + J^{-1} \nabla S^3\big((Y + y)/2\big), \tag{5.7}$$

where $Y = (P, Q)$, $y = (p, q)$, $S^3(w) = S^3(u, v)$ with $w = (u, v)$, and $J$ is the matrix of (2.3).

## VI.5.2 Generating Function for Symplectic Runge-Kutta Methods

We have just seen that all symplectic transformations can be written in terms of generating functions. What are these generating functions for symplectic Runge-Kutta methods? The following result, proved by Lasagni in an unpublished manuscript (with the same title as the note Lasagni (1988)), gives an alternative proof for Theorem 4.3.

## VI.5 Generating Functions

**Theorem 5.3.** *Suppose that*

$$b_i a_{ij} + b_j a_{ji} = b_i b_j \quad \text{for all } i,j \tag{5.8}$$

*(see Theorem 4.3). Then, the Runge-Kutta method*

$$\begin{aligned} P &= p - h\sum_{i=1}^{s} b_i H_q(P_i, Q_i), & P_i &= p - h\sum_{j=1}^{s} a_{ij} H_q(P_j, Q_j), \\ Q &= q + h\sum_{i=1}^{s} b_i H_p(P_i, Q_i), & Q_i &= q + h\sum_{j=1}^{s} a_{ij} H_p(P_j, Q_j) \end{aligned} \tag{5.9}$$

*can be written as (5.5) with*

$$S^1(P, q, h) = h\sum_{i=1}^{s} b_i H(P_i, Q_i) - h^2 \sum_{i,j=1}^{s} b_i a_{ij} H_q(P_i, Q_i)^T H_p(P_j, Q_j). \tag{5.10}$$

*Proof.* We first differentiate $S^1(P,q,h)$ with respect to $q$. Using the abbreviations $H[i] = H(P_i, Q_i)$, $H_p[i] = H_p(P_i, Q_i)$, ..., we obtain

$$\begin{aligned} \frac{\partial}{\partial q}\left(\sum_i b_i H[i]\right) &= \sum_i b_i H_p[i]^T \left(\frac{\partial p}{\partial q} - h\sum_j a_{ij}\frac{\partial}{\partial q} H_q[j]\right) \\ &+ \sum_i b_i H_q[i]^T \left(I + h\sum_j a_{ij}\frac{\partial}{\partial q} H_p[j]\right). \end{aligned}$$

With

$$0 = \frac{\partial p}{\partial q} - h\sum_j b_j \frac{\partial}{\partial q} H_q[j]$$

(this is obtained by differentiating the first relation of (5.9)), Leibniz' rule

$$\frac{\partial}{\partial q}\left(H_q[i]^T H_p[j]\right) = H_q[i]^T \frac{\partial}{\partial q} H_p[j] + H_p[j]^T \frac{\partial}{\partial q} H_q[i]$$

and the condition (5.8) therefore yield the first relation of

$$\frac{\partial S^1(P, q, h)}{\partial q} = h\sum_i b_i H_q[i], \qquad \frac{\partial S^1(P, q, h)}{\partial P} = h\sum_i b_i H_p[i].$$

The second relation is proved in the same way. This shows that the Runge-Kutta formulas (5.9) are equivalent to (5.5). □

It is interesting to note that, whereas Lemma 5.2 guarantees the *local* existence of a generating function $S^1$, the explicit formula (5.10) shows that for Runge-Kutta methods this generating function is *globally* defined. This means that it is well-defined in the same region where the Hamiltonian $H(p,q)$ is defined.

**Theorem 5.4.** *A partitioned Runge-Kutta method (II.2.2), satisfying the symplecticity conditions (4.3) and (4.4), is equivalent to (5.5) with*

$$S^1(P, q, h) = h \sum_{i=1}^{s} b_i H(P_i, Q_i) - h^2 \sum_{i,j=1}^{s} b_i \widehat{a}_{ij} H_q(P_i, Q_i)^T H_p(P_j, Q_j).$$

*If the Hamiltonian is of the form* $H(p, q) = T(p) + U(q)$, *i.e., it is separable, then the condition (4.3) alone implies that the method is of the form (5.5) with*

$$S^1(P, q, h) = h \sum_{i=1}^{s} \Big( b_i U(Q_i) + \widehat{b}_i T(P_i) \Big) - h^2 \sum_{i,j=1}^{s} b_i \widehat{a}_{ij} U_q(Q_i)^T T_p(P_j,).$$

*Proof.* This is a straightforward extension of the proof of the previous theorem. □

### VI.5.3 The Hamilton-Jacobi Partial Differential Equation

C.G.J. Jacobi[3]

We now return to the above construction of $S$ for a symplectic transformation $(p, q) \mapsto (P, Q)$ (see Theorem 5.1). This time, however, we imagine the point $P(t), Q(t)$ to move in the flow of the Hamiltonian system (1.7). We wish to determine a smooth generating function $S(q, Q, t)$, now also depending on $t$, which generates via (5.4) the symplectic map $(p, q) \mapsto (P(t), Q(t))$ of the *exact flow* of the Hamiltonian system.

In accordance with equation (5.4) we have to satisfy

$$P_i(t) = \frac{\partial S}{\partial Q_i}(q, Q(t), t),$$

$$p_i = -\frac{\partial S}{\partial q_i}(q, Q(t), t). \tag{5.11}$$

Differentiating the second relation with respect to $t$ yields

$$0 = \frac{\partial^2 S}{\partial q_i \partial t}(q, Q(t), t) + \sum_{j=1}^{d} \frac{\partial^2 S}{\partial q_i \partial Q_j}(q, Q(t), t) \cdot \dot{Q}_j(t) \tag{5.12}$$

$$= \frac{\partial^2 S}{\partial q_i \partial t}(q, Q(t), t) + \sum_{j=1}^{d} \frac{\partial^2 S}{\partial q_i \partial Q_j}(q, Q(t), t) \cdot \frac{\partial H}{\partial P_j}(P(t), Q(t)) \tag{5.13}$$

---

[3] Carl Gustav Jacob Jacobi, born: 10 December 1804 in Potsdam (near Berlin), died: 18 February 1851 in Berlin.

where we have inserted the second equation of (1.7) for $\dot{Q}_j$. Then, using the chain rule, this equation simplifies to

$$\frac{\partial}{\partial q_i}\left(\frac{\partial S}{\partial t} + H\left(\frac{\partial S}{\partial Q_1}, \ldots, \frac{\partial S}{\partial Q_d}, Q_1, \ldots, Q_d\right)\right) = 0. \tag{5.14}$$

This motivates the following surprisingly simple relation.

**Theorem 5.5.** *If $S(q, Q, t)$ is a smooth solution of*

$$\frac{\partial S}{\partial t} + H\left(\frac{\partial S}{\partial Q_1}, \ldots, \frac{\partial S}{\partial Q_d}, Q_1, \ldots, Q_d\right) = 0, \tag{5.15}$$

*and if the matrix $\left(\frac{\partial^2 S}{\partial q_i \partial Q_j}\right)$ is invertible, there is a map $(p,q) \mapsto (P(t), Q(t))$ defined by (5.11) which is the flow $\varphi_t(p,q)$ of the Hamiltonian system (1.7).*

Equation (5.15) is called the 'Hamilton-Jacobi partial differential equation'.

*Proof.* The invertibility of the matrix $\left(\frac{\partial^2 S}{\partial q_i \partial Q_j}\right)$ and the implicit function theorem imply that the mapping $(p,q) \mapsto (P(t), Q(t))$ is well-defined by (5.11), and, by differentiation, that (5.12) is true as well.

Since, by hypothesis, $S(q,Q,t)$ is a solution of (5.15), the equations (5.14) and hence also (5.13) are satisfied. Subtracting (5.12) and (5.13), and once again using the invertibility of the matrix $\left(\frac{\partial^2 S}{\partial q_i \partial Q_j}\right)$, we see that necessarily $\dot{Q}(t) = H_p(P(t), Q(t))$. This proves the validity of the second equation of the Hamiltonian system (1.7).

The first equation of (1.7) is obtained as follows: differentiate the first relation of (5.11) with respect to $t$ and the Hamilton-Jacobi equation (5.15) with respect to $Q_i$, then eliminate the term $\frac{\partial^2 S}{\partial Q_i \partial t}$. Using $\dot{Q}(t) = H_p(P(t), Q(t))$, this leads in a straightforward way to $\dot{P}(t) = -H_q(P(t), Q(t))$. □

In the hands of Jacobi (1842), this equation turned into a powerful tool for the analytic integration of many difficult problems. One has, in fact, to find a solution of (5.15) which contains sufficiently many parameters. This is often possible with the method of separation of variables. An example is presented in Exercise 12.

**Hamilton-Jacobi Equation for $S^1$, $S^2$, and $S^3$.** We now express the Hamilton-Jacobi differential equation in the coordinates used in Lemma 5.2. In these coordinates it is also possible to prescribe initial values for $S$ at $t = 0$.

From the proof of Lemma 5.2 we know that the generating functions in the variables $(q, Q)$ and $(P, q)$ are related by

$$S^1(P, q, t) = P^T(Q - q) - S(q, Q, t). \tag{5.16}$$

We consider $P, q, t$ as independent variables, and we differentiate this relation with respect to $t$. Using the first relation of (5.11) this gives

VI. Symplectic Integration of Hamiltonian Systems

$$\frac{\partial S^1}{\partial t}(P,q,t) = P^T \frac{\partial Q}{\partial t} - \frac{\partial S}{\partial Q}(q,Q,t)\frac{\partial Q}{\partial t} - \frac{\partial S}{\partial t}(q,Q,t) = -\frac{\partial S}{\partial t}(q,Q,t).$$

Differentiating (5.16) with respect to $P$ yields

$$\frac{\partial S^1}{\partial P}(P,q,t) = Q - q + P^T \frac{\partial Q}{\partial P} - \frac{\partial S}{\partial Q}(q,Q,t)\frac{\partial Q}{\partial P} = Q - q.$$

Inserting $\frac{\partial S}{\partial Q} = P$ and $Q = q + \frac{\partial S^1}{\partial P}$ into the Hamilton-Jacobi equation (5.15) we are led to the equation of the following theorem.

**Theorem 5.6.** *If $S^1(P,q,t)$ is a solution of the partial differential equation*

$$\frac{\partial S^1}{\partial t}(P,q,t) = H\Big(P, q + \frac{\partial S^1}{\partial P}(P,q,t)\Big), \qquad S^1(P,q,t_0) = 0, \qquad (5.17)$$

*then the mapping $(p,q) \mapsto (P(t), Q(t))$, defined by (5.5), is the exact flow of the Hamiltonian system (1.7).*

*Proof.* Whenever the mapping $(p,q) \mapsto (P(t), Q(t))$ can be written as (5.11) with a function $S(q,Q,t)$, and when the invertibility assumption of Theorem 5.5 holds, the proof is done by the above calculations. Since our mapping, for $t = t_0$, reduces to the identity and cannot be written as (5.11), we give a direct proof.

Let $S^1(P,q,t)$ be given by the Hamilton-Jacobi equation (5.17), and assume that $(p,q) \mapsto (P, Q) = (P(t), Q(t))$ is the transformation given by (5.5). Differentiation of the first relation of (5.5) with respect to time $t$ and using (5.17) yields[4]

$$\Big(I + \frac{\partial^2 S^1}{\partial P \partial q}(P,q,t)\Big)\dot{P} = -\frac{\partial^2 S^1}{\partial t \partial q}(P,q,t) = -\Big(I + \frac{\partial^2 S^1}{\partial P \partial q}(P,q,t)\Big)\frac{\partial H}{\partial Q}(P,Q).$$

Differentiation of the second relation of (5.5) gives

$$\dot{Q} = \frac{\partial^2 S^1}{\partial t \partial P}(P,q,t) + \frac{\partial^2 S^1}{\partial P^2}(P,q,t)\dot{P}$$

$$= \frac{\partial H}{\partial P}(P,Q) + \frac{\partial^2 S^1}{\partial P^2}(P,q,t)\Big(\frac{\partial H}{\partial Q}(P,Q) + \dot{P}\Big).$$

Consequently, $\dot{P} = -\frac{\partial H}{\partial Q}(P,Q)$ and $\dot{Q} = \frac{\partial H}{\partial P}(P,Q)$, so that $(P(t), Q(t)) = \varphi_t(p,q)$ is the exact flow of the Hamiltonian system. □

Writing the Hamilton-Jacobi differential equation in the variables $(P+p)/2$, $(Q+q)/2$ gives the following formula.

---

[4] Due to an inconsistent notation of the partial derivatives $\frac{\partial H}{\partial Q}$, $\frac{\partial S^1}{\partial q}$ as column or row vectors, this formula may be difficult to read. Use indices instead of matrices in order to check its correctness.

**Theorem 5.7.** *Assume that $S^3(u,v,t)$ is a solution of*

$$\frac{\partial S^3}{\partial t}(u,v,t) = H\left(u - \frac{1}{2}\frac{\partial S^3}{\partial v}(u,v,t), v + \frac{1}{2}\frac{\partial S^3}{\partial u}(u,v,t)\right) \quad (5.18)$$

*with initial condition $S^3(u,v,0) = 0$. Then, the exact flow $\varphi_t(p,q)$ of the Hamiltonian system (1.7) satisfies the system (5.6).*

*Proof.* As in the proof of Theorem 5.6, one considers the transformation $(p,q) \mapsto \bigl(P(t), Q(t)\bigr)$ defined by (5.6), and then checks by differentiation that $\bigl(P(t), Q(t)\bigr)$ is a solution of the Hamiltonian system (1.7). □

Writing $w = (u,v)$ and using the matrix $J$ of (2.3), the Hamilton-Jacobi equation (5.18) can also be written as

$$\frac{\partial S^3}{\partial t}(w,t) = H\left(w + \frac{1}{2}J^{-1}\nabla S^3(w,t)\right), \quad S^3(w,0) = 0. \quad (5.19)$$

The solution of (5.19) is anti-symmetric in $t$, i.e.,

$$S^3(w,-t) = -S^3(w,t). \quad (5.20)$$

This can be seen as follows: let $\varphi_t(w)$ be the exact flow of the Hamiltonian system $\dot{y} = J^{-1}\nabla H(y)$. Because of (5.7), $S^3(w,t)$ is defined by

$$\varphi_t(w) - w = J^{-1}\nabla S^3\bigl((\varphi_t(w) + w)/2, t\bigr).$$

Replacing $t$ with $-t$ and then $w$ with $\varphi_t(w)$ we get from $\varphi_{-t}\bigl(\varphi_t(t)\bigr) = w$ that

$$w - \varphi_t(w) = J^{-1}\nabla S^3\bigl((w + \varphi_t(w))/2, -t\bigr).$$

Hence $S^3(w,t)$ and $-S^3(w,-t)$ are generating functions of the same symplectic transformation. Since generating functions are unique up to an additive constant (because $dS = 0$ implies $S = Const$), the anti-symmetry (5.20) follows from the initial condition $S^3(w,0) = 0$.

## VI.5.4 Methods Based on Generating Functions

To construct symplectic numerical methods of high order, Feng Kang (1986), Feng Kang, Wu, Qin & Wang (1989) and Channell & Scovel (1990) proposed computing an approximate solution of the Hamilton-Jacobi equation. For this one inserts the ansatz

$$S^1(P,q,t) = tG_1(P,q) + t^2G_2(P,q) + t^3G_3(P,q) + \ldots$$

into (5.17), and compares like powers of $t$. This yields

$$G_1(P,q) = H(P,q),$$

$$G_2(P,q) = \frac{1}{2}\left(\frac{\partial H}{\partial P}\frac{\partial H}{\partial q}\right)(P,q),$$

$$G_3(P,q) = \frac{1}{6}\left(\frac{\partial^2 H}{\partial P^2}\left(\frac{\partial H}{\partial q}\right)^2 + \frac{\partial^2 H}{\partial P \partial q}\frac{\partial H}{\partial P}\frac{\partial H}{\partial q} + \frac{\partial^2 H}{\partial q^2}\left(\frac{\partial H}{\partial P}\right)^2\right)(P,q).$$

If we use the truncated series

$$S^1(P,q) = hG_1(P,q) + h^2 G_2(P,q) + \ldots + h^r G_r(P,q) \tag{5.21}$$

and insert it into (5.5), the transformation $(p,q) \mapsto (P,Q)$ defines a symplectic one-step method of order $r$. Symplecticity follows at once from Lemma 5.2 and order $r$ is a consequence of the fact that the truncation of $S^1(P,q)$ introduces a perturbation of size $\mathcal{O}(h^{r+1})$ in (5.17). We remark that for $r \geq 2$ the methods obtained require the computation of higher derivatives of $H(p,q)$, and for separable Hamiltonians $H(p,q) = T(p) + U(q)$ they are no longer explicit (compared to the symplectic Euler method (3.1)).

The same approach applied to the third characterization of Lemma 5.2 yields

$$S^3(w,t) = hG_1(w) + h^3 G_3(w) + \ldots + h^{2r-1} G_{2r-1}(w),$$

where $G_1(w) = H(w)$,

$$G_3(w) = \frac{1}{24}\nabla^2 H(w)\bigl(J^{-1}\nabla H(w), J^{-1}\nabla H(w)\bigr),$$

and further $G_j(w)$ can be obtained by comparing like powers of $h$ in (5.19). In this way we get symplectic methods of order $2r$. Since $S^3(w,h)$ has an expansion in odd powers of $h$, the resulting method is symmetric.

**The Approach of Miesbach & Pesch.** With the aim of avoiding higher derivatives of the Hamiltonian in the numerical method, Miesbach & Pesch (1992) propose considering generating functions of the form

$$S^3(w,h) = h\sum_{i=1}^{s} b_i H\bigl(w + hc_i J^{-1}\nabla H(w)\bigr), \tag{5.22}$$

and to determine the free parameters $b_i, c_i$ in such a way that the function of (5.22) agrees with the solution of the Hamilton-Jacobi equation (5.19) up to a certain order. For $b_{s+1-i} = b_i$ and $c_{s+1-i} = -c_i$ this function satisfies $S^3(w,-h) = -S^3(w,h)$, so that the resulting method is symmetric. A straightforward computation shows that it yields a method of order 4 if

$$\sum_{i=1}^{s} b_i = 1, \qquad \sum_{i=1}^{s} b_i c_i^2 = \frac{1}{12}.$$

For $s = 3$, these equations are fulfilled for $b_1 = b_3 = 5/18$, $b_2 = 4/9$, $c_1 = -c_3 = \sqrt{15}/10$, and $c_2 = 0$. Since the function $S^3$ of (5.22) has to be inserted into (5.19), these methods still need second derivatives of the Hamiltonian.

## VI.6 Variational Integrators

A third approach to symplectic integrators comes from using discretized versions of Hamilton's principle, which determines the equations of motion from a variational problem. This route has been taken by Suris (1990), MacKay (1992) and in a series of papers by Marsden and coauthors, see the review by Marsden & West (2001) and references therein. Basic theoretical properties were formulated by Maeda (1980,1982) and Veselov (1988,1991) in a non-numerical context.

### VI.6.1 Hamilton's Principle

> Ours, according to Leibniz, is the best of all possible worlds, and the laws of nature can therefore be described in terms of extremal principles.
> (C.L. Siegel & J.K. Moser 1971, p. 1)

> Man scheint dies Princip früher ... unbemerkt gelassen zu haben. *Hamilton* ist der erste, der von diesem Princip ausgegangen ist.
> (C.G.J. Jacobi 1842, p. 58)

> Hamilton gave an improved mathematical formulation of a principle which was well established by the fundamental investigations of Euler and Lagrange; the integration process employed by him was likewise known to Lagrange. The name "Hamilton's principle", coined by Jacobi, was not adopted by the scientists of the last century. It came into use, however, through the textbooks of more recent date.
> (C. Lanczos 1949, p. 114)

Lagrange's equations of motion (1.4) can be viewed as the Euler-Lagrange equations for the variational problem of extremizing the *action integral*

$$S(q) = \int_{t_0}^{t_1} L(q(t), \dot q(t))\, dt \tag{6.1}$$

among all curves $q(t)$ that connect two given points $q_0$ and $q_1$:

$$q(t_0) = q_0, \quad q(t_1) = q_1. \tag{6.2}$$

In fact, assuming $q(t)$ to be extremal and considering a variation $q(t) + \varepsilon\, \delta q(t)$ with the same end-points, i.e., with $\delta q(t_0) = \delta q(t_1) = 0$, gives, using a partial integration,

$$0 = \frac{d}{d\varepsilon}\Big|_{\varepsilon=0} S(q + \varepsilon\, \delta q) = \int_{t_0}^{t_1} \Big(\frac{\partial L}{\partial q}\delta q + \frac{\partial L}{\partial \dot q}\delta \dot q\Big) dt = \int_{t_0}^{t_1} \Big(\frac{\partial L}{\partial q} - \frac{d}{dt}\frac{\partial L}{\partial \dot q}\Big)\delta q\, dt,$$

which leads to (1.4). The principle that the motion extremizes the action integral is known as *Hamilton's principle*.

We now consider the action integral as a function of $(q_0, q_1)$, for the solution $q(t)$ of the Euler-Lagrange equations (1.4) with these boundary values (this exists uniquely locally at least if $q_0, q_1$ are sufficiently close),

$$S(q_0, q_1) = \int_{t_0}^{t_1} L(q(t), \dot{q}(t))\, dt\,. \tag{6.3}$$

The partial derivative of $S$ with respect to $q_0$ is, again using partial integration,

$$\begin{aligned}\frac{\partial S}{\partial q_0} &= \int_{t_0}^{t_1} \left(\frac{\partial L}{\partial q}\frac{\partial q}{\partial q_0} + \frac{\partial L}{\partial \dot{q}}\frac{\partial \dot{q}}{\partial q_0}\right) dt \\ &= \frac{\partial L}{\partial \dot{q}}\frac{\partial q}{\partial q_0}\Big|_{t_0}^{t_1} + \int_{t_0}^{t_1} \left(\frac{\partial L}{\partial q} - \frac{d}{dt}\frac{\partial L}{\partial \dot{q}}\right)\frac{\partial q}{\partial q_0}\, dt = -\frac{\partial L}{\partial \dot{q}}(q_0, \dot{q}_0)\end{aligned}$$

with $\dot{q}_0 = \dot{q}(t_0)$, where the last equality follows from (1.4) and (6.2). In view of the definition (1.5) of the conjugate momenta, $p = \partial L/\partial \dot{q}$, the last term is simply $-p_0$. Computing $\partial S/\partial q_1 = p_1$ in the same way, we thus obtain for the differential of $S$

$$dS = \frac{\partial S}{\partial q_1}\, dq_1 + \frac{\partial S}{\partial q_0}\, dq_0 = p_1\, dq_1 - p_0\, dq_0 \tag{6.4}$$

which is the basic formula for symplecticity generating functions (see (5.1) above), obtained here by working with the Lagrangian formalism.

## VI.6.2 Discretization of Hamilton's Principle

Discrete-time versions of Hamilton's principle are of mathematical interest in their own right, see Maeda (1980,1982), Veselov (1991) and references therein. Here they are considered with the aim of deriving or understanding numerical approximation schemes. The discretized Hamilton principle consists of extremizing, for given $q_0$ and $q_N$, the sum

$$S_h(\{q_n\}_0^N) = \sum_{n=0}^{N-1} L_h(q_n, q_{n+1})\,. \tag{6.5}$$

We think of the *discrete Lagrangian* $L_h$ as an approximation

$$L_h(q_n, q_{n+1}) \approx \int_{t_n}^{t_{n+1}} L(q(t), \dot{q}(t))\, dt\,, \tag{6.6}$$

where $q(t)$ is the solution of the Euler-Lagrange equations (1.4) with boundary values $q(t_n) = q_n$, $q(t_{n+1}) = q_{n+1}$. If equality holds in (6.6), then it is clear from the continuous Hamilton principle that the exact solution values $\{q(t_n)\}$ of the Euler-Lagrange equations (1.4) extremize the action sum $S_h$. Before we turn to concrete examples of approximations $L_h$, we continue with the general theory which is analogous to the continuous case.

The requirement $\partial S_h/\partial q_n = 0$ for an extremum yields the *discrete Euler-Lagrange equations*

$$\frac{\partial L_h}{\partial y}(q_{n-1}, q_n) + \frac{\partial L_h}{\partial x}(q_n, q_{n+1}) = 0 \tag{6.7}$$

for $n = 1, \ldots, N-1$, where the partial derivatives refer to $L_h = L_h(x, y)$. This gives a three-term difference scheme for determining $q_1, \ldots, q_{N-1}$.

We now set
$$S_h(q_0, q_N) = \sum_{n=0}^{N-1} L_h(q_n, q_{n+1})$$
where $\{q_n\}$ is a solution of the discrete Euler-Lagrange equations (6.7) with the boundary values $q_0$ and $q_N$. With (6.7) the partial derivatives reduce to
$$\frac{\partial S_h}{\partial q_0} = \frac{\partial L_h}{\partial x}(q_0, q_1), \quad \frac{\partial S_h}{\partial q_N} = \frac{\partial L_h}{\partial y}(q_{N-1}, q_N).$$

We introduce the *discrete momenta* via a discrete Legendre transformation,
$$p_n = -\frac{\partial L_h}{\partial x}(q_n, q_{n+1}). \tag{6.8}$$

The above formula and (6.7) for $n = N$ then yield
$$dS_h = p_N \, dq_N - p_0 \, dq_0. \tag{6.9}$$

If (6.8) defines a bijection between $p_n$ and $q_{n+1}$ for given $q_n$, then we obtain a one-step method $\Phi_h : (p_n, q_n) \mapsto (p_{n+1}, q_{n+1})$ by composing the inverse discrete Legendre transform, a step with the discrete Euler-Lagrange equations, and the discrete Legendre transformation as shown in the diagram:

$$\begin{array}{ccc} & (6.7) & \\ (q_n, q_{n+1}) & \longrightarrow & (q_{n+1}, q_{n+2}) \\ (6.8) \uparrow & & \downarrow (6.8) \\ (p_n, q_n) & & (p_{n+1}, q_{n+1}) \end{array}$$

The method is symplectic by (6.9) and Theorem 5.1. A short-cut in the computation is obtained by noting that (6.7) and (6.8) (for $n+1$ instead of $n$) imply
$$p_{n+1} = \frac{\partial L_h}{\partial y}(q_n, q_{n+1}), \tag{6.10}$$

which yields the scheme
$$(p_n, q_n) \xrightarrow{(6.8)} (q_n, q_{n+1}) \xrightarrow{(6.10)} (p_{n+1}, q_{n+1}).$$

Let us summarize these considerations, which can be found in Maeda (1980), Suris (1990), Veselov (1991) and MacKay (1992).

**Theorem 6.1.** *The discrete Hamilton principle for (6.5) gives the discrete Euler-Lagrange equations (6.7) and the symplectic method*

$$p_n = -\frac{\partial L_h}{\partial x}(q_n, q_{n+1}), \quad p_{n+1} = \frac{\partial L_h}{\partial y}(q_n, q_{n+1}). \quad (6.11)$$

These formulas also show that $L_h$ is a generating function (5.4) for the symplectic map $(p_n, q_n) \mapsto (p_{n+1}, q_{n+1})$. Conversely, since every symplectic method has a generating function (5.4), it can be interpreted as resulting from Hamilton's principle with the generating function (5.4) as the discrete Lagrangian. The classes of symplectic integrators and variational integrators are therefore identical.

We now turn to simple examples of variational integrators obtained by choosing a discrete Lagrangian $L_h$ with (6.6).

**Example 6.2 (MacKay 1992).** Choose $L_h(q_n, q_{n+1})$ by approximating $q(t)$ of (6.6) as the linear interpolant of $q_n$ and $q_{n+1}$ and approximating the integral by the trapezoidal rule. This gives

$$L_h(q_n, q_{n+1}) = \frac{h}{2} L\left(q_n, \frac{q_{n+1} - q_n}{h}\right) + \frac{h}{2} L\left(q_{n+1}, \frac{q_{n+1} - q_n}{h}\right) \quad (6.12)$$

and hence the symplectic scheme, with $v_{n+1/2} = (q_{n+1} - q_n)/h$ for brevity,

$$p_n = \frac{1}{2}\frac{\partial L}{\partial \dot q}(q_n, v_{n+1/2}) + \frac{1}{2}\frac{\partial L}{\partial \dot q}(q_{n+1}, v_{n+1/2}) - \frac{h}{2}\frac{\partial L}{\partial q}(q_n, v_{n+1/2})$$

$$p_{n+1} = \frac{1}{2}\frac{\partial L}{\partial \dot q}(q_n, v_{n+1/2}) + \frac{1}{2}\frac{\partial L}{\partial \dot q}(q_{n+1}, v_{n+1/2}) + \frac{h}{2}\frac{\partial L}{\partial q}(q_{n+1}, v_{n+1/2}).$$

For a mechanical Lagrangian $L(q, \dot q) = \frac{1}{2}\dot q^T M \dot q - U(q)$ this reduces to the Störmer/Verlet method

$$Mv_{n+1/2} = p_n + \frac{1}{2}hF_n$$

$$q_{n+1} = q_n + hv_{n+1/2}$$

$$p_{n+1} = Mv_{n+1/2} + \frac{1}{2}hF_{n+1}$$

where $F_n = -\nabla U(q_n)$. In this case, the discrete Euler-Lagrange equations (6.7) become the familiar second-difference formula $M(q_{n+1} - 2q_n + q_{n-1}) = h^2 F_n$.

**Example 6.3 (Wendlandt & Marsden 1997).** Approximating the integral in (6.6) instead by the midpoint rule gives

$$L_h(q_n, q_{n+1}) = hL\left(\frac{q_{n+1} + q_n}{2}, \frac{q_{n+1} - q_n}{h}\right). \quad (6.13)$$

This yields the symplectic scheme, with the abbreviations $q_{n+1/2} = (q_{n+1} + q_n)/2$ and $v_{n+1/2} = (q_{n+1} - q_n)/h$,

$$p_n = \frac{\partial L}{\partial \dot q}(q_{n+1/2}, v_{n+1/2}) - \frac{h}{2}\frac{\partial L}{\partial q}(q_{n+1/2}, v_{n+1/2})$$
$$p_{n+1} = \frac{\partial L}{\partial \dot q}(q_{n+1/2}, v_{n+1/2}) + \frac{h}{2}\frac{\partial L}{\partial q}(q_{n+1/2}, v_{n+1/2}).$$

For $L(q,\dot q) = \frac{1}{2}\dot q^T M \dot q - U(q)$ this becomes the implicit midpoint rule

$$Mv_{n+1/2} = p_n + \frac{1}{2}hF_{n+1/2}$$
$$q_{n+1} = q_n + hv_{n+1/2}$$
$$p_{n+1} = Mv_{n+1/2} + \frac{1}{2}hF_{n+1/2}$$

with $F_{n+1/2} = -\nabla U(\frac{1}{2}(q_{n+1}+q_n))$.

## VI.6.3 Symplectic Partitioned Runge-Kutta Methods Revisited

To obtain higher-order variational integrators, Marsden & West (2001) consider the discrete Lagrangian

$$L_h(q_0, q_1) = h \sum_{i=1}^{s} b_i L\big(u(c_i h), \dot u(c_i h)\big) \tag{6.14}$$

where $u(t)$ is the polynomial of degree $s$ with $u(0) = q_0$, $u(h) = q_1$ which extremizes the right-hand side. They then show that the corresponding variational integrator can be realized as a partitioned Runge-Kutta method. We here consider the slightly more general case

$$L_h(q_0, q_1) = h \sum_{i=1}^{s} b_i L(Q_i, \dot Q_i) \tag{6.15}$$

where
$$Q_i = q_0 + h \sum_{j=1}^{s} a_{ij} \dot Q_j$$

and the $\dot Q_i$ are chosen to extremize the above sum under the constraint

$$q_1 = q_0 + h \sum_{i=1}^{s} b_i \dot Q_i .$$

We assume that all the $b_i$ are non-zero and that their sum equals 1. Note that (6.14) is the special case of (6.15) where the $a_{ij}$ and $b_i$ are integrals (II.1.10) of Lagrange polynomials as for collocation methods.

With a Lagrange multiplier $\lambda = (\lambda_1, \ldots, \lambda_d)$ for the constraint, the extremality conditions obtained by differentiating (6.15) with respect to $\dot Q_j$ for $j = 1, \ldots, s$, read

$$\sum_{i=1}^{s} b_i \frac{\partial L}{\partial q}(Q_i, \dot{Q}_i) h a_{ij} + b_j \frac{\partial L}{\partial \dot{q}}(Q_j, \dot{Q}_j) = b_j \lambda.$$

With the notation

$$\dot{P}_i = \frac{\partial L}{\partial q}(Q_i, \dot{Q}_i), \quad P_i = \frac{\partial L}{\partial \dot{q}}(Q_i, \dot{Q}_i) \tag{6.16}$$

this simplifies to

$$b_j P_j = b_j \lambda - h \sum_{i=1}^{s} b_i a_{ij} \dot{P}_i. \tag{6.17}$$

The symplectic method of Theorem 6.1 now becomes

$$\begin{aligned}
p_0 &= -\frac{\partial L_h}{\partial x}(q_0, q_1) \\
&= -h \sum_{i=1}^{s} b_i \dot{P}_i \left( I + h \sum_{j=1}^{s} a_{ij} \frac{\partial \dot{Q}_j}{\partial q_0} \right) - h \sum_{j=1}^{s} b_j P_j \frac{\partial \dot{Q}_j}{\partial q_0} \\
&= -h \sum_{i=1}^{s} b_i \dot{P}_i + \lambda.
\end{aligned}$$

In the last equality we use (6.17) and $h \sum_j b_j \partial \dot{Q}_j / \partial q_0 = -I$, which follows from differentiating the constraint. In the same way we obtain

$$p_1 = \frac{\partial L_h}{\partial y}(q_0, q_1) = \lambda.$$

Putting these formulas together, we see that $(p_1, q_1)$ result from applying a partitioned Runge-Kutta method to the Lagrange equations (1.4) written as a differential-algebraic system

$$\dot{p} = \frac{\partial L}{\partial q}(q, \dot{q}), \quad p = \frac{\partial L}{\partial \dot{q}}(q, \dot{q}). \tag{6.18}$$

That is

$$\begin{aligned}
p_1 &= p_0 + h \sum_{i=1}^{s} b_i \dot{P}_i, & q_1 &= q_0 + h \sum_{i=1}^{s} b_i \dot{Q}_i, \\
P_i &= p_0 + h \sum_{j=1}^{s} \widehat{a}_{ij} \dot{P}_j, & Q_i &= q_0 + h \sum_{j=1}^{s} a_{ij} \dot{Q}_j,
\end{aligned} \tag{6.19}$$

with $\widehat{a}_{ij} = b_j - b_j a_{ji}/b_i$ so that the symplecticity condition (4.3) is fulfilled, and with $P_i, Q_i, \dot{P}_i, \dot{Q}_i$ related by (6.16). Since equations (6.16) are of the same form as (6.18), the proof of Theorem 1.3 shows that they are equivalent to

$$\dot{P}_i = -\frac{\partial H}{\partial q}(P_i, Q_i), \quad \dot{Q}_i = \frac{\partial H}{\partial p}(P_i, Q_i) \tag{6.20}$$

with the Hamiltonian $H = p^T \dot{q} - L(q, \dot{q})$ of (1.6). We have thus proved the following, which is similar in spirit to a result of Suris (1990).

**Theorem 6.4.** *The variational integrator with the discrete Lagrangian (6.15) is equivalent to the symplectic partitioned Runge-Kutta method (6.19), (6.20) applied to the Hamiltonian system with the Hamiltonian (1.6).* □

In particular, as noted by Marsden & West (2001), choosing Gaussian quadrature in (6.14) gives the Gauss collocation method applied to the Hamiltonian system, while Lobatto quadrature gives the Lobatto IIIA - IIIB pair.

## VI.6.4 Noether's Theorem

> ... enthält Satz I alle in Mechanik u.s.w. bekannten Sätze über erste Integrale.
> (E. Noether 1918)

We now return to the subject of Chap. IV, i.e., the existence of first integrals, but here in the context of Hamiltonian systems. E. Noether found the surprising result that continuous *symmetries* in the Lagrangian lead to such first integrals. We give in the following a version of her "Satz I", specialized to our needs, with a particularly short proof.

**Theorem 6.5 (Noether 1918).** *Consider a system with Hamiltonian $H(p,q)$ and Lagrangian $L(q,\dot{q})$. Suppose $\{g_s : s \in \mathbb{R}\}$ is a one-parameter group of transformations ($g_s \circ g_r = g_{s+r}$) which leaves the Lagrangian invariant:*

$$L(g_s(q), g'_s(q)\dot{q}) = L(q,\dot{q}) \quad \text{for all } s \text{ and all } (q,\dot{q}). \tag{6.21}$$

*Let $a(q) = (d/ds)|_{s=0}\, g_s(q)$ be defined as the vector field with flow $g_s(q)$. Then*

$$I(p,q) = p^T a(q) \tag{6.22}$$

*is a first integral of the Hamiltonian system.*

**Example 6.6.** Let $G$ be a matrix Lie group with Lie algebra $\mathfrak{g}$ (see Sect. IV.6). Suppose $L(Qq, Q\dot{q}) = L(q,\dot{q})$ for all $Q \in G$. Then $p^T Aq$ is a first integral for every $A \in \mathfrak{g}$. (Take $g_s(q) = \exp(sA)q$.) For example, $G = SO(n)$ yields conservation of angular momentum.

We prove Theorem 6.5 by using the discrete analogue, which reads as follows.

**Theorem 6.7.** *Suppose the one-parameter group of transformations $\{g_s : s \in \mathbb{R}\}$ leaves the discrete Lagrangian $L_h(q_0, q_1)$ invariant:*

$$L_h(g_s(q_0), g_s(q_1)) = L_h(q_0, q_1) \quad \text{for all } s \text{ and all } (q_0, q_1). \tag{6.23}$$

*Then (6.22) is a first integral of the method (6.11), i.e., $p_{n+1}^T a(q_{n+1}) = p_n^T a(q_n)$.*

*Proof.* Differentiating (6.23) with respect to $s$ gives

$$0 = \frac{d}{ds}\bigg|_{s=0} L_h(g_s(q_0), g_s(q_1)) = \frac{\partial L_h}{\partial x}(q_0, q_1)a(q_0) + \frac{\partial L_h}{\partial y}(q_0, q_1)a(q_1).$$

By (6.11) this becomes $0 = -p_0^T a(q_0) + p_1^T a(q_1)$. □

Theorem 6.5 now follows by choosing $L_h = S$ of (6.3) and noting (6.4) and

$$\begin{aligned}S(q(t_0), q(t_1)) &= \int_{t_0}^{t_1} L\big(q(t), \dot{q}(t)\big)\, dt \\ &= \int_{t_0}^{t_1} L\Big(g_s(q(t)), \frac{d}{dt} g_s(q(t))\Big)\, dt = S\Big(g_s(q(t_0)), g_s(q(t_1))\Big).\end{aligned}$$

Theorem 6.7 has the appearance of giving a rich source of first integrals for symplectic methods. However, it must be noted that, unlike the case of the exact flow map in the above formula, the invariance (6.21) of the Lagrangian $L$ does not in general imply the invariance (6.23) of the discrete Lagrangian $L_h$ of the numerical method. A noteworthy exception arises for linear transformations $g_s$ as in Example 6.6, for which Theorem 6.7 yields the conservation of quadratic first integrals $p^T A q$, such as angular momentum, by symplectic partitioned Runge-Kutta methods — a property we already know from Theorem IV.2.4. The Störmer/Verlet method plays an exceptional role in also preserving non-quadratic first integrals.

**Theorem 6.8.** *In the situation of Theorem 6.5 with a mechanical Lagrangian $L = \frac{1}{2}\dot{q}^T M \dot{q} - U(q)$, the Störmer/Verlet method conserves the first integral $I(p, q) = p^T a(q)$.*

*Proof.* The invariance condition (6.21) yields, with $\dot{q} = 0$,

$$U(g_s(q)) = U(q) \quad \text{for all } s \text{ and } q, \tag{6.24}$$

and then, for $T = \frac{1}{2}\dot{q}^T M \dot{q}$,

$$T(g'_s(q)\dot{q}) = T(\dot{q}) \quad \text{for all } s \text{ and all } q, \dot{q}. \tag{6.25}$$

This shows that $M^{1/2} g'_s(q) M^{-1/2}$ is an orthogonal matrix.

The transformed values $g_s(q)$ form the solution of the initial value problem $dy/ds = a(y)$, $y(0) = q$, and the derivative $g'_s(q)$ solves the variational equation $dY/ds = a'(y)Y$, $Y(0) = I$. The orthogonality of $M^{1/2} g'_s(q) M^{-1/2}$ therefore implies $0 = d/ds|_{s=0} Y^T M Y = a'(q)^T M + M a'(q)$, which shows that $M a'(q)$ is skew-symmetric for all $q$. Hence we obtain, with $y_0 = g_s(q_0)$ and $y_1 = g_s(q_1)$ for brevity,

$$\begin{aligned}\frac{d}{ds} T\Big(\frac{g_s(q_1) - g_s(q_0)}{h}\Big) &= (y_1 - y_0)^T M \big(a(y_1) - a(y_0)\big) \\ &= \int_0^1 (y_1 - y_0)^T M a'(\theta y_1 + (1-\theta) y_0)(y_1 - y_0)\, d\theta = 0,\end{aligned}$$

so that

$$T\Big(\frac{g_s(q_1) - g_s(q_0)}{h}\Big) = T\Big(\frac{q_1 - q_0}{h}\Big) \quad \text{for all } s \text{ and all } q_0, q_1. \tag{6.26}$$

Using (6.24) and (6.26) in the expression (6.12) of the discrete Lagrangian $L_h$ of the Störmer/Verlet method yields (6.23), and hence Theorem 6.7 gives the result. □

## VI.7 Characterization of Symplectic Methods

Up to now in this chapter, we have presented sufficient conditions for the symplecticity of numerical integrators (usually in terms of certain coefficients). Here, we will prove *necessary* conditions for symplecticity, i.e., answer the question as to which methods are *not* symplectic. It will turn out that the sufficient conditions of Sect. VI.4, under an irreducibility condition on the method, are also necessary. The main tool is the Taylor series expansion of the numerical flow $y_0 \mapsto \Phi_h(y_0)$, which we assume to be a P-series.

### VI.7.1 Symplectic P-Series (and B-Series)

The numerical solution of a partitioned Runge-Kutta method (II.2.2) can be written as a P-series

$$\begin{pmatrix} p_1 \\ q_1 \end{pmatrix} = \begin{pmatrix} p_0 \\ q_0 \end{pmatrix} + \begin{pmatrix} \sum_{u \in TP_p} \frac{h^{|u|}}{\sigma(u)} a(u) F(u)(p_0, q_0) \\ \sum_{v \in TP_q} \frac{h^{|v|}}{\sigma(v)} a(v) F(v)(p_0, q_0) \end{pmatrix} \qquad (7.1)$$

with coefficients $a(\tau)$ given by

$$a(\tau) = \begin{cases} \sum_{i=1}^{s} b_i \phi_i(\tau) & \text{for } \tau \in TP_p \\ \sum_{i=1}^{s} \widehat{b}_i \phi_i(\tau) & \text{for } \tau \in TP_q \end{cases} \qquad (7.2)$$

(see Theorem III.2.4). We assume here that the elementary differentials $F(t)(p,q)$ originate from a Hamiltonian system (1.7). Our aim is to express the symplecticity of the transformation (7.1) in terms of the coefficients $a(t)$. We first multiply Eq. (4.3) by $\phi_i(u) \cdot \phi_j(v)$ (where $u = [u_1, \ldots, u_m]_p \in TP_p$ and $v = [v_1, \ldots, v_l]_q \in TP_q$) and we sum over all $i$ and $j$. Using the recursion (III.2.7) this yields

$$\sum_{i=1}^{s} b_i \phi_i(u \circ v) + \sum_{j=1}^{s} \widehat{b}_j \phi_j(v \circ u) = \left( \sum_{i=1}^{s} b_i \phi_i(u) \right) \left( \sum_{j=1}^{s} \widehat{b}_j \phi_j(v) \right), \qquad (7.3)$$

where we have used the Butcher product (see, e.g., Butcher (1987), Sect. 143)

$$u \circ v = [u_1, \ldots, u_m, v]_p, \qquad v \circ u = [v_1, \ldots, v_l, u]_q \qquad (7.4)$$

(compare also Definition III.3.7 and Fig. 7.1 below). Because of (7.2), this implies

$$a(u \circ v) + a(v \circ u) = a(u) \cdot a(v) \qquad \text{for } u \in TP_p, \ v \in TP_q. \qquad (7.5)$$

Since $\phi_i(\tau)$ does not depend on the colour of the root of $\tau$, condition (4.4) implies that

$$a(\tau) \text{ is independent of the colour of the root of } \tau. \qquad (7.6)$$

For the special case of B-series, the following result was first obtained in Calvo & Sanz-Serna (1994).

**Theorem 7.1.** *If the P-series (7.1) defines a symplectic transformation for separable Hamiltonians $H(p, q) = T(p) + U(q)$, then (7.5) holds for trees where neighbouring vertices have different colours.*

*If it is symplectic for all (separable and non-separable) Hamiltonians, then (7.5) and (7.6) hold for all trees.*

*Proof.* a) We fix two trees $u \in TP_p$ and $v \in TP_q$, and we construct a (polynomial) Hamiltonian such that the transformation (7.1) satisfies

$$\left(\frac{\partial(p_1, q_1)}{\partial p_0^1}\right)^T J \left(\frac{\partial(p_1, q_1)}{\partial q_0^2}\right) = C\Big(a(u \circ v) + a(v \circ u) - a(u) \cdot a(v)\Big) \quad (7.7)$$

with $C \neq 0$ (here, $p_0^1$ denotes the first component of $p_0$, and $q_0^2$ the second component of $q_0$). The symplecticity of (7.1) implies that the expression in (7.7) vanishes. Both statements concerning condition (7.5) are then an immediate consequence. Observe that the constructed Hamiltonian will be separable, when $u$ and $v$ have no neighbouring vertices of the same colour.

For given $u \in TP_p$ and $v \in TP_q$ we define the Hamiltonian as follows: to the branches of $u \circ v$ we attach the numbers $3, \ldots, |u| + |v| + 1$ such that the branch between the roots of $u$ and $v$ is labelled by 3. Then, the Hamiltonian is a sum of as many terms as vertices in the tree. The summand corresponding to a vertex is a product containing the factor $p^j$ (resp. $q^j$) if an upward leaving branch "$j$" is directly connected with a black (resp. white) vertex, and the factor $q^i$ (resp. $p^i$) if the vertex itself is black (resp. white) and the downward leaving branch has label "$i$". Finally, the factors $q^2$ and $p^1$ are included in the terms corresponding to the roots of $u$ and $v$, respectively. For the example of Fig. 7.1 we have

$$H(p, q) = q^2 q^3 q^4 p^5 + p^1 p^3 p^7 p^8 + p^4 p^6 + q^5 + q^6 + q^7 + q^8. \quad (7.8)$$

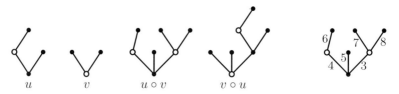

**Fig. 7.1.** Illustration of the Hamiltonian (7.8)

The components $F^i(\tau)(p, q)$ of the elementary differentials corresponding to the Hamiltonian system (with the Hamiltonian constructed above) satisfy

$$\begin{aligned} F^2(u \circ v)(p, q) &= (-1)^{\delta(u \circ v)} \sigma(u \circ v) \cdot p^1, \\ F^1(v \circ u)(p, q) &= (-1)^{\delta(v \circ u)} \sigma(v \circ u) \cdot q^2, \\ F^3(u)(p, q) &= (-1)^{\delta(u)} \sigma(u) \cdot q^2 \\ F^3(v)(p, q) &= (-1)^{\delta(v)} \sigma(v) \cdot p^1, \end{aligned} \quad (7.9)$$

and for all other trees $\tau \in TP$ and components $i$ we have

$$\frac{\partial F^i(\tau)}{\partial p^1}(0,0) = \frac{\partial F^i(\tau)}{\partial q^2}(0,0) = 0.$$

In (7.9), $\delta(\tau)$ counts the number of black vertices of $\tau$, and the *symmetry coefficient* $\sigma(\tau)$ is that of (III.2.3). For example, $\sigma(u) = 1$ and $\sigma(v) = 2$ for the trees of Fig. 7.1. The verification of (7.9) is straightforward. The coefficient $(-1)^{\delta(\tau)}$ is due to the minus sign in the first part of the Hamiltonian system (1.7), and the symmetry coefficient $\sigma(\tau)$ appears in exactly the same way as in the multidimensional Taylor formula. Due to the zero initial values, no elementary differential other than those of (7.9) give rise to non-vanishing expressions in (7.7). Consider for example the second component of $F(\tau)(p, q)$ for a tree $\tau \in TP_p$. Since we are concerned with the Hamiltonian system (1.7), this expression starts with a derivative of $H_{q^2}$. Therefore, it contributes to (7.7) at $p_0 = q_0 = 0$ only if it contains the factor $H_{q^2q^3q^4p^5}$ (for the example of Fig. 7.1). This in turn implies the presence of factors $H_{p^3\ldots}$, $H_{p^4\ldots}$ and $H_{q^5\ldots}$. Continuing this line of reasoning, we find that $F^2(\tau)(p, q)$ contributes to (7.7) at $p_0 = q_0 = 0$ only if $\tau = u \circ v$. With similar arguments we see that only the elementary differentials of (7.9) have to be considered. We now insert (7.9) into (7.1), and we compute its derivatives with respect to $p^1$ and $q^2$. This then yields (7.7) with $C = (-1)^{\delta(u)+\delta(v)} h^{|u|+|v|}$, and completes the proof concerning condition (7.5).

b) The necessity of condition (7.6) is seen similarly. We fix a tree $\tau \in TP_p$ and we let $\overline{\tau} \in TP_q$ be the tree obtained from $\tau$ by changing the colour of the root. We then attach the numbers $3, \ldots, |\tau| + 1$ to the branches of $\tau$, and we define a Hamiltonian as above but, different from adding the factors $q^2$ and $p^1$, we include the factor $p^1 q^2$ to the term corresponding to the root. For the tree $\tau = u$ of Fig. 7.1 this yields

$$H(p,q) = p^1 q^2 q^3 p^4 + p^3 p^5 + q^4 + q^5.$$

With this Hamiltonian we get

$$F^2(\tau)(p,q) = (-1)^{\delta(\tau)} \sigma(\tau) \cdot p^1,$$
$$F^1(\overline{\tau})(p,q) = (-1)^{\delta(\overline{\tau})} \sigma(\overline{\tau}) \cdot q^2,$$

and these are the only elementary differentials contributing to the left-hand expression of (7.7). We thus get

$$\left(\frac{\partial(p_1, q_1)}{\partial p_0^1}\right)^T J \left(\frac{\partial(p_1, q_1)}{\partial q_0^2}\right) = (-1)^{\delta(\tau)} h^{|\tau|} \big(a(\tau) - a(\overline{\tau})\big),$$

which completes the proof of Theorem 7.1. □

**Corollary 7.2.** *If a B-series (Definition II.1.8) defines a symplectic transformation for Hamiltonian systems (1.7), then*

$$a(u \circ v) + a(v \circ u) = a(u) \cdot a(v) \quad \text{for} \quad u, v \in T. \tag{7.10}$$

*Proof.* A B-series with coefficients $a(\tau), \tau \in T$, applied to a partitioned differential equation, can always be interpreted as a P-series (Definition II.2.1), where $a(\tau) := a(\varphi(\tau))$ for $\tau \in TP$ and $\varphi : TP \to T$ is the mapping that forgets the colouring of the vertices. This follows from the fact that

$$a(\tau)F(\tau)(y) = \begin{pmatrix} \sum_{u \in TP_p, \varphi(u)=\tau} a(u)\,F(u)(p,q) \\ \sum_{v \in TP_q, \varphi(v)=\tau} a(v)\,F(v)(p,q) \end{pmatrix}$$

for $\tau \in T$, because $\alpha(u) \cdot \sigma(u) = \alpha(v) \cdot \sigma(v) = \mathbf{e}(\tau) \cdot |\tau|!$. Here, $y = (p,q)$, the elementary differentials $F(\tau)(y)$ are those of Definition III.1.2, whereas $F(u)(p,q)$ and $F(v)(p,q)$ are those of Table III.2.1. □

## VI.7.2 Irreducible Runge-Kutta Methods

We are now able to study to what extent the conditions of Theorem 4.3 and Theorem 4.6 are also necessary for symplecticity. Consider first the 2-stage method

$$\begin{array}{c|cc} 1/2 & \alpha & 1/2 - \alpha \\ 1/2 & \beta & 1/2 - \beta \\ \hline & 1/2 & 1/2 \end{array}.$$

The solution of the corresponding Runge-Kutta system (II.1.4) is given by $k_1 = k_2 = k$, where $k = f(y_0 + k/2)$, and hence $y_1 = y_0 + hk$. Whatever the values of $\alpha$ and $\beta$ are, the numerical solution of the Runge-Kutta method is identical to that of the implicit midpoint rule, so that it defines a symplectic transformation. However, the condition (4.2) is only satisfied for $\alpha = \beta = 1/4$.

**Definition 7.3.** Two stages $i$ and $j$ of a Runge-Kutta method (II.1.4) are said to be *equivalent for a class (P)* of initial value problems, if for every problem in (P) and for every sufficiently small step size we have $k_i = k_j$ ($k_i = k_j$ and $\ell_i = \ell_j$ for partitioned Runge-Kutta methods (II.2.2)).

The method is called *irreducible for (P)* if it does not have equivalent stages. It is called *irreducible* if it is irreducible for all sufficiently smooth initial value problems.

For a more amenable characterization of irreducible Runge-Kutta methods, we introduce an ordering on $T$ (and on $TP$), and we consider the following $s \times \infty$ matrices

$\Phi_{\text{RK}} = \bigl(\phi(\tau); \tau \in T\bigr)$ with entries $\phi_i(\tau) = \mathbf{g}_i(\tau)$ given by (III.1.13),[5]
$\Phi_{\text{PRK}} = \bigl(\phi(\tau); \tau \in TP_p\bigr) = \bigl(\phi(\tau); \tau \in TP_q\bigr)$ with entries $\phi_i(\tau)$ given by (III.2.7); observe that $\phi_i(\tau)$ does not depend on the colour of the root,
$\Phi^*_{\text{PRK}} = \bigl(\phi(\tau); \tau \in TP^*_p\bigr) = \bigl(\phi(\tau); \tau \in TP^*_q\bigr)$ where $TP^*_p$ (resp. $TP^*_q$) is the set of trees in $TP_p$ (resp. $TP_q$) whose neighbouring vertices have different colours.

---

[5] In this section we let $\phi(\tau) \in \mathbb{R}^s$ denote the vector whose elements are $\phi_i(\tau), i = 1, \ldots, s$. This should not be mixed up with the value $\phi(\tau)$ of (III.1.16).

**Lemma 7.4 (Hairer 1994).** *A Runge-Kutta method is irreducible if and only if the matrix $\Phi_{RK}$ has full rank $s$.*

*A partitioned Runge-Kutta method is irreducible if and only if the matrix $\Phi_{PRK}$ has full rank $s$.*

*A partitioned Runge-Kutta method is irreducible for separable problems $\dot p = f(q), \dot q = g(p)$ if and only if the matrix $\Phi^*_{PRK}$ has full rank $s$.*

*Proof.* If the stages $i$ and $j$ are equivalent, it follows from the expansion

$$k_i = \sum_{\tau \in T} \frac{h^{|\tau|}}{\sigma(\tau)} \phi_i(\tau)\, F(\tau)(y_0)$$

(see the proof of Theorem III.1.4) and from the independency of the elementary differentials (Exercise III.3) that $\phi_i(\tau) = \phi_j(\tau)$ for all $\tau \in T$. Hence, the rows $i$ and $j$ of the matrix $\Phi_{RK}$ are identical. The analogous statement for partitioned Runge-Kutta methods follows from Theorem III.2.4 and Exercise III.6. This proves the sufficiency of the 'full rank' condition.

We prove its necessity only for partitioned Runge-Kutta methods applied to separable problems (the other situations can be treated similarly). For separable problems, only trees in $TP^*_p \cup TP^*_q$ give rise to non-vanishing elementary differentials. Irreducibility therefore implies that for every pair $(i, j)$ with $i \ne j$ there exists a tree $\tau \in TP^*_p$ such that $\phi_i(\tau) \ne \phi_j(\tau)$. Consequently, a certain finite linear combination of the columns of $\Phi^*_{PRK}$ has distinct elements, i.e., there exist vectors $\xi \in \mathbb{R}^\infty$ (only finitely many non zero elements) and $\eta \in \mathbb{R}^s$ with $\Phi^*_{PRK}\xi = \eta$ and $\eta_i \ne \eta_j$ for $i \ne j$. Due to the fact that $\phi_i([\tau_1,\ldots,\tau_m]) = \phi_i([\tau_1]) \cdot \ldots \cdot \phi_i([\tau_m])$, the componentwise product of two columns of $\Phi^*_{PRK}$ is again a column of $\Phi^*_{PRK}$. Continuing this argumentation and observing that $(1,\ldots,1)^T$ is a column of $\Phi^*_{PRK}$, we obtain a matrix $X$ such that $\Phi^*_{PRK} X = (\eta_i^{j-1})^s_{i,j=1}$ is a Vandermonde matrix. Since the $\eta_i$ are distinct, the matrix $\Phi^*_{PRK}$ has to be of full rank $s$. □

## VI.7.3 Characterization of Irreducible Symplectic Methods

The necessity of the condition (4.2) for symplectic Runge-Kutta methods was first stated by Lasagni (1988). Abia & Sanz-Serna (1993) extended his proof to partitioned methods. We follow here the ideas of Hairer (1994).

**Theorem 7.5.** *An irreducible Runge-Kutta method (II.1.4) is symplectic if and only if the condition (4.2) holds.*

*An irreducible partitioned Runge-Kutta method (II.2.2) is symplectic if and only if the conditions (4.3) and (4.4) hold.*

*A partitioned Runge-Kutta method, irreducible for separable problems, is symplectic for separable Hamiltonians $H(p,q) = T(p) + U(q)$ if and only if the condition (4.3) holds.*

*Proof.* The "if" part of all three statements has been proved in Theorem 4.3 and Theorem 4.6. We prove the "only if" part for partitioned Runge-Kutta methods applied to general Hamiltonian systems (the other two statements can be obtained in the same way).

We consider the $s \times s$ matrix $M$ with entries $m_{ij} = b_i \widehat{a}_{ij} + \widehat{b}_j a_{ji} - b_i \widehat{b}_j$. The computation at the beginning of Sect. VI.7.1 shows that for $u \in TP_p$ and $v \in TP_q$

$$\phi(u)^T M \, \phi(v) = a(u \circ v) + a(v \circ u) - a(u) \cdot a(v)$$

holds. Due to the symplecticity of the method, this expression vanishes and we obtain

$$\Phi_{\text{PRK}}^T M \, \Phi_{\text{PRK}} = 0,$$

where $\Phi_{\text{PRK}}$ is the matrix of Lemma 7.4. An application of this lemma then yields $M = 0$, which proves the necessity of (4.3).

For the vector $d$ with components $d_i = b_i - \widehat{b}_i$ we get $d^T \Phi_{\text{PRK}} = 0$, and we deduce from Lemma 7.4 that $d = 0$, so that (4.4) is also seen to be necessary. □

## VI.7.4 Conjugate Symplecticity

The symplecticity requirement may be too strong if we are interested in a correct long-time behaviour of a numerical integrator. Stoffer (1988) suggests considering methods that are not necessarily symplectic but conjugate to a symplectic method.

**Definition 7.6.** Two numerical methods $\Phi_h$ and $\Psi_h$ are mutually *conjugate*, if there exists a global change of coordinates $\chi_h$, such that

$$\Phi_h = \chi_h^{-1} \circ \Psi_h \circ \chi_h. \tag{7.11}$$

We assume that $\chi_h(y) = y + \mathcal{O}(h)$ uniformly for $y$ varying in a compact set.

For a numerical solution $y_{n+1} = \Phi_h(y_n)$, lying in a compact subset of the phase space, the transformed values $z_n = \chi_h(y_n)$ constitute a numerical solution $z_{n+1} = \Psi_h(z_n)$ of the second method. Since $y_n - z_n = \mathcal{O}(h)$, both numerical solutions have the same long-time behaviour, independently of whether one method shares certain properties (e.g., symplecticity) with the other.

**Trapezoidal Rule and Implicit Midpoint Rule.** The most prominent pair of conjugate methods are the trapezoidal and midpoint rules. Their conjugacy has been originally exploited by Dahlquist (1975) in an investigation on nonlinear stability.

If we denote by $\Phi_h^E$ and $\Phi_h^I$ the explicit and implicit Euler methods, respectively, then the trapezoidal rule $\Phi_h^T$ and the implicit midpoint rule $\Phi_h^M$ can be written as

$$\Phi_h^T = \Phi_{h/2}^I \circ \Phi_{h/2}^E, \qquad \Phi_h^M = \Phi_{h/2}^E \circ \Phi_{h/2}^I,$$

(see Fig. 7.2). This shows $\Phi_h^T = \chi_h^{-1} \Phi_h^M \chi_h$ with $\chi_h = \Phi_{h/2}^E$, implying that the trapezoidal and midpoint rules are mutually conjugate. The change of coordinates, which transforms the numerical solution of one method to that of the other, is $\mathcal{O}(h)$-close to the identity.

## VI.7 Characterization of Symplectic Methods

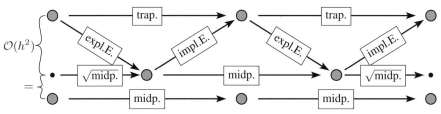

**Fig. 7.2.** Conjugacy of the trapezoidal rule and the implicit midpoint rule.

In fact, we can do even better. If we let $\Phi_{h/2}$ be the square root of $\Phi_h^M$ (i.e., $\Phi_{h/2} \circ \Phi_{h/2} = \Phi_h^M$, see Lemma III.3.2), then we have (Fig. 7.2)

$$\Phi_h^T = (\Phi_{h/2}^E)^{-1} \circ \Phi_h^M \circ \Phi_{h/2}^E = (\Phi_{h/2}^E)^{-1} \circ \Phi_{h/2} \circ \Phi_{h/2} \circ \Phi_{h/2}^{-1} \circ \Phi_{h/2}^E$$

so that the trapezoidal and the midpoint rules are conjugate via $\chi_h = \Phi_{h/2}^{-1} \circ \Phi_{h/2}^E$. Since $\Phi_{h/2}$ and $\Phi_{h/2}^E$ are both consistent with the same differential equation, the transformation $\chi_h$ is $\mathcal{O}(h^2)$-close to the identity. This shows that for every numerical solution of the trapezoidal rule there exists a numerical solution of the midpoint rule which remains $\mathcal{O}(h^2)$-close as long as it stays in a compact set. A single trajectory of the non-symplectic trapezoidal rule therefore behaves very much the same as a trajectory of the symplectic implicit midpoint rule.

**A Study via B-Series.** An investigation of Runge-Kutta methods, conjugate to a symplectic method, leads us to the following weaker requirement: we say that a numerical method $\Phi_h$ is *conjugate to a symplectic method* $\Psi_h$ *up to order* $r$, if there exists a transformation $\chi_h(y) = y + \mathcal{O}(h)$ such that

$$\Phi_h(h) = \left(\chi_h^{-1} \circ \Psi_h \circ \chi_h\right)(y) + \mathcal{O}(h^{r+1}). \tag{7.12}$$

This implies that the error of such a method behaves as the superposition of the error of a symplectic method of order $p$ with that of a non-symplectic method of order $r$.

In the following we assume that all methods considered as well as the conjugacy mapping $\chi_h$ can be represented as B-series

$$\Phi_h(y) = B(a, y), \qquad \Psi_h(y) = B(b, y), \qquad \chi_h(y) = B(c, y). \tag{7.13}$$

Using the composition formula (III.1.38) of B-series, condition (7.12) becomes

$$(ac)(\tau) = (cb)(\tau) \qquad \text{for} \quad |\tau| \leq r. \tag{7.14}$$

The following results are taken from the thesis of P. Leone (2000).

**Theorem 7.7.** *Let* $\Phi_h(y) = B(a, y)$ *represent a numerical method of order* 2.
a) *It is always conjugate to a symplectic method up to order* 3.
b) *It is conjugate to a symplectic method up to order* 4, *if and only if*

$$a(\bullet, \vee) - 2a(\bullet, \curlyvee) = 0, \qquad a(\curlyvee, \curlyvee) - 2a(\bullet, \curlyvee) = 0. \tag{7.15}$$

*Here, we use the abbreviation* $a(u, v) = a(u) \cdot a(v) - a(u \circ v) - a(v \circ u)$.

*Proof.* The condition (7.14) allows us to express $b(\tau)$ as a function of $a(u)$ for $|u| \leq |\tau|$ and of $c(v)$ for $|v| \leq |\tau| - 1$ (use the formulas of Example III.1.11). All we have to do is to check the symplecticity conditions $b(u,v) = 0$ for $|u| + |v| \leq r$ (see Corollary 7.2 and Theorem IX.10.11).

Since the method $\Phi_h$ is of order 2, we obtain $b(\bullet) = 1$ and $b(\mathfrak{f}) = 1/2$. We arbitrarily fix $c(\bullet) = 0$, so that the symplecticity condition $b(\bullet, \mathfrak{f}) = 0$ becomes $2c(\mathfrak{f}) = a(\bullet, \mathfrak{f})$. Defining $c(\mathfrak{f})$ by this relation proves statement (a).

For order 4, the three symplecticity conditions $b(\bullet, \mathsf{V}) = b(\bullet, [[\bullet]]) = b(\mathfrak{f}, \mathfrak{f}) = 0$ have to be fulfilled. One of them can be satisfied by defining suitably $c(\mathsf{V}) + c([[\bullet]])$; the other two conditions are then equivalent to (7.15). □

**Theorem 7.8.** *Let $\Phi_h(y) = B(a, y)$ represent a numerical method of order 4. It is conjugate to a symplectic method up to order 5, if and only if*

$$a(\bullet, \mathsf{Y}) - 2a(\bullet, \mathfrak{f}) = 0, \qquad a(\bullet, \mathsf{V}) - 3a(\bullet, \mathsf{V}) + 3a(\bullet, \mathfrak{f}) = 0,$$

$$a(\mathfrak{f}, \mathsf{V}) - a(\bullet, \mathsf{V}) - 2a(\mathfrak{f}, \mathfrak{f}) + 3a(\bullet, \mathfrak{f}) = 0.$$

*Proof.* The idea of the proof is the same as in the preceding theorem. The verification is left as an exercise for the reader. □

**Example 7.9.** A direct computation shows that for the Lobatto IIIB method with $s = 3$ we have $a(\mathfrak{f}, \mathsf{V}) = 1/144$, and $a(u, v) = 0$ for all other pairs with $|u| + |v| = 5$. Theorem 7.8 therefore proves that this method is not conjugate to a symplectic method up to order 5.

For the Lobatto IIIA method with $s = 3$ we obtain $a(\mathfrak{f}, \mathsf{V}) = -1/144$, $a(\mathfrak{f}, [[\bullet]]) = -1/288$, and $a(u, v) = 0$ for the remaining pairs with $|u| + |v| = 5$. This time the conditions of Theorem 7.8 are fulfilled, so that the Lobatto IIIA method with $s = 3$ is conjugate to a symplectic method up to order 5 at least.

# VI.8 Exercises

1. Prove that a linear transformation $A : \mathbb{R}^2 \to \mathbb{R}^2$ is symplectic, if and only if $\det A = 1$.
2. Prove that the flow of a Hamiltonian system satisfies $\det \varphi'_t(y) = 1$ for all $y$ and all $t$. Deduce from this result that the flow is *volume preserving*, i.e., for $B \subset \mathbb{R}^{2d}$ it holds that $\text{vol}(\varphi_t(B)) = \text{vol}(B)$ for all $t$.
3. Consider the Hamiltonian system $\dot{y} = J^{-1} \nabla H(y)$ and a variable transformation $y = \varphi(z)$. Prove that, for a symplectic transformation $\varphi(z)$, the system in the $z$-coordinates is again Hamiltonian with $\widetilde{H}(z) = H(\varphi(z))$.
4. Consider a Hamiltonian system with $H(p, q) = \frac{1}{2} p^T p + V(q)$. Let $q = \chi(Q)$ be a change of position coordinates. How has one to define the variable $P$ (as a function of $p$ and $q$) so that the system in the new variables $(P, Q)$ is again Hamiltonian? *Result.* $P = \chi'(Q)^T p$.

5. Let $\alpha$ and $\beta$ be the generalized coordinates of the double pendulum, whose kinetic and potential energies are

$$T = \frac{m_1}{2}(\dot{x}_1^2 + \dot{y}_1^2) + \frac{m_2}{2}(\dot{x}_2^2 + \dot{y}_2^2)$$
$$U = m_1 g y_1 + m_2 g y_2.$$

Determine the generalized momenta of the corresponding Hamiltonian system.

6. Consider the transformation $(r, \varphi) \mapsto (p, q)$, defined by

$$p = \psi(r) \cos \varphi, \qquad q = \psi(r) \sin \varphi.$$

For which function $\psi(r)$ is it a symplectic transformation?

7. On the set $U = \{(p, q) \,;\, p^2 + q^2 > 0\}$ consider the differential equation

$$\begin{pmatrix} \dot{p} \\ \dot{q} \end{pmatrix} = \frac{1}{p^2 + q^2} \begin{pmatrix} p \\ q \end{pmatrix}. \tag{8.1}$$

Prove that
a) its flow is symplectic everywhere on $U$;
b) on every simply-connected subset of $U$ the vector field (8.1) is Hamiltonian (with $H(p, q) = \text{Im} \log(p + iq) + Const$);
c) it is not possible to find a differentiable function $H : U \to \mathbb{R}$ such that (8.1) is equal to $J^{-1} \nabla H(p, q)$ for all $(p, q) \in U$.
*Remark.* The vector field (8.1) is locally (but not globally) Hamiltonian.

8. Prove that the definition (2.4) of $\Omega(M)$ does not depend on the parametrization $\varphi$, i.e., the parametrization $\psi = \varphi \circ \alpha$, where $\alpha$ is a diffeomorphism between suitable domains of $\mathbb{R}^2$, leads to the same result.

9. (Burnton & Scherer 1998). Prove that all members of the one-parameter family of Nyström methods of order $2s$, constructed in Exercise III.9, are symplectic and symmetric.

10. Compute the generating function $S^1(P, q, h)$ of a symplectic Nyström method applied to $\ddot{q} = U(q)$.

11. Find the Hamilton-Jacobi equation (cf. Theorem 5.6) for the generating function $S^2(p, Q)$ of Lemma 5.2.

12. (*Jacobi's method for exact integration*). Suppose we have a solution $S(q, Q, t, \alpha)$ of the Hamilton-Jacobi equation (5.15), depending on $d$ parameters $\alpha_1, \ldots, \alpha_d$ such that the matrix $\left(\frac{\partial^2 S}{\partial \alpha_i \partial Q_j}\right)$ is invertible. Since this matrix is the Jacobian of the system

$$\frac{\partial S}{\partial \alpha_i} = 0 \qquad i = 1, \ldots, d, \tag{8.2}$$

this system determines a solution path $Q_1, \ldots, Q_q$ which is locally unique. In possession of an additional parameter (and, including the partial derivatives with respect to $t$, an additional row and column in the Hessian matrix condition), we can also determine $Q_j(t)$ as function of $t$. Apply this method to the

Kepler problem (I.2.2) in polar coordinates, where, with the generalized momenta $p_r = \dot r$, $p_\varphi = r^2\dot\varphi$, the Hamiltonian becomes

$$H = \frac{1}{2}\left(p_r^2 + \frac{p_\varphi^2}{r^2}\right) - \frac{M}{r}$$

and the Hamilton-Jacobi differential equation (5.15) is

$$S_t + \frac{1}{2}(S_r)^2 + \frac{1}{2r^2}(S_\varphi)^2 - \frac{M}{r} = 0.$$

Solve this equation by the ansatz $S(t,r,\varphi) = \theta_1(t) + \theta_2(r) + \theta_3(\varphi)$ (separation of variables).

*Result.* One obtains

$$S = \int \sqrt{2\alpha_1 r^2 + 2Mr - \alpha_2^2}\,\frac{dr}{r} + \alpha_2 \varphi - \alpha_1 t.$$

Putting, e.g., $\partial S/\partial \alpha_2 = 0$, we obtain $\varphi = \arcsin \frac{Mr - \alpha_2^2}{\sqrt{M^2 + 2\alpha_1 \alpha_2^2}\,r}$ by evaluating an elementary integral. This, when resolved for $r$, leads to the elliptic movement of Kepler (Sect. I.2.1). This method turned out to be most effective for the exact integration of difficult problems. With the same ideas, just more complicated in the computations, Jacobi solves in "lectures" 24 through 30 of (Jacobi 1842) the Kepler motion in $\mathbb{R}^3$, the geodesics of ellipsoids (his greatest triumph), the motion with two centres of gravity, and proves a theorem of Abel.

13. (*Chan's Lobatto IIIS methods.*) Show that there exists a one-parameter family of symplectic, symmetric (and $A$-stable) Runge-Kutta methods of order $2s - 2$ based on Lobatto quadrature (Chan 1990). A special case of these methods can be obtained by taking the arithmetic mean of the Lobatto IIIA and Lobatto IIIB method coefficients (Sun 2000).

*Hint.* Use the $W$-transformation (see Hairer & Wanner (1996), p. 77) by putting $X_{s,s-1} = -X_{s-1,s}$ an arbitrary constant.

# Chapter VII.
# Further Topics in Structure Preservation

We discuss theoretical properties and the numerical treatment of three classes of problems that are closely related to Hamiltonian systems as considered in Chap. VI. In particular we study symmetric and symplectic methods for constrained Hamiltonian systems, we present Poisson integrators for Hamiltonian problems with a non-standard structure matrix, and we give volume-preserving algorithms for divergence-free differential equations that are not necessarily Hamiltonian systems.

## VII.1 Constrained Mechanical Systems

Constrained mechanical systems form an important class of differential equations on manifolds. Their numerical treatment has been extensively investigated in the context of *differential-algebraic equations* and is documented in monographs like that of Brenan, Campbell & Petzold (1996), Eich-Soellner & Führer (1998), Hairer, Lubich & Roche (1989), and Chap. VII of Hairer & Wanner (1996). We concentrate here on the symmetry and/or symplecticity of such numerical integrators.

### VII.1.1 Introduction and Examples

Consider a mechanical system described by position coordinates $q_1, \ldots, q_d$, and suppose that the motion is constrained to satisfy $g(q) = 0$ where $g : \mathbb{R}^d \to \mathbb{R}^m$ with $m < d$. Let $T(q, \dot q) = \frac{1}{2} \dot q^T M(q) \dot q$ be the kinetic energy of the system and $U(q)$ its potential energy, and put

$$L(q, \dot q) = T(q, \dot q) - U(q) - g(q)^T \lambda, \qquad (1.1)$$

where $\lambda = (\lambda_1, \ldots, \lambda_m)^T$ consists of Lagrange multipliers. The Euler-Lagrange equation of the variational problem for $\int_0^t L(q, \dot q)\, dt$ is then given by

$$\frac{d}{dt}\left(\frac{\partial L}{\partial \dot q}\right) - \frac{\partial L}{\partial q} = 0.$$

Written as a first order differential equation we get

$$\dot{q} = v$$
$$M(q)\dot{v} = f(q,v) - G(q)^T\lambda \qquad (1.2)$$
$$0 = g(q),$$

where $f(q,v) = \frac{\partial T}{\partial q}(q,v) - \nabla U(q)$ and $G(q) = g'(q)$.

**Example 1.1 (Spherical Pendulum).** We denote by $q_1, q_2, q_3$ the Cartesian coordinates of a point with mass $m$ that is connected with a massless rod of length $\ell$ to the origin. The kinetic and potential energies are $T = \frac{m}{2}(\dot{q}_1^2 + \dot{q}_2^2 + \dot{q}_3^2)$ and $U = mgq_3$, respectively, and the constraint is the fixed length of the rod. We thus get the system

$$\begin{aligned}
\dot{q}_1 &= v_1 & m\dot{v}_1 &= -2q_1\lambda \\
\dot{q}_2 &= v_2 & m\dot{v}_2 &= -2q_2\lambda \\
\dot{q}_3 &= v_3 & m\dot{v}_3 &= -mg - 2q_3\lambda \\
0 &= q_1^2 + q_2^2 + q_3^2 - \ell^2.
\end{aligned} \qquad (1.3)$$

The physical meaning of $\lambda$ is the tension in the rod which maintains the constant distance of the mass point from the origin.

**Existence and Uniqueness of the Solution.** A standard approach for studying the existence of solutions of differential-algebraic equations is to differentiate the constraints until an ordinary differential equation is obtained. Differentiating the constraint in (1.2) twice with respect to time yields

$$0 = G(q)v \quad \text{and} \quad 0 = g''(q)(v,v) + G(q)\dot{v}. \qquad (1.4)$$

The equation for $\dot{v}$ in (1.2) together with the second relation of (1.4) constitute a linear system for $\dot{v}$ and $\lambda$,

$$\begin{pmatrix} M(q) & G(q)^T \\ G(q) & 0 \end{pmatrix} \begin{pmatrix} \dot{v} \\ \lambda \end{pmatrix} = \begin{pmatrix} f(q,v) \\ -g''(q)(v,v) \end{pmatrix}. \qquad (1.5)$$

Throughout this chapter we require the matrix appearing in (1.5) to be invertible for $q$ close to the solution we are looking for. This then allows us to express $\dot{v}$ and $\lambda$ as functions of $(q,v)$. Notice that the matrix in (1.5) is invertible when $G(q)$ has full rank and $M(q)$ is invertible on $\ker G(q) = \{h \mid G(q)h = 0\}$.

We are now able to discuss the existence of a solution of (1.2). First of all, observe that the initial values $q_0, v_0, \lambda_0$ cannot be arbitrarily chosen. They have to satisfy the first relation of (1.4) and $\lambda_0 = \lambda(q_0, v_0)$, where $\lambda(q,v)$ is obtained from (1.5). In the case that $q_0, v_0, \lambda_0$ satisfy these conditions, we call them *consistent initial values*. Furthermore, every solution of (1.2) has to satisfy

$$\dot{q} = v, \qquad \dot{v} = \dot{v}(q,v), \qquad (1.6)$$

where $\dot{v}(q,v)$ is the function obtained from (1.5). It is known from standard theory of ordinary differential equations that (1.6) has locally a unique solution. This solution $(q(t), v(t))$ together with $\lambda(t) := \lambda(q(t), v(t))$ satisfies (1.5) by construction, and hence also the two differential equations of (1.2). Integrating the second relation of (1.4) twice and using the fact that the integration constants vanish for consistent initial values, proves also the remaining relation $0 = g(q)$ for this solution.

**Formulation as a Differential Equation on a Manifold.** The equations (1.6) define a differential equation on the manifold

$$\mathcal{M} = \{(q, v) \mid g(q) = 0,\ G(q)v = 0\}. \tag{1.7}$$

Indeed, we have just shown that for initial values $(q_0, v_0) \in \mathcal{M}$ (i.e., consistent initial values) the problems (1.6) and (1.2) are equivalent, so that the solutions of (1.6) stay on $\mathcal{M}$.

**Reversibility.** The system (1.2) and the corresponding differential equation (1.6) are reversible with respect to the involution $\rho(q, v) = (q, -v)$, if $f(q, -v) = f(q, v)$. This follows at once from Example V.1.3, because the solution $\dot{v}(q, v)$ of (1.5) satisfies $\dot{v}(q, -v) = \dot{v}(q, v)$

For the numerical solution of differential-algebraic equations "index reduction" is a very popular technique. This means that instead of directly treating the problem (1.2) one numerically solves the differential equation (1.6) on the manifold $\mathcal{M}$. Projection methods (Sect. IV.4) as well as methods based on local coordinates (Sect. IV.5) are much in use. If one is interested in a correct simulation of the reversible structure of the problem, the symmetric methods of Sect. V.4 can be applied. Here we do not repeat these approaches for this particular situation, instead we concentrate on the symplectic integration of constrained systems.

## VII.1.2 Hamiltonian Formulation

In Sect. VI.1 we have seen that, for unconstrained mechanical systems, the equations of motion become more structured if we use the momentum coordinates $p = \frac{\partial L}{\partial \dot{q}} = M(q)\dot{q}$ in place of the velocity coordinates $v = \dot{q}$. Let us do the same for the constrained system (1.2). As in the proof of Theorem VI.1.3 we obtain the equivalent system

$$\begin{aligned} \dot{q} &= H_p(p, q) \\ \dot{p} &= -H_q(p, q) - G(q)^T \lambda \\ 0 &= g(q), \end{aligned} \tag{1.8}$$

where

$$H(p, q) = \frac{1}{2} p^T M(q)^{-1} p + U(q) \tag{1.9}$$

is the total energy of the system; $H_p$ and $H_q$ denote the column vectors of partial derivatives. In the following we consider (1.8), where $H(p, q)$ is an arbitrary smooth function.

The existence and uniqueness of the solution of (1.8) can be discussed as before. Differentiating the constraint in (1.8) twice with respect to time, we get

$$0 = G(q)H_p(p,q), \tag{1.10}$$

$$0 = \frac{\partial}{\partial q}\Big(G(q)H_p(p,q)\Big)H_p(p,q) - G(q)H_{pp}(p,q)\Big(H_q(p,q) + G(q)^T\lambda\Big), \tag{1.11}$$

and assuming the matrix

$$G(q)H_{pp}(p,q)G(q)^T \quad \text{is invertible}, \tag{1.12}$$

equation (1.11) permits us to express $\lambda$ in terms of $(p,q)$. Inserting the so-obtained function $\lambda(p,q)$ into (1.8) gives a differential equation for $(p,q)$ on the manifold

$$\mathcal{M} = \{(p,q) \mid g(q) = 0,\ G(q)H_p(p,q) = 0\}. \tag{1.13}$$

It is reversible if $H(-p,q) = H(p,q)$.

**Preservation of the Hamiltonian.** Differentiation of $H(p(t),q(t))$ with respect to time yields

$$-H_p^T H_q - H_p^T G^T \lambda + H_q^T H_p$$

with all expressions evaluated at $(p(t),q(t))$. The first and the last terms cancel, and the central term vanishes because $GH_p = 0$ on the solution manifold. Consequently, the Hamiltonian $H(p,q)$ is constant along solutions of (1.8).

**Symplecticity of the Flow.** Since the flow of the system (1.8) is a transformation on $\mathcal{M}$, its derivative is a mapping between the corresponding tangent spaces. In agreement with Definition VI.2.2 we call a map $\varphi : \mathcal{M} \to \mathcal{M}$ symplectic if, for every $y = (p,q) \in \mathcal{M}$,

$$\xi^T \varphi'(y)^T J \varphi'(y)\eta = \xi^T J \eta \quad \text{for all } \xi,\eta \in T_y\mathcal{M}. \tag{1.14}$$

If $\varphi$ is actually defined and continuously differentiable in an open subset of $\mathbb{R}^{2d}$ that contains $\mathcal{M}$, then $\varphi'(y)$ in the above formula is just the usual Jacobian matrix. Otherwise, some care is necessary in the interpretation of (1.14): $\varphi'$ is the tangent map given by the directional derivative $\varphi'(y)\eta := (d/d\tau)|_{\tau=0}\,\varphi(\gamma(\tau))$ for $\eta \in T_y\mathcal{M}$, where $\gamma$ is a path on $\mathcal{M}$ with $\gamma(0) = y$, $\dot\gamma(0) = \eta$. The expression $\xi^T \varphi'(y)^T$ in (1.14) should then be interpreted as $(\varphi'(y)\xi)^T$.

**Theorem 1.2.** *Let $H(p,q)$ and $g(q)$ be twice continuously differentiable. The flow $\varphi_t : \mathcal{M} \to \mathcal{M}$ of the system (1.8) is then a symplectic transformation on $\mathcal{M}$, i.e., it satisfies (1.14).*

*Proof.* We let $y = (p,q)$, so that the system (1.8) becomes $\dot y = J^{-1}\big(\nabla H(y) + \sum_i \lambda_i(y)\nabla g_i(y)\big)$, where $\lambda_i(y)$ and $g_i(y)$ are the components of $\lambda(y)$ and $g(y)$, and $\lambda(y)$ is the function obtained from (1.11). The variational equation of this system, satisfied by the directional derivative $\Psi = \varphi'_t(y_0)\eta$, with $y_0 = (p_0,q_0)$, reads

$$\dot\Psi = J^{-1}\Big(\nabla^2 H(y) + \sum_{i=1}^m \lambda_i(y)\nabla^2 g_i(y) + \sum_{i=1}^m \nabla g_i(y)\nabla \lambda_i(y)^T\Big)\Psi.$$

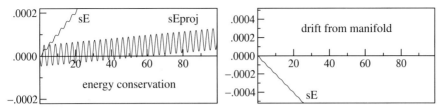

**Fig. 1.1.** Numerical solution of the symplectic Euler method applied to (1.16) with $H(p,q) = \frac{1}{2}(p_1^2 + p_2^2 + p_3^2) + q_3$, $g(q) = q_1^2 + q_2^2 + q_3^2 - 1$ (spherical pendulum); initial value $q_0 = (0, \sin(0.1), -\cos(0.1))$, $p_0 = (0.06, 0, 0)$, step size $h = 0.003$ for method 'sE' (without projection) and $h = 0.03$ for method 'sEproj' (with projection).

A direct computation, analogous to that in the proof of Theorem VI.2.4, yields for $\xi, \eta \in T_y \mathcal{M}$

$$\frac{d}{dt}\left(\xi^T \varphi_t'(y_0)^T J \varphi_t'(y_0)\eta\right) = \ldots = \sum_{i=1}^{m} \xi^T \varphi_t'(y_0)^T \nabla g_i(y) \nabla \lambda_i(y)^T \varphi_t'(y_0)\eta$$

$$- \sum_{i=1}^{m} \xi^T \varphi_t'(y_0)^T \nabla \lambda_i(y) \nabla g_i(y)^T \varphi_t'(y_0)\eta. \quad (1.15)$$

Since $g_i(\varphi_t(y_0)) = 0$ for $y_0 \in \mathcal{M}$, we have $\nabla g_i(y)^T \varphi_t'(y_0)\eta = 0$ and the same for $\xi$, so that the expression in (1.15) vanishes. This proves the symplecticity of the flow on $\mathcal{M}$. □

Applying the "index reduction" technique to (1.8) yields the differential equation

$$\dot{q} = H_p(p,q), \qquad \dot{p} = -H_q(p,q) - G(q)^T \lambda(p,q), \quad (1.16)$$

where $\lambda(p,q)$ is obtained from (1.11). If we solve this system with the symplectic Euler method (implicit in $p$, explicit in $q$), the qualitative behaviour of the numerical solution is not correct. As was observed by Leimkuhler & Reich (1994), there is a linear error growth in the Hamiltonian and also a drift from the manifold $\mathcal{M}$ (method 'sE' in Fig. 1.1). The explanation for this behaviour is the fact that (1.16) is no longer a Hamiltonian system. If we combine the symplectic Euler applied to (1.16) with an orthogonal projection onto $\mathcal{M}$ (method 'sEproj'), the result improves considerably but the linear error growth in the Hamiltonian is not eliminated. This numerical experiment illustrates that "index reduction" is not compatible with symplectic integration.

## VII.1.3 A Symplectic First Order Method

We extend the symplectic Euler method to Hamiltonian systems with constraints. We integrate the $q$-variable by the implicit and the $p$-variable by the explicit Euler method. This gives

$$\begin{aligned}
\widehat{p}_{n+1} &= p_n - h\left(H_q(\widehat{p}_{n+1}, q_n) + G(q_n)^T \lambda_{n+1}\right) \\
q_{n+1} &= q_n + h\, H_p(\widehat{p}_{n+1}, q_n) \\
0 &= g(q_{n+1}).
\end{aligned} \tag{1.17}$$

The numerical approximation $(\widehat{p}_{n+1}, q_{n+1})$ satisfies the constraint $g(q) = 0$, but not $G(q)H_p(p,q) = 0$. To get an approximation $(p_{n+1}, q_{n+1}) \in \mathcal{M}$, we append the projection

$$\begin{aligned}
p_{n+1} &= \widehat{p}_{n+1} - h\, G(q_{n+1})^T \mu_{n+1} \\
0 &= G(q_{n+1}) H_p(p_{n+1}, q_{n+1}).
\end{aligned} \tag{1.18}$$

Let us discuss some basic properties of this method.

**Existence and Uniqueness of the Numerical Solution.** Inserting the definition of $q_{n+1}$ from the second line of (1.17) into $0 = g(q_{n+1})$ gives a nonlinear system for $\widehat{p}_{n+1}$ and $h\lambda_{n+1}$. Due to the factor $h$ in front of $H_p(\widehat{p}_{n+1}, q_n)$, the implicit function theorem cannot be directly applied to prove existence and uniqueness of the numerical solution. We therefore write this equation as

$$0 = g(q_{n+1}) = g(q_n) + \int_0^1 G\big(q_n + \tau(q_{n+1} - q_n)\big)(q_{n+1} - q_n)\, d\tau.$$

We now use $g(q_n) = 0$, insert the definition of $q_{n+1}$ from the second line of (1.17) and divide by $h$. Together with the first line of (1.17) this yields the system $F(\widehat{p}_{n+1}, h\lambda_{n+1}, h) = 0$ with

$$F(p, \nu, h) = \begin{pmatrix} p - p_n + h H_q(p, q_n) + G(q_n)^T \nu \\ \int_0^1 G\big(q_n + \tau h H_p(p, q_n)\big) H_p(p, q_n)\, d\tau \end{pmatrix}.$$

Since $(p_n, q_n) \in \mathcal{M}$ with $\mathcal{M}$ from (1.13), we have $F(p_n, 0, 0) = 0$. Furthermore,

$$\frac{\partial F}{\partial(p, \nu)}(p_n, 0, 0) = \begin{pmatrix} I & G(q_n)^T \\ G(q_n) H_{pp}(p_n, q_n) & 0 \end{pmatrix},$$

and this matrix is invertible by (1.12). Consequently, an application of the implicit function theorem proves that the numerical solution $(\widehat{p}_{n+1}, h\lambda_{n+1})$ (and hence also $q_{n+1}$) exists and is locally unique for sufficiently small $h$.

The projection step (1.18) constitutes a nonlinear system for $p_{n+1}$ and $h\mu_{n+1}$, to which the implicit function theorem can be directly applied.

**Convergence of Order 1.** The above use of the implicit function theorem yields the rough estimates

$$\widehat{p}_{n+1} = p_n + \mathcal{O}(h), \quad h\lambda_{n+1} = \mathcal{O}(h), \quad h\mu_{n+1} = \mathcal{O}(h),$$

which, together with the equations (1.17) and (1.18), give

$$q_{n+1} = q(t_{n+1}) + \mathcal{O}(h^2), \quad p_{n+1} = p(t_{n+1}) - G\big(q(t_{n+1})\big)^T \nu + \mathcal{O}(h^2),$$

where $(p(t), q(t))$ is the solution of (1.8) passing through $(p_n, q_n) \in \mathcal{M}$ at $t = t_n$. Inserting these relations into the second equation of (1.18) we get

$$0 = G(q(t)) H_p(p(t), q(t)) + G(q(t)) H_{pp}(p(t), q(t)) G(q(t))^T \nu + \mathcal{O}(h^2)$$

at $t = t_{n+1}$. Since $G(q(t)) H_p(p(t), q(t)) = 0$, it follows from (1.12) that $\nu = \mathcal{O}(h^2)$. The local error is therefore of size $\mathcal{O}(h^2)$.

The convergence proof now follows standard arguments, because the method is a mapping $\Phi_h : \mathcal{M} \to \mathcal{M}$ on the solution manifold. We consider the solutions $(p_n(t), q_n(t))$ of (1.8) passing through the numerical values $(p_n, q_n) \in \mathcal{M}$ at $t = t_n$, we estimate the difference of two successive solutions in terms of the local error at $t_n$, and we sum up the propagated errors (see Fig. 3.2 of Sect. II.3 in Hairer, Nørsett & Wanner (1993)). This proves that the global error satisfies $p_n - p(t_n) = \mathcal{O}(h)$ and $q_n - q(t_n) = \mathcal{O}(h)$ as long as $t_n = nh \leq Const$.

**Symplecticity.** We first study the mapping $(p_n, q_n) \mapsto (\widehat{p}_{n+1}, q_{n+1})$ defined by (1.17), and we consider $\lambda_{n+1}$ as a function $\lambda(p_n, q_n)$. Differentiation with respect to $(p_n, q_n)$ yields

$$\begin{pmatrix} I + hH_{qp}^T & 0 \\ -hH_{pp} & I \end{pmatrix} \begin{pmatrix} \partial(\widehat{p}_{n+1}, q_{n+1}) \\ \partial(p_n, q_n) \end{pmatrix} = \begin{pmatrix} I - hG^T \lambda_p & S - hG^T \lambda_q \\ 0 & I + hH_{qp} \end{pmatrix}, \quad (1.19)$$

where $S = -hH_{qq} - h\lambda^T g_{qq}$ is a symmetric matrix, the expressions $H_{qp}, H_{pp}, H_{qq}, G$ are evaluated at $(\widehat{p}_{n+1}, q_n)$, and $\lambda, \lambda_p, \lambda_q$ at $(p_n, q_n)$. A computation, identical to that of the proof of Theorem VI.3.3, yields

$$\left(\frac{\partial(\widehat{p}_{n+1}, q_{n+1})}{\partial(p_n, q_n)}\right)^T J \left(\frac{\partial(\widehat{p}_{n+1}, q_{n+1})}{\partial(p_n, q_n)}\right) = \begin{pmatrix} 0 & -I + h\lambda_p^T G \\ I - hG\lambda_p & h(\lambda_q^T G - G^T \lambda_q) \end{pmatrix}.$$

We multiply this relation from the left by $\xi \in T_{(p_n, q_n)}\mathcal{M}$ and from the right by $\eta \in T_{(p_n, q_n)}\mathcal{M}$. With the partitioning $\xi = (\xi_p, \xi_q)$ we have $G(q_n)\xi_q = 0$ (and similarly also $G(q_n)\eta_q = 0$) so that the expression reduces to $\xi^T J \eta$. This proves the symplecticity condition (1.14) for the mapping $(p_n, q_n) \mapsto (\widehat{p}_{n+1}, q_{n+1})$.

Similarly, the projection step $(\widehat{p}_{n+1}, q_{n+1}) \mapsto (p_{n+1}, q_{n+1})$ of (1.18) gives

$$\frac{\partial(p_{n+1}, q_{n+1})}{\partial(\widehat{p}_{n+1}, q_{n+1})} = \begin{pmatrix} I - hG^T \mu_p & S - hG^T \mu_q \\ 0 & I \end{pmatrix},$$

where $\mu_{n+1}$ of (1.18) is considered as a function of $(\widehat{p}_{n+1}, q_{n+1})$, and $S = -h\mu^T g_{qq}$. This is formally the same as (1.19) with $H \equiv 0$. Consequently, the symplecticity condition is also satisfied for this mapping. As a composition of two symplectic transformations, the numerical flow of our first order method is therefore also symplectic.

**Numerical Experiment.** Consider the equations (1.3) for the spherical pendulum. For a mass $m = 1$ they coincide with the Hamiltonian formulation. Figure 1.2 (upper picture) shows the numerical solution (vertical coordinate $q_3$) over many

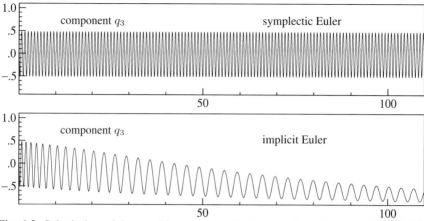
**Fig. 1.2.** Spherical pendulum problem solved with the symplectic Euler method (1.17)-(1.18) and with the implicit Euler method; initial value $q_0 = (\sin(1.3), 0, \cos(1.3))$, $p_0 = (3\cos(1.3), 6.5, 3\sin(1.3))$, step size $h = 0.01$.

periods obtained by method (1.17)-(1.18). We observe a regular qualitatively correct behaviour. For the implicit Euler method (i.e., the argument $q_n$ is replaced with $q_{n+1}$ in (1.17)) the numerical solution, obtained with the same step size and the same initial values, is less satisfactory. Already after one period the solution deteriorates and the pendulum loses energy.

## VII.1.4 SHAKE and RATTLE

The numerical method (1.17)-(1.18) is only of order 1 and it is not symmetric. An algorithm that is of order 2, symmetric and symplectic was originally considered for separable Hamiltonians

$$H(p,q) = \frac{1}{2} p^T M^{-1} p + U(q) \tag{1.20}$$

with constant mass matrix $M$. Notice that in this case we are concerned with a second order differential equation $M\ddot{q} = -U_q(q) - G(q)^T \lambda$ with $g(q) = 0$.

**SHAKE.** Ryckaert, Ciccotti & Berendsen (1977) propose the method

$$\begin{aligned} q_{n+1} - 2q_n + q_{n-1} &= -h^2 M^{-1} \big( U_q(q_n) + G(q_n)^T \lambda_n \big) \\ 0 &= g(q_{n+1}) \end{aligned} \tag{1.21}$$

for computations in molecular dynamics. It is a straightforward extension of the Störmer/Verlet scheme (I.3.4). The $p$-components, not used in the recursion, are approximated by $p_n = M(q_{n+1} - q_{n-1})/2h$.

**RATTLE.** The three-term recursion (1.21) may lead to an accumulation of round-off errors, and a reformulation as a one-step method is desirable. Using the same procedure as in (I.3.6) we formally get

$$\begin{aligned}
p_{n+1/2} &= p_n - \frac{h}{2}\left(U_q(q_n) + G(q_n)^T \lambda_n\right) \\
q_{n+1} &= q_n + hM^{-1}p_{n+1/2}, \qquad 0 = g(q_{n+1}) \\
p_{n+1} &= p_{n+1/2} - \frac{h}{2}\left(U_q(q_{n+1}) + G(q_{n+1})^T \lambda_{n+1}\right).
\end{aligned} \qquad (1.22)$$

The difficulty with this formulation is that $\lambda_{n+1}$ is not yet available at this step (it is computed together with $q_{n+2}$). As a remedy, Andersen (1983) suggests replacing the last line in (1.22) with a projection step similar to (1.18)

$$\begin{aligned}
p_{n+1} &= p_{n+1/2} - \frac{h}{2}\left(U_q(q_{n+1}) + G(q_{n+1})^T \mu_n\right) \\
0 &= G(q_{n+1}) M^{-1} p_{n+1}.
\end{aligned} \qquad (1.23)$$

This modification, called RATTLE, has the further advantage that the numerical approximation $(p_{n+1}, q_{n+1})$ lies on the solution manifold $\mathcal{M}$. The symplecticity of this algorithm has been established by Leimkuhler & Skeel (1994).

**Extension to General Hamiltonians.** As observed independently by Jay (1994) and Reich (1993), the RATTLE algorithm can be extended to general Hamiltonians as follows: for consistent values $(p_n, q_n) \in \mathcal{M}$ define

$$\begin{aligned}
p_{n+1/2} &= p_n - \frac{h}{2}\left(H_q(p_{n+1/2}, q_n) + G(q_n)^T \lambda_n\right) \\
q_{n+1} &= q_n + \frac{h}{2}\left(H_p(p_{n+1/2}, q_n) + H_p(p_{n+1/2}, q_{n+1})\right) \\
0 &= g(q_{n+1}) \\
p_{n+1} &= p_{n+1/2} - \frac{h}{2}\left(H_q(p_{n+1/2}, q_{n+1}) + G(q_{n+1})^T \mu_n\right) \\
0 &= G(q_{n+1}) H_p(p_{n+1}, q_{n+1}).
\end{aligned} \qquad (1.24)$$

The first three equations of (1.24) are very similar to (1.17) and the last two equations to (1.18). The existence of (locally) unique solutions $(p_{n+1/2}, q_{n+1}, \lambda_n)$ and $(p_{n+1}, \mu_n)$ can therefore be proved in the same way. Notice also that this method gives a numerical solution that stays exactly on the solution manifold $\mathcal{M}$.

**Theorem 1.3.** *The numerical method (1.24) is symmetric, symplectic, and convergent of order two.*

*Proof.* Although this theorem is the special case $s = 2$ of Theorem 1.4, we outline its proof. We will see that the convergence result is easier to obtain for $s = 2$ than for the general case.

If we add to (1.24) the consistency conditions $g(q_n) = 0$, $G(q_n)H_p(p_n, q_n) = 0$ of the initial values, the symmetry of the method follows at once by exchanging $h \leftrightarrow -h$, $p_{n+1} \leftrightarrow p_n$, $q_{n+1} \leftrightarrow q_n$, and $\lambda_n \leftrightarrow \mu_n$. The symplecticity can be proved as for (1.17)-(1.18) by computing the derivative of $(p_{n+1}, q_{n+1})$ with respect to $(p_n, q_n)$, and by verifying the condition (1.14). This does not seem to be simpler than the symplecticity proof of Theorem 1.4.

The implicit function theorem applied to the two subsystems of (1.24) shows

$$p_{n+1/2} = p_n + \mathcal{O}(h), \quad h\lambda = \mathcal{O}(h), \quad p_{n+1} = p_{n+1/2} + \mathcal{O}(h), \quad h\mu = \mathcal{O}(h),$$

and, inserted into (1.24), yields

$$q_{n+1} = q(t_{n+1}) + \mathcal{O}(h^2), \quad p_{n+1} = p(t_{n+1}) - G\bigl(q(t_{n+1})\bigr)^T \nu + \mathcal{O}(h^2).$$

Convergence of order one follows therefore in the same way as for method (1.17)-(1.18). Since the order of a symmetric method is always even, this implies convergence of order two. □

An easy way of obtaining high order methods for constrained Hamiltonian systems is by composition (Reich 1996a). Method (1.24) is an ideal candidate as basic integrator for compositions of the form (V.3.2). The resulting integrators are symmetric, symplectic, of high order, and yield a numerical solution that stays on the manifold $\mathcal{M}$.

### VII.1.5 The Lobatto IIIA - IIIB Pair

Another possibility for obtaining high order symplectic integrators for constrained Hamiltonian systems is by the use of partitioned Runge-Kutta or discontinuous collocation methods. We consider the system (1.8) and we search for polynomials $u(t)$ of degree $s$, $w(t)$ of degree $s-1$, and $v(t)$ of degree $s-2$ such that

$$u(t_n) = q_n, \quad v(t_n) = p_n - hb_1\delta(t_n) \tag{1.25}$$

with the defect

$$\delta(t) = \dot{v}(t) + H_q\bigl(v(t), u(t)\bigr) + G\bigl(u(t)\bigr)^T w(t) \tag{1.26}$$

and, using the abbreviation $t_{n,i} = t_n + c_i h$,

$$\dot{u}(t_{n,i}) = H_p\bigl(v(t_{n,i}), u(t_{n,i})\bigr), \quad i = 1, \ldots, s \tag{1.27}$$

$$\dot{v}(t_{n,i}) = -H_q\bigl(v(t_{n,i}), u(t_{n,i})\bigr) - G\bigl(u(t_{n,i})\bigr)^T w(t_{n,i}), \quad i = 2, \ldots, s-1$$

$$0 = g\bigl(u(t_{n,i})\bigr), \quad i = 1, \ldots, s.$$

If these polynomials exist, the numerical solution is defined by

$$q_{n+1} = u(t_n + h), \quad p_{n+1} = v(t_n + h) - hb_s \delta(t_n + h)$$
$$0 = G(q_{n+1}) H_p(p_{n+1}, q_{n+1}). \tag{1.28}$$

**Why Discontinuous Collocation Based on Lobatto Quadrature?** At a first glance (Theorem VI.4.2) it seems natural to consider collocation methods based on Gaussian quadrature for the entire system. This, however, has the disadvantage that the numerical solution does not satisfy $g(q_{n+1}) = 0$. To achieve this requirement, $t_n + h$ has to be one of the collocation points, i.e., we must have $c_s = 1$. Unfortunately, none of the collocation or discontinuous collocation methods with $c_s = 1$ is symplectic (see Exercise IV.6). We therefore turn our attention to partitioned methods, and we treat only the $q$-component by a collocation method satisfying $c_s = 1$. To satisfy the $s$ conditions $g(u(t_{n,i})) = 0$ of (1.27) there are only $s - 1$ free parameters $w(t_n), w(t_n + c_2 h), \ldots, w(t_n + c_{s-1} h)$ available. A remedy is to choose $c_1 = 0$ so that the first condition $g(u(t_n)) = 0$ is automatically verified. Encouraged by Theorem VI.4.5 we are thus led to consider the Lobatto nodes in the role of the $c_i$. The use of the partitioned Lobatto IIIA - IIIB pair for the treatment of constrained Hamiltonian systems has been suggested by Jay (1994, 1996).

**Existence and Uniqueness of the Numerical Solution.** The polynomial $u(t)$ of degree $s$ is uniquely determined by $u(t_n) = q_n$ and $\dot{u}(t_{n,i}) =: \dot{Q}_i$ $(i = 1, \ldots, s)$, the polynomial $v(t)$ of degree $s - 2$ is uniquely determined by $v(t_{n,i}) =: P_i$ $(i = 1, \ldots, s - 1)$, and the polynomial $w(t)$ of degree $s - 1$ is uniquely determined by $hw(t_{n,i}) =: \Lambda_i$ $(i = 1, \ldots, s)$. Notice that the value $\Lambda_s$ is only involved in (1.28) and not in (1.25)-(1.27). For the nonlinear system (1.25)-(1.27) we therefore consider

$$X = (\dot{Q}_1, \ldots, \dot{Q}_s, P_1, \ldots, P_{s-1}, \Lambda_1, \ldots, \Lambda_{s-1})$$

as independent variables, and we write the system as $F(X, h) = 0$. The function $F$ is composed of the $s$ conditions for $\dot{u}(t_{n,i})$, of the definition of $v(t_n)$ (divided by $b_1$) and the $s - 2$ conditions for $\dot{v}(t_{n,i})$ (multiplied by $h$), and finally of the $s - 1$ equations $0 = g(u(t_{n,i}))$ for $i = 2, \ldots, s$ (divided by $h$). Observe that $0 = g(u(t_n))$ is automatically satisfied by the consistency of $(p_n, q_n)$. We note that $P_s = v(t_n + h)$ and $\dot{P}_i = h\dot{v}(t_{n,i})$ are linear combinations of $P_1, \ldots, P_{s-1}$ with coefficients independent of the step size $h$.

The function $F(X, h)$ is well-defined for $h$ in a neighbourhood of 0. For the first two blocks this is evident, for the last one it follows from the identity

$$\frac{1}{h} g(u(t_{n,i})) = \int_0^{c_i} G(u(t_n + \theta h)) \dot{u}(t_n + \theta h) \, d\theta$$

using the fact that $\dot{u}(t_n + \theta h)$ is a linear combination of $\dot{Q}_i$ for $i = 1, \ldots, s$. With the values

$$X_0 = (H_p(p_n, q_n), \ldots, H_p(p_n, q_n), p_n, \ldots, p_n, 0, \ldots, 0)$$

we have that $F(X_0, 0) = 0$, because the values $(p_n, q_n)$ are assumed to be consistent. In view of an application of the implicit function theorem we compute

$$\frac{\partial F}{\partial X}(X_0, 0) = \begin{pmatrix} I \otimes I & -D \otimes H_{pp} & 0 \\ 0 & B \otimes I & I \otimes G^T \\ A \otimes G & 0 & 0 \end{pmatrix}, \qquad (1.29)$$

where $H_{pp}, G$ are evaluated at $(p_n, q_n)$, and $A, B, D$ are matrices of dimension $(s - 1) \times s$, $(s - 1) \times (s - 1)$ and $s \times (s - 1)$ respectively that depend only on the

Lobatto quadrature and not on the differential equation. For example, the matrix $B$ represents the linear mapping

$$(P_1, \ldots, P_{s-1}) \mapsto (\dot{P}_1 + b_1^{-1} P_1, \dot{P}_2, \ldots, \dot{P}_{s-1}).$$

This mapping is invertible, because the values on the right-hand side uniquely determine the polynomial $v(t)$ of degree $s - 2$.

Block Gaussian elimination then shows that (1.29) is invertible if and only if the matrix

$$ADB^{-1} \otimes GH_{pp}G^T \qquad \text{is invertible.}$$

Because of (1.12) it remains to show that $ADB^{-1}$ is invertible.

To achieve this without explicitly computing the matrices $A, B, D$, we apply the method to the problem where $p$ and $q$ are of dimension one, $H(p, q) = p^2/2$, and $g(q) = q$. Assuming $h = 1$ we get

$$\begin{aligned}
u(0) &= 0, & v(0) &= -b_1(\dot{v}(0) + w(0)) \\
\dot{u}(c_i) &= v(c_i) & \text{for } i &= 1, \ldots, s \\
\dot{v}(c_i) &= -w(c_i) & \text{for } i &= 2, \ldots, s-1 \\
0 &= u(c_i) & \text{for } i &= 1, \ldots, s,
\end{aligned} \qquad (1.30)$$

which is equivalent to

$$\begin{pmatrix} I & -D & 0 \\ 0 & B & I \\ A & 0 & 0 \end{pmatrix} \begin{pmatrix} (\dot{u}(c_i))_{i=1}^s \\ (v(c_i))_{i=1}^{s-1} \\ (w(c_i))_{i=1}^{s-1} \end{pmatrix} = \begin{pmatrix} 0 \\ 0 \\ 0 \end{pmatrix}, \qquad (1.31)$$

because $H_{pp}(p, q) = 1$ and $G(q) = 1$. Since $u(t)$ is a polynomial of degree $s$, the last equation of (1.30) implies that $u(t) = C \prod_{j=1}^s (t - c_j)$. By the second relation the polynomial $\dot{u}(t) - v(t)$, which is of degree $s - 1$, vanishes at $s$ points. Hence, $v(t) \equiv \dot{u}(t)$, which is possible only if $C = 0$, because the degree of $v(t)$ is $s - 2$. Consequently, the linear system (1.31) has only the trivial solution, so that the matrix in (1.31) and hence also $ADB^{-1}$ is invertible.

The implicit function theorem applied to $F(X, h) = 0$ shows that the nonlinear system (1.25)-(1.28) possesses a locally unique solution for sufficiently small step sizes $h$. Using the free parameter $\Lambda_s = hw(t_n + h)$, a further application of the implicit function theorem, this time to the small system (1.28), proves the existence and local uniqueness of $p_{n+1}$.

**Theorem 1.4.** *Let $(b_i, c_i)_{i=1}^s$ be the weights and nodes of the Lobatto quadrature (c.f. (II.1.17)). The method (1.25)-(1.27)-(1.28) is symmetric, symplectic, and super-convergent of order $2s - 2$.*

*Proof. Symmetry.* To the formulas (1.25)-(1.27)-(1.28) we add the consistency relations $g(q_n) = 0$, $G(q_n)H_p(p_n, q_n) = 0$. Then we exchange $(t_n, p_n, q_n) \leftrightarrow (t_{n+1}, p_{n+1}, q_{n+1})$ and $h \leftrightarrow -h$. Since $b_1 = b_s$ and $c_{s+1-i} = 1 - c_i$ for the Lobatto quadrature, the resulting formulas are equivalent to the original method (see also the proof of Theorem V.2.1).

**Symplecticity.** We fix $\xi, \eta \in T_{(p_n,q_n)}\mathcal{M}$, we put $y_n = (p_n, q_n)^T$, and we consider the bilinear mapping

$$Q\left(\frac{\partial p_{n+1}}{\partial y_n}, \frac{\partial q_{n+1}}{\partial y_n}\right) = \xi^T\left(\left(\frac{\partial q_{n+1}}{\partial y_n}\right)^T\left(\frac{\partial p_{n+1}}{\partial y_n}\right) - \left(\frac{\partial p_{n+1}}{\partial y_n}\right)^T\left(\frac{\partial q_{n+1}}{\partial y_n}\right)\right)\eta.$$

The symplecticity of the transformation $(p_n, q_n) \mapsto (p_{n+1}, q_{n+1})$ on the manifold $\mathcal{M}$ is then expressed by the relation

$$Q\left(\frac{\partial p_{n+1}}{\partial y_n}, \frac{\partial q_{n+1}}{\partial y_n}\right) = Q\left(\frac{\partial p_n}{\partial y_n}, \frac{\partial q_n}{\partial y_n}\right). \tag{1.32}$$

We now follow closely the proof of Theorem IV.2.3. We consider the polynomials $u(t), v(t), w(t)$ of the method (1.25)-(1.27)-(1.28) as functions of $t$ and $y_n = (p_n, q_n)$, and we compute

$$Q\left(\frac{\partial v(t_{n+1})}{\partial y_n}, \frac{\partial u(t_{n+1})}{\partial y_n}\right) - Q\left(\frac{\partial v(t_n)}{\partial y_n}, \frac{\partial u(t_n)}{\partial y_n}\right) \\ = \int_{t_n}^{t_{n+1}} \frac{dQ}{dt}\left(\frac{\partial v(t)}{\partial y_n}, \frac{\partial u(t)}{\partial y_n}\right) dt. \tag{1.33}$$

Since $u(t)$ is a polynomial of degree $s$ and $v(t)$ of degree $s - 2$, the integrand in (1.33) is a polynomial in $t$ of degree $2s - 3$. It is thus integrated without error by the Lobatto quadrature. By definition these polynomials satisfy the differential equation at the interior collocation points. Therefore, it follows from (1.15) that

$$\frac{dQ}{dt}\left(\frac{\partial v(t_{n,i})}{\partial y_n}, \frac{\partial u(t_{n,i})}{\partial y_n}\right) = 0 \qquad \text{for } i = 2, \ldots, s - 1,$$

and that

$$\frac{dQ}{dt}\left(\frac{\partial v(t_{n,i})}{\partial y_n}, \frac{\partial u(t_{n,i})}{\partial y_n}\right) = Q\left(\frac{\partial \delta(t_{n,i})}{\partial y_n}, \frac{\partial u(t_{n,i})}{\partial y_n}\right) \qquad \text{for } i = 1 \text{ and } i = s.$$

Applying the Lobatto quadrature to the integral in (1.33) thus yields

$$hb_1 Q\left(\frac{\partial \delta(t_n)}{\partial y_n}, \frac{\partial u(t_n)}{\partial y_n}\right) + hb_s Q\left(\frac{\partial \delta(t_{n+1})}{\partial y_n}, \frac{\partial u(t_{n+1})}{\partial y_n}\right),$$

and the symplecticity relation (1.32) follows in the same way as in the proof of Theorem IV.2.3.

**Superconvergence.** This is the most difficult part of the proof. We remark that superconvergence of Runge-Kutta methods for differential-algebraic systems of index 3 has been conjectured by Hairer, Lubich & Roche (1989), and a first proof has been obtained by Jay (1993) for collocation methods. In his thesis Jay (1994) proves superconvergence for a more general class of methods, including the Lobatto IIIA - IIIB pair, using a 'rooted-tree-type' theory. A sketch of that very elaborate proof is published in Jay (1996). Using the idea of discontinuous collocation, the elegant proof for collocation methods can now be extended to cover the Lobatto IIIA - IIIB pair. In the following we explain how the local error can be estimated.

We consider the polynomials $u(t), v(t), w(t)$ defined in (1.25)-(1.27)-(1.28), and we define defects $\mu(t), \delta(t), \theta(t)$ as follows:

$$\begin{aligned} \dot{u}(t) &= H_p\big(v(t), u(t)\big) + \mu(t) \\ \dot{v}(t) &= -H_q\big(v(t), u(t)\big) - G\big(u(t)\big)^T w(t) + \delta(t) \\ 0 &= g\big(u(t)\big) + \theta(t). \end{aligned} \qquad (1.34)$$

By definition of the method we have

$$\begin{aligned} \mu(t_n + c_i h) &= 0, & i &= 1, \ldots, s \\ \delta(t_n + c_i h) &= 0, & i &= 2, \ldots, s-1 \\ \theta(t_n + c_i h) &= 0, & i &= 1, \ldots, s. \end{aligned} \qquad (1.35)$$

We let $q(t), p(t), \lambda(t)$ be the exact solution of (1.8) satisfying $q(t_n) = q_n$, $p(t_n) = p_n$, and we consider the differences

$$\Delta u(t) = u(t) - q(t), \quad \Delta v(t) = v(t) - p(t), \quad \Delta w(t) = w(t) - \lambda(t).$$

Subtracting (1.8) from (1.34) we get by linearization that

$$\begin{aligned} \dot{\Delta u} &= a_{11}(t)\Delta u + a_{12}(t)\Delta v + \mu(t) \\ \dot{\Delta v} &= a_{21}(t)\Delta u + a_{22}(t)\Delta v + a_{23}(t)\Delta w + \delta(t), \end{aligned} \qquad (1.36)$$

where $a_{12}(t) = H_{pp}\big(p(t), q(t)\big)$, and where the other $a_{ij}(t)$ are given by similar expressions. We have suppressed quadratic and higher order terms to keep the presentation as simple as possible. They do not influence the convergence result. To eliminate $\Delta w$ in (1.36), we differentiate the algebraic relations in (1.8) and (1.34) twice, and we subtract them. This yields

$$\begin{aligned} 0 = {}& F\big(t, \mu(t)\big) + b_1(t)\Delta u + b_2(t)\Delta v + B(t)\Delta w \\ &+ G\big(u(t)\big) H_{pp}\big(v(t), u(t)\big)\delta(t) + G\big(u(t)\big)\dot{\mu}(t) + \ddot{\theta}(t), \end{aligned}$$

where $F(t, \mu)$, $B(t)$, $b_1(t)$, $b_2(t)$ are functions depending on $p(t), q(t), \lambda(t), u(t), v(t), w(t)$, and where $F(t, 0) = 0$ and $B(t) \approx G(q_n) H_{pp}(p_n, q_n) G(q_n)^T$. Because of our assumption (1.12) we can extract $\Delta w$ from this relation, and we insert it into (1.36). In this way we get a linear differential equation for $\Delta u, \Delta v$, which can be solved by the 'variation of constants' formula. Using $\Delta u(t_n) = 0$ (by (1.25)), the solution $\Delta v(t_n + h)$ is seen to be of the form

$$\Delta v(t_n + h) = R_{22}(t_n + h, t_n)\Delta v(t_n) + \int_{t_n}^{t_n+h} \Big( R_{21}(t_n + h, t)\mu(t)$$
$$+ R_{22}(t_n + h, t)\Big(\delta(t) + \widetilde{F}\big(t, \mu(t)\big) + c_1(t)\dot\mu(t)\Big) \quad (1.37)$$
$$+ C(t)\Big(G\big(u(t)\big) H_{pp}\big(v(t), u(t)\big)\delta(t) + \ddot\theta(t)\Big)\Big)\,dt,$$

where $R_{21}$ and $R_{22}$ are the lower blocks of the resolvent, and $\widetilde{F}, c_1, C$ are functions as before. To prove that the local error of the $p$-component

$$p_{n+1} - p(t_n + h) = \Delta v(t_n + h) - h b_s \delta(t_n + h) \quad (1.38)$$

is of size $\mathcal{O}(h^{2s-1})$, we first integrate by parts those expressions in (1.37) which contain a derivative. For example,

$$\int_{t_n}^{t_{n+1}} a(t)\dot\mu(t)\,dt = a(t)\mu(t)\Big|_{t_n}^{t_{n+1}} - \int_{t_n}^{t_{n+1}} \dot a(t)\mu(t)\,dt = \mathcal{O}(h^{2s-1}),$$

because $\mu(t_n) = \mu(t_n + h) = 0$ by (1.35) and an application of the Lobatto quadrature to the integral at the right-hand side gives zero as result with a quadrature error of size $\mathcal{O}(h^{2s-1})$. Similarly, integrating by parts twice yields

$$\int_{t_n}^{t_{n+1}} a(t)\ddot\theta(t)\,dt = a(t)\dot\theta(t)\Big|_{t_n}^{t_{n+1}} - \dot a(t)\theta(t)\Big|_{t_n}^{t_{n+1}} + \int_{t_n}^{t_{n+1}} \ddot a(t)\theta(t)\,dt$$
$$= a(t_{n+1})\dot\theta(t_{n+1}) - a(t_n)\dot\theta(t_n) + \mathcal{O}(h^{2s-1}).$$

To the other integrals in (1.37) we apply the Lobatto quadrature directly. Since $R_{22}(t_{n+1}, t_{n+1})$ is the identity, this gives

$$p_{n+1} - p(t_{n+1}) = R_{22}(t_{n+1}, t_n)\Big(\Delta v(t_n) + h b_1 \delta(t_n)\Big) \quad (1.39)$$
$$+ \widetilde{C}(t_{n+1})\Big(h b_s G\big(u(t_{n+1})\big) H_{pp}\big(v(t_{n+1}), u(t_{n+1})\big)\delta(t_{n+1}) + \dot\theta(t_{n+1})\Big)$$
$$+ \widetilde{C}(t_n)\Big(h b_1 G\big(u(t_n)\big) H_{pp}\big(v(t_n), u(t_n)\big)\delta(t_n) - \dot\theta(t_n)\Big) + \mathcal{O}(h^{2s-1}),$$

where $\widetilde{C}(t) = R(t_{n+1}, t)C(t)$. The term $\Delta v(t_n) + h b_1 \delta(t_n)$ vanishes by (1.25), and differentiation of the algebraic relation in (1.34) yields

$$0 = G\big(u(t)\big)\Big(H_p\big(v(t), u(t)\big) + \mu(t)\Big) + \dot\theta(t).$$

As a consequence of (1.25), (1.35) and the consistency of the initial values $(p_n, q_n)$, this gives

$$\dot\theta(t_n) = -G(q_n)H_p\big(p_n - h b_1 \delta(t_n), q_n\big)$$
$$= h b_1 G(q_n) H_{pp}(p_n, q_n)\delta(t_n) + \mathcal{O}\big(h^2 \delta(t_n)^2\big)$$
$$= h b_1 G\big(u(t_n)\big) H_{pp}\big(v(t_n), u(t_n)\big)\delta(t_n) + \mathcal{O}\big(h^2 \delta(t_n)^2\big).$$

Using (1.28) we get in the same way

$$\dot\theta(t_{n+1}) = -hb_s G\big(u(t_{n+1})\big) H_{pp}\big(v(t_{n+1}), u(t_{n+1})\big)\delta(t_{n+1}) + \mathcal{O}\big(h^2 \delta(t_{n+1})^2\big).$$

These estimates together show that the local error (1.39) is of size $\mathcal{O}(h^{2s-1}) + \mathcal{O}(h^2\delta(t)^2)$. The defect $\delta(t)$ vanishes at $s-2$ points in the interval $[t_n, t_{n+1}]$, so that $\delta(t) = \mathcal{O}(h^{s-2})$ for $t \in [t_n, t_{n+1}]$ (for a rigorous proof of this statement one has to apply the techniques of the proof of Theorem II.1.5). Therefore we obtain $p_{n+1} - p(t_{n+1}) = \mathcal{O}(h^{2s-2})$, and by the symmetry of the method also $\mathcal{O}(h^{2s-1})$.

In analogy to (1.37), the variation of constants formula yields also an expression for the local error $q_{n+1} - q(t_{n+1}) = \Delta u(t_{n+1})$. One only has to replace $R_{21}$ and $R_{22}$ with the upper blocks $R_{11}$ and $R_{12}$ of the resolvent. Using $R_{12}(t_{n+1}, t_{n+1}) = 0$, we prove in the same way that the local error of the $q$-component is of size $\mathcal{O}(h^{2s-1})$.

The estimation of the global error is obtained in the same way as for the first order method (1.17)-(1.18). Since the algorithm is a mapping $\Phi_h : \mathcal{M} \to \mathcal{M}$ on the solution manifold, it is not necessary to follow the technically difficult proofs in the context of differential-algebraic equations. Summing up the propagated local errors proves that the global error satisfies $p_n - p(t_n) = \mathcal{O}(h^{2s-2})$ and $q_n - q(t_n) = \mathcal{O}(h^{2s-2})$ as long as $t_n = nh \le Const$. □

## VII.1.6 Splitting Methods

When considering splitting methods for constrained mechanical systems, it should be borne in mind that such systems are differential equations on manifolds (see Sect. VII.1.2). Splitting methods should therefore be based on a decomposition $f(y) = f^{[1]}(y) + f^{[2]}(y)$, where both $f^{[i]}(y)$ are vector fields on the same manifold as $f(y)$. Let us consider here the Hamiltonian system (1.8) with Hamiltonian

$$H(p, q) = H^{[1]}(p, q) + H^{[2]}(p, q). \tag{1.40}$$

The manifold for this differential equation is

$$\mathcal{M} = \big\{(p, q) \mid g(q) = 0,\; G(q)H_p(p, q) = 0\big\}. \tag{1.41}$$

Notice that (1.8), when $H$ is simply replaced with $H^{[i]}$, is not a good candidate for splitting methods: the existence of a solution is not guaranteed, and if the solution exists it need not stay on the manifold $\mathcal{M}$. The following lemma indicates how splitting methods should be applied.

**Lemma 1.5.** *Consider a Hamiltonian (1.40), a function $g(q)$ with $G(q) = g'(q)$, and let the manifold $\mathcal{M}$ be given by (1.41). If (1.12) holds and if*

$$G(q)H_p^{[i]}(p, q) = 0 \quad \text{for all } (p, q) \in \mathcal{M}, \tag{1.42}$$

*then the system*

$$\begin{aligned}
\dot q &= H_p^{[i]}(p,q) \\
\dot p &= -H_q^{[i]}(p,q) - G(q)^T\lambda \qquad (1.43)\\
0 &= G(q)H_p(p,q)
\end{aligned}$$

*defines a differential equation on the manifold $\mathcal{M}$, and its flow is a symplectic transformation on $\mathcal{M}$.*

*Proof.* Differentiation of the algebraic relation in (1.43) with respect to time, and replacing $\dot q$ and $\dot p$ with their differential equations, yields an explicit relation for $\lambda = \lambda(p,q)$ (as a consequence of (1.12)). Hence, a unique solution of (1.43) exists locally if $G(q_0)H_p(p_0,q_0) = 0$. The assumption (1.42) implies $\frac{d}{dt}g(q(t)) = 0$. This together with the algebraic relation of (1.43) guarantees that for $(p_0,q_0) \in \mathcal{M}$ the solution stays on the manifold $\mathcal{M}$. The symplecticity of the flow is proved as for Theorem 1.2. □

Suppose now that the Hamiltonian $H(p,q)$ of (1.8) can be split as in (1.40), where both $H^{[i]}(p,q)$ satisfy (1.42). We denote by $\varphi_t^{[i]}$ the flow of the system (1.43). If these flows can be computed analytically, the Lie-Trotter splitting $\varphi_h^{[2]} \circ \varphi_h^{[1]}$ and the Strang splitting $\varphi_{h/2}^{[1]} \circ \varphi_h^{[2]} \circ \varphi_{h/2}^{[1]}$ yield first and second order numerical integrators, respectively. Considering more general compositions as in (II.5.6) and using the coefficients proposed in Sect. V.3, methods of high order are obtained. They give numerical approximations lying on the manifold $\mathcal{M}$, and they are symplectic (also symmetric if the splitting is well chosen). Such a use of splitting methods for differential equations on manifolds has been studied by Dullweber, Leimkuhler & McLachlan (1997) and by Benettin, Cherubini & Fassò (2001). In these papers, rigid body simulations are discussed in detail.

For the important special case where

$$H(p,q) = T(p,q) + U(q)$$

is the sum of the kinetic and potential energies, both summands satisfy assumption (1.42). This gives a natural splitting that is often used in practice.

**Example 1.6 (Spherical Pendulum).** We normalize all constants to 1 (cf. Example 1.1) and we consider the problem (1.8) with

$$H(p,q) = \frac{1}{2}(p_1^2 + p_2^2 + p_3^2) + q_3, \qquad g(q) = \frac{1}{2}(q_1^2 + q_2^2 + q_3^2 - 1).$$

We split the Hamiltonian as $H^{[1]}(p,q) = \frac{1}{2}(p_1^2 + p_2^2 + p_3^2)$ and $H^{[2]}(p,q) = q_3$, and we solve (1.43) with initial values on the manifold

$$\mathcal{M} = \{(p,q) \mid q_1^2 + q_2^2 + q_3^2 - 1 = 0,\ p_1q_1 + p_2q_2 + p_3q_3 = 0\}.$$

The kinetic energy $H^{[1]}(p,q)$ leads to the system

$$\dot q = p, \qquad \dot p = -q\lambda, \qquad q^T p = 0,$$

which gives $\lambda = p_0^T p_0$, so that the flow $\varphi_t^{[1]}$ is just a planar rotation around the origin. The potential energy $H^{[2]}(p,q)$ leads to

$$\dot{q} = 0, \qquad \dot{p} = -q_3(0,0,1)^T - q\lambda, \qquad q^T p = 0.$$

The flow $\varphi_t^{[2]}$ keeps $q(t)$ constant and changes $p(t)$ linearly with time. Splitting methods give simple, explicit and symplectic time integrators for this problem.

## VII.2 Poisson Systems

This section is devoted to an interesting generalization of Hamiltonian systems, where $J^{-1}$ in (VI.2.5) is replaced with a nonconstant matrix $B(y)$. Such structures were introduced by Sophus Lie (1888) and are today called *Poisson systems*. We investigate properties of the exact flow, and we present numerical methods that preserve them. In a first subsection, however, we discuss the Poisson structure of Hamiltonian systems in canonical form.

### VII.2.1 Canonical Poisson Structure

> ... quelques remarques sur la plus profonde découverte de M. Poisson, mais qui, je crois, n'a pas été bien comprise ni par Lagrange, ni par les nombreux géomètres qui l'ont citée, ni par son auteur lui-même.
>
> (C.G.J. Jacobi 1840, p. 350)

The derivative of a function $F(p,q)$ along the flow of a Hamiltonian system

$$\dot{p} = -\frac{\partial H}{\partial q}(p,q), \qquad \dot{q} = \frac{\partial H}{\partial p}(p,q), \tag{2.1}$$

is given by (Lie derivative, see (III.5.5))

$$\frac{d}{dt} F(p(t), q(t)) = \sum_{i=1}^{d} \left( \frac{\partial F}{\partial p_i} \dot{p}_i + \frac{\partial F}{\partial q_i} \dot{q}_i \right) = \sum_{i=1}^{d} \left( \frac{\partial F}{\partial q_i} \frac{\partial H}{\partial p_i} - \frac{\partial F}{\partial p_i} \frac{\partial H}{\partial q_i} \right). \tag{2.2}$$

This remarkably symmetric structure motivates the following definition.

**Definition 2.1.** The (canonical) *Poisson bracket* of two smooth functions $F(p,q)$ and $G(p,q)$ is the function

$$\{F, G\} = \sum_{i=1}^{d} \left( \frac{\partial F}{\partial q_i} \frac{\partial G}{\partial p_i} - \frac{\partial F}{\partial p_i} \frac{\partial G}{\partial q_i} \right), \tag{2.3}$$

or in vector notation $\{F, G\}(y) = \nabla F(y)^T J^{-1} \nabla G(y)$, where $y = (p,q)$ and $J$ is the matrix of (VI.2.3).

This Poisson bracket is bilinear, skew-symmetric ($\{F, G\} = -\{G, F\}$), it satisfies the *Jacobi identity* (Jacobi 1862, *Werke* 5, p. 46)

$$\{\{F, G\}, H\} + \{\{G, H\}, F\} + \{\{H, F\}, G\} = 0 \tag{2.4}$$

(notice the cyclic permutations among $F, G, H$), and *Leibniz'* rule

$$\{F \cdot G, H\} = F \cdot \{G, H\} + G \cdot \{F, H\}. \tag{2.5}$$

These formulas are obtained in a straightforward manner from standard rules of calculus (see also Exercise 1).

With this notation, the Lie derivative (2.2) becomes

$$\frac{d}{dt} F\bigl(y(t)\bigr) = \{F, H\}(y(t)). \tag{2.6}$$

It follows that a function $I(p, q)$ is a first integral of (2.1) if and only if

$$\{I, H\} = 0.$$

If we take $F(y) = y_i$, the mapping that selects the $i$th component of $y$, we see that the Hamiltonian system (2.1) or (VI.2.5) $\dot{y} = J^{-1} \nabla H(y)$ can be written as

$$\dot{y}_i = \{y_i, H\}, \quad i = 1, \ldots, 2d. \tag{2.7}$$

**Poisson's Discovery.** At the beginning of the 19th century, the hope of being able to integrate a given system of differential equations by analytic formulas faded more and more, and the energy of researchers went to the construction of, at least, first integrals. In this enthusiasm, Jacobi declared the subsequent result to be "Poisson's deepest discovery" (see citation) and his own identity, developed for its proof, a "gravissimum Theorema".

**Theorem 2.2 (Poisson 1809).** *If $I_1$ and $I_2$ are first integrals, then their Poisson bracket $\{I_1, I_2\}$ is again a first integral.*

*Proof.* This follows at once from the Jacobi identity with $F = I_1$ and $G = I_2$. □

Siméon Denis Poisson[1]

---

[1] Siméon Denis Poisson, born: 21 June 1781 in Pithiviers (France), died: 25 April 1840 in Sceaux (near Paris).

## VII.2.2 General Poisson Structures

> ... the general concept of the Poisson manifold should be credited to Sophus Lie in his treatise on transformation groups ...
> (J.E. Marsden & T.S. Ratiu 1994)

We now come to the announced generalization of Definition 2.1 of the canonical Poisson bracket, invented by Lie (1888). Indeed, many proofs of properties of Hamiltonian systems rely uniquely on the bilinearity, the skew-symmetry and the Jacobi identity of the Poisson bracket, but not on the special structure of (2.3). So the idea is, more generally, to start with a smooth matrix-valued function $B(y) = \bigl(b_{ij}(y)\bigr)$ and to set

$$\{F,G\}(y) = \sum_{i,j=1}^{n} \frac{\partial F(y)}{\partial y_i} b_{ij}(y) \frac{\partial G(y)}{\partial y_j} \tag{2.8}$$

(or more compactly $\{F,G\}(y) = \nabla F(y)^T B(y) \nabla G(y)$).

**Lemma 2.3.** *The bracket defined in (2.8) is bilinear, skew-symmetric and satisfies Leibniz' rule (2.5) as well as the Jacobi identity (2.4) if and only if*

$$b_{ij}(y) = -b_{ji}(y) \qquad \text{for all } i, j \tag{2.9}$$

*and for all $i, j, k$ (notice the cyclic permutations among $i, j, k$)*

$$\sum_{l=1}^{n} \left( \frac{\partial b_{ij}(y)}{\partial y_l} b_{lk}(y) + \frac{\partial b_{jk}(y)}{\partial y_l} b_{li}(y) + \frac{\partial b_{ki}(y)}{\partial y_l} b_{lj}(y) \right) = 0. \tag{2.10}$$

*Proof.* The main observation is that condition (2.10) is the Jacobi identity for the special choice of functions $F = y_i$, $G = y_j$, $H = y_k$ because of

$$\{y_i, y_j\} = b_{ij}(y). \tag{2.11}$$

If equation (2.4) is developed for the bracket (2.8), one obtains terms containing second order partial derivatives — these cancel due to the symmetry of the Jacobi identity — and terms containing first order partial derivatives; for the latter we may assume $F, G, H$ to be linear combinations of $y_i, y_j, y_k$, so we are back to (2.10). The details of this proof are left as an exercise (see Exercise 1). □

**Definition 2.4.** If the matrix $B(y)$ satisfies the properties of Lemma 2.3, formula (2.8) is said to represent a (general) *Poisson bracket*. The corresponding differential system, similar to (2.7),

$$\dot{y} = B(y)\nabla H(y), \tag{2.12}$$

is a *Poisson system*. We continue to call $H$ a Hamiltonian.

**Example 2.5.** The *Lotka-Volterra* equations of Sect. I.1.1 can be written as

$$\begin{pmatrix} \dot u \\ \dot v \end{pmatrix} = \begin{pmatrix} 0 & uv \\ -uv & 0 \end{pmatrix} \nabla H(u,v), \tag{2.13}$$

where $H(u,v) = u - \ln u + v - 2\ln v$ is the invariant (I.1.4). This is of the form (2.12) with a matrix that is skew-symmetric and satisfies the identity (2.10).

Higher dimensional Lotka-Volterra systems can also have a Poisson structure (see, e.g., Perelomov (1995) and Suris (1999)). For example, the system

$$\dot y_1 = y_1(y_2 + y_3), \quad \dot y_2 = y_2(y_1 - y_3 + 1), \quad \dot y_3 = y_3(y_1 + y_2 + 1)$$

can be written as

$$\begin{pmatrix} \dot y_1 \\ \dot y_2 \\ \dot y_3 \end{pmatrix} = \begin{pmatrix} 0 & y_1 y_2 & y_1 y_3 \\ -y_1 y_2 & 0 & -y_2 y_3 \\ -y_1 y_3 & y_2 y_3 & 0 \end{pmatrix} \nabla H(y) \tag{2.14}$$

with $H(y) = -y_1 + y_2 + y_3 + \ln y_2 - \ln y_3$. Again one can check by direct computation that (2.10) is satisfied.

In contrast to the structure matrix $J^{-1}$ of Hamiltonian systems in canonical form, the matrix $B(y)$ of (2.12) need not be invertible. All odd-dimensional skew-symmetric matrices are singular, and so is the matrix $B(y)$ of (2.14). In this case, the vector $v(y) = (-1/y_1, -1/y_2, 1/y_3)^T$ satisfies $v(y)^T B(y) = 0$. Since $v(y) = \nabla C(y)$ with $C(y) = -\ln y_1 - \ln y_2 + \ln y_3$, the function $C(y)$ is an invariant of (2.14) whatever the Hamiltonian $H(y)$ is.

An important motivation for studying Poisson systems is given by Hamiltonian problems expressed in non-canonical coordinates.

**Example 2.6 (Constrained Hamiltonian Systems).** Consider the system (1.8) written as the differential equation

$$\dot y = J^{-1}\Big(\nabla H(y) + \sum_{i=1}^m \lambda_i(y)\nabla g_i(y)\Big) \tag{2.15}$$

on the manifold $\mathcal{M} = \{y\,;\,c(y) = 0\}$ with $c(y) = \big(g(q), G(q)H_p(p,q)\big)^T$ and $y = (p,q)^T$ (see (1.13)). As in the proof of Theorem 1.2, $\lambda_i(y)$ and $g_i(y)$ are the components of $\lambda(y)$ and $g(y)$, and $\lambda(y)$ is the function obtained from (1.11). We use $z \in \mathbb{R}^{2(d-m)}$ as local coordinates of the manifold $\mathcal{M}$ via the transformation

$$y = \chi(z).$$

In these coordinates, the differential equation (2.15) becomes

$$\chi'(z)\,\dot z = J^{-1}\Big(\nabla H\big(\chi(z)\big) + \sum_{i=1}^m \lambda_i\big(\chi(z)\big)\nabla g_i\big(\chi(z)\big)\Big).$$

We multiply this equation from the left with $\chi'(z)^T J$ and note that the columns of $\chi'(z)$, which are tangent vectors, are orthogonal to the gradients $\nabla g_i$ of the constraints. This yields

$$\chi'(z)^T J \chi'(z) \dot z = \chi'(z)^T \nabla H(\chi(z)).$$

By assumption (1.12) the matrix $\chi'(z)^T J \chi'(z)$ is invertible. This is seen as follows: $\chi'(z)^T J \chi'(z) v = 0$ implies $J \chi'(z) v = c'(y)^T w$ for some $w$ ($y = \chi(z)$). By $c(\chi(z)) = 0$ and $c'(y)\chi'(z) = 0$ we get $c'(y) J^{-1} c'(y)^T w = 0$. It then follows from the structure of $c'(y)$ and from (1.12) that $w = 0$ and hence also $v = 0$. With

$$B(z) = \left(\chi'(z)^T J \chi'(z)\right)^{-1} \quad \text{and} \quad K(z) = H(\chi(z)), \tag{2.16}$$

the above equation for $\dot z$ thus becomes the Poisson system $\dot z = B(z) \nabla K(z)$. The matrix $B(z)$ is skew-symmetric and satisfies (2.10), see Exercise 10.

The above example also shows that a Hamiltonian system without constraints becomes a Poisson system in non-canonical coordinates. Interestingly, a converse also holds: every Poisson system can locally be written in canonical Hamiltonian form after a suitable change of coordinates. This result is a special case of the *Darboux-Lie Theorem*. Its proof was the result of several important papers: Jacobi's theory of simultaneous linear partial differential equations (Jacobi 1862), the works by Clebsch (1866) and Darboux (1882) on Pfaffian systems, and, finally, the paper of Lie (1888). We shall now retrace this development. Our first tool is a result on the commutativity of Poisson flows.

**Commutativity of Poisson Flows and Lie Brackets.** The elegant formula (2.6) for the Lie derivative also remains valid for general Poisson systems with the vector field $f(y) = B(y) \nabla H(y)$ of (2.12). Acting on a function $F : \mathbb{R}^n \to \mathbb{R}$, the Lie operator (III.5.4) becomes

$$DF = \nabla F^T f = \nabla F^T B(y) \nabla H = \{F, H\} \tag{2.17}$$

and is again the Poisson bracket. This observation is the key for the following lemma, which shows an interesting connection between the Lie bracket and the Poisson bracket.

**Lemma 2.7.** *Let two smooth Hamiltonians $H^{[1]}(y)$ and $H^{[2]}(y)$ be given.*

$$\begin{array}{lll} \text{If} & D_1 & \text{is the Lie operator of} \quad B(y) \nabla H^{[1]} \\ \text{and} & D_2 & \text{is the Lie operator of} \quad B(y) \nabla H^{[2]}, \\ \text{then} & [D_1, D_2] & \text{is the Lie operator of} \quad B(y) \nabla \{H^{[2]}, H^{[1]}\} \end{array} \tag{2.18}$$

*(notice, once again, that the indices 1 and 2 have been reversed).*

*Proof.* After some clever permutations, the Jacobi identity (2.4) can be written as

$$\{\{F, H^{[2]}\}, H^{[1]}\} - \{\{F, H^{[1]}\}, H^{[2]}\} = \{F, \{H^{[2]}, H^{[1]}\}\}. \tag{2.19}$$

By (2.17) this is nothing other than $D_1 D_2 F - D_2 D_1 F = [D_1, D_2] F$. □

**Lemma 2.8.** *Consider two smooth Hamiltonians $H^{[1]}(y)$ and $H^{[2]}(y)$ on an open connected set $U$, with $D_1$ and $D_2$ the corresponding Lie operators and $\varphi_s^{[1]}(y)$ and $\varphi_t^{[2]}(y)$ the corresponding flows. Then, if the matrix $B(y)$ is invertible, the following are equivalent in $U$:*

$$\begin{aligned}
&(i) \quad \{H^{[1]}, H^{[2]}\} = Const; \\
&(ii) \quad [D_1, D_2] = 0; \\
&(iii) \quad \varphi_t^{[2]} \circ \varphi_s^{[1]} = \varphi_s^{[1]} \circ \varphi_t^{[2]}.
\end{aligned}$$

*The conclusions "(i) $\Rightarrow$ (ii) $\Leftrightarrow$ (iii)" also hold for a non-invertible $B(y)$.*

*Proof.* This is obtained by combining Lemma III.5.4 and Lemma 2.7. We need the invertibility of $B(y)$ to conclude that $\{H^{[1]}, H^{[2]}\} = Const$ follows from $B(y)\nabla\{H^{[1]}, H^{[2]}\} = 0$. □

## VII.2.3 Simultaneous Linear Partial Differential Equations

If two functions $F(y)$ and $G(y)$ are given, formula (2.8) determines a function $h(y) = \{F, G\}(y)$ by differentiation. We now ask the *inverse* question: Given functions $G(y)$ and $h(y)$, can we find a function $F(y)$ such that $\{F, G\}(y) = h(y)$? This problem represents a first order linear partial differential equation for $F$. So we are led to the following problem, which we first discuss in two dimensions.

**One Equation.** Given functions $a(y_1, y_2)$, $b(y_1, y_2)$, $h(y_1, y_2)$, find all solutions $F(y_1, y_2)$ satisfying

$$a(y_1, y_2) \frac{\partial F}{\partial y_1} + b(y_1, y_2) \frac{\partial F}{\partial y_2} = h(y_1, y_2). \tag{2.20}$$

This equation is, for any point $(y_1, y_2)$, a linear relation between the partial derivatives of $F$, but does not determine them individually. There is *one* direction, however, where the derivative is uniquely determined, namely that of the vector $n = \bigl(a(y_1, y_2), b(y_1, y_2)\bigr)$, since the left-hand side of equation (2.20) is the directional derivative $\frac{\partial F}{\partial n}$. The lines, which everywhere respect this direction, are called *characteristic lines* (see left picture of Fig. 2.1). If we parametrize them with a parameter $t$, we can compute $y_1(t)$, $y_2(t)$ as well as $F(t) = F\bigl(y_1(t), y_2(t)\bigr)$ as solutions of the following ordinary differential equations

$$\dot{y}_1 = a(y_1, y_2), \qquad \dot{y}_2 = b(y_1, y_2), \qquad \dot{F} = h(y_1, y_2). \tag{2.21}$$

The *initial values* $\bigl(y_1(0), y_2(0)\bigr)$ can be chosen on an arbitrary curve $\gamma$ (which must be transversal to the characteristic lines) and the values $F|_\gamma$ can be arbitrarily prescribed. The solution $F(y_1, y_2)$ of (2.20) is then created by the curves (2.21) wherever the characteristic lines go (right picture of Fig. 2.1).

For one equation in $n$ dimensions, the initial values $\bigl(y_1(0), \ldots, y_n(0)\bigr)$ can be freely chosen on a manifold of dimension $n-1$ (e.g., the subspace orthogonal to the

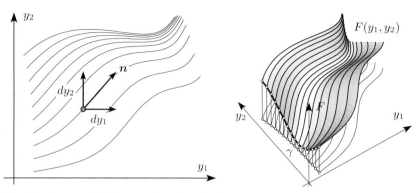

**Fig. 2.1.** Characteristic lines and solution of a first order linear partial differential equation.

characteristic line passing through a given point), and $F$ can be arbitrarily prescribed on this manifold. This guarantees the existence of $n-1$ independent solutions in the neighbourhood of a given point. Here, independent means that the gradients of these functions are linearly independent.

**Two Simultaneous Equations.** Two simultaneous equations of dimension two are trivial. We therefore suppose $y = (y_1, y_2, y_3)$ and two equations of the form

$$a_1^{[1]}(y)\frac{\partial F}{\partial y_1} + a_2^{[1]}(y)\frac{\partial F}{\partial y_2} + a_3^{[1]}(y)\frac{\partial F}{\partial y_3} = h_1(y),$$
$$a_1^{[2]}(y)\frac{\partial F}{\partial y_1} + a_2^{[2]}(y)\frac{\partial F}{\partial y_2} + a_3^{[2]}(y)\frac{\partial F}{\partial y_3} = h_2(y) \qquad (2.22)$$

for an unknown function $F(y_1, y_2, y_3)$. This system can also be written as $D_1 F = h_1$, $D_2 F = h_2$, where $D_i$ denotes the Lie operator corresponding to the vector field $a^{[i]}(y)$. Here, we have *two* directional derivatives prescribed, namely $\frac{\partial F}{\partial n_1}$ and $\frac{\partial F}{\partial n_2}$ where $n_i = a^{[i]}(y)$ (see Fig. 2.2). Therefore, we will have to follow both directions and, instead of (2.21), we will have *two* sets of ordinary differential equations

$$\dot{y}_1 = a_1^{[1]}(y), \quad \dot{y}_2 = a_2^{[1]}(y), \quad \dot{y}_3 = a_3^{[1]}(y), \quad \dot{F} = h_1(y)$$
$$\dot{y}_1 = a_1^{[2]}(y), \quad \dot{y}_2 = a_2^{[2]}(y), \quad \dot{y}_3 = a_3^{[2]}(y), \quad \dot{F} = h_2(y). \qquad (2.23)$$

If we prescribe $F$ on a curve that is orthogonal to $n_1$ and $n_2$, and if we follow the solutions of (2.23), we obtain the function $F$ on two 2-dimensional surfaces $S_1$ and $S_2$ containing the prescribed curve. Continuing from $S_1$ along the second flow and from $S_2$ along the first flow, we may be led to the same point, but nothing guarantees that the obtained values for $F$ are identical. To get a well-defined $F$, additional assumptions on the differential operators and on the inhomogeneities have to be made.

The following theorem, which is due to Jacobi (1862), has been extended by Clebsch (1866), who created the theory of *complete systems* ('vollständige Systeme'). These papers contained long analytic calculations with myriades of formulas. The wonderful geometric insight is mainly due to Sophus Lie.

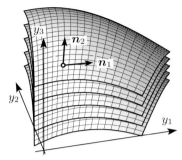

**Fig. 2.2.** Characteristic surfaces of two first order linear partial differential equations.

**Theorem 2.9.** *Let $D_1, \ldots, D_m$ be $m$ ($m < n$) linear differential operators in $\mathbb{R}^n$ corresponding to vector fields $a^{[1]}(y), \ldots, a^{[m]}(y)$ and suppose that these vectors are linearly independent for $y = y_0$. If*

$$[D_i, D_j] = 0 \qquad \text{for all } i, j, \tag{2.24}$$

*then the homogeneous system*

$$D_i F = 0 \qquad \text{for } i = 1, \ldots, m$$

*possesses (in a neighbourhood of $y_0$) $n - m$ solutions for which the gradients $\nabla F(y_0)$ are linearly independent.*

*Furthermore, the inhomogeneous system of partial differential equations*

$$D_i F = h_i \qquad \text{for } i = 1, \ldots, m$$

*possesses a particular solution in a neighbourhood of $y_0$, if and only if in addition to (2.24) the functions $h_1(y), \ldots, h_m(y)$ satisfy the integrability conditions*

$$D_i h_j = D_j h_i \qquad \text{for all } i, j. \tag{2.25}$$

*Proof.* (a) Let $V$ denote the space of vectors in $\mathbb{R}^n$ that are orthogonal to $a^{[1]}(y_0), \ldots, a^{[m]}(y_0)$, and consider the $(n-m)$-dimensional manifold $\mathcal{M} = y_0 + V$. We then extend an arbitrary smooth function $F : \mathcal{M} \to \mathbb{R}$ to a neighbourhood of $y_0$ by

$$F\left(\varphi_{t_m}^{[m]} \circ \ldots \circ \varphi_{t_1}^{[1]}(y_0 + v)\right) = F(y_0 + v). \tag{2.26}$$

Notice that $(t_1, \ldots, t_m, v) \mapsto y = \varphi_{t_m}^{[m]} \circ \ldots \circ \varphi_{t_1}^{[1]}(y_0 + v)$ defines a local diffeomorphism between neighbourhoods of $0$ and $y_0$. Since the application of the operator $D_m$ to (2.26) corresponds to a differentiation with respect to $t_m$ and the expression $F(\varphi_{t_m}^{[m]} \circ \ldots \circ \varphi_{t_1}^{[1]}(y_0 + v))$ is independent of $t_m$ by (2.26), we get $D_m F(y) = 0$. To prove $D_i F(y) = 0$ for $i < m$, we first have to change the order of the flows $\varphi_{t_j}^{[j]}$ in (2.26), which is permitted by Lemma III.5.4 and assumption (2.24), so that $\varphi_{t_i}^{[i]}$ is in the left-most position.

(b) The necessity of (2.25) follows immediately from $D_i h_j = D_i D_j F = D_j D_i F = D_j h_i$. For given $h_i$ satisfying (2.25) we define $F(y)$ in a neighbourhood of $y_0$ (i.e., for small $t_1, \ldots, t_m$ and small $v$) by

$$F\left(\varphi_{t_m}^{[m]} \circ \ldots \circ \varphi_{t_1}^{[1]}(y_0 + v)\right) = \int_0^{t_1} h_1\left(\varphi_t^{[1]}(y_0 + v)\right) dt$$
$$+ \ldots + \int_0^{t_m} h_m\left(\varphi_t^{[m]} \circ \varphi_{t_{m-1}}^{[m-1]} \circ \ldots \circ \varphi_{t_1}^{[1]}(y_0 + v)\right) dt,$$

and we prove that it is a solution of the system $D_i F = h_i$ for $i = 1, \ldots, m$. Since only the last integral depends on $t_m$, we immediately get by differentiation with respect to $t_m$ that $D_m F = h_m$. For the computation of $D_i F$ we differentiate with respect to $t_i$. The first $i-1$ integrals are independent of $t_i$. The derivative of the $i$th integral gives $h_i\left(\varphi_{t_i}^{[i]} \circ \ldots \circ \varphi_{t_1}^{[1]}(y_0 + v)\right)$, and the derivative of the remaining integrals gives

$$\int_0^{t_j} D_i h_j \left(\varphi_t^{[j]} \circ \ldots \circ \varphi_{t_1}^{[1]}(y_0 + v)\right) dt = \int_0^{t_j} D_j h_i \left(\varphi_t^{[j]} \circ \ldots \circ \varphi_{t_1}^{[1]}(y_0 + v)\right) dt$$
$$= h_i\left(\varphi_{t_j}^{[j]} \circ \ldots \circ \varphi_{t_1}^{[1]}(y_0 + v)\right) - h_i\left(\varphi_{t_{j-1}}^{[j-1]} \circ \ldots \circ \varphi_{t_1}^{[1]}(y_0 + v)\right)$$

for $j = i+1, \ldots, m$. Summing up, this proves $D_i F = h_i$. □

## VII.2.4 Coordinate Changes and the Darboux-Lie Theorem

The emphasis here is to simplify a given Poisson structure as much as possible by a coordinate transformation. We change from coordinates $y_1, \ldots, y_n$ to $\widetilde{y}_1(y), \ldots, \widetilde{y}_n(y)$ with continuously differentiable functions and an invertible Jacobian $A(y) = \partial \widetilde{y}/\partial y$,

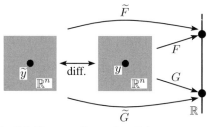

**Fig. 2.3.** New coordinates in a Poisson system

Jean Gaston Darboux[2]

and we denote $\widetilde{F}(\widetilde{y}) := F(y)$ and $\widetilde{G}(\widetilde{y}) := G(y)$ (see Fig. 2.3). The Poisson structure as well as the Poisson flow on one space

---

[2] Jean Gaston Darboux, born: 14 August 1842 in Nîmes (France), died: 23 February 1917 in Paris.

will become another Poisson structure and flow on the other space by simply applying the chain rule:

$$\sum_{i,j} \frac{\partial F(y)}{\partial y_i} b_{ij}(y) \frac{\partial G(y)}{\partial y_j} = \sum_{i,j,k,l} \frac{\partial \widetilde{F}(\widetilde{y})}{\partial \widetilde{y}_k} \frac{\partial \widetilde{y}_k}{\partial y_i} b_{ij}(y(\widetilde{y})) \frac{\partial \widetilde{y}_l}{\partial y_j} \frac{\partial \widetilde{G}(\widetilde{y})}{\partial \widetilde{y}_l}. \quad (2.27)$$

This is another Poisson structure with

$$\widetilde{b}_{kl} = \{\widetilde{y}_k, \widetilde{y}_l\} \quad \text{or} \quad \widetilde{B}(\widetilde{y}) = A(y)B(y)A(y)^T. \quad (2.28)$$

The same structure matrix is obtained if the Poisson system (2.12) is written in these new coordinates (Exercise 4).

Since $A$ is invertible, the structure matrices $B$ and $\widetilde{B}$ have the same rank. We now want to obtain the simplest possible form for $\widetilde{B}$.

**Theorem 2.10 (Darboux 1882, Lie 1888).** *Suppose that the matrix $B(y)$ defines a Poisson bracket and is of constant rank $n - q = 2m$ in a neighbourhood of $y_0 \in \mathbb{R}^n$. Then, there exist functions $P_1(y), \ldots, P_m(y)$, $Q_1(y), \ldots, Q_m(y)$, and $C_1(y), \ldots, C_q(y)$ satisfying*

$$\begin{aligned}
\{P_i, P_j\} &= 0 & \{P_i, Q_j\} &= -\delta_{ij} & \{P_i, C_l\} &= 0 \\
\{Q_i, P_j\} &= \delta_{ij} & \{Q_i, Q_j\} &= 0 & \{Q_i, C_l\} &= 0 \quad (2.29) \\
\{C_k, P_j\} &= 0 & \{C_k, Q_j\} &= 0 & \{C_k, C_l\} &= 0
\end{aligned}$$

*on a neighbourhood of $y_0$. The gradients of $P_i, Q_i, C_k$ are linearly independent, so that $y \mapsto (P_i(y), Q_i(y), C_k(y))$ constitutes a local change of coordinates to canonical form.*

The functions $C_1(y), \ldots, C_q(y)$, called *distinguished functions* (ausgezeichnete Funktionen) by Lie, are nowadays named *Casimirs*.

*Proof.* We follow Lie's original proof. Similar ideas, and the same notation, are also present in Darboux's paper. The proof proceeds in several steps, satisfying the conditions of (2.29), from one line to the next, by solving systems of linear partial differential equations.

(a) If all $b_{ij}(y_0) = 0$, the constant rank assumption implies $b_{ij}(y) = 0$ in a neighbourhood of $y_0$. We thus have $m = 0$ and all coordinates $C_i(y) = y_i$ are Casimirs.

(b) If there exist $i, j$ with $b_{ij}(y_0) \neq 0$, we set $Q_1(y) = y_i$ and we determine $P_1(y)$ as the solution of the linear partial differential equation

$$\{Q_1, P_1\} = 1. \quad (2.30)$$

Because of $b_{ij}(y_0) \neq 0$ the assumption of Theorem 2.9 is satisfied and this yields the existence of $P_1$. We next consider the homogeneous system

$$\{Q_1, F\} = 0 \quad \text{and} \quad \{P_1, F\} = 0 \quad (2.31)$$

of partial differential equations. By Lemma 2.8 and (2.30) the Lie operators corresponding to $Q_1$ and $P_1$ commute, so that by Theorem 2.9 the system (2.31) has $n-2$ independent solutions $F_3, \ldots, F_n$. Their gradients together with those of $Q_1$ and $P_1$ form a basis of $\mathbb{R}^n$. We therefore can change coordinates from $y_1, \ldots, y_n$ to $Q_1, P_1, F_3, \ldots, F_n$ (mapping $y_0$ to $\widetilde{y}_0$). In these coordinates the first two rows and the first two columns of the structure matrix $\widetilde{B}(\widetilde{y})$ have the required form.

(c) If $\widetilde{b}_{ij}(\widetilde{y}_0) = 0$ for all $i, j \geq 3$, we have $m = 1$ (similar to step (a)) and the coordinates $F_3, \ldots, F_n$ are Casimirs.

(d) If there exist $i \geq 3$ and $j \geq 3$ with $\widetilde{b}_{ij}(\widetilde{y}_0) \neq 0$, we set $Q_2 = F_i$ and we determine $P_2$ from the inhomogenous system

$$\{Q_1, P_2\} = 0, \quad \{P_1, P_2\} = 0, \quad \{Q_2, P_2\} = 1.$$

The inhomogeneities satisfy (2.25), and the Lie operators corresponding to $Q_1, P_1, Q_2$ commute (by Lemma 2.8). Theorem 2.9 proves the existence of such a $P_2$. We then consider the homogeneous system

$$\{Q_1, F\} = 0, \quad \{P_1, F\} = 0, \quad \{Q_2, F\} = 0, \quad \{P_2, F\} = 0$$

and apply once more Theorem 2.9. We get $n-4$ independent solutions, which we denote again $F_5, \ldots, F_n$. As in part (b) of the proof we get new coordinates $Q_1, P_1, Q_2, P_2, F_5, \ldots, F_n$, for which the first *four* rows and columns of the structure matrix are canonical.

(e) The proof now continues by repeating steps (c) and (d) until the structure matrix has the desired form. □

**Corollary 2.11 (Casimir Functions).** *In the situation of Theorem 2.10 the functions $C_1(y), \ldots, C_q(y)$ satisfy*

$$\{C_i, H\} = 0 \qquad \text{for all smooth } H. \tag{2.32}$$

*Proof.* Theorem 2.10 states that $\nabla C_i(y)^T B(y) \nabla H(y) = 0$, when $H(y)$ is one of the functions $P_j(y), Q_j(y)$ or $C_j(y)$. However, the gradients of these functions form a basis of $\mathbb{R}^n$. Consequently, $\nabla C_i(y)^T B(y) = 0$ and (2.32) is satisfied for all differentiable functions $H(y)$. □

This property implies that all Casimir functions are first integrals of (2.12) whatever $H(y)$ is. Consequently, (2.12) is (close to $y_0$) a differential equation on the manifold

$$\mathcal{M} = \{y \in U \mid C_i(y) = \text{Const}_i, \; i = 1, \ldots, m\}. \tag{2.33}$$

**Corollary 2.12 (Transformation to Canonical Form).** *Denote the transformation of Theorem 2.10 by $z = \chi(y) = \bigl(P_i(y), Q_i(y), C_k(y)\bigr)$. With this change of coordinates, the Poisson system $\dot{y} = B(y) \nabla H(y)$ becomes*

$$\dot{z} = B_0 \nabla K(z) \quad \text{with} \quad B_0 = \begin{pmatrix} J^{-1} & 0 \\ 0 & 0 \end{pmatrix}, \tag{2.34}$$

where $K(z) = H(y)$. Writing $z = (p, q, c)$, this system becomes

$$\dot{p} = -K_q(p, q, c), \qquad \dot{q} = K_p(p, q, c), \qquad \dot{c} = 0.$$

*Proof.* The transformed differential equation is

$$\dot{z} = \chi'(y) B(y) \chi'(y)^T \nabla K(z) \qquad \text{with} \qquad y = \chi^{-1}(z),$$

and Theorem 2.10 states that $\chi'(y) B(y) \chi'(y)^T = B_0$. □

## VII.2.5 Poisson Integrators

Before discussing geometric numerical integrators, we show that many important properties of Hamiltonian systems in canonical form remain valid for systems of the form

$$\dot{y} = B(y) \nabla H(y), \tag{2.35}$$

where $B(y)$ represents a Poisson bracket. We have already seen that the Hamiltonian $H(y)$ is a first integral of (2.35). We shall show here that the flow of (2.35) satisfies a property closely related to symplecticity.

**Definition 2.13 (Poisson Maps).** A transformation $\varphi : U \to \mathbb{R}^n$ (where $U$ is an open set in $\mathbb{R}^n$) is called a *Poisson map* with respect to the bracket (2.8), if its Jacobian matrix satisfies

$$\varphi'(y) B(y) \varphi'(y)^T = B(\varphi(y)). \tag{2.36}$$

For the canonical symplectic structure, where $B(y) = J^{-1}$, the condition (2.36) is equivalent to the symplecticity of the transformation $\varphi(y)$. This can be seen by taking the inverse of (2.36), and by multiplying the resulting equation with $\varphi'(y)$ from the right and with $\varphi'(y)^T$ from the left.

A comparison with (2.28) shows that Poisson maps leave the structure matrix invariant.

**Theorem 2.14.** *If $B(y)$ is the structure matrix of a Poisson bracket, the flow $\varphi_t(y)$ of the differential equation (2.12) is a Poisson map.*

*Proof.* (a) For $B(y) = J^{-1}$ this is exactly the statement of Theorem VI.2.4 on the symplecticity of the flow of Hamiltonian systems. This result can be extended in a straightforward way to the matrix $B_0$ of (2.34).

(b) For the general case consider the change of coordinates $z = \chi(y)$ which transforms (2.35) to canonical form (Theorem 2.10), i.e., $\chi'(y) B(y) \chi'(y)^T = B_0$ and $\dot{z} = B_0 \nabla K(z)$ with $K(z) = H(y)$ (Exercise 4). Denoting the flows of (2.35) and $\dot{z} = B_0 \nabla K(z)$ by $\varphi_t(y)$ and $\psi_t(z)$, respectively, we have $\psi_t(\chi(y)) = \chi(\varphi_t(y))$ and by the chain rule $\psi'_t(\chi(y)) \chi'(y) = \chi'(\varphi_t(y)) \varphi'_t(y)$. Inserting this relation into $\psi'_t(z) B_0 \psi'_t(z)^T = B_0$, which follows from (a), proves the statement.

A direct proof, avoiding the use of Theorem 2.10, is indicated in Exercise 5. □

The inverse of Theorem 2.14 is also true. It extends Theorem 2.6 from symplectic transformations to Poisson maps.

**Theorem 2.15.** *Let $f(y)$ and $B(y)$ be continuously differentiable on an open set $U \subset \mathbb{R}^{2d}$, and assume that $B(y)$ represents a Poisson bracket (Definition 2.4). Then, $\dot{y} = f(y)$ is locally of the form (2.35), if and only if*

- *its flow $\varphi_t(y)$ respects the Casimirs of $B(y)$, i.e., $C_i(\varphi_t(y)) = \text{Const}$, and*
- *its flow is a Poisson map for all $y \in U$ and for all sufficiently small $t$.*

*Proof.* The necessity follows from Corollary 2.11 and from Theorem 2.14. For the proof of sufficiency we apply the change of coordinates $(u, c) = \chi(y)$ of Theorem 2.10, which transforms $B(y)$ into canonical form (2.34). We write the differential equation $\dot{y} = f(y)$ in the new variables as

$$\dot{u} = g(u, c), \qquad \dot{c} = h(u, c). \tag{2.37}$$

Our first assumption expresses the fact that the Casimirs, which are the components of $c$, are first integrals of this system. Consequently, we have $h(u, c) \equiv 0$. The second assumption implies that the flow of (2.37) is a Poisson map for $B_0$ of (2.34). Writing down explicitly the blocks of condition (2.36), we see that this is equivalent to the symplecticity of the mapping $u_0 \mapsto u(t, u_0, c_0)$, with $c_0$ as a parameter. From Theorem VI.2.6 we thus obtain the existence of a function $K(u, c)$ such that $g(u, c) = J^{-1} \nabla_u K(u, c)$. Notice that for flows depending smoothly on a parameter, the Hamiltonian also depends smoothly on it. Consequently, the vector field (2.37) is of the form $B_0 \nabla K(u, c)$. Transforming back to the original variables we obtain $f(y) = B(y) \nabla H(y)$ with $H(y) = K(\chi(y))$ (see Corollary 2.12). □

The preceding theorem shows that "being a Poisson map and respecting the Casimirs" is characteristic for the flow of a Poisson system. This motivates the following definition.

**Definition 2.16.** *A numerical method $y_1 = \Phi_h(y_0)$ is a Poisson integrator for the structure matrix $B(y)$, if the transformation $y_0 \mapsto y_1$ respects the Casimirs and if it is a Poisson map whenever the method is applied to (2.35).*

Observe that for a Poisson integrator one has to specify the class of structure matrices $B(y)$. A method will never be a Poisson integrator for all possible $B(y)$.

**Example 2.17.** The symplectic Euler method reads

$$u_{n+1} = u_n + h u_{n+1} v_n H_v(u_{n+1}, v_n), \qquad v_{n+1} = v_n - h u_{n+1} v_n H_u(u_{n+1}, v_n)$$

for the Lotka-Volterra problem (2.13). It produces an excellent long-time behaviour (Fig. 2.4, left picture). We shall show that this is a Poisson integrator for all separable Hamiltonians $H(u, v) = K(u) + L(v)$. For this we compute the Jacobian of the map $(u_n, v_n) \mapsto (u_{n+1}, v_{n+1})$,

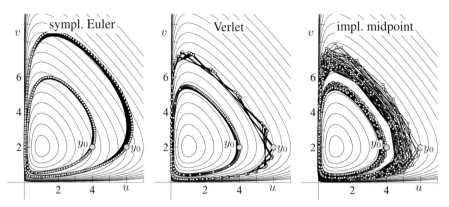

**Fig. 2.4.** Numerical solutions of the Lotka-Volterra equations (2.13) (step size $h = 0.25$, which is very large compared to the period of the solution; 1000 steps; initial values $(4, 2)$ and $(6, 2)$ for all methods.)

$$\begin{pmatrix} 1 - hv_n H_v & 0 \\ hv_n(H_u + u_{n+1} H_{uu}) & 1 \end{pmatrix} \begin{pmatrix} \partial(u_{n+1}, v_{n+1}) \\ \partial(u_n, v_n) \end{pmatrix} = \begin{pmatrix} 1 & hu_{n+1}(H_v + v_n H_{vv}) \\ 0 & 1 - hu_{n+1} H_u \end{pmatrix}$$

(the argument of the partial derivatives of $H$ is $(u_{n+1}, v_n)$ everywhere), and we check in a straightforward fashion the validity of (2.36). A different proof, using differential forms, is given in Sanz-Serna (1994) for a special choice of $H(u, v)$. Similarly, the adjoint of the symplectic Euler method is a Poisson integrator, and so is their composition - the Verlet scheme. Composition methods based on this scheme yield high order Poisson integrators, because the composition of Poisson maps is again a Poisson map.

The implicit midpoint rule, although symplectic, turns out not to be a Poisson map for the structure matrix $B(u, v)$ of (2.13). Figure 2.4 (right picture) shows that the numerical solution does not remain near a closed curve.

**Example 2.18 (Constrained Hamiltonian Systems).** Let $\Phi_h : \mathcal{M} \to \mathcal{M}$ be a symplectic integrator for the Hamiltonian system (1.8) on the manifold (1.13). By the definition (1.14) of symplecticity, we have

$$\eta_1^T \Phi_h'(y)^T J \Phi_h'(y) \eta_2 = \eta_1^T J \eta_2 \quad \text{for all} \quad \eta_1, \eta_2 \in T_y \mathcal{M}. \tag{2.38}$$

As in Example 2.6 we introduce local coordinates $z$ via the transformation $y = \chi(z)$. In the new variables the integrator becomes $\Psi_h$ where

$$\chi(\Psi_h(z)) = \Phi_h(\chi(z)). \tag{2.39}$$

The tangent vectors $\eta_i \in T_y \mathcal{M}$ for $y = \chi(z)$ can be written in the form $\eta_i = \chi'(z) \zeta_i$ with $\zeta_i \in \mathbb{R}^{2(d-m)}$. Upon inserting these vectors into (2.38) and using the differentiated equation (2.39), i.e., $\chi'(\Psi_h(z)) \Psi_h'(z) = \Phi_h'(\chi(z)) \chi'(z)$, the symplecticity condition becomes

$$\zeta_1^T \Psi_h'(z)^T \chi'(\Psi_h(z))^T J \chi'(\Psi_h(z)) \Psi_h'(z) \zeta_2 = \zeta_1^T \chi'(z)^T J \chi'(z) \zeta_2$$

for all $\zeta_1, \zeta_2 \in \mathbb{R}^{2(d-m)}$. With $B(z) = (\chi'(z)^T J \chi'(z))^{-1}$ as in Example 2.6, we obtain

$$\Psi_h'(z)^T B(\Psi_h(z))^{-1} \Psi_h'(z) = B(z)^{-1},$$

which is equivalent to $\Psi_h'(z) B(z) \Psi_h'(z)^T = B(\Psi_h(z))$. This shows that $\Psi_h$ is a Poisson integrator for the system $\dot{z} = B(z) \nabla K(z)$ of Example 2.6.

It is a difficult task to construct Poisson integrators for general Poisson systems. First of all, for non-constant $B(y)$ the condition (2.36) is no longer a quadratic first integral of the problem augmented by its variational equation (see Sect. VI.4.1). Secondly, the Casimir functions can be arbitrary and we know that only linear and quadratic first integrals can be conserved automatically (Chap. IV). Therefore, Poisson integrators will have to exploit special structures of the particular problem.

**Integrators Based on the Darboux-Lie Theorem.** If we explicitly know a transformation $z = \chi(y)$ that brings the system $\dot{y} = B(y) \nabla H(y)$ to canonical form (as in Corollary 2.12), we can proceed as follows: compute $z_n = \chi(y_n)$; apply a symplectic integrator to the transformed system $\dot{z} = B_0 \nabla K(z)$ ($B_0$ is the matrix (2.34) and $K(z) = H(y)$) which yields $z_{n+1} = \Psi_h(z_n)$; compute finally $y_{n+1}$ from $z_{n+1} = \chi(y_{n+1})$. This yields a Poisson integrator by the following lemma.

**Lemma 2.19.** *Let $z = (u, c) = \chi(y)$ be the transformation of Theorem 2.10. Suppose that the integrator $\Phi_h(y)$ takes the form*

$$\Psi_h(z) = \begin{pmatrix} \Psi_h^1(u, c) \\ c \end{pmatrix}$$

*in the new variables $z = (u, c)$. Then, $\Phi_h(y)$ is a Poisson integrator if and only if $u \mapsto \Psi_h^1(u, c)$ is a symplectic integrator for every $c$.*

*Proof.* The integrator $\Phi_h(y)$ is Poisson for the structure matrix $B(y)$ if and only if $\Psi_h(z)$ is Poisson for the matrix $B_0$ of (2.34); see Exercise 6. By assumption, $\Psi_h(z)$ preserves the Casimirs of $B_0$. The identity

$$\Psi_h'(z) B_0 \Psi_h'(z)^T = \begin{pmatrix} A J^{-1} A^T & 0 \\ 0 & 0 \end{pmatrix}$$

with $A = \partial \Psi_h^1 / \partial u$ proves the statement. □

Notice that the transformation $\chi$ has to be global in the sense that it has to be the same for all integration steps. Otherwise a degradation in performance, similar to that of the experiment in Example V.4.3, has to be expected.

**Example 2.20.** As a first illustration consider the Lotka-Volterra system (2.13). Applying the transformation $\chi(u, v) = (\ln u, \ln v) = (p, q)$, this system becomes Hamiltonian with

$$K(p, q) = -H(u, v) = -H(e^p, e^q).$$

If we apply the symplectic Euler method to this Hamiltonian system, and if we transform back to the original variables, we obtain the method

$$\begin{aligned} u_{n+1} &= u_n \exp\bigl(hv_n H_v(u_{n+1}, v_n)\bigr), \\ v_{n+1} &= v_n \exp\bigl(-hu_{n+1} H_u(u_{n+1}, v_n)\bigr). \end{aligned} \quad (2.40)$$

In contrast to the method of Example 2.17, (2.40) is also a Poisson integrator for (2.13) if $H(u, v)$ is not separable.

**Example 2.21 (Ablowitz-Ladik Discrete Nonlinear Schrödinger Equation).**
An interesting space discretization of the nonlinear Schrödinger equation is the Ablowitz-Ladik model

$$i\dot{y}_k + \frac{1}{\Delta x^2}(y_{k+1} - 2y_k + y_{k-1}) + |y_k|^2(y_{k+1} + y_{k-1}) = 0,$$

which we consider under periodic boundary conditions $y_{k+N} = y_k$ ($\Delta x = 1/N$). It is completely integrable (Ablowitz-Ladik 1976) and, as we shall see below, it is a Poisson system with noncanonical Poisson bracket. Splitting the variables into real and imaginary parts, $y_k = u_k + iv_k$, we obtain

$$\begin{aligned} \dot{u}_k &= -\frac{1}{\Delta x^2}(v_{k+1} - 2v_k + v_{k-1}) - (u_k^2 + v_k^2)(v_{k+1} + v_{k-1}) \\ \dot{v}_k &= \frac{1}{\Delta x^2}(u_{k+1} - 2u_k + u_{k-1}) + (u_k^2 + v_k^2)(u_{k+1} + u_{k-1}). \end{aligned}$$

With $u = (u_1, \ldots, u_N)$, $v = (v_1, \ldots, v_N)$ this system can be written as

$$\begin{pmatrix} \dot{u} \\ \dot{v} \end{pmatrix} = \begin{pmatrix} 0 & -D(u, v) \\ D(u, v) & 0 \end{pmatrix} \begin{pmatrix} \nabla_u H(u, v) \\ \nabla_v H(u, v) \end{pmatrix}, \quad (2.41)$$

where $D = \mathrm{diag}(d_1, \ldots, d_N)$ is the diagonal matrix with entries

$$d_k(u, v) = 1 + \Delta x^2 (u_k^2 + v_k^2),$$

and the Hamiltonian is

$$H(u, v) = \frac{1}{\Delta x^2} \sum_{l=1}^{N} (u_l u_{l-1} + v_l v_{l-1}) - \frac{1}{\Delta x^4} \sum_{l=1}^{N} \ln\bigl(1 + \Delta x^2 (u_l^2 + v_l^2)\bigr).$$

We thus get a Poisson system (the conditions of Lemma 2.3 are directly verified). There are many possibilities to transform this system to canonical form. Tang, Pérez-García & Vázquez (1997) propose the transformation

$$p_k = \frac{1}{\Delta x \sqrt{1 + \Delta x^2 v_k^2}} \arctan\left(\frac{\Delta x}{\sqrt{1 + \Delta x^2 v_k^2}} \cdot u_k\right), \qquad q_k = v_k,$$

for which the inverse can be computed in a straightforward way. Here, we suggest the transformation

$$p_k = u_k\, \sigma\!\left(\Delta x^2(u_k^2+v_k^2)\right)$$
$$q_k = v_k\, \sigma\!\left(\Delta x^2(u_k^2+v_k^2)\right) \qquad \text{with} \qquad \sigma(x)=\sqrt{\dfrac{\ln(1+x)}{x}}, \qquad (2.42)$$

which treats the variables more symmetrically. Its inverse is

$$u_k = p_k\, \tau\!\left(\Delta x^2(p_k^2+q_k^2)\right)$$
$$v_k = q_k\, \tau\!\left(\Delta x^2(p_k^2+q_k^2)\right) \qquad \text{with} \qquad \tau(x)=\dfrac{\exp x - 1}{x}.$$

Both transformations bring the system (2.41) to canonical form. For the transformation (2.42) the Hamiltonian in the new variables is

$$H(p,q) = \frac{1}{\Delta x^2}\sum_{l=1}^{N}\tau\!\left(\Delta x^2(p_l^2+q_l^2)\right)\tau\!\left(\Delta x^2(p_{l-1}^2+q_{l-1}^2)\right)\Big(p_l p_{l-1}+q_l q_{l-1}\Big)$$
$$-\frac{1}{\Delta x^2}\sum_{l=1}^{N}(p_l^2+q_l^2).$$

Applying standard symplectic schemes to this Hamiltonian yields Poisson integrators for (2.41).

## VII.2.6 Lie-Poisson Systems

We consider Poisson systems $\dot y = B(y)\nabla H(y)$ where the structure matrix $B(y)$ depends linearly on $y$, i.e.,

$$b_{ij}(y) = \sum_{k=1}^{n} C_{ij}^{k}\, y_k \qquad \text{for } i,j=1,\ldots,n. \qquad (2.43)$$

Such systems, called Lie-Poisson systems, are closely related to differential equations on Lie algebras.

Recall that a Lie algebra is a vector space with a bracket which is anti-symmetric and satisfies the Jacobi identity (Sect. IV.6). Let $E_1,\ldots,E_n$ be a basis of a vector space, and define a bracket by

$$[E_i, E_j] = \sum_{k=1}^{n} C_{ij}^{k} E_k \qquad (2.44)$$

with $C_{ij}^{k}$ from (2.43). If the structure matrix $B(y)$ of (2.43) is skew-symmetric and satisfies (2.10), then this bracket makes the vector space a Lie algebra and vice versa (the verification is left as an exercise).

In the following we assume that $\mathfrak{g}$ is the Lie algebra of a matrix Lie group $G$. Examples are given in Table IV.6.1. We assume that the bracket is the matrix commutator $[A,B]=AB-BA$, and that $E_1,\ldots,E_n$ is a basis of $\mathfrak{g}$. The coefficients $C_{ij}^{k}$, defined by (2.44), are called *structure constants* of the Lie algebra $\mathfrak{g}$. We let

$\mathfrak{g}^*$ be the dual Lie algebra, i.e., the set of all linear forms $Y : \mathfrak{g} \to \mathbb{R}$, we identify elements of $\mathfrak{g}^*$ with matrices. We denote by $F_1, \ldots, F_n$ the dual basis defined by $\langle F_i, E_j \rangle = \delta_{ij}$, the Kronecker $\delta$. Here, we consider the standard inner product which, in the case of matrix Lie algebras, is given by

$$\langle Y, X \rangle = \text{trace}\,(Y^T X).$$

For additional reading on the ideas of this section we refer the reader to Marsden & Ratiu (1994), in particular to Chap. 14 of this monograph.

**Theorem 2.22.** *Let $\mathfrak{g}$ be a matrix Lie algebra with basis $E_1, \ldots, E_n$ satisfying (2.44). To $y = (y_1, \ldots, y_n)^T \in \mathbb{R}^n$ we associate $Y = \sum_{j=1}^n y_j F_j \in \mathfrak{g}^*$, and we consider a Hamiltonian[3] $H(y) = H(Y)$.*

*Then, the Poisson system $\dot{y} = B(y)\nabla H(y)$ with $B(y)$ given by (2.43) is equivalent to the following differential equation on the Lie algebra $\mathfrak{g}^*$:*

$$\dot{Y} = -\text{ad}^*_{H'(Y)}(Y), \qquad (2.45)$$

*where $H'(Y)$ is the matrix $\left(\frac{\partial H(Y)}{\partial Y_{ij}}\right)$, and $\text{ad}^*_A$ is the adjoint of the operator $\text{ad}_A(X) = [A, X] = AX - XA$.*

*Proof.* Differentiating $H(y) = H(Y)$ with respect to $y_i$ gives

$$\frac{\partial H(y)}{\partial y_i} = \langle H'(Y), F_i \rangle \quad \text{and} \quad H'(Y) = \sum_{i=1}^n \frac{\partial H(y)}{\partial y_i} E_i.$$

Here we have used the identification $(\mathfrak{g}^*)^* = \mathfrak{g}$, because $H'(Y)$ is actually an element of $(\mathfrak{g}^*)^*$. With this formula for $H'(Y)$ we are able to compute

$$\langle \text{ad}^*_{H'(Y)}(Y), E_j \rangle = \langle Y, \text{ad}_{H'(Y)}(E_j) \rangle = \langle Y, [H'(Y), E_j] \rangle$$

$$= \left\langle Y, \sum_{i=1}^n \frac{\partial H(y)}{\partial y_i} [E_i, E_j] \right\rangle = \sum_{i=1}^n \sum_{k=1}^n \frac{\partial H(y)}{\partial y_i} C^k_{ij} \langle Y, E_k \rangle,$$

where we have used (2.44). Since $\langle \dot{Y}, E_j \rangle = \dot{y}_j$ and $\langle Y, E_k \rangle = y_k$, this shows that the differential equation (2.45) is equivalent to

$$\dot{y}_j = \sum_{i=1}^n \left( \sum_{k=1}^n C^k_{ji} y_k \right) \frac{\partial H(y)}{\partial y_i},$$

which is nothing more than $\dot{y} = B(y)\nabla H(y)$ with $B(y)$ from (2.43). □

**Connection with Constrained Hamiltonian Systems.** There is an interesting relationship between Lie-Poisson systems and Hamiltonian systems in canonical form

---

[3] We use the same symbol $H$ for the functions $H : \mathbb{R}^n \to \mathbb{R}$ and $H : \mathfrak{g}^* \to \mathbb{R}$.

on a Lie group (considered as a manifold). Let us assume that the Lie group $G$ is given by
$$G = \{Q \mid g_i(Q) = 0, \ i = 1, \ldots, m\}, \tag{2.46}$$
and consider a Hamiltonian system on $G$,
$$\dot{P} = -\frac{\partial H(P,Q)}{\partial Q} - \sum_{i=1}^{m} \lambda_i g_i'(Q), \qquad \dot{Q} = \frac{\partial H(P,Q)}{\partial P}$$
$$0 = g_i(Q), \quad i = 1, \ldots, m, \tag{2.47}$$

where $P, Q$ are square matrices, and $\frac{\partial H(P,Q)}{\partial Q} = \left(\frac{\partial H(P,Q)}{\partial Q_{ij}}\right)$. This is in canonical form as discussed in Sect. VII.1.2. In regions where the matrix
$$\left(\frac{\partial^2 H(P,Q)}{\partial P^2}\left(g_i'(Q), g_j'(Q)\right)\right)_{i,j=1}^{m} \quad \text{is invertible,} \tag{2.48}$$
the Lagrange parameters $\lambda_i$ can be expressed in terms of $P$ and $Q$ (cf. formula (1.12)). Hence, a unique solution exists locally provided the initial values $(P_0, Q_0)$ are consistent, i.e.,
$$g_i(Q_0) = 0, \qquad g_i'(Q_0)\left(\frac{\partial H(P_0, Q_0)}{\partial P}\right) = 0. \tag{2.49}$$

To relate the differential equation of Theorem 2.22 to the above system, the following lemma will be useful.

**Lemma 2.23.** *Let $G$ be a Lie group, $T_Q G$ the tangent space at $Q \in G$, $T_Q^* G$ its dual, and $\mathfrak{g} = T_I G$ the corresponding Lie algebra with $\mathfrak{g}^*$ as dual. Then we have:*
$$H \in T_Q G \quad \Longleftrightarrow \quad HQ^{-1} \in \mathfrak{g}, \tag{2.50}$$
$$P \in T_Q^* G \quad \Longleftrightarrow \quad PQ^T \in \mathfrak{g}^*. \tag{2.51}$$

*Proof.* The first equivalence follows from the fact that for a path $Q(t)$ in $G$ satisfying $Q(0) = Q$ and $\dot{Q}(0) = H$, the path $\Gamma(t) = Q(t)Q^{-1}$ lies in $G$ and satisfies $\Gamma(0) = I$ and $\dot{\Gamma}(0) = HQ^{-1}$. From
$$\langle PQ^T, E \rangle = \text{trace}\,(QP^T E) = \text{trace}\,(P^T EQ) = \langle P, EQ \rangle$$
for $E \in \mathfrak{g}$ we obtain the second equivalence. □

We call a Hamiltonian system (2.47) *right-invariant*, if $H(Pg^T, Qg^{-1}) = H(P,Q)$ for all $g \in G$. In this case $H(P,Q)$ only depends on the product $PQ^T$, and we write[4] $H(P,Q) = H(PQ^T)$.

---
[4] We again use the same letter for different functions. Since they have either one or two arguments, no confusion should arise.

**Theorem 2.24.** *For a function $H : \mathfrak{g}^* \to \mathbb{R}$ we consider $H(P, Q) := H(PQ^T)$, where $Q \in G$ and $P \in T_Q^*G$. If $(P(t), Q(t)) \in T_{Q(t)}^*G \times G$ is a solution of the system (2.47) with initial values $Q_0 \in G$, $P_0 \in T_{Q_0}^*G$, then $Y(t) := P(t)Q(t)^T$ solves the differential equation (2.45).*

*Proof.* We first compute the derivatives of $H(P, Q) = H(Y)$ (with $Y = PQ^T$),

$$\frac{\partial H(P,Q)}{\partial Q_{ij}} = \sum_{k,l=1}^{n} \frac{\partial H(PQ^T)}{\partial Y_{kl}} \frac{\partial Y_{kl}}{\partial Q_{ij}} = \sum_{k=1}^{n} \frac{\partial H(PQ^T)}{\partial Y_{ki}} P_{kj},$$

so that $\frac{\partial H(P,Q)}{\partial Q} = H'(PQ^T)^T P$. Similarly, we get $\frac{\partial H(P,Q)}{\partial P} = H'(PQ^T)Q$. Consequently, the differential equations (2.47) become

$$\dot{P} = -H'(PQ^T)^T P - \sum_{i=1}^{m} \lambda_i g_i'(Q), \qquad \dot{Q} = H'(PQ^T) Q. \qquad (2.52)$$

Leibniz' rule $\dot{Y} = \dot{P} Q^T + P \dot{Q}^T$ applied to $Y = PQ^T$ thus yields

$$\dot{Y} = Y H'(Y)^T - H'(Y)^T Y - \sum_{i=1}^{m} \lambda_i Q^T g_i'(Q). \qquad (2.53)$$

For $E \in \mathfrak{g}$, we now exploit the properties

$$\langle Q^T g_i'(Q), E \rangle = \langle g_i'(Q), QE \rangle = 0 \quad \text{(because } QE \in T_Q G \text{ by (2.50))}$$
$$\langle \mathrm{ad}^*_{H'(Y)}(Y), E \rangle = \langle Y, \mathrm{ad}_{H'(Y)}(E) \rangle = \mathrm{trace}\left(Y^T(H'(Y)E - EH'(Y))\right)$$
$$= \mathrm{trace}\left((Y^T H'(Y) - H'(Y)Y^T)E\right) = \langle [H'(Y)^T, Y], E \rangle.$$

Since $Y(t) \in \mathfrak{g}^*$ for all $t$ (by Lemma 2.23), $H'(Y(t)) \in \mathfrak{g}$, and the projection of formula (2.53) onto $\mathfrak{g}^*$ gives the differential equation (2.45). The projection onto the complement of $\mathfrak{g}^*$, which is spanned by $Q^T g_i'(Q)$ for $i = 1, \ldots, m$, defines the Lagrange multipliers $\lambda_i$. □

The above result has its analogue, if right-invariance is replaced with left-invariance, i.e., $H(g^T P, g^{-1}Q) = H(P, Q)$ for all $g \in G$, so that $H(P, Q)$ only depends on $Q^T P$.

**Example 2.25.** Consider the motion of a rigid body without any external forces. Relative to the centre of gravity this is a rotation described by an orthogonal 3-dimensional matrix $Q(t)$. The kinetic energy of the body is given by $\frac{1}{2} \sum_{k=1}^{3} I_k \omega_k^2$, where $(\omega_1, \omega_2, \omega_3)^T$, its angular velocity, is

$$\Omega = \begin{pmatrix} 0 & -\omega_3 & \omega_2 \\ \omega_3 & 0 & -\omega_1 \\ -\omega_2 & \omega_1 & 0 \end{pmatrix} = \dot{Q} Q^{-1}. \qquad (2.54)$$

In terms of the matrix $\Omega$, the kinetic energy of the body is $\frac{1}{2}\text{trace}\,(\Omega^T D\Omega)$, where $D = \text{diag}(d_1, d_2, d_3)$, and $I_1 = d_2 + d_3$, $I_2 = d_3 + d_1$, $I_3 = d_1 + d_2$. Taking into account the constraints $QQ^T = I$, the Lagrangian becomes

$$\mathcal{L}(Q, \dot{Q}) = \frac{1}{2}\text{trace}\,(Q^{-T}\dot{Q}^T D\dot{Q}Q^{-1}) + \text{trace}\,(\Lambda(QQ^T - I))$$

with $\Lambda$ composed of Lagrange multipliers. Introducing the conjugate variable $P = \partial \mathcal{L}/\partial \dot{Q} = D\dot{Q}Q^{-1}Q^{-T}$, we get a Hamiltonian system with

$$H(P,Q) = \frac{1}{2}\text{trace}\,(QP^T D^{-1} PQ^T) - \text{trace}\,(\Lambda(QQ^T - I)). \tag{2.55}$$

This Hamiltonian is right-invariant. We therefore introduce $Y = PQ^T$, and the computation of Theorem 2.24 gives

$$\dot{Y} = -Y^T D^{-1} Y + YY^T D^{-1} - (\Lambda^T + \Lambda).$$

The symmetric part of this formula yields $\Lambda^T + \Lambda$, and the skew-symmetric part is the differential equation (2.45). With the basis

$$E_1 = \begin{pmatrix} 0 & 0 & 0 \\ 0 & 0 & -1 \\ 0 & 1 & 0 \end{pmatrix}, \quad E_2 = \begin{pmatrix} 0 & 0 & 1 \\ 0 & 0 & 0 \\ -1 & 0 & 0 \end{pmatrix}, \quad E_3 = \begin{pmatrix} 0 & -1 & 0 \\ 1 & 0 & 0 \\ 0 & 0 & 0 \end{pmatrix}$$

for $\mathfrak{g} = \mathfrak{so}(3)$, the coordinates $y_k = \langle Y, E_k \rangle = I_k \omega_k$ become the angular momenta, and the differential equation for $Y$ is

$$\begin{pmatrix} \dot{y}_1 \\ \dot{y}_2 \\ \dot{y}_3 \end{pmatrix} = \begin{pmatrix} 0 & y_3 & -y_2 \\ -y_3 & 0 & y_1 \\ y_2 & -y_1 & 0 \end{pmatrix} \nabla H(y) \tag{2.56}$$

with $H(y) = \frac{1}{2}\sum_{k=1}^{3} I_k^{-1} y_k^2$. This is a system we have already encountered several times.

**Approach via the Hamiltonian System (2.47).** Realizing the close connection between a Lie-Poisson system and a constrained Hamiltonian system on a Lie group, the following algorithm suggests itself: apply any symplectic integrator of Sect. VII.1 to the system (2.52), and rewrite the formulas in terms of the variables $y$ of the Lie-Poisson system. By Example 2.18 this yields a Lie-Poisson integrator. This approach has been proposed and developed independently by McLachlan & Scovel (1995) and Reich (1994). We will explain this approach with an example.

For the rigid body problem of Example 2.25 $H(Y) = \frac{1}{2}\text{trace}\,(Y^T D^{-1} Y)$ holds, so that $H'(Y) = D^{-1} Y$. An application of the RATTLE algorithm (1.24) yields

$$\begin{aligned} P_{1/2} &= P_0 - \frac{h}{2} Q_0 P_{1/2}^T D^{-1} P_{1/2} - \frac{h}{2}(\Lambda + \Lambda^T) Q_0 \\ Q_1 &= Q_0 + h D^{-1} P_{1/2}, \quad Q_1 Q_1^T = I \\ P_1 &= P_{1/2} - \frac{h}{2} Q_1 P_{1/2}^T D^{-1} P_{1/2} - \frac{h}{2}(M + M^T) Q_1 \\ 0 &= Q_1^T D^{-1} P_1 + P_1^T D^{-1} Q_1. \end{aligned} \tag{2.57}$$

We let $Y_0 = P_0 Q_0^T, Y_{1/2} = P_{1/2} Q_0^T, Y_1 = P_1 Q_1^T$, and we recall that the connection between $Y \in \mathfrak{g}^*$ and $y = (y_1, y_2, y_3)^T$ is given by $y_k = I_k \omega_k$ and $Y = D\Omega$ with $\Omega$ from (2.54). This method can be reformulated as follows:

**Algorithm 2.26 (Rigid Body Integrator).** *The map $Y_0 \mapsto Y_1$ is defined by:*

- *determine a symmetric matrix $S$ such that $I + hD^{-1}Y_{1/2}$ is orthogonal, where $Y_{1/2} = Y_0 + hS$;*
- *determine a symmetric matrix $T$ such that $D^{-1}Y_1$ is skew-symmetric, where $Y_1 = Y_0 + hY_{1/2}Y_{1/2}^T D^{-1} + hT$.*

Since $D^{-1}Y_0$ is skew-symmetric by assumption, the first step is equivalent to solving the Ricatti equation

$$D^{-1}S + SD^{-1} = -(Y_0 + hS)^T D^{-2}(Y_0 + hS),$$

which can be done efficiently. The second step is a linear problem.

Numerical experiments with this algorithm revealed that it is of order 2, it preserves exactly the Casimir $y_1^2 + y_2^2 + y_3^2$ and, most surprisingly, it conserves exactly the Hamiltonian $y_1^2/I_1 + y_2^2/I_2 + y_3^2/I_3$.

**Further Approaches for Lie-Poisson Systems.** Historically the first Lie-Poisson integrators seem to be those of Ge & Marsden (1988). They extend the construction of symplectic methods by *generating functions* to Lie-Poisson systems. Channel & Scovel (1991) propose an implementation of these methods based on a coordinatization of the group by the exponential map. An adaptation of this approach to general Poisson structures is exemplified by Schober (1999) for the Ablowitz-Ladik equations of Example 2.21. A recent overview of Poisson systems and Poisson integrators is given by Karasözen (2001).

Whenever applicable, the technique of *splitting methods* gives powerful and simple geometric integrators. Consider a (general) Poisson system $\dot{y} = B(y)\nabla H(y)$ and suppose that the Hamiltonian permits a decomposition as $H(y) = H_1(y) + \ldots + H_m(y)$, such that the individual systems $\dot{y} = B(y)\nabla H_i(y)$ can be solved exactly. The flow of these subsystems is a Poisson map and automatically respects the Casimirs, and so does their composition. McLachlan (1993) and Reich (1993) present several interesting examples, where this approach is possible. An application of the splitting technique to the study of the rigid body dynamics in the solar system is given in Touma & Wisdom (1994).

Consider once more the rigid body equation (2.56). The Hamiltonian is already of the form $H(y) = H_1(y_1) + H_2(y_2) + H_3(y_3)$, and the system $\dot{y} = B(y)\nabla H_1(y)$ which is

$$\dot{y}_1 = 0, \quad \dot{y}_2 = -y_1 y_3 / I_1, \quad \dot{y}_3 = y_1 y_2 / I_1,$$

can be solved exactly. Similar systems are obtained for $H_2$ and $H_3$. Compositions such as those discussed in Sect. II.5 (see (II.5.11)) and Sect. V.3 lead to efficient integrators.

## VII.3 Volume Preservation

The flow $\varphi_t$ of a Hamiltonian system preserves volume in phase space: for every bounded open set $\Omega \subset \mathbb{R}^{2d}$ and for every $t$ for which $\varphi_t(y)$ exists for all $y \in \Omega$,

$$\text{vol}(\varphi_t(\Omega)) = \text{vol}(\Omega),$$

where $\text{vol}(\Omega) = \int_\Omega dy$. This identity is often referred to as *Liouville's theorem*. It is a consequence of the transformation formula for integrals and the fact that

$$\det \frac{\partial \varphi_t(y)}{\partial y} = 1 \qquad \text{for all } t \text{ and } y, \tag{3.1}$$

which follows directly from the symplecticity and $\varphi_0 = \text{id}$. The same argument shows that every symplectic transformation, and in particular every symplectic integrator applied to a Hamiltonian system, preserves volume in phase space.

More generally than for Hamiltonian systems, volume is preserved by the flow of differential equations with a divergence-free vector field:

**Lemma 3.1.** *The flow of a differential equation $\dot y = f(y)$ in $\mathbb{R}^n$ is volume-preserving if and only if $\text{div} f(y) = 0$ for all $y$.*

*Proof.* The derivative $Y(t) = \frac{\partial \varphi_t}{\partial y}(y_0)$ is the solution of the variational equation

$$\dot Y = A(t) Y, \qquad Y(0) = I,$$

with the Jacobian matrix $A(t) = f'(y(t))$ at $y(t) = \varphi_t(y_0)$. From the proof of Lemma IV.3.1 we obtain the *Abel-Liouville-Jacobi-Ostrogradskii identity*

$$\frac{d}{dt} \det Y = \text{trace } A(t) \cdot \det Y. \tag{3.2}$$

Note that here trace $A(t) = \text{div} f(y(t))$. Hence, $\det Y(t) = 1$ for all $t$ if and only if $\text{div} f(y(t)) = 0$ for all $t$. Since this is valid for all choices of initial values $y_0$, the result follows. □

**Example 3.2 (ABC Flow).** This flow, named after the three independent authors Arnold, Beltrami and Childress, is given by the equations

$$\begin{aligned} \dot x &= A \sin z + C \cos y \\ \dot y &= B \sin x + A \cos z \\ \dot z &= C \sin y + B \cos x \end{aligned} \tag{3.3}$$

and has all diagonal elements of $f'$ identically zero. It is therefore volume preserving. In Arnold (1966, p. 347) it appeared in a footnote as an example of a flow with $\text{rot} f$ parallel to $f$, thus violating Arnold's condition for the existence of invariant tori (Arnold 1966, p. 346). It was therefore expected to possess interesting chaotic properties and has since then been the object of many investigations showing their non-integrability (see e.g., Ziglin (1996)). We illustrate in Fig. 3.1 the action of this flow by transforming, in a volume preserving manner, a ball in $\mathbb{R}^3$. We see that, very soon, the set is strongly squeezed in one direction and dilated in two others. The solutions thus depend in a very sensitive way on the initial values.

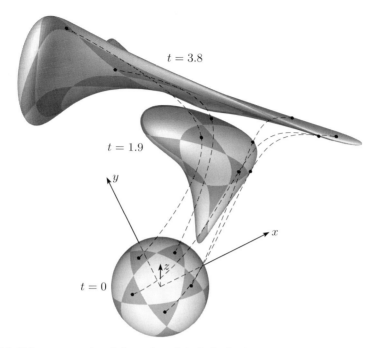

**Fig. 3.1.** Volume preserving deformation of the ball of radius 1, centred at the origin, by the ABC flow; $A = 1/2, B = C = 1$.

**Volume-Preserving Numerical Integrators.** The question arises as to whether volume-preserving integrators can be constructed for every differential equation with volume-preserving flow. Already for linear problems, Lemma IV.3.2 shows that no standard method can be volume-preserving for dimension $n \geq 3$. Nevertheless, positive answers were found by Qin & Zhu (1993), Shang (1994a,1994b), Feng & Shang (1995) and Quispel (1995). In the following we present the approach of Feng & Shang (1995). The key is the following result which generalizes and reinterprets a construction of H. Weyl (1940) for $n = 3$.

**Theorem 3.3 (Feng & Shang 1995).** *Every divergence-free vector field* $f : \mathbb{R}^n \to \mathbb{R}^n$ *can be written as the sum of* $n - 1$ *vector fields*

$$f = f_{1,2} + f_{2,3} + \ldots + f_{n-1,n}$$

*where each $f_{k,k+1}$ is Hamiltonian in the variables $(y_k, y_{k+1})$: there exist functions $H_{k,k+1} : \mathbb{R}^n \to \mathbb{R}$ such that*

$$f_{k,k+1} = \left(0, \ldots, 0, -\frac{\partial H_{k,k+1}}{\partial y_{k+1}}, \frac{\partial H_{k,k+1}}{\partial y_k}, 0, \ldots, 0\right)^T .$$

*Proof.* In terms of the components of $f = (f_1, \ldots, f_n)^T$, the functions $H_{k,k+1}$ must satisfy the equations

$$f_1 = -\frac{\partial H_{1,2}}{\partial y_2}, \quad f_2 = \frac{\partial H_{1,2}}{\partial y_1} - \frac{\partial H_{2,3}}{\partial y_3}, \ldots,$$

$$f_{n-1} = \frac{\partial H_{n-2,n-1}}{\partial y_{n-2}} - \frac{\partial H_{n-1,n}}{\partial y_n}, \quad f_n = \frac{\partial H_{n-1,n}}{\partial y_{n-1}}.$$

We thus set

$$H_{1,2} = -\int_0^{y_2} f_1 \, dy_2$$

and for $k = 2, \ldots, n-2$

$$H_{k,k+1} = \int_0^{y_{k+1}} \left( \frac{\partial H_{k-1,k}}{\partial y_{k-1}} - f_k \right) dy_{k+1}.$$

It remains to construct $H_{n-1,n}$ from the last two equations. We see by induction that for $k \le n-2$,

$$\frac{\partial^2 H_{k,k+1}}{\partial y_k \partial y_{k+1}} = -\left( \frac{\partial f_1}{\partial y_1} + \ldots + \frac{\partial f_k}{\partial y_k} \right),$$

and hence the integrability condition for $H_{n-1,n}$,

$$\frac{\partial}{\partial y_{n-1}} \left( \frac{\partial H_{n-2,n-1}}{\partial y_{n-2}} - f_{n-1} \right) = \frac{\partial f_n}{\partial y_n},$$

reduces to the condition $\operatorname{div} f = 0$, which is satisfied by assumption. $H_{n-1,n}$ can thus be constructed as

$$H_{n-1,n} = \int_0^{y_n} \left( \frac{\partial H_{n-2,n-1}}{\partial y_{n-2}} - f_{n-1} \right) dy_n + \int_0^{y_{n-1}} f_n|_{y_n=0} \, dy_{n-1},$$

which completes the proof. □

The above construction also shows that

$$f_{k,k+1} = (0, \ldots, 0, f_k + g_k, -g_{k+1}, 0, \ldots, 0)$$

with

$$g_{k+1} = \int_0^{y_{k+1}} \left( \frac{\partial f_1}{\partial y_1} + \ldots + \frac{\partial f_k}{\partial y_k} \right) dy_{k+1}$$

for $1 \le k \le n-2$, and $g_1 = 0$ and $g_n = -f_n$.

With the decomposition of Lemma 3.3 at hand, a volume-preserving algorithm is obtained by applying a splitting method with symplectic substeps. For example, as proposed by Feng & Shang (1995), a second-order volume-preserving method is obtained by Strang splitting with symplectic Euler substeps:

## VII.3 Volume Preservation

$$\varphi_h \approx \Phi_h = \Phi_{h/2}^{[1,2]*} \circ \ldots \circ \Phi_{h/2}^{[n-1,n]*} \circ \Phi_{h/2}^{[n-1,n]} \circ \ldots \circ \Phi_{h/2}^{[1,2]}$$

where $\Phi_{h/2}^{[k,k+1]}$ is a symplectic Euler step of length $h/2$ applied to the system with right-hand side $f_{k,k+1}$, and $*$ denotes the adjoint method. In this method, one step $\widehat{y} = \Phi_h(y)$ is computed component-wise, in a Gauss-Seidel-like manner, as

$$\overline{y}_1 = y_1 + \frac{h}{2} f_1(\overline{y}_1, y_2, \ldots, y_n)$$

$$\overline{y}_k = y_k + \frac{h}{2} f_k(\overline{y}_1, \ldots, \overline{y}_k, y_{k+1}, \ldots, y_n) + \frac{h}{2} g_k|_{y_k}^{\overline{y}_k} \quad \text{for } k = 2, \ldots, n-1$$

$$\overline{y}_n = y_n + \frac{h}{2} f_n(\overline{y}_1, \ldots, \overline{y}_{n-1}, y_n) \tag{3.4}$$

with $g_k|_{y_k}^{\overline{y}_k} = g_k(\overline{y}_1, \ldots, \overline{y}_k, y_{k+1}, \ldots, y_n) - g_k(\overline{y}_1, \ldots, \overline{y}_{k-1}, y_k, \ldots, y_n)$, and

$$\widehat{y}_n = \overline{y}_n + \frac{h}{2} f_n(\overline{y}_1, \ldots, \widehat{y}_n)$$

$$\widehat{y}_k = \overline{y}_k + \frac{h}{2} f_k(\overline{y}_1, \ldots, \overline{y}_k, \widehat{y}_{k+1} \ldots, \widehat{y}_n) - \frac{h}{2} g_k|_{\overline{y}_k}^{\widehat{y}_k} \quad \text{for } k = n-1, \ldots, 2$$

$$\widehat{y}_1 = \overline{y}_1 + \frac{h}{2} f_1(\overline{y}_1, \widehat{y}_2, \ldots, \widehat{y}_n) \tag{3.5}$$

with $\overline{g}_k|_{\overline{y}_k}^{\widehat{y}_k} = g_k(\overline{y}_1, \ldots, \overline{y}_{k-1}, \widehat{y}_k, \ldots, \widehat{y}_n) - g_k(\overline{y}_1, \ldots, \overline{y}_k, \widehat{y}_{k+1}, \ldots, \widehat{y}_n)$. The method is one-dimensionally implicit in general, but becomes explicit in the particular case where $\partial f_k / \partial y_k = 0$ for all $k$.

**Separable Partitioned Systems.** For problems of the form

$$\dot{y} = f(z), \qquad \dot{z} = g(y) \tag{3.6}$$

with $y \in \mathbb{R}^m$, $z \in \mathbb{R}^n$, the scheme (3.4) becomes the symplectic Euler method, (3.5) its adjoint, and its composition the Lobatto IIIA - IIIB extension of the Störmer/Verlet method. Since symplectic explicit partitioned Runge-Kutta methods are compositions of symplectic Euler steps (Theorem VI.4.7), this observation proves that such methods are volume-preserving for systems (3.6). This fact was obtained by Suris (1996) by a direct calculation, without interpreting the methods as composition methods. The question arises as to whether more symplectic partitioned Runge-Kutta methods are volume-preserving for systems (3.6).

**Theorem 3.4.** *Every symplectic Runge-Kutta method with at most two stages is volume-preserving for systems (3.6) of arbitrary dimension.*

*Proof.* (a) The idea is to consider the Hamiltonian system with

$$H(u, v, y, z) = u^T f(z) + v^T g(y),$$

where $(u, v)$ are the conjugate variables to $(y, z)$. This system is of the form

252    VII. Further Topics in Structure Preservation

$$\dot{y} = f(z) \qquad \dot{u} = -g'(y)^T v$$
$$\dot{z} = g(y) \qquad \dot{v} = -f'(z)^T u. \tag{3.7}$$

Applying the Runge-Kutta method to this augmented system does not change the numerical solution for $(y, z)$. For symplectic methods the matrix

$$\left(\frac{\partial(y_1, z_1, u_1, v_1)}{\partial(y_0, z_0, u_0, v_0)}\right) = M = \begin{pmatrix} R & 0 \\ S & T \end{pmatrix} \tag{3.8}$$

satisfies $M^T J M = J$ which implies $RT^T = I$. Below we shall show that $\det T = \det R$. This yields $\det R = 1$ which implies that the method is volume preserving.

(b) *One-stage methods.* The only symplectic one-stage method is the implicit midpoint rule for which $R$ and $T$ are computed as

$$\left(I - \frac{h}{2} E_1\right) R = I + \frac{h}{2} E_1 \tag{3.9}$$

$$\left(I + \frac{h}{2} E_1^T\right) T = I - \frac{h}{2} E_1^T, \tag{3.10}$$

where $E_1$ is the Jacobian of the system (3.6) evaluated at the internal stage value. Since

$$E_1 = \begin{pmatrix} 0 & f'(z_{1/2}) \\ g'(y_{1/2}) & 0 \end{pmatrix},$$

a similarity transformation with the matrix $D = \mathrm{diag}(I, -I)$ takes $E_1$ to $-E_1$. Hence, the transformed matrix satisfies

$$\left(I - \frac{h}{2} E_1^T\right)(D^{-1} T D) = I + \frac{h}{2} E_1^T.$$

A comparison with (3.9) and the use of $\det X^T = \det X$ proves $\det R = \det T$ for the midpoint rule.

(c) *Two-stage methods.* Applying a two-stage implicit Runge-Kutta method to (3.7) yields

$$\begin{pmatrix} I - h a_{11} E_1 & -h a_{12} E_2 \\ -h a_{21} E_1 & I - h a_{22} E_2 \end{pmatrix} \begin{pmatrix} R_1 \\ R_2 \end{pmatrix} = \begin{pmatrix} I \\ I \end{pmatrix},$$

where $R_i$ is the derivative of the $(y, z)$ components of the $i$th stage with respect to $(y_0, z_0)$, and $E_i$ is the Jacobian of the system (3.6) evaluated at the $i$th internal stage value. From the solution of this system the derivative $R$ of (3.8) is obtained as

$$R = I + (b_1 E_1, b_2 E_2) \begin{pmatrix} I - h a_{11} E_1 & -h a_{12} E_2 \\ -h a_{21} E_1 & I - h a_{22} E_2 \end{pmatrix}^{-1} \begin{pmatrix} I \\ I \end{pmatrix}.$$

With the determinant identity

$$\det(U) \det(X - W U^{-1} V) = \det \begin{pmatrix} U & V \\ W & X \end{pmatrix} = \det(X) \det(U - V X^{-1} W),$$

which is seen by Gaussian elimination, this yields

$$\det R = \frac{\det\bigl(I \otimes I - h((A - 1\!\!1 b^T) \otimes I)\, E\bigr)}{\det\bigl(I \otimes I - h(A \otimes I)\, E\bigr)},$$

where $A$ and $b$ collect the Runge-Kutta coefficients, and $E = \text{blockdiag}\,(E_1, E_2)$. For $D^{-1}TD$ we get the same formula with $E$ replaced by $E^T$. If $A$ is an arbitrary $2 \times 2$ matrix, it follows from block Gaussian elimination that

$$\det\bigl(I \otimes I - h(A \otimes I)\, E\bigr) = \det\bigl(I \otimes I - h(A \otimes I)\, E^T\bigr), \qquad (3.11)$$

which then proves $\det R = \det T$. Notice that the identity (3.11) is no longer true in general if $A$ is of dimension larger than two. □

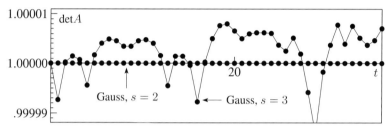

**Fig. 3.2.** Volume preservation of Gauss methods applied to (3.12) with $h = 0.8$.

We are curious to see whether Theorem 3.4 remains valid for symplectic Runge-Kutta methods with more than two stages. For this we apply the Gauss methods with $s = 2$ and $s = 3$ to the problem

$$\dot{x} = \sin z, \qquad \dot{y} = \cos z, \qquad \dot{z} = \sin y + \cos x \qquad (3.12)$$

with initial value $(0, 0, 0)$. We show in Fig. 3.2 the determinant of the derivative of the numerical flow as a function of time. Only the two-stage method is volume-preserving for this problem which is in agreement with Theorem 3.4.

## VII.4 Exercises

1. Prove that the Poisson bracket (2.8) satisfies the Jacobi identity (2.4) for all functions $F, G, H$, if and only if it satisfies (2.4) for the coordinate functions $y_i, y_j, y_k$.
   *Hint* (F. Engel, in Lie's *Gesammelte Abh.* vol. 5, p. 753). If the Jacobi identity is written as in (2.19), we see that there are no second partial derivatives of $F$ (the left hand side is a Lie bracket, the right hand side has no second derivatives of $F$ anyway). Other permutations show the same result for $G$ and $H$.

2. Solve the following first order partial differential equation:
$$3\frac{\partial F}{\partial y_1} + 2\frac{\partial F}{\partial y_2} - 5\frac{\partial F}{\partial y_3} = 0.$$
**Result.** $f(2y_1 - 3y_2, 5y_2 + 2y_3)$.

3. Find two solutions of the homogeneous system
$$3\frac{\partial F}{\partial y_1} + \frac{\partial F}{\partial y_2} - 2\frac{\partial F}{\partial y_3} - 5\frac{\partial F}{\partial y_4} = 0, \qquad 2\frac{\partial F}{\partial y_1} - \frac{\partial F}{\partial y_2} - 3\frac{\partial F}{\partial y_4} = 0,$$
such that their gradients are linearly independent.

4. Consider a Poisson system $\dot{y} = B(y)\nabla H(y)$ and a change of coordinates $z = \chi(y)$. Prove that in the new coordinates the system is of the form $\dot{z} = \widetilde{B}(z)\nabla K(z)$, where $\widetilde{B}(z) = \chi'(y)B(y)\chi'(y)^T$ (cf. formula (2.28)) and $K(z) = H(y)$.

5. Give an elementary proof of Theorem 2.14.
   *Hint.* Define $\delta(t) := \varphi_t'(y)B(y)\varphi_t'(y)^T - B(\varphi_t(y))$. Using the variational equation for (2.35) prove that $\delta(t)$ is the solution of a homogeneous linear differential equation. Therefore, $\delta(0) = 0$ implies $\delta(t) = 0$ for all $t$.

6. Let $z = \chi(y)$ be a transformation taking the Poisson system $\dot{y} = B(y)\nabla H(y)$ to $\dot{z} = \widetilde{B}(z)\nabla K(z)$. Prove that $\Phi_h(y)$ is a Poisson integrator for $B(y)$ if and only if $\Psi_h(z) = \chi \circ \Phi_h \circ \chi^{-1}(z)$ is a Poisson integrator for $\widetilde{B}(z)$.

7. Let $B$ be a skew-symmetric but otherwise arbitrary constant matrix, and consider the Poisson system $\dot{y} = B\nabla H(y)$. Prove that every symplectic Runge-Kutta method is a Poisson integrator for such a system.
   *Hint.* Transform $B$ to block-diagonal form.

8. (M.J. Gander 1994). Consider the Lotka-Volterra equation (2.13) with separable Hamiltonian $H(u,v) = K(u) + L(v)$. Prove that
$$u_{n+1} = u_n + hu_n v_n H_v(u_n, v_n), \qquad v_{n+1} = v_n - hu_{n+1}v_n H_u(u_{n+1}, v_n)$$
is a Poisson integrator for this system.

9. Find a change of coordinates that transforms the Lotka-Volterra system (2.14) into a Hamiltonian system (in canonical form). Following the approach of Example 2.20 construct Poisson integrators for this system.

10. Prove that the matrix $B(z)$ of (2.16) defines a Poisson bracket.
    *Hint.* Prove that
$$\{F, G\} = \{\widehat{F}, \widehat{G}\} - \sum_{i,j}\{\widehat{F}, c_i\}\gamma_{ij}\{c_j, \widehat{G}\}, \tag{4.1}$$
where $F$ and $G$ are functions of $z$, $\widehat{F}$ and $\widehat{G}$ are smooth functions of $y$ satisfying $\widehat{F}(\chi(z)) = F(z)$ and $\widehat{G}(\chi(z)) = G(z)$, $c_i(y)$ are the functions defining the manifold $\mathcal{M}$, and $\gamma_{ij}$ are the entries of the inverse of the matrix $(\{c_i, c_j\})$. The Poisson bracket to the left in (4.1) corresponds to $B(z)$ and those to the right are the canonical bracket evaluated at $y = \chi(z)$. Replacing $\widehat{F}(y)$ by $\widehat{F}(y) + \sum_k \mu_k(y)c_k(y)$ with $\mu_k(y)$ such that $\{\widehat{F}, c_k\} = 0$ on $\mathcal{M}$ eliminates the sum in (4.1) and proves the Jacobi identity for $B(z)$.

# Chapter VIII.
# Structure-Preserving Implementation

This chapter is devoted to practical aspects of an implementation of geometric integrators. We explain strategies for changing the step size which do not deteriorate the correct qualitative behaviour of the solution. We study multiple time stepping strategies, the effect of round-off in long-time integrations, and the efficient solution of nonlinear systems arising in implicit integration schemes.

## VIII.1 Dangers of Using Standard Step Size Control

> Another possible shortcoming of the method concerns its behavior when used with a variable step size ... The integrator completely loses its desirable qualities ... This can be understood at least qualitatively by realizing that by changing the time step one is in essence continually changing the nearby Hamiltonian ...     (B. Gladman, M. Duncan & J. Candy 1991)

In the previous chapters we have studied symmetric and symplectic integrators, and we have seen an enormous progress in long-time integrations of various problems. Decades ago, a similar enormous progress was the introduction of algorithms with automatic step size control. Naively, one would expect that the blind combination of both techniques leads to even better performances. We shall see by a numerical experiment that this is not the case, a phenomenon observed by Gladman, Duncan & Candy (1991) and Calvo & Sanz-Serna (1992).

We study the long-time behaviour of symplectic methods combined with the following standard step size selection strategy (see e.g., Hairer, Nørsett & Wanner (1993), Sect. II.4). We assume that an expression $err_n$ related to the local error is available for the current step computed with step size $h_n$ (usually obtained with an embedded method). Based on an asymptotic formula $err_n \approx Ch_n^r$ (for $h_n \to 0$) and on the requirement to get an error close to a user supplied tolerance $Tol$, we predict a new step size by

$$h_{new} = 0.85 \cdot h_n \left(\frac{Tol}{err_n}\right)^{1/r}, \tag{1.1}$$

where a safety factor $0.85$ is included. We then apply the method with step size $h_{n+1} = h_{new}$. If for the new step $err_{n+1} \leq Tol$, the step is accepted and the integration is continued. If $err_{n+1} > Tol$, it is rejected and recomputed with the step size $h_{new}$ obtained from (1.1) with $n+1$ instead of $n$. Similar step size strategies are implemented in most codes for solving ordinary differential equations.

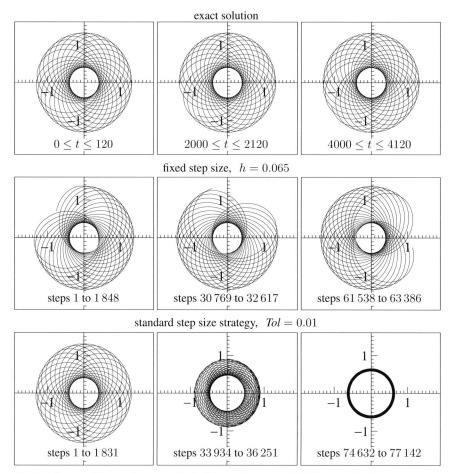

**Fig. 1.1.** Störmer/Verlet scheme applied with fixed step size (middle) or with the standard step size strategy (below) compared to the exact solution (above); solutions are for the interval $0 \leq t \leq 120$ (left), for $2000 \leq t \leq 2120$ (middle), and for $4000 \leq t \leq 4120$ (right).

**Numerical Experiment.** We consider the perturbed Kepler problem

$$\dot{q}_1 = p_1, \qquad \dot{p}_1 = -\frac{q_1}{(q_1^2 + q_2^2)^{3/2}} - \frac{\varepsilon q_1}{(q_1^2 + q_2^2)^{5/2}}$$

$$\dot{q}_2 = p_2, \qquad \dot{p}_2 = -\frac{q_2}{(q_1^2 + q_2^2)^{3/2}} - \frac{\varepsilon q_2}{(q_1^2 + q_2^2)^{5/2}} \qquad (1.2)$$

($\varepsilon = 0.015$) with initial values

$$q_1(0) = 1 - e, \quad q_2(0) = 0, \quad p_1(0) = 0, \quad p_2(0) = \sqrt{\frac{1+e}{1-e}}$$

(eccentricity $e = 0.6$). As a numerical method we take the *Störmer/Verlet scheme* (I.3.6) which is symmetric, symplectic, and of order 2. The fixed step size imple-

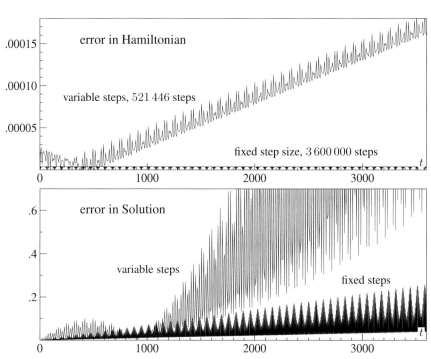

**Fig. 1.2.** Study of the error in the Hamiltonian and of the global error for the Störmer/Verlet scheme. Fixed step size implementation with $h = 10^{-3}$, variable step size with $Tol = 10^{-4}$.

mentation is straightforward. For the variable step size strategy we take for $err_n$ the Euclidean norm of the difference between the Störmer/Verlet solution and the symplectic Euler solution (which is available without any further function evaluation). Since $err_n = \mathcal{O}(h_n^2)$, we take $r = 2$ in (1.1).

The numerical solution in the $(q_1, q_2)$-plane is presented in Fig. 1.1. To make the long-time behaviour of the two implementations visible, we show the numerical solution on three different parts of the integration interval. We have included the numbers of steps needed for the integration to reach $t = 120$, $2120$, and $4120$, respectively. We see that the qualitative behaviour of the variable step size implementation is not correct, although it is more precise on short intervals. Moreover, the near-preservation of the Hamiltonian is lost (see Fig. 1.2) as is the linear error growth. Apparently, the error in the Hamiltonian behaves like $|a - bt|$ for the variable step size implementation, and that for the solution like $|ct - dt^2|$ (with constants $a, b, c, d$ depending on $Tol$). Due to the relatively large eccentricity of the problem, the variable step size implementation needs fewer function evaluations for a given accuracy on a short time interval, but the opposite is true for long-time integrations.

The aim of the next two sections is to present approaches which permit the use of variable step sizes for symmetric or symplectic methods without losing the qualitatively correct long-time behaviour.

## VIII.2 Reversible Adaptive Step Size Selection

The disappointing long-time behaviour in Fig. 1.1 of the variable step size implementation of the Störmer/Verlet scheme can be understood from the fact that it is certainly symmetric, but not reversible. Indeed, consider a $\rho$-reversible differential equation and denote by $y_{n+1}$ the numerical solution obtained from $y_n$ with step size $h_n$. From a reversible method (cf. Fig. V.1.1) we expect that starting from $\rho y_{n+1}$ we obtain $\rho y_n$ as result. The step size strategy of Sect. VIII.1, for which $h_n$ depends on information from the preceding step, cannot guarantee such a property.

Following Stoffer (1988) we therefore consider step sizes

$$h_n = \varepsilon\, s(y_n, \varepsilon), \tag{2.1}$$

where $h_n$ only depends on $y_n$ and on a small parameter $\varepsilon$ that is usually related to the tolerance *Tol*. The numerical flow of a method $\Phi_h$ applied with step size (2.1),

$$\Psi_\varepsilon(y) := \Phi_{\varepsilon s(y,\varepsilon)}(y), \tag{2.2}$$

can be considered as a transformation on the phase space, and the definitions of symmetry and reversibility can be extended in a straightforward way.

**Symmetry.** We call the numerical method $\Phi_h$ with step size strategy (2.1) symmetric, if $\Psi_\varepsilon = \Psi_{-\varepsilon}^{-1}$. In the case of a symmetric $\Phi_h$ this is equivalent to

$$s(\widehat{y}, -\varepsilon) = s(y, \varepsilon) \qquad \text{with} \qquad \widehat{y} = \Phi_{\varepsilon s(y,\varepsilon)}(y). \tag{2.3}$$

Hut, Makino & McMillan (1995) propose the use of $s(y, \varepsilon) = \frac{1}{2}\bigl(\sigma(y) + \sigma(\widehat{y})\bigr)$, where $\sigma(y)$ is some function that uses an a priori knowledge of the solution of the differential equation.

**Reversibility.** The method $\Phi_h$ with step size strategy (2.1) is called $\rho$-reversible if, when applied to a $\rho$-reversible differential equation, $\rho \circ \Psi_\varepsilon = \Psi_\varepsilon^{-1} \circ \rho$ holds (cf. Definition V.1.2). If the method $\Phi_h$ is symmetric and if $\rho \circ \Phi_h = \Phi_{-h} \circ \rho$ (cf. Theorem V.1.5) then this is equivalent to

$$s(\rho\widehat{y}, \varepsilon) = s(y, \varepsilon) \qquad \text{with} \qquad \widehat{y} = \Phi_{\varepsilon s(y,\varepsilon)}(y). \tag{2.4}$$

Together with (2.3) this implies that $s(\rho y, \varepsilon) = s(y, -\varepsilon)$ for all $y$.

How can we find suitable step size functions $s(y, \varepsilon)$ which satisfy all these properties, and which do not require any a priori knowledge of the solution? In a remarkable publication, Stoffer (1995) gives the key to the answer of this question. He simply proposes to choose the step size $h$ in such a way that the local error estimate satisfies $err = Tol$ (in contrast to $err \leq Tol$ for the standard strategy). Let us explain this idea in some more detail for Runge-Kutta methods.

**Example 2.1 (Symmetric, Variable Step Size Runge-Kutta Methods).** For the numerical solution of $\dot{y} = f(y)$ we consider Runge-Kutta methods

$$Y_i = y_n + h \sum_{j=1}^{s} a_{ij} f(Y_j), \qquad y_{n+1} = y_n + h \sum_{i=1}^{s} b_i f(Y_i), \qquad (2.5)$$

with coefficients satisfying $a_{s+1-i,s+1-j} + a_{ij} = b_j$ for all $i, j$. By Theorem V.2.3 such methods are symmetric. A common approach for step size control is to consider an embedded method $\widehat{y}_{n+1} = y_n + h \sum_{i=1}^{s} \widehat{b}_i f(Y_i)$ (which is a method with the same internal stages $Y_i$) and to take the difference $y_{n+1} - \widehat{y}_{n+1}$, i.e.,

$$D(y_n, h) = h \sum_{i=1}^{s} e_i f(Y_i) \qquad (2.6)$$

with $e_i = b_i - \widehat{b}_i$, as indicator of the local error. For methods where $Y_i \approx y(t_n + c_i h)$ (e.g., collocation or discontinuous collocation) one usually computes the coefficients $e_i$ from a nontrivial solution of the homogeneous linear system

$$\sum_{i=1}^{s} e_i c_i^{k-1} = 0 \qquad \text{for} \quad k = 1, \ldots, s-1. \qquad (2.7)$$

This yields $D(y_n, h) = \mathcal{O}(h^r)$ with $r$ close to $s$. According to the suggestion of Stoffer (1995) we determine the step size $h_n$ such that

$$\|D(y_n, h_n)\| = Tol. \qquad (2.8)$$

A Taylor expansion around $h = 0$ shows that $D(y, h) = d_r(y) h^r + \mathcal{O}(h^{r+1})$ with some $r \geq 1$. We assume $\|d_r(y)\| \neq 0$ and we put $\varepsilon = Tol^{1/r}$, so that $h_n$ from (2.8) can be expressed by a smooth function $s(y, \varepsilon)$ as (2.1).

In order to satisfy the *symmetry* relation (2.3) we determine the $e_i$ such that

$$e_{s+1-i} = e_i \quad \text{for all } i \qquad \text{or} \qquad e_{s+1-i} = -e_i \quad \text{for all } i \qquad (2.9)$$

(Hairer & Stoffer 1997). If the Runge-Kutta method is symmetric, this then implies

$$\|D(y_n, h)\| = \|D(y_{n+1}, -h)\| \qquad \text{with} \qquad y_{n+1} = \Phi_h(y_n). \qquad (2.10)$$

This follows from the fact that the internal stage vectors $Y_i$ of the step from $y_n$ to $y_{n+1}$ and the stage vectors $\overline{Y}_i$ of the step from $y_{n+1}$ to $y_n$ (negative step size $-h$) are related by $\overline{Y}_i = Y_{s+1-i}$. The step size determined by (2.8) is thus the same for both steps and, consequently, condition (2.3) holds.

The *reversibility* requirement (2.4) is a consequence of

$$\|D(y_n, h)\| = \|D(\rho y_{n+1}, h)\| \qquad \text{with} \qquad y_{n+1} = \Phi_h(y_n) \qquad (2.11)$$

which is satisfied for orthogonal mappings $\rho$ (i.e., $\rho^T \rho = I$). This is seen as follows: applying $\Phi_h$ to $\rho y_{n+1}$ gives $\rho y_n$, and the internal stages are $\overline{Y}_i = \rho Y_{s+1-i}$. Hence, we have from (2.9) that $D(\rho y_{n+1}, h) = \pm \rho D(y_n, h)$, and (2.11) follows from the orthogonality of $\rho$.

A simple special case is the trapezoidal rule

$$y_{n+1} = y_n + \frac{h_n}{2}\Big(f(y_n) + f(y_{n+1})\Big) \tag{2.12}$$

combined with

$$D(y_n, h) = \frac{h}{2}\Big(f(y_{n+1}) - f(y_n)\Big).$$

The scalar nonlinear equation (2.8) for $h_n$ can be solved in tandem with the nonlinear system (2.12).

**Example 2.2 (Variable Step Size Verlet Scheme).** The strategy of Example 2.1 can be extended in a straightforward way to partitioned Runge-Kutta methods. Let us illustrate this for the second order symmetric Verlet scheme which, for a problem $\dot{q} = p$, $\dot{p} = -\nabla U(q)$, is given by

$$\begin{aligned} p_{n+1/2} &= p_n - \frac{h}{2}\nabla U(q_n) \\ q_{n+1} &= q_n + h\, p_{n+1/2} \\ p_{n+1} &= v_{n+1/2} - \frac{h}{2}\nabla U(q_{n+1}). \end{aligned} \tag{2.13}$$

As an error indicator we propose

$$D(p_n, q_n, h) = \frac{h}{2}\begin{pmatrix} \nabla U(q_{n+1}) - \nabla U(q_n) \\ h\big(\nabla U(q_{n+1}) + \nabla U(q_n)\big) \end{pmatrix}$$

The first component is just the difference of the Verlet solution and the numerical approximation obtained by the symplectic Euler method. The second component is a symmetrized version of it.

We apply this method with $h_n$ determined by (2.8) and $Tol = 0.01$ to the perturbed Kepler problem (1.2) with initial values as in Fig. 1.1. The result is given in Fig. 2.1. We identify a correct qualitative behaviour (compared to the wrong behaviour for the standard step size strategy in Fig. 1.1). It should be mentioned that

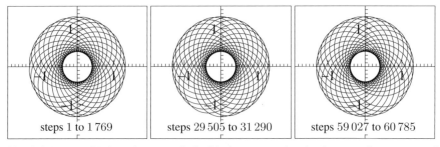

**Fig. 2.1.** Störmer/Verlet scheme applied with the symmetric adaptive step size strategy of Example 2.2 ($Tol = 0.01$); the three pictures have the same meaning as in Fig. 1.1

the work for solving the scalar equation (2.8) for $h_n$ is not negligible, because the Störmer/Verlet scheme is explicit. Solving this equation iteratively, every iteration requires one force evaluation $\nabla U(q)$. An efficient solver for this scalar nonlinear equation should be used.

## VIII.3 Time Transformations

A variable step size implementation produces approximations $y_n$ on a (non-equidistant) grid $\{t_n\}$. The same effect can be achieved by performing in advance a time transformation $t \leftrightarrow \tau$ and by applying a constant step size implementation to the transformed system. If the time transformation is given as the solution of a differential equation, it follows from the chain rule $\frac{dy}{d\tau} = \frac{dy}{dt}\frac{dt}{d\tau}$ that the transformed system is

$$y' = s(y)f(y), \qquad t' = s(y). \tag{3.1}$$

Here, prime indicates a derivative with respect to $\tau$, and we use the same letter $y$ for the solutions $y(t)$ of $\dot{y} = f(y)$ and $y(\tau)$ of (3.1). If $s(y) > 0$, the correspondence $t \leftrightarrow \tau$ is bijective.

Applying a numerical method with constant step size $\varepsilon$ to (3.1) yields approximations $y_n \approx y(\tau_n) = y(t_n)$, where $\tau_n = n\varepsilon$ and

$$t_{n+1} - t_n = \int_{n\varepsilon}^{(n+1)\varepsilon} s\bigl(y(\tau)\bigr)\, d\tau \approx \varepsilon s(y_n).$$

Approximations to $t_n$ are obtained by integrating numerically the differential equation $t' = s(y)$ together with $y' = s(y)f(y)$.

### VIII.3.1 Symplectic Integration

For a Hamiltonian system $\dot{y} = f(y) = J^{-1}\nabla H(y)$ it is natural to investigate for which step size functions $s(y)$ the problem (3.1) is again Hamiltonian. Applying a symplectic integrator with constant step size to (3.1) would give numerical approximations on a non-equidistant grid, which have the same good long-time behaviour as constant step size implementations of symplectic methods.

By the Integrability Lemma VI.2.7 we have to check whether the Jacobian of $s(y)\nabla H(y)$ is symmetric. This is the case only if $\nabla H(y)\nabla s(y)^T$ is symmetric, i.e., $\nabla H(y)$ and $\nabla s(y)$ are collinear, so that $\frac{d}{dt}s\bigl(y(t)\bigr) = \nabla s\bigl(y(t)\bigr)^T J \nabla H\bigl(y(t)\bigr) = 0$. Consequently, $s(y) = Const$ along solutions of the Hamiltonian system which is what makes this approach unattractive for a variable step size integration. This disappointing fact has been observed by Stoffer (1988, 1995) and Skeel & Gear (1992).

The main idea for circumventing this difficulty is the following: suppose we want to integrate the Hamiltonian system with steps of size $h \approx \varepsilon\, s(y)$ (cf. (2.1)), where $s(y) > 0$ is a state-dependent given function and $\varepsilon > 0$ is a small parameter.

Instead of multiplying the vector field $f(y) = J^{-1}\nabla H(y)$ by $s(y)$, we consider the *new Hamiltonian*

$$K(y) = s(y)\Big(H(y) - H_0\Big), \tag{3.2}$$

where $H_0 = H(y_0)$ for a fixed initial value $y_0$. The corresponding Hamiltonian system is

$$y' = s(y)J^{-1}\nabla H(y) + \big(H(y) - H_0\big)J^{-1}\nabla s(y). \tag{3.3}$$

Compared to (3.1) we have introduced a perturbation, which vanishes along the solution of the Hamiltonian system passing through $y_0$, but which makes the system Hamiltonian.

Time transformations such as in (3.2) are used in classical mechanics for an analytic treatment of Hamiltonian systems (Levi-Civita (1906, 1920), where (3.2) is called the "Darboux-Sundman transformation", see Sundman (1912)). Zare & Szebehely (1975) consider such time transformations for numerical purposes (without taking care of symplecticity). Waldvogel & Spirig (1995) apply the transformations proposed by Levi-Civita to Hill's lunar problem and solve the transformed equations by composition methods in order to preserve the symplectic structure. The following general procedure was proposed independently by Hairer (1997) and Reich (1999).

**Algorithm 3.1.** *Apply an arbitrary symplectic one-step method with constant step size $\varepsilon$ to the Hamiltonian system (3.3), augmented by $t' = s(y)$. This yields numerical approximations $(y_n, t_n)$ with $y_n \approx y(t_n)$.*

Although this algorithm yields numerical approximations on a non-equidistant grid, it can be considered as a fixed step size, symplectic method applied to a different Hamiltonian system. This interpretation allows us to apply the standard techniques for the study of its long-time behaviour.

A disadvantage of this algorithm is that for separable Hamiltonians $H(p, q) = T(p) + U(q)$ the transformed Hamiltonian (3.2) is no longer separable. Hence, methods that are explicit for separable Hamiltonians are not explicit in the implementation of Algorithm 3.1. The following examples illustrate that this disadvantage can be partially overcome for the important case of Hamiltonian functions

$$H(p, q) = \frac{1}{2}p^T M^{-1} p + U(q), \tag{3.4}$$

where $M$ is a constant symmetric matrix.

**Example 3.2 (Symplectic Euler with $p$-Independent Step Size Function).** For step size functions $s(q, \varepsilon)$ the symplectic Euler method, applied with constant step size $\varepsilon$ to (3.3), reads

$$p_{n+1} = p_n - \varepsilon s(q_n)\nabla U(q_n) - \varepsilon s'(q_n)\Big(\frac{1}{2}p_{n+1}^T M^{-1} p_{n+1} + U(q_n) - H_0\Big)$$

$$q_{n+1} = q_n + \varepsilon s(q_n) M^{-1} p_{n+1}$$

and yields an approximation at $t_{n+1} = t_n + \varepsilon s(q_n)$. The first equation is non-linear (quadratic) in $p_{n+1}$. Introducing the scalar quantity $\beta := \|p_{n+1}\|_M^2 := p_{n+1}^T M^{-1} p_{n+1}$, it reduces to the scalar quadratic equation

$$\beta = \left\| p_n - \varepsilon s(q_n)\nabla U(q_n) - \varepsilon s'(q_n)\left(\frac{\beta}{2} + U(q_n) - H_0\right) \right\|_M^2$$

which can be solved directly. The numerical solution $(p_{n+1}, q_{n+1})$ is then given explicitly.

**Choices of Step Size Functions.** For certain problems, suitable functions $s(p,q)$ are known a priori. For example, for the two-body problem one can take $s(p,q) = \|q\|^r$ (e.g., $r=2$), so that smaller step sizes are taken when the two bodies are close.

An interesting choice, which does not require any a priori knowledge of the solution, is $s(y) = \|f(y)\|^{-1}$. The solution of (3.1) then satisfies $\|y'(\tau)\| = 1$ (arclength parameterization) and we get approximations $y_n$ that are nearly equidistant in the phase space. Such time transformations have been proposed by McLeod & Sanz-Serna (1982) for graphical reasons and by Huang & Leimkuhler (1997). For a Hamiltonian system with $H(p,q)$ given by (3.4), it is thus natural to consider

$$s(p,q) = \left(\frac{1}{2}p^T M^{-1} p + \nabla U(q)^T M^{-1} \nabla U(q)\right)^{-1/2}. \tag{3.5}$$

We have chosen this particular norm, because it leaves the expression (3.5) invariant with respect to linear coordinate changes $q \mapsto Aq$ (implying $p \mapsto A^{-T}p$). Exploiting the fact that the Hamiltonian (3.4) is constant along solutions, the step size function (3.5) can be replaced by the $p$-independent function

$$s(q) = \left((H_0 - U(q)) + \nabla U(q)^T M^{-1} \nabla U(q)\right)^{-1/2}. \tag{3.6}$$

The use of (3.5) and (3.6) gives nearly identical results, but (3.6) is easier to implement. If we are interested in an output that is approximatively equidistant in the $q$-space, we can take

$$s(q) = \left(H_0 - U(q)\right)^{-1/2}. \tag{3.7}$$

**Example 3.3 (Störmer/Verlet Scheme with $p$-Independent Step Size Function).** For a step size function $s(q, \varepsilon)$ the Störmer/Verlet scheme gives

$$\begin{aligned} p_{n+1/2} &= p_n - \frac{\varepsilon}{2}s(q_n)\nabla U(q_n) - \frac{\varepsilon}{2}s'(q_n)\Big(H(p_{n+1/2}, q_n) - H_0\Big) \\ q_{n+1} &= q_n + \frac{\varepsilon}{2}\big(s(q_n) + s(q_{n+1})\big)M^{-1}p_{n+1/2} \\ p_{n+1} &= p_{n+1/2} - \frac{\varepsilon}{2}s(q_{n+1})\nabla U(q_{n+1}) \\ &\quad - \frac{\varepsilon}{2}s'(q_{n+1})\Big(H(p_{n+1/2}, q_{n+1}) - H_0\Big). \end{aligned} \tag{3.8}$$

The first equation is essentially the same as that for the symplectic Euler method, and it can be solved for $p_{n+1/2}$ as explained in Example 3.2. The second equation is implicit in $q_{n+1}$, but it is sufficient to solve the scalar equation

$$\gamma = s\Big(q_n + \frac{\varepsilon}{2}\big(s(q_n) + \gamma\big)M^{-1}p_{n+1/2}\Big) \tag{3.9}$$

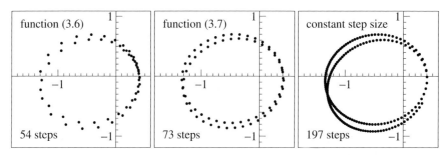

**Fig. 3.1.** Various step size strategies for the Störmer/Verlet scheme (Example 3.3) applied to the perturbed Kepler problem (1.2) on the interval $[0, 10]$ (approximately two periods).

for $\gamma = s(q_{n+1})$. Newton iterations can be efficiently applied, because $s'(q)$ is available already. The last equation (for $p_{n+1}$) is explicit. This variable step size Störmer/Verlet scheme gives approximations at $t_n$, where

$$t_{n+1} = t_n + \frac{\varepsilon}{2}\big(s(q_n) + s(q_{n+1})\big).$$

In Fig. 3.1 we illustrate how the different step size functions influence the position of the output points. We apply the Störmer/Verlet method of Example 3.3 to the perturbed Kepler problem (1.2) with initial values, perturbation parameter, and eccentricity as in Sect. VIII.1. As step size functions we use (3.6), (3.7), and constant step size $s(q) \equiv 1$. For all three choices of $s(q)$ we have adjusted the parameter $\varepsilon$ in such a way that the maximal error in the Hamiltonian is close to 0.01. The step size strategy (3.6) is apparently the most efficient one. For this strategy, we observe that the output points in the $q$-plane concentrate in regions where the velocity is large, while the constant step size implementation shows the opposite behaviour.

### VIII.3.2 Reversible Integration

For $\rho$-reversible differential equations $\dot{y} = f(y)$, i.e., $f(\rho y) = -\rho f(y)$ for all $y$, the time transformed problem (3.1) remains $\rho$-reversible if

$$s(\rho y) = s(y). \tag{3.10}$$

This condition is not very restrictive and is satisfied by many important time transformations. In particular, (3.10) holds for the arc length parameterization $s(y) = \|f(y)\|^{-1}$ if $\rho$ is orthogonal. Consequently, it makes sense to apply symmetric numerical methods with constant step size $\varepsilon$ directly to the system (3.1).

However, similar to the symplectic integration of Sect. VIII.3.1, there is a serious disadvantage. For separable differential equations (i.e., problems that can be split as $\dot{p} = f_1(q)$, $\dot{q} = f_2(p)$) and for non-constant $s(p, q)$ the transformed system (3.1) is no longer separable. Hence, methods that are explicit for separable problems are not explicit for (3.1).

**Example 3.4 (Adaptive Verlet Method).** We consider a Hamiltonian system with separable Hamiltonian (3.4), and we apply the Störmer/Verlet scheme to the time transformed differential equation. This yields

$$p_{n+1/2} = p_n - \frac{\varepsilon}{2} s_n \nabla U(q_n)$$
$$q_{n+1} = q_n + \frac{\varepsilon}{2}(s_n + s_{n+1}) M^{-1} p_{n+1/2} \qquad (3.11)$$
$$p_{n+1} = p_{n+1/2} - \frac{\varepsilon}{2} s_{n+1} \nabla U(q_{n+1}),$$

where $s_n = s(p_{n+1/2}, q_n)$ and $s_{n+1} = s(p_{n+1/2}, q_{n+1})$ (notice that the $s_{n+1}$ of the current step is not the same as the $s_n$ of the subsequent step, if $s(p, q)$ depends on $p$). The values $(p_{n+1}, q_{n+1})$ are approximations to the solution at $t_n$, where

$$t_{n+1} = t_n + \frac{\varepsilon}{2}(s_n + s_{n+1}).$$

For a $p$-independent step size function $s$, method (3.11) corresponds to that of Example 3.3, where the terms involving $s'(q)$ are removed. The implicitness of (3.11) is comparable to the method of Example 3.3.

With the aim of obtaining a completely explicit integrator, Huang & Leimkuhler (1997) propose the use of two-term recurrence relations for $s_n$. An improved modification of their approach is presented in Holder, Leimkuhler & Reich (2001), where the values $s_n$ in (3.11) are determined by (starting with $s_0 = s(p_0, q_0)$)

$$\frac{1}{s_{n+1}} + \frac{1}{s_n} = \frac{2}{s(p_{n+1/2}, q_{n+1/2})}. \qquad (3.12)$$

Here, $p_{n+1/2}$ is given by (3.11) and $q_{n+1/2} = q_n + \frac{\varepsilon}{2} s_n M^{-1} p_{n+1/2}$. The resulting method is easy to implement, and these authors report an excellent performance for realistic problems.

A rigorous analysis of the long-time behaviour of this variable step size Störmer/Verlet method is still not available. The results of Chapters IX and XI cannot be applied, because (3.11)-(3.12) is not a mapping $(p_n, q_n) \mapsto (p_{n+1}, q_{n+1})$. A first analysis is given in Cirilli, Hairer & Leimkuhler (1999). There it is shown that, similar to weakly stable multistep methods (Chap. XIV), the numerical solution contains oscillatory terms. Although these oscillations are usually very small (and not visible), it seems difficult to get rigorous estimates for them.

We conclude this section with a brief comparison of the variable step size Störmer/Verlet methods of Examples 3.3 and 3.4. Method (3.11) is easier to implement and more efficient when the step size function $s(p, q)$ is expensive to evaluate. In a few numerical comparisons we observed, however, that the error in the Hamiltonian and in the solution is in general larger for method (3.11), and that the method (3.8) becomes competitive when $s(p, q)$ is $p$-independent and easy to evaluate.

For the perturbed Kepler problem of Sect. VIII.1 with the Sundman time transformation $s(q) = \|q\|^2$ the error in the Hamiltonian is about four times larger, so that method (3.11) has to be applied with halved step size in order to get the same accuracy. For this choice of $s(q)$ the nonlinear equation in (3.9) is quadratic and so can be solved immediately. This means that method (3.8) does not need any iterative procedures and allows a straightforward implementation.

## VIII.4 Multiple Time Stepping

A completely different approach to variable step sizes will be described in this section. We are interested in situations where:

- many solution components of the differential equation vary slowly and only a few components have fast dynamics; or
- computationally expensive parts of the right-hand side do not contribute much to the dynamics of the solution.

In the first case it is tempting to use large step sizes for the slow components and small step sizes for the fast ones. Such integrators, called *multirate methods*, were first formulated by Rice (1960) and Gear & Wells (1984). They were further developed by Günther & Rentrop (1993) in view of applications in electric circuit simulation, and by Engstler & Lubich (1997) with applications in astrophysics. Symmetric multirate methods are obtained from the approaches described below and are specially constructed by Leimkuhler & Reich (2001).

The second case suggests the use of methods that evaluate the expensive part of the vector field less often than the rest. This approach is called *multiple time stepping*. It was originally proposed for astronomy by Hayli (1967) and has become very popular in molecular dynamics simulations (Streett, Tildesley & Saville 1978, Grubmüller, Heller, Windemuth & Schulten 1991, Tuckerman, Berne & Martyna 1992). As noticed by Biesiadecki & Skeel (1993), one approach to such methods is within the framework of splitting and composition methods, which yields symmetric and symplectic methods. A second family of symmetric multiple time stepping methods results from the concept of using averaged force evaluations.

### VIII.4.1 Fast-Slow Splitting: the Impulse Method

Consider a differential equation

$$\dot{y} = f(y), \qquad f(y) = f^{[\text{slow}]}(y) + f^{[\text{fast}]}(y), \qquad (4.1)$$

where the vector field is split into summands contributing to slow and fast dynamics, respectively, and where $f^{[\text{slow}]}(y)$ is more expensive to evaluate than $f^{[\text{fast}]}(y)$. Multirate methods can often be cast into this framework by collecting in $f^{[\text{slow}]}(y)$ those components of $f(y)$ which produce slow dynamics and in $f^{[\text{fast}]}(y)$ the remaining components.

**Algorithm 4.1.** *For a given $N \geq 1$ and for the differential equation (4.1) a multiple time stepping method is obtained from*

$$\left(\Phi_{h/2}^{[\text{slow}]}\right)^* \circ \left(\Phi_{h/N}^{[\text{fast}]}\right)^N \circ \Phi_{h/2}^{[\text{slow}]}, \tag{4.2}$$

where $\Phi_h^{[\text{slow}]}$ and $\Phi_h^{[\text{fast}]}$ are numerical integrators consistent with $\dot{y} = f^{[\text{slow}]}(y)$ and $\dot{y} = f^{[\text{fast}]}(y)$, respectively.

The method of Algorithm 4.1 is already stated in symmetrized form ($\Phi_h^*$ denotes the adjoint of $\Phi_h$). It is often called the *impulse method*, because the slow part $f^{[\text{slow}]}$ of the vector field is used – impulse-like – only at the beginning and at the end of the step, whereas the many small substeps in between are concerned solely through integrating the fast system $\dot{y} = f^{[\text{fast}]}(y)$.

**Lemma 4.2.** *Let $\Phi_h^{[\text{slow}]}$ be an arbitrary method of order 1, and $\Phi_h^{[\text{fast}]}$ a symmetric method of order 2. Then, the multiple time stepping algorithm (4.2) is symmetric and of order 2.*

*If $f^{[\text{slow}]}(y)$ and $f^{[\text{fast}]}(y)$ are Hamiltonian and if $\Phi_h^{[\text{slow}]}$ and $\Phi_h^{[\text{fast}]}$ are both symplectic, then the multiple time stepping method is also symplectic.*

*Proof.* Due to the interpretation of multiple time stepping as composition methods the proof of these statements is obvious. □

The order statement of Lemma 4.2 is valid for $h \to 0$, but should be taken with caution if the product of the step size $h$ with a Lipschitz constant of the problem is not small (see Chap. XIII for a detailed analysis): it is *not* stated, and is not true in general for large $N$, that if $h$ and $h/N$ are the step sizes needed to integrate the slow and fast system, respectively, with an error bounded by $\varepsilon$, then the error of the combined scheme is $\mathcal{O}(\varepsilon)$.

The most important application of multiple time stepping is in Hamiltonian systems with a separable Hamiltonian

$$H(p,q) = T(p) + U(q), \qquad U(q) = U^{[\text{slow}]}(q) + U^{[\text{fast}]}(q). \tag{4.3}$$

If we let the fast vector field correspond to $T(p) + U^{[\text{fast}]}(q)$ and the slow vector field to $U^{[\text{slow}]}(q)$, and if we apply the Störmer/Verlet method and exact integration, respectively, Algorithm 4.1 reads

$$\varphi_{h/2}^{[\text{slow}]} \circ \left(\varphi_{h/2N}^{[\text{fast}]} \circ \varphi_{h/N}^T \circ \varphi_{h/2N}^{[\text{fast}]}\right)^N \circ \varphi_{h/2}^{[\text{slow}]}, \tag{4.4}$$

where $\varphi_t^T, \varphi_t^{[\text{slow}]}, \varphi_t^{[\text{fast}]}$ are the exact flows corresponding to the Hamiltonian systems for $T(p), U^{[\text{slow}]}(q), U^{[\text{fast}]}(q)$, respectively. Notice that for $N = 1$ the method (4.4) reduces to the Störmer/Verlet scheme applied to the Hamiltonian system with $H(p,q)$. This is a consequence of the fact that $\varphi_t^{[\text{fast}]} \circ \varphi_t^{[\text{slow}]} = \varphi_t^U$ is the exact flow of the Hamiltonian system corresponding to $U(q)$ of (4.3). In the molecular dynamics literature, the method (4.4) is known as the Verlet-I method (Grubmüller et al. 1991, who consider the method with little enthusiasm) or r-RESPA method (Tuckerman et al. 1992, with much more enthusiasm).

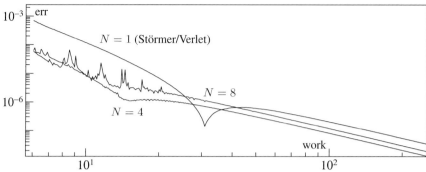

**Fig. 4.1.** Maximal error in the Hamiltonian as a function of computational work.

**Example 4.3.** In order to illustrate the effect of multiple time stepping we choose a 'solar system' with two planets, i.e., with a Hamiltonian

$$H(p,q) = \frac{1}{2}\left(\frac{p_0^T p_0}{m_0} + \frac{p_1^T p_1}{m_1} + \frac{p_2^T p_2}{m_2}\right) - \frac{m_0 m_1}{\|q_0 - q_1\|} - \frac{m_0 m_2}{\|q_0 - q_2\|} - \frac{m_1 m_2}{\|q_1 - q_2\|},$$

where $m_0 = 1, m_1 = m_2 = 10^{-2}$ and initial values $q_0 = (0,0), \dot{q}_0 = (0,0)$, $q_1 = (1,0), \dot{q}_1 = (0,1), q_2 = (4,0), \dot{q}_2 = (0,0.5)$. With these data, the motion of the two planets is nearly circular with periods close to $2\pi$ and $14\pi$, respectively.

We split the potential as

$$U^{[\text{fast}]}(q) = -\frac{m_0 m_1}{\|q_0 - q_1\|}, \qquad U^{[\text{slow}]}(q) = -\frac{m_0 m_2}{\|q_0 - q_2\|} - \frac{m_1 m_2}{\|q_1 - q_2\|},$$

and we apply the algorithm of (4.4) with $N = 1$ (Störmer/Verlet), $N = 4$, and $N = 8$. Since the evaluation of $\varphi_t^{[\text{slow}]}$ is about twice as expensive as $\varphi_t^{[\text{fast}]}$ and that of $\varphi_t^T$ is of negligible cost, the computational work of applying (4.4) on a fixed interval is proportional to

$$\frac{2\pi}{h} \cdot \frac{(2+N)}{3}. \tag{4.5}$$

Our computations have shown that this measure of work corresponds very well to the actual cpu time.

We have solved this problem with many different step sizes $h$. Figure 4.1 shows the maximal error in the Hamiltonian (over the interval $[0, 200\pi]$) as a function of the computational work (4.5). We notice that the value $N = 4$ yields excellent results for relatively large as well as small step sizes. It noticeably improves the performance of the Störmer/Verlet method. If $N$ becomes too large, an irregular behaviour for large step sizes is observed. Such "artificial resonances" are notorious for this method and have been discussed by Biesiadecki & Skeel (1993) for a similar experiment; also see Chap. XIII. For large $N$ we also note a loss of accuracy for small step sizes. The optimal choice of $N$ (which here is close to 4) depends on the problem and on the splitting into fast and slow parts, and has to be determined by experiment.

The multiple time stepping technique can be iteratively extended to problems with more than two different time scales. The idea is to split the 'fast' vector field of (4.1) into $f^{[\text{fast}]}(y) = f^{[ff]}(y) + f^{[fs]}(y)$, and to replace the method $\Phi_h^{[\text{fast}]}$ in Algorithm 4.1 with a multiple time stepping method. Depending on the problem, a significant gain in computer time may be achieved in this way.

Many more multiple time stepping methods that extend the above Verlet-I/r-RESPA/impulse method, have been proposed in the literature, most notably the mollified impulse method of García-Archilla, Sanz-Serna & Skeel (1999); see Sect. XIII.1.

## VIII.4.2 Averaged Forces

A different approach to multiple time stepping arises from the idea of advancing the step with *averaged force evaluations*. We describe such a method for the second-order equation

$$\ddot{y} = f(y), \qquad f(y) = f^{[\text{slow}]}(y) + f^{[\text{fast}]}(y). \tag{4.6}$$

The exact solution satisfies

$$y(t+h) - 2y(t) + y(t-h) = h^2 \int_{-1}^{1} (1 - |\theta|) f\big(y(t+\theta h)\big) d\theta ,$$

where the integral on the right-hand side represents a weighted average of the force along the solution, which is now going to be approximated. At $t = t_n$, we replace

$$f\big(y(t_n + \theta h)\big) \approx f^{[\text{slow}]}(y_n) + f^{[\text{fast}]}\big(u(\theta h)\big)$$

where $u(\tau)$ is a solution of the differential equation

$$\ddot{u} = f^{[\text{slow}]}(y_n) + f^{[\text{fast}]}(u) . \tag{4.7}$$

We then have

$$h^2 \int_{-1}^{1} (1 - |\theta|) \Big( f^{[\text{slow}]}(y_n) + f^{[\text{fast}]}\big(u(\theta h)\big) \Big) d\theta = u(h) - 2u(0) + u(-h) .$$

The velocities are treated similarly, starting from the identity

$$\dot{y}(t+h) - \dot{y}(t-h) = h \int_{-1}^{1} f\big(y(t+\theta h)\big) d\theta .$$

**A Symmetric Two-Step Method.** For the differential equation (4.7) we assume the initial values

$$u(0) = y_n, \quad \dot{u}(0) = \dot{y}_n . \tag{4.8}$$

This initial value problem is solved numerically, e.g., by the Störmer/Verlet method with a smaller step size $\pm h/N$ on the interval $[-h, h]$, yielding numerical approximations $u_N(\pm h)$ and $v_N(\pm h)$ to $u(\pm h)$ and $\dot{u}(\pm h)$, respectively. Note that no further evaluations of $f^{[\text{slow}]}$ are needed for the computation of $u_N(\pm h)$ and $v_N(\pm h)$. This finally gives the symmetric two-step method (Hochbruck & Lubich 1999a)

$$\begin{aligned} y_{n+1} - 2y_n + y_{n-1} &= u_N(h) - 2u_N(0) + u_N(-h) \\ \dot{y}_{n+1} - \dot{y}_{n-1} &= v_N(h) - v_N(-h) \,. \end{aligned} \quad (4.9)$$

The starting values $y_1$ and $\dot{y}_1$ are chosen as $u_N(h)$ and $v_N(h)$ which correspond to (4.7) and (4.8) for $n = 0$.

**A Symmetric One-step Method.** An explicit one-step method with similar averaged forces is obtained when the initial values for (4.7) are chosen as

$$u(0) = y_n, \quad \dot{u}(0) = 0 \,. \quad (4.10)$$

It may appear crude to take zero initial values for the velocity, but we remark that for linear $f^{[\text{fast}]}$ the averaged force $(u(h) - 2u(0) + u(-h))/h^2$ does not depend on the choice of $\dot{u}(0)$. Moreover the solution then satisfies $u(-t) = u(t)$, so that the computational cost is halved. We again denote by $u_N(h) = u_N(-h)$ the numerical approximation to $u(h)$ obtained with step size $\pm h/N$ from a one-step method (e.g., from the Störmer/Verlet scheme). Because of (4.10) the averaged forces

$$F_n = \frac{1}{h^2}\bigl(u_N(h) - 2u_N(0) + u_N(-h)\bigr) = \frac{2}{h^2}\bigl(u_N(h) - u_N(0)\bigr)$$

now depend only on $y_n$ and not on the velocity $\dot{y}_n$. In trustworthy Verlet manner, the scheme $y_{n+1} - 2y_n + y_{n-1} = h^2 F_n$ can be written as the one-step method

$$\begin{aligned} v_{n+1/2} &= v_n + \frac{h}{2} F_n \\ y_{n+1} &= y_n + h v_{n+1/2} \\ v_{n+1} &= v_{n+1/2} + \frac{h}{2} F_{n+1} \,. \end{aligned} \quad (4.11)$$

The auxiliary variables $v_n$ can be interpreted as averaged velocities: we have

$$v_n = \frac{y_{n+1} - y_{n-1}}{2h} \approx \frac{y(t_{n+1}) - y(t_{n-1})}{2h} = \frac{1}{2}\int_{-1}^{1} \dot{y}(t_n + \theta h)\,d\theta \,.$$

This average may differ substantially from $\dot{y}(t_n)$ if the solution is highly oscillatory in $[-h, h]$. In the experiments of this section it turned out that the choice $v_0 = \dot{y}_0$ and $\dot{y}_n = v_n$ as velocity approximations gives excellent results.

In a multirate context, symmetric one-step schemes using averaged forces were studied by Hochbruck & Lubich (1999b), Nettesheim & Reich (1999), and Leimkuhler & Reich (2001).

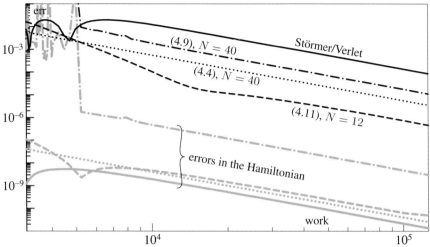

**Fig. 4.2.** Errors in position and in the Hamiltonian as a function of the computational work; the classical Störmer/Verlet method, the impulse method (4.4), and the averaged force methods (4.11) and (4.9). The errors in the Hamiltonian are indicated by grey lines (same linestyle).

**Example 4.4.** We add a satellite of mass $m_3 = 10^{-4}$ to the three body-problem of Example 4.3. It moves rapidly around the planet number one. The initial positions and velocities are $q_3 = (1.01, 0)$ and $p_3 = (0, 0)$. We split the potential as

$$U^{[\text{fast}]}(q) = -\frac{m_1 m_3}{\|q_1 - q_3\|}, \quad U^{[\text{slow}]}(q) = -\sum_{\substack{i<j \\ (i,j)\neq(1,3)}} \frac{m_i m_j}{\|q_i - q_j\|},$$

and we apply the methods (4.9), (4.11), and the impulse method (4.4). Since the sum in $U^{[\text{slow}]}$ contains 5 terms, the computational work is proportional to

$$\frac{5+N}{6h} \quad \text{for methods (4.11) and (4.4)}$$

$$\frac{6+2N}{6h} \quad \text{for method (4.9)}.$$

For each of the methods we have optimized the number $N$ of small steps. We obtained a flat minimum near $N = 40$ for (4.9) and (4.4), and a more pronounced minimum at $N = 12$ for (4.11). Figure 4.2 shows the errors at $t = 10$ in the positions and in the Hamiltonian as a function of the computational work.

The error in the position is largest for the Störmer/Verlet method and significantly smallest for the one-step averaged-force method (4.11). The errors in the velocities are about a factor 100 larger for all methods. They are not included in the figure. The error in the Hamiltonian is very similar for all methods with the exception of the two-step averaged-force method (4.9), for which it is much larger.

## VIII.5 Reducing Rounding Errors

> ... the idea is to capture the rounding errors and feed them back into the summation.
> (N.J. Higham 1993)

All numerical methods for solving ordinary differential equations require the computation of a recursion of the form

$$y_{n+1} = y_n + \delta_n, \tag{5.1}$$

where $\delta_n$, the increment, is usually smaller in magnitude than the approximation $y_n$ to the solution. In this situation the rounding errors caused by the computation of $\delta_n$ are in general smaller than those due to the addition in (5.1).

A first attempt at reducing the accumulation of rounding errors (in fixed-point arithmetic for his Runge-Kutta code) was due to Gill (1951). Kahan (1965) and Möller (1965) both extended this idea to floating point arithmetic. The resulting algorithm is nowadays called 'compensated summation', and a particularly nice presentation and analysis is given by N. Higham (1993). In the following algorithm we assume that $y_n$ is a scalar; vector valued recursions have to be treated component-wise.

**Algorithm 5.1 (Compensated Summation).** *Let $y_0$ and $\{\delta_n\}_{n \geq 0}$ be given and put $e = 0$. Compute $y_1, y_2, \ldots$ from (5.1) as follows:*

$$
\begin{aligned}
&\text{for } n = 0, 1, 2, \ldots \text{ do} \\
&\quad a = y_n \\
&\quad e = e + \delta_n \\
&\quad y_{n+1} = a + e \\
&\quad e = e + (a - y_{n+1}) \\
&\text{end do}
\end{aligned}
$$

This algorithm can best be understood with the help of Fig. 5.1 (following the presentation of N. Higham (1993)). We present the mantissas of floating point numbers by boxes, for which the horizontal position indicates the exponent (for a large exponent the box is more to the left). The mantissas of $y_n$ and $e$ together represent the accurate value of $y_n$ (notice that in the beginning $e = 0$). The operations of Algorithm 5.1 yield $y_{n+1}$ and a new $e$, which together represent $y_{n+1} = y_n + \delta_n$.

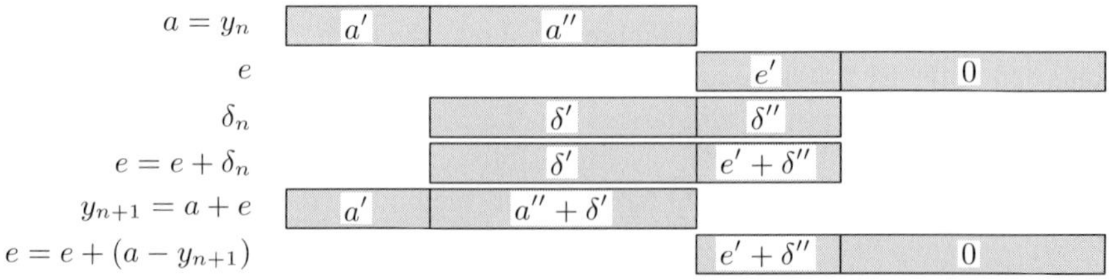

**Fig. 5.1.** Illustration of the technique of 'compensated summation'.

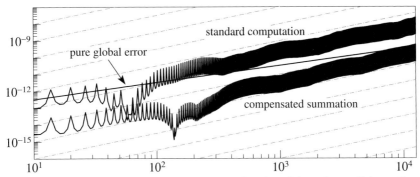

**Fig. 5.2.** Rounding errors and pure global error as a function of time; the parallel grey lines indicate a growth of $\mathcal{O}(t^{3/2})$.

No digit of $\delta_n$ is lost in this way. With a standard summation the last digits of $\delta_n$ (those indicated by $\delta''$ in Fig. 5.1) would have been missed.

**Numerical Experiment.** We study the effect of compensated summation on the Kepler problem (I.2.2) (written as a first order system) with eccentricity $e = 0.6$ and initial values as in (I.2.11), so that the period of the elliptic orbit is exactly $2\pi$. As the numerical integrator we take the composition method (V.3.13) of order 8 with the Störmer/Verlet scheme as basic integrator. We compute the numerical solution with step size $h = 2\pi/500$ once with standard update of the increment, once with compensated summation (both in double precision) and, in order to get a reference solution, we also perform the whole computation in quadruple precision. The difference between the double and quadruple precision computations gives us the rounding errors. Their Euclidean norms as a function of time are displayed in Fig. 5.2.

We see that throughout the whole integration interval the rounding errors of the standard implementation are nearly a factor of $100$ larger than those of the implementation with compensated summation. This corresponds to the inverse of the step size or, more precisely, to the mean quotient between $y_n$ and $\delta_n$ in (5.1). In Fig. 5.2 we have also included the pure global error of the method (without rounding errors) at integral multiples of the period $2\pi$ (hence no oscillations are visible). This is obtained as the difference of the numerical solution computed with quadruple precision and the exact solution. We observe a linear growth of the pure global error (this will be explained in Sect. X.3) and a growth like $\mathcal{O}(t^{3/2})$ due to the rounding errors. Thus, eventually the rounding errors will surpass the truncation errors, but this happens for the compensated summation only after some $1000$ periods.

**Probabilistic Explanation of the Error Growth.** Our aim is to explain the growth rate of rounding errors observed in Fig. 5.2. Denote by $\varepsilon_k$ the vector of rounding errors produced during the computations in the $k$th step. Since the derivative of the flow $\varphi_t(y)$ describes the propagation of these errors, the accumulated rounding error at time $t = t_N$ ($t_k = kh$) is

$$\eta_t = \sum_{k=1}^{N} \varphi'_{t-t_k}(y_k)\varepsilon_k. \tag{5.2}$$

For the Kepler problem and, in fact, for all completely integrable differential equations (cf. Sect. X.1) the flow and its derivative grow at most linearly with time, i.e.,

$$\|\varphi'_{t-t_k}(y)\| \le a + b(t - t_k) \quad \text{for} \quad t \ge t_k. \tag{5.3}$$

Using $\varepsilon_k = \mathcal{O}(eps)$, where $eps$ denotes the roundoff unit of the computer, an application of the triangle inequality to (5.2) yields $\eta_t = \mathcal{O}(t^2 eps)$. From our experiment of Fig. 5.2 we see that such an estimate is too pessimistic.

For a better understanding of accumulated rounding errors over long time intervals we make use of probability theory. Such an approach has been developed in the classical book of Henrici (1962). We assume that the components $\varepsilon_{ki}$ of $\varepsilon_k$ are *random variables* with mean and variance

$$E(\varepsilon_{ki}) = 0, \qquad \text{Var}(\varepsilon_{ki}) = C_{ki} \cdot eps^2,$$

and uniformly bounded $C_{ki} \le C$. For simplicity we assume that all $\varepsilon_{ki}$ are independent random variables. Replacing the matrix $\varphi_{t-t_k}(y_k)$ in (5.2) with $\varphi_{t-t_k}(y(t_k))$ and denoting its entries by $w_{ijk}$, the $i$th component of the accumulated rounding error (5.2) becomes

$$\eta_{ti} = \sum_{k=1}^{N} \sum_{j=1}^{n} w_{ijk}\varepsilon_{kj},$$

a linear combination of the random variables $\varepsilon_{kj}$. Elementary probability theory thus implies that

$$E(\eta_{ti}) = 0 \quad \text{and} \quad \text{Var}(\eta_{ti}) = \sum_{k=1}^{N}\sum_{j=1}^{n} w_{ijk}^2 \text{Var}(\varepsilon_{kj}).$$

Inserting the estimate (5.3) for $w_{ijk}$ we get

$$\text{Var}(\eta_{ti}) \le \sum_{k=1}^{N} (a + b(t - t_k))^2 \max_{j=1,\ldots,n} \text{Var}(\varepsilon_{kj}) = \mathcal{O}\left(\frac{C}{h} t^3 eps^2\right).$$

Consequently, the Euclidean norm of the expected rounding error $\eta_t$ is

$$\left(\sum_{i=1}^{n} \text{Var}(\eta_{ti})\right)^{1/2} = \mathcal{O}\left(\sqrt{\frac{C}{h}}\, t^{3/2}\, eps\right).$$

This is in excellent agreement with the results displayed in Fig. 5.2.

## VIII.6 Implementation of Implicit Methods

Symplectic methods for general Hamiltonian equations are implicit, and so are symmetric methods for general reversible systems. Also, when we consider variable step size extensions as described in Sections VIII.2 and VIII.3, we are led to nonlinear equations. The efficient numerical solution of such nonlinear equations is the main difficulty in an implementation of implicit methods. Notice that in the context of geometric integration there is no need of ad-hoc strategies for step size and order selection, so that the remaining parts of a computer code are more or less straightforward.

In the following we discuss the numerical solution of the nonlinear system defined by an implicit Runge-Kutta method. We have the Gauss methods of Sect. II.1.3 in mind which are symplectic and symmetric. An extension of the ideas to partitioned Runge-Kutta methods and to Nyström methods is obvious. For simplicity of notation we consider autonomous differential equations $\dot{y} = f(y)$, and we write the nonlinear system of Definition II.1.1 in the form

$$Z_{in} - h \sum_{j=1}^{s} a_{ij} f(y_n + Z_{jn}) = 0, \qquad i = 1, \ldots, s. \tag{6.1}$$

The unknown variables are $Z_{1n}, \ldots, Z_{sn}$, and the equivalence of the two formulations is via the relation $k_i = f(y_n + Z_{in})$. The numerical solution after one step can be expressed as

$$y_{n+1} = y_n + h \sum_{i=1}^{s} b_i f(y_n + Z_{in}). \tag{6.2}$$

For implicit Runge-Kutta methods the equations (6.1) represent a nonlinear system that has to be solved iteratively. We discuss the choice of good starting approximations for $Z_{in}$ as well as different nonlinear equation solvers (fixed-point iteration, modified Newton methods).

### VIII.6.1 Starting Approximations

The most simple approximations to the solution $Z_{in}$ of (6.1) are $Z_{in}^0 = 0$ or $Z_{in}^0 = hc_i f(y_n)$ where $c_i = \sum_{j=1}^{s} a_{ij}$. They are, however, not very accurate and we will try to exploit the information of previous steps for improving them. There are essentially two possibilities: either use only the information of the last step $y_{n-1} \mapsto y_n$ (methods (A) and (B) below), or consider a fixed $i$ and use the interpolation polynomial that passes through $Z_{i,n-l}$ for $l = 1, 2, \ldots$ (method (C)). Let us separately discuss these two approaches.

**(A) Use of Continuous Output.** Consider the polynomial $w_{n-1}(t)$ of degree $s$ that interpolates the values $(t_{n-1}, y_{n-1})$ and $(t_{n-1}+c_i h, Y_{i,n-1})$ for $i = 1, \ldots, s$, where $Y_{i,n-1} = y_{n-1} + Z_{i,n-1}$ is the argument in (6.1) of the previous step. For collocation methods (such as Gauss methods) $w_{n-1}(t)$ is the collocation polynomial, and we know from Lemma II.1.6 that on compact intervals

$$w_{n-1}(t) - y(t) = \mathcal{O}(h^{q+1}) \tag{6.3}$$

with $q = s$, where $y(t)$ denotes the solution of $\dot{y} = f(y)$ satisfying $y(t_{n-1}) = y_{n-1}$. For Runge-Kutta methods that are not collocation methods, (6.3) holds with $q$ defined by the condition $C(q)$ of (II.1.11). Since the solution of $\dot{y} = f(y)$ passing through $y(t_n) = y_n$ is $\mathcal{O}(h^{p+1})$ close to $y(t)$ with $p \geq q$, we have $w_n(t) = w_{n-1}(t) + \mathcal{O}(h^{q+1})$ and the computable value

$$Z_{in}^0 = Y_{in}^0 - y_n, \qquad Y_{in}^0 = w_{n-1}(t_n + c_i h) \tag{6.4}$$

serves as starting approximation for (6.1) with an error of size $\mathcal{O}(h^{q+1})$. This approach is standard in variable step size implementations of implicit Runge-Kutta methods (cf. Sect. IV.8 of Hairer & Wanner (1996)). Since $w_{n-1}(t) - y_{n-1}$ is a linear combination of the $Z_{i,n-1} = Y_{i,n-1} - y_{n-1}$, it follows from (6.1) that it is also a linear combination of $hf(Y_{i,n-1})$, so that

$$Y_{in}^0 = y_{n-1} + h \sum_{j=1}^{s} \beta_{ij} f(Y_{j,n-1}). \tag{6.5}$$

For a constant step size implementation, the $\beta_{ij}$ depend only on the method coefficients and can be computed in advance as the solution of the linear Vandermonde type system

$$\sum_{j=1}^{s} \beta_{ij} c_j^{k-1} = \frac{(1+c_i)^k}{k}, \qquad k = 1, \ldots, s \tag{6.6}$$

(see Exercise 2). For collocation methods and for methods with $q \geq s - 1$ the coefficients $\beta_{ij}$ from (6.6) are optimal in the sense that they are the only ones making (6.5) an $s$th order approximation to the solution of (6.1). For $q < s - 1$, more complicated order conditions have to be considered (Sand 1992).

**(B) Starting Algorithms Using Additional Function Evaluations.** In particular for high order methods where $s$ is relatively large, a much more accurate starting approximation can be constructed we the aid of a few additional function evaluations. Such starting algorithms have been investigated by Laburta (1997), who presents coefficients for the Gauss methods up to order 8 in Laburta (1998).

The idea is to use starting approximations of the form

$$Y_{in}^0 = y_{n-1} + h \sum_{j=1}^{s} \beta_{ij} f(Y_{j,n-1}) + h \sum_{j=1}^{m} \nu_{ij} f(Y_{s+j,n-1}), \tag{6.7}$$

where $Y_{1,n-1}, \ldots, Y_{s,n-1}$ are the internal stages of the basic implicit Runge-Kutta method (with coefficients $c_i, a_{ij}, b_j$), and the additional internal stages are computed from

$$Y_{s+i,n-1} = y_{n-1} + h \sum_{j=1}^{s+i-1} \mu_{ij} f(Y_{j,n-1}).$$

For a fixed $i$, we interpret $Y_{in}^0$ as the result of the explicit Runge-Kutta method with coefficients of the right tableau of

$$\begin{array}{c|cc} \text{exact } i\text{th stage} \\ c & A \\ \mathbb{1}+c & B & A \\ \hline & b^T & a_i^T \end{array} \qquad \begin{array}{c|cc} \text{approximate} \\ c & A \\ \mu & M_1 & M_2 \\ \hline & \beta_i^T & \nu_i^T \end{array} \qquad (6.8)$$

Here, $(M_1, M_2) = M = (\mu_{jk})$, $\mu_j = \sum_{k=1}^{s+j-1} \mu_{jk}$, and $c, \mu, \beta_i, \nu_i$ are the vectors composed of $c_j, \mu_j, \beta_{ij}, \nu_{ij}$, respectively. The exact stage values $Y_{in}$ are interpreted as the result of the Runge-Kutta method with coefficients given in the left tableau of (6.8). The entries of the vectors $\mathbb{1}, b$ and $a_i$ are $1, b_j$ and $a_{ij}$, respectively, and $B$ is the matrix whose rows are all equal to $b^T$.

If the order conditions (see Sect. III.1) for the two Runge-Kutta methods of (6.8) give the same result for all trees with $\leq r$ vertices, we get an approximation of order $r$, i.e., $Y_{in}^0 - Y_{in} = \mathcal{O}(h^{r+1})$. For the bushy tree $\tau_k = [\bullet, \ldots, \bullet]$ with $k$ vertices we have

$$\sum_{j=1}^s \beta_{ij} c_j^{k-1} + \sum_{j=1}^m \nu_{ij} \mu_j^{k-1} = \sum_{j=1}^s b_j c_j^{k-1} + \sum_{j=1}^s a_{ij}(1+c_j)^{k-1}. \qquad (6.9)$$

Notice that for collocation methods (such as the Gauss methods) the condition $C(s)$ reduces the right-hand expression of this equation to $(1+c_i)^k/k$ for $k \leq s$. For $m=0$, these conditions are thus equivalent to (6.6).

For the tree $[\tau_k] = [[\bullet, \ldots, \bullet]]$ with $k+1$ vertices we get the condition

$$\sum_{j,l=1}^s \beta_{ij} a_{jl} c_l^{k-1} + \sum_j^s \nu_{ij} \left( \sum_{l=1}^s \mu_{jl} c_l^{k-1} + \sum_{l=1}^m \mu_{j,s+l} \mu_l^{k-1} \right)$$
$$= \sum_{j,l=1}^s b_j a_{jl} c_l^{k-1} + \sum_{j,l=1}^s a_{ij} \left( b_l c_l^{k-1} + a_{jl}(1+c_l)^{k-1} \right). \qquad (6.10)$$

We now assume that the Runge-Kutta method corresponding to the right tableau of (6.8) satisfies condition $C(s)$. This means that the method $(c, A, b)$ is a collocation method, and that the coefficients $\mu_{ij}$ have to be computed from the linear system

$$\sum_{j=1}^{s+i-1} \mu_{ij} c_j^{k-1} = \frac{\mu_i^k}{k}, \qquad k = 1, \ldots, s. \qquad (6.11)$$

The method corresponding to the left tableau of (6.8) then also satisfies $C(s)$. Consequently, the order conditions are simplified considerably, and it follows from

Sect. III.1 that $Y_{in}^0$ is an approximation to the exact stage value $Y_{in}$ of order $s+1$ or $s+2$ if the following conditions hold:

$$\begin{array}{ll} \text{order } s+1 & \text{if (6.9) for } k=1,\ldots,s+1; \\ \text{order } s+2 & \text{if (6.9) for } k=1,\ldots,s+2, \text{ and (6.10) for } k=s+1. \end{array} \quad (6.12)$$

For an approximation of *order* $s+1$ we put $m=1$, we arbitrarily choose $\mu_1$, we compute $\mu_{1j}$ from (6.11), and the coefficients $\beta_{ij}$ and $\nu_{i1}$ from (6.9) with $k=1,\ldots,s+1$. A reasonable choice for the free parameter is $\mu_1 \in [1,2]$ (in our computations we take $\mu_1 = 1.75$ for $s=2,4$, and $\mu_1 = 1.8$ for $s=6$.[1]

For an approximation of *order* $s+2$ we put $m=3$. One of the three additional function evaluations can be saved if we put $\mu_1 = 0$ and $\mu_2 = 1$. This implies $Y_{s+1,n-1} = y_{n-1}$ and $Y_{s+2,n-1} = y_n$, so that the evaluation of $f(Y_{s+1,n-1})$ is already available from computations for the preceding step (FSAL technique, "first same as last"). In our experiments we take $\mu_3 = 1.6$ for $s=2$, $\mu_3 = 1.65$ for $s=4$, and $\mu_3 = 1.75$ for $s=6$. The coefficients $\mu_{ij}, \beta_{ij}, \nu_{ij}$ are then obtained as the solution of Vandermonde like linear systems.

For an implementation it is more convenient to work with the quantities $Z_{in}^0 = Y_{in}^0 - y_n$ and to write (6.7) in the form

$$Z_{in}^0 = h \sum_{j=1}^{s} \alpha_{ij} f(Y_{j,n-1}) + h \sum_{j=1}^{m} \nu_{ij} f(Y_{s+j,n-1}) \quad (6.13)$$

with $\alpha_{ij} = \beta_{ij} - b_j$.

**(C) Equistage Approximation.** From the implicit function theorem, applied to the nonlinear system (6.1), we know that $Z_{in} = z(y_n, h)$, where the function $z(y, h)$ is as smooth as $f(y)$. Furthermore, since on compact intervals the global error of a one-step method permits an asymptotic expansion in powers of $h$, we have $y_{n-l} = y_N(t_{n-l}, h) + \mathcal{O}(h^{N+1})$ with $y_N(t, h) = y(t) + h^p e_p(t) + \ldots + h^N e_N(t)$ (the value of $N$ can be chosen arbitrarily large if $f(y)$ is sufficiently smooth). Consequently, $Z_{i,n-l}$ is $\mathcal{O}(h^{N+1})$ close to the smooth function $z(y_N(t, h), h)$ at $t = t_n - lh$. Let $\zeta_i(t)$ be the polynomial of degree $k-1$ defined by $\zeta_i(t_{n-l}) = Z_{i,n-l}$ for $l = 1, \ldots, k$. Then, the value

$$Z_{in}^0 = \zeta_i(t_n) \quad (6.14)$$

yields a $\mathcal{O}(h^{k+1})$ approximation to the solution of (6.1). This interpolation procedure was first proposed by In't Hout (1992) for the numerical solution of delay differential equations. For the iterative solution of the nonlinear Runge-Kutta equations (6.1), the starting approximation (6.14) is proposed and analyzed by Calvo (2002).

The implementation of this approach is very simple. Using Newton's interpolation formula we have

---

[1] Laburta (1997) proposes to consider $m=2$, $\mu_1 = 0$, $\mu_2 = 1$ (apart from the first step this also needs only one additional function evaluation per step), and to optimize free parameters by satisfying the order conditions for some trees with one order higher.

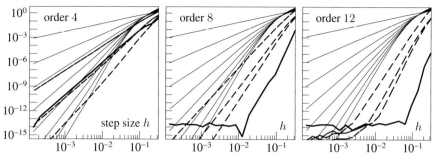

**Fig. 6.1.** Errors of starting approximations for Gauss methods as functions of the step size $h$: thick dashed lines for the extrapolated continuous output (6.4) and for the approximations (6.7) of order $s+1$ and $s+2$; thin solid lines for the equistage approximation (6.15) with $k = 0, 1, \ldots, 7$; the thick solid line represents the global error of the method after one period.

$$Z_{in}^0 = Z_{i,n-1} + \nabla Z_{i,n-1} + \ldots + \nabla^{k-1} Z_{i,n-1} \qquad (6.15)$$

with backward differences given by $\nabla Z_{i,n} = Z_{i,n} - Z_{i,n-1}$, $\nabla^2 Z_{i,n} = \nabla Z_{i,n} - \nabla Z_{i,n-1}$, etc.

**Numerical Study of Starting Approximations.** We consider the Kepler problem with eccentricity $e = 0.6$ and initial values such that the period is $2\pi$. With many different step sizes $h = 2\pi/N$ we compute $N+1$ steps with the Gauss method of order $p = 2s$ ($p = 4, 8, 12$). In the last step we compute the different starting approximations and their error $(\sum_{i=1}^s \|Z_{in} - Z_{in}^0\|^2)^{1/2}$ as a function of the step size $h$. The result is plotted in Fig. 6.1. There, the pictures also contain the global errors after one period. They allow us to localize the values of $h$, which are of practical interest.

We observe that the equistage approximation (6.15) also behaves like $\mathcal{O}(h^{k+1})$ when $k+1$ is larger than the order of the integrator. However, due to the increasing error constants, the accuracy is improved only for small step sizes. An optimal $k$ could be estimated by checking the decrease of the backward differences $\|\nabla^j Z_{i,n-1}\|$. The error of the starting approximation obtained from the continuous output behaves like $\mathcal{O}(h^{s+1})$ (for the Gauss methods) and, in contrast to the equistage approximation, improves with increasing order. The approximations (6.7) of order $s+1$ and $s+2$ are a clear improvement. As a conclusion we find that for this example the equistage approximation (which is free from additional function evaluations) is preferable only for $s = 2$ (order 4). For higher order, the approximation obtained from (6.7) is significantly more accurate and so it is worthwhile to spend these two additional function evaluations per step.

## VIII.6.2 Fixed-Point Versus Newton Iteration

Finally we investigate the iterative solution of the nonlinear Runge-Kutta system (6.1). We discuss fixed-point and Newton-like iterations, and we compare their efficiency to the use of composition methods.

**Fixed-Point Iteration.** This is the most simple and most natural iteration for the solution of (6.1). With any starting approximation $Z_{in}^0$ from Sect. VIII.6.1 it reads

$$Z_{in}^{k+1} = h \sum_{j=1}^{s} a_{ij} f(y_n + Z_{jn}^k), \qquad i = 1, \ldots, s. \qquad (6.16)$$

In the case where the entries of the Jacobian matrix $f'(y)$ are not excessively large (nonstiff problems) and that the step size is sufficiently small, this iteration converges for $k \to \infty$ to the solution of (6.1). Usually, the iteration is stopped if a certain norm of the differences $Z_{in}^{k+1} - Z_{in}^k$ is sufficiently small. We then use $Z_{in}^k$ in the update formula (6.2) so that no additional function evaluation is required.

For a numerical study of the convergence of this iteration, we consider the Kepler problem with eccentricity $e = 0.6$ and initial values as in the preceding experiments (period of the solution is $2\pi$). We apply the Gauss methods of order 4, 8, and 12 with various step sizes. For the integration over one period we show in Table 6.1 the total number of function evaluations, the mean number of required iterations per step, and the global error at the endpoint of integration. As a stopping criterion for the fixed-point iteration we check whether the norm of the difference of two successive approximations is smaller than $10^{-16}$ (roundoff unit in double precision). As a starting approximation $Z_{in}^0$ we use (6.15) with $k = 8$ for the method of order 4, and the approximation (6.7) of order $s + 2$ for the methods of orders 8 and 12. The coefficients are those presented after equation (6.12).

Since the starting approximations are more accurate for small $h$, the number of necessary iterations decreases drastically. In particular, for the 4th order method we need about 16 iterations per step for $h = 2\pi/25$, but at most 2 iterations when $h \le 2\pi/800$. If one is interested in high accuracy computations (e.g., long-time simulations in astronomy), for which the error over one period is not larger than $10^{-10}$, Table 6.1 illustrates that high order methods ($p \ge 12$) are most efficient.

**Table 6.1.** Statistics of Gauss methods (total number of function evaluations, number of fixed-point iterations per step, and the global error at the endpoint) for computations of the Kepler problem over one period with $e = 0.6$.

Fixed-point iteration (general problems)

| Gauss | $h = 2\pi/25$ | $h = 2\pi/50$ | $h = 2\pi/100$ | $h = 2\pi/200$ | $h = 2\pi/400$ |
|---|---|---|---|---|---|
| order 4 | 803<br>16.1<br>$9.2 \cdot 10^{-2}$ | 1 043<br>10.4<br>$1.7 \cdot 10^{-2}$ | 1 393<br>7.0<br>$1.3 \cdot 10^{-3}$ | 1 825<br>4.6<br>$8.4 \cdot 10^{-5}$ | 2 319<br>2.9<br>$5.3 \cdot 10^{-6}$ |
| order 8 | 1 021<br>9.7<br>$1.1 \cdot 10^{-3}$ | 1 455<br>6.8<br>$6.9 \cdot 10^{-7}$ | 2 091<br>4.7<br>$3.6 \cdot 10^{-9}$ | 3 007<br>3.3<br>$1.8 \cdot 10^{-11}$ | 4 183<br>2.1<br>$6.9 \cdot 10^{-14}$ |
| order 12 | 1 297<br>8.3<br>$2.7 \cdot 10^{-6}$ | 1 731<br>5.4<br>$8.0 \cdot 10^{-11}$ | 2 311<br>3.5<br>$2.7 \cdot 10^{-14}$ | 3 441<br>2.5<br>$\le$ roundoff | 5 917<br>2.1<br>$\le$ roundoff |

**Newton-Type Iterations.** A standard technique for solving nonlinear equations is Newton's method or some modification of it. Writing the nonlinear system (6.1) of an implicit Runge-Kutta method as $F(Z) = 0$ with $Z = (Z_{1n}, \ldots, Z_{sn})^T$, the Newton iteration is

$$Z^{k+1} = Z^k - M^{-1} F(Z^k), \qquad (6.17)$$

where $M$ is some approximation to the Jacobian matrix $F'(Z^k)$. Since the solution $Z$ of the nonlinear system is $\mathcal{O}(h)$ close to zero, it is common to use $M = F'(0)$ so that the matrix $M$ is independent of the iteration index $k$. In our special situation we get

$$M = I \otimes I - hA \otimes J \qquad (6.18)$$

with $J = f'(y_n)$. Here, $I$ denotes the identity matrix of suitable dimension, and $A$ is the Runge-Kutta matrix.

We repeat the experiment of Table 6.1 with modified Newton iterations instead of fixed-point iterations. The result is shown in Table 6.2. We have suppressed the error at the end of the period, because it is the same as in Table 6.1. As expected, the convergence is faster (i.e., the number of iterations per step is smaller) so that the total number of function evaluations is reduced. However, we do not see in this table that we computed at every step the Jacobian $f'(y_n)$ and an $LR$-decomposition of the matrix $M$. Even if we exploit the tensor product structure in (6.18) as explained in Hairer & Wanner (1996, Sect. IV.8), the cpu time is now considerably larger. Further improvements are possible, if the Jacobian of $f$ and hence also the $LR$-decomposition of $M$ is frozen over a couple of steps. But all these efforts can hardly beat (in cpu time) the straightforward fixed-point iterations. In accordance with the experience of Sanz-Serna & Calvo (1994, Sect. 5.5) we recommend in general the use of fixed-point iterations.

**Separable Systems and Second Order Differential Equations.** Many interesting differential equations are of the form

$$\dot{\eta} = f(y), \qquad \dot{y} = g(\eta). \qquad (6.19)$$

**Table 6.2.** Statistics of Gauss methods (total number of function evaluations, number of iterations per step) for computations of the Kepler problem over one period with $e = 0.6$.

Modified Newton iteration (general problems)

| Gauss | $h = 2\pi/25$ | $h = 2\pi/50$ | $h = 2\pi/100$ | $h = 2\pi/200$ | $h = 2\pi/400$ |
|---|---|---|---|---|---|
| order 4 | 383<br>7.7 | 511<br>5.1 | 765<br>3.8 | 1 125<br>2.8 | 1 677<br>2.1 |
| order 8 | 597<br>5.5 | 883<br>3.9 | 1 387<br>3.0 | 2 307<br>2.4 | 3 667<br>1.8 |
| order 12 | 763<br>4.7 | 1 095<br>3.3 | 1 717<br>2.5 | 3 003<br>2.2 | 5 689<br>2.0 |

For example, the second order differential equation $\ddot{y} = f(y)$ is obtained by putting $g(\eta) = \eta$. Also Hamiltonian systems with separable Hamiltonian $H(p,q) = T(p) + U(q)$ are of the form (6.19).

For this particular system the Runge-Kutta equations (6.1) become

$$\zeta_{in} - h\sum_{j=1}^{s} a_{ij} f(y_n + Z_{jn}) = 0, \qquad Z_{in} - h\sum_{j=1}^{s} a_{ij} g(\eta_n + \zeta_{jn}) = 0.$$

In this case we can still do better: instead of the standard fixed-point iteration (6.16) we apply a Gauss-Seidel like iteration

$$\zeta_{in}^{k+1} = h\sum_{j=1}^{s} a_{ij} f(y_n + Z_{jn}^{k}), \qquad Z_{in}^{k+1} = h\sum_{j=1}^{s} a_{ij} g(\eta_n + \zeta_{jn}^{k+1}), \qquad (6.20)$$

which is explicit for separable systems (6.19). Notice that the starting approximations have to be computed only for $\zeta_{in}$. Those for $Z_{in}$ are then obtained by (6.20) with $k + 1 = 0$.

For second order differential equations $\ddot{y} = f(y)$, where $g(\eta) = \eta$, this iteration becomes

$$Z_{in}^{k+1} = hc_i \eta_n + h^2 \sum_{j=1}^{s} \widehat{a}_{ij} f(y_n + Z_{jn}^{k}), \qquad (6.21)$$

where $c_i = \sum_{j=1}^{s} a_{ij}$ and $\widehat{a}_{ij}$ are the entries of the square $A^2$ of the Runge-Kutta matrix (any Nyström method could be applied as well). Due to the factor $h^2$ in (6.21) we expect this iteration to converge about twice as fast as the standard fixed-point iteration.

The Kepler problem is a second order differential equation, so that the iteration (6.21) can be applied. In analogy to the previous tables we present in Table 6.3 the statistics of such an implementation of the Gauss methods. We observe that for relatively large step sizes the number of iterations required per step in nearly halved

**Table 6.3.** Statistics of iterations (6.20) for Gauss methods (total number of function evaluations, number of iterations per step) for computations of the Kepler problem over one period with $e = 0.6$.

| Gauss | Fixed-point iteration (separable problems) | | | | |
|---|---|---|---|---|---|
| | $h = 2\pi/25$ | $h = 2\pi/50$ | $h = 2\pi/100$ | $h = 2\pi/200$ | $h = 2\pi/400$ |
| order 4 | 437<br>8.7 | 603<br>6.0 | 857<br>4.3 | 1 201<br>3.0 | 1 717<br>2.1 |
| order 8 | 613<br>5.6 | 923<br>4.1 | 1 427<br>3.1 | 2 339<br>2.4 | 3 647<br>1.8 |
| order 12 | 781<br>4.9 | 1 131<br>3.4 | 1 741<br>2.6 | 3 027<br>2.2 | 5 677<br>2.0 |

(compared to Table 6.1). For high accuracy requirements the number of necessary iterations is surprisingly small, and the question arises whether such an implementation can compete with high order explicit composition methods.

**Comparison Between Implicit Runge-Kutta and Composition Methods.** We consider second order differential equations $\ddot{y} = f(y)$, so that composition methods based on the explicit Störmer/Verlet scheme can be applied. We use the coefficients of method (V.3.14) which has turned out to be excellent in the experiments of Sect. V.3.2. It is a method of order 8 and uses 17 function evaluations per integration step.

We compare it with the Gauss methods of order 8 and 12 (i.e., $s = 4$ and $s = 6$). As a starting approximation for the solution of the nonlinear system (6.1) we use (6.7) with $m = 3$, $\mu_1 = 0$, $\mu_2 = 1$, $\mu_3 = 1.75$, $\mu_{ij}$ chosen such that (6.11) holds for $k = 1, \ldots, s + i - 1$, and $\beta_{ij}, \nu_{ij}$ such that order $s + 2$ is obtained. Since we are concerned with second order differential equations, we apply the iterations (6.20) until the norm of the difference of two successive approximations is below $10^{-17}$.

For both classes of methods we use compensated summation (Algorithm 5.1), which permits us to reduce rounding errors. For composition methods we apply this technique for all updates of the basic integrator. For Runge-Kutta methods, we use it for adding the increment to $y_n$ and also for computing the sum $\sum_{i=1}^{s} b_i k_i$.

The work–precision diagrams of the comparison are given in Fig. 6.2. The upper pictures correspond to the Kepler problem with $e = 0.6$ and an integration over 100 periods; the lower pictures correspond to the outer solar system with data given in Sect. I.2.3 and an integration over 500 000 earth days. The left pictures show the Euclidean norm of the error at the end of the integration interval as a function of total numbers of function evaluations required for the integration; the pictures to the right present the same error as a function of the cpu times (with optimizing compiler on a SunBlade 100 workstation). We can draw the following conclusions from this experiment:

- the implementation of composition methods based on the Störmer/Verlet scheme is extremely easy; that of implicit Runge-Kutta methods is slightly more involved because it requires a stopping criterion for the fixed-point iterations;
- the overhead (total cpu time minus that used for the function evaluations) is much higher for the implicit Runge-Kutta methods; this is seen from the fact that implicit Runge-Kutta methods require less function evaluations for a given accuracy, but often more cpu time;
- among the two Gauss methods, the higher order method is more efficient for all precisions of practical interest;
- for very accurate computations (say, in quadruple precision), high order Runge-Kutta methods are more efficient than composition methods;
- much of the computation in the Runge-Kutta code can be done in parallel (e.g., the $s$ function evaluations of a fixed-point iteration); composition methods do not have this potential;
- implicit Runge-Kutta methods can be applied to general (non-separable) differential equations, and the cost of the implementation is at most twice as large; if one

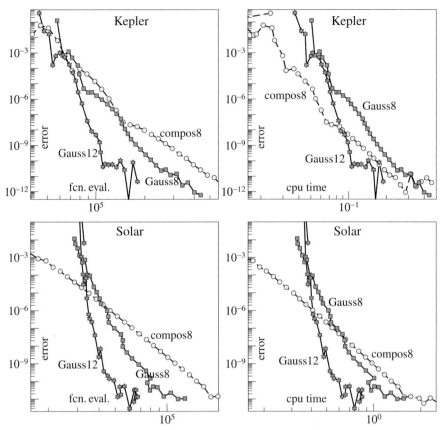

**Fig. 6.2.** Work–precision diagrams for two problems (Kepler and outer solar system) and three numerical integrators (composition method with coefficients of method (V.3.14) based on the explicit Störmer/Verlet scheme and the Gauss methods of orders 8 and 12).

is obliged to use an implicit method as the basic method for composition, many advantages of composition methods are lost.

Both classes of methods (composition and implicit Runge-Kutta) are of interest in the geometric integration of differential equations. Each one has its advantages and disadvantages.

Codes of these computations will be made available on the Internet under the homepage <http://www.math.unige.ch/folks/hairer>.

## VIII.7 Exercises

1. Consider a one-step method applied to a Hamiltonian system. Give a probabilistic proof of the property that the error of the numerical Hamiltonian due to roundoff grows like $\mathcal{O}(\sqrt{t}\,eps)$.

2. Prove that the collocation polynomial can be written as
$$w_n(t) = y_n + h \sum_{i=1}^{s} \beta_i(t) f(Y_{in}),$$
where the polynomials $\beta_i(t)$ are a solution of
$$\sum_{j=1}^{s} \beta_j(t) c_j^{k-1} = \frac{t^k}{k}.$$

3. Apply your favourite code to the Kepler problem and to the outer solar system with data as in Fig. 6.2. Plot a work-precision diagram.
   *Remark.* Figure 7.1 shows our results obtained with the 8th order Runge-Kutta code Dop853 (Hairer, Nørsett & Wanner 1993) compared to an 8th order composition method. Rounding errors are more pronounced for Dop853, because compensated summation is not applied. Computations on shorter time intervals and comparisons of required function evaluations would be more in favour for Dop853. It is also of interest to consider high order Runge-Kutta Nyström methods.

4. Consider starting approximations
$$Y_{in}^0 = y_{n-2} + h \sum_{j=1}^{s} \beta_{ij}^{(2)} f(Y_{j,n-2}) + h \sum_{j=1}^{s} \beta_{ij}^{(1)} f(Y_{j,n-1}) \qquad (7.1)$$

which use the internal stages of two consecutive steps without any additional function evaluation. What are the conditions such that (7.1) is of order $s+1$, of order $s+2$?
Compare the efficiency of these formulas with the algorithms (A) and (B) of Sect. VIII.6.1.

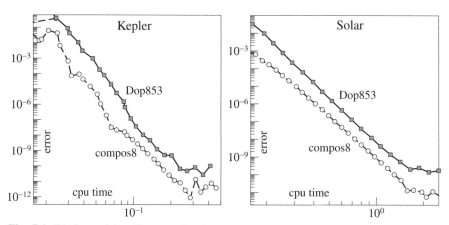

**Fig. 7.1.** Work–precision diagrams for the explicit, variable step size Runge-Kutta code Dop853 applied to two problems (Kepler and outer solar system). For a comparison, the results of Fig. 6.2 for the composition method are included.

5. Prove that for a second order differential equation $\ddot{y} = f(y)$ (more precisely, for $\dot{y} = z, \dot{z} = f(y)$) the application of the $s$-stage Gauss method gives

$$y_{n+1} = y_n + h\dot{y}_n + h^2 \sum_{i=1}^{s} b_i(1 - c_i)f(y_n + Z_{in})$$

$$\dot{y}_{n+1} = \dot{y}_n + h \sum_{i=1}^{s} b_i f(y_n + Z_{in}),$$

where $Z_{in}$ is obtained from the iteration (6.21).
*Hint.* The coefficients of the Gauss methods satisfy $\sum_j b_j a_{ji} = b_i(1 - c_i)$ for all $i$.

# Chapter IX.
# Backward Error Analysis and Structure Preservation

> One of the greatest virtues of backward analysis ... is that when it *is* the appropriate form of analysis it tends to be very markedly superior to forward analysis. Invariably in such cases it has remarkable formal simplicity and gives deep insight into the stability (or lack of it) of the algorithm.
> (J.H. Wilkinson, IMA Bulletin 1986)

The origin of backward error analysis dates back to the work of Wilkinson (1960) in numerical linear algebra. For the study of integration methods for ordinary differential equations, its importance was seen much later. The present chapter is devoted to this theory. It is very useful, when the qualitative behaviour of numerical methods is of interest, and when statements over very long time intervals are needed. The formal analysis (construction of the modified equation, study of its properties) gives already a lot of insight into numerical methods. For a rigorous treatment, the modified equation, which is a formal series in powers of the step size, has to be truncated. The error, induced by such a truncation, can be made exponentially small, and the results remain valid on exponentially long time intervals.

## IX.1 Modified Differential Equation – Examples

Consider an ordinary differential equation

$$\dot{y} = f(y),$$

and a numerical method $\Phi_h(y)$ which produces the approximations

$$y_0, y_1, y_2, \ldots .$$

A forward error analysis consists of the study of the errors $y_1 - \varphi_h(y_0)$ (local error) and $y_n - \varphi_{nh}(y_0)$ (global error) in the solution space. The idea of backward error analysis is to search for a *modified differential equation* $\dot{\widetilde{y}} = f_h(\widetilde{y})$ of the form

$$\dot{\widetilde{y}} = f(\widetilde{y}) + h f_2(\widetilde{y}) + h^2 f_3(\widetilde{y}) + \ldots, \qquad (1.1)$$

such that $y_n = \widetilde{y}(nh)$, and in studying the difference of the vector fields $f(y)$ and $f_h(y)$. This then gives much insight into the qualitative behaviour of the numerical solution and into the global error $y_n - y(nh) = \widetilde{y}(nh) - y(nh)$. We remark that the series in (1.1) usually diverges and that one has to truncate it suitably. The effect of such a truncation will be studied in Sect. IX.7. For the moment we content ourselves with a formal analysis without taking care of convergence issues. The idea of interpreting the numerical solution as the exact solution of a modified equation is common to many numerical analysts ("... This is possible since the map is the solution of some physical Hamiltonian problem which, in some sense, is close to the original problem", Ruth (1983), or "... the symplectic integrator creates a numerical Hamiltonian system that is close to the original ...", Gladman, Duncan & Candy 1991). A systematic study started with the work of Griffiths & Sanz-Serna (1986), Feng (1991), Sanz-Serna (1992), Yoshida (1993), Eirola (1993), Fiedler & Scheurle (1996), and many others.

For the computation of the modified equation (1.1) we put $y := \widetilde{y}(t)$ for a fixed $t$, and we expand the solution of (1.1) into a Taylor series

$$\begin{aligned} \widetilde{y}(t+h) &= y + h\big(f(y) + hf_2(y) + h^2 f_3(y) + \ldots\big) \\ &+ \frac{h^2}{2!}\big(f'(y) + hf'_2(y) + \ldots\big)\big(f(y) + hf_2(y) + \ldots\big) + \ldots \,. \end{aligned} \quad (1.2)$$

We assume that the numerical method $\Phi_h(y)$ can be expanded as

$$\Phi_h(y) = y + hf(y) + h^2 d_2(y) + h^3 d_3(y) + \ldots \quad (1.3)$$

(the coefficient of $h$ is $f(y)$ for consistent methods). The functions $d_j(y)$ are known and are typically composed of $f(y)$ and its derivatives. For the explicit Euler method we simply have $d_j(y) = 0$ for all $j \geq 2$. In order to get $\widetilde{y}(nh) = y_n$ for all $n$, we must have $\widetilde{y}(t+h) = \Phi_h(y)$. Comparing like powers of $h$ in the expressions (1.2) and (1.3) yields recurrence relations for the functions $f_j(y)$, namely,

$$f_2(y) = d_2(y) - \frac{1}{2!}f'f(y) \quad (1.4)$$

$$f_3(y) = d_3(y) - \frac{1}{3!}\Big(f''(f,f)(y) + f'f'f(y)\Big) - \frac{1}{2!}\Big(f'f_2(y) + f'_2 f(y)\Big).$$

**Example 1.1.** Consider the scalar differential equation

$$\dot{y} = y^2, \qquad y(0) = 1 \quad (1.5)$$

with exact solution $y(t) = 1/(1-t)$. It has a singularity at $t = 1$. We apply the explicit Euler method $y_{n+1} = y_n + hf(y_n)$ with step size $h = 0.02$. The picture in Fig. 1.1 presents the exact solution (dashed curve) together with the numerical solution (bullets). The above procedure for the computation of the modified equation, implemented as a Maple program (see Hairer & Lubich 2000) gives

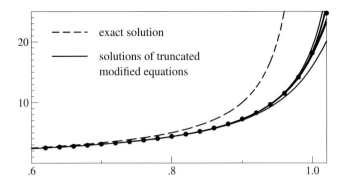

**Fig. 1.1.** Solutions of the modified equation for the problem (1.5)

```
> fcn := y -> y^2:
> nn := 6:
> fcoe[1] := fcn(y):
> for n from 2 by 1 to nn do
>     modeq := sum(h^j*fcoe[j+1], j=0..n-2):
>     diffy[0] := y:
>     for i from 1 by 1 to n do
>         diffy[i] := diff(diffy[i-1],y)*modeq:
>     od:
>     ytilde := sum(h^k*diffy[k]/k!, k=0..n):
>     res := ytilde-y-h*fcn(y):
>     tay := convert(series(res,h=0,n+1),polynom):
>     fcoe[n] := -coeff(tay,h,n):
> od:
> simplify(sum(h^j*fcoe[j+1], j=0..nn-1));
```

Its output is

$$\dot{\tilde{y}} = \tilde{y}^2 - h\tilde{y}^3 + h^2 \frac{3}{2}\tilde{y}^4 - h^3 \frac{8}{3}\tilde{y}^5 + h^4 \frac{31}{6}\tilde{y}^6 - h^5 \frac{157}{15}\tilde{y}^7 \pm \dots . \quad (1.6)$$

The above picture also presents the solution of the modified equation, when truncated after 1,2,3, and 4 terms. We observe an excellent agreement of the numerical solution with the exact solution of the modified equation.

A similar program for the implicit midpoint rule (I.1.7) computes the modified equation

$$\dot{\tilde{y}} = \tilde{y}^2 + h^2 \frac{1}{4}\tilde{y}^4 + h^4 \frac{1}{8}\tilde{y}^6 + h^6 \frac{11}{192}\tilde{y}^8 + h^8 \frac{3}{128}\tilde{y}^{10} \pm \dots , \quad (1.7)$$

and for the classical Runge-Kutta method of order 4 (left tableau of (II.1.8))

$$\dot{\tilde{y}} = \tilde{y}^2 - h^4 \frac{1}{24}\tilde{y}^6 + h^6 \frac{65}{576}\tilde{y}^8 - h^7 \frac{17}{96}\tilde{y}^9 + h^8 \frac{19}{144}\tilde{y}^{10} \pm \dots . \quad (1.8)$$

We observe that the perturbation terms in the modified equation are of size $\mathcal{O}(h^p)$, where $p$ is the order of the method. This is true in general.

**Theorem 1.2.** *Suppose that the method* $y_{n+1} = \Phi_h(y_n)$ *is of order p, i.e.,*

$$\Phi_h(y) = \varphi_h(y) + h^{p+1}\delta_{p+1}(y) + \mathcal{O}(h^{p+2}),$$

*where $\varphi_t(y)$ denotes the exact flow of $\dot{y} = f(y)$, and $h^{p+1}\delta_{p+1}(y)$ the leading term of the local truncation error. The modified equation then satisfies*

$$\dot{\widetilde{y}} = f(\widetilde{y}) + h^p f_{p+1}(\widetilde{y}) + h^{p+1} f_{p+2}(\widetilde{y}) + \ldots, \qquad \widetilde{y}(0) = y_0 \qquad (1.9)$$

*with $f_{p+1}(y) = \delta_{p+1}(y)$.*

*Proof.* The construction of the functions $f_j(y)$ (see the beginning of this section) shows that $f_j(y) = 0$ for $2 \leq j \leq p$ if and only if $\Phi_h(y) - \varphi_h(y) = \mathcal{O}(h^{p+1})$. □

A first application of the modified equation (1.1) is the existence of an *asymptotic expansion of the global error*. Indeed, by the nonlinear variation of constants formula, the difference between its solution $\widetilde{y}(t)$ and the solution $y(t)$ of $\dot{y} = f(y)$ satisfies

$$\widetilde{y}(t) - y(t) = h^p e_p(t) + h^{p+1} e_{p+1}(t) + \ldots . \qquad (1.10)$$

Since $y_n = \widetilde{y}(nh) + \mathcal{O}(h^N)$ for the solution of a truncated modified equation, this proves the existence of an asymptotic expansion in powers of $h$ for the global error $y_n - y(nh)$.

A large part of this chapter studies properties of the modified differential equation, and the question of the extent to which structures (such as conservation of invariants, Hamiltonian structure) in the problem $\dot{y} = f(y)$ can carry over to the modified equation.

**Example 1.3.** We next consider the Lotka-Volterra equations

$$\dot{q} = q(p-1), \qquad \dot{p} = p(2-q),$$

and we apply (a) the explicit Euler method, and (b) the symplectic Euler method, both with constant step size $h = 0.1$. The first terms of their modified equations are

(a) $\quad \dot{q} = q(p-1) - \dfrac{h}{2} q(p^2 - pq + 1) + \mathcal{O}(h^2),$

$\quad \dot{p} = -p(q-2) - \dfrac{h}{2} p(q^2 - pq - 3q + 4) + \mathcal{O}(h^2),$

(b) $\quad \dot{q} = q(p-1) - \dfrac{h}{2} q(p^2 + pq - 4p + 1) + \mathcal{O}(h^2),$

$\quad \dot{p} = -p(q-2) + \dfrac{h}{2} p(q^2 + pq - 5q + 4) + \mathcal{O}(h^2).$

Figure 1.2 shows the numerical solutions for initial values indicated by a thick dot. In the pictures to the left they are embedded in the exact flow of the differential equation, whereas in those to the right they are embedded in the flow of the modified differential equation, truncated after the $h^2$ terms. As in the first example, we observe an excellent agreement of the numerical solution with the exact solution of

IX.1 Modified Differential Equation – Examples    291

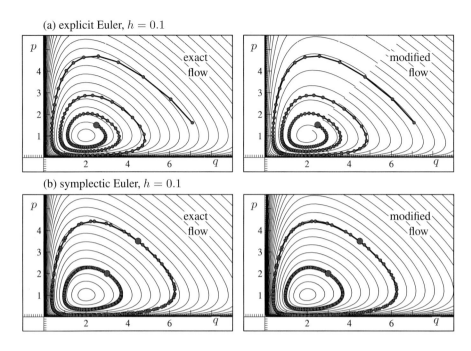

**Fig. 1.2.** Numerical solution compared to the exact and modified flows

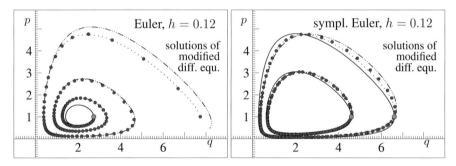

**Fig. 1.3.** Study of the truncation in the modified equation

the modified equation. For the symplectic Euler method, the solutions of the truncated modified equation are periodic, as is the case for the unperturbed problem (Exercise 5).

In Fig. 1.3 we present the numerical solution and the exact solution of the modified equation, once truncated after the $h$ terms (dashed-dotted), and once truncated after the $h^2$ terms (dotted). The exact solution of the problem is included as a solid curve. This shows that taking more terms in the modified equation usually improves the agreement of its solution with the numerical approximation of the method.

**Example 1.4.** For a linear differential equation with constant coefficients
$$\dot{y} = Ay, \qquad y(0) = y_0.$$
we consider numerical methods which yield $y_{n+1} = R(hA)y_n$, where $R(z)$ is the stability function (VI.4.9) of the method. In this case we get $y_n = R(hA)^n y_0$, so that $y_n = \widetilde{y}(nh)$, where $\widetilde{y}(t) = R(hA)^{t/h} y_0 = \exp\bigl(\frac{t}{h} \ln R(hA)\bigr) y_0$ is the solution of the modified differential equation
$$\dot{\widetilde{y}} = \frac{1}{h} \ln R(hA)\, \widetilde{y} = (A + hb_2 A^2 + h^2 b_3 A^3 + \ldots)\, \widetilde{y} \tag{1.11}$$
with suitable constants $b_2, b_3, \ldots$. Since $R(z) = 1 + z + \mathcal{O}(z^2)$ and $\ln(1+x) = x - x^2/2 + \mathcal{O}(x^3)$ both have a positive radius of convergence, the series (1.11) converges for $|h| < h_0$ with some $h_0 > 0$. We shall see later that this is an exceptional situation. In general, the modified equation is a formal divergent series.

## IX.2 Modified Equations of Symmetric Methods

In this and the following sections we investigate how the structure of the differential equation and geometric properties of the method are reflected in the modified differential equation. Here we begin by studying this question for symmetric/reversible methods.

Consider a numerical method $\Phi_h$. Recall that its adjoint $y_{n+1} = \Phi_h^*(y_n)$ is defined by the relation $y_n = \Phi_{-h}(y_{n+1})$ (see Definition II.1.4).

**Theorem 2.1 (Adjoint Methods).** *Let $f_j(y)$ be the coefficient functions of the modified equation for the method $\Phi_h$. Then, the coefficient functions $f_j^*(y)$ of the modified equation for the adjoint method $\Phi_h^*$ satisfy*
$$f_j^*(y) = (-1)^{j+1} f_j(y). \tag{2.1}$$
*Proof.* The solution $\widetilde{y}(t)$ of the modified equation for $\Phi_h^*$ has to satisfy $\widetilde{y}(t) = \Phi_{-h}\bigl(\widetilde{y}(t+h)\bigr)$ or, equivalently, $\widetilde{y}(t-h) = \Phi_{-h}(y)$ with $y := \widetilde{y}(t)$. We get (2.1) if we replace $h$ with $-h$ in the formulas (1.1), (1.2) and (1.3). □

For symmetric methods we have $\Phi_h^* = \Phi_h$, implying $f_j^*(y) = f_j(y)$. We therefore get the following corollary to Theorem 2.1.

**Theorem 2.2 (Symmetric Methods).** *The coefficient functions of the modified equation of a symmetric method satisfy $f_j(y) = 0$ whenever $j$ is even, so that (1.1) has an expansion in even powers of $h$.* □

This theorem explains the $h^2$-expansion in the modified equation (1.7) of the midpoint rule.

As a consequence of Theorem 2.2, the asymptotic expansion (1.10) of the global error is also in even powers of $h$. This property is responsible for the success of $h^2$-extrapolation methods.

Consider now a numerical method applied to a $\rho$-reversible differential equation as studied in Sect. V.1. Recall from Theorem V.1.5 that a symmetric method is $\rho$-reversible under the $\rho$-compatibility condition (V.1.4), which is satisfied for most numerical methods.

**Theorem 2.3 (Reversible Methods).** *Consider a $\rho$-reversible differential equation $\dot{y} = f(y)$ and a $\rho$-reversible numerical method $\Phi_h(y)$. Then, every truncation of the modified differential equation is again $\rho$-reversible.*

*Proof.* Let $f_j(y)$ be the $j$th coefficient of the modified equation (1.1) for $\Phi_h$. The proof is by induction on $j$. So assume that for $j = 1, \ldots, r$, the vector field $f_j(y)$ is $\rho$-reversible, i.e.,
$$\rho \circ f_j = -f_j \circ \rho.$$
We show that the same relation holds also for $j = r + 1$. By assumption, the truncated modified equation
$$\dot{\widetilde{y}} = f(\widetilde{y}) + h f_2(\widetilde{y}) + \ldots + h^{r-1} f_r(\widetilde{y})$$
is $\rho$-reversible, so that by (V.1.2), it has a $\rho$-reversible flow $\varphi_{r,t}(y)$, that is, $\rho \circ \varphi_{r,t} = \varphi_{r,t}^{-1} \circ \rho$. By construction of the modified equation, we have
$$\Phi_h(y) = \varphi_{r,h}(y) + h^{r+1} f_{r+1}(y) + \mathcal{O}(h^{r+2}).$$
Since $\varphi_{r,h}(y) = y + \mathcal{O}(h)$, this implies
$$\Phi_h^{-1}(y) = \varphi_{r,h}^{-1}(y) - h^{r+1} f_{r+1}(y) + \mathcal{O}(h^{r+2}).$$
Since both $\Phi_h$ and $\varphi_{r,h}$ are $\rho$-reversible maps, these two relations yield $\rho \circ f_{r+1} = -f_{r+1} \circ \rho$ as desired. □

## IX.3 Modified Equations of Symplectic Methods

We now present one of the most important results of this chapter. We consider a Hamiltonian system $\dot{y} = J^{-1} \nabla H(y)$ with an infinitely differentiable Hamiltonian $H(y)$, and we show that the modified equation of symplectic methods is also Hamiltonian.

### IX.3.1 Existence of a Local Modified Hamiltonian

> ... if we neglect convergence questions then one can always find a formal integral ...
> (J. Moser 1968)

**Theorem 3.1.** *If a symplectic method $\Phi_h(y)$ is applied to a Hamiltonian system with a smooth Hamiltonian $H : \mathbb{R}^{2d} \to \mathbb{R}$, then the modified equation (1.1) is also Hamiltonian. More precisely, there exist smooth functions $H_j : \mathbb{R}^{2d} \to \mathbb{R}$ for $j = 2, 3, \ldots$, such that $f_j(y) = J^{-1} \nabla H_j(y)$.*

The following proof by induction, whose ideas can be traced back to Moser (1968), was given by Benettin & Giorgilli (1994) and Tang (1994). It can be extended to many other situations. We have already encountered its reversible version in the proof of Theorem 2.3.

*Proof.* Assume that $f_j(y) = J^{-1}\nabla H_j(y)$ for $j = 1, 2, \ldots, r$ (this is satisfied for $r = 1$, because $f_1(y) = f(y) = J^{-1}\nabla H(y)$). We have to prove the existence of a Hamiltonian $H_{r+1}(y)$. The idea is to consider the truncated modified equation

$$\dot{\widetilde{y}} = f(\widetilde{y}) + hf_2(\widetilde{y}) + \ldots + h^{r-1}f_r(\widetilde{y}), \tag{3.1}$$

which is a Hamiltonian system with Hamiltonian $H(y)+hH_2(y)+\ldots+h^{r-1}H_r(y)$. Its flow $\varphi_{r,t}(y_0)$, compared to that of (1.1), satisfies

$$\Phi_h(y_0) = \varphi_{r,h}(y_0) + h^{r+1}f_{r+1}(y_0) + \mathcal{O}(h^{r+2}),$$

and also

$$\Phi_h'(y_0) = \varphi_{r,h}'(y_0) + h^{r+1}f_{r+1}'(y_0) + \mathcal{O}(h^{r+2}).$$

By our assumption on the method and by the induction hypothesis, $\Phi_h$ and $\varphi_{r,h}$ are symplectic transformations. This, together with $\varphi_{r,h}'(y_0) = I + \mathcal{O}(h)$, therefore implies

$$J = \Phi_h'(y_0)^T J \Phi_h'(y_0) = J + h^{r+1}\left(f_{r+1}'(y_0)^T J + J f_{r+1}'(y_0)\right) + \mathcal{O}(h^{r+2}).$$

Consequently, the matrix $J f_{r+1}'(y)$ is symmetric and the existence of $H_{r+1}(y)$ satisfying $f_{r+1}(y) = J^{-1}\nabla H_{r+1}(y)$ follows from the Integrability Lemma VI.2.7. This part of the proof is similar to that of Theorem VI.2.6. □

For Hamiltonians $H : D \to \mathbb{R}$ the statement of the above theorem remains valid with $H_j : D \to \mathbb{R}$ on domains $D \subset \mathbb{R}^{2d}$ on which the Integrability Lemma VI.2.7 is applicable. This is the case for simply connected domains $D$, but not in general (see the discussion after the proof of Lemma VI.2.7).

## IX.3.2 Existence of a Global Modified Hamiltonian

By Lemma VI.5.2 every symplectic one-step method $\Phi_h : (p, q) \mapsto (P, Q)$ can be locally expressed in terms of a generating function $S(P, q, h)$ as

$$p = P + \frac{\partial S}{\partial q}(P, q, h), \qquad Q = q + \frac{\partial S}{\partial P}(P, q, h). \tag{3.2}$$

This property allows us to give an independent proof of Theorem 3.1 and in addition to show that the modified equation is Hamiltonian with $\widetilde{H}(p, q)$ defined on the same domain as the generating function. The following result is mentioned in Benettin & Giorgilli (1994) and in the thesis of Murua (1994), p. 100.

## IX.3 Modified Equations of Symplectic Methods

**Theorem 3.2.** *Assume that the symplectic method $\Phi_h$ has a generating function*
$$S(P, q, h) = h\, S_1(P, q) + h^2 S_2(P, q) + h^3 S_3(P, q) + \ldots \tag{3.3}$$
*with smooth $S_j(P, q)$ defined on an open set $D$. Then, the modified differential equation is a Hamiltonian system with*
$$\widetilde{H}(p, q) = H(p, q) + h\, H_2(p, q) + h^2 H_3(p, q) + \ldots, \tag{3.4}$$
*where the functions $H_j(p, q)$ are defined and smooth on the whole of $D$.*

*Proof.* By Theorem VI.5.6, the exact solution $(P, Q) = \big(\widetilde{p}(t), \widetilde{q}(t)\big)$ of the Hamiltonian system corresponding to $\widetilde{H}(p, q)$ is given by
$$p = P + \frac{\partial \widetilde{S}}{\partial q}(P, q, t), \qquad Q = q + \frac{\partial \widetilde{S}}{\partial P}(P, q, t),$$
where $\widetilde{S}$ is the solution of the Hamilton-Jacobi differential equation
$$\frac{\partial \widetilde{S}}{\partial t}(P, q, t) = \widetilde{H}\Big(P, q + \frac{\partial \widetilde{S}}{\partial P}(P, q, t)\Big), \qquad \widetilde{S}(P, q, 0) = 0. \tag{3.5}$$

Since $\widetilde{H}$ depends on the parameter $h$, this is also the case for $\widetilde{S}$. Our aim is to determine the functions $H_j(p, q)$ such that the solution $\widetilde{S}(P, q, t)$ of (3.5) coincides for $t = h$ with (3.3).

We first express $\widetilde{S}(P, q, t)$ as a series
$$\widetilde{S}(P, q, t) = t\, \widetilde{S}_1(P, q, h) + t^2 \widetilde{S}_2(P, q, h) + t^3 \widetilde{S}_3(P, q, h) + \ldots,$$
insert it into (3.5) and compare powers of $t$. This allows us to obtain the functions $\widetilde{S}_j(p, q, h)$ recursively in terms of derivatives of $\widetilde{H}$:
$$\begin{aligned}
\widetilde{S}_1(p, q, h) &= \widetilde{H}(p, q) \\
2\widetilde{S}_2(p, q, h) &= \Big(\frac{\partial \widetilde{H}}{\partial q} \cdot \frac{\partial \widetilde{S}_1}{\partial P}\Big)(p, q, h) \\
3\widetilde{S}_3(p, q, h) &= \Big(\frac{\partial \widetilde{H}}{\partial q} \cdot \frac{\partial \widetilde{S}_2}{\partial P}\Big)(p, q, h) + \frac{1}{2}\Big(\frac{\partial^2 \widetilde{H}}{\partial q^2}\Big(\frac{\partial \widetilde{S}_1}{\partial P}, \frac{\partial \widetilde{S}_1}{\partial P}\Big)\Big)(p, q, h).
\end{aligned} \tag{3.6}$$

We then write $\widetilde{S}_j$ as a series
$$\widetilde{S}_j(p, q, h) = \widetilde{S}_{j1}(p, q) + h\, \widetilde{S}_{j2}(p, q) + h^2 \widetilde{S}_{j3}(p, q) + \ldots,$$
insert it and the expansion (3.4) for $\widetilde{H}$ into (3.6), and compare powers of $h$. This yields $\widetilde{S}_{1k}(p, q) = H_k(p, q)$ and for $j > 1$ we see that $\widetilde{S}_{jk}(p, q)$ is a function of derivatives of $H_l$ with $l < k$.

The requirement $S(p, q, h) = \widetilde{S}(p, q, h)$ finally shows $S_1(p, q) = \widetilde{S}_{11}(p, q)$, $S_2(p, q) = \widetilde{S}_{12}(p, q) + \widetilde{S}_{21}(p, q)$, etc., so that

$$S_j(p, q) = H_j(p, q) + \text{"function of derivatives of } H_k(p, q) \text{ with } k < j\text{"}.$$

For a given generating function $S(P, q, h)$, this recurrence relation allows us to determine successively the $H_j(p, q)$. We see from these explicit formulas that the functions $H_j$ are defined on the same domain as the $S_j$. □

As a consequence of Theorem 3.2 and Theorems VI.5.3 and VI.5.4 we obtain the following result.

**Theorem 3.3.** *A symplectic (partitioned) Runge-Kutta method applied to a system with smooth Hamiltonian $H : D \to \mathbb{R}$ (with $D \subset \mathbb{R}^{2d}$ an arbitrary open set) has a modified Hamiltonian (3.4) with smooth functions $H_j : D \to \mathbb{R}$.* □

**Example 3.4 (Symplectic Euler Method).** The symplectic Euler method is nothing other than (3.2) with $S(P, q, h) = h H(P, q)$. We therefore have (3.3) with $S_1(p, q) = H(p, q)$ and $S_j(p, q) = 0$ for $j > 1$. Following the constructive proof of Theorem 3.2 we obtain

$$\widetilde{H} = H - \frac{h}{2} H_p H_q + \frac{h^2}{12} \left( H_{pp} H_q^2 + H_{qq} H_p^2 + 4 H_{pq} H_q H_p \right) + \ldots . \quad (3.7)$$

as the modified Hamiltonian of the symplectic Euler method. For vector-valued $p$ and $q$, the expression $H_p H_q$ is the scalar product of the vectors $H_p$ and $H_q$, and $H_{pp} H_q^2 = H_{pp}(H_q, H_q)$ with the second derivative interpreted as a bilinear mapping.

As a particular example consider the pendulum problem (I.1.13), which is Hamiltonian with $H(p, q) = p^2/2 - \cos q$, and apply the symplectic Euler method. By (3.7), the modified Hamiltonian is

$$\widetilde{H}(p, q) = H(p, q) - \frac{h}{2} p \sin q + \frac{h^2}{12} \left( \sin^2 q + p^2 \cos q \right) + \ldots .$$

This example illustrates that the modified equation corresponding to a separable Hamiltonian (i.e., $H(p, q) = T(p) + U(q)$) is in general not separable. Moreover, it shows that the modified equation of a second order differential equation $\ddot{q} = -\nabla U(q)$ (or equivalently, $\dot{q} = p, \dot{p} = -\nabla U(q)$) is in general not a second order equation.

In principle, the constructive proof of Theorem 3.2 allows us to explicitly compute the modified equation of every symplectic (partitioned) Runge-Kutta method. In Sect. IX.10 below we shall, however, give explicit formulas for the modified Hamiltonian in terms of trees. This also yields an alternative proof of Theorem 3.3.

## IX.3.3 Poisson Integrators

Consider a Poisson system, i.e., a differential equation

$$\dot{y} = B(y)\nabla H(y), \tag{3.8}$$

where the structure matrix $B(y)$ satisfies the conditions of Lemma VII.2.3, and apply a Poisson integrator (Definition VII.2.16).

**Theorem 3.5.** *If a Poisson integrator $\Phi_h(y)$ is applied to the Poisson system (3.8), then the modified equation is locally a Poisson system. More precisely, for every $y_0 \in \mathbb{R}^n$ there exist a neighbourhood $U$ and smooth functions $H_j : U \to \mathbb{R}$ such that on $U$, the modified equation is of the form*

$$\dot{\widetilde{y}} = B(\widetilde{y})\Big(\nabla H(\widetilde{y}) + h\,\nabla H_2(\widetilde{y}) + h^2 \nabla H_3(\widetilde{y}) + \ldots\Big). \tag{3.9}$$

*Proof.* We use the local change of coordinates $(u, c) = \chi(y)$ of the Darboux-Lie Theorem. By Corollary VII.2.12, this transforms (3.8) to

$$\dot{u} = J^{-1}\nabla_u K(u, c), \qquad \dot{c} = 0,$$

where $K(u, c) = H(y)$ and $\nabla_u$ is the gradient with respect to $u$. The same transformation takes $\Phi_h(y)$ to $\chi \circ \Phi_h \circ \chi^{-1}(u, c) = \big(\Psi_h^1(u, c), c\big)$, where by Lemma VII.2.19 $u \mapsto \Psi_h^1(u, c)$ is a symplectic transformation for every $c$. By Theorem 3.1, the modified equation in the $(u, c)$ variables is of the form

$$\dot{\widetilde{u}} = J^{-1}\nabla_u \widetilde{K}(\widetilde{u}, \widetilde{c}), \qquad \dot{\widetilde{c}} = 0$$

with $\widetilde{K}(u, c) = K(u, c) + h\,K_2(u, c) + h^2 K_3(u, c) + \ldots$ . Transforming back to the $y$-variables gives the modified equation (3.9) with $H_j(y) = K_j(u, c)$. □

The above result is purely local in that it relies on the local transformation of the Darboux-Lie Theorem. It can be made more global under additional conditions on the differential equation.

**Theorem 3.6.** *If $H(y)$ and $B(y)$ are defined and smooth on a simply connected domain $D$, and if $B(y)$ is invertible on $D$, then a Poisson integrator $\Phi_h(y)$ has a modified equation (3.9) with smooth functions $H_j(y)$ defined on all of $D$.*

*Proof.* By the construction of Sect. IX.1, the coefficient functions $f_j(y)$ of the modified equation (1.1) are defined and smooth on $D$. Since $B(y)$ is assumed invertible, there exist unique smooth functions $g_j(y)$ such that $f_j(y) = B(y)g_j(y)$. It remains to show that $g_j(y) = \nabla H_j(y)$ for a function $H_j(y)$ defined on $D$.

By the local result of Theorem 3.5, we know that for every $y_0 \in D$ there exist functions $H_j^0(y)$ such that $g_j(y) = \nabla H_j^0(y)$ in a neighbourhood of $y_0$. This implies that the Jacobian of $g_j(y)$ is symmetric on $D$. The Integrability Lemma VI.2.7 thus proves the existence of functions $H_j(y)$ defined on all of $D$ such that $g_j(y) = \nabla H_j(y)$. □

## IX.4 Modified Equations of Splitting Methods

For splitting methods applied to a differential equation

$$\dot{y} = f^{[1]}(y) + f^{[2]}(y), \tag{4.1}$$

the modified differential equation is obtained directly with the calculus of Lie derivatives and the Baker-Campbell-Hausdorff formula. This approach is due to Yoshida (1993) who considered the case of separable Hamiltonian systems.

**First-Order Splitting.** Consider the splitting method

$$\Phi_h = \varphi_h^{[1]} \circ \varphi_h^{[2]},$$

where $\varphi_h^{[i]}$ is the time-$h$ flow of $\dot{y} = f^{[i]}(y)$. In terms of the Lie derivatives $D_i$ defined by $D_i g(y) = g'(y) f^{[i]}(y)$, this method becomes, using Lemma III.5.1,

$$\Phi_h = \exp(hD_2)\exp(hD_1)\mathrm{Id},$$

and with the BCH formula (III.4.11), (III.4.12) this reads

$$\Phi_h = \exp(h\widetilde{D})\mathrm{Id}$$

with

$$\widetilde{D} = D_1 + D_2 + \frac{h}{2}[D_2, D_1] + \frac{h^2}{12}\Big([D_2, [D_2, D_1]] + [D_1, [D_1, D_2]]\Big) + \dots. \tag{4.2}$$

It follows that $\Phi_h$ is formally the exact time-$h$ flow of the modified equation

$$\dot{\widetilde{y}} = \widetilde{f}(\widetilde{y}) \qquad \text{with} \qquad \widetilde{f} = \widetilde{D}\,\mathrm{Id}. \tag{4.3}$$

This gives

$$\widetilde{f}(y) = f(y) + h f_2(y) + h^2 f_3(y) + \dots$$

with $f = f^{[1]} + f^{[2]}$ and

$$f_2 = \frac{1}{2}\Big(f^{[1]'} f^{[2]} - f^{[2]'} f^{[1]}\Big)$$

$$f_3 = \frac{1}{12}\Big(f^{[1]''}(f^{[2]}, f^{[2]}) + f^{[1]'} f^{[2]'} f^{[2]} - f^{[2]''}(f^{[1]}, f^{[2]}) - f^{[2]'} f^{[1]'} f^{[2]}$$
$$+ f^{[2]''}(f^{[1]}, f^{[1]}) + f^{[2]'} f^{[1]'} f^{[1]} - f^{[1]''}(f^{[2]}, f^{[1]}) - f^{[1]'} f^{[2]'} f^{[1]}\Big).$$

**Strang Splitting.** For the symmetric splitting

$$\Phi_h^{[S]} = \varphi_{h/2}^{[1]} \circ \varphi_h^{[2]} \circ \varphi_{h/2}^{[1]}$$

the symmetric BCH formula (III.4.14), (III.4.15) yields

$$\Phi_h^{[S]} = \exp(\frac{h}{2}D_1)\exp(hD_2)\exp(\frac{h}{2}D_1)\,\text{Id} = \exp(h\widetilde{D}^{[S]})\,\text{Id}$$

with

$$\widetilde{D}^{[S]} = D_1 + D_2 + h^2\left(-\frac{1}{24}[D_1,[D_1,D_2]] + \frac{1}{12}[D_2,[D_2,D_1]]\right) + \dots . \quad (4.4)$$

Hence, $\Phi_h^{[S]}$ is the formally exact flow of the modified equation

$$\dot{\widetilde{y}} = \widetilde{f}^{[S]}(\widetilde{y}) \quad \text{with} \quad \widetilde{f}^{[S]} = \widetilde{D}^{[S]}\,\text{Id}. \quad (4.5)$$

This gives

$$\widetilde{f}^{[S]}(y) = f(y) + h^2 f_3^{[S]}(y) + h^4 f_5^{[S]}(y) + \dots$$

with $f = f^{[1]} + f^{[2]}$ and

$$\begin{aligned}
f_3^{[S]} &= \Big(\frac{1}{12}\big(f^{[1]''}(f^{[2]},f^{[2]}) + f^{[1]'}f^{[2]'}f^{[2]} - f^{[2]''}(f^{[1]},f^{[2]}) - f^{[2]'}f^{[1]'}f^{[2]}\big) \\
&\quad -\frac{1}{24}\big(f^{[2]''}(f^{[1]},f^{[1]}) + f^{[2]'}f^{[1]'}f^{[1]} - f^{[1]''}(f^{[2]},f^{[1]}) - f^{[1]'}f^{[2]'}f^{[1]}\big)\Big).
\end{aligned}$$

The modified equations for general splitting methods (III.5.15) are obtained in the same way, using Lemma III.5.5.

**Hamiltonian Splittings.** Consider a differential equation (4.1) where the vector fields $f^{[i]}(y) = J^{-1}\nabla H^{[i]}(y)$ are Hamiltonian. Lemma VII.2.7 shows that the commutator of the Lie derivatives of two Hamiltonian vector fields is the Lie derivative of another Hamiltonian vector field which corresponds to the Poisson bracket of the two Hamiltonians: $[D_F, D_G] = D_{\{G,F\}}$. This implies in particular that the modified differential equations (4.3) and (4.5) are again Hamiltonian. For the first-order splitting, we thus get $f_j(y) = J^{-1}\nabla H_j(y)$, where by (4.2) and (4.3),

$$\begin{aligned}
H_2 &= \tfrac{1}{2}\{H^{[1]},H^{[2]}\} \\
H_3 &= \tfrac{1}{12}\Big(\{\{H^{[1]},H^{[2]}\},H^{[2]}\} + \{\{H^{[2]},H^{[1]}\},H^{[1]}\}\Big),
\end{aligned}$$

and for the Strang splitting, by (4.4) and (4.5),

$$H_3^{[S]} = -\frac{1}{24}\{\{H^{[2]},H^{[1]}\},H^{[1]}\} + \frac{1}{12}\{\{H^{[1]},H^{[2]}\},H^{[2]}\}.$$

The explicit expressions from the BCH-formula show that the modified Hamiltonian is defined on the same open set as the smooth Hamiltonians $H^{[i]}$.

For the splitting $H(p,q) = T(p) + U(q)$ of a separable Hamiltonian, this approach gives an alternative derivation of the modified equation (3.7) of the symplectic Euler method, and a simple construction of the modified equation of the Störmer/Verlet method (Yoshida 1993). Here, the formula simplifies to

$$\widetilde{H}^{[S]} = H + h^2\left(-\frac{1}{24}U_{qq}(T_p,T_p) + \frac{1}{12}T_{pp}(U_q,U_q)\right) + \dots . \quad (4.6)$$

## IX.5 Modified Equations of Methods on Manifolds

Consider a differential equation on a smooth manifold $\mathcal{M}$,

$$\dot{y} = f(y) \quad \text{with} \quad f(y) \in T_y\mathcal{M}, \tag{5.1}$$

with a smooth vector field $f(y)$ defined on $\mathcal{M}$.

**Theorem 5.1.** *Let $\Phi_h : \mathcal{M} \to \mathcal{M}$ be an integrator on the manifold $\mathcal{M}$, with $\Phi_h(y)$ depending smoothly on $(y, h)$. Then, there exists a modified differential equation on $\mathcal{M}$,*

$$\dot{\widetilde{y}} = f(\widetilde{y}) + h f_2(\widetilde{y}) + h^2 f_3(\widetilde{y}) + \ldots \tag{5.2}$$

*with smooth $f_j(y) \in T_y\mathcal{M}$, such that $\varphi_{r,h}(y) = \Phi_h(y) + \mathcal{O}(h^{r+1})$, where $\varphi_{r,t}(y)$ denotes the flow of the truncation of (5.2) after $r$ terms.*

*For symmetric methods, the expansion (5.2) contains only even powers of $h$.*

*Proof.* We choose a local parametrization $y = \chi(z)$ of the manifold $\mathcal{M}$. In the coordinates $z$ the differential equation (5.1) reads

$$\dot{z} = g(z) \quad \text{with } g(z) \text{ defined by} \quad \chi'(z)g(z) = f(\chi(z)),$$

and the numerical integrator becomes

$$\Psi_h(z) = \chi^{-1} \circ \Phi_h \circ \chi(z).$$

Since $g(z)$ and $\Psi_h(z)$ are smooth, the standard backward error analysis on $\mathbb{R}^n$ of Sect. IX.1 yields a modified equation for the integrator $\Psi_h(z)$,

$$\dot{\widetilde{z}} = g(\widetilde{z}) + h g_2(\widetilde{z}) + h^2 g_3(\widetilde{z}) + \ldots .$$

Defining

$$f_j(y) = \chi'(z) \, g_j(z) \quad \text{for} \quad y = \chi(z)$$

gives the desired vector fields $f_j(y)$ on $\mathcal{M}$. It follows from the uniqueness of the modified equation in the parameter space that $f_j(y)$ is independent of the choice of the local parametrization.

The additional statement on symmetric methods follows from Theorem 2.2, because $\Psi_h$ is symmetric if and only if $\Phi_h$ is symmetric. □

Theorem 5.1 applies to many situations treated in Chap. IV.

**First Integrals.** The following result was obtained by Gonzalez, Higham & Stuart (1999) and Reich (1999) with different arguments.

**Corollary 5.2.** *Consider a differential equation $\dot{y} = f(y)$ with a first integral $I(y)$, i.e., $I'(y)f(y) = 0$ for all $y$. If the numerical method preserves this first integral, then every truncation of the modified equation has $I(y)$ as a first integral.*

*Proof.* This follows from Theorem 5.1 by considering $\dot y = f(y)$ as a differential equation on the manifold $\mathcal{M} = \{y \mid I(y) = Const\}$, for which the tangent space is $T_y\mathcal{M} = \{v \mid I'(y)v = 0\}$. □

**Projection Methods.** Algorithm IV.4.2 defines a smooth mapping on the manifold if the direction of projection depends smoothly on the position. This is satisfied by orthogonal projection, but is not fulfilled if switching coordinate projections are used (as in Example 4.3). The symmetric orthogonal projection method of Algorithm V.4.1 gives a symmetric method on the manifold to which Theorem 5.1 can be applied.

**Methods Based on Local Coordinates.** If the parametrization of the manifold employed in Algorithms IV.5.3 and V.4.5 depends smoothly on the position, then again Theorem 5.1 applies. This is the case for the tangent space parametrization, but not for the generalized coordinate partitioning considered in Sect. IV.5.3.

**Corollary 5.3 (Lie Group Methods).** *Consider a differential equation on a matrix Lie group $G$,*
$$\dot Y = A(Y)Y,$$
*where $A(Y)$ is in the associated Lie algebra $\mathfrak{g}$. A Lie group integrator $\Phi_h : G \to G$ has the modified equation*
$$\dot{\widetilde Y} = \big(A(\widetilde Y) + hA_2(\widetilde Y) + h^2 A_3(\widetilde Y) + \ldots\big)\widetilde Y \tag{5.3}$$
*with $A_j(Y) \in \mathfrak{g}$ for $Y \in G$.*

*Proof.* This is a direct consequence of Theorem 5.1 and (IV.6.3), viz., $T_Y G = \{AY \mid A \in \mathfrak{g}\}$. □

**Constrained Hamiltonian Systems.** In Sect. VII.1 we studied symplectic numerical integrators for constrained Hamiltonian systems
$$\begin{aligned} \dot q &= H_p(p,q) \\ \dot p &= -H_q(p,q) - G(q)^T \lambda \\ 0 &= g(q). \end{aligned} \tag{5.4}$$

Assuming the regularity condition (VII.1.12), the Lagrange parameter $\lambda = \lambda(p,q)$ is given by (VII.1.11). This system can be interpreted as a differential equation on the manifold
$$\mathcal{M} = \{(p,q) \mid g(q) = 0, \; G(q)H_p(p,q) = 0\}, \tag{5.5}$$
where $G(q) = g'(q)$. The symplectic Euler method (VII.1.17)–(VII.1.18), the RATTLE scheme (VII.1.24), and the Lobatto IIIA-IIIB pair (VII.1.25)–(VII.1.28) were found to be symplectic integrators $\Phi_h$ on the manifold $\mathcal{M}$.

**Theorem 5.4.** *A symplectic integrator* $\Phi_h : \mathcal{M} \to \mathcal{M}$ *for the constrained Hamiltonian system (5.4) has a modified equation which is locally of the form*

$$\begin{aligned} \dot{\widetilde{q}} &= \widetilde{H}_p(\widetilde{p}, \widetilde{q}) \\ \dot{\widetilde{p}} &= -\widetilde{H}_q(\widetilde{p}, \widetilde{q}) - G(\widetilde{q})^T \widetilde{\lambda} \\ 0 &= g(\widetilde{q}), \end{aligned} \quad (5.6)$$

*where* $\widetilde{\lambda} = \widetilde{\lambda}(\widetilde{p}, \widetilde{q})$ *is given by (VII.1.11) with $H$ replaced by $\widetilde{H}$, and*

$$\widetilde{H}(p, q) = H(p, q) + h\, H_2(p, q) + h^2 H_3(p, q) + \ldots \quad (5.7)$$

*with $H_j(p, q)$ satisfying $G(q)\nabla_p H_j(p, q) = 0$ for $(p, q) \in \mathcal{M}$ and all $j$.*

*Proof.* As explained in Example VII.2.6, a local parametrization $(p, q) = \chi(z)$ of the manifold $\mathcal{M}$ transforms (5.4) to the Poisson system

$$\dot{z} = B(z)\nabla K(z) \quad (5.8)$$

with $B(z) = (\chi'(z)^T J \chi'(z))^{-1}$ and $K(z) = H(\chi(z))$. It was shown in Example VII.2.18 that the numerical method $\Phi_h(p, q)$ on $\mathcal{M}$ becomes a Poisson integrator $\Psi_h(z)$ for (5.8). By Theorem 3.5, $\Psi_h(z)$ has the modified equation

$$\dot{\widetilde{z}} = B(\widetilde{z})\Big(\nabla K(\widetilde{z}) + h\,\nabla K_2(\widetilde{z}) + h^2 \nabla K_3(\widetilde{z}) + \ldots\Big). \quad (5.9)$$

Let $\pi$ be a smooth projection onto the manifold $\mathcal{M}$, defined on a neighbourhood of $\mathcal{M}$ in $\mathbb{R}^{2d}$. We then define

$$H_j(p, q) = K_j\big(\chi^{-1}(\pi(p, q))\big) + \mu(p, q)^T G(q) \nabla_p H(p, q)$$

where we choose $\mu(p, q)$ such that

$$G(q)\nabla_p H_j(p, q) = 0 \quad \text{for} \quad (p, q) \in \mathcal{M}. \quad (5.10)$$

This is possible because of the regularity assumption (VII.1.12), and because $G(q)\nabla_p H(p, q) = 0$ on $\mathcal{M}$. The condition (5.10) implies that the system (5.6) can be viewed as a differential equation on the original manifold $\mathcal{M}$. Using the same parametrization $(p, q) = \chi(z)$ as before shows that (5.6) is equivalent to (5.9). □

We note that, due to the arbitrary choice of the projection $\pi$, the functions $H_j(p, q)$ of the modified equation are uniquely defined only on $\mathcal{M}$.

A different approach to the backward error analysis of symplectic integrators for constrained Hamiltonian systems is given by Reich (1996a) and by Hairer & Wanner (1996). This approach applies to the constrained symplectic Euler and the RATTLE method, and it yields a modified Hamiltonian that is globally defined. It is based on an extension of the numerical integrator to a neighbourhood of the manifold $\mathcal{M}$. The existence of a globally defined modified Hamiltonian for the Lobatto IIIA - IIIB pair is still an open problem.

## IX.6 Modified Equations for Variable Step Sizes

The modified differential equation of a numerical integrator depends on the step size employed. Therefore, if the step size is changed arbitrarily, a different modified equation occurs at every step. This is the reason for the poor longtime behaviour observed in Sect. VIII.1. On the other hand, a satisfactory backward error analysis is possible for the variable-step approaches of Sects. VIII.2 and VIII.3.

**Reversible Adaptive Step Size Selection.** As in Sect. VIII.2 we consider the choice of step size
$$h_n = \varepsilon\, s(y_n, \varepsilon), \tag{6.1}$$
where $\varepsilon$ is a small accuracy parameter. The step size is not allowed to use information from previous steps. The idea is to consider expansions in powers of the fixed parameter $\varepsilon$ instead of the step sizes. The following development is given in Hairer & Stoffer (1997).

**Theorem 6.1.** *Let $\Phi_h(y)$ be a smooth one-step method. Then the variable-step method $y \mapsto \Phi_{\varepsilon s(y,\varepsilon)}(y)$ has a modified differential equation*
$$\dot{\widetilde{y}} = f(\widetilde{y}) + \varepsilon f_2(\widetilde{y}) + \varepsilon^2 f_3(\widetilde{y}) + \dots, \tag{6.2}$$
*with smooth vector fields $f_j(y)$, such that*
$$\varphi_{r,\varepsilon s(y,\varepsilon)}(y) = \Phi_{\varepsilon s(y,\varepsilon)}(y) + \mathcal{O}(\varepsilon^{r+1}),$$
*where $\varphi_{r,t}(y)$ denotes the flow of the truncation of (6.2) after $r$ terms.*

*Proof.* The modified equation is derived by Taylor expansion in the same way as (1.1), using $\varepsilon$-expansions instead of $h$-expansions. □

Theorems 2.1, 2.2, and 2.3 extend directly to this situation. In particular, Theorem 2.3 yields the following.

**Theorem 6.2.** *Consider a $\rho$-reversible differential equation and a numerical method $y \mapsto \Phi_{\varepsilon s(y,\varepsilon)}(y)$, where $\Phi_h(y)$ satisfies (V.1.4) and where $s(\rho y, \varepsilon) = s(y, -\varepsilon)$. If the method $\Phi_h$ is symmetric and if (VIII.2.3) holds, then the modified differential equation (6.2) is $\rho$-reversible.*

*Proof.* The assumptions imply that $\Psi_\varepsilon(y) = \Phi_{\varepsilon s(y,\varepsilon)}(y)$ satisfies $\rho \circ \Psi_\varepsilon = \Psi_{-\varepsilon} \circ \rho$ and the symmetry relation $\Psi_\varepsilon = \Psi_{-\varepsilon}^{-1}$. Theorem 2.3 applies to $\Psi_\varepsilon$ and yields the result. □

**Time Transformations.** The adaptive approaches of Sect. VIII.3 amount to applying a fixed step size method to a transformed differential equation. Hence the backward error analysis applies and yields modified equations for the transformed problem. These modified equations are Hamiltonian for Algorithm VIII.3.1 and reversible for method (VIII.3.11).

## IX.7 Rigorous Estimates – Local Error

> Wherefore it is highly desirable that it be clearly and rigorously shown why series of this kind, which at first converge very rapidly and then ever more slowly, and at length diverge more and more, nevertheless give a sum close to the true one if not too many terms are taken, and to what degree such a sum can safely be considered as exact.
> (a footnote in Gauss' thesis, 1799)

Up to now we have considered the modified equation (1.1) as a formal series without taking care of convergence issues. Here,

- we show that already in very simple situations the modified differential equation does not converge;
- we give bounds on the coefficient functions $f_j(y)$ of the modified equation (1.1), so that an optimal truncation index can be determined;
- we estimate the difference between the numerical solution $y_1 = \Phi_h(y_0)$ and the exact solution $\widetilde{y}(h)$ of the truncated modified equation.

These estimates will be the basis for rigorous statements concerning the long-time behaviour of numerical solutions. The rigorous estimates of the present section have been given in the articles Benettin & Giorgilli (1994), Hairer & Lubich (1997) and Reich (1999). We mainly follow the approach of Benettin & Giorgilli, but we also use ideas of the other two papers.

**Example 7.1.** We consider the differential equation[1] $\dot{y} = f(t)$, $y(0) = 0$, and we apply the trapezoidal rule $y_1 = h\bigl(f(0) + f(h)\bigr)/2$. In this case, the numerical solution has an expansion $\Phi_h(t, y) = y + h\bigl(f(t) + f(t+h)\bigr)/2 = y + hf(t) + h^2 f'(t)/2 + h^3 f''(t)/4 + \ldots$, so that the modified equation is necessarily of the form
$$\dot{\widetilde{y}} = f(t) + hb_1 f'(t) + h^2 b_2 f''(t) + h^3 b_3 f'''(t) + \ldots \,. \tag{7.1}$$
The real coefficients $b_k$ can be computed by putting $f(t) = e^t$. The relation $\Phi_h(t, y) = \widetilde{y}(t+h)$ (with initial value $\widetilde{y}(t) = y$) yields after division by $e^t$
$$\frac{h}{2}\bigl(e^h + 1\bigr) = \bigl(1 + b_1 h + b_2 h^2 + b_3 h^3 + \ldots\bigr)\bigl(e^h - 1\bigr).$$

This proves that $b_1 = 0$, and $b_k = B_k/k!$, where $B_k$ are the Bernoulli numbers (see for example Hairer & Wanner (1997), Sect. II.10). Since these numbers behave like $B_k/k! \approx \text{Const} \cdot (2\pi)^{-k}$ for $k \to \infty$, the series (7.1) diverges for all $h \neq 0$, as soon as the derivatives of $f(t)$ grow like $f^{(k)}(t) \approx k! \, MR^{-k}$. This is typically the case for analytic functions $f(t)$ with finite poles.

It is interesting to remark that the relation $\Phi_h(t, y) = \widetilde{y}(t+h)$ is nothing other than the Euler-MacLaurin summation formula.

---

[1] Observe that after adding the equation $\dot{t} = 1$, $t(0) = 0$, we get for $Y = (t, y)^T$ the autonomous differential equation $\dot{Y} = F(Y)$ with $F(Y) = (1, f(t))^T$. Hence, all results of this chapter are applicable.

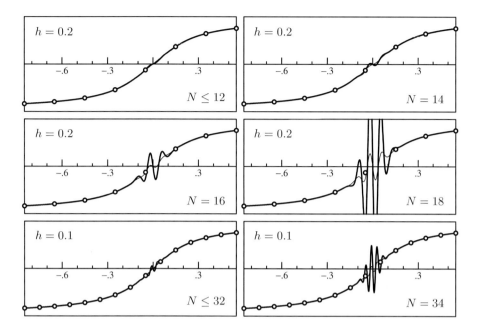

**Fig. 7.1.** Numerical solution with the trapezoidal rule compared to the solution of the truncated modified equation for $h = 0.2$ (upper four pictures), and for $h = 0.1$ (lower two pictures).

As a particular example we choose the function

$$f(t) = \frac{5}{1 + 25t^2}.$$

Figure 7.1 shows the numerical solution and the exact solution of the modified equation truncated at different values of $N$. For $h = 0.2$, there is an excellent agreement for $N \leq 12$, whereas oscillations begin to appear from $N = 14$ onwards. For the halved step size $h = 0.1$, the oscillations become visible for $N$ twice as large.

The main ingredient of a rigorous backward error analysis is an analyticity assumption on the differential equation $\dot{y} = f(y)$ and on the method. Throughout this section we assume that $f(y)$ is analytic in a complex neighbourhood of $y_0$ and that

$$\|f(y)\| \leq M \quad \text{for} \quad \|y - y_0\| \leq 2R \tag{7.2}$$

i.e., for all $y$ of $B_{2R}(y_0) := \{y \in \mathbb{C}^d\,;\, \|y - y_0\| \leq 2R\}$. Our strategy is the following: using (7.2) and Cauchy's estimates we derive bounds for the coefficient functions $d_j(y)$ of (1.3) on $B_R(y_0)$ (Sect. IX.7.1), then we estimate the functions $f_j(y)$ of the modified differential equation on $B_{R/2}(y_0)$ (Sect. IX.7.2), and finally we search for a suitable truncation for the formal series (1.1) and we prove the closeness of the numerical solution to the exact solution of the truncated modified equation (Sect. IX.7.3).

## IX.7.1 Estimation of the Derivatives of the Numerical Solution

If we apply a numerical method to $\dot y = f(y)$ with analytic $f(y)$, the expression $\Phi_h(y)$ will usually be analytic in a neighbourhood of $h = 0$ and $y \in B_R(y_0)$. Consequently, the coefficients $d_j(y)$ of the Taylor series expansion

$$\Phi_h(y) = y + hf(y) + h^2 d_2(y) + h^3 d_3(y) + \ldots \tag{7.3}$$

are also analytic and the functions $d_j(y)$ can be estimated by the use of Cauchy's inequalities. Let us demonstrate this for Runge-Kutta methods.

**Theorem 7.2.** *For a Runge-Kutta method (II.1.4) let*

$$\mu = \sum_{i=1}^{s} |b_i|, \qquad \kappa = \max_{i=1,\ldots,s} \sum_{j=1}^{s} |a_{ij}|. \tag{7.4}$$

*If $f(y)$ is analytic in the complex ball $B_{2R}(y_0)$ and satisfies (7.2), then the coefficient functions $d_j(y)$ of (7.3) are analytic in $B_R(y_0)$ and satisfy*

$$\|d_j(y)\| \leq \mu M \left(\frac{2\kappa M}{R}\right)^{j-1} \qquad \text{for} \qquad \|y - y_0\| \leq R. \tag{7.5}$$

*Proof.* For $y \in B_{3R/2}(y_0)$ and $\|\Delta y\| \leq 1$ the function $\alpha(z) = f(y + z\Delta y)$ is analytic for $|z| \leq R/2$ and bounded by $M$. Cauchy's estimate therefore yields

$$\|f'(y)\Delta y\| = \|\alpha'(0)\| \leq 2M/R.$$

Consequently, $\|f'(y)\| \leq 2M/R$ for $y \in B_{3R/2}(y_0)$ in the operator norm.

For $y \in B_R(y_0)$, the Runge-Kutta method (II.1.4) requires the solution of the nonlinear system $g_i = y + h \sum_{j=1}^{s} a_{ij} f(g_j)$, which can be solved by fixed point iteration. If $|h| 2\kappa M/R \leq \gamma < 1$, it represents a contraction on the closed set $\{(g_1, \ldots, g_s); \|g_i - y\| \leq R/2\}$ and possesses a unique solution. Consequently, the method is analytic for $|h| \leq \gamma R/(2\kappa M)$ and $y \in B_R(y_0)$. This implies that the functions $d_j(y)$ of (7.3) are also analytic. Furthermore, $\|\Phi_h(y) - y\| \leq |h|\mu M$ for $y \in B_R(y_0)$ so that, again by Cauchy's estimate,

$$\|d_j(y)\| = \frac{1}{j!} \left\| \frac{d^j}{dh^j}\big(\Phi_h(y) - y\big) \right\|_{h=0} \leq \mu M \left(\frac{2\kappa M}{\gamma R}\right)^{j-1}$$

for $j \geq 1$. The statement is then obtained by considering the limit $\gamma \to 1$. $\square$

**Table 7.1.** The constants $\mu$ and $\kappa$ of formula (7.4)

| method | $\mu$ | $\kappa$ | method | $\mu$ | $\kappa$ |
|---|---|---|---|---|---|
| explicit Euler | 1 | 0 | implicit Euler | 1 | 1 |
| implicit midpoint | 1 | 1/2 | trapezoidal rule | 1 | 1 |
| Gauss methods | 1 | $c_s$ | Lobatto IIIA | 1 | 1 |

Due to the consistency condition $\sum_{i=1}^{s} b_i = 1$, methods with positive weights $b_i$ all satisfy $\mu = 1$. The values $\mu, \kappa$ of some classes of Runge-Kutta methods are given in Table 7.1 (those for the Gauss methods and for the Lobatto IIIA methods have been checked for $s \le 9$ and $s \le 5$, respectively).

Estimates of the type (7.5), possibly with a different interpretation of $M$ and $R$, hold for all one-step methods which are analytic in $h$ and $y$, e.g., partitioned Runge-Kutta methods, splitting and composition methods, projection methods, Lie group methods, ... .

## IX.7.2 Estimation of the Coefficients of the Modified Equation

At the beginning of this chapter we gave an explicit formula for the first coefficient functions of the modified differential equation (see (1.4)). Using the Lie derivative

$$(D_i g)(y) = g'(y) f_i(y) \qquad (7.6)$$

(cf. (VI.5.4)) and $f_1(y) := f(y)$, these formulas can be written as

$$f_2(y) = d_2(y) - \frac{1}{2!}(D_1 f_1)(y)$$

$$f_3(y) = d_3(y) - \frac{1}{3!}(D_1^2 f_1)(y) - \frac{1}{2!}(D_2 f_1 + D_1 f_2)(y).$$

We have the following recurrence relation for the general case.

**Lemma 7.3.** *If the numerical method has an expansion of the form (7.3), then the functions $f_j(y)$ of the modified differential equation (1.1) satisfy*

$$f_j(y) = d_j(y) - \sum_{i=2}^{j} \frac{1}{i!} \sum_{k_1 + \ldots + k_i = j} \left( D_{k_1} \ldots D_{k_{i-1}} f_{k_i} \right)(y),$$

*where $k_m \ge 1$ for all $m$. Observe that the right-hand expression only involves $f_k(y)$ with $k < j$.*

*Proof.* The solution of the modified equation (1.1) with initial value $y(t) = y$ can be formally written as (cf. (1.2))

$$\widetilde{y}(t+h) = y + \sum_{i \ge 1} \frac{h^i}{i!} D^{i-1} F(y),$$

where $F(y) = f_1(y) + h f_2(y) + h^2 f_3(y) + \ldots$ stands for the modified equation, and $hD = hD_1 + h^2 D_2 + h^3 D_3 + \ldots$ for the corresponding Lie operator. We expand the formal sums and obtain

$$\widetilde{y}(t+h) = y + \sum_{i \ge 1} \frac{1}{i!} \sum_{k_1, \ldots, k_i} h^{k_1 + \ldots + k_i} \left( D_{k_1} \ldots D_{k_{i-1}} f_{k_i} \right)(y), \qquad (7.7)$$

where all $k_m \ge 1$. Comparing like powers of $h$ in (7.3) and (7.7) yields the desired recurrence relations for the functions $f_j(y)$. □

To get bounds for $\|f_j(y)\|$, we have to estimate repeatedly expressions like $\|(D_i g)(y)\|$. The following variant of Cauchy's estimate will be extremely useful.

**Lemma 7.4.** *For analytic functions $f_i(y)$ and $g(y)$ we have for $0 \leq \sigma < \rho$ the estimate*
$$\|D_i g\|_\sigma \leq \frac{1}{\rho - \sigma} \cdot \|f_i\|_\sigma \cdot \|g\|_\rho.$$

*Here, $\|g\|_\rho := \max\{\|g(y)\|\,;\, y \in B_\rho(y_0)\}$ and $\|f_i\|_\sigma$, $\|D_i g\|_\sigma$ are defined similarly.*

*Proof.* For a fixed $y \in B_\sigma(y_0)$ the function $\alpha(z) = g(y + z f_i(y))$ is analytic for $\|z\| \leq \varepsilon := (\rho - \sigma)/M$ with $M := \|f_i\|_\sigma$. Since $\alpha'(0) = g'(y) f_i(y) = (D_i g)(y)$, we get from Cauchy's estimate that
$$\|(D_i g)(y)\| = \|\alpha'(0)\| \leq \frac{1}{\varepsilon} \sup_{|z| \leq \varepsilon} \|\alpha(z)\| \leq \frac{M}{\rho - \sigma} \cdot \|g\|_\rho.$$

This proves the statement. □

We are now able to estimate the coefficients $f_j(y)$ of the modified differential equation.

**Theorem 7.5.** *Let $f(y)$ be analytic in $B_{2R}(y_0)$, let the Taylor series coefficients of the numerical method (7.3) be analytic in $B_R(y_0)$, and assume that (7.2) and (7.5) are satisfied. Then, we have for the coefficients of the modified differential equation*
$$\|f_j(y)\| \leq \ln 2 \, \eta \, M \left(\frac{\eta M j}{R}\right)^{j-1} \quad \text{for} \quad \|y - y_0\| \leq R/2, \tag{7.8}$$

*where $\eta = 2 \max\bigl(\kappa, \mu/(2\ln 2 - 1)\bigr)$.*

*Proof.* We fix an index, say $J$, and we estimate (in the notation of Lemma 7.4)
$$\|f_j\|_{R - (j-1)\delta} \quad \text{for} \quad j = 1, 2, \ldots, J,$$

where $\delta = R/(2(J-1))$. This will then lead to the desired estimate for $\|f_J\|_{R/2}$.

In the following we abbreviate $\|\cdot\|_{R-(j-1)\delta}$ by $\|\cdot\|_j$. Using repeatedly Cauchy's estimate of Lemma 7.4 we get for $k_1 + \ldots + k_i = j$ that

$$\|D_{k_1} \ldots D_{k_{i-1}} f_{k_i}\|_j \leq \frac{1}{\delta} \|f_{k_1}\|_j \|D_{k_2} \ldots D_{k_{i-1}} f_{k_i}\|_{j-1}$$
$$\leq \ldots \leq \frac{1}{\delta^{i-1}} \|f_{k_1}\|_j \|f_{k_2}\|_{j-1} \cdot \ldots \cdot \|f_{k_i}\|_{j-i+1}$$
$$\leq \frac{1}{\delta^{i-1}} \|f_{k_1}\|_{k_1} \|f_{k_2}\|_{k_2} \cdot \ldots \cdot \|f_{k_i}\|_{k_i}.$$

The last inequality follows from $\|g\|_j \leq \|g\|_l$ for $l \leq j$, which is an immediate

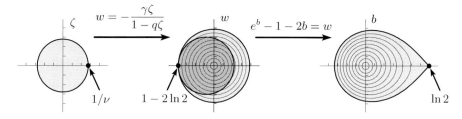

**Fig. 7.2.** Complex functions of the proof of Theorem 7.5 ($\gamma = q = 1$)

consequence of $B_{R-(j-1)\delta}(y_0) \subset B_{R-(l-1)\delta}(y_0)$. It therefore follows from Lemma 7.3 that

$$\|f_j\|_j \le \|d_j\|_j + \sum_{i=2}^{j} \frac{1}{i!} \sum_{k_1+\ldots+k_i=j} \frac{1}{\delta^{i-1}} \|f_{k_1}\|_{k_1} \|f_{k_2}\|_{k_2} \cdot \ldots \cdot \|f_{k_i}\|_{k_i}.$$

By induction on $j$ ($1 \le j \le J$) we obtain that $\|f_j\|_j \le \delta \beta_j$, where $\beta_j$ is defined by

$$\beta_j = \frac{\mu M}{\delta} \left(\frac{2\kappa M}{R}\right)^{j-1} + \sum_{i=2}^{j} \frac{1}{i!} \sum_{k_1+\ldots+k_i=j} \beta_{k_1} \beta_{k_2} \cdot \ldots \cdot \beta_{k_i}. \quad (7.9)$$

Observe that $\beta_j$ is defined for all $j \ge 1$. We let $b(\zeta) = \sum_{j \ge 1} \beta_j \zeta^j$ be its generating function and we obtain (by multiplying (7.9) with $\zeta^j$ and summing over $j \ge 1$)

$$b(\zeta) = \frac{\gamma \zeta}{1 - q\zeta} + \sum_{j \ge 2} \frac{1}{j!} b(\zeta)^j = \frac{\gamma \zeta}{1 - q\zeta} + e^{b(\zeta)} - 1 - b(\zeta), \quad (7.10)$$

where we have used the abbreviations $\gamma := \mu M/\delta$ and $q := 2\kappa M/R$.

Whenever $e^{b(\zeta)} \ne 2$ (i.e., for $\zeta \ne (2b-1)/(\gamma+q(2b-1))$ with $b = \ln 2 + 2k\pi i$) the implicit function theorem can be applied to (7.10). This implies that $b(\zeta)$ is analytic in a disc with radius $1/\nu = (2\ln 2 - 1)/(\gamma + q(2\ln 2 - 1))$ and centre at the origin. On the disc $|\zeta| \le 1/\nu$, the solution $b(\zeta)$ of (7.10) with $b(0) = 0$ is bounded by $\ln 2$. This is seen as follows (Fig. 7.2): with the function $w = -\gamma\zeta/(1 - q\zeta)$ the disc $|\zeta| \le 1/\nu$ is mapped into a disc which, for all possible choices of $\gamma \ge 0$ and $q \ge 0$, lies in $|w| \le 2\ln 2 - 1$. The image of this disc under the mapping $b(w)$ defined by $e^b - 1 - 2b = w$ and $b(0) = 0$ is completely contained in the disc $|b| \le \ln 2$. Cauchy's inequalities therefore imply $|\beta_j| \le \ln 2 \cdot \nu^j$, and we get

$$\|f_J\|_{R/2} = \|f_J\|_J \le \delta\beta_J \le \ln 2 \cdot \delta \cdot \nu^J.$$

Since $\nu = q + \gamma/(2\ln 2 - 1) \le \eta M J/R$ with $\eta$ given by $\eta = 2\max(\kappa, \mu/(2\ln 2 - 1))$ and $\delta\nu \le \eta M$, this proves the statement for $J$. □

## IX.7.3 Choice of $N$ and the Estimation of the Local Error

To get rigorous estimates, we truncate the modified differential equation (1.1), and we consider

$$\dot{\widetilde{y}} = F_N(\widetilde{y}), \qquad F_N(\widetilde{y}) = f(\widetilde{y}) + hf_2(\widetilde{y}) + \ldots + h^{N-1}f_N(\widetilde{y}) \qquad (7.11)$$

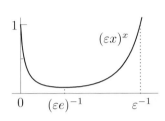

with initial value $\widetilde{y}(0) = y_0$. It is common in the theory of asymptotic expansions to truncate the series at the index where the corresponding term is minimal. Motivated by the bound (7.8) and by the fact that $(\varepsilon x)^x$ admits a minimum for $x = (\varepsilon e)^{-1}$ (see the picture to the left with $\varepsilon = 0.15$), we suppose that the truncation index $N$ satisfies

$$hN \le h_0 \qquad \text{with} \qquad h_0 = \frac{R}{e\eta M}. \qquad (7.12)$$

Under the less restrictive assumption $hN \le eh_0$, the estimates (7.2) and (7.8) imply for $\|y - y_0\| \le R/2$ that

$$\begin{aligned}\|F_N(y)\| &\le M\left(1 + \eta \ln 2 \sum_{j=2}^{N} \left(\frac{\eta Mjh}{R}\right)^{j-1}\right) \\ &\le M\left(1 + \eta \ln 2 \sum_{j=2}^{N} \left(\frac{j}{N}\right)^{j-1}\right) \le M(1 + 1.65\,\eta).\end{aligned} \qquad (7.13)$$

One can check that the sum in the lower formula of (7.13) is maximal for $N = 7$ and bounded by $2.38$. For a $p$th order method we obtain under the same assumptions

$$\|F_N(y) - f(y)\| \le cMh^p, \qquad (7.14)$$

where $c$ depends only on the method.

**Theorem 7.6.** *Let $f(y)$ be analytic in $B_{2R}(y_0)$, let the coefficients $d_j(y)$ of the method (7.3) be analytic in $B_R(y_0)$, and assume that (7.2) and (7.5) hold. If $h \le h_0/4$ with $h_0 = R/(e\eta M)$, then there exists $N = N(h)$ (namely $N$ equal to the largest integer satisfying $hN \le h_0$) such that the difference between the numerical solution $y_1 = \Phi_h(y_0)$ and the exact solution $\widetilde{\varphi}_{N,t}(y_0)$ of the truncated modified equation (7.11) satisfies*

$$\|\Phi_h(y_0) - \widetilde{\varphi}_{N,h}(y_0)\| \le h\gamma M e^{-h_0/h},$$

*where $\gamma = e(2 + 1.65\eta + \mu)$ depends only on the method (we have $5 \le \eta \le 5.18$ and $\gamma \le 31.4$ for the methods of Table 7.1).*

The quotient $L = M/R$ is an upper bound of the first derivative $f'(y)$ and can be interpreted as a Lipschitz constant for $f(y)$. The condition $h \le h_0/4$ is therefore equivalent to $hL \le Const$, where $Const$ depends only on the method. Because of this condition, Theorem 7.6 requires unreasonably small step sizes for the numerical solution of stiff differential equations.

*Proof of Theorem 7.6.* We follow here the elegant proof of Benettin & Giorgilli (1994). It is based on the fact that $\Phi_h(y_0)$ (as a convergent series (7.3)) and $\widetilde{\varphi}_{N,h}(y_0)$ (as the solution of an analytic differential equation) are both analytic functions of $h$. Hence,

$$g(h) := \Phi_h(y_0) - \widetilde{\varphi}_{N,h}(y_0) \tag{7.15}$$

is analytic in a complex neighbourhood of $h = 0$. By definition of the functions $f_j(y)$ of the modified equation (1.1), the coefficients of the Taylor series for $\Phi_h(y_0)$ and $\widetilde{\varphi}_{N,h}(y_0)$ are the same up to the $h^N$ term, but not further due to the truncation of the modified equation. Consequently, the function $g(h)$ contains the factor $h^{N+1}$, and the maximum principle for analytic functions, applied to $g(h)/h^{N+1}$, implies that

$$\|g(h)\| \le \left(\frac{h}{\varepsilon}\right)^{N+1} \max_{|z| \le \varepsilon} \|g(z)\| \quad \text{for } 0 \le h \le \varepsilon, \tag{7.16}$$

if $g(z)$ is analytic for $|z| \le \varepsilon$. We shall show that we can take $\varepsilon = eh_0/N$, and we compute an upper bound for $\|g(z)\|$ by estimating separately $\|\Phi_h(y_0) - y_0\|$ and $\|\widetilde{\varphi}_{N,h}(y_0) - y_0\|$.

The function $\Phi_z(y_0)$ is given by the series (7.3) which, due to the bounds of Theorem 7.2, converges certainly for $|z| \le R/(4\kappa M)$, and therefore also for $|z| \le \varepsilon$ (because $2\kappa \le \eta$ and $N \ge 4$, which is a consequence of $h_0/h \ge 4$). Hence, it is analytic in $|z| \le \varepsilon$. Moreover, we have from Theorem 7.2 that $\|\Phi_z(y_0) - y_0\| \le |z|M(1+\mu)$ for $|z| \le \varepsilon$.

Because of the bound (7.13) on $F_N(y)$, which is valid for $y \in B_{R/2}(y_0)$ and for $|h| \le \varepsilon$, we have $\|\widetilde{\varphi}_{N,z}(y_0) - y_0\| \le |z|M(1+1.65\eta)$ as long as the solution $\widetilde{\varphi}_{N,z}(y_0)$ stays in the ball $B_{R/2}(y_0)$. Because of $\varepsilon M(1+1.65\eta) \le R/2$, which is a consequence of the definition of $\varepsilon$, of $N \ge 4$, and of $(1+1.65\eta) \le 1.85\eta$ (because for consistent methods $\mu \ge 1$ holds and therefore also $\eta \ge 2/(2\ln 2 - 1) \ge 5$), this is the case for all $|z| \le \varepsilon$. In particular, the solution $\widetilde{\varphi}_{N,z}(y_0)$ is analytic in $|z| \le \varepsilon$.

Inserting $\varepsilon = eh_0/N$ and the bound on $\|g(z)\| \le \|\Phi_z(y_0) - y_0\| + \|\widetilde{\varphi}_{N,z}(y_0) - y_0\|$ into (7.16) yields (with $C = 2 + 1.65\eta + \mu$)

$$\|g(h)\| \le \varepsilon MC\left(\frac{h}{\varepsilon}\right)^{N+1} \le hMC\left(\frac{h}{\varepsilon}\right)^{N} = hMC\left(\frac{hN}{eh_0}\right)^{N} \le hMCe^{-N},$$

because $hN \le h_0$. The statement now follows from the fact that $N \le h_0/h < N+1$, so that $e^{-N} \le e \cdot e^{-h_0/h}$. □

## IX.8 Long-Time Energy Conservation

> In particular, one easily explains in this way why symplectic algorithms give rise to a good energy conservation, with essentially no accumulation of errors in time.
> (G. Benettin & A. Giorgilli 1994)

As a first application of Theorem 7.6 we study the long-time energy conservation of symplectic numerical schemes applied to Hamiltonian systems $\dot{y} = J^{-1}\nabla H(y)$. It follows from Theorem 3.1 that the corresponding modified differential equation is also Hamiltonian. After truncation we thus get a modified Hamiltonian

$$\widetilde{H}(y) = H(y) + h^p H_{p+1}(y) + \ldots + h^{N-1} H_N(y), \tag{8.1}$$

which we assume to be defined on the same open set as the original Hamiltonian $H$; see Theorem 3.2 and Sect. IX.4. We also assume that the numerical method satisfies the analyticity bounds (7.5), so that Theorem 7.6 can be applied. The following result is given by Benettin & Giorgilli (1994).

**Theorem 8.1.** *Consider a Hamiltonian system with analytic $H : D \to \mathbb{R}$ (where $D \subset \mathbb{R}^{2d}$), and apply a symplectic numerical method $\Phi_h(y)$ with step size $h$. If the numerical solution stays in the compact set $K \subset D$, then there exist $h_0$ and $N = N(h)$ (as in Theorem 7.6) such that*

$$\widetilde{H}(y_n) = \widetilde{H}(y_0) + \mathcal{O}(e^{-h_0/2h})$$
$$H(y_n) = H(y_0) + \mathcal{O}(h^p)$$

*over exponentially long time intervals $nh \leq e^{h_0/2h}$.*

*Proof.* We let $\widetilde{\varphi}_{N,t}(y_0)$ be the flow of the truncated modified equation. Since this differential equation is Hamiltonian with $\widetilde{H}$ of (8.1), $\widetilde{H}(\widetilde{\varphi}_{N,t}(y_0)) = \widetilde{H}(y_0)$ holds for all times $t$. From Theorem 7.6 we know that $\|y_{n+1} - \widetilde{\varphi}_{N,h}(y_n)\| \leq h\gamma M e^{-h_0/h}$ and, by using a global $h$-independent Lipschitz constant for $\widetilde{H}$ (which exists by Theorem 7.5), we also get $\widetilde{H}(y_{n+1}) - \widetilde{H}(\widetilde{\varphi}_{N,h}(y_n)) = \mathcal{O}(he^{-h_0/h})$. From the identity

$$\widetilde{H}(y_n) - \widetilde{H}(y_0) = \sum_{j=1}^{n}\left(\widetilde{H}(y_j) - \widetilde{H}(y_{j-1})\right) = \sum_{j=1}^{n}\left(\widetilde{H}(y_j) - \widetilde{H}(\widetilde{\varphi}_{N,h}(y_{j-1}))\right)$$

we thus get $\widetilde{H}(y_n) - \widetilde{H}(y_0) = \mathcal{O}(nhe^{-h_0/h})$, and the statement on the long-time conservation of $\widetilde{H}$ is an immediate consequence. The statement for the Hamiltonian $H$ follows from (8.1), because $H_{p+1}(y) + hH_{p+2}(y) + \ldots + h^{N-p-1}H_N(y)$ is uniformly bounded on $K$ independently of $h$ and $N$. This follows from the proof of Lemma VI.2.7 and from the estimates of Theorem 7.5. □

**Example 8.2.** Let us check explicitly the assumptions of Theorem 8.1 for the pendulum problem $\dot{q} = p$, $\dot{p} = -\sin q$. The vector field $f(p,q) = (p, -\sin q)^T$ is also

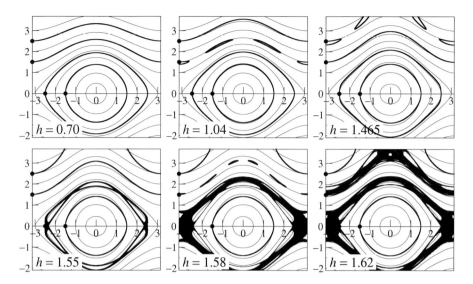

**Fig. 8.1.** Numerical solutions of the implicit midpoint rule with large step sizes

well-defined for complex $p$ and $q$, and it is analytic everywhere on $\mathbb{C}^2$. We let $K$ be a compact subset of $\{(p,q) \in \mathbb{R}^2 \; ; \; |p| \leq c\}$. As a consequence of $|\sin q| \leq e^{|\Im q|}$, we get the bound

$$\|f(p,q)\| \leq \sqrt{c^2 + 4R^2 + e^{2R}}$$

for $\|(p,q) - (p_0, q_0)\| \leq 2R$ and $(p_0, q_0) \in K$. If we choose $c \leq 2$, $R = 1$, and $M = 4$, the value $h_0$ of Theorem 7.6 is given by $h_0 = 1/4e\eta \approx 0.018$ for the methods of Table 7.1. For step sizes that are smaller than $h_0/20$, Theorem 8.1 guarantees that the numerical Hamiltonian is well conserved on intervals $[0, T]$ with $T \approx e^{10} \approx 2 \cdot 10^4$.

The numerical experiment of Fig. 8.1 shows that the estimates for $h_0$ are often too pessimistic. We have drawn $200\,000$ steps of the numerical solution of the implicit midpoint rule for various step sizes $h$ and for initial values $(p_0, q_0) = (0, -1.5)$, $(p_0, q_0) = (0, -2.5)$, $(p_0, q_0) = (1.5, -\pi)$, and $(p_0, q_0) = (2.5, -\pi)$. They are compared to the contour lines of the truncated modified Hamiltonian

$$\widetilde{H}(p,q) = \frac{p^2}{2} - \cos q + \frac{h^2}{48}\Big(\cos(2q) - 2p^2 \cos q\Big).$$

This shows that for step sizes as large as $h \leq 0.7$ the Hamiltonian $\widetilde{H}$ is extremely well conserved. Beyond this value, the dynamics of the numerical method soon turns into chaotic behaviour (see also Yoshida (1993) and Hairer, Nørsett & Wanner (1993), page 336).

Theorem 8.1 explains the near conservation of the Hamiltonian with the symplectic Euler method, the implicit midpoint rule and the Störmer/Verlet method as observed in the numerical experiments of Chap. I: in Fig. I.1.3 for the pendulum problem, in Fig. I.2.2 for the Kepler problem, and in Fig. I.3.2 for the frozen argon crystal.

The linear drift of the numerical Hamiltonian for non-symplectic methods can be explained by a computation similar to that of the proof of Theorem 8.1. From a Lipschitz condition of the Hamiltonian and from the standard local error estimate, we obtain $H(y_{n+1}) - H(\varphi_h(y_n)) = \mathcal{O}(h^{p+1})$. Since $H(\varphi_h(y_n)) = H(y_n)$, a summation of these terms leads to

$$H(y_n) - H(y_0) = \mathcal{O}(th^p) \qquad \text{for } t = nh. \tag{8.2}$$

This explains the linear growth in the error of the Hamiltonian observed in Fig. I.2.2 and in Fig. I.3.2 for the explicit Euler method.

## IX.9 Modified Equation in Terms of Trees

By Theorem III.1.4 the numerical solution $y_1 = \Phi_h(y_0)$ of a Runge-Kutta method can be written as a B-series

$$\Phi_h(y) = y + hf(y) + h^2 a(\mathbf{\jmath})(f'f)(y) \\ + h^3\left(\frac{1}{2} a(\mathbf{V}) f''(f,f)(y) + a(\mathbf{\jmath})f'f'f(y)\right) + \ldots . \tag{9.1}$$

For consistent methods, i.e., methods of order at least 1, we always have $a(\bullet) = 1$, so that the coefficient of $h$ is equal to $f(y)$. In this section we exploit this special structure of $\Phi_h(y)$ in order to get practical formulas for the coefficient functions of the modified differential equation. Using (9.1) instead of (1.3), the equations (1.4) yield

$$\begin{aligned} f_2(y) &= \left(a(\mathbf{\jmath}) - \frac{1}{2}\right)(f'f)(y) \\ f_3(y) &= \frac{1}{2}\left(a(\mathbf{V}) - a(\mathbf{\jmath}) + \frac{1}{6}\right)f''(f,f)(y) \\ &\quad + \left(a(\mathbf{\jmath}) - a(\mathbf{\jmath}) + \frac{1}{3}\right)f'f'f(y). \end{aligned} \tag{9.2}$$

Continuing this computation, one is quickly convinced of the general formula

$$f_j(y) = \sum_{|\tau|=j} \frac{b(\tau)}{\sigma(\tau)} F(\tau)(y), \tag{9.3}$$

so that the modified equation (1.1) becomes

$$\dot{\tilde{y}} = \sum_{\tau \in T} \frac{h^{|\tau|-1}}{\sigma(\tau)} b(\tau) F(\tau)(\tilde{y}) \tag{9.4}$$

with $b(\bullet) = 1$, $b(\mathbf{\jmath}) = a(\mathbf{\jmath}) - \frac{1}{2}$, etc. Since the coefficients $\sigma(\tau)$ are known from Definition III.1.7, all we have to do is to find suitable recursion formulas for the real coefficients $b(\tau)$.

## IX.9.1 B-Series of the Modified Equation

Recurrence formulas for the coefficients $b(\tau)$ in (9.4) were first given by Hairer (1994) and by Calvo, Murua & Sanz-Serna (1994). We follow here the approach of Hairer (1999), which uses the Lie-derivative of B-series and thus simplifies the construction of the coefficients.

We make use of the notion of ordered trees introduced in Sect. III.1.3. For a given tree $\tau$ we define the set of all *splittings* as

$$SP(\tau) = \{\theta \in OST(\tau) \,;\, \tau \setminus \theta \text{ consists of only one element}\}. \quad (9.5)$$

Here, $OST(\tau) = OST(\omega(\tau))$ is the set of ordered subtrees as defined in (III.1.33).

**Lemma 9.1 (Lie-Derivative of B-series).** *Let $b(\tau)$ (with $b(\emptyset) = 0$) and $c(\tau)$ be the coefficients of two B-series, and let $y(t)$ be a formal solution of the differential equation $h\dot{y}(t) = B(b, y(t))$. The Lie derivative of the function $B(c, y)$ with respect to the vector field $B(b, y)$ is again a B-series*

$$h\frac{d}{dt} B(c, y(t)) = B(\partial_b c, y(t)).$$

*Its coefficients are given by $\partial_b c(\emptyset) = 0$ and for $|\tau| \geq 1$ by*

$$\partial_b c(\tau) = \sum_{\theta \in SP(\tau)} c(\theta)\, b(\tau \setminus \theta). \quad (9.6)$$

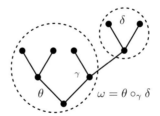

**Fig. 9.1.** Splitting of an ordered tree $\omega$ into a subtree $\theta$ and $\{\delta\} = \omega \setminus \theta$.

*Proof.* For the proof of this lemma it is convenient to work with ordered trees $\omega \in OT$. Since $\nu(\tau)$ of (III.1.31) denotes the number of possible orderings of a tree $\tau \in T$, a sum $\sum_{\tau \in T} \cdot/\cdot$ becomes $\sum_{\omega \in OT} \nu(\omega)^{-1} \cdot/\cdot$.

For the computation of the Lie derivative of $B(c, y)$ we have to differentiate the elementary differential $F(\theta)(y(t))$ with respect to $t$. Using Leibniz' rule, this yields $|\theta|$ terms, one for every vertex of $\theta$. Then we insert the series $B(b, y(t))$ for $h\dot{y}(t)$. This means that all the trees $\delta$ appearing in $B(b, y(t))$ are attached with a new branch to the distinguished vertex. Written out as formulas, this gives

$$h\frac{d}{dt} B(c, y(t)) = \sum_{\theta \in OT \cup \{\emptyset\}} \frac{h^{|\theta|} c(\theta)}{\nu(\theta)\, \sigma(\theta)} \sum_{\gamma} \sum_{\delta \in OT} \frac{h^{|\delta|} b(\delta)}{\nu(\delta)\, \sigma(\delta)} F(\theta \circ_\gamma \delta)(y(t)),$$

where $\sum_\gamma$ is a sum over all vertices of $\theta$, and $\theta \circ_\gamma \delta$ is the ordered tree obtained when attaching the root of $\delta$ with a new branch to $\gamma$ (see Fig. 9.1). We choose one of the $n(\gamma)+1$ possibilities of doing this, where $n(\gamma)$ denotes the number of upwards leaving branches of $\theta$ at the vertex $\gamma$. We now collect the terms with equal ordered tree $\omega = \theta \circ_\gamma \delta$, and notice that $\nu(\theta)\sigma(\theta) = \kappa(\theta)$ with $\kappa(\theta)$ given by (III.1.32). This gives

$$h\frac{d}{dt}B(c,y(t)) = \sum_{\omega \in OT} h^{|\omega|} \Big( \sum_{\theta \circ_\gamma \delta = \omega} \frac{c(\theta)\,b(\delta)}{(n(\gamma)+1)\,\kappa(\theta)\,\kappa(\delta)} \Big) F(\omega)(y(t)),$$

where $\sum_{\theta \circ_\gamma \delta = \omega}$ is over all triplets $(\theta, \gamma, \delta)$ such that $\theta \circ_\gamma \delta = \omega$. Because of $\kappa(\omega) = \kappa(\theta)\kappa(\delta)(n(\gamma)+1)$, we obtain

$$h\frac{d}{dt}B(c,y(t)) = \sum_{\omega \in OT} \frac{h^{|\omega|}}{\kappa(\omega)} \Big( \sum_{\theta \circ_\gamma \delta = \omega} c(\theta)\,b(\delta) \Big) F(\omega)(y(t))$$

$$= \sum_{\tau \in T} \frac{h^{|\tau|}}{\sigma(\tau)} \Big( \sum_{\theta \in SP(\tau)} c(\theta)\,b(\tau \setminus \delta) \Big) F(\tau)(y(t)),$$

which proves the statement. $\square$

**Fig. 9.2.** Illustration of the formula (9.6) for an ordered tree with 5 vertices.

Let us illustrate this proof and the formula (9.6) with an ordered tree having 5 vertices. All possible splittings $\omega = \theta \circ_\gamma \delta$ are given in Fig. 9.2. Notice that $\theta$ may be the empty tree $\emptyset$, and that always $|\delta| \geq 1$. We see that the tree $\omega$ is obtained in several ways: (i) differentiation of $F(\emptyset)(y) = y$ and adding $F(\omega)(y)$ as argument, (ii) differentiation of the factor corresponding to the root in $F(\theta)(y) = f''(f,f)(y)$ and adding $F(\mathbf{\int})(y) = (f'f)(y)$, (iii) differentiation of all $f$'s in $F(\theta)(y) = f'''(f,f,f)(y)$ and adding $F(\bullet)(y) = f(y)$, and finally, (iv) differentiation of the factor for the root in $F(\theta)(y) = f''(f'f,f)(y)$ and adding $F(\bullet)(y) = f(y)$. This proves that

$$\partial_b c(\mathbf{V}) = c(\emptyset)b(\mathbf{V}) + c(\mathbf{V})b(\mathbf{\int}) + c(\mathbf{V})b(\bullet) + 2\,c(\mathbf{V})b(\bullet).$$

For the trees up to order 3 the formulas for $\partial_b c$ are:

$$\partial_b c(\bullet) = c(\emptyset)\,b(\bullet)$$
$$\partial_b c(\mathbf{\int}) = c(\emptyset)\,b(\mathbf{\int}) + c(\bullet)\,b(\bullet)$$
$$\partial_b c(\mathbf{V}) = c(\emptyset)\,b(\mathbf{V}) + 2\,c(\mathbf{\int})\,b(\bullet)$$
$$\partial_b c(\mathbf{\int}) = c(\emptyset)\,b(\mathbf{\int}) + c(\bullet)\,b(\mathbf{\int}) + c(\mathbf{\int})\,b(\bullet).$$

The above lemma permits us to get recursion formulas for the coefficients $b(\tau)$ of the modified differential equation (9.4).

**Theorem 9.2.** *If the method $\Phi_h(y)$ is given by (9.1), the functions $f_j(y)$ of the modified differential equation (1.1) satisfy (9.3), where the real coefficients $b(\tau)$ are recursively defined by $b(\emptyset) = 0$, $b(\bullet) = 1$ and*

$$b(\tau) = a(\tau) - \sum_{j=2}^{|\tau|} \frac{1}{j!} \partial_b^{j-1} b(\tau). \qquad (9.7)$$

Here, $\partial_b^{j-1}$ is the $(j-1)$-th iterate of the Lie-derivative $\partial_b$ defined in Lemma 9.1.

*Proof.* The right-hand side of the modified equation (9.4) is the B-series $B(b, \widetilde{y}(t))$ divided by $h$. It therefore follows from an iterative application of Lemma 9.1 that

$$h^j \widetilde{y}^{(j)}(t) = B(\partial_b^{j-1} b, \widetilde{y}(t)),$$

so that by Taylor series expansion $\widetilde{y}(t+h) = y + B\left(\sum_{j\geq 1} \frac{1}{j!} \partial_b^{j-1} b, y\right)$, where $y := \widetilde{y}(t)$. Since we have to determine the coefficients $b(\tau)$ in such a way that $\widetilde{y}(t+h) = \Phi_h(y) = B(a, y)$, a comparison of the two B-series gives $\sum_{j\geq 1} \frac{1}{j!} \partial_b^{j-1} b(\tau) = a(\tau)$. This proves the statement, because $\partial_b^0 b(\tau) = b(\tau)$ for $\tau \in T$, and $\partial_b^{j-1} b(\tau) = 0$ for $j > |\tau|$ (as a consequence of $b(\emptyset) = 0$). $\square$

We present in Table 9.1 the formula (9.7) for trees up to order 3.

**Table 9.1.** Examples of formula (9.7)

| $\tau = \bullet$ | $b(\bullet) = a(\bullet)$ |
|---|---|
| $\tau = \mathord{\vcenter{\hbox{$\scriptstyle\mathrm{I}$}}}$ | $b(\mathrm{I}) = a(\mathrm{I}) - \frac{1}{2} b(\bullet)^2$ |
| $\tau = \mathrm{V}$ | $b(\mathrm{V}) = a(\mathrm{V}) - b(\mathrm{I}) b(\bullet) - \frac{1}{3} b(\bullet)^3$ |
| $\tau = \mathrm{Y}$ | $b(\mathrm{Y}) = a(\mathrm{Y}) - b(\mathrm{I}) b(\bullet) - \frac{1}{6} b(\bullet)^3$ |

## IX.9.2 Extension to Partitioned Systems

Many interesting numerical methods are partitioned (e.g., symplectic Euler method, Störmer/Verlet scheme). We therefore present the results of Lemma 9.1 and Theorem 9.2 also in the context of P-series. We consider the partitioned system

$$\dot{p} = f(p, q), \qquad \dot{q} = g(p, q), \qquad (9.8)$$

where, in view of an application to Hamiltonian systems, we use $(p, q)$ instead of $(y, z)$ for the variables. By Theorem III.2.4 all consistent partitioned Runge-Kutta methods can be written as P-series (cf. Definition III.2.1)

$$\begin{pmatrix} p_1 \\ q_1 \end{pmatrix} = \begin{pmatrix} p_0 \\ q_0 \end{pmatrix} + h \begin{pmatrix} f \\ g \end{pmatrix}_0 + h^2 \begin{pmatrix} a(\mathbf{\mathit{l}})(f_p f) + a(\mathbf{\mathit{f}})(f_q g) \\ a(\mathbf{\mathit{l}})(g_p f) + a(\mathbf{\mathit{f}})(g_q g) \end{pmatrix}_0 + \ldots, \quad (9.9)$$

where the subscript 0 indicates an evaluation at the initial value $(p_0, q_0)$. The first perturbation term of the modified equation (1.1) can therefore be written as

$$\begin{pmatrix} f_2(p,q) \\ g_2(p,q) \end{pmatrix} = \begin{pmatrix} \left(a(\mathbf{\mathit{l}}) - \tfrac{1}{2}\right)(f_p f)(p,q) + \left(a(\mathbf{\mathit{f}}) - \tfrac{1}{2}\right)(f_q g)(p,q) \\ \left(a(\mathbf{\mathit{l}}) - \tfrac{1}{2}\right)(g_p f)(p,q) + \left(a(\mathbf{\mathit{f}}) - \tfrac{1}{2}\right)(g_q g)(p,q) \end{pmatrix}$$

and, in general, one finds

$$\begin{pmatrix} f_j(p,q) \\ g_j(p,q) \end{pmatrix} = \begin{pmatrix} \sum_{\tau \in TP_p, |\tau|=j} \frac{b(\tau)}{\sigma(\tau)} F(\tau)(p,q) \\ \sum_{\tau \in TP_q, |\tau|=j} \frac{b(\tau)}{\sigma(\tau)} F(\tau)(p,q) \end{pmatrix}. \quad (9.10)$$

Hence, the modified equation (1.1) is of the form

$$\begin{pmatrix} \dot{\widetilde{p}} \\ \dot{\widetilde{q}} \end{pmatrix} = \begin{pmatrix} \sum_{\tau \in TP_p} \frac{h^{|\tau|-1}}{\sigma(\tau)} b(\tau) F(\tau)(\widetilde{p}, \widetilde{q}) \\ \sum_{\tau \in TP_q} \frac{h^{|\tau|-1}}{\sigma(\tau)} b(\tau) F(\tau)(\widetilde{p}, \widetilde{q}) \end{pmatrix}, \quad (9.11)$$

where $b(\tau) = 1$ for $|\tau| = 1$, $b(\tau) = a(\tau) - \tfrac{1}{2}$ for $|\tau| = 2$. For $|\tau| > 2$, the coefficients $b(\tau)$ can be obtained recursively from Theorem 9.4 below. The proofs of the following two results are straightforward extensions of those for Lemma 9.1 and Theorem 9.2, and are therefore omitted.

**Lemma 9.3 (Lie-Derivative of P-series).** *Let $b(\tau)$ (with $b(\emptyset_p) = b(\emptyset_q) = 0$) and $c(\tau)$ be the coefficients of two P-series, and let $\bigl(p(t), q(t)\bigr)$ be a formal solution of the differential equation $h(\dot{p}(t), \dot{q}(t))^T = P\bigl(b, (p(t), q(t))\bigr)$, i.e., (9.11). The Lie derivative of the function $P\bigl(c, (p,q)\bigr)$ with respect to the vector field $P\bigl(b, (p,q)\bigr)$ is again a P-series*

$$h \frac{d}{dt} P\bigl(c, (p(t), q(t))\bigr) = P\bigl(\partial_b c, (p(t), q(t))\bigr).$$

*Its coefficients are given by $\partial_b c(\emptyset_p) = \partial_b c(\emptyset_q) = 0$, and for $|\tau| \geq 1$ by*

$$\partial_b c(\tau) = \sum_{\theta \in SP(\tau)} c(\theta)\, b(\tau \setminus \theta), \quad (9.12)$$

*where, analogously to (9.5), $SP(\tau)$ denotes the set of splittings of $\tau \in TP$.* □

In formula (9.12), $\emptyset_p \in SP(\tau)$ defines a splitting only if $\tau \in TP_p$, and $\emptyset_q \in SP(\tau)$ only if $\tau \in TP_q$. We therefore have $\partial_b c(\bullet) = c(\emptyset_p) b(\bullet)$, $\partial_b c(\circ) = c(\emptyset_q) b(\circ)$, and as examples for trees of order 3

$$\begin{aligned}
\partial_b c(\mathbf{\mathit{V}}) &= c(\emptyset_p) b(\mathbf{\mathit{V}}) + 2\, c(\mathbf{\mathit{f}}) b(\circ), \\
\partial_b c(\mathbf{\mathit{V}}) &= c(\emptyset_p) b(\mathbf{\mathit{V}}) + c(\mathbf{\mathit{l}}) b(\circ) + c(\mathbf{\mathit{f}}) b(\bullet).
\end{aligned}$$

**Theorem 9.4.** *If the method* $(p_1, q_1) = \Phi_h(p_0, q_0)$ *can be written as (9.9), the modified differential equation is given by (9.11), where the real coefficients $b(\tau)$ are recursively defined by $b(\emptyset_p) = b(\emptyset_q) = 0$, $b(\tau) = 1$ for $|\tau| = 1$, and*

$$b(\tau) = a(\tau) - \sum_{j=2}^{|\tau|} \frac{1}{j!} \partial_b^{j-1} b(\tau) \quad \text{for} \quad \tau \in TP. \tag{9.13}$$

Here, $\partial_b^{j-1}$ denotes the iterate of the Lie derivative $\partial_b$ defined in Lemma 9.3. □

**Example 9.5.** The symplectic Euler method

$$p_{n+1} = p_n + hf(p_{n+1}, q_n), \qquad q_{n+1} = q_n + hg(p_{n+1}, q_n) \tag{9.14}$$

is a partitioned Runge-Kutta method ($a_{11} = 1$, $\widehat{a}_{11} = 0$, $b_1 = \widehat{b}_1 = 1$) and can therefore be expressed as a P-series (9.9). From Theorem III.2.4 we get its coefficients:

$$a(\tau) = \begin{cases} 1 & \text{if all vertices (different from the root) are black,} \\ 0 & \text{otherwise.} \end{cases}$$

From Theorem 9.4 we can compute the coefficients $b(\tau)$ of the modified equation (9.11). They are given in Table 9.2 for the trees a with black root. Since $a(\tau)$ does not depend on the colour of the root of $\tau$, the same holds for the coefficients $b(\tau)$. Hence, we do not include the values of $b(\tau)$ for trees with a white root.

**Table 9.2.** Coefficients $b(\tau)$ of the modified equation for symplectic Euler (9.14)

| $\tau$ | • | ╱ | ╱ | V | V | V | ⋎ | ⋎ | ⋎ | ⋎ |
|---|---|---|---|---|---|---|---|---|---|---|
| $b(\tau)$ | 1 | 1/2 | −1/2 | 1/6 | −1/3 | 1/6 | 1/3 | −1/6 | −1/6 | 1/3 |

# IX.10 Modified Hamiltonian

We know from Theorem 3.1 that the modified equation of a symplectic method applied to a Hamiltonian system

$$\dot{p} = -H_q(p, q), \qquad \dot{q} = H_p(p, q) \tag{10.1}$$

is again Hamiltonian. Can we get explicit expressions of this modified Hamiltonian? Since (10.1) is a special case of (9.8), all formulas of Sect. IX.9.2 apply.

For methods, whose modified equation is of the form (9.11), we have the following result, which will be the key for obtaining explicit formulas for the modified Hamiltonian. This and the subsequent results of this section were originally obtained by Hairer (1994).

**Theorem 10.1.** *Suppose that for all separable Hamiltonians $H(p,q) = T(p) + U(q)$ the modified vector field (9.11), truncated after an arbitrary power of $h$, is (locally) Hamiltonian. Then, we have*

$$b(u \circ v) + b(v \circ u) = 0 \qquad u \in TP_p, \ v \in TP_q \qquad (10.2)$$

*for trees, where neighbouring vertices have different colours.*

*If it is (locally) Hamiltonian for all $H(p,q)$, then (10.2) holds for all $u \in TP_p$, $v \in TP_q$, and additionally we have*

$$b(\tau) \text{ is independent of the colour of the root of } \tau \in TP. \qquad (10.3)$$

*Proof.* Let $\widetilde{\varphi}_{N,t}(p_0, q_0)$ be the flow of the modified differential equation (9.11), truncated after the $h^{N-1}$ terms. It is symplectic for all $t$, and in particular for $t = h$. As a consequence of Theorem 9.4 (cf. proof of Theorem 9.2) we obtain that $\widetilde{\varphi}_{N,h}(p_0, q_0)$ is a symplectic P-series $P(a_N, (p_0, q_0))$. The coefficients $a_N(\tau)$ are given by (9.13), where $b(\tau)$ is replaced with 0 for $|\tau| > N$. For $u, v \in TP$ with $|u| + |v| = N$ we therefore have

$$b(u \circ v) = a_N(u \circ v) - a_{N-1}(u \circ v).$$

Since $a_N(\tau) = a_{N-1}(\tau)$ for $|\tau| < N$, formula (10.2) is an immediate consequence of Theorem VI.7.1 and of formula (VI.7.5). The remaining statements are obtained in the same way by an additional use of (VI.7.6). □

**Remark 10.2.** Let $G = \{a : TP \to \mathbb{R}\}$ be the analogue of the Butcher group (see Sect. III.1.5) for partitioned methods, and consider the mapping $S : G \to \mathbb{R}$ defined by

$$S(a) = a(u \circ v) + a(v \circ u) - a(u) \cdot a(v).$$

If we denote by $e \in G$ the element corresponding to the identity (i.e., $e(\emptyset_p) = e(\emptyset_q) = 1$ and $e(\tau) = 0$ for $|\tau| \geq 1$), we have for its derivative

$$S'(e)b = b(u \circ v) + b(v \circ u).$$

Hence, coefficient mappings $b(\tau)$ satisfying (10.2) lie in the tangent space at $e(\tau)$ of the symplectic subgroup of $G$ (i.e., $a \in G$ satisfying (VI.7.5) and (VI.7.6)). This is in complete analogy to the fact that Hamiltonian vector fields can be considered as elements of the tangent space at the identity of the group of symplectic diffeomorphisms.

The conditions (10.2) and (10.3) define relations between the coefficients $b(\tau)$ of a Hamiltonian vector field (9.11). Trees, which are related by these conditions, will be called equivalent throughout this section (Hairer 1994).

**Definition 10.3.** We denote by $\sim$ the smallest *equivalence relation* on $TP$ which satisfies the two properties

- $u \sim v$ if $u$ and $v$ are identical with the exception of the colour of the root;
- $u \circ v \sim v \circ u$ for $u \in TP_p$ and $v \in TP_q$.

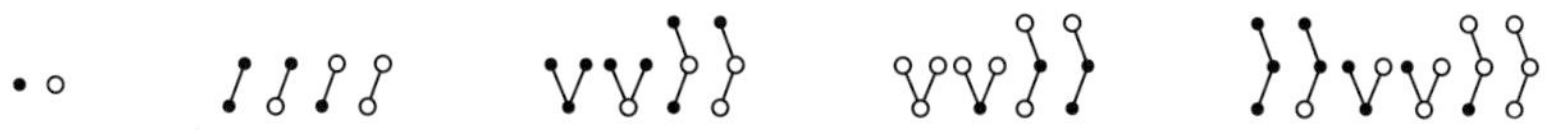

**Fig. 10.1.** Groups of equivalent trees of orders up to three

Equivalent trees of orders up to three are grouped together in Fig. 10.1. We can change the colour of the root, and we can move the root to a neighbouring vertex if it has the opposite colour. In the case of separable Hamiltonians, one has to consider only trees for which neighbouring vertices have different colours. This implies that the first condition of Definition 10.3 is empty. The second condition means that the root can be moved arbitrarily in the tree without changing the equivalence class. For this special situation, equivalence classes have been considered already by Abia & Sanz-Serna (1993) and are named "bicolour (unrooted) trees".

## IX.10.1 Elementary Hamiltonians

Example 3.4 shows that the modified Hamiltonian of the symplectic Euler method is composed of expressions such as $H_p H_q$, $H_{pq}(H_q, H_p)$, etc. The same will be true for general partitioned Runge-Kutta methods. This justifies the introduction of the name "elementary Hamiltonians" for these expressions (cf. Hairer (1994); closely related functions have been considered in the last section of Sanz-Serna & Abia (1991)). In the following definition, the elementary differentials $F(\tau)(p,q)$ correspond to the partitioned system $f(p,q) = -H_q(p,q)$, $g(p,q) = H_p(p,q)$.

**Definition 10.4.** For a given function $H : D \to \mathbb{R}$ (with open $D \subset \mathbb{R}^d \times \mathbb{R}^d$) and for $\tau \in TP$ we define the *elementary Hamiltonian* $H(\tau) : D \to \mathbb{R}$ by

$$H(\bullet)(p,q) = H(\circ)(p,q) = H(p,q)$$

$$H(\tau)(p,q) = (-1)^\delta \frac{\partial^{m+l} H(p,q)}{\partial^m p \, \partial^l q} \Big( F(u_1)(p,q), ..., F(v_1)(p,q), ... \Big)$$

where $\tau = [u_1, \ldots, u_m, v_1, \ldots, v_l]_p$ or $\tau = [u_1, \ldots, u_m, v_1, \ldots, v_l]_q$ with $u_i \in TP_p$ and $v_i \in TP_q$. Furthermore, $\delta = \delta(u_1) + \ldots + \delta(u_m) + \delta(v_1) + \ldots + \delta(v_l)$, where the integers $\delta(u_i)$ and $\delta(v_i)$ count the number of black vertices of $u_i$ and $v_i$, respectively.

The factor $(-1)^\delta$ in the definition of $H(\tau)$ compensates for the minus sign which is present in the function $f(p,q) = -H_q(p,q)$. Examples of elementary Hamiltonians are

$$H(\bullet) = H, \qquad H(\textit{\textbf{f}}) = H_q H_p,$$

$$H(\textit{\textbf{V}}) = H_{pp}(H_q, H_q), \qquad H(\textit{\textbf{V}}) = H_{pq}(H_q, H_p), \qquad H(\textit{\textbf{V}}) = H_{qq}(H_p, H_p).$$

**Lemma 10.5.** *For equivalent trees the elementary Hamiltonians are identical.*

*Proof.* By definition, the elementary Hamiltonians do not depend on the colour of the root. Therefore, all we have to show is that $H(u \circ v)(p,q) = H(v \circ u)(p,q)$ for $u \in TP_p$ and $v \in TP_q$. Let $u = [u_1, \ldots, u_m]_p$, $v = [v_1, \ldots, v_l]_q$, and denote by $\mu$ and $\lambda$ the numbers of trees in $TP_p$ among $u_1, \ldots, u_m$ and $v_1, \ldots, v_l$, respectively. Omitting the obvious argument $(p,q)$ we then have

$$H(u \circ v) = \pm \sum_{j=1}^{d} \frac{\partial^{m+1} H}{\partial^\mu p\, \partial^{m-\mu} q\, \partial q^j}\Big(F(u_1), \ldots, F(u_m)\Big) \cdot F^j(v) \qquad (10.4)$$

where the superscript $j$ denotes the $j$th component. Inserting the expression

$$F^j(v) = \frac{\partial^{l+1} H}{\partial^\lambda p\, \partial^{l-\lambda} q\, \partial p^j}\Big(F(v_1), \ldots, F(v_l)\Big)$$

into (10.4) yields a formula which is completely symmetric with respect to $u$ and $v$. Since the sign of $H(\tau)$ is always positive, we have $H(u \circ v) = H(v \circ u)$. □

In the following we denote the equivalence class of trees $\tau \in TP$ by

$$\langle \tau \rangle = \{u \in TP \mid u \sim \tau\}. \qquad (10.5)$$

**Lemma 10.6.** *For a fixed tree $\tau \in TP$ we have*

$$\begin{aligned}
-\frac{\partial H(\tau)}{\partial q}(p,q) &= \sum_{u \in \langle t \rangle \cap TP_p} \frac{\beta(u)}{\sigma(u)} F(u)(p,q), \\
\frac{\partial H(\tau)}{\partial p}(p,q) &= \sum_{v \in \langle t \rangle \cap TP_q} \frac{\beta(v)}{\sigma(v)} F(v)(p,q),
\end{aligned} \qquad (10.6)$$

*where the coefficients $\beta : TP \to \mathbb{R}$ are defined by the relations (10.2) and (10.3) (b replaced with $\beta$) and by the normalization (cf. Table 10.1)*

$$|\tau| = \sum_{u \in \langle \tau \rangle \cap TP_p} (-1)^{\delta(u)+1} \frac{\beta(u)}{\sigma(u)} = \sum_{v \in \langle \tau \rangle \cap TP_q} (-1)^{\delta(v)} \frac{\beta(v)}{\sigma(v)}. \qquad (10.7)$$

**Table 10.1.** Coefficients $\sigma(\tau)$ and $\beta(\tau)$

| $t$ | • | ○ | ∫∫ | ∫♀ | ∨∨ | ⁀⁀ | ♀∨♀ | ⁀⁀ | ⁀⁀ | ∨∨ | ⁀⁀ |
|---|---|---|---|---|---|---|---|---|---|---|---|
| $\sigma(\tau)$ | 1 | 1 | 1 | 1 | 2 | 1 | 2 | 1 | 1 | 1 | 1 |
| $\beta(\tau)$ | 1 | 1 | $-1$ | 1 | 2 | $-2$ | 2 | $-2$ | 1 | $-1$ | 1 |

*Proof.* We first prove that the derivative of $H(\tau)(p,q)$ with respect to $p$ is a linear combination of elementary differentials $F(u)(p,q)$ with $u \in \langle \tau \rangle \cap TP_p$. In fact, $H(\tau)(p,q)$ consists of $|\tau|$ factors corresponding to the vertices of $\tau$, each of which has to be differentiated by Leibniz' rule. Differentiation of $\frac{\partial^{m+l} H(p,q)}{\partial^m p \, \partial^l q}$ (cf. Definition 10.4) yields $F(\tau)(p,q)$ (eventually, one has to change the colour of the root, so that $\tau \in TP_p$). Before differentiating the other factors, we bring the corresponding vertex down to the root (by adjusting the colour of the root, and by moving the root to a neighbouring vertex with opposite colour). In view of Lemma 10.5 this does not change $H(\tau)(p,q)$, and it shows that a differentiation of the corresponding factor yields $F(u)(p,q)$ with a $u \in \langle \tau \rangle \cap TP_p$. Hence we obtain the formulas stated in (10.6).

Since (10.6) is clearly a Hamiltonian vector field, it follows from Theorem 10.1 that the coefficients $\beta(\tau)$ satisfy (10.2) and (10.3). The condition (10.7) is a consequence of the fact that the above application of Leibniz' rule requires exactly $|\tau|$ differentiations. □

The relations (10.2) and (10.3) together with the normalization (10.7) uniquely define the coefficients $\beta(\tau)$. Their values (and those of $\sigma(\tau)$) are presented in Table 10.1 for all trees up to order 3. We are now able to give the main result of this section.

**Theorem 10.7.** *Consider a numerical method that can be written as a P-series (9.9). If it is symplectic for all $H(p,q)$, then there exists a mapping $c : TP/\sim \, \to \mathbb{R}$, such that the corresponding modified equation is Hamiltonian with*

$$\widetilde{H}(p,q) = H_1(p,q) + h H_2(p,q) + h^2 H_3(p,q) + \ldots,$$

*where*

$$H_k(p,q) = \sum_{\langle \tau \rangle \in TP/\sim, \, |\tau|=k} c(\tau) \, H(\tau)(p,q). \tag{10.8}$$

*Proof.* We apply the method (9.9) to the Hamiltonian system, so that by Theorem 3.1 the modified differential equation is (locally) Hamiltonian. It therefore follows from Theorem 10.1 that the coefficients $b(\tau)$ of (9.11) satisfy (10.2) and (10.3). We now define

$$c(\tau) = b(\tau)/\beta(\tau) \qquad \tau \in TP. \tag{10.9}$$

Since $b(\tau)$ and $\beta(\tau)$ satisfy both the relations (10.2) and (10.3), the coefficient $c(\tau)$ is independent of the representative in the equivalence class $\langle \tau \rangle$, and the sum in (10.8) is well defined. We also get

$$\sum_{u \in TP_p} \frac{h^{|u|-1}}{\sigma(u)} b(u) F(u) = \sum_{\langle \tau \rangle \in TP/\sim} h^{|\tau|-1} c(\tau) \sum_{u \in \langle \tau \rangle \cap TP_p} \frac{\beta(u)}{\sigma(u)} F(\tau)$$

and a similar formula for $TP_p$ replaced with $TP_q$. An application of Lemma 10.6 finally proves the statement. □

**Table 10.2.** Coefficients $a(\tau)$ and $b(\tau)$ for the Störmer/Verlet scheme (Table II.2.1)

| $\tau$ | • | ! | ! | V | V | V | } | } | } | } |
|---|---|---|---|---|---|---|---|---|---|---|
| $a(\tau)$ | 1 | 1/2 | 1/2 | 1/2 | 1/4 | 1/4 | 1/4 | 1/4 | 0 | 1/4 |
| $b(\tau)$ | 1 | 0 | 0 | 1/6 | −1/12 | −1/12 | 1/12 | 1/12 | −1/6 | 1/12 |

If the method (9.9) is known to be symplectic for separable Hamiltonians only, and if it is applied to $H(p,q) = T(p) + U(q)$, the statement of Theorem 10.7 is still valid. In this situation $H(\tau)(p,q)$ is non-vanishing only for trees whose neighbouring vertices have different colours (otherwise it would contain a factor $H_{pq...} = 0$), and the relation (10.9) has to be considered only for these trees.

**Remark 10.8.** Theorem 3.1 proves that the modified differential equation of every symplectic method applied to a Hamiltonian system is *locally Hamiltonian*. Theorem 10.7 gives explicit formulas for the modified Hamiltonian (for methods expressed as P-series). Since they depend only on the derivatives of $H(p,q)$, this modified Hamiltonian is *globally* defined (similar to Theorem 3.2)

**Example 10.9.** Consider the 2-stage Lobatto IIIA - IIIB pair (cf. Table II.2.1), which is the natural extension of the Störmer/Verlet scheme to non-separable problems. We compute the coefficients $a(\tau)$ from Theorem III.2.4, and $b(\tau)$ from Theorem 9.4. The result is given in Table 10.2. Notice that $a(\tau)$ and $b(\tau)$ are both independent of the colour of the root. With $\beta(\tau)$ from Table 10.1 we then compute the coefficients $c(\tau)$ given by (10.9). Theorem 10.7 finally yields

$$\widetilde{H} = H + \frac{h^2}{24}\left(2H_{pp}H_q^2 - H_{qq}H_p^2 + 2H_{pq}H_qH_p\right) + \ldots \tag{10.10}$$

for the modified Hamiltonian. Since the method is symmetric, $\widetilde{H}$ is in even powers of $h$. The next non-vanishing term requires the consideration of trees up to order 5.

### IX.10.2 Characterization of Symplectic P-Series

For the sake of completeness we give a characterization of the property that the vector field (9.11) is Hamiltonian.

**Theorem 10.10.** *For separable Hamiltonians $H(p,q) = T(p) + U(q)$, the (arbitrarily) truncated differential equation (9.11) is Hamiltonian if and only if (10.2) holds for trees where neighbouring vertices have different colours.*

*For general Hamiltonians $H(p,q)$, (9.11) is Hamiltonian if and only if (10.2) and (10.3) hold for all trees.*

*Proof.* The "only if" part follows from Theorem 10.1. We prove the "if" part for general Hamiltonians. The case of separable Hamiltonians can be treated in the same way. If $b(\tau)$ satisfies (10.2) and (10.3), the coefficients $c(\tau)$ of (10.9) are independent of the representative of $\langle\tau\rangle$. The proof of Theorem 10.7 therefore implies that (9.11) is Hamiltonian. □

A complete characterization of symplectic P-series in terms of their coefficients $a(\tau)$ is the content of the next theorem.

**Theorem 10.11.** *A P-series (9.9) defines a symplectic transformation for separable Hamiltonians $H(p,q) = T(p) + U(q)$, if and only if (VI.7.5) holds for trees where neighbouring vertices have different colours.*

*It is symplectic for all (separable and non separable) Hamiltonians, if and only if (VI.7.5) and (VI.7.6) hold for all trees.*

*Proof.* The "only if" part is nothing other than Theorem VI.7.1. For the proof of the "if" part, assume that (VI.7.5) and (VI.7.6) hold. We shall prove below that this implies (10.2) and (10.3) for the coefficients $b(\tau)$ obtained from the corresponding modified differential equation. By Theorem 10.10, this then implies that (9.11) is Hamiltonian, so that the P-series is symplectic.

To prove (10.2) and (10.3), we proceed by induction on the order of the trees. Suppose that both relations hold, whenever the trees involved are of order $\leq N$. This implies that the differential equation (9.11), truncated at $|\tau| \leq N$, is Hamiltonian (cf. Theorem 10.10). Hence, the corresponding P-series (9.9), denoted by $P(a_N, (p_0, q_0))$, is symplectic and satisfies (VI.7.5) and (VI.7.6) by Theorem VI.7.1. The coefficients $a_N(\tau)$ are given by (9.13), where $b(\tau)$ is replaced with 0 for $|\tau| > N$ (cf. proof of Theorem 10.1). We therefore have

$$b(\tau) = a(\tau) - a_N(\tau) \quad \text{for} \quad |\tau| = N+1.$$

Since $a_N(\tau) = a(\tau)$ for $|\tau| \leq N$, the relations (10.2) and (10.3) follow from the corresponding formulas (VI.7.5) and (VI.7.6) for $a(\tau)$ and $a_N(\tau)$. □

# IX.11 Exercises

1. Change the Maple program of Example 1.1 in such a way that the modified equations for the implicit Euler method, the implicit midpoint rule, or the trapezoidal rule are obtained. Observe that for symmetric methods one gets expansions in even powers of $h$.
2. Write a short Maple program which, for simple methods such as the symplectic Euler method, computes some terms of the modified equation for a two-dimensional system $\dot{p} = f(p,q)$, $\dot{q} = g(p,q)$. Check the modified equations of Example 1.3.

3. Prove that the modified equation of the Störmer/Verlet scheme (I.3.4) applied to $\ddot{y} = g(y)$ is a second order differential equation of the form $\ddot{\widetilde{y}} = g_h(\widetilde{y}, \dot{\widetilde{y}})$ with initial values given by $\widetilde{y}(0) = y_0$ and $\dot{\widetilde{y}}(0)$ such that $\widetilde{y}(h) = y_1$ holds.
   *Hint.* Taylor expansion shows that for a smooth function $\widetilde{y}(t)$ satisfying $\widetilde{y}(t) = y_n$ we have
   $$\left(1 + \frac{h^2}{12} D^2 + \frac{h^4}{360} D^4 + \ldots\right) \ddot{\widetilde{y}}(t) = g(\widetilde{y}(t)),$$
   where $D$ represents differentiation with respect to time.
   *Warning.* In general, we do not have that $\dot{\widetilde{y}}(t_n) = \dot{y}_n$.

4. Prove that for $\rho$-reversible differential equations the elementary differentials satisfy
   $$F(\tau)(\rho y) = (-1)^{|\tau|} \rho F(\tau)(y).$$
   Use this to give an alternative proof of Theorem 2.3 for the case that the method is symmetric and can be expressed as a $B$-series.

5. Find a first integral of the truncated modified equation for the symplectic Euler method and the Lotka-Volterra problem (Example 1.3).
   *Hint.* With the transformation $p = \exp P$, $q = \exp Q$ you will get a Hamiltonian system.
   *Result.* $\widetilde{I}(p,q) = I(p,q) - h\big((p+q)^2 - 8p - 10q + 2\ln p + 8\ln q\big)/4$.

6. Compute $\partial_b c(\tau)$ for the tree $\tau = [[\tau], \tau]$ of order 4.

7. For the implicit mid-point rule compute the coefficients $a(\tau)$ of the expansion (9.1), and also a few coefficients $b(\tau)$ of the modified equation.
   *Result.* $a(\tau) = 2^{1-|\tau|}$, $b(\bullet) = 1$, $b(\mathcal{J}) = 0$, $b(\tau) = a(\tau) - 1/\gamma(\tau)$ for $|\tau| = 3$.

8. Check the formulas of Table 9.1.

9. Consider a differential equation $\dot{y} = f(y)$ with a divergence-free vector field, and apply a volume-preserving integrator. Show that every truncation of the modified equation has again a divergence-free vector field.
   *Hint.* Adapt the proof by induction of Theorems 2.3 and 3.1.

10. Consider explicit 2-stage Runge-Kutta methods of order 2, applied to the pendulum problem $\dot{q} = p$, $\dot{p} = -\sin q$. With the help of Exercise 2 compute $f_3(p,q)$ of the modified differential equation. Is there a choice of the free parameter $c_2$, such that $f_3(p,q)$ is a Hamiltonian vector field?

11. Find at least two linear transformations $\rho$ for which the Kepler problem (I.2.2), written as a first order system, is $\rho$-reversible.

12. Consider the Kepler problem (I.2.2), written as a Hamiltonian system (I.1.10). Find constants $M$ and $R$ such that (7.2) holds for all $(p,q) \in \mathbb{R}^4$ satisfying
    $$\|p\| \leq 2 \quad \text{and} \quad 0.8 \leq \|q\| \leq 1.2.$$

13. (Ge & Marsden 1988). Consider a Hamiltonian system with one degree of freedom, i.e., $d = 1$. Prove that, if a numerical method $\Phi_h(y_0)$ is symplectic and if it preserves the Hamiltonian exactly, then it satisfies $\Phi_h(y) = \varphi_{\alpha(h,y)}(y)$, where $\alpha(h,y) = h + \mathcal{O}(h^2)$ is a formal series in $h$.

# Chapter X.
# Hamiltonian Perturbation Theory and Symplectic Integrators

> Perturbation theory is in fact an outgrowth of the necessity to determine the orbits with ever greater accuracy. This problem can be solved today, but in what is for the theoretician a rather disappointing way. With modern calculating machines, one is now able to compute directly results even more accurately than those provided by perturbation theory.
> (J. Moser 1978)

> ... allows computer prediction of planetary positions far more accurate (by brute computation) than anything provided by classical perturbation theory. In a very real sense, one of the most exalted of human endeavors, going back to the priests of Babylon and before, has been taken over by the machine. (S. Sternberg 1969)

In this chapter we study the long-time behaviour of symplectic integrators, combining backward error analysis and the perturbation theory of integrable Hamiltonian systems.

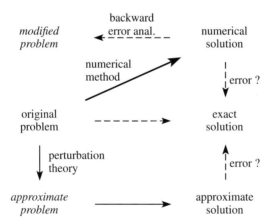

During the 18th and 19th centuries, scientists struggled for the integration of complicated problems of dynamics, with the main aim of solving them analytically by "quadrature". But only few problems could be treated successfully in this way. In cases where the original problem could not be solved, much effort was put into replacing it by an integrable *approximate problem*, by using and developing perturbation theory. Thereby, a rich arsenal of very ingenuous theories has been discovered since the 19th century.

In the 1960s and 1970s, the enormous progress of "calculating machines" and numerical software allowed many of the original problems to be solved with extreme accuracy, so that for the first time numerical integration methods superseded analytical perturbation methods in the computations of celestial mechanics (see the above citations). Since then, the further increase in computing speed has allowed problems to be treated on larger and larger time scales, where huge amounts of errors are accumulated and need to be understood and controlled. In the spirit of backward error analysis, these numerical errors are interpreted as those of a *modified problem*, for the study of which perturbation theory is once again the appropriate tool.

## X.1 Completely Integrable Hamiltonian Systems

Integrable Hamiltonian systems were originally of interest because their equations of motion can be solved analytically. Their interest in the present context lies in the fact that their flow is simply uniform motion on a Cartesian product of circles and straight lines in suitable coordinates, and that many physical systems can be viewed as perturbations of integrable systems.

### X.1.1 Local Integration by Quadrature

> M. Liouville a fait voir qu'il fallait que toutes les combinaisons $(\alpha, \beta)$ des intégrales trouvées fussent nulles. (E. Bour 1855)

One of the great dreams of 18th and 19th century analytical mechanics was to solve the equations of motion of mechanical systems by "quadrature", that is, using only evaluations and inversions of functions and calculating integrals of known functions. In this spirit, Newton's (1687) equations of motion of Kepler's two-body problem were solved by Joh. Bernoulli (1710) and Newton (1713), see Sect. I.2.1. Euler's (1760) solution of the problem of the attraction of a particle by two fixed centres, and Lagrange's (1766) study of motion of a particle in a field with one attracting centre and under an additional constant force were among the important achievements of the 18th century. The three-body problem, however, resisted all efforts aiming at an integration by quadrature, and though it continued to do so, this problem spurred the development of extremely useful mathematical theories of a much wider scope throughout the 19th century, from Poisson to Poincaré via Hamilton, Jacobi, Liouville, to name but a few of the most eminent mathematicians contributing to analytical mechanics.

Consider the Hamiltonian system

$$\dot{p} = -\frac{\partial H}{\partial q}(p, q), \quad \dot{q} = \frac{\partial H}{\partial p}(p, q), \tag{1.1}$$

with $d$ degrees of freedom: $(p,q) \in \mathbb{R}^d \times \mathbb{R}^d$. We try to find a symplectic transformation $(p,q) \mapsto (x,y)$, such that the system has a more amenable form in the new coordinates. In particular, this is the case if the Hamiltonian expressed in the new variables,

$$H(p,q) = K(x),  \qquad (1.2)$$

does not depend on $y$. Since $\frac{\partial K}{\partial y} \equiv 0$, the transformed system then becomes (recall the conservation of the Hamiltonian form of the differential equations under symplectic transformations, Theorem VI.2.8)

$$\dot{x} = 0, \quad \dot{y} = \omega(x), \qquad (1.3)$$

with $\omega(x) = \frac{\partial K}{\partial x}(x)$. This is readily integrated:

$$x(t) = x_0, \quad y(t) = y_0 + \omega(x_0)t.$$

As we recall from Sect. VI.5, a symplectic transformation $(p,q) \mapsto (x,y)$ can be constructed via a *generating function* $S(x,q)$ by the equations

$$y = \frac{\partial S}{\partial x}(x,q), \quad p = \frac{\partial S}{\partial q}(x,q). \qquad (1.4)$$

If $(p_0, q_0)$ and $(x_0, y_0)$ are related by (1.4), and if $\partial^2 S / \partial x \partial q$ is invertible at $(x_0, q_0)$, then the equations (1.4) define a symplectic transformation between neighbourhoods of $(p_0, q_0)$ and $(x_0, y_0)$.

The equation (1.2) together with the second equation of (1.4) give a partial differential equation for $S$, the *Hamilton-Jacobi equation*

$$H\left(\frac{\partial S}{\partial q}(x,q), q\right) = K(x).$$

If $S(x,q)$ is a solution of such an equation (for some function $K$), then (1.3) shows that $x_i = F_i(p,q)$ ($i = 1, \ldots, d$) as given implicitly by the second equation of (1.4), are first integrals of the Hamiltonian system (1.1). Moreover, these functions $F_i$ are *in involution*, which means that their Poisson brackets vanish pairwise:

$$\{F_i, F_j\} = 0, \quad i,j = 1, \ldots, d.$$

This is an immediate consequence of the definition $\{F, G\} = \nabla F^T J^{-1} \nabla G$ of the Poisson bracket and of the symplecticity of the transformation (the left upper block of $J^{-1}$ is 0).

Conversely, it was realized by Bour (1855) and Liouville (1855) that a Hamiltonian system having $d$ first integrals in involution can *locally* be transformed to the form (1.3) by "quadrature". This observation is based on the following completion result and its proof.

**Lemma 1.1 (Liouville Lemma).** *Let $F_1, \ldots, F_d$ be smooth real-valued functions, defined in a neighbourhood of $(p_0, q_0) \in \mathbb{R}^d \times \mathbb{R}^d$. Suppose that these functions are in involution (i.e., all Poisson brackets $\{F_i, F_j\} = 0$), and that their gradients are linearly independent at $(p_0, q_0)$. Then, there exist smooth functions $G_1, \ldots, G_d$, defined on some neighbourhood of $(p_0, q_0)$, such that*

$$(F_1, \ldots, F_d, G_1, \ldots, G_d) : (p, q) \mapsto (x, y) \quad \text{is a symplectic transformation.}$$

*Proof.* Let $F = (F_1, \ldots, F_d)^T$. The linear independence of the gradients $\nabla F_i$ implies that there are $d$ columns of the $d \times 2d$ Jacobian $\partial F/\partial(p,q)$ that form an invertible $d \times d$ submatrix. After some suitable symplectic transformations (see Exercise 1) we may assume without loss of generality that $F_p = \partial F/\partial p$ is invertible. By the implicit function theorem, we can then locally solve $x = F(p, q)$ for $p$:

$$p = P(x, q) \quad \text{with partial derivatives} \quad P_x = F_p^{-1}, \ P_q = -F_p^{-1} F_q \ .$$

The condition that the $F_i$ are in involution, reads in matrix notation

$$F_p F_q^T - F_q F_p^T = 0 \ .$$

Multiplying this equation with $F_p^{-1}$ from the left and with $F_p^{-T}$ from the right, we obtain

$$-P_q^T + P_q = 0 \ ,$$

so that $P_q = \partial P/\partial q$ is symmetric. By the Integrability Lemma VI.2.7, $P(x, q)$ is thus locally the gradient with respect to $q$ of some function $S(x, q)$ (which is constructed by quadrature). Moreover, $\frac{\partial^2 S}{\partial x \partial q} = P_x = F_p^{-1}$ is invertible. The equations (1.4) define a symplectic transformation $(p, q) \mapsto (x, y)$, and by construction $x = F(p, q)$. □

If, in a Hamiltonian system with $d$ degrees of freedom, we can find $d$ independent first integrals in involution $H = F_1, F_2, \ldots, F_d$, then Lemma 1.1 yields a symplectic change of coordinates, constructed by quadrature, which transforms (1.1) *locally* to (1.2) with $K(x_1, \ldots, x_d) = x_1$.

**Example 1.2.** Consider the Hamiltonian of motion in a central field,

$$H = \tfrac{1}{2}(p_1^2 + p_2^2) + V(r) \quad \text{for} \quad r = \sqrt{q_1^2 + q_2^2} \ ,$$

with a potential $V(r)$ that is defined and smooth for $r > 0$. The Kepler problem corresponds to the special case $V(r) = -1/r$, and the perturbed Kepler problem to $V(r) = -1/r - \mu/(3r^3)$. Changing to polar coordinates (see Exercise VI.4)

$$\begin{pmatrix} q_1 \\ q_2 \end{pmatrix} = \begin{pmatrix} r \cos \varphi \\ r \sin \varphi \end{pmatrix}, \quad \begin{pmatrix} p_r \\ p_\varphi \end{pmatrix} = \begin{pmatrix} \cos \varphi & \sin \varphi \\ -r \sin \varphi & r \cos \varphi \end{pmatrix} \begin{pmatrix} p_1 \\ p_2 \end{pmatrix}, \quad (1.5)$$

this becomes

$$H(p_r, p_\varphi, r, \varphi) = \frac{1}{2}\left(p_r^2 + \frac{p_\varphi^2}{r^2}\right) + V(r).$$

The system has the angular momentum $L = p_\varphi$ as a first integral, since $H$ does not depend on $\varphi$. Clearly, $\{H, L\} = 0$ everywhere. The gradients of $H$ and $L$ are linearly independent unless both $p_r = 0$ and $p_\varphi^2 = r^3 V'(r)$. By inserting $p_\varphi^2 = 2r^2(H - V(r))$ and eliminating $r$ this becomes a condition of the form $\alpha(H, L) = 0$, which for the Kepler problem reads explicitly $L^2(1 + 2HL^2) = 0$. The conditions of Lemma 1.1 are thus satisfied on the domain

$$M = \{(p_r, p_\varphi, r, \varphi)\,;\, r > 0,\; \alpha(H, L) \neq 0\}.$$

The equations $x_1 = H = \frac{1}{2}(p_r^2 + p_\varphi^2/r^2) + V(r)$, $x_2 = L = p_\varphi$ can be solved for

$$p_r = \pm\sqrt{2(H - V(r)) - L^2/r^2},\quad p_\varphi = L,$$

and $p_r = \partial S/\partial r$, $p_\varphi = \partial S/\partial \varphi$ with

$$S(H, L, r, \varphi) = L\varphi \pm \int_{r_0}^{r}\sqrt{2(H - V(\rho)) - L^2/\rho^2}\, d\rho.$$

The conjugate variables are

$$y_1 = \frac{\partial S}{\partial H} = \pm\int_{r_0}^{r}\frac{1}{\sqrt{2(H - V(\rho)) - L^2/\rho^2}}\, d\rho, \tag{1.6}$$

$$y_2 = \frac{\partial S}{\partial L} = \varphi \mp \int_{r_0}^{r}\frac{L/\rho^2}{\sqrt{2(H - V(\rho)) - L^2/\rho^2}}\, d\rho. \tag{1.7}$$

This defines (locally) the transformation $(p_r, p_\varphi, r, \varphi) \mapsto (x_1, x_2, y_1, y_2)$. In these variables, the equations of motion read $\dot{x}_1 = 0$, $\dot{x}_2 = 0$, $\dot{y}_1 = 1$, $\dot{y}_2 = 0$. Over any time interval where $p_r(t)$ does not change sign, solutions therefore satisfy

$$t_1 - t_0 = \pm\int_{r(t_0)}^{r(t_1)}\frac{1}{\sqrt{2(H - V(\rho)) - L^2/\rho^2}}\, d\rho, \tag{1.8}$$

$$\varphi(t_1) - \varphi(t_0) = \pm\int_{r(t_0)}^{r(t_1)}\frac{L/\rho^2}{\sqrt{2(H - V(\rho)) - L^2/\rho^2}}\, d\rho. \tag{1.9}$$

## X.1.2 Completely Integrable Systems

Lemma 1.1 appears as a powerful tool for an explicit solution by quadrature. However, because of its purely local nature this lemma does not tell us anything about the dynamics of the system. This was not a concern at Liouville's time, but the first rigorous non-integrability results by Poincaré (1892) put a definite end to the hope of being eventually able to construct explicit analytic solutions of most equations of motion by quadrature, and shifted the interest to understanding the *global*, qualitative behaviour of dynamical systems.

Lemma 1.1 can be globalized by a procedure similar to analytic continuation if the conditions of the following definition are satisfied.

**Definition 1.3.** A Hamiltonian system with Hamiltonian $H : M \to \mathbb{R}$ ($M$ an open subset of $\mathbb{R}^d \times \mathbb{R}^d$) is called *completely integrable* if there exist smooth functions $F_1 = H, F_2, \ldots, F_d : M \to \mathbb{R}$ with the following properties:
1) $F_1, \ldots, F_d$ are in involution (i.e., all $\{F_i, F_j\} = 0$) on $M$.
2) The gradients of $F_1, \ldots, F_d$ are linearly independent at every point of $M$.
3) The solution trajectories of the Hamiltonian systems with Hamiltonian $F_i$ ($i = 1, \ldots, d$) exist for all times and remain in $M$.

Obviously, all the Hamiltonian systems with Hamiltonian $F_i$ ($i = 1, \ldots, d$) are then completely integrable, and so there will be no mathematical reason to further distinguish $H = F_1$. We note that condition (1) of Definition 1.3 implies that all $F_j$ are first integrals of the Hamiltonian system with Hamiltonian $F_i$, and that the flows $\varphi_t^{[i]}$ of these Hamiltonian systems commute: $\varphi_t^{[i]} \circ \varphi_s^{[j]} = \varphi_s^{[j]} \circ \varphi_t^{[i]}$ for all $i, j$ and all $t, s \in \mathbb{R}$; see Lemma VII.2.8.

For $x = (x_i) \in \mathbb{R}^d$ we define the level set

$$M_x = \{(p, q) \in M \,;\, F_i(p, q) = x_i \text{ for } i = 1, \ldots, d\}. \tag{1.10}$$

**Theorem 1.4.** *Suppose that $F_1, \ldots, F_d : M \to \mathbb{R}$ satisfy the conditions of Definition 1.3. Assume that $M_x$ is connected (and non-empty) for all $x$ in a neighbourhood of $x_0 \in \mathbb{R}^d$. Then, on some neighbourhood $B$ of $x_0$, there exists a symplectic and surjective mapping*

$$e : B \times \mathbb{R}^d \to \bigcup_{x \in B} M_x : (x, y) \mapsto (p, q) \in M_x$$

*that linearizes, for all $i = 1, \ldots, d$, the flow $\varphi_t^{[i]}$ of the system with Hamiltonian $F_i$:*

$$\text{if } (p, q) = e(x, y), \quad \text{then} \quad \varphi_t^{[i]}(p, q) = e(x, y + te_i), \tag{1.11}$$

*where $e_i = (0, \ldots, 1, \ldots, 0)^T$ is the $i$th unit vector of $\mathbb{R}^d$.*

Since $e$ is symplectic, $e$ is a local diffeomorphism. Its local inverse is a transformation as constructed in Lemma 1.1. However, $(p, q)$ can have countably many discretely lying pre-images $(x, y)$, so that $e^{-1}$ becomes a multi-valued function. The situation is analogous to that of the complex exponential and logarithm. The following example illustrates that this analogy is not incidental.

**Example 1.5.** Consider the harmonic oscillator, i.e., $d = 1$ and $H(p, q) = \frac{1}{2}(p^2 + q^2)$. For $x = \frac{1}{2}r^2$, we have $e(x, y) = (r \cos y, r \sin y)$.

*Proof of Theorem 1.4.* We fix $(p_0, q_0) \in M_{x_0}$, and in a neighbourhood $U$ of $(p_0, q_0)$ we consider a symplectic transformation

$$\ell = (F_1, \ldots, F_d, G_1, \ldots, G_d) : (p, q) \mapsto (x, y)$$

as constructed in Lemma 1.1. We have $\ell(p_0, q_0) = (x_0, y_0)$ where we may assume $y_0 = 0$. To every $v = (v_i) \in \mathbb{R}^d$ we associate the Hamiltonian

$$F_v = v_1 F_1 + \ldots + v_d F_d$$

and note that, because of the commutativity of the flows $\varphi_t^{[i]}$, the flow of the system with Hamiltonian $F_v$ equals

$$\varphi_{tv} = \varphi_{tv_1}^{[1]} \circ \ldots \circ \varphi_{tv_d}^{[d]}.$$

In the neighbourhood $U$ of $(p_0, q_0)$, the system with Hamiltonian $F_v$ is transformed under the symplectic mapping $\ell$ to

$$\dot{x} = 0, \quad \dot{y} = v.$$

Hence, the following diagram commutes for $(p, q) \in U$ and for sufficiently small $tv$:

$$\begin{array}{ccc} (p,q) & \longrightarrow & \varphi_{tv}(p,q) \\ \downarrow \ell & & \uparrow \ell^{-1} \\ (x,y) & \longrightarrow & (x, y+tv) \end{array} \qquad (1.12)$$

We now construct $e$ by extending this diagram to arbitrary $tv$:

$$\begin{array}{ccc} (p,q) & \longrightarrow & \varphi_y(p,q) \\ & & \uparrow \ell^{-1} \\ (x,0) & \longleftarrow & (x,y) \end{array} \qquad (1.13)$$

That is, we define on $B \times \mathbb{R}^d$ (with $B$ a neighbourhood of $x_0$ on which $\ell^{-1}(x, 0)$ is defined)

$$e(x, y) = \varphi_y(\ell^{-1}(x, 0)).$$

For $(x, y)$ near some fixed $(\widehat{x}, \widehat{y})$, we have by (1.12) with $0$ and $y - \widehat{y}$ instead of $y$ and $tv$ that

$$e(x, y) = \varphi_{\widehat{y}}(\ell^{-1}(x, y - \widehat{y})),$$

which shows that $e$ is symplectic, being locally the composition of symplectic transformations. The property (1.11) is obvious from the definition of $e$ and from the commutativity of the flows $\varphi_t^{[i]}$. Since $\ell^{-1}(x, 0) \in M_x$ and $M_x$ is invariant under the flows $\varphi_t^{[i]}$, we have $e(x, y) \in M_x$ for all $(x, y)$.

It remains to show that $e : \{x\} \times \mathbb{R}^d \to M_x$ is surjective for every $x$ near $x_0$. Let $(\widehat{p}, \widehat{q})$ be an arbitrary point on $M_x$. By assumption, there exists a path on $M_x$ connecting $\ell^{-1}(x, 0)$ and $(\widehat{p}, \widehat{q})$. Moreover, by (1.12) and by the compactness of the path, there is a $\delta > 0$ such that, for every $(p, q)$ on this path, the mapping $y \mapsto \varphi_y(p, q)$ is a diffeomorphism between the ball $\|y\| < \delta$ and a neighbourhood of $(p, q)$ on $M_x$. Therefore, $(\widehat{p}, \widehat{q})$ can be reached from $\ell^{-1}(x, 0)$ by a finite composition of maps:

$$(\widehat{p}, \widehat{q}) = \varphi_{y^{(m)}} \circ \ldots \circ \varphi_{y^{(1)}}(\ell^{-1}(x, 0)) = \varphi_{\widehat{y}}(\ell^{-1}(x, 0)) = e(x, \widehat{y}),$$

where $\widehat{y} = y^{(1)} + \ldots + y^{(m)}$ once again by the commutativity of the flows $\varphi_t^{[i]}$. □

**Illustration of the Liouville Transform.** We illustrate the above construction at a simple example, the pendulum (I.1.12) with Hamiltonian $H = p^2/2 - \cos q$. The first coordinate is $x = H(p,q)$, a first integral. The second coordinate $y$ is, following (1.13), the time $t$ which is necessary to reach the point $(p,q)$ from an initial line, which we assume at $q = 0$. Then we have (Fig. 1.1 left) $dp\, dq = dH\, dt$ (because

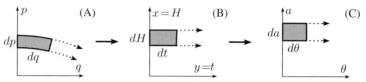

**Fig. 1.1.** Liouville and action-angle coordinate transforms

of $dq = H_p\, dt$ and $dH = H_p\, dp$). We see again that we have area preservation, because the symplecticity of the flow preserves this property for all times. This symplectic change of coordinates $(p,q) \mapsto (x,y)$ is illustrated in Fig. 1.2, which transforms the problem (A) to a much simpler form (B) with uniform horizontal movement.

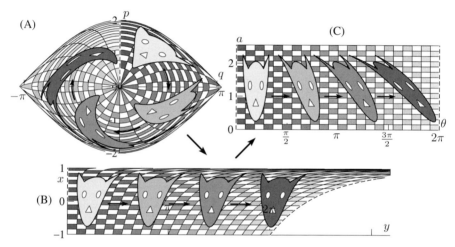

**Fig. 1.2.** Liouville and action-angle coordinates illustrated at the pendulum problem.

We are not yet completely satisfied, however, because the orbits have periods $g = g(H)$ which are not all the same. We therefore append a *second* transform by putting $\theta = \frac{2\pi}{g} \cdot t$ (see picture (C) in Fig. 1.1 and Fig. 1.2), which forces all periods into a Procrustean bed of length $2\pi$. Area preservation $da\, d\theta = dH\, dt$ now requires that $2\pi\, da = g(H)\, dH$, which is a differential equation between $a$ and $H$. The new coordinates $(a, \theta)$ are the *action-angle variables* and we see that they transform the phase space into $D \times \mathbb{T}^1$ where $D \subset \mathbb{R}^1$. We again have horizontal movement, but this time the speed depends on $a$. The general existence for completely integrable systems will be proved in Theorem 1.6 below.

## X.1.3 Action-Angle Variables

> We show here that, under the hypotheses of Liouville's theorem, we can find symplectic coordinates $(\mathbf{I}, \boldsymbol{\varphi})$ such that the first integrals $\mathbf{F}$ depend only on $\mathbf{I}$, and $\boldsymbol{\varphi}$ are angular coordinates on the torus $M_{\mathbf{f}}$.
>
> (V.I. Arnold 1989, p. 279)

We are now in the position to prove the main result of this section, which establishes a symplectic change of coordinates to the so-called *action-angle variables*, such that $d$ first integrals of a completely integrable system depend only on the actions, and the angles are defined globally mod $2\pi$ (provided the level sets of the first integrals are compact). This is known as the Arnold-Liouville theorem; cf. Arnold (1963, 1989), Arnold, Kozlov & Neishtadt (1997; Ch. 4, Sect. 2.1), Jost (1968). Here and in the following,

$$\mathbb{T}^d = \mathbb{R}^d/2\pi\mathbb{Z}^d = \{(\theta_1 \bmod 2\pi, \ldots, \theta_d \bmod 2\pi)\, ;\, \theta_i \in \mathbb{R}\}$$

denotes the standard $d$-dimensional torus.

**Theorem 1.6 (Arnold-Liouville Theorem).** *Let $F_1, \ldots, F_d : M \to \mathbb{R}$ be first integrals of a completely integrable system as in Definition 1.3. Suppose that the level sets $M_x$ (see (1.10)) are compact and connected for all $x$ in a neighbourhood of $x_0 \in \mathbb{R}^d$. Then, there are neighbourhoods $B$ of $x_0$ and $D$ of $0$ in $\mathbb{R}^d$ such that the following holds:*
*(i) For every $x \in B$, the level set $M_x$ is a $d$-dimensional torus that is invariant under the flow of the system with Hamiltonian $F_i$ ($i = 1, \ldots, d$).*
*(ii) There exists a bijective symplectic transformation*

$$\psi : D \times \mathbb{T}^d \to \bigcup_{x \in B} M_x \subset \mathbb{R}^d \times \mathbb{R}^d : (a, \theta) \mapsto (p, q)$$

*such that $(F_i \circ \psi)(a, \theta)$ depends only on $a$, i.e.,*

$$F_i(p, q) = f_i(a) \quad \text{for } (p, q) = \psi(a, \theta) \qquad (i = 1, \ldots, d)$$

*with functions $f_i : D \to \mathbb{R}$.*

The variables $(a, \theta) = (a_1, \ldots, a_d, \theta_1 \bmod 2\pi, \ldots, \theta_d \bmod 2\pi)$ are called *action-angle variables*.

**Remark 1.7.** If the level sets $M_x$ are not compact, then the proof of Theorem 1.6 shows that $M_x$ is diffeomorphic to a Cartesian product of circles and straight lines $\mathbb{T}^k \times \mathbb{R}^{d-k}$ for some $k < d$, and there is a bijective symplectic transformation $(a, \theta) \mapsto (p, q)$ between $D \times (\mathbb{T}^k \times \mathbb{R}^{d-k})$ and a neighbourhood $\bigcup \{M_x : x \in B\}$ of $M_{x_0}$ such that the first integrals again depend only on $a$.

**Remark 1.8.** If the Hamiltonian is real-analytic, then the proof shows that also the transformation to action-angle variables is real-analytic.

**Proof of Theorem 1.6.** (a) We return to Theorem 1.4. For $x \in B$, we consider the set
$$\Gamma_x = \{y \in \mathbb{R}^d\ ;\ e(x,y) = e(x,0)\}\ .$$
Since $e$ is locally a diffeomorphism, for every fixed $y_0 \in \Gamma_{x_0}$ there exists a unique smooth function $\eta$ defined on a neighbourhood of $x_0$, such that $\eta(x_0) = y_0$ and $\eta(x) \in \Gamma_x$ for $x$ near $x_0$. In particular, $\Gamma_x$ is a discrete subset of $\mathbb{R}^d$. By (1.11), for $y \in \Gamma_x$ we have $e(x, y+v) = e(x,v)$ for all $v \in \mathbb{R}^d$. Therefore, $\Gamma_x$ is a subgroup of $\mathbb{R}^d$, i.e., with $y, v \in \Gamma_x$ also $y + v \in \Gamma_x$ and $-y \in \Gamma_x$. It then follows (see Exercise 4) that $\Gamma_x$ is a grid, generated by $k \leq d$ linearly independent vectors $g_1(x), \ldots, g_k(x) \in \mathbb{R}^d$:
$$\Gamma_x = \{m_1 g_1(x) + \ldots + m_k g_k(x)\ ;\ m_i \in \mathbb{Z}\}\ .$$
We extend $g_1(x), \ldots, g_k(x)$ to a basis $g_1(x), \ldots, g_d(x)$ of $\mathbb{R}^d$. Then, $e$ induces a diffeomorphism
$$\mathbb{T}^k \times \mathbb{R}^{d-k} \to M_x$$
$$(\theta_1, \ldots, \theta_k, \tau_{k+1}, \ldots, \tau_d) \mapsto e\left(x, \sum_{i=1}^{k} \frac{\theta_i}{2\pi} g_i(x) + \sum_{j=k+1}^{d} \tau_j g_j(x)\right).$$
If $M_x$ is compact, then necessarily $k = d$ and $M_x$ is a torus. The above map then becomes the bijection
$$\mathbb{T}^d \to M_x : \theta \mapsto e\left(x, \sum_{i=1}^{d} \frac{\theta_i}{2\pi} g_i(x)\right)\ .$$

(b) Next we show that $g_i(x)$ is the gradient of some function $U_i(x)$. For notational convenience, we omit the subscript $i$ and consider a differentiable function $g$ with
$$e(x, g(x)) = e(x, 0)\ , \qquad x \in B\ ,$$
or equivalently,
$$\ell \circ e(x, g(x)) = (x, 0)\ , \qquad x \in B\ .$$
Differentiating this relation gives (with $I$ the $d$-dimensional identity)
$$A \begin{pmatrix} I \\ g'(x) \end{pmatrix} = \begin{pmatrix} I \\ 0 \end{pmatrix}$$
where $A$ is the Jacobian matrix of $\ell \circ e$ at $(x, g(x))$. We thus have
$$(I\ \ g'(x)^T) A^T J A \begin{pmatrix} I \\ g'(x) \end{pmatrix} = (I\ \ 0) J \begin{pmatrix} I \\ 0 \end{pmatrix} = 0\ .$$
Since $\ell \circ e$ is a symplectic transformation, we have $A^T J A = J$, and hence the above equation reduces to

$$g'(x)^T - g'(x) = 0 .$$

By the Integrability Lemma VI.2.7, there is a function $U$ such that $g(x) = \nabla U(x)$. We may assume $U(x_0) = 0$.

(c) The result of (b) allows us to extend the bijection of (a) to a symplectic transformation. For this, we consider the generating function

$$S(x, \theta) = \sum_{i=1}^{d} \frac{\theta_i}{2\pi} U_i(x) .$$

With $u(x) = (U_1(x), \ldots, U_d(x))$, the mixed second derivative of $S$ is

$$S_{x\theta}(x, \theta) = \frac{1}{2\pi} u_x(x) = \frac{1}{2\pi} \big(g_1(x), \ldots, g_d(x)\big) ,$$

which is invertible because of the linear independence of the $g_i$. The equations

$$a = \frac{\partial S}{\partial \theta} = \frac{1}{2\pi} u(x) , \quad y = \frac{\partial S}{\partial x} = \sum_{i=1}^{d} \frac{\theta_i}{2\pi} g_i(x)$$

define a bijective symplectic transformation (for some neighbourhood $D$ of 0, and possibly with a reduced neighbourhood $B$ of $x_0$)

$$\beta : D \times \mathbb{R}^d \to B \times \mathbb{R}^d : (a, \theta) \mapsto (x, y) = \Big(f(a), \sum_{i=1}^{d} \frac{\theta_i}{2\pi} g_i(f(a))\Big)$$

where $x = f(a)$ is the inverse map of $a = \frac{1}{2\pi} u(x)$. We now define

$$\widehat{\psi} = e \circ \beta : D \times \mathbb{R}^d \to \bigcup_{x \in B} M_x .$$

By construction, this map is smooth and symplectic, and such that $f_i(a) = x_i = F_i(p, q)$ for $(p, q) = \widehat{\psi}(a, \theta)$. It is surjective by Theorem 1.4. By part (a) of this proof, it becomes injective when the $\theta_i$ are taken mod $2\pi$, thus yielding a transformation $\psi$ defined on $D \times \mathbb{T}^d$ with the stated properties. □

## X.1.4 Conditionally Periodic Flows

An immediate and important consequence of Theorem 1.6 is the following.

**Corollary 1.9.** *In the situation of Theorem 1.6, consider the completely integrable system with Hamiltonian $H = F_1$. In the action-angle variables $(a, \theta)$, the Hamiltonian equations become*

$$\dot{a}_i = 0, \quad \dot{\theta}_i = \omega_i(a) \quad (i = 1, \ldots, d)$$

*with $\omega_i(a) = \partial K / \partial a_i(a)$, where $K(a) = H(p, q)$ for $(p, q) = \psi(a, \theta)$.*

The flow of a differential system

$$\dot{\theta} = \omega, \quad \omega = (\omega_i) \in \mathbb{R}^d$$

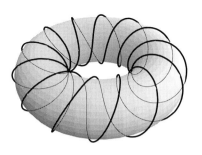

on the torus $\mathbb{T}^d$ is called *conditionally periodic* with *frequencies* $\omega_i$. The flow is periodic if there exist integers $k_i$ such that for any two frequencies the relation $\omega_i/\omega_j = k_i/k_j$ holds. Otherwise, the flow is called *quasi-periodic*. In particular, the latter occurs when the frequencies are rationally independent, or *non-resonant*: the only integers $k_i$ with $k_1\omega_1 + \ldots + k_d\omega_d = 0$ are $k_1 = \ldots = k_d = 0$. For non-resonant frequencies, it is well known (see Arnold (1989), p. 287) that every trajectory $\{\theta(t) : t \in \mathbb{R}\}$ is dense on the torus $\mathbb{T}^d$ and uniformly distributed.

**Example 1.10.** We take up again the example of motion in a central field, Exercise 1.2. For given $H$ and $L$, we now assume that

$$\{r > 0 \,;\, 2(H - V(r)) - L^2/r^2 > 0\} = [r_0, r_1]$$

is a non-empty interval and the derivatives of $2(H - V(r)) - L^2/r^2$ are non-vanishing at $r_0, r_1$. By (1.8) and (1.9), the motion from $r_0$ to $r_1$ and back again takes a time $T$ and runs through an angle $\Phi$ which are given by

$$T = 2 \int_{r_0}^{r_1} \frac{1}{\sqrt{2(H - V(\rho)) - L^2/\rho^2}} \, d\rho, \quad (1.14)$$

$$\Phi = 2 \int_{r_0}^{r_1} \frac{L/\rho^2}{\sqrt{2(H - V(\rho)) - L^2/\rho^2}} \, d\rho. \quad (1.15)$$

Note that $r_0, r_1, T, \Phi$ are functions of $H$ and $L$. The solution is periodic if $\Phi$ is a rational multiple of $2\pi$. This occurs for the Kepler problem, where $\Phi = 2\pi$ and where $T = 2\pi/(-2H)^{3/2}$ (for $H < 0$) depends only on $H$; see Exercise I.5.

We now construct action-angle variables and compute the frequencies of the system. We begin by constructing the mapping $e(x, y)$ as defined by (1.13) for the variables $x = (x_1, x_2) = (H, L)$ and $y = (y_1, y_2)$ of (1.6), (1.7). For a given $(x, y)$, we consider $(x, 0)$ and we fix $(p, q)$ with $p = (p_r, p_\varphi)$ and $q = (r, \varphi)$ such that $\ell(p, q) = (x, 0)$, e.g., by choosing $r = r_0$, $\varphi = 0$, $p_r = 0$, $p_\varphi = L$. The mapping $e(x, y)$ is defined by the flow at time $t = 1$ corresponding to the Hamiltonian

$$F_y = y_1 H + y_2 L = y_1 \left(\tfrac{1}{2}(p_r^2 + p_\varphi^2/r^2) + V(r)\right) + y_2 p_\varphi,$$

i.e., by the solution at $t = 1$ of

$$\begin{aligned} \dot{p}_r &= -y_1 \frac{p_\varphi^2}{r^3} - y_1 V'(r), & \dot{p}_\varphi &= 0 \\ \dot{r} &= y_1 p_r, & \dot{\varphi} &= y_1 \frac{p_\varphi}{r^2} + y_2. \end{aligned} \quad (1.16)$$

## X.1 Completely Integrable Hamiltonian Systems

If we denote the flow of the original system with Hamiltonian $H(p_r, p_\varphi, r, \varphi)$ by $\varphi_t$, then we have

$$e(x, y) = \varphi_{y_1}(0, L, r_0, 0) + (0, 0, 0, y_2)^T$$

with the last component taken modulo $2\pi$. Hence, the values of $y$ satisfying $e(x, y) = e(x, 0)$ are

$$y = m_1 g_1(x) + m_2 g_2(x)$$

with integers $m_1, m_2$ and

$$g_1 = \begin{pmatrix} T \\ -\Phi \end{pmatrix}, \quad g_2 = \begin{pmatrix} 0 \\ 2\pi \end{pmatrix}.$$

We know from the proof of Theorem 1.6 that $g_1$ and $g_2$ are the gradients of functions $U_1(H, L)$ and $U_2(H, L)$, respectively. Clearly, $U_2 = 2\pi L$. The expression for $U_1$ is less explicit. With the construction of the Integrability Lemma VI.2.7, this function is obtained by quadrature, in a neighbourhood of $(H_0, L_0)$, as

$$U_1(H, L) = \int_0^1 \Big( (H - H_0) T(H_0 + s(H - H_0), L_0 + s(L - L_0)) - (L - L_0) \Phi(H_0 + s(H - H_0), L_0 + s(L - L_0)) \Big) ds.$$

(For the Kepler problem, $T = 2\pi/(-2H)^{3/2}$, $\Phi = 0 \mod 2\pi$, and hence $U_1 = 2\pi/\sqrt{-2H}$.) For the action variables we thus obtain

$$a_1 = \frac{1}{2\pi} U_1(H, L), \quad a_2 = L.$$

The angle variables are given by $y = \frac{1}{2\pi}(\theta_1 g_1 + \theta_2 g_2)$, i.e.,

$$\theta_1 = y_1 \frac{2\pi}{T}, \quad \theta_2 = y_2 + y_1 \frac{\Phi}{T}. \tag{1.17}$$

Writing the total energy $H = K(a_1, L)$ if $a_1$ is given by the above formula, we obtain, by differentiation of the identity $2\pi a_1 = U_1(K(a_1, L), L)$,

$$2\pi = \frac{\partial U_1}{\partial H} \frac{\partial K}{\partial a_1}, \quad 0 = \frac{\partial U_1}{\partial H} \frac{\partial K}{\partial a_2} + \frac{\partial U_1}{\partial L}$$

and hence the frequencies

$$\omega_1 = \frac{\partial K}{\partial a_1} = \frac{2\pi}{T}, \quad \omega_2 = \frac{\partial K}{\partial a_2} = \frac{\Phi}{T}. \tag{1.18}$$

## X.1.5 The Toda Lattice – an Integrable System

> Our method is based on the realization that the Toda lattice belongs to a class of evolution equations which can be studied, and in some cases solved, by utilization of a certain associated eigenvalue problem.
>
> (H. Flaschka 1974)

Classical examples of integrable systems from mechanics include Kepler's problem (Newton 1687/1713, Joh. Bernoulli 1710), the planar motion of a point mass attracted by two fixed centres (Euler 1760), Kepler's problem in a homogeneous force field (Lagrange 1766 solved this as the limit of the previous problem when one centre is at infinity), various spinning tops (Euler 1758, Lagrange 1788, Kovalevskaya 1889, Goryachev 1899 and Chaplygin 1901), a number of integrable cases of the motion of a rigid body in a fluid, the motion of point vortices in the plane. We refer to Arnold, Kozlov & Neishtadt (1997) and Kozlov (1983) for interesting accounts of these problems and for further references.

Here we consider the celebrated example of the Toda lattice which was the starting point for a huge amount of work on integrable systems in the last few decades, with fascinating relationships to soliton theory in partial differential equations (most notably the Korteweg-de Vries equation) and to eigenvalue algorithms of Numerical Analysis; see Deift (1996) for an account of these developments.

The Toda lattice (or chain) is a system of particles on a line interacting pairwise with exponential forces. Such systems were studied by Toda (1970) as discrete models for nonlinear wave propagation. The motion is determined by the Hamiltonian

$$H(p,q) = \sum_{k=1}^{n} \left( \frac{1}{2} p_k^2 + \exp(q_k - q_{k+1}) \right). \tag{1.19}$$

Two types of boundary conditions have found particular attention in the literature:

(i) periodic boundary conditions: $q_{n+1} = q_1$;

(ii) put formally $q_{n+1} = +\infty$, so that the term $\exp(q_n - q_{n+1})$ does not appear.

It was found by Hénon, Flaschka and independently Manakov in 1974 that the periodic Toda system is integrable. Moser (1975) then gave a detailed study of the non-periodic case (ii).

Flaschka (1974) introduced new variables

$$a_k = -\tfrac{1}{2} p_k, \qquad b_k = \tfrac{1}{2} \exp\left(\tfrac{1}{2}(q_k - q_{k+1})\right).$$

(Take $b_n = 0$ in case (ii)). Along a solution $(p(t), q(t))$ of the Toda system, the corresponding functions $(a(t), b(t))$ satisfy the differential equations

$$\dot{a}_k = 2(b_k^2 - b_{k-1}^2), \qquad \dot{b}_k = b_k(a_{k+1} - a_k)$$

(with $a_{n+1} = a_1$ in case (i), $b_n = 0$ in case (ii)). With the matrices

## X.1 Completely Integrable Hamiltonian Systems

$$L = \begin{pmatrix} a_1 & b_1 & & & & & b_n \\ b_1 & a_2 & b_2 & & & 0 & \\ & b_2 & a_3 & b_3 & & & \\ & & \ddots & \ddots & \ddots & & \\ & 0 & & b_{n-2} & a_{n-1} & b_{n-1} & \\ b_n & & & & b_{n-1} & a_n \end{pmatrix},$$

$$B = B(L) = \begin{pmatrix} 0 & b_1 & & & & & -b_n \\ -b_1 & 0 & b_2 & & & 0 & \\ & -b_2 & 0 & b_3 & & & \\ & & \ddots & \ddots & \ddots & & \\ & 0 & & -b_{n-2} & 0 & b_{n-1} \\ b_n & & & & -b_{n-1} & 0 \end{pmatrix},$$

the differential equations can be written in the *Lax pair* form

$$\dot{L} = BL - LB. \tag{1.20}$$

This system has an *isospectral flow*, that is, along any solution $L(t)$ of (1.20) the eigenvalues do not depend on $t$; see Lemma IV.3.4. The eigenvalues $\lambda_1, \ldots, \lambda_n$ of $L$ are therefore first integrals of the Toda system. They are independent and turn out to be in involution, in a neighbourhood of every point where the $\lambda_i$ are all different; see Exercise 6. Hence, the Toda lattice is a completely integrable system. Its Hamiltonian can be written as

$$H = \sum_{k=1}^{n} \left(2a_k^2 + 4b_k^2\right) = 2\,\text{trace}\,L^2 = 2\sum_{i=1}^{n} \lambda_i^2.$$

We conclude this section with a numerical example for the periodic Toda lattice. We choose $n = 3$ and the initial conditions $p_1 = -1.5$, $p_2 = 1$, $p_3 = 0.5$ and $q_1 = 1$, $q_2 = 2$, $q_3 = -1$. We apply to the system with Hamiltonian (1.19) the symplectic second-order Störmer/Verlet method and the non-symplectic classical fourth-order Runge-Kutta method with two different step sizes. The left pictures of Fig. 1.3 show the numerical approximations to the eigenvalues, and the right pictures the deviations of the eigenvalues $\lambda_1, \lambda_2, \lambda_3$ along the numerical solution from their initial values. Clearly, the eigenvalues are not invariants of the numerical schemes. However, Fig. 1.3 illustrates that the eigenvalues along the numerical solution remain close to their correct values over very long time intervals for the symplectic method, whereas they drift off for the non-symplectic method.

An explanation of the long-time near-preservation of the first integrals of completely integrable systems by symplectic methods will be given in the following sections, using backward error analysis and the perturbation theory for integrable Hamiltonian systems.

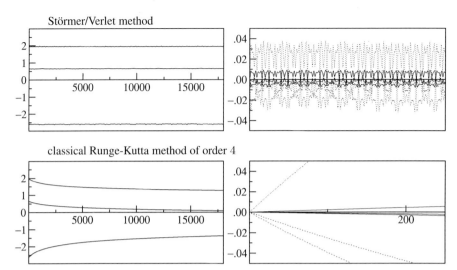

**Fig. 1.3.** Numerically obtained eigenvalues (left pictures) and errors in the eigenvalues (right pictures) for the step sizes $h = 0.1$ (dotted) and $h = 0.05$ (solid line).

## X.2 Transformations in the Perturbation Theory for Integrable Systems

> **Problème général de la Dynamique.** Nous sommes donc conduit à nous proposer le problème suivant: Étudier les équations canoniques
> $$\frac{dx_i}{dt} = \frac{dF}{dy_i}, \qquad \frac{dy_i}{dt} = -\frac{dF}{dx_i},$$
> en supposant que la fonction F peut se développer suivant les puissances d'un paramétre très petit $\mu$ de la manière suivante:
> $$F = F_0 + \mu F_1 + \mu^2 F_2 + \ldots,$$
> en supposant de plus que $F_0$ ne dépend que des $x$ et est indépendant des $y$; et que $F_1, F_2, \ldots$ sont des fonctions périodiques de période $2\pi$ par rapport aux $y$. (H. Poincaré 1892, p. 32f.)

Consider a small perturbation of a completely integrable Hamiltonian. In action-angle variables $(a, \theta)$ on $D \times \mathbb{T}^d$ ($D$ an open subset of $\mathbb{R}^d$), this takes the form

$$H(a, \theta) = H_0(a) + \varepsilon H_1(a, \theta), \tag{2.1}$$

where $\varepsilon$ is a small parameter. We assume that $H_0$ and $H_1$ are real-analytic, and that the perturbation $H_1$ (which may depend also on $\varepsilon$) is bounded by a constant on a complex neighbourhood of $D \times \mathbb{T}^d$ that is independent of $\varepsilon$. No other restriction shall be imposed on the perturbation.

For the unperturbed system ($\varepsilon = 0$) we have seen that the motion is conditionally periodic on invariant tori $\{a = \text{const.}, \theta \in \mathbb{T}^d\}$. Perturbation theory aims at an understanding of the flow of the perturbed system. The basic tools are symplectic coordinate transformations which take the system to a form that allows the long-time behaviour (perpetually, or over time scales large compared to $\varepsilon^{-1}$) of solutions of the system (certain solutions, or all solutions with initial values in some ball) to be read off. There are different transformations that provide answers to these problems. The emphasis in this section will be on the construction of suitable transformations, not on the technical but equally important aspects of obtaining estimates for them.

The methods in Poincaré's *Méthodes Nouvelles* form the now classical part of perturbation theory, but the theories of Birkhoff, Siegel, Kolmogorov/Arnold/Moser (KAM) and Nekhoroshev in the 20th century have become "classics" in their own right.

## X.2.1 The Basic Scheme of Classical Perturbation Theory

In the spirit of the preceding section, one might search for a symplectic change of coordinates $(a, \theta) \mapsto (b, \varphi)$ close to the identity such that the perturbed Hamiltonian written in the new variables $(b, \varphi)$ depends only on $b$, or more modestly, depends only on $b$ up to a remainder term of order $\mathcal{O}(\varepsilon^N)$ with a large $N > 1$, or to begin even more modestly, with $N = 2$. We search for a generating function

$$S(b, \theta) = b \cdot \theta + \varepsilon S_1(b, \theta)$$

where $\cdot$ symbolizes the Euclidean product of vectors in $\mathbb{R}^d$ and $S_1$ is $2\pi$-periodic in $\theta$. Naively, we require that the symplectic transformation defined by

$$a = \frac{\partial S}{\partial \theta}(b, \theta), \quad \varphi = \frac{\partial S}{\partial b}(b, \theta)$$

be such that the order-$\varepsilon$ term in the expansion of the Hamiltonian in the new variables, $K(b, \varphi) = H(a, \theta)$, $K(b, \varphi) = H_0(b) + \varepsilon K_1(b, \varphi) + \ldots$ depends only on $b$. Since

$$H(a, \theta) = H\left(b + \varepsilon \frac{\partial S_1}{\partial \theta}(b, \theta), \theta\right) = H_0(b) + \varepsilon \left\{\omega(b) \cdot \frac{\partial S_1}{\partial \theta}(b, \theta) + H_1(b, \theta)\right\} + \ldots$$

with the vector of frequencies

$$\omega(b) = \frac{\partial H_0}{\partial b}(b),$$

the function $S_1$ must satisfy the partial differential equation

$$\omega(b) \cdot \frac{\partial S_1}{\partial \theta}(b, \theta) + H_1(b, \theta) = \overline{H}_1(b) \tag{2.2}$$

for a function $\overline{H}_1$ that does not depend on $\theta$. Since $S_1$ is required to be $2\pi$-periodic in $\theta$, the function $\overline{H}_1$ must equal the average of $H_1$ over the angles:

$$\overline{H}_1(b) = \frac{1}{(2\pi)^d} \int_{\mathbb{T}^d} H_1(b, \theta) \, d\theta.$$

Equation (2.2) is the basic equation of Hamiltonian perturbation theory. From the Fourier series of $S_1$ and $H_1$,

$$S_1(b,\theta) = \sum_{k \in \mathbb{Z}^d} s_k(b)\, e^{ik \cdot \theta}, \qquad H_1(b,\theta) = \sum_{k \in \mathbb{Z}^d} h_k(b)\, e^{ik \cdot \theta}$$

we obtain a formal solution of (2.2) by comparing Fourier coefficients: $s_0(b)$ is arbitrary and

$$s_k(b) = -\frac{h_k(b)}{ik \cdot \omega(b)}, \qquad k \neq 0. \tag{2.3}$$

At this point, however, we are struck by the *problem of small denominators*. For any values of the frequencies $\omega_j(b)$, the denominator $k \cdot \omega(b) = k_1 \omega_1(b) + \ldots + k_d \omega_d(b)$ becomes arbitrarily small for some $k = (k_1, \ldots, k_d) \in \mathbb{Z}^d$, and even vanishes if the frequencies are rationally dependent.

For a perturbation where only finitely many Fourier coefficients $h_k$ are nonzero, the construction above excludes only a finite number of resonant frequencies (i.e., those with $k \cdot \omega(b) = 0$ for a $k \in \mathbb{Z}^d$ with $h_k \neq 0$) and small neighbourhoods around them. For $\omega(b)$ outside these neighbourhoods and for $\varphi$ on a complex neighbourhood of $\mathbb{T}^d$, we obtain for the Hamiltonian in the new variables

$$K(b, \varphi) = H_0(b) + \varepsilon \overline{H}_1(b) + \mathcal{O}(\varepsilon^2).$$

In the general case, we can approximate the perturbation $H_1$ up to $\mathcal{O}(\varepsilon^2)$ by a trigonometric polynomial. For analytic $H_1$, the Fourier coefficients $h_k$ decay exponentially with $|k| = \sum_i |k_i|$, and hence the required degree $m$ of the approximating trigonometric polynomial grows logarithmically with $\varepsilon$, i.e., $m \sim |\log \varepsilon|$.

As $\varepsilon \to 0$, the remainder term is under control only for those frequencies $\omega = \omega(b)$ for which the exponentially decaying Fourier coefficients $h_k$ of the perturbation decay faster than the denominators $ik \cdot \omega$ with growing $|k|$. This is certainly the case for frequencies satisfying *Siegel's diophantine condition* (or *strong non-resonance condition*, as it is sometimes called)

$$|k \cdot \omega| \geq \gamma |k|^{-\nu}, \qquad k \in \mathbb{Z}^d, k \neq 0 \tag{2.4}$$

for some positive constants $\gamma, \nu$. (Here again, $|k| = \sum_i |k_i|$). If $\nu > d-1$, the set of frequencies in a fixed ball that do *not* satisfy (2.4) has Lebesgue measure bounded by $Const \cdot \gamma$ (Exercise 5). Therefore, almost all frequencies satisfy (2.4) for some $\gamma > 0$. However, for any $\gamma$ and $\nu$, the complementary set is open and dense in $\mathbb{R}^d$.

## X.2.2 Lindstedt-Poincaré Series

> ... pour que la méthode de M. Lindstedt soit applicable, soit sous sa forme primitive, sois sous celle que je lui ai ensuite donnée, il faut qu'en première approximation les moyens mouvements ne soient liés par aucune relation linéaire à coefficients entières; ...
>
> Il semble donc permis de conclure que les séries (...) ne convergent pas. Toutefois le raisonnement qui précède ne suffit pas pour établir ce point avec une rigueur complète. (H. Poincaré 1893, pp. vi, 103.)

**Fig. 2.1.** Henri Poincaré (left), born: 29 April 1854 in Nancy (France), died: 17 July 1912 in Paris; Anders Lindstedt (right), born: 27 June 1854 in Sundborn (Sweden), died: 1939. Reproduced with permission of Bibl. Math. Univ. Genève.

The above construction is extended without any additional difficulty to arbitrary finite order in $\varepsilon$. The generating function is now sought in the form

$$S(b,\theta) = b \cdot \theta + \varepsilon S_1(b,\theta) + \varepsilon^2 S_2(b,\theta) + \ldots + \varepsilon^{N-1} S_{N-1}(b,\theta) \qquad (2.5)$$

and, as before, the requirement that the first $N$ terms in the $\varepsilon$-expansion of the Hamiltonian in the new variables be independent of the angles, leads via a Taylor expansion of the Hamiltonian to equations of the form (2.2) for $S_1, \ldots, S_{N-1}$:

$$\omega(b) \cdot \frac{\partial S_j}{\partial \theta} + K_j(b,\theta) = \overline{K}_j(b) \qquad (2.6)$$

where $K_1 = H_1$,

$$K_2 = \frac{1}{2} \frac{\partial^2 H_0}{\partial a^2} \left( \frac{\partial S_1}{\partial \theta}, \frac{\partial S_1}{\partial \theta} \right) + \frac{\partial H_1}{\partial a} \cdot \frac{\partial S_1}{\partial \theta},$$

and in general, $K_j$ is a sum of terms

$$\frac{1}{i!} \frac{\partial^i H_{k_0}}{\partial a^i} \left( \frac{\partial S_{k_1}}{\partial \theta}, \ldots, \frac{\partial S_{k_i}}{\partial \theta} \right) \quad \text{with} \quad k_0 + k_1 + \ldots + k_i = j \, .$$

The function $\overline{K}_j$ denotes again the angular average of $K_j$. These equations can be formally solved in the case of rationally independent frequencies. The Hamiltonian in the new variables is then

$$K(b,\varphi) = H_0(b)+\varepsilon\overline{K}_1(b)+\varepsilon^2\overline{K}_2(b)+\ldots+\varepsilon^{N-1}\overline{K}_{N-1}(b)+\varepsilon^N R_N(b,\theta). \quad (2.7)$$

The possible convergence of the series for $N \to \infty$ is a delicate issue that was not resolved conclusively by Poincaré (1893) in his chapter on "Divergence des séries de M. Lindstedt". If for some $b^*$, the series (2.5) together with its partial derivatives converged as $N \to \infty$, then $\{b = b^*, \varphi \in \mathbb{T}^d\}$ would be an invariant torus of the perturbed Hamiltonian system. However, it was not until Kolmogorov (1954) that the existence of invariant tori – for diophantine frequencies – was found, using a different construction. A direct proof of the convergence of the series of classical perturbation theory for diophantine frequencies was obtained only in 1988 by Eliasson (published in 1996); also see Giorgilli & Locatelli (1997) and references therein.

Nevertheless, already the truncated series (2.5) leads in a rather simple way to strong conclusions about the flow over long time scales when it is combined with the idea of approximating the Hamiltonian by a trigonometric polynomial: the "ultraviolet cut-off", an idea briefly addressed by Poincaré (1893), p. 98f., and taken to its full bearing by Arnold (1963) in his proof of the KAM theorem. We formulate a lemma for a fixed truncation index $N$. Here, $\omega_{\varepsilon,N}(b)$ denotes the derivative of the truncated series (2.7) with respect to $b$.

**Lemma 2.1.** *Suppose that $\omega(b^*)$ satisfies the diophantine condition (2.4). For any fixed $N \geq 2$, there are positive constants $\varepsilon_0, c, C$ such that the following holds for $\varepsilon \leq \varepsilon_0$: there exists a real-analytic symplectic change of coordinates $(a, \theta) \mapsto (b, \varphi)$ such that every solution $(b(t), \varphi(t))$ of the perturbed system in the new coordinates, starting with $\|b(0) - b^*\| \leq c|\log \varepsilon|^{-\nu-1}$, satisfies*

$$\|b(t) - b(0)\| \leq C t \varepsilon^N \quad \text{for } t \leq \varepsilon^{-N+1},$$

$$\|\varphi(t) - \omega_{\varepsilon,N}(b(0))t - \varphi(0)\| \leq C\left(t^2 + t|\log \varepsilon|^{\nu+1}\right)\varepsilon^N \quad \text{for } t^2 \leq \varepsilon^{-N+1}.$$

*Moreover, the transformation is $\mathcal{O}(\varepsilon)$-close to the identity: $\|(a,\theta) - (b,\varphi)\| \leq C\varepsilon$ holds for $(a,\theta)$ and $(b,\varphi)$ related by the above coordinate transform, for $\|b - b^*\| \leq c|\log\varepsilon|^{-\nu-1}$ and for $\varphi$ in an $\varepsilon$-independent complex neighbourhood of $\mathbb{T}^d$.*

*The constants $\varepsilon_0, c, C$ depend on $N, d, \gamma, \nu$ and on bounds of $H_0$ and $H_1$ on a complex neighbourhood of $\{b^*\} \times \mathbb{T}^d$.*

*Proof.* Using the relations (2.3) and their analogues for (2.6), it is a straightforward but somewhat tedious exercise to show that at the given particular $b^*$, the functions $K_j(b^*, \cdot), S_j(b^*, \cdot)$ are all analytic on the same complex neighbourhood of $\mathbb{T}^d$, and that the remainder term is bounded by

$$|R_N(b^*, \theta)| \leq C = C(N, d, \gamma, \nu)$$

for all $\theta$ in a complex neighbourhood of $\mathbb{T}^d$ which is independent of $\varepsilon$. Here, $C$ depends in addition on the bound of $H_1$ on a complex neighbourhood of $\{b^*\} \times \mathbb{T}^d$, or what amounts to the same by Cauchy's estimates, on bounds of the exponential decay of the Fourier coefficients $h_k$ of $H_1$. (In case of doubt, see also Sect. X.4 for explicit estimates.)

Assume first that $H_1(b,\theta)$ is a trigonometric polynomial in $\theta$ of degree $m$. Then $K_j$, $S_j$ are trigonometric polynomials of degree $jm$. Since $|k\cdot\omega(b)| \geq |k\cdot\omega(b^*)| - |k|(\max\|\omega'\|)\|b-b^*\|$, there is a $\delta > 0$ such that

$$|k\cdot\omega(b)| \geq \tfrac{1}{2}\gamma|k|^{-\nu} \quad \text{for} \quad \|b-b^*\| \leq \delta, \ |k| \leq Nm.$$

This number $\delta$ is proportional to $\gamma(Nm)^{-\nu-1}$. Consequently, since the construction involves only the trigonometric polynomials $K_j$, $S_j$ of degree up to $Nm$, the above estimate for the remainder term $R_N$ holds also for $\|b-b^*\| \leq \delta$. To approximate a general analytic $H_1$ by trigonometric polynomials up to $\mathcal{O}(\varepsilon^N)$, we must choose the degree $m$ proportional to $|\log\varepsilon^N|$. With the choice $\delta = c(N^2|\log\varepsilon|)^{-\nu-1}$, for a sufficiently small $c > 0$ independent of $\varepsilon$ (and $N$), the above bound for the remainder $R_N(b,\theta)$ is then valid for $b$ in the complex ball $\|b-b^*\| \leq 2\delta$ and for $\varphi$ in a complex neighbourhood of $\mathbb{T}^d$ (which depends only on $N$). By Cauchy's estimates, this implies

$$\left\|\frac{\partial R_N}{\partial\theta}(b,\theta)\right\| \leq C, \quad \left\|\frac{\partial R_N}{\partial b}(b,\theta)\right\| \leq \frac{C}{\delta}$$

for $\|b-b^*\| \leq \delta$ and $\theta \in \mathbb{T}^d$. Hence, as long as $\|b(t)-b^*\| \leq \delta$, the Hamiltonian differential equations are of the form

$$\dot b = -\frac{\partial K}{\partial\varphi} = -\varepsilon^N\frac{\partial R_N}{\partial\theta}\frac{\partial\theta}{\partial\varphi} = \mathcal{O}(\varepsilon^N), \quad \dot\varphi = \frac{\partial K}{\partial b} = \omega_{\varepsilon,N}(b) + \mathcal{O}(\varepsilon^N/\delta).$$

This implies the result. □

Hence, the tori $\{b = b(0), \ \varphi \in \mathbb{T}^d\}$ are nearly invariant over a time scale $\varepsilon^{-N+1}$, and the flow is close to a quasiperiodic flow over times bounded by the square root of $\varepsilon^{-N+1}$. Lemma 2.1 is just a preliminary to more substantial results (which hold under appropriate additional conditions): invariant tori carrying a quasi-periodic flow with diophantine frequencies persist under small Hamiltonian perturbations (Kolmogorov 1954); every solution of the perturbed system remains close, within a positive power of $\varepsilon$, to some torus over times that are exponentially long in a negative power of $\varepsilon$ (Nekhoroshev 1977); solutions starting close to an invariant torus with diophantine frequencies stay within twice the initial distance over time intervals that are exponentially long in a negative power of the distance (Perry & Wiggins 1994) or even exponentially long in the exponential of the inverse of the distance (Morbidelli & Giorgilli 1995).

The symplectic transformations of this subsection were constructed using the mixed-variable generating function $S(b,\theta)$. As was pointed out for example by Benettin, Galgani & Giorgilli (1985), rigorous estimates for the remainder terms are often obtained in a simpler way using the *Lie method*, which involves constructing the near-identity symplectic transformation as the time-$\varepsilon$ flow of some auxiliary Hamiltonian system with a suitably defined Hamiltonian $\chi(b,\varphi)$. As before, the condition that the Hamiltonian $H(a,\theta) = K(b,\varphi)$ should depend on $\varphi$ only in higher-order terms, leads to equations of the form (2.2), now for $\chi$ instead of $S_1$. We will use such a construction in the following subsection.

## X.2.3 Kolmogorov's Iteration

> It is easy to grasp the meaning of Theorem 1 for mechanics. It indicates that an $s$-parametric family of conditionally periodic motions [...] cannot, under conditions (3) and (4) [here: (2.4) and (2.9)], disappear as a result of a small change in the Hamilton function $H$.
>
> In this note we confine ourselves to the construction of the transformation. (A.N. Kolmogorov 1954)

For the completely integrable Hamiltonian $H_0(a)$, the phase space is foliated into invariant tori parametrized by $a$. We now fix one such torus $\{a = a^*,\ \theta \in \mathbb{T}^d\}$ with strongly diophantine frequencies $\omega = \omega(a^*)$. Without loss of generality, we may assume $a^* = 0$. This particular torus is invariant under the flow of every Hamiltonian $H(a, \theta)$ for which the linear terms in the Taylor expansion with respect to $a$ at $0$ are independent of $\theta$:

$$H(a, \theta) = c + \omega \cdot a + \tfrac{1}{2} a^T M(a, \theta) a \tag{2.8}$$

with $c \in \mathbb{R}$, $\omega \in \mathbb{R}^d$, and a real symmetric $d \times d$-matrix $M(a, \theta)$ analytic in its arguments. Since the Hamiltonian equations are of the form

$$\dot{a} = \mathcal{O}(\|a\|^2), \quad \dot{\theta} = \omega + \mathcal{O}(\|a\|),$$

the torus $\{a = 0,\ \theta \in \mathbb{T}^d\}$ is invariant and the flow on it is quasi-periodic with frequencies $\omega$.

Consider now an analytic perturbation of such a Hamiltonian: $H(a, \theta) + \varepsilon G(a, \theta)$ with a small $\varepsilon$. Kolmogorov (1954) found a near-identity symplectic transformation $(a, \theta) \mapsto (\widetilde{a}, \widetilde{\theta})$, constructed by an iterative procedure, such that the perturbed Hamiltonian in the new variables is again of the form (2.8) with the same $\omega$, and hence has the invariant torus $\{\widetilde{a} = 0,\ \widetilde{\theta} \in \mathbb{T}^d\}$ carrying a quasi-periodic flow with the frequencies of the unperturbed system. This holds under the conditions that $\omega$ satisfies the diophantine condition (2.4), and that the angular average

$$\overline{M}_0 := \frac{1}{(2\pi)^d} \int_{\mathbb{T}^d} M(0, \theta)\, d\theta \quad \text{is an invertible matrix.} \tag{2.9}$$

Here we describe the iterative construction of this symplectic transformation. The proof of convergence of the iteration will be given in Sect. X.5.

We construct a symplectic transformation $(a, \theta) \mapsto (b, \varphi)$ as the time-$\varepsilon$ flow of an auxiliary Hamiltonian of the form

$$\chi(b, \varphi) = \xi \cdot \varphi + \chi_0(\varphi) + \sum_{i=1}^{d} b_i \chi_i(\varphi), \tag{2.10}$$

where $\xi \in \mathbb{R}^d$ is a constant vector, and $\chi_0, \chi_1, \ldots, \chi_d$ are $2\pi$-periodic functions. (Quadratic and higher-order terms in $b$ play no role in the construction and are therefore omitted right at the outset.) The old and new coordinates are then related by

$$a = b + \varepsilon \frac{\partial \chi}{\partial \varphi}(b, \varphi) + \mathcal{O}(\varepsilon^2), \quad \theta = \varphi - \varepsilon \frac{\partial \chi}{\partial b}(b, \varphi) + \mathcal{O}(\varepsilon^2).$$

We insert this into

$$H(a,\theta) + \varepsilon G(a,\theta) = c + \omega \cdot b + \tfrac{1}{2} b^T M(b,\varphi) b$$
$$+ \varepsilon \left\{ \omega \cdot \frac{\partial \chi}{\partial \varphi}(b,\varphi) + b^T M(b,\varphi) \frac{\partial \chi}{\partial \varphi}(b,\varphi) + G(b,\varphi) \right\} + \mathcal{O}(\varepsilon \|b\|^2) + \mathcal{O}(\varepsilon^2).$$

We now require that the term in curly brackets be $Const + \mathcal{O}(\|b\|^2)$. Writing down the Taylor expansion

$$G(b,\varphi) = G_0(\varphi) + \sum_{i=1}^{d} b_i G_i(\varphi) + b^T Q(b,\varphi) b \tag{2.11}$$

and inserting the above ansatz for $\chi$, this condition becomes

$$\omega \cdot \frac{\partial \chi_0}{\partial \varphi}(\varphi) + \sum_{i=1}^{d} b_i \left( \omega \cdot \frac{\partial \chi_i}{\partial \varphi}(\varphi) + u_i(\varphi) + v_i(\varphi) \right)$$
$$+ G_0(\varphi) + \sum_{i=1}^{d} b_i G_i(\varphi) = Const.,$$

where $u = (u_1, \ldots, u_d)^T$ and $v = (v_1, \ldots, v_d)^T$ are defined by

$$u(\varphi) = M(0,\varphi) \xi, \tag{2.12}$$
$$v(\varphi) = M(0,\varphi) \frac{\partial \chi_0}{\partial \varphi}(\varphi). \tag{2.13}$$

The condition is fulfilled if

$$\omega \cdot \frac{\partial \chi_0}{\partial \varphi}(\varphi) + G_0(\varphi) = \overline{G}_0 \tag{2.14}$$

$$\omega \cdot \frac{\partial \chi_i}{\partial \varphi}(\varphi) + u_i(\varphi) + v_i(\varphi) + G_i(\varphi) = \overline{u}_i + \overline{v}_i + \overline{G}_i \tag{2.15}$$

$$\overline{u}_i + \overline{v}_i + \overline{G}_i = 0 \qquad (i = 1, \ldots, d). \tag{2.16}$$

Here the bars again denote angular averages. Note that equations (2.14), (2.15) are of the form (2.2). Equation (2.14) determines $\chi_0$ and hence $v = (v_1, \ldots, v_d)^T$ by (2.13). Equations (2.16) then give $\overline{u} = (\overline{u}_1, \ldots, \overline{u}_d)^T$. By (2.12), we need

$$\overline{u} = \overline{M}_0 \xi,$$

which determines $\xi$ uniquely because $\overline{M}_0$ is assumed to be invertible. Equation (2.12) then yields $u = (u_1, \ldots, u_d)^T$. Finally, (2.15) determines $\chi_1, \ldots, \chi_d$, and the construction of $\chi(b,\varphi)$ is complete. In the new variables $(b,\varphi)$, the perturbed Hamiltonian then takes the form

$$H(a,\theta) + \varepsilon G(a,\theta) = \widehat{c} + \omega \cdot b + \tfrac{1}{2} b^T \widehat{M}(b,\varphi) b + \varepsilon^2 \widehat{G}(b,\varphi) \tag{2.17}$$

with unchanged frequencies $\omega$ and with $\widehat{M}(b,\varphi) = M(b,\varphi) + \mathcal{O}(\varepsilon)$. The perturbation to the form (2.8) is thus reduced from $\mathcal{O}(\varepsilon)$ to $\mathcal{O}(\varepsilon^2)$. The iteration of this procedure turns out to be convergent, see Sect. X.5. This finally yields a symplectic change of coordinates that transforms the perturbed Hamiltonian to the form (2.8). The perturbed system thus has an invariant torus carrying a quasi-periodic flow with frequencies $\omega$ — a KAM torus, as it is named after Kolmogorov, Arnold and Moser.

## X.2.4 Birkhoff Normalization Near an Invariant Torus

> KAM tori are very sticky.
> (A.D. Perry & S. Wiggins 1994)

In this subsection we describe a transformation studied by Pöschel (1993) and Perry & Wiggins (1994) for systems with Hamiltonian in the Kolmogorov form (2.8) in a neighbourhood of the invariant torus $\{a = 0, \theta \in \mathbb{T}^d\}$. This transformation is an analogue of a transformation of Birkhoff (1927) for Hamiltonian systems near an elliptic stationary point.

The symplectic change of coordinates $(a, \theta) \mapsto (b, \varphi)$ considered here transforms a Hamiltonian (2.8) with diophantine frequencies $\omega$ to the form $H(a, \theta) = K_N(b) + \mathcal{O}(\|b\|^N)$ for arbitrary $N$, or more precisely, the Hamiltonian in the new variables, $H_N(b, \varphi) = H(a, \theta)$, is of the form

$$H_N(b, \varphi) = \omega \cdot b + Z_N(b) + R_N(b, \varphi) \tag{2.18}$$

with $Z_N(b) = \mathcal{O}(\|b\|^2)$ and $R_N(b, \varphi) = \mathcal{O}(\|b\|^N)$. (We have taken the irrelevant constant term in (2.8) $c = 0$.) The equations of motion then take the form

$$\dot{b} = \mathcal{O}(\|b\|^N), \quad \dot{\varphi} = \omega + \mathcal{O}(\|b\|).$$

Therefore, in these variables $\{b = 0, \varphi \in \mathbb{T}^d\}$ is an invariant torus, and for sufficiently small $r$,

$$\|b(0)\| \leq r \quad \text{implies} \quad \|b(t)\| \leq 2r \text{ for } t \leq C_N\, r^{-N+1}.$$

A judicious choice of $N$ even yields time intervals that are exponentially long in a negative power of $r$ on which solutions starting at a distance $r$ stay within twice the initial distance (Perry & Wiggins 1994). Motion away from the torus can thus be only very slow.

The normal form (2.18) is constructed iteratively. Each iteration step is very similar to the procedure in Sect. X.2.1, where now the distance to the torus plays the role of the small parameter. Consider a Hamiltonian

$$H(a, \theta) = \omega \cdot a + Z(a) + R(a, \theta)$$

where $Z(a) = \mathcal{O}(\|a\|^2)$ and $R(a, \theta) = \mathcal{O}(\|a\|^k)$ for some $k \geq 2$ in a complex neighbourhood of $\{0\} \times \mathbb{T}^d$. We construct a symplectic change of coordinates $(a, \theta) \mapsto (b, \varphi)$ via a generating function $b \cdot \theta + S(b, \theta)$ as

$$a = b + \frac{\partial S}{\partial \theta}(b,\theta), \quad \varphi = \theta + \frac{\partial S}{\partial b}(b,\theta).$$

We expand (omitting the arguments $(b,\theta)$ in $\partial S/\partial \theta$ and $\partial H/\partial a$)

$$\begin{aligned}
H\left(b + \frac{\partial S}{\partial \theta}, \theta\right) &= H(b,\theta) + \frac{\partial H}{\partial a} \cdot \frac{\partial S}{\partial \theta} + Q(b,\theta) \\
&= \omega \cdot b + Z(b) + \left\{ R(b,\theta) + \frac{\partial H}{\partial a} \cdot \frac{\partial S}{\partial \theta} \right\} + Q(b,\theta),
\end{aligned}$$

where $|Q(b,\theta)| \leq \text{Const.} \, \|\partial S/\partial \theta\|^2$. Since $\partial H/\partial b = \omega + \mathcal{O}(\|b\|)$, we can make the expression in curly brackets independent of $\theta$ up to $\mathcal{O}(\|b\|^{k+1})$ by determining $S$ from the equation of the form (2.2):

$$\omega \cdot \frac{\partial S}{\partial \theta}(b,\theta) + R(b,\theta) = \overline{R}(b).$$

For diophantine frequencies $\omega$, we obtain $S(b,\theta) = \mathcal{O}(\|b\|^k)$ on a (reduced) complex neighbourhood of $\{0\} \times \mathbb{T}^d$ from the corresponding estimate for $R(b,\theta)$. It follows that the above symplectic transformation with generating function $b \cdot \theta + S(b,\theta)$ is well-defined for small $\|b\|$, and the Hamiltonian in the new variables, $\widehat{H}(b,\varphi) = H(a,\theta)$, becomes

$$\widehat{H}(b,\varphi) = \omega \cdot b + \widehat{Z}(b) + \widehat{R}(b,\varphi)$$

with $\widehat{Z}(b) = Z(b) + \overline{R}(b)$ and

$$\widehat{R}(b,\varphi) = \left(\frac{\partial H}{\partial a}(b,\theta) - \omega\right) \cdot \frac{\partial S}{\partial \theta}(b,\theta) + Q(b,\theta) = \mathcal{O}(\|b\|^{k+1}),$$

so that the order in $b$ of the remainder term is augmented by 1. The procedure can be iterated, but unlike the iteration of the preceding subsection, this iteration is in general divergent. Nevertheless, a suitable finite termination yields remainder terms that are exponentially small in a positive power of $r$ for $\|b\| \leq r$, by arguments similar to those of Sect. X.4.

# X.3 Linear Error Growth and Near-Preservation of First Integrals

In this section we study the error growth of symplectic numerical methods when they are applied to integrable or near-integrable systems. A preliminary analysis of linear error growth for the Kepler problem was first given by Calvo & Sanz-Serna (1993). Using backward error analysis and KAM theory, Calvo & Hairer (1995a) then showed linear error growth of symplectic methods applied to integrable systems when the frequencies at the initial value satisfy a diophantine condition (2.4). Here we give such a result under milder conditions on the initial values, combining backward error analysis and Lemma 2.1. We derive also a first result on the long-time near-preservation of all first integrals, which will be extended to exponentially long times in Sections X.4.3 and X.5.2 (under stronger assumptions on the starting values), and perpetually in Sect. X.6 (only for a Cantor set of step sizes).

Figure 3.1 illustrates the linear error growth of the symplectic Störmer/Verlet method, as opposed to the quadratic error growth for the classical fourth-order Runge-Kutta method, on the example of the Toda lattice. The same number of function evaluations was used for both methods.

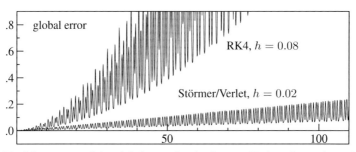

**Fig. 3.1.** Euclidean norm of the global error for the Störmer/Verlet scheme (step size $h = 0.02$) and the classical Runge-Kutta method of order 4 (step size $h = 0.08$) applied to the Toda lattice with $n = 3$ and initial values as in Fig. 1.3.

We consider a completely integrable Hamiltonian system (usually not given in action-angle variables)

$$\dot{p} = -\frac{\partial H}{\partial q}(p, q), \quad \dot{q} = \frac{\partial H}{\partial p}(p, q) \tag{3.1}$$

and apply to it a symplectic numerical method with step size $h$, yielding a numerical solution sequence $(p_n, q_n)$. We assume that the Hamiltonian is real-analytic and that the conditions of the Arnold-Liouville theorem, Theorem 1.6, are fulfilled. Consider the symplectic transformation $(p, q) = \psi(a, \theta)$ to action-angle variables. We denote the inverse transformation as

$$(a, \theta) = \bigl(I(p, q), \Theta(p, q)\bigr). \tag{3.2}$$

We recall that the components $I_1, \ldots, I_d$ of $I = (I_i)$ are first integrals of the system: $I(p(t), q(t)) = I(p_0, q_0)$ for all $t$. In the action-angle variables, the Hamiltonian is $\mathcal{H}(a) = H(p, q)$, and we denote the frequencies

$$\omega(a) = \frac{\partial \mathcal{H}}{\partial a}(a). \tag{3.3}$$

We consider this in a neighbourhood of some $a^* \in \mathbb{R}^d$.

**Theorem 3.1.** *Consider applying a symplectic numerical integrator of order $p$ to the completely integrable Hamiltonian system (3.1). Suppose that $\omega(a^*)$ satisfies the diophantine condition (2.4). Then, there exist positive constants $C, c$ and $h_0$ such that the following holds for all step sizes $h \leq h_0$: every numerical solution starting with $\|I(p_0, q_0) - a^*\| \leq c |\log h|^{-\nu-1}$ satisfies*

$$\begin{aligned} \|(p_n, q_n) - (p(t), q(t))\| &\leq C t h^p \\ \|I(p_n, q_n) - I(p_0, q_0)\| &\leq C h^p \end{aligned} \quad \text{for } t = nh \leq h^{-p}.$$

*The constants $h_0, c, C$ depend on $d, \gamma, \nu$, on bounds of the real-analytic Hamiltonian $H$ on a complex neighbourhood of the torus $\{(p, q)\,;\, I(p, q) = a^*\}$, and on the numerical method.*

*Proof.* (a) In the action-angle variables $(a, \theta)$, the exact flow is given as

$$a(t) = a(0), \quad \theta(t) = \omega(a(0))\, t + \theta(0). \tag{3.4}$$

By Theorem IX.3.1 (and Theorem IX.1.2), the truncated modified equation of the numerical method is Hamiltonian with[1]

$$\widetilde{H}(p, q) = H(p, q) + h^p H_{p+1}(p, q) + \ldots + h^r H_{r+1}(p, q).$$

We choose $r = 2p$, and we denote by $(\widetilde{p}(t), \widetilde{q}(t))$ the solution of the modified equations with initial values $(p_0, q_0)$. In the variables $(a, \theta)$, the modified Hamiltonian becomes $\widetilde{H}(p, q) = \widetilde{\mathcal{H}}(a, \theta)$ with

$$\widetilde{\mathcal{H}}(a, \theta) = \mathcal{H}(a) + \varepsilon\, \mathcal{G}_h(a, \theta), \tag{3.5}$$

where $\varepsilon = h^p$ and the perturbation function $\mathcal{G}_h$ is bounded independently of $h$ on a complex neighbourhood of $\{a^*\} \times \mathbb{T}^d$. By Lemma 2.1 with $\varepsilon = h^p$ and $N \geq 3$, there is a symplectic change of coordinates $\mathcal{O}(h^p)$-close to the identity, such that the solution of the modified equation in the new variables $(b, \varphi)$ is of the form

$$\begin{aligned} \widetilde{b}(t) &= \widetilde{b}(0) + \mathcal{O}(th^{pN}), \\ \widetilde{\varphi}(t) &= \omega_h(\widetilde{b}(0))\, t + \widetilde{\varphi}(0) + \mathcal{O}(th^{pN-1} + t^2 h^{pN}) \end{aligned} \quad \text{for } t \leq h^{-p}, \tag{3.6}$$

with $\omega_h(b) = \omega(b) + \mathcal{O}(h^p)$. The constants symbolized by the $\mathcal{O}$-notation are independent of $h$, of $t \leq h^{-p}$ and of $(\widetilde{b}(0), \widetilde{\varphi}(0))$ with $|\widetilde{b}(0) - a^*| \leq c |\log h|^{-\nu-1}$. Since the transformation between the variables $(a, \theta)$ and $(b, \varphi)$ is $\mathcal{O}(h^p)$ close to the identity, it follows that the flow of the modified equations in the variables $(a, \theta)$ satisfies

---

[1] We always assume, without further mention, that the modified Hamiltonian is well-defined on the same open set $D$ as the original Hamiltonian. This is true for arbitrary symplectic methods if $D$ is simply connected; on general domains it is satisfied for (partitioned) Runge-Kutta methods and for splitting methods; see Sections IX.3 and IX.4.

$$\widetilde{a}(t) = \widetilde{a}(0) + \mathcal{O}(h^p),$$
$$\widetilde{\theta}(t) = \omega(\widetilde{a}(0))\, t + \widetilde{\theta}(0) + t e_h + \mathcal{O}(h^p) \qquad \text{for } 1 \le t \le h^{-p},$$

where $e_h = \omega_h(\widetilde{b}(0)) - \omega(\widetilde{a}(0)) = \mathcal{O}(h^p)$ yields the dominant contribution to the error. By comparison with (3.4) and since $\widetilde{a}(t) = I(\widetilde{p}(t), \widetilde{q}(t))$, the difference between the exact solution and the solution of the modified equation therefore satisfies

$$(\widetilde{p}(t), \widetilde{q}(t)) - (p(t), q(t)) = \mathcal{O}(t h^p)$$
$$I(\widetilde{p}(t), \widetilde{q}(t)) - I(p_0, q_0) = \mathcal{O}(h^p) \qquad \text{for } 1 \le t \le h^{-p}.$$

The same bounds for $t \le 1$ follow by standard error estimates.

(b) It remains to bound the difference between the solution of the modified equation and the numerical solution. By construction of the modified equation with $r = 2p$ and by comparison with (3.6), one step of the method is of the form

$$b_{n+1} = b_n + \mathcal{O}(h^{r+1}), \qquad \varphi_{n+1} = \omega_h(b_n)\, h + \varphi_n + \mathcal{O}(h^{r+1}).$$

It follows that for $t = nh$,

$$b_n = \widetilde{b}(t) + \mathcal{O}(t h^r), \qquad \varphi_n = \widetilde{\varphi}(t) + \mathcal{O}(t^2 h^r).$$

For $t \le h^{-p}$ and $r = 2p$, we have $t h^r \le h^p$. Hence the difference between the numerical solution and the solution of the modified equations in the original variables $(p, q)$ is bounded by

$$(p_n, q_n) - (\widetilde{p}(t), \widetilde{q}(t)) = \mathcal{O}(t h^p)$$
$$I(p_n, q_n) - I(\widetilde{p}(t), \widetilde{q}(t)) = \mathcal{O}(h^p) \qquad \text{for } t = nh \le h^{-p}.$$

Together with the bound of part (a) this gives the result. $\square$

**Remark 3.2.** The linear error growth holds also when the symplectic method is applied to a perturbed integrable system with a perturbation parameter $\varepsilon$ bounded by a positive power of the step size: $\varepsilon \le K h^\alpha$ for some $\alpha > 0$. The proof of this generalization is the same as above, except that possibly a larger $N$ is required in using Lemma 2.1.

**Example 3.3 (Linear Error Growth for the Kepler Problem).** From Example 1.10 we know that for the Kepler problem the frequencies (1.18) do not satisfy the diophantine condition (2.4). Nevertheless we observed a linear error growth for symplectic methods in the experiments of Fig. I.2.2 (see also Table I.2.1). This can be explained as follows: in action-angle variables the Hamiltonian of the Kepler problem is $\mathcal{H}(a_1, a_2)$, where $a_2 = L$ is the angular momentum. Since the angular momentum is a quadratic invariant that is exactly conserved by symplectic integrators such as symplectic partitioned Runge-Kutta methods, the modified Hamiltonian

$$\widetilde{\mathcal{H}}(a, \theta) = \mathcal{H}(a_1, a_2) + \varepsilon\, \mathcal{G}_h(a_1, a_2, \theta_1)$$

does not depend on the angle variable $\theta_2$ (see Corollary IX.5.2). As in the proof of Lemma 2.1 we average out the angle $\theta_1$ up to a certain power of $\varepsilon$. Since we are concerned here with one degree of freedom, the diophantine condition is trivially satisfied, and we can conclude as in Theorem 3.1.

# X.4 Near-Invariant Tori on Exponentially Long Times

We refine the results for the classical perturbation series of Sect. X.2.2 to yield locally integrable behaviour, up to exponentially small deviations, over time intervals that are exponentially long in a power of the small perturbation parameter. We then combine this result with backward error analysis to show the near-preservation of invariant tori over exponentially long times in a negative power of the step size for symplectic integrators. We begin with the necessary technical estimates.

## X.4.1 Estimates of Perturbation Series

We will estimate the coefficients of the perturbation series (2.5), which requires a bound for the solution of (2.6). We use the following notation: for $\rho > 0$ and with $\|\cdot\|$ the maximum norm on $\mathbb{R}^d$,

$$U_\rho = \{\theta \in \mathbb{T}^d + i\mathbb{R}^d \,;\, \|\operatorname{Im}\theta\| < \rho\}$$

denotes the complex extension of the $d$-dimensional torus $\mathbb{T}^d$ of width $\rho$. For a bounded analytic function $F$ on $U_\rho$, we write

$$\|F\|_\rho = \sup_{\theta \in U_\rho} |F(\theta)|, \qquad \left\|\frac{\partial F}{\partial \theta}\right\|_\rho = \sum_{j=1}^d \left\|\frac{\partial F}{\partial \theta_j}\right\|_\rho.$$

Following Arnold (1963), we prove the following bounds for the solution of the basic partial differential equation (2.2).

**Lemma 4.1.** *Suppose $\omega \in \mathbb{R}^d$ satisfies the diophantine condition (2.4). Let $G$ be a bounded real-analytic function on $U_\rho$, and let $\overline{G}$ denote the average of $G$ over $\mathbb{T}^d$. Then, the equation*

$$\omega \cdot \frac{\partial F}{\partial \theta} + G = \overline{G}$$

*has a unique real-analytic solution $F$ on $U_\rho$ with zero average $\overline{F} = 0$. For every positive $\delta < \min(\rho, 1)$, $F$ is bounded on $U_{\rho-\delta}$ by*

$$\|F\|_{\rho-\delta} \leq \kappa_0 \, \delta^{-\alpha+1} \|G\|_\rho, \qquad \left\|\frac{\partial F}{\partial \theta}\right\|_{\rho-\delta} \leq \kappa_1 \, \delta^{-\alpha} \|G\|_\rho,$$

*where $\alpha = \nu + d + 1$ and $\kappa_0 = \gamma^{-1} 8^d \, 2^\nu \nu!$, $\kappa_1 = \gamma^{-1} 8^d \, 2^{\nu+1} (\nu+1)!$.*

Rüssmann (1975, 1976) has shown that the estimates hold with the optimal exponent $\alpha = \nu + 1$ and with $\kappa_0 = 2^{d+1-\nu}\sqrt{(2\nu)!}$ and $\kappa_1 = 2^{d-\nu}\sqrt{(2\nu+2)!}$. This optimal value of $\alpha$ would yield slightly more favourable estimates in the following, but here we content ourselves with the simpler result given above.

*Proof of Lemma 4.1.* We have the Fourier series, convergent on the complex extension $\|\operatorname{Im} \theta\| < \rho$,

$$G(\theta) - \overline{G} = \sum_{k \neq 0} g_k\, e^{ik \cdot \theta}, \qquad F(\theta) = \sum_k f_k\, e^{ik \cdot \theta}$$

with Fourier coefficients $f_0 = \overline{F} = 0$ and

$$f_k = -\frac{g_k}{ik \cdot \omega} \qquad \text{for } k \in \mathbb{Z}^d,\ k \neq 0.$$

By Cauchy's estimates, $|g_k| \leq M e^{-|k|\rho}$ with $M = \|G - \overline{G}\|_\rho \leq 2\|G\|_\rho$ and $|k| = \sum |k_i|$. It follows with (2.4) that

$$\|F\|_{\rho-\delta} \leq \sum_k |f_k|\, e^{|k|(\rho-\delta)} \leq \frac{M}{\gamma} \sum_k |k|^\nu\, e^{-|k|\delta},$$

$$\left\|\frac{\partial F}{\partial \theta}\right\|_{\rho-\delta} \leq \sum_k |f_k| \cdot |k|\, e^{|k|(\rho-\delta)} \leq \frac{M}{\gamma} \sum_k |k|^{\nu+1}\, e^{-|k|\delta}.$$

It remains to bound the right-hand sums. We use the inequality $x^\nu/\nu! \leq e^x$ with $x = |k|\delta/2$ to obtain

$$\sum_k |k|^\nu\, e^{-|k|\delta} \leq 2^\nu \delta^{-\nu} \nu! \sum_k e^{-|k|\delta/2}.$$

The last sum is bounded by

$$\sum_k e^{-|k|\delta/2} = \left(1 + 2\sum_{j=1}^\infty e^{-j\delta/2}\right)^d = \left(\frac{1 + e^{-\delta/2}}{1 - e^{-\delta/2}}\right)^d \leq (8\delta^{-1})^d.$$

Taken together, the above inequalities yield the stated bound for $\|F\|_{\rho-\delta}$. The bound for the derivative is obtained in the same way, with $\nu$ replaced by $\nu + 1$. $\square$

The coefficients of the perturbation series (2.5) are bounded as follows.

**Lemma 4.2.** *Let $H_0$, $H_1$ be real-analytic and bounded by $M$ on the complex $r$-neighbourhood $B_r(b^*)$ of $b^* \in \mathbb{R}^d$ and on $B_r(b^*) \times U_\rho$, respectively. Suppose that $\omega(b^*) = (\partial H_0/\partial a)(b^*)$ satisfies the diophantine condition (2.4). Then, the coefficients of the perturbation series (2.5) are bounded by*

$$\left\|\frac{\partial S_j}{\partial \theta}(b^*, \cdot)\right\|_{\rho/2} \leq C_0 (C_1 j^\alpha)^{j-1}$$

*for all $j \geq 0$. Here $C_0 = 2r$, and $C_1 = 128(\kappa_1 M/r\rho^\alpha)^2$ with $\alpha$ and $\kappa_1$ of Lemma 4.1.*

*Proof.* We recall from Sect. X.2.2 that $S_j$ is determined by (2.6), where $K_1 = H_1$ and for $j \geq 2$,

$$K_j = \sum_{i=2}^{j} \sum_{k_1+\ldots+k_i=j} \frac{1}{i!} \frac{\partial^i H_0}{\partial a^i} \left( \frac{\partial S_{k_1}}{\partial \theta}, \ldots, \frac{\partial S_{k_i}}{\partial \theta} \right)$$
$$+ \sum_{i=1}^{j-1} \sum_{k_1+\ldots+k_i=j-1} \frac{1}{i!} \frac{\partial^i H_1}{\partial a^i} \left( \frac{\partial S_{k_1}}{\partial \theta}, \ldots, \frac{\partial S_{k_i}}{\partial \theta} \right).$$

We fix an index, say $J$, set $\delta = \rho/(2J)$ and abbreviate

$$\|K_k\|_j = \|K_k(b^*, \cdot)\|_{\rho - j\delta}$$

and similarly for $\partial S_k / \partial \theta$. By (2.6) and Lemma 4.1, we have

$$\left\| \frac{\partial S_j}{\partial \theta} \right\|_j \leq \kappa_1 \delta^{-\alpha} \|K_j\|_{j-1}.$$

We use the Cauchy estimate

$$\left| \frac{1}{i!} \frac{\partial^i H_0}{\partial a^i}(v_1, \ldots, v_i) \right| \leq \frac{M}{r^i} |v_1| \cdot \ldots \cdot |v_i|,$$

where $|\cdot|$ denotes the sum norm on $\mathbb{C}^d$, and bound $\|\cdot\|_{j-1}$ by $\|\cdot\|_k$ for $k \leq j-1$. We thus obtain from the above formula for $K_j$

$$\|K_j\|_{j-1} \leq \sum_{i=2}^{j} \sum_{k_1+\ldots+k_i=j} \frac{M}{r^i} \left\| \frac{\partial S_{k_1}}{\partial \theta} \right\|_{k_1} \cdot \ldots \cdot \left\| \frac{\partial S_{k_i}}{\partial \theta} \right\|_{k_i}$$
$$+ \sum_{i=1}^{j-1} \sum_{k_1+\ldots+k_i=j-1} \frac{M}{r^i} \left\| \frac{\partial S_{k_1}}{\partial \theta} \right\|_{k_1} \cdot \ldots \cdot \left\| \frac{\partial S_{k_i}}{\partial \theta} \right\|_{k_i}.$$

Combining the two bounds yields

$$\frac{1}{r} \left\| \frac{\partial S_j}{\partial \theta} \right\|_j \leq \beta_j,$$

where, with $\mu = (M/r)(\kappa_1/\delta^\alpha)$, we have $\beta_1 = \mu$ and recursively for $j \geq 2$,

$$\beta_j = \mu \sum_{i=2}^{j} \sum_{k_1+\ldots+k_i=j} \beta_{k_1} \cdot \ldots \cdot \beta_{k_i} + \mu \sum_{i=1}^{j-1} \sum_{k_1+\ldots+k_i=j-1} \beta_{k_1} \cdot \ldots \cdot \beta_{k_i}.$$

Multiplying this equation with $\zeta^j$ and summing over $j$, we see that the generating function $b(\zeta) = \sum_{j=1}^{\infty} \beta_j \zeta^j$ is given implicitly by

$$b(\zeta) - \mu \zeta = \mu \left( \frac{1}{1 - b(\zeta)} - 1 - b(\zeta) \right) + \mu \zeta \left( \frac{1}{1 - b(\zeta)} - 1 \right),$$

or explicitly, after solving the quadratic equation, by

$$b(\zeta) = \frac{1}{2}\frac{1}{1+\mu} - \sqrt{\frac{1}{4}\left(\frac{1}{1+\mu}\right)^2 - \frac{\mu}{1+\mu}\zeta}\,.$$

Hence, $b(\zeta)$ is analytic on the disc $|\zeta| < 1/(4\mu(1+\mu))$, and is there bounded by $1/(2(1+\mu))$. For $\mu \geq 1$, Cauchy's estimate yields

$$\|\partial S_j/\partial\theta\|_j \leq r\beta_j \leq 2r\,(8\mu^2)^{j-1}\,.$$

(For the uninteresting case $\mu \leq 1$ the bound is $2r \cdot 8^{j-1}$.) For $j = J$ this almost gives the stated result upon inserting the definition of $\mu$, but with an exponent $2\alpha$ instead of $\alpha$. This can be reduced to $\alpha$ if in the above proof $\delta$ is chosen as $\delta_1 = \rho/4$ in the first step and in the other steps as $\delta_j = \rho/(4J)$. This leads to a more complicated quadratic equation where now $b(\zeta)$ is analytic for $|\zeta| \leq (C_1 J^\alpha)^{-1}$. We omit the details of this refinement of the proof. □

For the remainder term in (2.7) we then obtain the following bound.

**Lemma 4.3.** *In the situation of Lemma 4.2, with $r \leq 1$ and for $C_1 N^\alpha \leq 1/(2\varepsilon)$,*

$$\|R_N(b^*,\cdot)\|_{\rho/2} \leq 4Mr\left(\frac{4C_1}{r}N^\alpha\right)^N\,.$$

*Proof.* The remainder term $R_N$ in (2.7) is a sum of terms

$$\frac{1}{i!}\frac{\partial^i H_{k_0}}{\partial a^i}(Q_{k_1},\ldots,Q_{k_i}) \quad \text{for } k_0 + k_1 + \ldots + k_i = N,$$

where

$$Q_k = \frac{\partial S_k}{\partial \theta} + \varepsilon\frac{\partial S_{k+1}}{\partial \theta} + \ldots + \varepsilon^{N-k-1}\frac{\partial S_{N-1}}{\partial \theta}\,.$$

As long as $C_1 N^\alpha \leq 1/(2\varepsilon)$, we have, by Lemma 4.2,

$$\|Q_k(b^*,\cdot)\|_{\rho/2} \leq \sum_{j=k}^{N-1} \varepsilon^{(j-k)} C_0 (C_1 j^\alpha)^j$$

$$\leq C_0 \sum_{j=k}^{N-1} 2^{-(j-k)}\left(\frac{j}{N}\right)^{\alpha j}(C_1 N^\alpha)^k \leq 2C_0\,(C_1 N^\alpha)^k\,.$$

This implies

$$\left\|\frac{1}{i!}\frac{\partial^i H_{k_0}}{\partial a^i}(Q_{k_1},\ldots,Q_{k_i})(b^*,\cdot)\right\|_{\rho/2} \leq \frac{M}{r^i}\,2C_0(C_1 N^\alpha)^N$$

for $k_0 + k_1 + \ldots + k_i = N$. (This bound is also valid when an argument different from $b^*$ appears in the derivatives of $H_0$ and $H_1$, as is needed for the remainder terms in the Taylor expansion.) Estimating the number of such expressions by

$$2\sum_{i=1}^{N}\binom{N+i-1}{i} \leq 2\sum_{i=0}^{2N-1}\binom{2N-1}{i} = 2^{2N}$$

yields the result. □

## X.4.2 Near-Invariant Tori of Perturbed Integrable Systems

The following result extends Lemma 2.1 to exponentially long times for sufficiently small values of the perturbation parameter.

**Theorem 4.4.** *Let $H_0$, $H_1$ be real-analytic on the complex $r$-neighbourhood $B_r(b^*)$ of $b^* \in \mathbb{R}^d$ and on $B_r(b^*) \times U_\rho$, respectively, with $r \leq 1$ and $\rho \leq 1$. Suppose that $\omega(b^*) = (\partial H_0/\partial a)(b^*)$ satisfies the diophantine condition (2.4). There are positive constants $\varepsilon_0$, $c_0$, $C$ such that the following holds for every positive $\beta \leq 1$ and for $\varepsilon \leq \varepsilon_0$: there exists a real-analytic symplectic change of coordinates $(a, \theta) \mapsto (b, \varphi)$ such that every solution $(b(t), \varphi(t))$ of the perturbed system in the new coordinates, starting with $\|b(0) - b^*\| \leq c_0 \varepsilon^{2\beta}$, satisfies*

$$\|b(t) - b(0)\| \leq Ct \exp(-c\varepsilon^{-\beta/\alpha}) \quad \text{for} \quad t \leq \exp(\tfrac{1}{2}c\varepsilon^{-\beta/\alpha}).$$

*Here, $\alpha = \nu + d + 1$ and $c = (16\, C_1 e/r)^{-1/\alpha}$ with $C_1$ of Lemma 4.2. Moreover, the transformation is such that, for $(a, \theta)$ and $(b, \varphi)$ related by the above coordinate transform,*

$$\|a - b\| \leq C\varepsilon \quad \text{for} \quad \|b - b^*\| \leq c_0\, \varepsilon^{2\beta}, \ \varphi \in U_{\rho/2}.$$

*The thresholds $\varepsilon_0$ and $c_0$ are such that $\varepsilon_0^{2\beta}$ is inversely proportional to $\gamma C_1^2$, and $c_0$ is proportional to $\gamma C_1^2$.*

**Remark 4.5.** Theorem 4.4 is a *local* result, showing that for $b_0$ near $b^*$ the tori $\{b = b_0,\ \varphi \in \mathbb{T}^d\}$ are nearly invariant, up to exponentially small deviations, over exponentially long times. Nekhoroshev (1977, 1979) has shown the *global* result, under a "steepness condition" which is in particular satisfied for convex Hamiltonians, that for sufficiently small $\varepsilon$ *every* solution of the perturbed Hamiltonian system satisfies, for some positive constants $A, B < 1$ (proportional to the inverse of the square of the dimension),

$$\|a(t) - a(0)\| \leq \varepsilon^B \quad \text{for} \quad t \leq \exp(\varepsilon^{-A}).$$

**Remark 4.6.** The constant $C_1$ in Lemma 4.2 and constants in similar estimates of Hamiltonian perturbation theory are very large, with the consequence that the results on the long-time behaviour derived from them are meaningful, in a rigorous sense, only for extremely small values of the perturbation parameter $\varepsilon$. Nevertheless, apart from their pure theoretical interest these results are of value as they describe the behaviour to be expected if one presupposes that the constants obtained from the worst-case estimations are unduly pessimistic for a given problem, as is typically the case.

*Proof of Theorem 4.4.* The proof combines Lemmas 4.2 and 4.3 with the proof of Lemma 2.1. An appropriate choice of the truncation indices $N$ and $m$ then gives the exponential estimates.

As in the proof of Lemma 2.1, we approximate $H_1(b,\theta)$ by a trigonometric polynomial of order $m$ in $\theta$. The error of this approximation is bounded by $\mathcal{O}(e^{-m\rho/2})$ on $B_r(b^*) \times U_{\rho/2}$, which is $\mathcal{O}(e^{-N})$ for the choice $m = 2N/\rho$ made below. By the arguments of the proof of Lemma 2.1, the estimates of Lemmas 4.2 and 4.3 (for $\gamma$ replaced by $\gamma/2$, which increases $C_1$ to $4C_1$) are then valid in $\mathcal{O}((jm)^{-\alpha})$ and $\mathcal{O}((Nm)^{-\alpha})$ neighbourhoods of $b^*$: for a sufficiently small constant $c^*$ and with $C_2 = 16\, C_1/r$,

$$\left\|\frac{\partial S_j}{\partial \theta}(b,\theta)\right\| \le C_0 (4C_1 j^\alpha)^{j-1} \quad \text{for} \quad \|b-b^*\| \le c^*(jm)^{-\alpha},\; \theta \in U_{\rho/2},$$

$$|R_N(b,\theta)| \le 4Mr\,(C_2 N^\alpha)^N \quad \text{for} \quad \|b-b^*\| \le c^*(Nm)^{-\alpha},\; \theta \in U_{\rho/2}.$$

We now consider the symplectic change of variables $(a,\theta) \mapsto (b,\varphi)$ defined by the generating function $S(b,\theta)$. The Hamiltonian equations in the variables $(b,\varphi)$ are then of the form, for $\|b-b^*\| \le c^*(Nm)^{-\alpha}$,

$$\begin{aligned}
\dot b &= -\frac{\partial K}{\partial \varphi}(b,\varphi) = -\varepsilon^N \frac{\partial R_N}{\partial \theta}\frac{\partial \theta}{\partial \varphi} = \mathcal{O}(\varepsilon^N (C_2 N^\alpha)^N) \\
\dot \varphi &= \frac{\partial K}{\partial b}(b,\varphi) = \omega_{\varepsilon,N}(b) + \mathcal{O}((Nm)^\alpha \cdot \varepsilon^N (C_2 N^\alpha)^N).
\end{aligned} \qquad (4.1)$$

Choosing $m = 2N/\rho$ and $N$ such that $C_2 N^\alpha = 1/(e\varepsilon^\beta)$ gives

$$\begin{aligned}
\dot b &= \mathcal{O}(\exp(-c\varepsilon^{-\beta/\alpha})) \\
\dot \varphi &= \omega_{\varepsilon,N}(b) + \mathcal{O}(\varepsilon^{-2\beta} \exp(-c\varepsilon^{-\beta/\alpha}))
\end{aligned} \qquad \text{for} \quad \|b-b^*\| \le c_0\,\varepsilon^{2\beta} \quad (4.2)$$

with $c = (C_2 e)^{-\alpha}$, which yields the result. $\square$

### X.4.3 Near-Invariant Tori of Symplectic Integrators

We return to the situation of Sect. X.3 and apply a symplectic numerical method to the integrable Hamiltonian system (3.1) with (3.2) and (3.3).

**Theorem 4.7.** *Consider applying a symplectic numerical integrator of order $p$ to the real-analytic completely integrable Hamiltonian system (3.1). Suppose that $\omega(a^*)$ satisfies the diophantine condition (2.4). Then, there exist positive constants $c_0, c, C$ and $h_0$ such that the following holds for all step sizes $h \le h_0$ and for all $\mu \le \min(p,\alpha)$ with $\alpha = \nu + d + 1$: every numerical solution starting with $\|I(p_0,q_0) - a^*\| \le c_0 h^{2\mu}$ satisfies*

$$\|I(p_n,q_n) - I(p_0,q_0)\| \le C h^p \quad \text{for} \quad nh \le \exp(c\,h^{-\mu/\alpha}).$$

*The constants $h_0, c_0, c, C$ depend on $d, \gamma, \nu$, on bounds of the real-analytic Hamiltonian $H$ on a complex neighbourhood of the torus $\{(p,q)\,;\, I(p,q) = a^*\}$, and on the numerical method.*

*Proof.* The proof is obtained by following the arguments of the proof of Theorem 3.1. Instead of Lemma 2.1, now Theorem 4.4 is applied to the modified Hamiltonian system (3.5) with $\varepsilon = h^p$. This gives a change of coordinates $(a, \theta) \mapsto (b, \varphi)$ $\mathcal{O}(h^p)$-close to the identity, such that in the new variables, the solution $(\widetilde{b}(t), \widetilde{\varphi}(t))$ of (3.5) satisfies

$$\widetilde{b}(t) = b_0 + \mathcal{O}(\exp(-ch^{-\mu/\alpha})) \quad \text{for} \quad t \leq \exp(ch^{-\mu/\alpha}).$$

On the other hand, using the exponentially small bound of Theorem IX.7.6, together with Theorem 4.4 and the arguments of part (b) of the proof of Theorem 3.1, yields for the numerical solution in the new variables

$$b_n = \widetilde{b}(t) + \mathcal{O}(\exp(-ch^{-\mu/\alpha})) \quad \text{for} \quad t = nh \leq \exp(ch^{-\mu/\alpha}).$$

Together with $a_n - b_n = \mathcal{O}(h^p)$ this gives the result. □

**Remark 4.8.** When the symplectic method is applied to a perturbed integrable system as in Theorem 4.4, then the same argument yields for $\|I(p_0, q_0) - a^*\| \leq c_0 \eta^{2\beta}$ with $\eta = \max(\varepsilon, h^p)$ and $\beta \leq 1$ the bound

$$\|I(p_n, q_n) - I(p_0, q_0)\| \leq C\eta \quad \text{for} \quad t \leq \exp(c\eta^{-\beta/\alpha}).$$

## X.5 Kolmogorov's Theorem on Invariant Tori

> (The proof of this theorem was published in Dokl. Akad. Nauk SSSR **98** (1954), 527–530 [MR **16**, 924], but the convergence discussion does not seem convincing to the reviewer.) This very interesting theorem would imply that for an analytic canonical system which is close to an integrable one, all solutions but a set of small measure lie on invariant tori.
> 
> (J. Moser 1959)

It was a celebrated discovery by Kolmogorov (1954) that invariant tori carrying a conditionally periodic flow with diophantine frequencies persist under small perturbations of the Hamiltonian. Together with the extensions and refinements by Arnold (1963), Moser (1962) and later authors, Kolmogorov's result forms what is now known as KAM theory. Here we give a proof of Kolmogorov's theorem and use it in studying the long-time behaviour of symplectic numerical methods applied to perturbed integrable systems.

### X.5.1 Kolmogorov's Theorem

In Sect. X.2.3 we have already given Kolmogorov's transformation which reduces the size of a perturbation to a Hamiltonian of the form (2.8) from $\mathcal{O}(\varepsilon)$ to $\mathcal{O}(\varepsilon^2)$, at least formally. The iteration of that procedure is convergent and yields the following result.

A.N. Kolmogorov[2]  V.I. Arnold[3]  J.K. Moser[4]

**Theorem 5.1 (Kolmogorov 1954).** *Consider a real-analytic Hamiltonian $H(a,\theta)$, defined for $a$ in a neighbourhood of $0 \in \mathbb{R}^d$ and $\theta \in \mathbb{T}^d$, for which the linearization at $a^* = 0$ does not depend on the angles:*

$$H(a,\theta) = c + \omega \cdot a + \tfrac{1}{2} a^T M(a,\theta) a \,. \tag{5.1}$$

*Suppose that $\omega \in \mathbb{R}^d$ satisfies the diophantine condition (2.4), viz.,*

$$|k \cdot \omega| \geq \gamma |k|^{-\nu} \quad \text{for} \quad k \in \mathbb{Z}^d, k \neq 0, \tag{5.2}$$

*and that the angular average $\overline{M}_0$ of $M(0,\cdot)$ is an invertible $d \times d$ matrix:*

$$\|\overline{M}_0 v\| \geq \mu \|v\| \quad \text{for} \quad v \in \mathbb{R}^d, \tag{5.3}$$

*with positive constants $\gamma, \nu, \mu$. Let $H_\varepsilon(a,\theta) = H(a,\theta) + \varepsilon G(a,\theta)$ be a real-analytic perturbation of $H(a,\theta)$. Then, there exists $\varepsilon_0 > 0$ such that for every $\varepsilon$ with $|\varepsilon| \leq \varepsilon_0$, there is an analytic symplectic transformation $\psi_\varepsilon : (b,\varphi) \mapsto (a,\theta)$, $\mathcal{O}(\varepsilon)$ close to the identity and depending analytically on $\varepsilon$, which puts the perturbed Hamiltonian back to the form*

$$H_\varepsilon(a,\theta) = c_\varepsilon + \omega \cdot b + \tfrac{1}{2} b^T M_\varepsilon(b,\varphi) b \quad \text{for} \quad (a,\theta) = \psi_\varepsilon(b,\varphi). \tag{5.4}$$

*The perturbed system therefore has the invariant torus $\{b = 0, \varphi \in \mathbb{T}^d\}$ carrying a quasi-periodic flow with the same frequencies $\omega$ as the unperturbed system.*
*(The threshold $\varepsilon_0$ depends on $d, \nu, \gamma, \mu$ and on bounds of $H$ and $G$ on a complex neighbourhood of $\{0\} \times \mathbb{T}^d$.)*

---

[2] Andrei Nikolaevich Kolmogorov, born: 25 April 1903 in Tambov (Russia), died: 20 October 1987 in Moscow.
[3] Vladimir Igorevich Arnold, born: 12 June 1937 in Odessa (USSR).
[4] Jürgen K. Moser, born: 4 July 1928 in Königsberg, now Kaliningrad, died: 17 December 1999 in Zürich (Switzerland).

Of particular interest is the case when $H(a, \theta) = H_0(a)$ is independent of $\theta$, so that we are considering perturbations of an integrable system. In this case, the theorem shows that all invariant tori with frequencies $\omega(a) = \partial H_0/\partial a(a)$ satisfying (5.2) and with invertible Hessian $\partial^2 H_0/\partial a^2(a)$ persist under small perturbations and are only slightly deformed. in some neighbourhood

Kolmogorov (1954) stated the theorem and formulated the iteration of Section X.2.3, but did not give the details of the convergence estimates. Arnold (1963) gave a first complete proof of the theorem for perturbed integrable systems, using a construction based on the "ultra-violet cutoff" (cf. Lemma 2.1) which yields a single transformation simultaneously for all frequencies satisfying the diophantine condition (2.4), in contrast to Kolmogorov's iteration which yields a different transformation for every choice of diophantine frequencies. However, Arnold's transformation is no longer analytic in the perturbation parameter $\varepsilon$. Moser (1962) showed that the analyticity of the Hamiltonian can be replaced by differentiability of sufficiently high order. Full proofs of Kolmogorov's theorem along his original construction were published by Thirring (1977) (for a reduced model problem) and by Benettin, Galgani, Giorgilli & Strelcyn (1984).

As in Remark 4.6, a practical difficulty with Theorem 5.1 is that the theoretically obtained threshold $\varepsilon_0$ is very small. The proof below requires $\varepsilon_0 \le \delta_0^{5\alpha}$ with $\alpha = \nu + d + 1$ of Lemma 4.1, where $\delta_0$ is inversely proportional to $\nu$. This pessimistic estimate of the threshold can be somewhat improved by first reducing the perturbation of an integrable Hamiltonian system via a perturbation series expansion as in the proof of Theorem 4.4 and then applying Kolmogorov's theorem to the remainder of the truncated perturbation series.

The proof of Theorem 5.1 uses iteratively the following lemma, which refers to the transformation constructed in Sect. X.2.3. Similar to Sect. X.4 we use the notation
$$\|G\|_\rho = \sup\{|G(a,\theta)|\,;\, \|a\| < \rho,\, \|\mathrm{Im}\,\theta\| < \rho\}$$
for a bounded analytic function $G$ on $W_\rho := B_\rho(0) \times U_\rho$, where again $B_\rho(0)$ is the complex ball of radius $\rho$ around 0 and $U_\rho$ is the complex extension of $\mathbb{T}^d$ of width $\rho$. The same notation is used for vector- and matrix-valued functions, in which case the underlying norm on $\mathbb{C}^d$ or $\mathbb{C}^{d \times d}$ is the maximum norm or its induced matrix norm, respectively.

**Lemma 5.2.** *In the situation of Sect. X.2.3 and under the conditions of Theorem 5.1, suppose that $H$ and $G$ are real-analytic and bounded on $W_\rho$. Then, there exists $\delta_0 > 0$ such that the following bounds hold for Kolmogorov's transformation whenever $0 < \delta \le \delta_0$:*

*if* $\|\varepsilon G\|_\rho \le \delta^{5\alpha}$, *then* $\|\varepsilon^2 \widehat{G}\|_{\rho-\delta} \le (\tfrac{1}{2}\delta)^{5\alpha}$

*and* $\|\varepsilon \nabla \chi\|_{\rho-\delta} \le \delta^{3\alpha}$, $\|\widehat{M} - M\|_{\rho-\delta} \le \delta^{2\alpha}$,

*where $\alpha = \nu + d + 1$. The threshold $\delta_0$ depends only on $d, \nu, \gamma, \mu$ and on $\|H\|_\rho$.*

*Proof.* We estimate the terms arising in the construction of Kolmogorov's transformation of Sect. X.2.3. For brevity we denote $\|\cdot\|_j = \|\cdot\|_{\rho - j\delta/4}$ for $j = 0, 1, 2, 3, 4$.

(a) The transformation $(b, \varphi) \mapsto (a, \theta)$ is constructed such that $(a, \theta) = y(\varepsilon)$, where $y(t)$ is the solution of $\dot{y} = J^{-1} \nabla \chi(y)$ with $y(0) = (b, \varphi)$. Suppose for the moment that

$$\|\varepsilon \nabla \chi\|_3 \leq \tfrac{1}{4}\delta . \tag{5.5}$$

Let $(b, \varphi) \in W_{\rho - \delta}$. Then, $\|y(t) - y(0)\| \leq \tfrac{1}{4}\delta$ for $0 \leq t \leq \varepsilon$, and in particular $\|(a, \theta) - (b, \varphi)\| \leq \tfrac{1}{4}\delta$. We define

$$\begin{aligned} \varepsilon^2 R(b, \varphi) &:= \left( a - b + \varepsilon \frac{\partial \chi}{\partial \varphi}(b, \varphi),\; \theta - \varphi - \varepsilon \frac{\partial \chi}{\partial b}(b, \varphi) \right) \\ &= y(\varepsilon) - y(0) - \varepsilon J^{-1} \nabla \chi(y(0)) \end{aligned}$$

and note

$$\|R(b, \varphi)\| \leq \tfrac{1}{2} \max_{0 \leq t \leq \varepsilon} \|\ddot{y}(t)\| \leq \tfrac{1}{2} \| J^{-1} \nabla^2 \chi \, J^{-1} \nabla \chi \|_3$$

so that

$$\|R\|_4 \leq \tfrac{1}{2} \|\nabla^2 \chi\|_3 \|\nabla \chi\|_3 . \tag{5.6}$$

(b) Tracing the construction of Sect. X.2.3, we find by Taylor expansion of $H(a, \theta)$ that the new matrix is

$$\widehat{M}(b, \varphi) = M(b, \varphi) + \varepsilon L(b, \varphi)$$

with

$$L(b, \varphi) = \sum_{i=1}^{d} \left( \frac{\partial M}{\partial a_i} \frac{\partial \chi}{\partial \varphi_i} - \frac{\partial M}{\partial \theta_i} \frac{\partial \chi}{\partial b_i} \right)(b, \varphi) + P(b, \varphi) + Q(b, \varphi)$$

where $P(b, \varphi)$ is symmetric with

$$b^T P(b, \varphi) b = b^T \bigl( M(b, \varphi) - M(0, \varphi) \bigr) \frac{\partial \chi}{\partial \varphi}$$

and where $Q(b, \varphi)$ is given by (2.11). It follows that

$$\|\widehat{M} - M\|_4 \leq 2\varepsilon \bigl( \|\nabla M\|_4 \|\nabla \chi\|_4 + \|\nabla^2 G\|_4 \bigr) . \tag{5.7}$$

From the construction of $\widehat{G}$ we also find by simple estimates of Taylor remainders

$$\|\widehat{G}\|_4 \leq \|\nabla H\|_3 \|R\|_4 + \|\nabla G\|_3 \|\nabla \chi\|_4 + \|\nabla^2 H\|_3 \|\nabla \chi\|_4^2 . \tag{5.8}$$

(c) Using Lemma 4.1 in the equations (2.12)–(2.16) defining $\chi$ of (2.10), we obtain first

$$\|\chi_0\|_1 \leq \kappa_0 \delta^{-\alpha+1} \|G_0\|_0 , \qquad \left\| \frac{\partial \chi_0}{\partial \varphi} \right\|_1 \leq \kappa_1 \delta^{-\alpha} \|G_0\|_0$$

and by a second application of that lemma, for $i = 1, \ldots, d$,

$$\|\chi_i\|_2 \leq \kappa_0 \delta^{-\alpha+1} (\|u\|_1 + \|v\|_1 + \|G_i\|_1)$$

where, by construction of $u$ and $v$,

$$\|v\|_1 \leq \|M\|_1 \left\|\frac{\partial \chi_0}{\partial \varphi}\right\|_1, \quad \|u\|_1 \leq \|M\|_1 \mu^{-1}\left(\|v\|_1 + \sum_{j=1}^{d} \|G_j\|_1\right).$$

It then follows by Cauchy's estimates that

$$\|\nabla\chi\|_3 \leq C\delta^{-2\alpha} \|G\|_0, \quad \|\nabla^2\chi\|_3 \leq C\delta^{-2\alpha-1} \|G\|_0. \quad (5.9)$$

(d) Combining the estimates (5.6)–(5.9) and using once more Cauchy's estimates to bound derivatives of $H$ and $G$ yields

$$\|\varepsilon^2 \widehat{G}\|_{\rho-\delta} \leq C\delta^{-4\alpha-1}\|\varepsilon G\|_\rho^2$$
$$\|\varepsilon \nabla \chi\|_{\rho-\delta} \leq C\delta^{-2\alpha}\|\varepsilon G\|_\rho$$
$$\|\widehat{M} - M\|_{\rho-\delta} \leq C\delta^{-2\alpha-3}\|\varepsilon G\|_\rho.$$

All this holds under the condition (5.5). By (5.9), this condition is satisfied if $\|\varepsilon G\|_\rho \leq \delta^{5\alpha}$ and $\delta \leq \delta_0$ with a sufficiently small $\delta_0$. (Tracing the above constants shows that $\delta_0$ needs to be inversely proportional to $\kappa_1^{1/\alpha}$, or inversely proportional to $\nu$.) This yields the stated bounds. □

*Proof of Theorem 5.1.* Kolmogorov's iteration yields sequences

$$G^{(0)} = G, G^{(1)}, G^{(2)}, \ldots$$
$$M^{(0)} = M, M^{(1)}, M^{(2)}, \ldots$$
$$\chi^{(0)}, \chi^{(1)}, \chi^{(2)}, \ldots.$$

By Lemma 5.2 they satisfy, provided that $\|\varepsilon G\|_\rho = \delta^{5\alpha}$ with $\delta \leq \delta_0$,

$$\|\varepsilon^{2^j} G^{(j)}\|_{\rho^{(j)}} \leq (2^{-j}\delta)^{5\alpha} \quad (5.10)$$
$$\|M^{(j+1)} - M^{(j)}\|_{\rho^{(j)}} \leq (2^{-j}\delta)^{2\alpha} \quad (5.11)$$
$$\|\varepsilon^{2^j} \nabla \chi^{(j)}\|_{\rho^{(j)}} \leq (2^{-j}\delta)^{3\alpha} \quad (5.12)$$

where $\rho^{(j)} = \rho - (1 + \frac{1}{2} + \ldots + 2^{-j})\delta > \frac{1}{2}\rho$ for all $j$. Note that (5.11) implies that the inverse of $M^{(j)}$ is bounded by $2\mu^{-1}$ for all $j$, so that the iterative use of Lemma 5.2 is justified. The time-$\varepsilon^{2^j}$ flow of $\chi^{(j)}$ is a symplectic transformation $\sigma_\varepsilon^{(j)}$, which by (5.12) satisfies

$$\|\sigma_\varepsilon^{(j)} - \mathrm{id}\|_{\rho/2} \leq (2^{-j}\delta)^{3\alpha}. \quad (5.13)$$

The composed transformation

$$\psi_\varepsilon^{(j)} := \sigma_\varepsilon^{(0)} \circ \sigma_\varepsilon^{(1)} \circ \ldots \circ \sigma_\varepsilon^{(j)}$$

is constructed such that

$$H(\psi_\varepsilon^{(j-1)}(b,\varphi)) = c^{(j)} + \omega \cdot b + b^T M^{(j)}(b,\varphi) b + \varepsilon^{2^j} G^{(j)}(b,\varphi) . \tag{5.14}$$

By (5.13), the sequence $\psi_\varepsilon^{(j)}(b,\varphi)$ converges uniformly on $W_{\rho/2} \times (-\varepsilon_0, \varepsilon_0)$ to a limit $\psi_\varepsilon(b,\varphi)$. By Weierstrass' theorem, $\psi_\varepsilon(b,\varphi)$ is analytic in $(b,\varphi,\varepsilon)$ (and in any further parameters on which $M$ and $G$ might possibly depend analytically). Since $\psi_\varepsilon$ depends analytically on $\varepsilon$ and $\psi_0 = \mathrm{id}$, it follows that $\psi_\varepsilon$ is $\mathcal{O}(\varepsilon)$-close to the identity on $W_{\rho/2}$. By (5.10) and (5.14), the transformed Hamiltonian $H \circ \psi_\varepsilon$ is of the desired form (5.4). $\square$

## X.5.2 KAM Tori under Symplectic Discretization

Consider a Hamiltonian system

$$\dot{p} = -\frac{\partial \mathcal{H}}{\partial q}(p,q) , \quad \dot{q} = \frac{\partial \mathcal{H}}{\partial p}(p,q) , \tag{5.15}$$

for which, in suitable coordinates $(a,\theta)$, the Hamiltonian $\mathcal{H}(p,q) = H(a,\theta) + \varepsilon G(a,\theta)$ satisfies the conditions of Theorem 5.1. Kolmogorov's theorem yields a transformation to variables $(b,\varphi)$ in terms of which

$$\mathcal{H}(p,q) = \omega \cdot b + \tfrac{1}{2} b^T M_\varepsilon(b,\varphi) b ,$$

so that the torus $\mathcal{T}_\omega = \{b = 0, \varphi \in \mathbb{T}^d\}$ is invariant and the flow on it is quasi-periodic with frequencies $\omega$.

For a symplectic integrator of order $p$ applied to (5.15), backward analysis gives a modified Hamiltonian $\widetilde{\mathcal{H}}(p,q)$ which is an $\mathcal{O}(h^p)$ perturbation of $\mathcal{H}(p,q)$:

$$\widetilde{\mathcal{H}}(p,q) = \omega \cdot b + \tfrac{1}{2} b^T M_\varepsilon(b,\varphi) b + h^p \widetilde{G}(b,\varphi) . \tag{5.16}$$

Kolmogorov's theorem can be applied once more, yielding an invariant torus $\widetilde{\mathcal{T}}_\omega$ of the modified Hamiltonian $\widetilde{\mathcal{H}}(p,q)$ which again carries a quasi-periodic flow with frequencies $\omega$. Combined with the exponentially small estimates of backward analysis for the difference between numerical solutions and the flow of the modified Hamiltonian system, this gives the following result of Hairer & Lubich (1997).

**Theorem 5.3.** *In the above situation, for a symplectic integrator of order $p$ used with sufficiently small step size $h$, there is a modified Hamiltonian $\widetilde{\mathcal{H}}$ with an invariant torus $\widetilde{\mathcal{T}}_\omega$ carrying a quasi-periodic flow with frequencies $\omega$, $\mathcal{O}(h^p)$ close to the invariant torus $\mathcal{T}_\omega$ of the original Hamiltonian $\mathcal{H}$, such that the difference between any numerical solution $(p_n, q_n)$ starting on the torus $\widetilde{\mathcal{T}}_\omega$ and the solution*

$(\widetilde{p}(t), \widetilde{q}(t))$ of the modified Hamiltonian system with the same starting values remains exponentially small in $1/h$ over exponentially long times:

$$\|(p_n, q_n) - (\widetilde{p}(t), \widetilde{q}(t))\| \leq C e^{-\kappa/h} \quad \text{for} \quad t = nh \leq e^{\kappa/h} .$$

*The constants $C$ and $\kappa$ are independent of $n, h, \varepsilon$ (for $h, \varepsilon$ sufficiently small) and of the initial value $(p_0, q_0) \in \widetilde{\mathcal{T}}_\omega$.*

*Proof.* (a) For sufficiently small $h$, Kolmogorov's theorem applied to (5.16) yields a change of coordinates $(b, \varphi) \mapsto (c, \psi)$, $\mathcal{O}(h^p)$ close to the identity, which transforms the modified Hamiltonian to the form

$$\widetilde{\mathcal{H}}(p, q) = \omega \cdot c + \tfrac{1}{2} c^T M_{\varepsilon, h}(c, \psi) c ,$$

with the invariant torus $\widetilde{\mathcal{T}}_\omega = \{c = 0, \ \psi \in \mathbb{T}^d\}$. The corresponding differential equations read in these coordinates

$$\dot{c} = u(c, \psi) , \qquad \dot{\psi} = \omega + v(c, \psi) \tag{5.17}$$

where $u(c, \psi) = \mathcal{O}(\|c\|^2)$ and $v(c, \psi) = \mathcal{O}(\|c\|)$, and similarly for the derivatives $\partial u / \partial c = \mathcal{O}(\|c\|)$, $\partial u / \partial \psi = \mathcal{O}(\|c\|^2)$, and $\partial v / \partial c = \mathcal{O}(1)$, $\partial v / \partial \psi = \mathcal{O}(\|c\|)$. The constants in these $\mathcal{O}$-terms are independent of $h$ and $\varepsilon$. Let $(c(t), \psi(t))$ and $(\widehat{c}(t), \widehat{\psi}(t))$ be two solutions of (5.17) such that $\|c(t)\| \leq \beta$, $\|\widehat{c}(t)\| \leq \beta$ ($\beta$ sufficiently small) for all $t$ under consideration. Then, an argument based on Gronwall's lemma shows that their difference is bounded over a time interval $0 \leq t \leq 1/\beta$ by

$$\begin{aligned}\|c(t) - \widehat{c}(t)\| &\leq C \left( \|c(0) - \widehat{c}(0)\| + \beta \|\psi(0) - \widehat{\psi}(0)\| \right) \\ \|\psi(t) - \widehat{\psi}(t)\| &\leq C \left( t \|c(0) - \widehat{c}(0)\| + \|\psi(0) - \widehat{\psi}(0)\| \right) ,\end{aligned} \tag{5.18}$$

for some constant $C$ that does not depend on $\beta$, $h$ or $\varepsilon$.

(b) In the following we denote $y = (p, q)$ for brevity, and more specifically, $y_n$ denotes the numerical solution starting from any $y_0$ on the torus $\widetilde{\mathcal{T}}_\omega$, i.e., the $c$-coordinate of $y_0$ vanishes: $c_0 = 0$. We denote by $\widetilde{y}(t, s, z)$ the solution of the modified Hamiltonian system with initial value $\widetilde{y}(s, s, z) = z$, and more briefly $\widetilde{y}(t) = \widetilde{y}(t, 0, y_0)$ the solution starting from $y_0$. By Theorem IX.7.6, the local error of backward error analysis at $t_j = jh$ is bounded by

$$\|y_j - \widetilde{y}(t_j, t_{j-1}, y_{j-1})\| \leq \delta := Const. \, h \, e^{-3\kappa/h}$$

for some constant $\kappa$, as long as $y_j$ remains in a compact subset of the domain of analyticity of $\mathcal{H}$. We further denote the $c$-coordinates of $y_n$, $\widetilde{y}(t)$ and $\widetilde{y}(t, t_j, y_j)$ by $c_n$, $\widetilde{c}(t)$ and $\widetilde{c}(t, t_j, y_j)$, respectively. To apply the error propagation estimate (5.18), we assume that

$$\|\widetilde{c}(t, t_j, y_j)\| \leq \beta \quad \text{for} \quad t_j \leq t \leq 1/\beta \tag{5.19}$$

and for all $j$ satisfying $t_j = jh \leq 1/\beta$. This assumption will be justified by induction later, and the value of $\beta$ will be specified in (5.21) below. By (5.18) we thus obtain the bound

$$\|\widetilde{y}(t,t_j,y_j) - \widetilde{y}(t,t_{j-1},y_{j-1})\| \leq C\left(1 + (t-t_j)\right)\delta \quad \text{for} \quad t_j \leq t \leq 1/\beta.$$

Summing up from $j=1$ to $n$ gives for $t_n \leq t \leq 1/\beta$ (and $t > 2$)

$$\begin{aligned}\|\widetilde{y}(t,t_n,y_n) - \widetilde{y}(t)\| &\leq \sum_{j=1}^{n} C\left(1 + (t-t_j)\right)\delta \leq Ch^{-1}\delta\left(t_n + tt_n - t_n^2/2\right) \\ &< Ch^{-1}\delta t^2 \leq Ch^{-1}\delta/\beta^2.\end{aligned} \quad (5.20)$$

We now set
$$\beta = (2Ch^{-1}\delta)^{1/3}, \quad (5.21)$$
so that $Ch^{-1}\delta/\beta^2 = \beta/2$, and we obtain the desired estimate from (5.20) by putting $t = t_n$.

(c) We still have to justify the assumption (5.19). This will be done by induction. For $j = 0$ nothing needs to be shown, because $\widetilde{c}(t,0,y_0) = \widetilde{c}(t) \equiv 0$ as a consequence of the fact that $\widetilde{y}(t)$ stays on the invariant torus $\mathcal{T}_\omega = \{c = 0, \psi \in \mathbb{T}^d\}$. Suppose now that (5.19) holds for $j \leq n$. It then follows from (5.20) that

$$\|\widetilde{c}(t,t_n,y_n)\| < Ch^{-1}\delta/\beta^2 = \beta/2 \quad \text{for} \quad t_n \leq t \leq 1/\beta$$

(again because of $\widetilde{c}(t) \equiv 0$). Consequently we also have

$$\|c_{n+1}\| \leq \|c_{n+1} - \widetilde{c}(t_{n+1},t_n,y_n)\| + \|\widetilde{c}(t_{n+1},t_n,y_n)\| < \delta + \beta/2 \leq \beta,$$

provided that $h$ is sufficiently small so that $\delta \leq \beta/2$. By continuity, $\widetilde{c}(t,t_{n+1},y_{n+1})$ is bounded by $\beta$ on a non-empty interval $[t_{n+1},T_{n+1}]$. The computation of part (b) shows that $\|\widetilde{c}(t,t_{n+1},y_{n+1})\| \leq \beta/2$ on this interval. Hence, $T_{n+1}$ can be increased until $T_{n+1} \geq 1/\beta$. This proves the estimate (5.19) for $j = n+1$. □

## X.6 Invariant Tori of Symplectic Maps

In the preceding section, backward error analysis combined with Kolmogorov's theorem has shown that a symplectic integrator applied to a Hamiltonian system with KAM tori possesses tori that are near-invariant, up to exponentially small terms, over exponentially long times in the inverse of the step size. To obtain truly invariant tori, we need a discrete KAM theorem for perturbations of integrable near-identity maps depending on a small parameter, the step size. Such a result was recently obtained by Shang (1999, 2000), who gave a discrete Arnold-type construction. Here, we use instead a discrete-time version of Kolmogorov's iteration. This establishes the existence of invariant tori of symplectic integrators applied to integrable Hamiltonian systems or to near-integrable systems with KAM tori, for a Cantor set of non-resonant step sizes.

## X.6.1 A KAM Theorem for Symplectic Near-Identity Maps

We consider a discrete-time analogue of the situation in Sections X.2.3 and X.5.1 and construct the corresponding version of Kolmogorov's iteration. Consider the symplectic map $\sigma_h : (a, \theta) \mapsto (\widehat{a}, \widehat{\theta})$ for $a$ near $0 \in \mathbb{R}^d$, $\theta \in \mathbb{T}^d$ defined by

$$\widehat{a} = a - h\frac{\partial S}{\partial \widehat{\theta}}(a, \widehat{\theta}), \quad \widehat{\theta} = \theta + h\frac{\partial S}{\partial a}(a, \widehat{\theta}) \tag{6.1}$$

where $h$ is a small parameter (the step size), and $S : B_r(0) \times \mathbb{T}^d \to \mathbb{R}$ is a real-analytic generating function. If $S(a, \widehat{\theta})$ has the form (cf. (2.8))

$$S(a, \widehat{\theta}) = c + \omega \cdot a + \tfrac{1}{2} a^T M(a, \widehat{\theta}) a, \tag{6.2}$$

then the associated symplectic map is of the form

$$\widehat{a} = a + \mathcal{O}(h\|a\|^2), \quad \widehat{\theta} = \theta + h\omega + \mathcal{O}(h\|a\|).$$

Hence, the torus $\{a = 0, \theta \in \mathbb{T}^d\}$ is invariant, and on it the map $\sigma_h$ reduces to rotation by $h\omega$.

Consider now an analytic perturbation of such a generating function: $S(a, \widehat{\theta}) + \varepsilon R(a, \widehat{\theta})$ with a small $\varepsilon$. We construct a near-identity symplectic change of coordinates, via an iterative procedure similar to Kolmogorov's iteration of Sect. X.2.3, such that the generating function of the perturbed symplectic map in the new variables is again of the form (6.2) with the same $\omega$, and hence the perturbed map has an invariant torus on which it is conjugate to rotation by $h\omega$. This holds if $h\omega$ satisfies the following diophantine condition (cf. (2.4)):

$$\left|\frac{1 - e^{-ik \cdot h\omega}}{h}\right| \geq \gamma^* |k|^{-\nu^*} \quad \text{for} \quad k \in \mathbb{Z}^d, k \neq 0, \tag{6.3}$$

for some positive constants $\gamma^*, \nu^*$; and if the angular average $\overline{M}_0$ of $M(0, \cdot)$ is invertible:

$$\|\overline{M}_0 v\| \geq \mu^* \|v\| \quad \text{for} \quad v \in \mathbb{R}^d \tag{6.4}$$

for a positive constant $\mu^*$. As in Sect. X.2.3, we construct a symplectic transformation $(a, \theta) \mapsto (b, \varphi)$ as the time-$\varepsilon$ flow of an auxiliary Hamiltonian of the form (2.10), viz.,

$$\chi(b, \varphi) = \xi \cdot \varphi + \chi_0(\varphi) + \sum_{i=1}^{d} b_i \chi_i(\varphi)$$

where $\xi \in \mathbb{R}^d$ is a constant vector, and $\chi_0, \chi_1, \ldots, \chi_d$ are $2\pi$-periodic functions. We then consider the map conjugate to the perturbed map $(a, \theta) \mapsto (\widehat{a}, \widehat{\theta})$ generated by $S(a, \widehat{\theta}) + \varepsilon R(a, \widehat{\theta})$:

$$\begin{array}{ccc} (a, \theta) & \longrightarrow & (\widehat{a}, \widehat{\theta}) \\ \uparrow & & \downarrow \\ (b, \varphi) & & (\widehat{b}, \widehat{\varphi}) \end{array}$$

We construct $\chi$ in such a way that the above composed symplectic map is generated by $\widetilde{S}(b\widehat{\varphi}) + \varepsilon^2 \widetilde{R}(b, \widehat{\varphi})$ with $\widetilde{S}$ of the form (6.2) and both $\widetilde{S}$ and $\widetilde{R}$ real-analytic and bounded independently of $\varepsilon$ and of $h$ with (6.3). The map $(b, \varphi) \mapsto (\widehat{b}, \widehat{\varphi})$ is then of the form

$$\widehat{b} = b + \mathcal{O}(h\|b\|^2) + \mathcal{O}(h\varepsilon^2), \quad \widehat{\varphi} = \varphi + h\omega + \mathcal{O}(h\|b\|) + \mathcal{O}(h\varepsilon^2).$$

As an elementary calculation shows, this holds if $\chi$ satisfies for all $(b, \widehat{\varphi})$ with $b$ near $0$, $\widehat{\varphi} \in \mathbb{T}^d$

$$\frac{\chi(b,\widehat{\varphi}) - \chi(b, \widehat{\varphi} - h\omega)}{h} + b^T M(b, \widehat{\varphi}) \frac{\partial \chi}{\partial \varphi}(b, \widehat{\varphi} - h\omega) + R(b, \widehat{\varphi}) = C_h + \mathcal{O}(\|b\|^2)$$

where $C_h$ does not depend on $(b, \widehat{\varphi})$ and $\varepsilon$. Writing down the Taylor expansion

$$R(b, \widehat{\varphi}) = R_0(\widehat{\varphi}) + \sum_{i=1}^{d} b_i R_i(\widehat{\varphi}) + \mathcal{O}(\|b\|^2)$$

and inserting the above ansatz for $\chi$, this condition becomes fulfilled if, with $u(\widehat{\varphi}) = M(0, \widehat{\varphi})\xi$ and $v(\widehat{\varphi}) = M(0, \widehat{\varphi})(\partial \chi_0/\partial\varphi)(\widehat{\varphi} - h\omega)$,

$$\frac{\chi_0(\widehat{\varphi}) - \chi_0(\widehat{\varphi} - h\omega)}{h} + R_0(\widehat{\varphi}) = \overline{R}_0 \quad (6.5)$$

$$\frac{\chi_i(\widehat{\varphi}) - \chi_i(\widehat{\varphi} - h\omega)}{h} + u_i(\widehat{\varphi}) + v_i(\widehat{\varphi}) + R_i(\widehat{\varphi}) = \overline{u}_i + \overline{v}_i + \overline{R}_i \quad (6.6)$$

$$\overline{u}_i + \overline{v}_i + \overline{R}_i = 0 \quad (i = 1, \ldots, d) \quad (6.7)$$

where the bars again denote angular averages. We note

$$\frac{\chi_0(\widehat{\varphi}) - \chi_0(\widehat{\varphi} - h\omega)}{h} = \sum_k \frac{1 - e^{-ik \cdot h\omega}}{h} \chi_{0,k} e^{ik \cdot \widehat{\varphi}},$$

where $\chi_{0,k}$ are the Fourier coefficients of $\chi_0$. Under the diophantine condition (6.3), Equation (6.5) is thus solved like (2.14) under condition (2.4). Equations (6.6) are of the same type. The above system is then solved in the same way as (2.12)–(2.16), yielding that the perturbed map in the new coordinates, $(b, \varphi) \mapsto (\widehat{b}, \widehat{\varphi})$, is generated by

$$S^{(1)}(b, \widehat{\varphi}) = c^{(1)} + \omega \cdot b + \tfrac{1}{2} b^T M^{(1)}(b, \widehat{\varphi}) b + \varepsilon^2 R^{(1)}(b, \widehat{\varphi})$$

with unchanged frequencies $\omega$ and with $M^{(1)}(b, \widehat{\varphi}) = M(b, \widehat{\varphi}) + \mathcal{O}(\varepsilon)$. The perturbation to the form (6.2) is thus reduced from $\mathcal{O}(\varepsilon)$ to $\mathcal{O}(\varepsilon^2)$. By the same arguments as in the proof of Theorem 5.1 it is shown that the iteration of this procedure converges. This proves the following discrete-time version of Kolmogorov's theorem.

**Theorem 6.1.** *Consider a real-analytic function $S(a, \widehat{\theta})$ of the form (6.2) with (6.4), defined on a neighbourhood of $\{0\} \times \mathbb{T}^d$. Let $|h| < h_0$ ($h_0$ so small that (6.1) is a well-defined map) and suppose that $h\omega$ satisfies (6.3).*

Let $S_\varepsilon(a,\widehat\theta) = S(a,\theta) + \varepsilon R(a,\widehat\theta)$ be an analytic perturbation of $S(a,\theta)$, generating a symplectic map $\sigma_{h,\varepsilon} : (a,\theta) \mapsto (\widehat a, \widehat\theta)$ via (6.1) with $S_\varepsilon$ in place of $S$.

Then, there exists $\varepsilon_0 > 0$ such that for every $\varepsilon$ with $|\varepsilon| < \varepsilon_0$, there is an analytic symplectic transformation $\psi_{h,\varepsilon} : (b,\varphi) \mapsto (a,\theta)$, $\mathcal{O}(\varepsilon)$ close to the identity uniformly in $h$ satisfying (6.3) and analytic in $\varepsilon$, such that $\psi_{h,\varepsilon}^{-1} \circ \sigma_{h,\varepsilon} \circ \psi_{h,\varepsilon} : (b,\varphi) \mapsto (\widehat b, \widehat\varphi)$ is generated, via (6.1), by a function $S^*_{h,\varepsilon}(b,\widehat\varphi)$ which is again of the form (6.2), i.e.,

$$S^*_{h,\varepsilon}(b,\widehat\varphi) = c_{h,\varepsilon} + \omega \cdot b + \tfrac{1}{2} b^T M_{h,\varepsilon}(b,\widehat\varphi) b .$$

The perturbed map $\sigma_{h,\varepsilon}$ therefore has an invariant torus on which it is conjugate to rotation by $h\omega$.
(The threshold $\varepsilon_0$ depends only on $d, \nu^*, \gamma^*, \mu^*$ and on bounds of $S$ and $R$ on a complex neighbourhood of $\{0\} \times \mathbb{T}^d$.) □

### X.6.2 Invariant Tori of Symplectic Integrators

As a direct consequence of Theorem 6.1 we obtain the following result on invariant tori of symplectic integrators applied to KAM systems.

**Theorem 6.2.** *Apply a symplectic integrator of order $p$ to a perturbed integrable system with a KAM torus $\mathcal{T}_\omega$ which carries a quasi-periodic flow with diophantine frequencies $\omega$. Then, if the step size $h$ is sufficiently small and satisfies the strong non-resonance condition (6.3), the numerical method has an invariant torus $\mathcal{T}_{\omega,h}$ $\mathcal{O}(h^p)$-close to $\mathcal{T}_\omega$, on which it is conjugate to rotation by $h\omega$.*

*Proof.* Theorem 6.1 applies directly, with $\varepsilon = h^p$, to the above situation. Here, the generating function $S(a,\widehat\theta)$ of the time-$h$ flow $\varphi_h$ of the Hamiltonian system with the KAM torus $\mathcal{T}_\omega$ is of the form (6.2) in the variables $(a,\theta)$ obtained by Kolmogorov's theorem. The matrix $M(a,\widehat\theta)$ in (6.2) then differs from the corresponding matrix of (2.8) by $\mathcal{O}(h)$, so that (5.3) implies (6.4). Finally, the generating function of the numerical one-step map $\Phi_h$ is an $\mathcal{O}(h^p)$-perturbation $S(a,\widehat\theta) + h^p R(a,\widehat\theta)$. □

### X.6.3 Strongly Non-Resonant Step Sizes

Theorem 6.2 leaves us with an interesting question: if $\omega \in \mathbb{R}^d$ is a vector of frequencies that satisfies the diophantine condition (2.4), then which step sizes $h$ satisfy the non-resonance condition (6.3)? Here we give a lemma in the spirit of results by Shang (2000). It shows that the probability of picking an $h \in (0, h_0)$ satisfying (6.3) tends to 1 as $h_0 \to 0$.

**Lemma 6.3.** *Suppose $\omega \in \mathbb{R}^d$ satisfies (2.4), and let $h_0 > 0$. For any choice of positive $\gamma^*$ and $\nu^*$, the set*

$$Z(h_0) = \{h \in (0, h_0) \;;\; h \text{ does not satisfy (6.3)}\}$$

is open and dense in $(0, h_0)$. If $\gamma^* \leq \gamma$ and $\nu^* > \nu + d + r$ with $r > 1$, then the Lebesgue measure of $Z(h_0)$ is bounded by

$$\text{measure}\left(Z(h_0)\right) \leq C \frac{\gamma^*}{\gamma} h_0^{r+1}$$

where $C$ depends only on $d, \nu, \nu^*$ and $\|\omega\|$.

*Proof.* It is clear from the definition that $Z(h_0)$ is open and dense in $(0, h_0)$. It remains to prove the estimate of the Lebesgue measure. For every $k \in \mathbb{Z}^d$ and $|h| \leq h_0$, there exists an integer $l = l(k, h)$ such that

$$|1 - e^{-ik \cdot h\omega}| \geq \frac{2}{\pi} |k \cdot h\omega - 2\pi l| = \frac{2}{\pi} |k \cdot \omega| \cdot \left| h - \frac{2\pi l}{|k \cdot \omega|} \right|.$$

For this $l$ we must have, by the triangle inequality,

$$2\pi |l| \leq \pi + |k| h_0 \|\omega\|,$$

so that in case $l \neq 0$

$$\frac{1}{|k|} \leq \frac{h_0 \|\omega\|}{2\pi(|l| - \frac{1}{2})}.$$

On the other hand, $l = 0$ yields

$$\left| \frac{1 - e^{-ik \cdot h\omega}}{h} \right| \geq \frac{2}{\pi} |k \cdot \omega| \geq \frac{2}{\pi} \gamma |k|^{-\nu}$$

which implies $h \notin Z(h_0)$. Hence, $h$ can be in $Z(h_0)$ only if there exist $k \in \mathbb{Z}^d$, $k \neq 0$ and an integer $l \neq 0$ such that

$$\left| h - \frac{2\pi l}{|k \cdot \omega|} \right| \leq \frac{\pi}{2} \frac{|h|}{|k \cdot \omega|} \frac{\gamma^*}{|k|^{\nu^*}} \leq \frac{\pi}{2} |h| \frac{|k|^\nu}{\gamma} \frac{\gamma^*}{|k|^{\nu^*}}$$

$$\leq \frac{\pi}{2} \frac{\gamma^*}{\gamma} |k|^{\nu + r - \nu^*} \left( \frac{\|\omega\|}{2\pi} \frac{1}{|l| - \frac{1}{2}} \right)^r h_0^{r+1}.$$

It follows that

$$\text{measure}\left(Z(h_0)\right) \leq 2 \sum_{k \neq 0} \sum_{l \neq 0} \frac{\pi}{2} \frac{\gamma^*}{\gamma} |k|^{\nu + r - \nu^*} \left( \frac{\|\omega\|}{2\pi} \frac{1}{|l| - \frac{1}{2}} \right)^r h_0^{r+1},$$

which yields the stated result. □

## X.7 Exercises

1. Let $R$ be a $d \times 2d$ matrix of rank $d$. Show that there exists a symplectic $2d \times 2d$ matrix $A$ such that $RA = (P, Q)$ with an invertible $d \times d$ matrix $P$.
   *Hint.* Consider first the case $d = 2$ and then reduce the general situation to a sequence of transformations for that case.

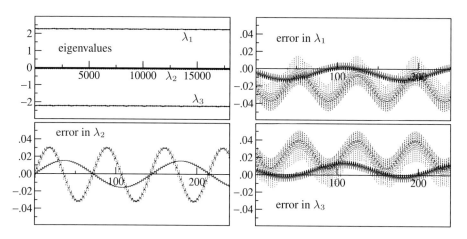

**Fig. 7.1.** Numerically obtained eigenvalues (left pictures) and errors in the eigenvalues (right pictures) for the step sizes $h = 0.1$ (dotted) and $h = 0.05$ (solid line).

2. The transformation $(x, y) \mapsto (x, y + d(x, y))$ is symplectic if and only if the partial derivatives of $d$ satisfy $d_x = d_x^T$, $d_y = 0$.

3. In the situation of Lemma 1.1, if $(F_1, \ldots, F_d, \widetilde{G}_1, \ldots, \widetilde{G}_d)^T$ is another such symplectic transformation, then there exists a smooth function $W$ depending only on $x = (x_1, \ldots, x_d)$ such that, for $x_j = F_j(p, q)$,

$$\widetilde{G}_i(p, q) - G_i(p, q) = \frac{\partial W}{\partial x_i}(x).$$

*Hint.* Use the previous exercise.

4. Show that every discrete subgroup of $\mathbb{R}^d$ is a grid, generated by $k \leq d$ linearly independent vectors.
*Solution.* See e.g. Arnold (1989), Sect. 10D.

5. Show the following bound of the Lebesgue measure of non-diophantine frequencies (Arnold 1963): for any bounded domain $\Omega \subset \mathbb{R}^d$,

$$\text{measure}\{\omega \in \Omega \,;\, \omega \text{ does not satisfy (2.4) with } \nu \geq d\} \leq C(d, \Omega)\gamma.$$

*Hint.* For a fixed $k$, decompose $\omega = \omega_0 + \alpha k/|k|$ with $\omega_0 \cdot k = 0$.

6. Show that the eigenvalues $\lambda_j$ of the matrix $L$ of the Toda system are first integrals in involution.
*Hint.* For $P_\lambda = \det(\lambda I - L)$, show that $\{P_\lambda, P_\mu\} = 0$ for all $\lambda, \mu$.

7. We repeat the experiment of Fig. 1.3 with the Störmer/Verlet scheme, where we keep the initial values for the $q$-variables, but change the initial values for the $p$-variables to $p_1 = p_2 = p_3 = 0$. The numerical results, given in Fig. 7.1, are qualitatively different from those in Fig. 1.3. The errors behave more like $hc(th)$ rather than $h^2c(t)$. We do not understand this behaviour; do you?

8. Show that for a non-symplectic numerical method, there is at worst quadratic error growth in time when it is applied to an integrable Hamiltonian system.

9. Consider a numerical integrator of order $p$ (i.e., $\Phi_h(y) = \varphi_h(y) + \mathcal{O}(h^{p+1})$), and assume that
$$\Phi_h'(y)^T J \Phi_h'(y) = J + \mathcal{O}(h^{q+1})$$
with $q > p$, when the method is applied to a Hamiltonian system. Prove that under the assumptions of Theorem 3.1 the global error behaves for $t = nh$ like
$$y_n - y(t) = \mathcal{O}(th^p) + \mathcal{O}(t^2 h^q),$$
and the action variables like
$$I(y_n) - I(y_0) = \mathcal{O}(h^p) + \mathcal{O}(th^q).$$

*Remark.* Methods satisfying the assumptions of this exercise are called *pseudo-symplectic* of order $(p, q)$ (Aubry & Chartier 1998). Pseudo-symplectic methods behave like symplectic methods on time intervals of length $\mathcal{O}(h^{p-q})$.

10. Using the theory of B-series, in particular Theorem VI.7.1, derive the conditions for the coefficients of a Runge-Kutta method such that it is pseudo-symplectic of order $p(q)$. Prove that there exist explicit, pseudo-symplectic Runge-Kutta methods of order $(2, 4)$ with 3 stages.

# Chapter XI.
# Reversible Perturbation Theory and Symmetric Integrators

> There is a very close similarity between the behaviour of solutions of reversible systems and that of Hamiltonian ones.
>
> (M.B. Sevryuk 1986, p. 3)

Numerical experiments indicate that symmetric methods applied to integrable and near-integrable reversible systems share similar properties to symplectic methods applied to (near-)integrable Hamiltonian systems: linear error growth, long-time near-conservation of first integrals, existence of invariant tori. The present chapter gives a theoretical explanation of the good long-time behaviour of symmetric methods. The results and techniques are largely analogous to those of the previous chapter — the extent of the analogy may indeed be seen as the most surprising feature of this chapter.

## XI.1 Integrable Reversible Systems

We consider a system of differential equations on a domain of $\mathbb{R}^m \times \mathbb{R}^n$,

$$\begin{aligned} \dot{u} &= f(u,v) \\ \dot{v} &= g(u,v) \,, \end{aligned} \qquad (1.1)$$

which is *reversible* with respect to the involution $(u,v) \mapsto (u,-v)$: for all $(u,v)$,

$$\begin{aligned} f(u,-v) &= -f(u,v) \\ g(u,-v) &= g(u,v) \,. \end{aligned} \qquad (1.2)$$

From Sect. V.1 we recall that the time-$t$ flow $\varphi_t$ of a reversible system is a *reversible map*:

$$\varphi_t(u,v) = (\widehat{u},\widehat{v}) \qquad \text{implies} \qquad \varphi_t^{-1}(u,-v) = (\widehat{u},-\widehat{v}) \,.$$

A coordinate transform $u = \mu(x,y)$, $v = \nu(x,y)$ is said to *preserve reversibility* if the relations

$$\begin{aligned} \mu(x,-y) &= \mu(x,y) \\ \nu(x,-y) &= -\nu(x,y) \end{aligned} \qquad (1.3)$$

hold for all $(x,y)$. This implies that every reversible system (1.1) written in the new variables $(x,y)$ is again reversible, and that every reversible map $(u,v) \mapsto (\widehat{u},\widehat{v})$

expressed in the variables $(x, y)$ again becomes a reversible map $(x, y) \mapsto (\widehat{x}, \widehat{y})$. Conversely, (1.3) is necessary for these properties.

For Hamiltonian systems, complete integrability is tied to the existence of a symplectic transformation to action-angle variables; see Sect. X.1. For reversible systems, we take the existence of a reversibility-preserving transformation to such variables as the definition of integrability.

**Definition 1.1.** The system (1.1) is called an *integrable reversible system* if, for every point $(u_0, v_0) \in \mathbb{R}^m \times \mathbb{R}^n$ in the domain of $(f, g)$, there exist a function $\omega = (\omega_1, \ldots, \omega_n) : D \to \mathbb{R}^n$ and a diffeomorphism

$$\psi = (\mu, \nu) : D \times \mathbb{T}^n \to U \subset \mathbb{R}^m \times \mathbb{R}^n : (a, \theta) \mapsto (u, v)$$

(with $D$ and $U$ open sets in $\mathbb{R}^m$ and $\mathbb{R}^m \times \mathbb{R}^n$, respectively, and $(u_0, v_0) \in U$), which preserves reversibility and transforms the system (1.1) to the form

$$\begin{aligned} \dot{a} &= 0 \\ \dot{\theta} &= \omega(a) \,. \end{aligned} \tag{1.4}$$

We speak of a *real-analytic integrable reversible system* if all the functions appearing in the above definition are real-analytic.

**Example 1.2 (Motion in a Central Field).** In Examples X.1.2 and X.1.10 we constructed action-angle variables via a series of transformations

$$\begin{pmatrix} q_1, p_2 \\ p_1, q_2 \end{pmatrix} \xrightarrow{\text{(X.1.5)}} \begin{pmatrix} r, p_\varphi \\ \varphi, p_r \end{pmatrix} \xrightarrow{\text{(X.1.16)}} \begin{pmatrix} H, L \\ y_1, y_2 \end{pmatrix} \xrightarrow{\text{(X.1.17)}} \begin{pmatrix} H, L \\ \theta_1, \theta_2 \end{pmatrix}.$$

It is easily verified that all these transformations preserve reversibility. They transform the reversible system

$$\begin{aligned} \dot{q}_1 &= p_1, & \dot{p}_2 &= -q_2 V'(r)/r \\ \dot{q}_2 &= p_2, & \dot{p}_1 &= -q_1 V'(r)/r \end{aligned} \tag{1.5}$$

(with $r = \sqrt{q_1^2 + q_2^2}$) to the form

$$\begin{aligned} \dot{H} &= 0, \quad \dot{L} = 0 \\ \dot{\theta}_1 &= \frac{2\pi}{T}, \quad \dot{\theta}_2 = \frac{\Phi}{T} \end{aligned} \tag{1.6}$$

with $T = T(H, L)$ and $\Phi = \Phi(H, L)$ given by (X.1.14) and (X.1.15).

As the following result shows, it is not incidental that the above transformations preserve reversibility.

**Theorem 1.3.** *In the situation of the Arnold-Liouville theorem, Theorem X.1.6, let the first integrals $F_1, \ldots, F_d$ of the completely integrable Hamiltonian system be such that all $F_i$ are even functions of the second half of the arguments:*

$$F_i(u, v) = F_i(u, -v) \qquad (i = 1, \ldots, d). \tag{1.7}$$

*Suppose further that $\partial F_1/\partial u, \ldots, \partial F_d/\partial u$ are linearly independent everywhere (on $\bigcup \{M_x : x \in B\}$) except possibly on a set that has no interior points. Then, the transformation $\psi : (a, \theta) \mapsto (u, v)$ to action-angle variables as given by Theorem X.1.6 preserves reversibility.*

*Proof.* The result follows by tracing the proofs of Lemma X.1.1, Theorem X.1.4 and Theorem X.1.6.

(a) For $F_i$ satisfying (1.7) and at points where the Jacobian matrix $\partial F/\partial u$ is invertible, the construction of the local symplectic transformation $\ell = (F_1, \ldots, F_d, G_1, \ldots, G_d) : (u, v) \mapsto (x, y)$ shows that the generating function $S(x, v)$ becomes odd in $v$ when the integration constant is chosen such that $S(x, 0) = 0$. By (X.1.4), this implies that $\ell$ preserves reversibility. A continuity argument used together with the essential uniqueness of the transformation $\ell$ (see Exercise X.3) does away with the exceptional points where $\partial F/\partial u$ is singular.

(b) In Theorem X.1.4, the construction of the mapping

$$e(x, y) = \varphi_y(\ell^{-1}(x, 0)) =: (u, v)$$

is such that

$$e(x, -y) = \varphi_{-y}(\ell^{-1}(x, 0)) = (u, -v).$$

This holds because by (a), $\ell^{-1}(x, 0) = (u_0, 0)$ for some $u_0$, and because $\varphi_{\pm y}$ is the time $\pm 1$ flow of the Hamiltonian system with Hamiltonian $y_1 F_1 + \ldots + y_d F_d$. Condition (1.7) implies that this is a reversible system, which in turn yields that $e$ preserves reversibility as stated above.

(c) The transformation in the proof of Theorem X.1.6 is of the form

$$a = w(x), \quad y = W(x)\theta$$

(with invertible $W(x) = w'(x)$) and hence preserves reversibility. $\square$

**Example 1.4 (Motion in a Central Field, Continued).** The condition (1.7) is satisfied for $F_1 = H$, $F_2 = L = p_1 q_2 - p_2 q_1$ if we take the symplectic coordinates $u = (q_1, p_2)$ and $v = (-p_1, q_2)$. With $F_1 = H$, $F_2 = L^2$ ($L \neq 0$ as always) we can also take $u = (p_1, p_2)$ and $v = (q_1, q_2)$, or $u = (q_1, q_2)$ and $v = (-p_1, -p_2)$.

**Example 1.5 (Toda Lattice).** Consider the Toda lattice of Sect. X.1.5. The eigenvalues of the matrix $L$ are first integrals in involution. They are odd functions of the momenta $p$, as is seen by a similarity transformation with the matrix that contains $+1$ and $-1$ as alternating values on the diagonal. (This requires even $n$ or case (ii): $b_n = 0$.) The squares of the eigenvalues therefore satisfy (1.7) for the symplectic coordinates $(u, v) = (q, -p)$. Theorem 1.3 is applicable when the eigenvalues are different from each other and different from 0.

For a different choice of invariants Theorem 1.3 also becomes applicable when $n$ is odd or in cases where an eigenvalue is 0. We illustrate this for $n = 3$ and leave the general case to Exercise 2. The characteristic polynomial of the matrix $L$ is

$$\chi(\lambda) = -\lambda^3 + (a_1 + a_2 + a_3)\lambda^2 - (a_1 a_2 + a_2 a_3 + a_3 a_1 - b_1^2 - b_2^2 - b_3^2)\lambda + (a_1 a_2 a_3 - a_1 b_2^2 - a_2 b_3^2 - a_3 b_1^2 + 2 b_1 b_2 b_3).$$

We know that the coefficients of this polynomial are first integrals in involution. The coefficient of $\lambda^2$ is an odd function of $p = (a_1, a_2, a_3)$ and hence its square is even in $p$. The coefficient of $\lambda$ is already even in $p$. Since $b_1 b_2 b_3 = \exp(\frac{1}{2}(q_1 - q_2)) \exp(\frac{1}{2}(q_2 - q_3)) \exp(\frac{1}{2}(q_3 - q_1)) \equiv 1$, also $a_1 a_2 a_3 - a_1 b_2^2 - a_2 b_3^2 - a_3 b_1^2$ is a first integral, which is again odd in $p$ and hence its square is even in $p$.

**Example 1.6 (Rigid Body Equations on the Unit Sphere).** We reconsider an example that has accompanied us all the way through Chapters IV and V: the rigid body equations (IV.1.4), here considered as differential equations on the unit sphere. We assume $I_3 < I_1, I_2$ for the inertia, which implies that any solution starting with $y_3(0) > 0$ will have $y_3(t) > 0$ for all $t$. We consider the equations in the neighbourhood of such a solution. We can then choose $u = y_1$, $v = y_2$ as coordinates on the upper half-sphere $\{y_1^2 + y_2^2 + y_3^2 = 1, y_3 > 0\}$. This gives the reversible system

$$\begin{aligned} \dot{u} &= a_1 v \sqrt{1 - u^2 - v^2} \\ \dot{v} &= a_2 u \sqrt{1 - u^2 - v^2} \end{aligned} \quad (1.8)$$

with $a_1 = (I_2 - I_3)/I_2 I_3 > 0$ and $a_2 = (I_3 - I_1)/I_3 I_1 < 0$, which has $H = u^2/I_1 + v^2/I_2 + (1 - u^2 - v^2)/I_3 = a_2 u^2 - a_1 v^2 + I_3^{-1}$ as an invariant. We introduce polar coordinates $u = r \cos \varphi$, $v = r \sin \varphi$ and express $r$ as a function of $H$ and $\varphi$:

$$r = \sqrt{\frac{I_3^{-1} - H}{a_1 \sin^2 \varphi - a_2 \cos^2 \varphi}}.$$

This leaves us with differential equations

$$\dot{H} = 0, \quad \dot{\varphi} = \gamma(H, \varphi),$$

where $\gamma$ is even in $\varphi$ and has no zeros. The time needed to run through an angle $\varphi$ is

$$\tau(H, \varphi) = \int_0^\varphi \frac{1}{\gamma(H, \phi)} d\phi, \quad \text{and} \quad \omega(H) = \frac{2\pi}{\tau(H, 2\pi)}$$

is the frequency. With $\theta = \omega(H)\tau(H, \varphi)$ we then have

$$\dot{H} = 0, \quad \dot{\theta} = \omega(H).$$

The transformation from $(u, v)$ in the open unit disc (except the origin) to $(H, \theta) \in (0, I_3^{-1}) \times \mathbb{T}$ is a diffeomorphism that preserves reversibility. This shows that the rigid body equations (1.8) are an integrable reversible system.

**Example 1.7 (Rigid Body Equations in $\mathbb{R}^3$).** We now consider the rigid body equations (IV.1.4) in the ambient space $\mathbb{R}^3$, rather than on the unit sphere. The system then has the invariants $H = y_1^2/I_1 + y_2^2/I_2 + y_3^2/I_3$ and $K = y_1^2 + y_2^2 + y_3^2$, and it is reversible with respect to the partition $u = (y_1, y_3)$ and $v = y_2$. In the case $I_3 < I_1, I_2$ we can again restrict our attention to $y_3 > 0$. We then write $y_3 = \sqrt{K - y_1^2 - y_2^2}$ and introduce polar coordinates $y_1 = r\cos\varphi$, $y_2 = r\sin\varphi$. As above, we express $r$ as a function of $H, K$ and $\varphi$ (this just requires replacing $I_3^{-1}$ with $K/I_3$ in the above formula for $r$) and we obtain differential equations

$$\dot{H} = 0, \quad \dot{K} = 0, \quad \dot{\varphi} = \gamma(H, K, \varphi)$$

with $\gamma$ even in $\varphi$ and without zeros. In the same way as above, this is transformed to

$$\dot{H} = 0, \quad \dot{K} = 0, \quad \dot{\theta} = \omega(H, K).$$

The transformation $((y_1, y_3), y_2) \mapsto ((H, K), \theta)$ preserves reversibility. The rigid body equations (IV.1.4) are thus an integrable reversible system. Note that this time the dimensions differ.

## XI.2 Transformations in Reversible Perturbation Theory

We consider perturbations of an integrable reversible system such that the perturbed system is still reversible. This takes the form

$$\begin{aligned} \dot{a} &= \varepsilon r(a, \theta) \\ \dot{\theta} &= \omega(a) + \varepsilon \rho(a, \theta) \end{aligned} \quad (2.1)$$

where $\varepsilon$ is a small parameter, and $r$ is an odd function of $\theta$ and $\rho$ is even function of $\theta$:

$$\begin{aligned} r(a, -\theta) &= -r(a, \theta) \\ \rho(a, -\theta) &= \rho(a, \theta). \end{aligned} \quad (2.2)$$

Similar to Sect. X.2 for Hamiltonian perturbation theory, we study coordinate transformations that change (2.1) to reversible systems which – in various ways – look closer to an integrable system in action-angle variables than (2.1).

### XI.2.1 The Basic Scheme of Reversible Perturbation Theory

We look for a transformation between neighbourhoods of $\{a_0\} \times \mathbb{T}^n$,

$$\begin{aligned} a &= b + \varepsilon s(b, \varphi) \\ \theta &= \varphi + \varepsilon \sigma(b, \varphi), \end{aligned} \quad (2.3)$$

which preserves reversibility and hence has $s$ even in $\varphi$ and $\sigma$ odd in $\varphi$, such that the transformed system is of the form

$$\dot b = \mathcal{O}(\varepsilon^2)$$
$$\dot\varphi = \omega(b) + \varepsilon\mu(b) + \mathcal{O}(\varepsilon^2) \,. \tag{2.4}$$

Inserting (2.3) into (2.1) gives the system

$$\left\{\begin{pmatrix} I & 0 \\ 0 & I \end{pmatrix} + \varepsilon \begin{pmatrix} \partial s/\partial b & \partial s/\partial\varphi \\ \partial\sigma/\partial b & \partial\sigma/\partial\varphi \end{pmatrix}\right\}\begin{pmatrix}\dot b \\ \dot\varphi\end{pmatrix} = \begin{pmatrix}\varepsilon r(a,\theta) \\ \omega(a)+\varepsilon\rho(a,\theta)\end{pmatrix}$$

with $(a,\theta)$ from (2.3). Inverting the matrix on the left-hand side and expanding in powers of $\varepsilon$, it is seen that (2.4) requires that $s, \sigma$ satisfy the equations

$$\frac{\partial s}{\partial\varphi}(b,\varphi)\,\omega(b) = r(b,\varphi) \tag{2.5}$$

$$\frac{\partial\sigma}{\partial\varphi}(b,\varphi)\,\omega(b) = \rho(b,\varphi) + \omega'(b)\,s(b,\varphi) - \mu(b) \,. \tag{2.6}$$

A necessary condition for the solvability of (2.5) is that the angular average of $r$ vanishes:

$$\overline{r}(b) = 0\,, \qquad \text{where} \quad \overline{r}(b) = \frac{1}{(2\pi)^n}\int_{\mathbb{T}^n} r(b,\varphi)\,d\varphi \,. \tag{2.7}$$

In the Hamiltonian case this condition was satisfied because $r$ was a gradient with respect to $\varphi$. Here, in the reversible case, this is satisfied because $r$ is an odd function of $\varphi$.

If (2.7) holds, then (2.5) can be solved by Fourier series expansion in the same way as we solved (X.2.2), provided that the frequencies $\omega_1(b), \ldots, \omega_n(b)$ are non-resonant. Of course, there is again the same problem of small denominators as in the Hamiltonian case. Equations (2.6) are solved in the same way as (2.5), upon setting

$$\mu(b) = \overline{\rho}(b) + \omega'(b)\,\overline{s}(b) \,. \tag{2.8}$$

Since $r$ is odd in $\varphi$, the solution $s$ of (2.5) becomes even in $\varphi$. It is determined uniquely only up to a constant: we are still free to choose the angular average $\overline{s}(b)$. If $\omega'(b)$ has rank $n$, we may actually choose $\overline{s}(b)$ such that $\mu(b) = 0$ results from (2.8). Since the right-hand side of (2.6) is even in $\varphi$, the solution $\sigma$ of (2.6) becomes odd in $\varphi$ if we choose $\overline{\sigma}(b) = 0$.

## XI.2.2 Reversible Perturbation Series

The above construction extends to arbitrary finite order in $\varepsilon$. The transformation is now sought for in the form

$$a = b + \varepsilon s_1(b,\varphi) + \varepsilon^2 s_2(b,\varphi) + \ldots + \varepsilon^{N-1} s_{N-1}(b,\varphi) \tag{2.9}$$
$$\theta = \varphi + \varepsilon\sigma_1(b,\varphi) + \varepsilon^2\sigma_2(b,\varphi) + \ldots + \varepsilon^{N-1}\sigma_{N-1}(b,\varphi) \tag{2.10}$$

with $s_j$ even in $\varphi$ and $\sigma_j$ odd in $\varphi$ to preserve reversibility. This transformation is to be chosen such that the system in the new variables is of the form

$$\dot{b} = \varepsilon^N r_N(b,\varphi)$$
$$\dot{\varphi} = \omega_{\varepsilon,N}(b) + \varepsilon^N \rho_N(b,\varphi)$$

with $\omega_{\varepsilon,N}(b) = \omega(b) + \varepsilon\mu_1(b) + \ldots + \varepsilon^{N-1}\mu_{N-1}(b)$, and with $r_N(b,\varphi)$ odd in $\varphi$ and $\rho_N(b,\varphi)$ even in $\varphi$, and with all these functions bounded independently of $\varepsilon$.

Inserting the transformation into (2.1) and expanding in powers of $\varepsilon$, it is seen that the functions $s_j$ and $\sigma_j$ must satisfy equations of the form of (2.5), (2.6):

$$\frac{\partial s_j}{\partial \varphi}(b,\varphi)\omega(b) = p_j(b,\varphi) \tag{2.11}$$

$$\frac{\partial \sigma_j}{\partial \varphi}(b,\varphi)\omega(b) = \pi_j(b,\varphi) + \omega'(b)s_j(b,\varphi) - \mu_j(b) \tag{2.12}$$

where $p_j, \pi_j$ are given by expressions that depend linearly on higher-order derivatives of $r, \rho$ and polynomially on the functions $s_i, \sigma_i$ with $i < j$ and on their first-order derivatives. Using the rules

$$\begin{pmatrix} \text{even} & \text{odd} \\ \text{odd} & \text{even} \end{pmatrix} \begin{pmatrix} \text{odd} \\ \text{even} \end{pmatrix} = \begin{pmatrix} \text{odd} \\ \text{even} \end{pmatrix}$$

and

$$\frac{\partial \text{ even}}{\partial \varphi} = \text{odd}, \qquad \frac{\partial \text{ odd}}{\partial \varphi} = \text{even},$$

it is found that $p_j$ is odd in $\varphi$ and $\pi_j$ is even in $\varphi$ for all $j$. For non-resonant frequencies $\omega(b)$, the equations (2.11), (2.12) can therefore be solved with $s_j$ even in $\varphi$, $\sigma_j$ odd in $\varphi$. If $\omega'(b)$ is invertible, we can obtain $\mu_j(b) = 0$ for all $j$.

Beyond these formal calculations, there is the following reversible analogue of Lemma X.2.1 in the Hamiltonian case. This result is obtained by the same "ultraviolet cut-off" argument as the earlier result.

**Lemma 2.1.** *Let the right-hand side functions of (2.1) be real-analytic in a neighbourhood of $\{b^*\} \times \mathbb{T}^n$ and satisfy (2.2). Suppose that $\omega(b^*)$ satisfies the diophantine condition (X.2.4). For any fixed $N \geq 2$, there are positive constants $\varepsilon_0, c, C$ such that the following holds for $\varepsilon \leq \varepsilon_0$: there exists a real-analytic reversibility-preserving change of coordinates $(a,\theta) \mapsto (b,\varphi)$ such that every solution $(b(t),\varphi(t))$ of the perturbed system in the new coordinates, starting with $\|b(0) - b^*\| \leq c|\log \varepsilon|^{-\nu-1}$, satisfies*

$$\|b(t) - b(0)\| \leq Ct\varepsilon^N \qquad \text{for } t \leq \varepsilon^{-N+1},$$

$$\|\varphi(t) - \omega_{\varepsilon,N}(b(0))t - \varphi(0)\| \leq C\left(t^2 + t|\log \varepsilon|^{\nu+1}\right)\varepsilon^N \quad \text{for } t^2 \leq \varepsilon^{-N+1}.$$

*Moreover, the transformation is $\mathcal{O}(\varepsilon)$-close to the identity: $\|(a,\theta) - (b,\varphi)\| \leq C\varepsilon$ holds for $(a,\theta)$ and $(b,\varphi)$ related by the above coordinate transform, for $\|b - b^*\| \leq c|\log \varepsilon|^{-\nu-1}$ and for $\varphi$ in an $\varepsilon$-independent complex neighbourhood of $\mathbb{T}^n$.*

*The constants $\varepsilon_0, c, C$ depend on $N, n, \gamma, \nu$ and on bounds of $\omega, r, \rho$ on a complex neighbourhood of $\{b^*\} \times \mathbb{T}^n$.* □

The equations determining the coefficient functions of the perturbation series are of the form to which Lemma X.4.1 applies. Therefore, that lemma is again the tool for estimating the terms in the perturbation series, similar to Sect. X.4.1. This yields a reversible analogue of Theorem X.4.4 showing near-invariance of tori (up to exponentially small terms in a negative power of $\varepsilon$) over time intervals that are exponentially large in a negative power of $\varepsilon$, with the same exponents $\alpha, \beta$ as in Theorem X.4.4.

### XI.2.3 Reversible KAM Theory

For an integrable reversible system, just as for an integrable Hamiltonian system, the phase space is foliated into invariant tori on which the flow is conditionally periodic. We fix one such torus $\{a = a^*, \theta \in \mathbb{T}^n\}$ with diophantine frequencies $\omega_1, \ldots, \omega_n$. For convenience we may assume $a^* = 0 \in \mathbb{R}^m$. This torus is invariant under the flow of systems of the form $\dot{a} = \mathcal{O}(\|a\|^2)$, $\dot{\theta} = \omega + \mathcal{O}(\|a\|)$, or written more explicitly,

$$\begin{aligned} \dot{a} &= \tfrac{1}{2} a^T K(a, \theta) a \\ \dot{\theta} &= \omega + M(a, \theta) a \,. \end{aligned} \quad (2.13)$$

Here, $K = [K_1, \ldots, K_m]$ where each $K_i(a, \theta)$ is a symmetric $m \times m$ matrix, and $M(a, \theta)$ is an $n \times m$ matrix. The first equation is to be interpreted as $\dot{a}_i = \tfrac{1}{2} a^T K_i(a, \theta) a$ for the components $i = 1, \ldots, m$. Consider now a perturbation of this system:

$$\begin{aligned} \dot{a} &= \tfrac{1}{2} a^T K(a, \theta) a + \varepsilon r(a, \theta) \\ \dot{\theta} &= \omega + M(a, \theta) a + \varepsilon \rho(a, \theta) \,. \end{aligned} \quad (2.14)$$

For the reversible case, i.e., for $K$ and $r$ odd in $\theta$ and for $M$ and $\rho$ even in $\theta$, we construct a sequence of reversibility-preserving transformations in the spirit of Kolmogorov's transformation of Sect. X.2.3, which transform (2.14) back to the form (2.13) in the new variables, showing the persistence of an invariant torus with frequencies $\omega_i$ under small reversible perturbations of the system. This holds again under the diophantine condition (X.2.4) on $\omega$ and additionally under the condition that the angular average $\overline{M}_0$ of $M$ at $a = 0$ has rank $n$. A result of this type – a reversible KAM theorem – was shown by Moser (1973), Chap. V, in a different setting. See also Sevryuk (1986) for further results in that direction.

We look for a transformation of the form

$$\begin{aligned} a &= b + \varepsilon \Big( s(\varphi) + S(\varphi) b \Big) \\ \theta &= \varphi + \varepsilon \sigma(\varphi) \end{aligned} \quad (2.15)$$

with an $m \times m$ matrix $S(\varphi)$. Preserving reversibility requires that $s$ and $S$ are even functions and $\sigma$ is odd. Higher-order terms in $b$ play no role and are therefore omitted from the beginning. We insert this into (2.14) and obtain

$$\dot{b} = \frac{1}{2} b^T K(b,\varphi)b + \varepsilon \left\{ r(0,\varphi) - \frac{\partial s}{\partial \varphi}(\varphi)w \right.$$
$$\left. + \frac{\partial r}{\partial b}(0,\varphi)b - \frac{\partial s}{\partial \varphi}(\varphi)M(0,\varphi)b - \frac{\partial}{\partial \varphi}\Big(S(\varphi)b\Big)w + s(\varphi)^T K(0,\varphi)b \right\}$$
$$+ \mathcal{O}(\varepsilon^2) + \mathcal{O}(\varepsilon \|b\|^2)$$

$$\dot{\varphi} = w + M(b,\varphi)b$$
$$+ \varepsilon \left\{ \rho(0,\varphi) - \frac{\partial \sigma}{\partial \varphi}(\varphi)w + M(0,\varphi)s(\varphi) \right\} + \mathcal{O}(\varepsilon^2) + \mathcal{O}(\varepsilon \|b\|) .$$

We require that the terms in curly brackets vanish. This holds if the following equations are satisfied (the last equation is written component-wise for notational clarity):

$$\frac{\partial s}{\partial \varphi}(\varphi)w = r(0,\varphi)$$
$$\frac{\partial \sigma}{\partial \varphi}(\varphi)w = \rho(0,\varphi) + M(0,\varphi)s(\varphi) \qquad (2.16)$$
$$\frac{\partial S_{ij}}{\partial \varphi}(\varphi)w = \frac{\partial r_i}{\partial b_j}(\varphi) - \sum_k \frac{\partial s_i}{\partial \varphi_k}(\varphi)M_{kj}(0,\varphi) + \sum_k s_k(\varphi)K_{i,kj}(0,\varphi) .$$

Since $r$ is odd in $\varphi$, the first equation can be solved for $s$ even in $\varphi$, uniquely up to a constant, the angular average $\bar{s}$. Since the angular average of $M$ is assumed to be of full rank $n$, $\bar{s}$ can be chosen such that the angular average of the right-hand side of the equation for $\sigma$ becomes zero. Since the right-hand side is even, the equation can then be solved uniquely for an odd $\sigma$. The equations for $S$ have an odd right-hand side and can therefore be solved for an even $S$.

In this way, the perturbation to the form (2.13) is reduced from $\mathcal{O}(\varepsilon)$ to $\mathcal{O}(\varepsilon^2)$. By the same arguments as in the Hamiltonian case (see Sect. X.5), the iteration of this procedure is seen to be convergent. This finally yields a change of coordinates that preserves reversibility and transforms the perturbed system (2.14) back to the form (2.13). We summarize this in the following theorem, which is the reversible analogue of Kolmogorov's Theorem X.5.1.

**Theorem 2.2.** *Consider a real-analytic reversible system (2.13). Suppose that $w \in \mathbb{R}^n$ satisfies the diophantine condition (X.2.4), and that the angular average of $M(0,\cdot)$ is an $n \times m$ matrix of rank $n$. Let (2.14) be a real-analytic reversible perturbation of the system (2.13). Then, there exists $\varepsilon_0 > 0$ (which depends on the perturbation functions only through a bound of their norms on a complex neighbourhood of $\{0\} \times \mathbb{T}^n$) such that for every $\varepsilon$ with $|\varepsilon| \leq \varepsilon_0$, there is a real-analytic transformation $\psi_\varepsilon : (b,\varphi) \mapsto (a,\theta)$, $\mathcal{O}(\varepsilon)$ close to the identity and depending analytically on $\varepsilon$, which preserves reversibility and puts the perturbed system back to the form (2.13) in the new variables: $\dot{b} = \mathcal{O}(\|b\|^2)$, $\dot{\varphi} = w + \mathcal{O}(\|b\|)$. The perturbed system therefore has the invariant torus $\{b=0,\ \varphi \in \mathbb{T}^n\}$ carrying a quasi-periodic flow with the same frequencies $w$ as the unperturbed system.* □

## XI.2.4 Reversible Birkhoff-Type Normalization

We show that, in the situation of diophantine frequencies $\omega$, there is a reversibility-preserving transformation that takes a reversible system of the form (2.13) to the form

$$\begin{aligned} \dot{b} &= r_k(b, \varphi) \\ \dot{\varphi} &= \omega + \zeta_k(b) + \rho_k(b, \varphi) \end{aligned} \quad \text{with} \quad r_k, \rho_k = \mathcal{O}(\|b\|^k) \quad (2.17)$$

for arbitrary $k \geq 2$, where $\zeta_k = \overline{\rho}_1 + \ldots + \overline{\rho}_{k-1}$ with the bars denoting angular averages and with $\rho_1(b, \varphi) = M(b, \varphi)b$. This implies again that the invariant torus is "very sticky": $\|b(0)\| \leq \delta$ implies $\|b(t)\| \leq 2\delta$ for $t \leq C_k \delta^{-k+1}$. As in the Hamiltonian case, a suitable choice of $k$ would even yield time intervals exponentially long in a negative power of $\delta$ during which solutions stay within twice the initial distance $\delta$.

The transformation to the normal form (2.17) is constructed recursively. Suppose that in some variables $(a, \theta)$ we have, for some $k \geq 2$,

$$\begin{aligned} \dot{a} &= r_{k-1}(a, \theta) \\ \dot{\theta} &= \omega + \zeta_{k-1}(a) + \rho_{k-1}(a, \theta) \end{aligned} \quad \text{with} \quad r_{k-1}, \rho_{k-1} = \mathcal{O}(\|a\|^{k-1}).$$

Note, for $k = 2$ we have $r_1 = \mathcal{O}(\|a\|^2)$ by (2.13). We search for a transformation

$$\begin{aligned} a &= b + s(b, \varphi) \\ \theta &= \varphi + \sigma(b, \varphi) \end{aligned} \quad \text{with} \quad s, \sigma = \mathcal{O}(\|b\|^{k-1}),$$

(and $s = \mathcal{O}(\|b\|^2)$ for $k = 2$) that preserves reversibility, i.e., has $s$ even in $\varphi$ and $\sigma$ odd in $\varphi$, and is such that (2.17) holds. Inserting the transformation into the above differential equation shows that this is indeed achieved if $s, \sigma$ solve the following system of the form (2.5), (2.6):

$$\begin{aligned} \frac{\partial s}{\partial \varphi}(b, \varphi) \omega &= r_{k-1}(b, \varphi) \\ \frac{\partial \sigma}{\partial \varphi}(b, \varphi) \omega &= \rho_{k-1}(b, \varphi) + \zeta'_{k-1}(b) s(b, \varphi) - \mu_k(b). \end{aligned}$$

Choosing $\overline{s}(b) = 0$ leads to $\mu_k = \overline{\rho}_{k-1}$ and gives (2.17) with $\zeta_k = \zeta_{k-1} + \overline{\rho}_{k-1}$.

## XI.3 Linear Error Growth and Near-Preservation of First Integrals

We now study the error behaviour of reversible methods applied to integrable reversible systems. Recall from Theorem V.1.5 that symmetric methods are reversible under the compatibility condition (V.1.4). We give an analogue of Theorem X.3.1

on the error behaviour of symplectic methods applied to integrable Hamiltonian systems. We consider an integrable reversible system (1.1) (usually not given in action-angle variables) and let $(u, v) = \psi(a, \theta)$ be the reversibility-preserving transformation to action-angle variables. The inverse transformation is denoted as

$$(a, \theta) = (I(u, v), \Theta(u, v)).$$

The following is the reversible analogue of Theorem X.3.1.

**Theorem 3.1.** *Consider applying a reversible numerical integrator of order $p$ to the integrable reversible system (1.1) with real-analytic right-hand side. Suppose that $\omega(a^*)$ satisfies the diophantine condition (X.2.4). Then, there exist positive constants $C, c$ and $h_0$ such that the following holds for all step sizes $h \leq h_0$: every numerical solution starting with $\|I(u_0, v_0) - a^*\| \leq c |\log h|^{-\nu-1}$ satisfies*

$$\begin{aligned} \|(u_n, v_n) - (u(t), v(t))\| &\leq C t h^p \\ \|I(u_n, v_n) - I(u_0, v_0)\| &\leq C h^p \end{aligned} \quad \text{for } t = nh \leq h^{-p}.$$

*The constants $h_0, c, C$ depend on $\gamma, \nu$ of (X.2.4), on the dimensions, on bounds of the real-analytic functions $f, g$ on a complex neighbourhood of the torus $\{(u, v) : I(u, v) = a^*\}$, and on the numerical method.*

*Proof.* The proof of Theorem X.3.1 relied on Theorem IX.3.1 and Lemma X.2.1. Using their reversible analogues Theorem IX.2.3 and Lemma 2.1 with the same arguments gives the above result for the reversible case. □

**Remark 3.2.** As in the analogous remark for the Hamiltonian case, the error bounds of Theorem 3.1 also hold when the reversible method is applied to a perturbed integrable system with a perturbation parameter $\varepsilon$ bounded by a positive power of the step size: $\varepsilon \leq K h^\alpha$ for some $\alpha > 0$.

Figures 3.1 and 3.2 illustrate the long-time conservation of the first integrals and the linear error growth, respectively, of the symmetric (but not symplectic) Lobatto IIIB method on the Toda lattice example.

Theorem 3.1 together with Examples 1.6 and 1.7 also explains the good behaviour of symmetric (in fact, reversible) integrators on the rigid body equations which we observed in Chap. V (Figs. V.4.2 and V.4.6).

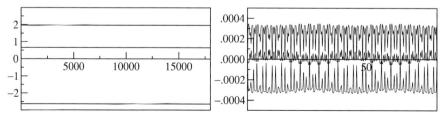

**Fig. 3.1.** Numerically obtained eigenvalues (left picture) and errors in the eigenvalues (right picture) of the 3-stage Lobatto IIIA scheme (step size $h = 0.1$) applied to the Toda lattice with the data of Sect. X.1.5.

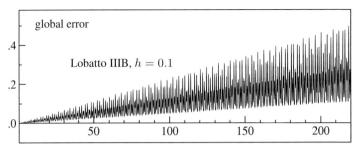

**Fig. 3.2.** Euclidean norm of the global error for the 3-stage Lobatto IIIB scheme (step size $h = 0.1$) applied to the Toda lattice with $n = 3$ and initial values as in Fig. 3.1.

## XI.4 Invariant Tori under Reversible Discretization

In this section we study the question as to how invariant tori of reversible systems are preserved under discretization of the system by reversible numerical methods. We give reversible analogues of Theorems X.5.3 and X.6.1.

### XI.4.1 Near-Invariant Tori over Exponentially Long Times

We consider a reversible system (1.1) which in suitable coordinates takes the perturbed form (2.14). Under the conditions of the reversible KAM theorem, Theorem 2.2, this system has an invariant torus carrying a quasi-periodic flow with frequencies $\omega$ for sufficiently small $\varepsilon$. Consider now a reversible numerical integrator applied to this system. By the same arguments as in Sect. X.5.2, using the reversible KAM theorem 2.2 in place of Kolmogorov's Theorem X.5.1, we obtain the following analogue of Theorem X.5.3, which states the existence of a torus such that numerical solutions starting on this torus remain exponentially close to a quasi-periodic flow on that torus over exponentially long times in $1/h$.

**Theorem 4.1.** *In the above situation, for a reversible numerical method of order $p$ used with sufficiently small step size $h$, there is a modified reversible system with an invariant torus $\widetilde{\mathcal{T}}_\omega$ carrying a quasi-periodic flow with frequencies $\omega$, $\mathcal{O}(h^p)$ close to the invariant torus $\mathcal{T}_\omega$ of the original reversible system, such that the difference between any numerical solution $(u_n, v_n)$ starting on the torus $\widetilde{\mathcal{T}}_\omega$ and the solution $(\widetilde{u}(t), \widetilde{v}(t))$ of the modified Hamiltonian system with the same starting values remains exponentially small in $1/h$ over exponentially long times:*

$$\|(u_n, v_n) - (\widetilde{u}(t), \widetilde{v}(t))\| \leq C e^{-\kappa/h} \quad \text{for} \quad t = nh \leq e^{\kappa/h}.$$

*The constants $C$ and $\kappa$ are independent of $h, \varepsilon$ (for $h, \varepsilon$ sufficiently small) and of the initial value $(u_0, v_0) \in \widetilde{\mathcal{T}}_\omega$.* □

The case of initial values lying close to, but not on $\widetilde{\mathcal{T}}_\omega$, can again be treated by a reversible analogue of Theorem X.4.7.

## XI.4.2 A KAM Theorem for Reversible Near-Identity Maps

To obtain truly invariant tori, we need a discrete analogue of the reversible KAM theorem, which is derived in this subsection. This result can also be viewed as the reversible analogue of Theorem X.6.1. It establishes the existence of invariant tori of reversible integrators, but as in the symplectic case, only for a Cantor set of non-resonant step sizes.

A map $\Phi : (a, \theta) \mapsto (\widehat{a}, \widehat{\theta})$ has the invariant torus $\{a = 0, \ \theta \in \mathbb{T}^n\}$, and reduces on this torus to rotation by $h\omega$ ($h$ a real parameter and $\omega \in \mathbb{R}^n$), when it is of the form (cf. (2.13))

$$\begin{aligned} \widehat{a} &= a + \tfrac{1}{2} h a^T K(a, \theta) a \\ \widehat{\theta} &= \theta + h\omega + h M(a, \theta) a \ . \end{aligned} \quad (4.1)$$

Here, $K = [K_1, \ldots, K_m]$ where each $K_i(a, \theta)$ is a symmetric $m \times m$ matrix, and $M(a, \theta)$ is an $n \times m$ matrix. The expression in the first equation is again to be interpreted as $a^T K_i(a, \theta) a$ for the components $i = 1, \ldots, m$.

A necessary condition for the above map $\Phi$ to be *reversible* with respect to the involution $(a, \theta) \mapsto (a, -\theta)$, cf. Definition V.1.2, is seen to be

$$\begin{aligned} K(0, -\theta) &= -K(0, \theta - h\omega) \\ M(0, -\theta) &= M(0, \theta - h\omega) \ . \end{aligned} \quad (4.2)$$

Consider now a perturbed map

$$\begin{aligned} \widehat{a} &= a + \tfrac{1}{2} h a^T K(a, \theta) a + h\varepsilon r(a, \theta) \\ \widehat{\theta} &= \theta + h\omega + h M(a, \theta) a + h\varepsilon \rho(a, \theta) \end{aligned} \quad (4.3)$$

where $r$ and $\rho$, which like $K$ and $M$ are assumed real-analytic, might depend analytically also on $h$ and $\varepsilon$. Reversibility of this map implies, by direct computation, that in addition to (4.2), the following equations are satisfied up to an error $\mathcal{O}(h\varepsilon)$:

$$\begin{aligned} r(0, -\theta) &= -r(0, \theta - h\omega) \\ \frac{\partial r}{\partial a}(0, -\theta) &= -\frac{\partial r}{\partial a}(0, \theta) \\ \rho(0, -\theta) &= \rho(0, \theta - h\omega) - h M(0, \theta - h\omega) r(0, \theta - h\omega) \ . \end{aligned} \quad (4.4)$$

Similar to Sect. XI.2.3, we construct a reversibility-preserving near-identity transformation of coordinates $(a, \theta) \mapsto (b, \varphi)$ such that the above map $\Phi_{h,\varepsilon}$ in the new variables is of the form (4.3) with the perturbation terms reduced from $\mathcal{O}(\varepsilon)$ to $\mathcal{O}(\varepsilon^2)$. Similar to Sect. X.6.1, this is possible if $h\omega$ satisfies the diophantine condition (X.6.3) and if the angular average $\overline{M}_0$ of $M(0, \cdot)$ has rank $n$.

We look for the transformation in the form (2.15). The functions defining this transformation must satisfy the following equations, cf. (2.16):

$$\frac{s(\varphi + h\omega) - s(\varphi)}{h} = r(0, \varphi)$$

$$\frac{\sigma(\varphi + h\omega) - \sigma(\varphi)}{h} = \rho(0, \varphi) + M(0, \varphi)s(\varphi)$$

$$\frac{S_{ij}(\varphi + h\omega) - S_{ij}(\varphi)}{h} = \frac{\partial r_i}{\partial b_j}(\varphi) - \sum_k \frac{\partial s_i}{\partial \varphi_k}(\varphi) M_{kj}(0, \varphi) + \sum_k s_k(\varphi) K_{i,kj}(0, \varphi) \,.$$
(4.5)

Under the conditions (X.6.3), (X.6.4) these equations can be solved by Fourier expansion, in the same way as the analogous equations in Sections X.6.1 and XI.2.3, and the map in the variables $(b, \varphi)$ becomes of the form

$$\begin{aligned}\widehat{b} &= b + \tfrac{1}{2} h b^T K(b, \varphi) b + \mathcal{O}(h\varepsilon \|b\|^2) + \mathcal{O}(h\varepsilon^2) \\ \widehat{\varphi} &= \varphi + h\omega + h M(b, \varphi) b + \mathcal{O}(h\varepsilon \|b\|) + \mathcal{O}(h\varepsilon^2) \,.\end{aligned}$$
(4.6)

We still need to know that the change of variables $(a, \theta) \mapsto (b, \varphi)$ preserves reversibility, i.e., that $s$ and $S$ are even functions of $\varphi$ and $\sigma$ is an odd function of $\varphi$. This is indeed a consequence of (4.2) and (4.4). (We may modify $r$ and $\rho$ such that (4.4) holds exactly, at the expense of introducing additional $\mathcal{O}(h^2\varepsilon^2)$ perturbations in (4.3).) Let us show this property for $s$. The Fourier coefficients $s_k$ of $s$ must satisfy

$$\frac{e^{ik \cdot h\omega} - 1}{h} s_k = r_k \,.$$

Since (4.4) implies $r_{-k} = -r_k e^{-ik \cdot h\omega}$ for all $k$, it follows that $s_{-k} = s_k$, and hence $s$ is an even function of $\varphi$. Similarly it is shown that $S$ is even and $\sigma$ is odd.

In summary, we have found a transformation $\mathcal{O}(\varepsilon)$ close to the identity, which transforms the reversible map (4.3) to a reversible map (4.6), thus reducing the perturbation terms from $\mathcal{O}(\varepsilon)$ to $\mathcal{O}(\varepsilon^2)$. The iteration of this procedure can again be shown to be convergent. This finally yields a transformation to coordinates in terms of which the perturbed map is back in the form (2.13). In this way we obtain the following discrete analogue of Theorem 2.2 or reversible analogue of Theorem X.6.1.

**Theorem 4.2.** *Consider a real-analytic reversible map $\Phi_{h,\varepsilon}$ of the form (4.3), defined on a neighbourhood of $\{0\} \times \mathbb{T}^n$, with $0 \in \mathbb{R}^m$. Suppose that $h\omega$ satisfies the diophantine condition (X.6.3), and that the angular average of $M(0, \cdot)$ has rank $n$. Then, there exists $\varepsilon_0 > 0$ such that for every $\varepsilon$ with $|\varepsilon| < \varepsilon_0$, there is a real-analytic transformation $\psi_{h,\varepsilon} : (b, \varphi) \mapsto (a, \theta)$, which preserves reversibility and is $\mathcal{O}(\varepsilon)$ close to the identity uniformly in $h$ satisfying (X.6.3) and is analytic in $\varepsilon$, such that $\psi_{h,\varepsilon}^{-1} \circ \Phi_{h,\varepsilon} \circ \psi_{h,\varepsilon} : (b, \varphi) \mapsto (\widehat{b}, \widehat{\varphi})$ is again of the form (4.1): $\widehat{b} = b + \mathcal{O}(\|b\|^2)$, $\widehat{\varphi} = \varphi + h\omega + \mathcal{O}(\|b\|)$. The perturbed map $\Phi_{h,\varepsilon}$ therefore has an invariant torus on which it is conjugate to rotation by $h\omega$.* □

As in the analogous situation of Sect. X.6.2, Theorem 4.2 applies directly, with $\varepsilon = h^p$, to the situation where a reversible numerical method of order $p$ is used to discretize an integrable reversible system, or more generally, a reversible system with a KAM torus with diophantine frequencies $\omega$. Here (4.1) corresponds to the time-$h$ flow of the reversible system, and (4.3) represents the numerical map. This establishes the existence of invariant tori for reversible integrators, in perfect analogy to the symplectic counterpart Theorem X.6.2.

Concerning condition (X.6.3) we refer back to Sect. X.6.3, where it is shown that this condition is satisfied for a Cantor set of step sizes $h$ if $\omega$ satisfies the diophantine condition (X.2.4).

## XI.5 Exercises

1. This exercise shows that reversibility with respect to the particular involution $(u, v) \mapsto (u, -v)$ is not as special as it might seem at first glance.
   (a) If the system $\dot y = f(y)$ is $\rho$-reversible (i.e., $f(\rho y) = -\rho f(y)$), then the transformed system $\dot z = T^{-1} f(Tz)$ is $\sigma$-reversible with $\sigma = T^{-1} \rho T$.
   (b) Every linear involution ($\rho^2 = I$) is similar to a diagonal matrix with entries $\pm 1$.

2. Show that the Toda system with an arbitrary number $n$ of degrees of freedom has $n$ independent first integrals in involution which are even functions of the momenta.
   *Hint.* Generalize the discussion for $n = 3$ in the text.

3. A reversible system of the form
$$\dot a = 0$$
$$\dot \theta = \omega(a, \theta)$$
with $\omega$ an even function of $\theta \in \mathbb{T}^n$, also has a foliation of invariant tori. Consider reversible perturbations of such systems like in (2.1) and search for a reversibility-preserving transformation (2.3) that takes the perturbed system to the form
$$\dot b = \mathcal{O}(\varepsilon^2)$$
$$\dot \varphi = \omega(b, \varphi) + \varepsilon\mu(b, \varphi) + \mathcal{O}(\varepsilon^2)$$
with $\mu$ even in $\varphi$. Write down the partial differential equations that the transformation must satisfy and discuss (sufficient) conditions for their solvability.

4. The torus $\{a = 0, \theta \in \mathbb{T}^n\}$ is invariant and carries a conditionally periodic flow with frequencies $\omega$ for reversible systems of the form $\dot a = \mathcal{O}(\|a\|)$, $\dot \theta = \omega + \mathcal{O}(\|a\|)$, which is more general than (2.13) in the differential equation for $a$. Discuss the difficulties that arise in trying to transform a reversible perturbation of such a system back to this form.

5. Apply an arbitrary (non-symmetric) Runge-Kutta method of even order $p = 2k$ to an integrable reversible system. Prove that under the assumptions of Theorem 3.1 the global error behaves for $t = nh$ like

$$y_n - y(t) = \mathcal{O}(th^p) + \mathcal{O}(t^2 h^{p+1}),$$

and the action variables like

$$I(y_n) - I(y_0) = \mathcal{O}(h^p) + \mathcal{O}(th^{p+1}).$$

# Chapter XII.
# Dissipatively Perturbed Hamiltonian and Reversible Systems

Symplectic integrators also show a favourable long-time behaviour when they are applied to non-Hamiltonian perturbations of Hamiltonian systems. The same is true for symmetric methods applied to non-reversible perturbations of reversible systems. In this chapter we study the behaviour of numerical integrators when they are applied to dissipative perturbations of integrable systems, where only one invariant torus persists under the perturbation and becomes weakly attractive. The simplest example of such a system is Van der Pol's equation with small parameter, which has a single limit cycle in contrast to the infinitely many periodic orbits of the unperturbed harmonic oscillator.

## XII.1 Numerical Experiments with Van der Pol's Equation

> One of the first such methods is the method of Van-der-Pol. [...] It should, however, be noted that in the formulation given by Van-der-Pol, approximation was effected by simple intuitive reasonings.
> (N.N. Bogoliubov & Y.A. Mitropolski 1961, p. 10f.)

Consider Van der Pol's equation

$$\begin{aligned} \dot{p} &= -q + \varepsilon(1-q^2)p \\ \dot{q} &= p \end{aligned} \quad (1.1)$$

with small positive $\varepsilon$, which is a perturbation of the harmonic oscillator. A symplectic change to polar coordinates $p = \sqrt{2a}\cos\theta$, $q = \sqrt{2a}\sin\theta$ puts the system into the form

$$\begin{aligned} \dot{a} &= \varepsilon\, 2a\cos^2\theta(1-2a\sin^2\theta) \\ \dot{\theta} &= 1 + \varepsilon\,\cos\theta\sin\theta(1-2a\sin^2\theta)\,. \end{aligned}$$

Since the angle $\theta$ evolves much faster than $a$, we may expect that the *averaged system*, which replaces the right-hand side functions by their angular averages, gives a good approximation:

$$\dot{a} = \varepsilon a(1 - \tfrac{1}{2}a)$$
$$\dot{\theta} = 1.$$

Approximating by the averaged equation is the "method of Van-der-Pol" cited above, and the belief in the long-time validity of such an approximation is the *averaging principle*. The averaged differential equation for $a$ has an unstable equilibrium at zero, and an asymptotically stable equilibrium at $a^* = 2$. The averaged system therefore has the circle $\{a^* = 2,\ \theta \in \mathbb{R} \bmod 2\pi\}$ as an attractive limit cycle. This suggests that the original Van der Pol equation has a nearby limit cycle, which is indeed the case.

Following the numerical experiment of Hairer & Lubich (1999), we solve the equation (1.1) with two initial values, $(p_0, q_0) = (0, 1.3)$ and $(p_0, q_0) = (0, 2.7)$, and with three numerical methods: the non-symplectic explicit and implicit Euler methods, and the symplectic Euler method. All of them have order 1. The numerical results are displayed in Fig. 1.1. For large step sizes (compared to the perturbation parameter $\varepsilon$), the non-symplectic methods give a completely wrong numerical solution, whereas that of the symplectic method is qualitatively correct. For smaller step sizes, the numerical solutions of the non-symplectic methods also show a limit cycle.

For the moment we explain these observations by "simple intuitive reasonings", that is, by the averaging principle and formal backward error analysis. The rigorous treatment is developed in the course of this chapter in a more general framework of perturbed integrable systems.

For a differential equation
$$\dot{y} = f(y) + \varepsilon g(y),$$
the numerical solution $y_n$ obtained by the explicit Euler method is the (formally) exact solution of a modified differential equation
$$\dot{\tilde{y}} = f(\tilde{y}) + \varepsilon g(\tilde{y}) - \tfrac{1}{2}hf'(\tilde{y})f(\tilde{y}) + \mathcal{O}(h^2 + \varepsilon h).$$

For the Van der Pol equation in the above coordinates, the averaged modified equation becomes
$$\dot{\tilde{a}} = h\tilde{a} + \varepsilon \tilde{a}(1 - \tfrac{1}{2}\tilde{a}) + \dots$$
which has approximately $\tilde{a} = 2 + 2h/\varepsilon$ as an equilibrium. Hence, the limit cycle of the numerical solution of the explicit Euler method has approximate radius $2\sqrt{1 + h/\varepsilon}$ (Fig. 1.1) which is far from the correct value unless $h \ll \varepsilon$.

The implicit Euler discretization is adjoint to the explicit Euler method. Therefore, its modified differential equation is as above with $h$ replaced by $-h$. In this case, the radius of the limit cycle is approximately $2\sqrt{1 - h/\varepsilon}$ (for $h < \varepsilon$), which again agrees very well with the pictures of Fig. 1.1.

For the symplectic Euler method, the modified differential equation for Van der Pol's equation is

## XII.1 Numerical Experiments with Van der Pol's Equation

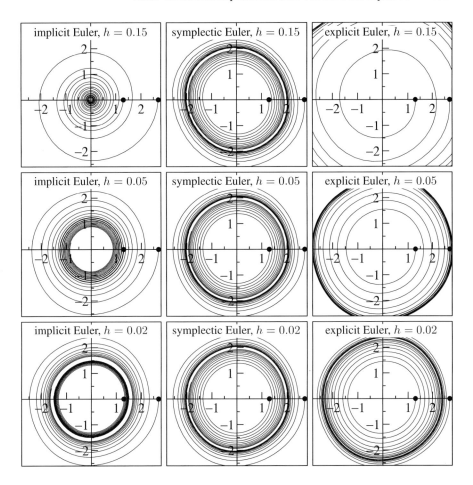

**Fig. 1.1.** Numerical experiments with Van der Pol's equation (1.1), $\varepsilon = 0.05$

$$\dot{\widetilde{p}} = \widetilde{q} + \varepsilon(1 - \widetilde{q}^2)\widetilde{p} + \tfrac{1}{2}h\widetilde{p} + \mathcal{O}(h^2 + \varepsilon h)$$
$$\dot{\widetilde{q}} = \widetilde{p} - \tfrac{1}{2}h\widetilde{q} + \mathcal{O}(h^2 + \varepsilon h).$$

Here, the modified differential equation for the unperturbed harmonic oscillator is Hamiltonian (Theorem IX.3.1), and so all $\varepsilon$-independent terms in the averaged modified equation vanish:

$$\int_0^{2\pi} \frac{\partial H_j}{\partial \theta}(a, \theta)\, d\theta = 0.$$

Therefore, the radius of the limit cycle is of size $2 + \mathcal{O}(h)$ in accordance with Fig. 1.1.

## XII.2 Averaging Transformations

> Le problème des oscillations non linéaires a actuellement une grande importance dans les domaines les plus divers de la technique et de la physique. Parmi les méthodes analytiques d'étude des oscillations non linéaires, la méthode asymptotique de développement en série par rapport à un paramètre petit est particulièrement efficace. Toute une série de monographies publiées en 1930–1938 par N. Krylov et N. Bogolioubov tant en russe qu'en français ont été consacrées à cette question, malheureusement ces ouvrages sont devenus aujourd'hui des raretés bibliographiques. Par ailleurs les méthodes exposées ont été largement développées depuis.
> (N. Bogolioubov & I. Mitropolski 1962, préface à la traduction française)

In this section we consider rather general perturbations of integrable systems. We study transformations that eliminate the dependence on the angles in the perturbation functions, up to arbitrary powers of the small perturbation parameter. The construction and properties of these "averaging" transformations are obtained by a slight extension of the arguments in Sections X.2 and XI.2.

### XII.2.1 The Basic Scheme of Averaging

As in Sections X.2.1 and XI.2.1, we consider perturbations of an integrable system written in action-angle variables:

$$\begin{aligned} \dot{a} &= \varepsilon\, r(a, \theta) \\ \dot{\theta} &= \omega(a) + \varepsilon\, \rho(a, \theta) \end{aligned} \quad (2.1)$$

where $\varepsilon$ is a small parameter and $r, \rho$ are real-analytic in a neighbourhood of $\{a^*\} \times \mathbb{T}^d$. Unlike the situation of the previous chapters, we do not impose conditions that make the angular average

$$\bar{r}(a) = \frac{1}{(2\pi)^d} \int_{\mathbb{T}^d} r(a, \theta)\, d\theta \quad (2.2)$$

vanish identically. We look for a transformation to new variables $(b, \varphi)$, of the form

$$\begin{aligned} a &= b + \varepsilon s(b, \varphi) \\ \theta &= \varphi + \varepsilon \sigma(b, \varphi)\,, \end{aligned} \quad (2.3)$$

which eliminates the dependence on the angles in the $\mathcal{O}(\varepsilon)$ terms of (2.1):

$$\begin{aligned} \dot{b} &= \varepsilon m(b) + \mathcal{O}(\varepsilon^2) \\ \dot{\varphi} &= \omega(b) + \varepsilon \mu(b) + \mathcal{O}(\varepsilon^2)\,. \end{aligned} \quad (2.4)$$

This is just a minor modification of the problem in Sect. XI.2.1. The equations that $s$ and $\sigma$ must satisfy, differ from (XI.2.5) and (XI.2.6) only in that the right-hand side $r(b, \varphi)$ of (XI.2.5) is replaced by $r(b, \varphi) - m(b)$, viz.,

$$\frac{\partial s}{\partial \varphi}(b,\varphi)\,\omega(b) = r(b,\varphi) - m(b) \tag{2.5}$$

$$\frac{\partial \sigma}{\partial \varphi}(b,\varphi)\,\omega(b) = \rho(b,\varphi) + \omega'(b)\,s(b,\varphi) - \mu(b). \tag{2.6}$$

Necessary conditions for solvability are now

$$m(b) = \overline{r}(b), \quad \mu(b) = \overline{\rho}(b), \tag{2.7}$$

where the second equation corresponds to the choice $\overline{s}(b) = 0$. In other words, the leading terms in (2.4) are the angular averages of the perturbations in (2.1).

The equations (2.5), (2.6) are solvable for $b = b^*$ if $\omega(b^*)$ satisfies the diophantine condition (X.2.4). The "ultraviolet cutoff" argument of the proof of Lemma X.2.1 then shows that (2.4) holds uniformly as long as the solution remains in the ball $\|b-b^*\| \le c|\log\varepsilon|^{-\nu-1}$, with a sufficiently small constant $c$. This may hold over a very long time interval if the equation $\dot{b} = \varepsilon m(b)$ has a stable equilibrium in that ball.

## XII.2.2 Perturbation Series

As in Sections X.2.2 and XI.2.2, the above construction extends to arbitrary finite order in $\varepsilon$. A transformation of the form (XI.2.9), which eliminates the angles in all terms up to order $\varepsilon^{N-1}$, is sought for:

$$\begin{aligned}\dot{b} &= \varepsilon m_1(b) + \varepsilon^2 m_2(b) + \ldots + \varepsilon^{N-1} m_{N-1}(b) + \varepsilon^N r_N(b,\varphi) \\ \dot{\varphi} &= \omega(b) + \varepsilon\mu_1(b) + \varepsilon^2\mu_2(b) + \ldots + \varepsilon^{N-1}\mu_{N-1}(b) + \varepsilon^N \rho_N(b,\varphi).\end{aligned} \tag{2.8}$$

The equations determining the transformation are a slight modification of (XI.2.11) and (XI.2.12): on the right-hand side of (XI.2.11), $p_j(b,\varphi)$ is replaced by the difference $p_j(b,\varphi) - m_j(b)$, with $m_j(b) = \overline{p}_j(b)$. We then have the following variant of Lemmas X.2.1 and XI.2.1.

**Lemma 2.1.** *Let the right-hand side functions of (2.1) be real-analytic in a neighbourhood of $\{b^*\} \times \mathbb{T}^d$. Suppose that $\omega(b^*)$ satisfies the diophantine condition (X.2.4) with exponent $\nu$. For any fixed $N \ge 2$, there are positive constants $\varepsilon_0, c, C$ such that the following holds for $|\varepsilon| \le \varepsilon_0$: there exists a real-analytic change of coordinates $(a,\theta) \mapsto (b,\varphi)$ which transforms (2.1) to (2.8) with*

$$\begin{array}{ll} \|m_j(b)\| \le C/\delta^{j-1}, & \|\mu_j(b)\| \le C/\delta^{j-1} \\ \|r_N(b,\varphi)\| \le C/\delta^{N-1}, & \|\rho_N(b,\varphi)\| \le C/\delta^{N-1} \end{array} \quad \text{for } \|b-b^*\| \le \delta,$$

*where*

$$\delta = c|\log\varepsilon|^{-\nu-1}. \tag{2.9}$$

*Moreover, the transformation is $\mathcal{O}(\varepsilon)$-close to the identity: $\|(a,\theta) - (b,\varphi)\| \le C\varepsilon$ holds for $(a,\theta)$ and $(b,\varphi)$ related by the above coordinate transform, for $\|b-b^*\| \le \delta$ and for $\varphi$ in an $\varepsilon$-independent complex neighbourhood of $\mathbb{T}^d$.*

*The constants $\varepsilon_0, c, C$ depend on $N, d, \gamma, \nu$ and on bounds of $\omega, r, \rho$ on a complex neighbourhood of $\{b^*\} \times \mathbb{T}^d$.*

*Proof.* The proof uses again the ultraviolet cutoff argument of the proof of Lemma X.2.1. This makes all the functions $s_i, \sigma_i, m_i, \mu_i$ real-analytic in $b$ for $\|b-b^*\| \leq 2\delta$ and of $\varphi$ in an $\varepsilon$-independent complex neighbourhood of $\mathbb{T}^d$. The powers of $\delta$ in the denominators of the estimates come from the presence of terms $\partial s_j/\partial b$, $\partial \sigma_j/\partial b$ in $p_i(b, \varphi)$ and $\pi_i(b, \varphi)$ of (XI.2.11) and (XI.2.12) and from Cauchy's estimates applied to $s_j, \sigma_j$ on $\|b - b^*\| \leq 2\delta$. □

## XII.3 Attractive Invariant Manifolds

> Theorems on invariant manifolds for maps have been proved many times for many different settings. The first results were obtained by Hadamard (1901) and Perron (1929). [...] Our aim was to derive a global invariant manifold result with conditions that are easy to verify for the applications in mind.
> (K. Nipp & D. Stoffer 1992)

In this section we give results on the existence and properties of attractive invariant manifolds of maps, with a very explicit handling of constants. These results are due to Kirchgraber, Lasagni, Nipp & Stoffer (1991) and Nipp & Stoffer (1992). They will allow us to understand the weakly attractive closed curves that we observed in Sect. XII.1. Beyond that particular example, these results are extremely useful for studying the long-time behaviour of numerical discretizations in a great variety of applications; see Nipp & Stoffer (1995, 1996) and Lubich (2001) and references therein, and also Stuart & Humphries (1996) for a related invariant manifold theorem and its use in analyzing the dynamics of numerical integrators for non-conservative problems.

Consider a map $\Phi : X \times Y \to X \times Y$ defined on the Cartesian product of a Banach space $X$ and a closed bounded subset $Y$ of another Banach space. We write $\Phi(x, y) = (\widehat{x}, \widehat{y})$ with

$$\begin{aligned} \widehat{x} &= x + f(x,y) \\ \widehat{y} &= g(x,y) \,. \end{aligned} \tag{3.1}$$

We assume that $f$ and $g$ are Lipschitz bounded, with Lipschitz constants $L_{xx}, L_{xy}$ and $L_{yx}, L_{yy}$ with respect to $x, y$. If these Lipschitz constants are sufficiently small, then the map $\Phi$ has an attractive invariant manifold. More precisely, there is the following result, stated without proof by Kirchgraber, Lasagni, Nipp & Stoffer (1991) and proved in a more general setting by Nipp & Stoffer (1992).

**Theorem 3.1.** *In the above situation, if*

$$L_{xx} + L_{yy} + 2\sqrt{L_{xy}L_{yx}} < 1 \,, \tag{3.2}$$

*then there exists a function $s : X \to Y$, which is Lipschitz bounded with the constant $\lambda = 2L_{yx}/(1 - L_{xx} - L_{yy})$, such that*

$$\mathcal{M} = \{(x, s(x)) : x \in X\} \text{ is invariant under } \Phi.$$

$\mathcal{M}$ attracts orbits of $\Phi$ with the attractivity factor $\rho = \lambda L_{xy} + L_{yy} < 1$, that is, $\|\widehat{y} - s(\widehat{x})\| \le \rho \|y - s(x)\|$ holds for all $(x, y) \in X \times Y$.

*Proof.* (a) We search for a function $s : X \to Y$ such that for $(\widehat{x}, \widehat{y}) = \Phi(x, y)$, the relation $y = s(x)$ implies also $\widehat{y} = s(\widehat{x})$. For an arbitrary function $\sigma : X \to Y$, we first study which relation holds between $\widehat{x}$ and $\widehat{y}$ if $y = \sigma(x)$. To write $\widehat{y}$ as a function of $\widehat{x}$, we need a bijective correspondence between $x$ and $\widehat{x}$ via the first equation of (3.1). By the Banach fixed-point theorem, the equation

$$\widehat{x} = x + f(x, \sigma(x)) \text{ has a unique solution } x = u_\sigma(\widehat{x})$$

for every $\widehat{x} \in X$ if $x \mapsto f(x, \sigma(x))$ is a contraction. This is the case if $\sigma$ has the Lipschitz constant $\lambda$ and

$$L_{xx} + L_{xy}\lambda < 1 . \qquad (3.3)$$

We then obtain $\widehat{y} = \widehat{\sigma}(\widehat{x})$ from the following scheme:

$$\begin{array}{ccc} x = u_\sigma(\widehat{x}) & \longleftarrow & \widehat{x} \\ \downarrow \sigma & & \\ y = \sigma(x) & \longrightarrow & \widehat{y} = g(x, y) \end{array}$$

That is, we set $\widehat{y} = \widehat{\sigma}(\widehat{x}) = g(u_\sigma(\widehat{x}), \sigma(u_\sigma(\widehat{x})))$. By construction, $(\widehat{x}, \widehat{y}) = \Phi(x, y)$. Under condition (3.3), the function $u_\sigma : X \to X$ is Lipschitz bounded by $\mu = 1/(1 - L_{xx} - L_{xy}\lambda)$. Consequently, the function $\widehat{\sigma} : X \to Y$ is Lipschitz bounded by $(L_{yx} + L_{yy}\lambda)\mu$. The condition that the transformed function $\widehat{\sigma}$ is again Lipschitz bounded by the same $\lambda$ as $\sigma$, therefore reads

$$\frac{L_{yx} + L_{yy}\lambda}{1 - L_{xx} - L_{xy}\lambda} \le \lambda , \qquad (3.4)$$

or equivalently,

$$L_{xy}\lambda^2 - (1 - L_{xx} - L_{yy})\lambda + L_{yx} \le 0 .$$

Under condition (3.2), there exists a non-empty real interval of values $\lambda$ satisfying this quadratic inequality. In particular, (3.4) then holds for

$$\lambda = \frac{2L_{yx}}{1 - L_{xx} - L_{yy}} . \qquad (3.5)$$

(This is close to the smallest possible value of $\lambda$ if $2\sqrt{L_{xy}L_{yx}} \ll 1 - L_{xx} - L_{yy}$.) It is easily checked that (3.2) and (3.5) imply (3.3).

Under conditions (3.3) and (3.4), the transformation $H : \sigma \mapsto \widehat{\sigma}$, which is called a *Hadamard graph transform*, maps the set of functions

$$S = \{\sigma : X \to Y \mid \sigma \text{ is Lipschitz bounded by } \lambda\}$$

into itself, i.e.,

$$H : S \to S : \sigma \mapsto \widehat{\sigma} .$$

$S$ is a closed subset of $C(X, Y)$, the Banach space of continuous functions from $X$ to the bounded closed set $Y$, equipped with the supremum norm $\|\sigma\|_\infty = \sup_{x \in X} \|\sigma(x)\|$. If $H$ is a contraction, then the Banach fixed-point theorem tells us that there is a unique function $s \in S$ with $\widehat{s} = s$. By construction, this means that if $(\widehat{x}, \widehat{y}) = \Phi(x, y)$ and $y = s(x)$, then also $\widehat{y} = s(\widehat{x})$. The graph $\mathcal{M} = \{(x, s(x)) : x \in X\}$ is then an invariant manifold for the map $\Phi$.

(b) We now show that $H$ is already a contraction under condition (3.2). Let $\sigma_0, \sigma_1$ be two arbitrary functions in $S$, and $\widehat{x} \in X$. With $x_i = u_{\sigma_i}(\widehat{x})$,

$$\|H\sigma_1(\widehat{x}) - H\sigma_0(\widehat{x})\| = \|g(x_1, \sigma_1(x_1)) - g(x_0, \sigma_0(x_0))\|$$
$$\leq \|g(x_1, \sigma_1(x_1)) - g(x_1, \sigma_0(x_1))\| + \|g(x_1, \sigma_0(x_1)) - g(x_0, \sigma_0(x_0))\|$$
$$\leq L_{yy} \|\sigma_1 - \sigma_0\|_\infty + (L_{yx} + L_{yy}\lambda) \|x_1 - x_0\| .$$

By definition, $\widehat{x} = x_i + f(x_i, \sigma_i(x_i))$ for $i = 0, 1$. Subtracting these two equations yields similarly

$$\|x_1 - x_0\| \leq \|f(x_1, \sigma_1(x_1)) - f(x_0, \sigma_0(x_0))\|$$
$$\leq \|f(x_1, \sigma_1(x_1)) - f(x_1, \sigma_0(x_1))\| + \|f(x_1, \sigma_0(x_1)) - f(x_0, \sigma_0(x_0))\|$$
$$\leq L_{xy} \|\sigma_1 - \sigma_0\|_\infty + (L_{xx} + L_{xy}\lambda) \|x_1 - x_0\| .$$

Hence,

$$\|x_1 - x_0\| \leq \frac{L_{xy}}{1 - L_{xx} - L_{xy}\lambda} \|\sigma_1 - \sigma_0\|_\infty .$$

Combining both inequalities and recalling (3.4), we obtain

$$\|H\sigma_1 - H\sigma_0\|_\infty \leq (L_{yy} + \lambda L_{xy}) \|\sigma_1 - \sigma_0\|_\infty .$$

Since the inequality

$$L_{yy} + \lambda L_{xy} < 1 \tag{3.6}$$

is satisfied by the $\lambda$ of (3.5) under condition (3.2), $H$ is indeed a contraction.

(c) It remains to show that the invariant manifold $\mathcal{M}$ is attractive. With $(\widehat{x}, \widehat{y}) = \Phi(x, y)$, we write

$$\widehat{y} - s(\widehat{x}) = g(x, y) - s(x + f(x, y))$$
$$= \Big(g(x, y) - g(x, s(x))\Big) + \Big(s(x + f(x, s(x))) - s(x + f(x, y))\Big) .$$

Here we used the identity

$$s(x + f(x, s(x))) = \widehat{s}(x + f(x, s(x))) = g(x, s(x)) ,$$

which holds because $\widehat{s} = s$ and by construction of the Hadamard transform. It follows that

$$\|\widehat{y} - s(\widehat{x})\| \leq (L_{yy} + \lambda L_{xy}) \|y - s(x)\| ,$$

which together with (3.6) yields the result. □

Next we study the effect of a perturbation of the map on the invariant manifold.

**Theorem 3.2.** *Consider maps $\Phi_0, \Phi_1 : X \times Y \to X \times Y$ both of which satisfy the conditions of Theorem 3.1 with the same Lipschitz constants $L_{xx}, L_{xy}, L_{yx}, L_{yy}$. Let $s_0$ and $s_1$ be the functions defining the attractive invariant manifolds $\mathcal{M}_0$ and $\mathcal{M}_1$, respectively. If the bound*

$$\|\Phi_1(x,y) - \Phi_0(x,y)\| \leq \delta \quad \text{for} \quad (x,y) \in \mathcal{M}_0$$

*holds in the norm $\|(x,y)\| = \lambda \|x\| + \|y\|$ on $X \times Y$, then*

$$\|s_1(x) - s_0(x)\| \leq \frac{\delta}{1-\rho} \quad \text{for} \quad x \in X.$$

*(Here $\lambda$ and $\rho$ are defined as in Theorem 3.1.)*

*Proof.* The proof is similar to part (b) of the previous proof. Let $\widehat{x} \in X$. For $i = 0, 1$, we have $s_i(\widehat{x}) = g_i(x_i, s_i(x_i))$ with $x_i$ defined by the equation $\widehat{x} = x_i + f_i(x_i, s_i(x_i))$. We estimate

$$\begin{aligned}
\|s_1(\widehat{x}) - s_0(\widehat{x})\| &\leq \|g_1(x_1, s_1(x_1)) - g_1(x_1, s_0(x_1))\| \\
&+ \|g_1(x_1, s_0(x_1)) - g_1(x_0, s_0(x_0))\| \\
&+ \|g_1(x_0, s_0(x_0)) - g_0(x_0, s_0(x_0))\| \\
&\leq L_{yy}\|s_1 - s_0\|_\infty + (L_{yx} + L_{yy}\lambda)\|x_1 - x_0\| \\
&+ \|g_1(x_0, s_0(x_0)) - g_0(x_0, s_0(x_0))\|
\end{aligned}$$

and in the same way

$$\begin{aligned}
\|x_1 - x_0\| &\leq \|f_1(x_1, s_1(x_1)) - f_1(x_1, s_0(x_1))\| \\
&+ \|f_1(x_1, s_0(x_1)) - f_1(x_0, s_0(x_0))\| \\
&+ \|f_1(x_0, s_0(x_0)) - f_0(x_0, s_0(x_0))\| \\
&\leq L_{xy}\|s_1 - s_0\|_\infty + (L_{xx} + L_{xy}\lambda)\|x_1 - x_0\| \\
&+ \|f_1(x_0, s_0(x_0)) - f_0(x_0, s_0(x_0))\|.
\end{aligned}$$

Inserting the second bound into the first one and using (3.4) and the assumed bound on $\Phi_1 - \Phi_0$ gives

$$\|s_1 - s_0\|_\infty \leq (L_{yy} + \lambda L_{xy})\|s_1 - s_0\|_\infty + \delta,$$

which implies the result. ∎

## XII.4 Weakly Attractive Invariant Tori of Perturbed Integrable Systems

> We assume that the perturbation is dissipative such that one torus persists under the perturbation and gets attractive.
>
> Our analysis is done by the method of averaging. The problem of this section is classical, see e.g. Bogoliubov & Mitropolski (1961), Kirchgraber & Stiefel (1978). (D. Stoffer 1998)

In the example of the Van der Pol equation, we have seen that only one of the periodic orbits of the harmonic oscillator persists under the small nonlinear perturbation and becomes an attractive limit cycle. More generally, we consider perturbations of integrable systems

$$\begin{aligned} \dot a &= \varepsilon\, r(a,\theta) \\ \dot\theta &= \omega(a) + \varepsilon\, \rho(a,\theta) \end{aligned} \quad (4.1)$$

where (locally) just one invariant torus survives the perturbation and attracts nearby solutions. Using the results of the two previous sections, it will be shown that this situation occurs if, at some point $a^*$ where the frequencies $\omega_i(a^*)$ are diophantine, the angular average $\bar r(a^*)$ is small and its Jacobian matrix

$$A = \bar r\,'(a^*)$$

has all eigenvalues with negative real part.

The following theorem is a slight modification of a result of Stoffer (1998). Early versions of it are much older; see the citations above. The origins of the problem can be traced back to the work of Van der Pol (1927) and Krylov & Bogoliubov (1934).

Here we assume the following: $\omega(a^*)$ satisfies the diophantine condition (X.2.4) with exponent $\nu$. The perturbation functions $r(a,\theta)$ and $\rho(a,\theta)$ are real-analytic on a fixed complex neighbourhood of $\{a^*\} \times \mathbb{T}^d$ and bounded independently of $\varepsilon$ (though they may depend on $\varepsilon$). In some norm $\|\cdot\|$ on $\mathbb{R}^d$ and its induced matrix norm, the bounds

$$\|\bar r(a^*)\| \le C|\log\varepsilon|^{-2(\nu+1)} \quad (4.2)$$

$$\|e^{tA}\| \le e^{-t\alpha} \quad \text{for } t > 0 \quad (4.3)$$

hold with some constants $C$ and $\alpha > 0$.

**Theorem 4.1.** *Under the above conditions, for sufficiently small $\varepsilon > 0$, the system (4.1) has an invariant torus $\mathcal{T}_\varepsilon$ which attracts an $\mathcal{O}(|\log\varepsilon|^{-\nu-1})$-neighbourhood of $\{a^*\} \times \mathbb{T}^d$ with an exponential rate proportional to $\varepsilon$.*

*Proof.* The proof combines Lemma 2.1 and Theorem 3.1. For convenience we assume $a^* = 0$ in the following. Lemma 2.1 (with $N = 3$) gives us a change of coordinates $(a,\theta) \mapsto (b,\varphi)$, $\mathcal{O}(\varepsilon)$-close to the identity, such that for $\|b\| \le \delta$ with $\delta = c|\log\varepsilon|^{-\nu-1}$ of (2.9),

$$\begin{aligned}
\dot b &= \varepsilon m_1(b) + \varepsilon^2 m_2(b) + \mathcal{O}(\varepsilon^3/\delta^2) \\
\dot\varphi &= \omega(b) + \varepsilon\mu_1(b) + \varepsilon^2\mu_2(b) + \mathcal{O}(\varepsilon^3/\delta^2) \,.
\end{aligned} \quad (4.4)$$

Since $m_1(b) = \bar r(b) = Ab + \mathcal{O}(\delta^2)$ by (4.2), this system is of the form

$$\begin{aligned}
\dot b &= \varepsilon Ab + \mathcal{O}(\varepsilon\delta^2) \\
\dot\varphi &= \omega(b) + \mathcal{O}(\varepsilon) \,.
\end{aligned}$$

Similarly, the corresponding variational equation is of the form

$$\begin{pmatrix} \dot B \\ \dot\Phi \end{pmatrix} = \begin{pmatrix} \varepsilon A + \mathcal{O}(\varepsilon\delta) & \mathcal{O}(\varepsilon^3/\delta^2) \\ \mathcal{O}(1) & \mathcal{O}(\varepsilon^3/\delta^2) \end{pmatrix} \begin{pmatrix} B \\ \Phi \end{pmatrix} \,.$$

These relations and condition (4.3) imply that, for sufficiently small $\varepsilon$ and for any fixed $\tau > 0$, the time-$\tau$ flow of (4.1) maps the strip $D = \{(b, \varphi) : \|b\| \leq \frac{1}{2}\delta, \varphi \in \mathbb{T}^d\}$ into itself, and the following bounds hold for the derivatives of the solution with respect to the initial values:

$$\begin{aligned}
\left\|\frac{\partial b(\tau)}{\partial b(0)}\right\| &\leq L_{bb} = e^{-\tau\varepsilon\alpha} + \mathcal{O}(\varepsilon\delta) \,, & \left\|\frac{\partial b(\tau)}{\partial\varphi(0)}\right\| &\leq L_{b\varphi} = \mathcal{O}(\varepsilon^3/\delta^2) \\
\left\|\frac{\partial\varphi(\tau)}{\partial b(0)}\right\| &\leq L_{\varphi b} = \mathcal{O}(1) \,, & \left\|\frac{\partial\varphi(\tau)}{\partial\varphi(0)} - I\right\| &\leq L_{\varphi\varphi} = \mathcal{O}(\varepsilon^3/\delta^2) \,.
\end{aligned} \quad (4.5)$$

Hence, for sufficiently small $\varepsilon$,

$$L_{\varphi\varphi} + L_{bb} + 2\sqrt{L_{\varphi b}L_{b\varphi}} \leq e^{-\tau\varepsilon\alpha/2} < 1 \,.$$

Theorem 3.1 (and Exercise 1) used with $\varphi, b$ in the roles of $x, y$ now shows that the time-$\tau$ flow has an attractive invariant torus $\{(s(\varphi), \varphi) : \varphi \in \mathbb{T}^d\}$, where $s : \mathbb{T}^d \to \{\|b\| \leq \frac{1}{2}\delta\}$ is Lipschitz bounded by $\lambda = 2L_{b\varphi}/(1 - L_{\varphi\varphi} - L_{bb}) = \mathcal{O}(\varepsilon^3/\delta^2)$. This invariant torus attracts orbits of the time-$\tau$ flow map in the strip $D$ with the attractivity factor $\lambda L_{\varphi b} + L_{bb} \leq e^{-\tau\varepsilon\alpha/2}$. As Exercise 2 shows, the torus is actually invariant for the differential equation (4.1). □

## XII.5 Weakly Attractive Invariant Tori of Numerical Integrators

Does the attractive invariant torus of Theorem 4.1 persist under numerical discretization of the perturbed integrable system? This question was first studied by Stoffer (1998) who worked directly with the discrete equations in his analysis. Here we take up the approach of Hairer & Lubich (1999) where the problem was studied by combining backward error analysis and perturbation theory, similar to what was done in the two preceding chapters.

## XII.5.1 Modified Equations of Perturbed Differential Equations

Below we need to use backward error analysis for the numerical solution of a perturbed differential equation

$$\dot{y} = f(y) + \varepsilon g(y, \varepsilon), \qquad y(0) = y_0 \qquad (5.1)$$

with real-analytic functions $f$ and $g$ and small parameter $\varepsilon$. We consider applying a one-step method $y_1 = \Phi_h^\varepsilon(y_0)$ of order $p \geq 1$ with step size $h > 0$. The associated modified differential equations constructed in Chap. IX are then of the form

$$\dot{\widetilde{y}} = \widetilde{f}(\widetilde{y}) + \varepsilon \widetilde{g}(\widetilde{y}, \varepsilon), \qquad \widetilde{y}(0) = y_0 \qquad (5.2)$$

with suitably truncated series

$$\begin{aligned}
\widetilde{f}(y) &= f(y) + h^p f_{p+1}(y) + \ldots + h^{N-1} f_N(y) \\
\widetilde{g}(y, \varepsilon) &= g(y, \varepsilon) + h^p g_{p+1}(y, \varepsilon) + \ldots + h^{N-1} g_N(y, \varepsilon),
\end{aligned} \qquad (5.3)$$

where the functions $f_j$ are independent of $\varepsilon, h, N$, whereas the functions $g_j$ are allowed to depend on $\varepsilon$. The following adapts Theorem IX.7.6 to the above situation.

**Theorem 5.1.** *Let $f(y) + \varepsilon g(y, \varepsilon)$ be real-analytic (in $y$ and $\varepsilon$) and bounded by $M$ for $y \in B_{2R}(y_0)$ and for all complex $\varepsilon$ with $|\varepsilon| \leq \varepsilon_0$. Let the coefficients of the Taylor series (in $h$) of the numerical method be analytic in $B_R(y_0)$ with bounds (IX.7.5) for $|\varepsilon| \leq \varepsilon_0$. Then, there exists $h_0 > 0$ (proportional to $R/M$), such that for $h \leq h_0/4$ and for $N = N(h)$ the largest integer with $hN \leq h_0$, the difference between the numerical solution $y_1 = \Phi_h^\varepsilon(y_0)$ and the exact solution $\widetilde{\varphi}_{N,t}^\varepsilon(y_0)$ of the truncated modified equation (5.2)-(5.3) satisfies*

$$\|\Phi_h^\varepsilon(y_0) - \widetilde{\varphi}_{N,h}^\varepsilon(y_0)\| \leq Ch\, e^{-h_0/h}.$$

*The functions $\widetilde{f}$ and $\widetilde{g}$ of (5.3) are real-analytic in $B_R(y_0)$ with*

$$\|\widetilde{f}(y) - f(y)\| \leq Ch^p, \qquad \|\widetilde{g}(y, \varepsilon) - g(y, \varepsilon)\| \leq Ch^p$$

*for $y \in B_{R/2}(y_0)$ and $|\varepsilon| \leq \varepsilon_0$. The constants $C$ are independent of $h \leq h_0/4$ and $|\varepsilon| \leq \varepsilon_0$.*

*Proof.* The exponentially small estimate for $\Phi_h^\varepsilon(y_0) - \widetilde{\varphi}_{N,h}^\varepsilon(y_0)$ is that of Theorem IX.7.6 applied to the differential equation (5.1). The $\mathcal{O}(h^p)$ bound for $\widetilde{f}(y) - f(y)$ is the estimate (IX.7.14) applied to $\dot{y} = f(y)$. By applying that estimate to (5.1), a bound of the same type is obtained for $(\widetilde{f}(y) + \varepsilon \widetilde{g}(y, \varepsilon)) - (f(y) + \varepsilon g(y))$, uniformly for all complex $\varepsilon$ in the complex disk $|\varepsilon| \leq \varepsilon_0$. For any fixed $y \in B_{R/2}(y_0)$, the difference

$$\widetilde{g}(y, \varepsilon) - g(y, \varepsilon) = \frac{1}{\varepsilon} \Big( [(\widetilde{f}(y) + \varepsilon \widetilde{g}(y, \varepsilon)) - (f(y) + \varepsilon g(y, \varepsilon))] - [\widetilde{f}(y) - f(y)] \Big)$$

is an analytic function of $\varepsilon$ in the complex disk $|\varepsilon| \leq \varepsilon_0$, which is bounded by $\mathcal{O}(h^p)$ for $|\varepsilon| = \varepsilon_0$. By the maximum principle, the same bound then holds for $|\varepsilon| \leq \varepsilon_0$. □

## XII.5.2 Symplectic Methods

We apply a symplectic integrator with step size $h$ to a real-analytic perturbed integrable Hamiltonian system in coordinates $(p, q)$,

$$\dot{p} = -\frac{\partial H}{\partial q}(p, q) + \varepsilon k(p, q)$$
$$\dot{q} = \frac{\partial H}{\partial p}(p, q) + \varepsilon \ell(p, q) \,. \tag{5.4}$$

We assume that the unperturbed system ($\varepsilon = 0$) is a completely integrable system which satisfies the conditions of the Arnold-Liouville theorem, Theorem X.1.6. Hence, there exists a transformation to action-angle variables for the

*integrable system:* $(p, q) \mapsto (a, \theta)$ by Theorem X.1.6.

This change of coordinates transforms the integrable system to the equations $\dot{a} = 0$, $\dot{\theta} = \omega(a)$, and it transforms (5.4) to a system (4.1), for which we assume (4.2), (4.3) and the diophantine condition (X.2.4) with exponent $\nu$ for $\omega(a^*)$. The following theorem is a variant of results in Stoffer (1998) and Hairer & Lubich (1999). It shows that for symplectic methods, the invariant torus persists under a very mild restriction on the step size. For non-symplectic methods, this would require step sizes $h$ with $h^p \ll \varepsilon$ (see Exercise 5).

**Theorem 5.2.** *Let a symplectic numerical integrator of order $p$ be applied to a perturbed integrable Hamiltonian system (5.4) which satisfies the conditions stated above. Then, there exist $\varepsilon_0 > 0$ and $c_0 > 0$ such that, for $0 < \varepsilon \leq \varepsilon_0$ and for step sizes $h > 0$ satisfying*

$$h^p \leq c_0 |\log \varepsilon|^{-\kappa} \tag{5.5}$$

*with $\kappa = \max(\nu + d + 1, p)$, the numerical method has an attractive invariant torus $\mathcal{T}_{\varepsilon,h}$. This torus is $\mathcal{O}(h^p)$ close to the invariant torus $\mathcal{T}_\varepsilon$ of (5.4). It attracts an $\mathcal{O}(|\log \varepsilon|^{-2\kappa})$ neighbourhood with an exponential rate proportional to $\varepsilon$, uniformly in $h$.*

**Remark 5.3.** The exponent $\nu + d + 1$ comes from Lemma X.4.1. It could be reduced to $\nu + 1$ by using Rüssmann's estimates in place of that lemma; cf. the remark after Lemma X.4.1.

*Proof of Theorem 5.2.* The proof combines backward error analysis (Theorem IX.3.1 and Theorem 5.1), perturbation theory (Theorem X.4.4 and Lemma 2.1), and the invariant manifold theorem (Theorem 3.1).

(a) We begin by considering the symplectic method applied to the integrable Hamiltonian system (5.4) with $\varepsilon = 0$. This leads us back to the questions of Chap. X. We use backward error analysis and recall (Theorem IX.3.1) that the modified equation is again Hamiltonian and an $\mathcal{O}(h^p)$ perturbation of the integrable system, both in the $(p, q)$ and the $(a, \theta)$ variables. We transform variables for the *modified equation of the integrable system:* $(a, \theta) \mapsto (\widetilde{a}, \widetilde{\theta})$ by Theorem X.4.4,

with $h^p$ in the role of the perturbation parameter. By (X.4.1) with $N$ proportional to $|\log \varepsilon|$, and by condition (5.5) with a sufficiently small $c_0$, the modified equations in these variables become

$$\begin{aligned} \dot{\widetilde{a}} &= \mathcal{O}(\varepsilon^3) \\ \dot{\widetilde{\theta}} &= \widetilde{\omega}(\widetilde{a}) + \mathcal{O}(\varepsilon^3) \end{aligned} \quad \text{for} \quad \|\widetilde{a} - a^*\| \leq c^*|\log \varepsilon|^{-2\kappa},$$

with $\widetilde{\omega}(\widetilde{a}) = \omega(\widetilde{a}) + \mathcal{O}(h^p)$. Moreover, the transformation $(a, \theta) \mapsto (\widetilde{a}, \widetilde{\theta})$ is $\mathcal{O}(h^p)$ close to the identity.

(b) The modified equations of the perturbed system, written in the $(\widetilde{a}, \widetilde{\theta})$ variables, become

$$\begin{aligned} \dot{\widetilde{a}} &= \varepsilon \widetilde{r}(\widetilde{a}, \widetilde{\theta}) + \mathcal{O}(\varepsilon^3) \\ \dot{\widetilde{\theta}} &= \widetilde{\omega}(\widetilde{a}) + \varepsilon \widetilde{\rho}(\widetilde{a}, \widetilde{\theta}) + \mathcal{O}(\varepsilon^3) \end{aligned} \quad \text{for} \quad \|\widetilde{a} - a^*\| \leq c^*|\log \varepsilon|^{-2\kappa}, \quad (5.6)$$

where $\widetilde{r}(\widetilde{a}, \widetilde{\theta}) = r(\widetilde{a}, \widetilde{\theta}) + \mathcal{O}(h^p)$ and $\widetilde{\rho}(\widetilde{a}, \widetilde{\theta}) = \rho(\widetilde{a}, \widetilde{\theta}) + \mathcal{O}(h^p)$ by Theorem 5.1. Consider now these equations with the $\mathcal{O}(\varepsilon^3)$ terms dropped. We change variables for the

*modified equation of the perturbed system:* $(\widetilde{a}, \widetilde{\theta}) \mapsto (\widetilde{b}, \widetilde{\varphi})$ by Lemma 2.1.

(Note Exercise 4 with $\widetilde{\omega}(a^*) = \omega(a^*) + \mathcal{O}(h^p)$ and (5.5).) The system (5.6) is transformed to the form of (4.4),

$$\begin{aligned} \dot{\widetilde{b}} &= \varepsilon \widetilde{m}(\widetilde{b}) + \mathcal{O}(\varepsilon^3/\delta^2) \\ \dot{\widetilde{\varphi}} &= \widetilde{\omega}(\widetilde{b}) + \varepsilon \widetilde{\mu}(\widetilde{b}) + \mathcal{O}(\varepsilon^3/\delta^2) \end{aligned} \quad (5.7)$$

with $\delta = c^*|\log \varepsilon|^{-2\kappa}$, and where $\widetilde{m}(\widetilde{b}) = \overline{\widetilde{r}}(\widetilde{b}) + \mathcal{O}(\varepsilon/\delta) = \overline{r}(\widetilde{b}) + \mathcal{O}(h^p) + \mathcal{O}(\varepsilon/\delta)$, and also the Jacobian of $\widetilde{m}$ at $a^*$ is close to that of $\overline{r}$, so that it satisfies again (4.3), at least with $\alpha$ replaced by $\alpha/2$. In the same way as in the proof of Theorem 4.1 and with the same Lipschitz constants as in (4.5), we now obtain an attractive invariant torus of the modified equation of the perturbed system. The time-$h$ flow of this equation is an exponentially small (in $1/h$) Lipschitz perturbation of the numerical one-step map, so that under condition (5.5) it is an $\mathcal{O}(\varepsilon^3)$ perturbation. Therefore, Theorem 3.1 yields an invariant torus $\mathcal{T}_{\varepsilon,h}$ of the numerical method.

(c) It remains to bound the distance between the tori $\mathcal{T}_{\varepsilon,h}$ and $\mathcal{T}_\varepsilon$. We recall that $\mathcal{T}_\varepsilon$ was obtained by a transformation of the

*perturbed system:* $(a, \theta) \mapsto (b, \varphi)$ by Lemma 2.1,

which puts (4.1) into the form (4.4). We thus have the transformations

$$
\begin{array}{ccc}
(a,\theta) & \xrightarrow{\varepsilon} & (b,\varphi) \\
{\scriptstyle h^p}\downarrow & & \\
(\tilde{a},\tilde{\theta}) & \xrightarrow{\varepsilon} & (\tilde{b},\tilde{\varphi})
\end{array}
$$

where the symbols $h^p$ and $\varepsilon$ indicate that the transformation is $\mathcal{O}(h^p)$ or $\mathcal{O}(\varepsilon)$ close to the identity. By the construction of Lemma 2.1, the composed transformation $(b,\varphi) \mapsto (\tilde{b},\tilde{\varphi})$ is $\mathcal{O}(h^p)$ close to the identity and moreover, the right-hand sides of (4.4) and (5.7) differ by $\mathcal{O}(\varepsilon h^p)$. Theorem 3.2 (with $\rho = e^{-\varepsilon \tau \alpha/2}$) now shows that the functions $s_{\varepsilon,h}$ and $s_\varepsilon$ defining $\mathcal{T}_{\varepsilon,h}$ and $\mathcal{T}_\varepsilon$, respectively, differ by $\mathcal{O}(h^p)$. This yields the desired distance bound. □

### XII.5.3 Symmetric Methods

A result analogous to the theorem of the previous subsection holds for reversible methods applied to perturbed reversible systems

$$
\begin{aligned}
\dot{u} &= f(u,v) + \varepsilon k(u,v) \\
\dot{v} &= g(u,v) + \varepsilon \ell(u,v)
\end{aligned}
$$

where the unperturbed system ($\varepsilon = 0$) is a real-analytic integrable reversible system. If the perturbed system, written in action-angle variables of the unperturbed system, satisfies the conditions of Theorem 4.1, then a reversible analogue of Theorem 5.2 holds, where the terms "symplectic" and "Hamiltonian" are simply replaced by "reversible". The proof remains the same, working with the reversible analogues of the results used for the Hamiltonian case.

## XII.6 Exercises

1. In the situation of the invariant manifold theorem, Theorem 3.1, suppose in addition that $f$ and $g$ are $\alpha$-periodic in $x$: $f(x+\alpha,y) = f(x,y)$, $g(x+\alpha,y) = g(x,y)$ for all $x \in X$, $y \in Y$. Show that in this case the function $s$ defining the invariant manifold is also $\alpha$-periodic.
   *Hint.* The Hadamard transform maps $\alpha$-periodic functions to $\alpha$-periodic functions.
2. Show that if the time-$\tau$ flow map $\Phi = \varphi_\tau$ of a differential equation has an attractive invariant manifold $\mathcal{M}$, and if the flow $\varphi_t$ maps a domain of attractivity of $\mathcal{M}$ under $\Phi$ into itself for every real $t$, then $\mathcal{M}$ is also invariant under the flow $\varphi_t$ for every real $t$.
   *Hint.* Write $\varphi_t = \Phi^n \circ \varphi_t \circ \Phi^{-n}$ and use the attractivity of $\mathcal{M}$ for $n \to \infty$.

3. Prove that in the situation of Theorem 3.1, iterates $(x_{n+1}, y_{n+1}) = \Phi(x_n, y_n)$ have the *property of asymptotic phase* (Nipp & Stoffer 1992): there exists a sequence $(\widetilde{x}_n, \widetilde{y}_n)$ of iterates on the invariant manifold, i.e., with $(\widetilde{x}_{n+1}, \widetilde{y}_{n+1}) = \Phi(\widetilde{x}_n, \widetilde{y}_n)$ and $\widetilde{y}_n = s(\widetilde{x}_n)$, such that for all $n \geq 0$,

$$\|x_n - \widetilde{x}_n\| \leq c \|y_n - s(x_n)\|$$
$$\|y_n - \widetilde{y}_n\| \leq (1 + \lambda c) \|y_n - s(x_n)\|,$$

where $c = \lambda/(1 - \lambda\lambda^*)$ with $\lambda = 2L_{yx}/(1 - L_{xx} - L_{yy})$ of (3.5) and $\lambda^* = 2L_{xy}/(1 - L_{xx} - L_{yy})$. Note that $\|y_n - s(x_n)\| \leq \rho^n \|y_0 - s(x_0)\|$ by Theorem 3.1.

*Hint.* Consider the sequences $(\widetilde{x}_n^{(k)}, \widetilde{y}_n^{(k)})$ defined by $\widetilde{x}_k^{(k)} = x_k$, $\widetilde{y}_k^{(k)} = s(x_k)$ and $(\widetilde{x}_{n+1}^{(k)}, \widetilde{y}_{n+1}^{(k)}) = \Phi(\widetilde{x}_n^{(k)}, \widetilde{y}_n^{(k)})$ for $n = k-1, \ldots, 1, 0$. Show that, for fixed $n$, the sequence $(x_n^{(k)})$ ($k \geq n$) is a Cauchy sequence.

4. Show that Lemma 2.1 holds unchanged if the diophantine condition (X.2.4) for $\omega(a^*)$ is weakened to $\omega(a^*) = \omega^* + \mathcal{O}(\delta^2)$ with $\omega^*$ satisfying (X.2.4).

5. In the situation of Theorem 5.2, show that every numerical integrator of order $p$ has an attractive invariant torus if $h^p \ll \varepsilon$. This torus is $\mathcal{O}(h^p/\varepsilon)$ close to the invariant torus of the continuous system.

# Chapter XIII.
# Highly Oscillatory Differential Equations

This chapter deals with numerical methods for second-order differential equations with oscillatory solutions. These methods are designed to require a new complete function evaluation only after a time step over one or many periods of the fastest oscillations in the system. Various such methods have been proposed in the literature – some of them decades ago, some very recently, motivated by problems from molecular dynamics, astrophysics and nonlinear wave equations. For these methods it is not obvious what implications geometric properties like symplecticity or reversibility have on the long-time behaviour, e.g., on energy conservation. The backward error analysis of Chap. IX, which was the backbone of the results of the three preceding chapters, is no longer applicable when the product of the step size with the highest frequency is not small, which is the situation of interest here. The "exponentially small" remainder terms are now only $\mathcal{O}(1)$! At least for a class of nonlinear model problems, which includes the Fermi-Pasta-Ulam problem of Sect. I.4.1, a substitute for the backward error analysis of Chap. IX is given by the *modulated Fourier expansions* of the exact and the numerical solutions. Among other properties, they permit us to understand the numerical long-time conservation of energy (or the failure of conserving energy in certain cases). It turns out, symmetry of the methods is still essential, but symplecticity plays no role in the analysis and in the numerical experiments, and new conditions of an apparently non-geometric nature come into play.

## XIII.1 Towards Longer Time Steps in Solving Oscillatory Differential Equations

> Dynamical systems with multiple time scales pose a major problem in simulations because the small time steps required for stable integration of the fast motions lead to large numbers of time steps required for the observation of slow degrees of freedom and thus to the need to compute a large number of forces.
>
> (M. Tuckerman, B.J. Berne & G.J. Martyna 1992)

We describe numerical methods that have been proposed for solving highly oscillatory second-order differential equations with fewer force evaluations than are needed by standard integrators like the Störmer/Verlet method. We present the ideas underlying the construction of the methods and leave numerical comparisons to Sect. XIII.2 and the analysis of the methods to Sections XIII.3–XIII.6. We consider only methods that are symmetric or symplectic. The presentation in this section follows roughly the chronological order.

### XIII.1.1 The Störmer/Verlet Method vs. Multiple Time Scales

> Perhaps the most widely used method of integrating the equations of motion is that initially adopted by Verlet (1967) and attributed to Störmer.
> (M.P. Allen & D.J. Tildesley 1987, p. 78)

The Newtonian equations of motion of particle systems (in molecular dynamics, astrophysics and elsewhere) are second-order differential equations

$$\ddot{q} = -\nabla V(q) . \tag{1.1}$$

To simplify the presentation, we omit the positive definite mass matrix $M$ which would usually multiply $\ddot{q}$. This entails no loss of generality, since a transformation $q \to M^{1/2}q$ and $V(q) \to V(M^{-1/2}q)$ gives the very form (1.1).

The standard numerical integrator of molecular dynamics is the Störmer/Verlet scheme; see Chap. I. We recall that this method computes the new positions $q_{n+1}$ at time $t_{n+1}$ from

$$q_{n+1} - 2q_n + q_{n-1} = h^2 f_n \tag{1.2}$$

with the force $f_n = -\nabla V(q_n)$. Velocity approximations are given by

$$\dot{q}_n = \frac{q_{n+1} - q_{n-1}}{2h} .$$

In its one-step formulation (see (I.3.6)) the method reads[1]

$$\begin{aligned} p_{n+1/2} &= p_n + \tfrac{1}{2} h f_n \\ q_{n+1} &= q_n + h p_{n+1/2} \\ p_{n+1} &= p_{n+1/2} + \tfrac{1}{2} h f_{n+1} . \end{aligned} \tag{1.3}$$

We recall that this is a symmetric and symplectic method of order 2. For linear stability, i.e., for bounded error propagation in linearized equations, the step size must be restricted to

$$h\omega < 2$$

where $\omega$ is the largest eigenfrequency (i.e., square root of an eigenvalue) of the Hessian matrix $\nabla^2 V(q)$ along the numerical solution; see Sect. I.4.1. Good energy conservation requires an even stronger restriction on the step size. Values of $h\omega \approx \tfrac{1}{2}$ are frequently used in molecular dynamics simulations.

The potential $V(q)$ is often a sum of potentials that act on different time scales,

---

[1] We write $p$ when the Hamiltonian structure and symplecticity are an issue, and $\dot{q}$ otherwise.

$$V(q) = W(q) + U(q) \quad \text{with} \quad \begin{array}{l} \nabla^2 W(q) \text{ positive semi-definite and} \\ \|\nabla^2 W(q)\| \gg \|\nabla^2 U(q)\| \, . \end{array} \qquad (1.4)$$

In this situation, solutions are in general highly oscillatory on the slow time scale $\tau \sim 1/\|\nabla^2 U(q)\|^{1/2}$.

In particular when the *fast* forces $-\nabla W(q)$ are cheaper to evaluate than the *slow* forces $-\nabla U(q)$, it is of interest to devise methods where the required number of slow-force evaluations is not (or not severely) affected by the presence of the fast forces which are responsible for the oscillatory behaviour and which restrict the step size of standard integrators like the Störmer/Verlet scheme. This situation occurs in molecular dynamics, where $W(q)$ corresponds to short-range molecular bonds, whereas $U(q)$ includes *inter alia* long-range electrostatic potentials.

In some approaches to this computational problem, the differential model is modified: highly oscillatory components are replaced by constraints (Ryckaert, Ciccotti & Berendsen 1977), or stochastic and dissipative terms are added to the model (see Schlick 1999). Such modifications may prove highly successful in some applications. In the following, however, we restrict our attention to methods which aim at long time steps directly for the problem (1.1) with (1.4).

Spatial semi-discretizations of nonlinear wave equations, such as the sine-Gordon equation

$$u_{tt} = u_{xx} - \sin u \, ,$$

form another important class of equations (1.1) with (1.4). Here $W(q) = \frac{1}{2} q^T A q$, where $A$ is the discretization matrix of the differential operator $-\partial^2/\partial x^2$.

## XIII.1.2 Gautschi's and Deuflhard's Trigonometric Methods

> It is anticipated that trigonometric methods can be applied, with similar success, also to nonlinear differential equations describing oscillation phenomena.
> (W. Gautschi 1961)

The oldest methods allowing the use of long time steps in oscillatory problems concern the particular case of a quadratic potential $W(q) = \frac{1}{2}\omega^2 q^T q$ with $\omega \gg 1$, for which the equations take the form

$$\ddot{q} = -\omega^2 q + g(q) \, . \qquad (1.5)$$

For such equations, Gautschi (1961) proposed a number of methods of multistep type which are constructed to be exact if the solution is a trigonometric polynomial in $\omega t$ of a prescribed degree. The simplest of these methods (and the only symmetric one) reads

$$q_{n+1} - 2q_n + q_{n-1} = h^2 \operatorname{sinc}^2(\tfrac{1}{2} h\omega) \, \ddot{q}_n \, , \qquad (1.6)$$

where $\operatorname{sinc} \xi = \sin \xi / \xi$ and $\ddot{q}_n = -\omega^2 q_n + g_n$ with $g_n = g(q_n)$, or equivalently

$$q_{n+1} - 2\cos(h\omega) q_n + q_{n-1} = h^2 \operatorname{sinc}^2(\tfrac{1}{2} h\omega) \, g_n \, . \qquad (1.7)$$

The method gives the exact solution for equations (1.5) with $g = Const$ and arbitrary $\omega$ (see also Hersch (1958) for such a construction principle). This property is readily verified with the variation-of-constants formula

$$\begin{pmatrix} q(t) \\ \dot{q}(t) \end{pmatrix} = \begin{pmatrix} \cos t\omega & \omega^{-1} \sin t\omega \\ -\omega \sin t\omega & \cos t\omega \end{pmatrix} \begin{pmatrix} q_0 \\ \dot{q}_0 \end{pmatrix} \qquad (1.8)$$
$$+ \int_0^t \begin{pmatrix} \omega^{-1} \sin(t-s)\omega \\ \cos(t-s)\omega \end{pmatrix} g(q(s))\, ds\,.$$

This formula also shows that the following scheme for a velocity approximation becomes exact for $g = Const$:

$$\dot{q}_{n+1} - \dot{q}_{n-1} = 2h \operatorname{sinc}(h\omega)\, \ddot{q}_n\,. \qquad (1.9)$$

Starting values $q_1$ and $\dot{q}_1$ are also obtained from (1.8) with $g(q_0)$ in place of $g(q(s))$.

Deuflhard (1979) considered $h^2$-extrapolation based on the explicit symmetric method that is obtained by replacing the integral term in (1.8) by its trapezoidal rule approximation:

$$\begin{pmatrix} q_{n+1} \\ h\dot{q}_{n+1} \end{pmatrix} = \begin{pmatrix} \cos h\omega & \operatorname{sinc} h\omega \\ -h\omega \sin h\omega & \cos h\omega \end{pmatrix} \begin{pmatrix} q_n \\ h\dot{q}_n \end{pmatrix} + \frac{h^2}{2} \begin{pmatrix} \operatorname{sinc}(h\omega)\, g_n \\ g_{n+1} + \cos(h\omega)\, g_n \end{pmatrix}. \qquad (1.10)$$

Eliminating the velocities yields the two-step formulation

$$q_{n+1} - 2\cos(h\omega) q_n + q_{n-1} = h^2 \operatorname{sinc}(h\omega)\, g_n\,. \qquad (1.11)$$

The velocity approximation is obtained back from

$$2h \operatorname{sinc}(h\omega)\, \dot{q}_n = q_{n+1} - q_{n-1} \qquad (1.12)$$

or alternatively from

$$\dot{q}_{n+1} - 2\cos(h\omega)\dot{q}_n + \dot{q}_{n-1} = h^2 \frac{g_{n+1} - g_{n-1}}{2h}\,.$$

Both Gautschi's and Deuflhard's method reduce to the Störmer/Verlet scheme for $\omega = 0$. Both methods extend in a straightforward way to systems

$$\ddot{q} = -Aq + g(q) \qquad (1.13)$$

with a symmetric positive semi-definite matrix $A$, by formally replacing $\omega$ by $\Omega = A^{1/2}$ in the above formulas. The methods then require the computation of products of entire functions of the matrix $h^2 A$ with vectors. This can be done by diagonalizing $A$, which is efficient for problems of small dimension or in spectral methods for nonlinear wave equations. In high-dimensional problems where a diagonalization is not feasible, these matrix function times vector products can be efficiently computed by superlinearly convergent Krylov subspace methods, see Druskin & Knizhnerman (1995) and Hochbruck & Lubich (1997).

The above methods permit extensions to more general problems (1.1) with (1.4), but this requires a reinterpretation to which we turn next.

## XIII.1.3 The Impulse Method

> Integrators based on r-RESPA [...] have led to considerable speed-up in the CPU time for large scale simulations of biomacromolecular solutions. Since r-RESPA is symplectic such integrators are very stable.
>
> (B.J. Berne 1999)

The Störmer/Verlet method (1.3) can be interpreted as approximating the flow $\varphi_h^H$ of the system with Hamiltonian $H(p,q) = T(p) + V(q)$ with $T(p) = \frac{1}{2}p^T p$ by the symmetric splitting

$$\varphi_{h/2}^V \circ \varphi_h^T \circ \varphi_{h/2}^V,$$

which involves only the flows of the systems with Hamiltonians $T(p)$ and $V(q)$, which are trivial to compute; see Sect. II.5.

In the situation (1.4) of a potential $V = W + U$, we may instead use a different splitting of $H = (T + W) + U$ and approximate the flow $\varphi_h^H$ of the system by

$$\varphi_{h/2}^U \circ \varphi_h^{T+W} \circ \varphi_{h/2}^U.$$

This gives a method that was proposed in the context of molecular dynamics by Grubmüller, Heller, Windemuth & Schulten (1991) (their Verlet-I scheme) and by Tuckerman, Berne & Martyna (1992) (their r-RESPA scheme). Following the terminology of García-Archilla, Sanz-Serna & Skeel (1999) we here refer to this method as the *impulse method*:

1. kick: set $p_n^+ = p_n - \frac{1}{2}h\,\nabla U(q_n)$
2. oscillate: solve $\ddot{q} = -\nabla W(q)$ with initial values $(q_n, p_n^+)$ over a time step $h$ to obtain $(q_{n+1}, p_{n+1}^-)$
3. kick: set $p_{n+1} = p_{n+1}^- - \frac{1}{2}h\nabla U(q_{n+1})$

(1.14)

Step 2 must in general be computed approximately by a numerical integrator with a smaller time step, which results in the multiple time stepping method that we encountered in Sect. VIII.4. If the inner integrator is symplectic and symmetric, as it would be for the natural choice of the Störmer/Verlet method, then also the overall method is symplectic – as a composition of symplectic transformations, and it is symmetric – as a symmetric composition of symmetric steps.

It is interesting to note that the impulse method (with exact solution of step 2) reduces to Deuflhard's method in the case of a quadratic potential $W(q) = \frac{1}{2}q^T A q$ (Exercise 1).

Though the method does allow larger step sizes than the Störmer/Verlet method in molecular dynamics simulations, it is not free from numerical difficulties. Biesadecki & Skeel (1993) and García-Archilla et al. (1999) report and in linear model problems analyze instabilities and numerical resonance phenomena when the product of the step size $h$ with an eigenfrequency $\omega$ of $\nabla^2 W$ is near an integral multiple of $\pi$.

## XIII.1.4 The Mollified Impulse Method

> We also propose a nontrivial improvement of the impulse method that we call the *mollified impulse method*, for which superior stability and accuracy is demonstrated.
>
> (B.García-Archilla, J.M. Sanz-Serna & R.D. Skeel 1999)

Difficulties with the impulse method can be intuitively seen to come from two sources: the slow force $-\nabla U(q)$ has an effect only at the ends of a time step, but it does not enter into the oscillations in between; the slow force is evaluated, somewhat arbitrarily, at isolated points of the oscillatory solution.

García-Archilla et al. (1999) propose to evaluate the slow force at an *averaged* value $\overline{q}_n = a(q_n)$. They replace the potential $U(q)$ by $\overline{U}(q) = U(a(q))$ and hence the slow force $-\nabla U(q)$ in the impulse method by the *mollified force*

$$-\nabla \overline{U}(q) = -a'(q)^T \nabla U(a(q)) . \tag{1.15}$$

Since this *mollified impulse method* is the impulse method for a modified potential, it is again symplectic and symmetric.

There are numerous possibilities to choose the average $a(q_n)$, but care should be taken that it is only a function of the position $q_n$ and thus independent of $p_n$, in order to obtain a symplectic and symmetric method. This precludes taking averages of the solution of the problem in the oscillation step (Step 2) of the algorithm. Instead, one solves the auxiliary initial value problem

$$\ddot{x} = -\nabla W(x) \quad \text{with} \quad x(0) = q, \ \dot{x}(0) = 0 \tag{1.16}$$

together with the variational equation (using the same method and the same step size)

$$\ddot{X} = -\nabla^2 W(x(t))X \quad \text{with} \quad X(0) = I, \ \dot{X}(0) = 0 \tag{1.17}$$

and computes the time average over an interval of length $ch$ for some $c > 0$:

$$a(q) = \frac{1}{ch}\int_0^{ch} x(t)\,dt, \quad a'(q) = \frac{1}{ch}\int_0^{ch} X(t)\,dt . \tag{1.18}$$

García-Archilla et al. (1999) found that the choice $c = 1$ gives the best results. Weighted averages instead of the simple average used above give no improvement.

Izaguirre, Reich & Skeel (1999) propose to take $a(q)$ as a projection of $q$ to the manifold $\nabla W(q) = 0$ of rest positions of the fast forces, for situations where all non-zero eigenfrequencies of $\nabla^2 W(q)$ are much larger than those of $\nabla^2 U(q)$. This choice is motivated by the fact that solutions oscillate about this manifold.

We now turn to the interesting special case of a quadratic $W(q) = \frac{1}{2}q^T A q$ with a symmetric positive semi-definite matrix $A$. In this case, the above average can be computed analytically. It becomes

$$a(q) = \phi(h\Omega)q$$

with $\Omega = A^{1/2}$ and the function $\phi(\xi) = \mathrm{sinc}(c\xi)$. For $a(q)$ defined by the orthogonal projection to $Aq = 0$ we have $\phi(0) = 1$ and $\phi(\xi) = 0$ for $\xi$ away from 0. With $g_n = -\phi(h\Omega)\nabla U(\phi(h\Omega)q_n)$, the mollified impulse method reduces to

$$p_n^+ = p_n + \tfrac{1}{2}hg_n$$

$$\begin{pmatrix} q_{n+1} \\ p_{n+1}^- \end{pmatrix} = \begin{pmatrix} \cos h\Omega & h\,\mathrm{sinc}\,h\Omega \\ -\Omega\sin h\Omega & \cos h\Omega \end{pmatrix} \begin{pmatrix} q_n \\ p_n^+ \end{pmatrix} \quad (1.19)$$

$$p_{n+1} = p_{n+1}^- + \tfrac{1}{2}hg_{n+1}\,.$$

This can equivalently be written as (1.10) with the same $g_n$ (and $\Omega$ in place of $\omega$), or in the two-step form (1.11) with (1.12).

## XIII.1.5 Gautschi's Method Revisited

We recall that Gautschi's method (1.7) (with $\Omega = A^{1/2}$ in place of $\omega$) integrates equations $\ddot{q} = -Aq + g(q)$ exactly in the case of a constant inhomogeneity $g(q) = Const$. This property is obviously kept if the argument of $g$ in the algorithm is modified to

$$g_n = g(\phi(h\Omega)q_n)$$

similar to the previous subsection. Such Gautschi-type methods were analyzed by Hochbruck & Lubich (1999a). Functions $\phi$ with $\phi(0) = 1$ that vanish at integral multiples of $\pi$ give a substantial improvement over the original Gautschi method. The choice

$$\phi(\xi) = \mathrm{sinc}\,\xi\left(1 + \tfrac{1}{3}\sin^2\tfrac{1}{2}\xi\right) \quad (1.20)$$

was found to give particularly good accuracy. The methods are symmetric but not symplectic.

The following symmetric method for general problems (1.1) with (1.4) was proposed by Hochbruck & Lubich (1999a). The method reduces to Gautschi-type methods for quadratic $W(q) = \tfrac{1}{2}q^T Aq$. Given $q_n$ and $\dot{q}_n$, one computes an averaged value $\bar{q}_n = a(q_n)$ and the solution of

$$\ddot{u} = -\nabla W(u) - \nabla U(\bar{q}_n) \quad \text{with} \quad u(0) = q_n,\ \dot{u}(0) = \dot{q}_n \quad (1.21)$$

backwards and forwards on the intervals from 0 to $-h$ and 0 to $h$. Note that this requires only one evaluation of the slow force $-\nabla U$. Then, $q_{n+1}$ and $\dot{q}_{n+1}$ are computed from

$$\begin{aligned} q_{n+1} - 2q_n + q_{n-1} &= u(h) - 2u(0) + u(-h) \\ \dot{q}_{n+1} - \dot{q}_{n-1} &= \dot{u}(h) - \dot{u}(-h)\,. \end{aligned} \quad (1.22)$$

When the differential equation for $u$ is solved approximately by a symmetric numerical method with smaller time steps, then this becomes a symmetric multiple time-stepping method. For the interpretation as an averaged-force method and for the corresponding one-step version, where the initial value for the velocity in (1.21) is replaced by $\dot{u}(0) = 0$, we refer back to Sect. VIII.4 (where $q_n$ instead of the average $\bar{q}_n = a(q_n)$ was taken as the argument of the slow force $-\nabla U$).

## XIII.1.6 Two-Force Methods

Hairer & Lubich (2000a) compare the analytical solution and the numerical solutions given by the above methods in the Fermi-Pasta-Ulam model of Sect. I.4.1, using the tool of modulated Fourier expansions (see Sections XIII.3 and XIII.5 below). Their analysis of the slow energy exchange between stiff springs leads them to propose the following method for equations $\ddot{q} = -Aq + g(q)$, which requires two evaluations of the slow force per time step: with $\Omega = A^{1/2}$, set

$$q_{n+1} - 2\cos(h\Omega)\, q_n + q_{n-1} = h^2 \operatorname{sinc}(h\Omega)\, g(q_n) + h^2 d_n \qquad (1.23)$$

with

$$d_n = \operatorname{sinc}^2(h\Omega)\, g(q_n) - \operatorname{sinc}(h\Omega)\, g\!\left(\operatorname{sinc}(h\Omega) q_n\right). \qquad (1.24)$$

This method gives the correct slow energy exchange between stiff components in the model problem and has better energy conservation than the Deuflhard/impulse method. With the velocity approximation (1.12) the method can equivalently be written in the one-step forms (1.19) or (1.10). The method extends again to a symmetric method for general problems (1.1) with (1.4), giving a correction to the impulse method: let $g(q) = -\nabla U(q)$ and let $a(q)$ be defined by (1.18) with $c = 1$. Set $\overline{q}_n = a(q_n)$ and

$$\overline{g}(q_n) = \frac{2}{h^2}\left( a\!\left(q_n + \tfrac{1}{2} h^2 g(q_n)\right) - a(q_n) \right).$$

The method then consists of taking

$$g_n = g(q_n) + \overline{g}(q_n) - g(\overline{q}_n)$$

instead of $g(q_n) = -\nabla U(q_n)$ in the impulse method (1.14).

A two-force method with interesting properties, for situations where all non-zero eigenfrequencies of $A$ are much larger than those of $\nabla^2 U(q)$, is given by (1.23) with

$$d_n = \operatorname{sinc}^2(\tfrac{1}{2} h\Omega)\, g\!\left(\chi(h\Omega) q_n\right) - \operatorname{sinc}(h\Omega)\, g\!\left(\chi(h\Omega) q_n\right), \qquad (1.25)$$

where $\chi(0) = 1$ and $\chi(\xi) = 0$ for $\xi$ away from 0.

## XIII.2 A Nonlinear Model Problem and Numerical Phenomena

To gain insight into the properties of the various numerical methods described in the previous section, it is helpful to study the methods when they are applied to suitably chosen, rather simple model problems which show characteristic features but are still accessible to an analysis. Such an approach has traditionally been very successful for stiff differential equations (see, e.g., Hairer & Wanner 1996). For the present stiff-oscillatory case we investigate the behaviour of the numerical methods on nonlinear systems

$$\ddot{x} + \Omega^2 x = g(x) \tag{2.1}$$

with a smooth gradient nonlinearity $g(x) = -\nabla U(x)$ and with the square matrix

$$\Omega = \begin{pmatrix} 0 & 0 \\ 0 & \omega I \end{pmatrix}, \quad \omega \gg 1, \tag{2.2}$$

with blocks of arbitrary dimension. The Fermi-Pasta-Ulam (FPU) problem of Section I.4.1 belongs precisely to this class, and we will present numerical experiments with this example. In the model problem (2.1) we clearly impose strong restrictions in that the high frequencies are confined to the linear part (unlike molecular dynamics) and that the linear part has a single high frequency (unlike nonlinear wave equations). In any case, satisfactory behaviour of a method on the model problem can be anticipated to be necessary for a successful treatment of more general situations.

We consider only solutions whose energy is bounded independently of $\omega$, so that in particular the initial values satisfy

$$\tfrac{1}{2}\|\dot{x}_0\|^2 + \tfrac{1}{2}\|\Omega x_0\|^2 \leq E \tag{2.3}$$

with $E$ independent of $\omega$.

## XIII.2.1 Time Scales in the Fermi-Pasta-Ulam Problem

The FPU model shows different behaviour on different time scales: almost-harmonic motion of the stiff springs on the time scale $\omega^{-1}$, motion of the soft springs on the scale $\omega^0$, energy exchange between stiff springs on the time scale $\omega$, and almost-preservation of the oscillatory energy over intervals that are exponentially long in $\omega$. This is illustrated in the following.

We consider the FPU problem with three stiff springs with the data of Sect. I.4.1. The four pictures of Fig. 2.1 show the evolution of the following quantities: the total energy

$$H(x, \dot{x}) = \tfrac{1}{2}\dot{x}^T \dot{x} + \tfrac{1}{2}x^T \Omega^2 x + U(x), \tag{2.4}$$

(or rather $H - 0.8$ for graphical reasons), which is a conserved quantity; the oscillatory energy

$$I = I_1 + I_2 + I_3 \quad \text{with} \quad I_j = \tfrac{1}{2}\dot{x}_{2,j}^2 + \tfrac{1}{2}\omega^2 x_{2,j}^2, \tag{2.5}$$

where $x_{2,j}$, the $j$th component of the lower half $x_2$ of $x$, represents the elongation of the $j$th stiff spring. Further quantities shown are the kinetic energy of the mass centre motion and of the relative motion of masses joined by a stiff spring,

$$T_1 = \tfrac{1}{2}\|\dot{x}_1\|^2, \quad T_2 = \tfrac{1}{2}\|\dot{x}_2\|^2.$$

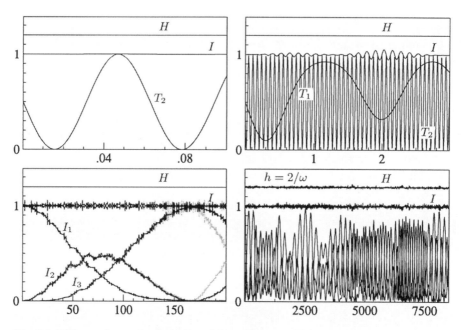

**Fig. 2.1.** Different time scales in the Fermi-Pasta-Ulam model ($\omega = 50$).

**Time Scale $\omega^{-1}$.** The vibration of the stiff linear springs is nearly harmonic with almost-period $\pi/\omega$. This is illustrated by the plot of $T_2$ in the first picture.

**Time Scale $\omega^0$.** This is the time scale of the motion of the soft nonlinear springs, as is exemplified by the plot of $T_1$ in the second picture of Fig. 2.1.

**Time Scale $\omega$.** A slow energy exchange among the stiff springs takes place on the scale $\omega$. In the third picture, the initially excited first stiff spring passes energy to the second one, and then also the third stiff spring begins to vibrate. The picture also illustrates that the problem is very sensitive to perturbations of the initial data: the grey curves of each of $I_1, I_2, I_3$ correspond to initial data where $10^{-5}$ has been added to $x_1(0)$, $\dot{x}_1(0)$ and $\dot{x}_4(0)$. The displayed solutions of the first three pictures have been computed very accurately by an adaptive integrator.

**Time Scale $\omega^N$, $N \geq 2$.** The oscillatory energy $I$ has only $\mathcal{O}(\omega^{-1})$ deviations from the initial value over very long time intervals. The fourth picture of Fig. 2.1 shows the total energy $H$ and the oscillatory energy $I$ as computed by the method (1.10)-(1.11) of Sect. XIII.1.2 with the step size $h = 2/\omega$, which is nearly as large as the length of the time interval of the first picture. No drift is seen for $H$ or $I$.

## XIII.2.2 Numerical Methods

The methods described in Sect. XIII.1 all have in common that they reduce to the Störmer/Verlet method when they are applied to (2.1) with $\Omega = 0$, and they become exact solvers for the linear homogeneous problem with $g(x) \equiv 0$. They can be formulated as one-step or two-step schemes.

**Two-Step Formulation.** All the methods of Sections XIII.1.2–XIII.1.5, when applied to the system (2.1), can be written in the two-step form

$$x_{n+1} - 2\cos(h\Omega)\, x_n + x_{n-1} = h^2 \Psi g(\Phi x_n). \tag{2.6}$$

Here $\Psi = \psi(h\Omega)$ and $\Phi = \phi(h\Omega)$, where the *filter functions* $\psi$ and $\phi$ are even, real-valued functions with $\psi(0) = \phi(0) = 1$. In our numerical experiments we will consider the following choices of $\psi$ and $\phi$, where again $\mathrm{sinc}(\xi) = \sin\xi/\xi$:

| | | | |
|---|---|---|---|
| (A) | $\psi(\xi) = \mathrm{sinc}^2(\tfrac{1}{2}\xi)$ | $\phi(\xi) = 1$ | Gautschi (1961) |
| (B) | $\psi(\xi) = \mathrm{sinc}(\xi)$ | $\phi(\xi) = 1$ | Deuflhard (1979) |
| (C) | $\psi(\xi) = \mathrm{sinc}(\xi)\phi(\xi)$ | $\phi(\xi) = \mathrm{sinc}(\xi)$ | García-Archilla & al. (1999) |
| (D) | $\psi(\xi) = \mathrm{sinc}^2(\tfrac{1}{2}\xi)$ | $\phi(\xi)$ of (1.20) | Hochbruck & Lubich (1999a) |
| (E) | $\psi(\xi) = \mathrm{sinc}^2(\xi)$ | $\phi(\xi) = 1$ | Hairer & Lubich (2000a) |

**One-Step Formulation.** The method (2.6) can be written as a symmetric one-step method of a form that is motivated by the variation-of-constants formula (1.8). This now also includes a velocity approximation $\dot{x}_n$:

$$x_{n+1} = \cos h\Omega\, x_n + \Omega^{-1} \sin h\Omega\, \dot{x}_n + \tfrac{1}{2} h^2 \Psi\, g_n \tag{2.7}$$

$$\dot{x}_{n+1} = -\Omega \sin h\Omega\, x_n + \cos h\Omega\, \dot{x}_n + \tfrac{1}{2} h\big(\Psi_0\, g_n + \Psi_1\, g_{n+1}\big) \tag{2.8}$$

where $g_n = g(\Phi x_n)$ and $\Psi_0 = \psi_0(h\Omega)$, $\Psi_1 = \psi_1(h\Omega)$ with even functions $\psi_0$, $\psi_1$ satisfying $\psi_0(0) = 1$, $\psi_1(0) = 1$. Exchanging $n \leftrightarrow n+1$ and $h \leftrightarrow -h$ in the method, it is seen that the method is symmetric if and only if

$$\psi(\xi) = \mathrm{sinc}(\xi)\, \psi_1(\xi), \qquad \psi_0(\xi) = \cos(\xi)\, \psi_1(\xi). \tag{2.9}$$

The method is then symplectic if and only if (Exercise 2)

$$\psi(\xi) = \mathrm{sinc}(\xi)\, \phi(\xi). \tag{2.10}$$

**Two-Step Velocity Schemes.** For a symmetric method (2.7)–(2.8) the velocity approximation can be equivalently obtained from

$$2h\, \mathrm{sinc}(h\omega)\, \dot{x}_n = x_{n+1} - x_{n-1} \tag{2.11}$$

(for $\sin(h\omega) \neq 0$) or from

$$\dot{x}_{n+1} - 2\cos(h\Omega)\, \dot{x}_n + \dot{x}_{n-1} = \tfrac{1}{2} h \Psi_1 (g_{n+1} - g_{n-1}). \tag{2.12}$$

The latter formula gives a symmetric two-step method for arbitrary even functions $\psi_1$ with $\psi_1(0) = 1$, which do not necessarily satisfy (2.9).

**Multi-Force Methods.** The methods of Sect. XIII.1.6 belong to the class of multi-force methods, which generalize the right-hand side of (2.6) to a linear combination of such terms:

$$x_{n+1} - 2\cos(h\Omega)\,x_n + x_{n-1} = h^2 \sum_{j=1}^{k} \Psi_j\, g(\Phi_j x_n) \tag{2.13}$$

with $\Psi_j = \psi_j(h\Omega)$, $\Phi_j = \phi_j(h\Omega)$, where $\psi_j$, $\phi_j$ are even functions with

$$\sum_{j=1}^{k} \psi_j(0) = 1, \qquad \phi_j(0) = 1 \qquad \text{for } j = 1, \ldots, k.$$

In our numerical experiments we include the method

$(F)$   two-force method (1.23) with (1.24).

### XIII.2.3 Accuracy Comparisons

The accuracy of the methods (A)-(E) and the Störmer/Verlet method on a short time interval is shown in Fig. 2.2, where the errors at $t = 1$ of the different solution components in the FPU problem (with $\omega = 50$) are plotted as a function of the step size $h$. Here and in all the following numerical experiments, the methods were

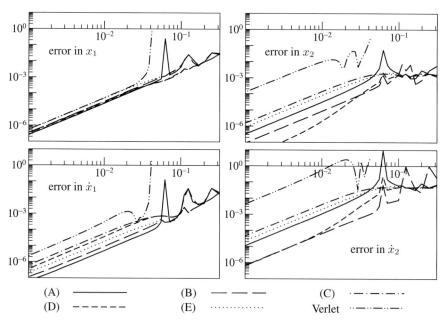

**Fig. 2.2.** Global error at $t = 1$ for the different components and for the five methods (A) - (E) and the Störmer/Verlet method as a function of the step size $h$.

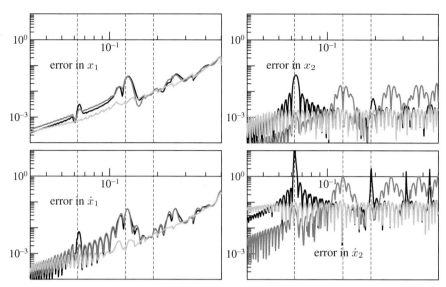

**Fig. 2.3.** Global error at the first grid point after $t = 1$ for the different components as a function of the step size $h$. The error for method (A) is drawn in black, for method (B) in dark grey, and for method (C) in light grey. The vertical lines indicate step sizes for which $h\omega$ equals $\pi$, $2\pi$, or $3\pi$.

implemented in the one-step formulation (2.7)-(2.8) with (2.9). The errors in the $x_1$-components are nearly identical for all the methods in the stability range of the Störmer/Verlet method ($h\omega < 2$). Differences between the methods are however visible for larger step sizes. For the other solution components $x_2$, $\dot{x}_1$, $\dot{x}_2$ there are pronounced differences in the error behaviour of the methods. All five methods (A)-(E) are considerably more accurate than the Störmer/Verlet method. Figure 2.3 shows the errors of methods (A)-(C) for step sizes beyond the stability range of the Störmer/Verlet method. Methods (A) and (B) lose accuracy when $h\omega$ is near integral multiples of $\pi$, a phenomenon that does not occur with method (C).

## XIII.2.4 Energy Exchange between Stiff Components

Figure 2.4 shows the energy exchange of the six methods (A)-(F) applied to the Fermi-Pasta-Ulam problem with the same data as in Fig. 2.1. The figures show again the oscillatory energies $I_1$, $I_2$, $I_3$ of the stiff springs, their sum $I = I_1 + I_2 + I_3$ and the total energy $H - 0.8$ as functions of time on the interval $0 \le t \le 200$. Only the methods (B), (D) and (F) give a good approximation of the energy exchange between the stiff springs. It will turn out in Sect. XIII.4.2 that a necessary condition for a correct approximation of the energy exchange is $\psi(h\omega)\phi(h\omega) = \mathrm{sinc}(h\omega)$, which is satisfied for method (B). The two-force method (F) satisfies an analogous condition for multi-force methods. The good behaviour of method (D) comes from the fact that here $\psi(h\omega)\phi(h\omega) \approx 0.95 \,\mathrm{sinc}(h\omega)$ for $h\omega = 1.5$.

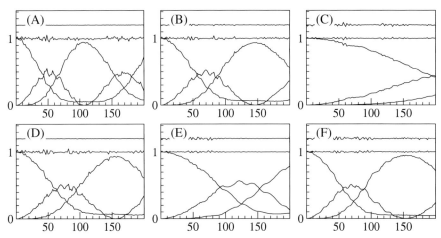

**Fig. 2.4.** Energy exchange between stiff springs for methods (A)-(F) ($h = 0.03$, $\omega = 50$).

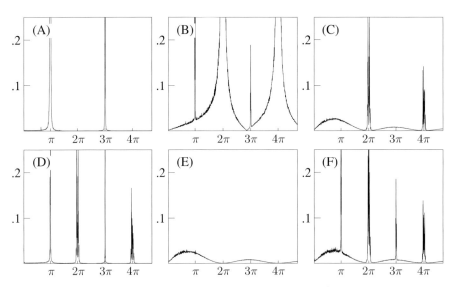

**Fig. 2.5.** Maximum error of the total energy on the interval $[0, 1000]$ for methods (A) - (F) as a function of $h\omega$ (step size $h = 0.02$).

## XIII.2.5 Near-Conservation of Total and Oscillatory Energy

Figure 2.5 shows the maximum error of the total energy $H$ as a function of the scaled frequency $h\omega$ (step size $h = 0.02$). We consider the long time interval $[0, 1000]$. The pictures for the different methods show that in general the total energy is well conserved. Exceptions are near integral multiples of $\pi$. Certain methods show a bad energy conservation close to odd multiples of $\pi$, other methods close to even multiples of $\pi$. Only method (E) shows a uniformly good behaviour for all frequencies. In

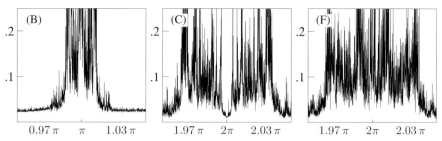

**Fig. 2.6.** Zoom (close to $\pi$ or $2\pi$) of the maximum error of the total energy on the interval $[0, 1000]$ for three methods as a function of $h\omega$ (step size $h = 0.02$).

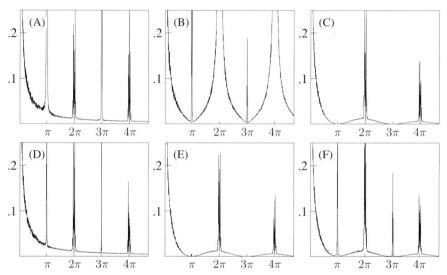

**Fig. 2.7.** Maximum deviation of the oscillatory energy on the interval $[0, 1000]$ for methods (A) - (F) as a function of $h\omega$ (step size $h = 0.02$).

Fig. 2.6 we show in more detail what happens close to such integral multiples of $\pi$. If there is a difficulty close to $\pi$, it is typically in an entire neighbourhood. Close to $2\pi$, the picture is different. Method (C) has good energy conservation for values of $h\omega$ that are very close to $2\pi$, but there are small intervals to the left and to the right, where the error in the total energy is large. Unlike the other methods shown, method (B) has poor energy conservation in rather large intervals around even multiples of $\pi$. Methods (A) and (D) conserve the total energy particularly well, for $h\omega$ away from integral multiples of $\pi$.

Figure 2.7 shows similar pictures where the total energy $H$ is replaced by the oscillatory energy $I$ (cf. Sect. XIII.2.1). For the exact solution we have $I(t) = Const + \mathcal{O}(\omega^{-1})$. It is therefore not surprising that this quantity is not well conserved for small values of $\omega$. For larger values of $\omega$, we observe that the methods have difficulties in conserving the oscillatory energy when $h\omega$ is near integral multiples of $\pi$. None of the considered methods conserves both quantities $H$ and $I$ uniformly for all values of $h\omega$.

## XIII.3 Principal Terms of the Modulated Fourier Expansion

The analytical tool for understanding the above numerical phenomena is provided by *modulated Fourier expansions*, which decompose both the exact and the numerical solution into a slowly varying part and into oscillatory components built up of trigonometric functions multiplied with slowly varying coefficient functions. A comparison of these expansions will serve as a partial substitute for the backward error analysis of Chap. IX, which yields results only for $h\omega \to 0$ and is not applicable to the situation of $h\omega \geq c > 0$ that is of interest here. In this section we derive the first terms of the modulated Fourier expansion.

### XIII.3.1 Decomposition of the Exact Solution

Every solution of the linear equation $\ddot{x} + \Omega^2 x = g(t)$ with $\Omega$ of (2.2) can be written as $y(t) + \cos(\omega t)\, u(t) + \sin(\omega t)\, v(t) + \mathcal{O}(\omega^{-N})$ (for $\omega \to \infty$), where $y(t)$, $u(t)$, $v(t)$ are truncated asymptotic expansions in powers of $\omega^{-1}$ (see Exercise 3). These functions have the property that all their derivatives are bounded independently of the parameter $\omega \gg 1$. Here and in the following, a *smooth* function is understood to be a function with this property. We may hope to find a similar decomposition for solutions of the nonlinear problem (2.1). So we look for a smooth real-valued function $y(t)$ and a smooth complex-valued function $z(t) = u(t) + iv(t)$ such that the function

$$x_*(t) = y(t) + e^{i\omega t} z(t) + e^{-i\omega t} \overline{z}(t) \tag{3.1}$$

gives a small defect when it is inserted into the differential equation (2.1) and has the given initial values

$$x_*(0) = x_0, \quad \dot{x}_*(0) = \dot{x}_0. \tag{3.2}$$

Under the condition (2.3) the exact solution $x(t)$ has bounded energy, and we may expect the same of the approximation $x_*(t)$, which would then imply $z(t) = \mathcal{O}(\omega^{-1})$. We therefore insert the ansatz (3.1) into the differential equation (2.1) and expand the nonlinearity around the smooth part $y(t)$. With the variables $y = (y_1, y_2)$, $z = (z_1, z_2)$ partitioned according to the blocks of $\Omega$, this gives the expressions

$$\ddot{x}_* + \Omega^2 x_* = \begin{pmatrix} \ddot{y}_1 \\ \ddot{y}_2 + \omega^2 y_2 \end{pmatrix} + e^{i\omega t} \begin{pmatrix} -\omega^2 z_1 + 2i\omega \dot{z}_1 + \ddot{z}_1 \\ 2i\omega \dot{z}_2 + \ddot{z}_2 \end{pmatrix}$$
$$+ e^{-i\omega t} \begin{pmatrix} -\omega^2 \overline{z}_1 - 2i\omega \dot{\overline{z}}_1 + \ddot{\overline{z}}_1 \\ -2i\omega \dot{\overline{z}}_2 + \ddot{\overline{z}}_2 \end{pmatrix}$$

and, as long as $z(t) = \mathcal{O}(\omega^{-1})$,

$$\begin{aligned}
g(x_*) &= g(y) + g''(y)(z,\overline{z}) + e^{i\omega t} g'(y) z + e^{-i\omega t} g'(y) \overline{z} \\
&\quad + e^{2i\omega t} \tfrac{1}{2} g''(y)(z,z) + e^{-2i\omega t} \tfrac{1}{2} g''(y)(\overline{z},\overline{z}) + \mathcal{O}(\omega^{-3}).
\end{aligned}$$

**Equations for the Coefficient Functions.** We now compare the coefficients of $1, e^{i\omega t}, e^{-i\omega t}$ and require that the dominant terms in these expressions be equal:

$$\begin{aligned}
\ddot{y}_1 &= g_1(y) + g_1''(y)(z,\overline{z}) \\
\omega^2 y_2 &= g_2(y) \\
-\omega^2 z_1 &= g_1'(y) z \\
2i\omega \dot{z}_2 &= g_2'(y) z .
\end{aligned} \qquad (3.3)$$

This gives a system of differential equations for $y_1, z_2$ and expresses $y_2, z_1$ as functions of $y_1, z_2$. We note that $y_1$ evolves on the time scale 1, whereas $z_2$ changes on the slow time scale $\omega$. As long as $y_1(t)$ stays in a bounded domain and $z_2(t) = \mathcal{O}(\omega^{-1})$, (3.3) implies the bounds

$$y_2(t) = \mathcal{O}(\omega^{-2}), \quad z_1(t) = \mathcal{O}(\omega^{-3}), \quad \dot{z}_2(t) = \mathcal{O}(\omega^{-2}). \qquad (3.4)$$

**Initial Values.** The initial values for $y_1, \dot{y}_1$ and $z_2$ are obtained from condition (3.2), which gives a system that can be solved by fixed point iteration to yield

$$\begin{aligned}
y_1(0) &= x_{0,1} + \mathcal{O}(\omega^{-3}), & \dot{y}_1(0) &= \dot{x}_{0,1} + \mathcal{O}(\omega^{-2}) \\
2\,\mathrm{Re}\, z_2(0) &= x_{0,2} + \mathcal{O}(\omega^{-2}), & -\omega\, 2\,\mathrm{Im}\, z_2(0) &= \dot{x}_{0,2} + \mathcal{O}(\omega^{-2}).
\end{aligned} \qquad (3.5)$$

**Defect.** As long as $z_2(t) = \mathcal{O}(\omega^{-1})$, the above equations show that the defect

$$d(t) = \ddot{x}_*(t) + \Omega^2 x_*(t) - g(x_*(t))$$

is of the form

$$d(t) = \mathrm{Re}\begin{pmatrix} \omega^{-2} e^{i\omega t} a(t) + \omega^{-2} e^{2i\omega t} b(t) + \mathcal{O}(\omega^{-3}) \\ \mathcal{O}(\omega^{-2}) \end{pmatrix} \qquad (3.6)$$

with smooth functions $a, b$. Together with (3.3) this also shows that the smooth $\mathcal{O}(\omega^{-2})$-term $g''(y)(z,\overline{z})$ is the principal term describing the influence of oscillatory solution components on the evolution of smooth components.

**Example.** To illustrate the approximation of the solution $x(t)$ by $x_*(t)$ of (3.1), we have solved numerically, with high accuracy, the system (3.3) for the FPU problem with the data of Sect. I.4.1. In Figure 3.1 we plot the oscillatory energy $I = I_1 + I_2 + I_3$ with $x$ replaced by the approximation $x_*$ in the definition (2.5) of these quantities. The figure agrees rather well with Figure I.4.2.

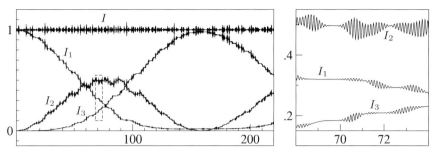

**Fig. 3.1.** Same experiment as in Fig. I.4.2 for the solution (3.1) of (3.3).

## XIII.3.2 Decomposition of the Numerical Solution

For the numerical method (2.6), which solves linear equations $\ddot{x} = -\Omega^2 x$ exactly, we look similarly to the above for a function of the form

$$x_h(t) = y_h(t) + e^{i\omega t} z_h(t) + e^{-i\omega t} \overline{z}_h(t) \tag{3.7}$$

with coefficient functions $y_h(t)$, $z_h(t)$ which are smooth in the sense that all their derivatives are bounded independently of $h$ and $\omega$, such that $x_h(t)$ gives a small defect when inserted into the difference scheme (2.6) and has the correct starting values:

$$x_h(0) = x_0, \quad x_h(h) = x_1. \tag{3.8}$$

Taylor expansion of $z_h(t \pm h)$ at the point $t$ shows, after some calculation,

$$\frac{1}{h^2}\Big(x_h(t+h) - 2\cos(h\Omega)x_h(t) + x_h(t-h)\Big) = \begin{pmatrix} \delta_h^2 y_{h,1}(t) \\ \sigma_1^2 \omega^2 y_{h,2}(t) + \delta_h^2 y_{h,2}(t) \end{pmatrix}$$
$$+ e^{i\omega t} \begin{pmatrix} -\sigma_1^2 \omega^2 z_{h,1}(t) + \sigma_2\, 2i\omega \dot{z}_{h,1}(t) + \cos(h\omega)\ddot{z}_{h,1}(t) + \ldots \\ \sigma_2\, 2i\omega \dot{z}_{h,2}(t) + \cos(h\omega)\ddot{z}_{h,2}(t) + \ldots \end{pmatrix} \tag{3.9}$$

$+$ *the complex conjugate of the expression in the previous line,*

where $y_h(t) = \big(y_{h,1}(t), y_{h,2}(t)\big)$ and $z_h(t) = \big(z_{h,1}(t), z_{h,2}(t)\big)$ according to the partitioning in (2.2),

$$\delta_h^2 y_h(t) = \frac{1}{h^2}\Big(y_h(t+h) - 2y_h(t) + y_h(t-h)\Big)$$

is the symmetric second-order difference quotient, $\sigma_k = \operatorname{sinc}(\tfrac{1}{2}kh\omega)$, and the dots stand for higher powers of $h$ multiplied by derivatives of $z_{h,1}$ or $z_{h,2}$. Taylor expansion of the nonlinearity now gives

$$\begin{aligned}\Psi g(\Phi x_h) &= \Psi g(\Phi y_h) + \Psi g''(\Phi y_h)(\Phi z_h, \Phi \overline{z}_h) \\ &\quad + e^{i\omega t} \Psi g'(\Phi y_h)\Phi z_h + e^{-i\omega t}\Psi g'(\Phi y_h)\Phi \overline{z}_h + \ldots \end{aligned} \tag{3.10}$$

**Modified Equations for the Numerical Coefficient Functions.** For the moment we consider the case where the absolute values of $\sigma_1$ and $\sigma_2$ are bounded from

XIII.3 Principal Terms of the Modulated Fourier Expansion    425

below by a positive constant, so that $h\omega$ is assumed bounded and bounded away from a non-zero integral multiple of $\pi$. We also assume $h\omega$ to be bounded away from zero, which is the computational situation of interest. In this case, the first term in each line of each bracket in (3.9) can be considered as the dominant one. We therefore require that the functions $y_h$, $z_h$ satisfy

$$\delta_h^2 y_{h,1} = g_1(\Phi y_h) + g_1''(\Phi y_h)(\Phi z_h, \Phi \bar{z}_h)$$

$$\operatorname{sinc}^2(\tfrac{1}{2}h\omega)\,\omega^2 y_{h,2} = \psi(h\omega)\, g_2(\Phi y_h) \qquad (3.11)$$

$$-\operatorname{sinc}^2(\tfrac{1}{2}h\omega)\,\omega^2 z_{h,1} = \frac{\partial g_1}{\partial x_1}(\Phi y_h) z_{h,1} + \frac{\partial g_1}{\partial x_2}(\Phi y_h)\phi(h\omega) z_{h,2}$$

$$\operatorname{sinc}(h\omega)\, 2i\omega\, \dot{z}_{h,2} = \psi(h\omega)\frac{\partial g_2}{\partial x_1}(\Phi y_h) z_{h,1} + \psi(h\omega)\frac{\partial g_2}{\partial x_2}(\Phi y_h)\phi(h\omega) z_{h,2}\,.$$

The first equation should be stated more precisely as $y_{h,1}$ being a solution of a modified equation for the Störmer/Verlet method (see Exercise IX.3) applied to the corresponding differential equation:

$$\ddot{y}_{h,1} = \left(1 - \frac{h^2}{12}\frac{d^2}{dt^2}\right)\Big(g_1(\Phi y_h) + g_1''(\Phi y_h)(\Phi z_h, \Phi \bar{z}_h)\Big),$$

where the time derivatives of $y_{h,2}$, $z_h$ that result from applying the chain rule are replaced by using the expressions in (3.11). As long as $y_{h,1}(t)$ remains in a bounded domain and $z_{h,2}(t) = \mathcal{O}(\omega^{-1})$, we have again bounds of the same type as for the coefficients of the exact solution:

$$y_{h,2}(t) = \mathcal{O}(\omega^{-2}),\quad z_{h,1}(t) = \mathcal{O}(\omega^{-3}),\quad \dot{z}_{h,2}(t) = \mathcal{O}(\omega^{-2})\,. \qquad (3.12)$$

**Initial Values.** We next determine the initial values $y_{h,1}(0)$, $\dot{y}_{h,1}(0)$ and $z_{h,2}(0)$ such that $x_h(0)$ and $x_h(h)$ coincide with the starting values $x_0$ and $x_1$ of the numerical method. We let $x_1$ be computed from $x_0$ and $\dot{x}_0$ via the formula (2.7) with $n = 0$, and we still assume that $\sigma_1$ and $\sigma_2$ are bounded away from zero. Using (3.11), the condition $x_h(0) = x_0 = (x_{0,1}, x_{0,2})$ then becomes

$$\begin{aligned} x_{0,1} &= y_{h,1}(0) + \mathcal{O}\big(\omega^{-2} z_{h,2}(0)\big) \\ x_{0,2} &= z_{h,2}(0) + \bar{z}_{h,2}(0) + \mathcal{O}(\omega^{-2})\,. \end{aligned} \qquad (3.13)$$

The formula for the first component of (2.7), $x_{1,1} - x_{0,1} = h\dot{x}_{0,1} + \tfrac{1}{2}h^2 g_1(\Phi x_0)$, together with $x_{h,1}(h) - x_{h,1}(0) = h\dot{y}_{h,1}(0) + \tfrac{1}{2}h^2 g_1(\Phi x_0) + \mathcal{O}(h^3) + \mathcal{O}(\omega^{-2} z_{h,2}(0))$ implies that

$$\dot{x}_{0,1} = \dot{y}_{h,1}(0) + \mathcal{O}(h^2) + \mathcal{O}\big(\omega^{-1} z_{h,2}(0)\big). \qquad (3.14)$$

For the second component we have from (2.7)

$$x_{1,2} - \cos(h\omega) x_{0,2} = h\operatorname{sinc}(h\omega)\dot{x}_{0,2} + \tfrac{1}{2}h^2 \psi(h\omega)\, g_2(\Phi x_0),$$

and by Taylor expansion and (3.11),

$$x_{h,2}(h) - \cos(h\omega)x_{h,2}(0) = (1 - \cos(h\omega))y_{h,2}(0) + \mathcal{O}(h\omega^{-2})$$
$$+ i\sin(h\omega)(z_{h,2}(0) - \overline{z}_{h,2}(0)) + \mathcal{O}(h\omega^{-1}z_{h,2}(0)),$$

where we note the relation $(1 - \cos(h\omega))y_{h,2}(0) = \tfrac{1}{2}h^2\psi(h\omega)g_2(\Phi y_h(0))$ by (3.11) and a trigonometric identity. After division by $h\operatorname{sinc} h\omega = \omega^{-1}\sin h\omega$ the above formulas yield

$$\dot{x}_{0,2} = i\omega(z_{h,2}(0) - \overline{z}_{h,2}(0)) + \mathcal{O}(\omega^{-2}) + \mathcal{O}(\omega^{-1}z_{h,2}(0)). \tag{3.15}$$

The four equations (3.13), (3.14), (3.15) constitute a nonlinear system for the four quantities $y_1(0), \dot{y}_1(0), \omega(z_{h,2}(0) + \overline{z}_{h,2}(0))$, and $\omega(z_{h,2}(0) - \overline{z}_{h,2}(0))$. By fixed-point iteration and using the bounded-energy assumption (2.3), we get a locally unique solution for sufficiently small $h$, with $z_{h,2}(0) = \mathcal{O}(\omega^{-1})$ and hence

$$y_{h,1}(0) = x_{0,1} + \mathcal{O}(\omega^{-3}), \qquad \dot{y}_{h,1}(0) = \dot{x}_{0,1} + \mathcal{O}(h^2)$$
$$2\operatorname{Re} z_{h,2}(0) = x_{0,2} + \mathcal{O}(\omega^{-2}), \quad -\omega\, 2\operatorname{Im} z_{h,2}(0) = \dot{x}_{0,2} + \mathcal{O}(h\omega^{-1}). \tag{3.16}$$

**Defect.** As long as $z_{h,2}(t) = \mathcal{O}(\omega^{-1})$, the defect

$$d_h(t) = \frac{1}{h^2}\Big(x_h(t+h) - 2\cos(h\Omega)x_h(t) + x_h(t-h)\Big) - \Psi g(\Phi x_h(t)) \tag{3.17}$$

is of size $\mathcal{O}(h^2)$ by (3.9)–(3.10) and the very construction (3.11) of the coefficient functions. This estimate refers again to the non-resonant case where $\sigma_1, \sigma_2$ are bounded away from zero and hence $h\omega$ is bounded away from non-zero integral multiples of $\pi$. The case of $h\omega$ near a multiple of $\pi$ requires a special treatment and will be considered in the next subsection.

## XIII.4 Accuracy and Slow Exchange

A comparison of the principal terms of the modulated Fourier expansions of the numerical and the exact solution gives much insight into the behaviour of the numerical method and the role of the filter functions $\psi$ and $\phi$. From this comparison we obtain error bounds over finite time intervals, and we discuss the slow energy exchange between oscillatory components and the slow energy transfer from oscillatory to smooth components which take place on the time scale $\omega$.

### XIII.4.1 Convergence Properties on Bounded Time Intervals

As a first application of the modulated Fourier expansion we consider error bounds on bounded time intervals. Second-order convergence estimates for more general equations $\ddot{x} = -Ax + g(x)$ with symmetric positive semi-definite matrix $A$, uniformly in the (arbitrarily large) eigenfrequencies of $A$, were given by García-Archilla, Sanz-Serna & Skeel (1999) for the mollified impulse method, and by

Hochbruck & Lubich (1999a) for Gautschi-type methods. Those results were proved with different techniques. The following bounds on the filter functions $\psi$ and $\phi$ are needed for second-order error bounds of the method (2.6):

$$
\begin{aligned}
|\psi(h\omega)| &\leq C_1 \operatorname{sinc}^2(\tfrac{1}{2}h\omega), \\
|\phi(h\omega)| &\leq C_2 |\operatorname{sinc}(\tfrac{1}{2}h\omega)|, \\
|\psi(h\omega)\phi(h\omega)| &\leq C_3 |\operatorname{sinc}(h\omega)|.
\end{aligned}
\quad (4.1)
$$

**Theorem 4.1.** *Consider the numerical solution of the system (2.1) – (2.3) by the method (2.6) with a step size $h \leq h_0$ (with a sufficiently small $h_0$ independent of $\omega$) for which $h\omega \geq c_0 > 0$. Let the starting value $x_1$ be given by (2.7) with $n = 0$. If the conditions (4.1) are satisfied, then the error is bounded by*

$$\|x_n - x(nh)\| \leq C h^2 \quad \text{for} \quad nh \leq T.$$

*If only $|\psi(h\omega)| \leq C_0 |\operatorname{sinc}(\tfrac{1}{2}h\omega)|$ holds instead of (4.1), then the order of convergence reduces to one: $\|x_n - x(nh)\| \leq C h$ for $nh \leq T$. In both cases, $C$ is independent of $\omega$, $h$ and $n$ with $nh \leq T$ and of bounds of solution derivatives, but depends on $T$, on $E$ of (2.3), on bounds of derivatives of the nonlinearity $g$, and on $C_1, C_2, C_3$ or $C_0$.*

To obtain second-order error bounds uniformly in $h\omega$, condition (4.1) requires a double zero of $\psi$ and a zero of $\phi$ at even multiples of $\pi$, and a zero of $\psi$ or $\phi$ at odd multiples of $\pi$. This is satisfied for the mollified impulse method with $\phi(\xi) = \operatorname{sinc}(\xi)$, for which $\psi(\xi) = \operatorname{sinc}^2(\xi)$. Gautschi-type methods have $\psi(\xi) = \operatorname{sinc}^2(\tfrac{1}{2}\xi)$, so that the first condition on $\psi$ in (4.1) is trivially satisfied. The conditions on $\phi$ hold, for example, for $\phi = \operatorname{sinc}$ or for $\phi$ of (1.20). The original Gautschi method has $\phi = 1$, which does not satisfy the second condition of (4.1), and the Deuflhard/impulse method ($\psi = \operatorname{sinc}, \phi = 1$) satisfies only the third condition of (4.1). These latter methods are not of second order uniformly in $h\omega$.

*Proof of Theorem 4.1.* (a) First we consider the case where $h\omega$ is bounded away from integral multiples of $\pi$, so that condition (4.1) is not needed. Comparing the equations (3.3) and (3.11), which determine the modulated Fourier expansion coefficients, shows

$$y_h(t) - y(t) = \mathcal{O}(h^2), \quad z_h(t) - z(t) = \mathcal{O}(h^2)$$

on bounded intervals, and hence

$$x_h(t) - x_*(t) = \mathcal{O}(h^2). \quad (4.2)$$

The variation-of-constants formula (1.8) and a Gronwall-type inequality show that, on bounded intervals, the error $x_*(t) - x(t)$ is of the same magnitude as the defect: by (3.6),

$$x_*(t) - x(t) = \mathcal{O}(\omega^{-2}).$$

The errors $e_n = x_n - x_h(t_n)$ satisfy

$$e_{n+1} - 2\cos(h\Omega)e_n + e_{n-1} = b_n \tag{4.3}$$

with $b_n = h^2\big(\Psi g(\Phi x_n) - \Psi g(\Phi x_h(t_n)) - d_h(t_n)\big)$. This recurrence relation can be solved to yield (Exercise 4)

$$e_{n+1} = -W_{n-1}e_0 + W_n e_1 + \sum_{j=1}^{n} W_{n-j} b_j \tag{4.4}$$

with

$$W_n = \begin{pmatrix} (n+1)I & 0 \\ 0 & \dfrac{\sin(n+1)h\omega}{\sin h\omega} I \end{pmatrix}.$$

A discrete Gronwall inequality now yields that on bounded intervals, $e_n$ is of the same magnitude as the defect $d_h(t)$ of (3.17), which is $\mathcal{O}(h^2)$ by the construction of (3.11) and by $z_{h,2} = \mathcal{O}(\omega^{-1})$. Hence,

$$x_n - x_h(t_n) = \mathcal{O}(h^2).$$

Combining these estimates yields the desired second-order error bound.

(b) We now consider the case where $\omega|\text{sinc}(\tfrac{1}{2}h\omega)| \geq c$ with a sufficiently large constant $c$, which depends only on bounds of derivatives of $g$. This condition means that $h\omega$ is outside of an $\mathcal{O}(h)$ neighbourhood of integral multiples of $2\pi$. Under conditions (4.1), the equations (3.11) still give

$$y_{h,2}(t) = \mathcal{O}(\omega^{-2}), \quad z_{h,1}(t) = \mathcal{O}(\omega^{-2}), \quad \dot{z}_{h,2}(t) = \mathcal{O}(\omega^{-2}) \tag{4.5}$$

as long as $z_{h,2}(t) = \mathcal{O}(\omega^{-1})$. Here the first condition of (4.1) gives the bound of $y_{h,2}$, the second one the bound of $z_{h,1}$, and the third one the bound of $\dot{z}_{h,2}$. As in Sect. XIII.3.2, we determine the initial values $y_{h,1}(0)$, $\dot{y}_{h,1}(0)$ and $z_{h,2}(0)$ such that $x_h(0)$ and $x_h(h)$ coincide with the starting values $x_0$ and $x_1$ of the numerical method. Using once more (4.1), we obtain a system for the initial values similar to (3.13)–(3.15):

$$\begin{aligned} x_{0,1} &= y_{h,1}(0) + \mathcal{O}\big(\omega^{-1} z_{h,2}(0)\big) \\ x_{0,2} &= z_{h,2}(0) + \overline{z}_{h,2}(0) + \mathcal{O}(\omega^{-2}) \\ \dot{x}_{0,1} &= \dot{y}_{h,1}(0) + \mathcal{O}(h) + \mathcal{O}\big(\omega^{-1} z_{h,2}(0)\big) \\ \dot{x}_{0,2} &= i\omega\big(z_{h,2}(0) - \overline{z}_{h,2}(0)\big) + \mathcal{O}(\omega^{-1}) + \mathcal{O}\big(z_{h,2}(0)\big). \end{aligned} \tag{4.6}$$

With the weaker estimates for $z_{h,1}(t)$ and in (4.6) we still obtain estimates for the initial values of the type (3.16) with at most one factor $\omega^{-1}$ or $h$ less in the remainder terms. Condition (2.3) implies again $z_2(0) = \mathcal{O}(\omega^{-1})$, which ensures that (4.5) holds for $0 \leq t \leq T$. The defect is then $d_h(t) = \mathcal{O}(h^2)$, and as in part (a) we get the second-order error bound.

(c) Now let $\omega|\text{sinc}(\tfrac{1}{2}h\omega)| \leq c$, so that $h\omega$ is $\mathcal{O}(h)$ close to a multiple of $2\pi$. In this case we replace the third equation in (3.11) simply by

## XIII.4 Accuracy and Slow Exchange

$$z_{h,1} = 0.$$

Under condition (4.1) we still obtain the bounds (4.5). The initial values are now chosen to satisfy

$$\begin{aligned}
x_{0,1} &= y_{h,1}(0) \\
x_{0,2} &= z_{h,2}(0) + \bar{z}_{h,2}(0) + \omega^{-2} \frac{\psi(h\omega)}{\text{sinc}^2(\tfrac{1}{2}h\omega)} g_2(\Phi x_0) \\
\dot{x}_{0,1} &= \dot{y}_{h,1}(0) \\
\dot{x}_{0,2} &= i\omega\bigl(z_{h,2}(0) - \bar{z}_{h,2}(0)\bigr).
\end{aligned} \qquad (4.7)$$

They are then bounded as in (b) and, by the arguments used in the determination of the initial values of Sect. XIII.3.2, yield the estimates $x_h(0) = x_0 + \mathcal{O}(h^3)$ and $x_h(h) = x_1 + \mathcal{O}(h^3)$, and again $z_{h,2}(t) = \mathcal{O}(\omega^{-1})$. Since (4.1) implies $\phi(h\omega) z_{h,2} = \mathcal{O}(\omega^{-2})$ in the present situation of $|\text{sinc}(\tfrac{1}{2}h\omega)| \leq c\omega^{-1}$, the defect is still $d_h(t) = \mathcal{O}(h^2)$. The bound (4.2) is also seen to hold. Therefore the second-order error bound remains valid in this case.

(d) If only $|\psi(h\omega)| \leq |\text{sinc}(\tfrac{1}{2}h\omega)|$ holds, then we replace the third equation in (3.11) by $z_{h,1} = 0$. If $\omega|\text{sinc}(\tfrac{1}{2}h\omega)| \leq 1$, we also set $y_{h,2} = 0$. The defect is then only $d_h(t) = \mathcal{O}(h)$, which yields the first-order error bound. □

For the velocity approximation, we obtain the following for the method (2.12) or its equivalent formulations.

**Theorem 4.2.** *Under the conditions of Theorem 4.1, consider the velocity approximation scheme (2.12) with a function $\psi_1$ satisfying $\psi_1(0) = 1$ and*

$$|\psi_1(h\omega)| \leq C_1' \, |\text{sinc}(\tfrac{1}{2}h\omega)| \, . \qquad (4.8)$$

*Let the starting values satisfy $\dot{x}_0 = \dot{x}(0)$ and $\dot{x}_1 = \dot{x}(h) + h\sin(h\Omega)a_1 + \mathcal{O}(h^2)$ with $a_1 = \mathcal{O}(1)$. Then, the error in the velocities is bounded by*

$$\|\dot{x}_n - \dot{x}(nh)\| \leq Ch \quad \text{for} \quad nh \leq T \, ,$$

*where $C$ is independent of $\omega$, $h$ and $n$ with $nh \leq T$ and of bounds of solution derivatives, but depends on $T$, on $E$ of (2.3), on bounds of derivatives of the nonlinearity $g$, and on $C_1, C_2, C_3$ and $C_1'$.*

*Proof.* (a) By the variation-of-constants formula (1.8), the exact solution satisfies

$$\begin{aligned}
\dot{x}(t+h) &- 2\cos(h\Omega)\dot{x}(t) + \dot{x}(t-h) \\
&= \int_0^h \cos((h-s)\Omega)\Bigl(g(x(t+s)) - g(x(t-s))\Bigr) ds \, .
\end{aligned}$$

With the modulated Fourier expansion, we write the exact solution as

$$x(t) = y(t) + e^{i\omega t} z(t) + e^{-i\omega t}\bar{z}(t) + \mathcal{O}(\omega^{-2})$$

to obtain

$$g(x(t+s)) - g(x(t-s))$$
$$= g'(y(t))\Big(2s\,\dot{y}(t) - 4\sin(\omega s)\,\text{Im}\,(e^{i\omega t}z(t)) + \mathcal{O}(s^2) + \mathcal{O}(\omega^{-2})\Big).$$

Using the bounds (3.4), abbreviating $g_{i,j} = \partial g_i/\partial x_j$ and omitting the arguments $t$ and $y(t)$ on the right-hand side, we therefore have

$$\dot{x}(t+h) - 2\cos(h\Omega)\dot{x}(t) + \dot{x}(t-h)$$
$$= \begin{pmatrix} h^2\,g_{1,1}\,\dot{y}_1 - 2h^2\,\text{sinc}^2(\tfrac{1}{2}h\omega)\,\omega\,g_{1,2}\,\text{Im}\,(e^{i\omega t}z_2) + \mathcal{O}(h^3) \\ h^2\,\text{sinc}^2(\tfrac{1}{2}h\omega)\,g_{2,1}\dot{y}_1 - 2h^2\,\text{sinc}(h\omega)\,\omega\,g_{2,2}\,\text{Im}\,(e^{i\omega t}z_2) + \mathcal{O}(h^3) \end{pmatrix}.$$

We now use the discrete variation-of-constants formula (4.4) and partial summation. For example, the expression

$$\sum_{j=1}^{n} \frac{\sin(n+1-j)h\omega}{\sin h\omega}\,\tfrac{1}{2}h^2\text{sinc}^2(\tfrac{1}{2}h\omega)\,g_{2,1}\big(y(jh)\big)\,\dot{y}_1(jh)$$

is seen to be $\mathcal{O}(h)$ uniformly in $h\omega$ and for $nh \le T$ by partial summation, using that the function $g_{2,1}(y(t))\dot{y}_1(t)$ has a bounded derivative and that

$$\frac{\sin(\tfrac{1}{2}h\omega)}{\sin(h\omega)}\sum_{j=1}^{k}\sin(jh\omega) = \mathcal{O}(k).$$

In this way we obtain

$$\dot{x}(nh) = -W_{n-1}\dot{x}(0) + W_n\dot{x}(h) \qquad (4.9)$$
$$+ \begin{pmatrix} h\sum_{j=1}^{n}(n+1-j)h\,g_{1,1}(y(jh))\,\dot{y}_1(jh) \\ 0 \end{pmatrix} + \mathcal{O}(h).$$

(b) For the numerical approximation we proceed similarly. Inserting the modulated Fourier expansion of the numerical solution,

$$x_n = y_h(t) + e^{i\omega t}z_h(t) + e^{-i\omega t}\bar{z}_h(t) + \mathcal{O}(h^2) \qquad \text{for } t = nh \le T,$$

into the numerical scheme, we have with (3.12) or (4.5)

$$\dot{x}_{n+1} - 2\cos(h\omega)\dot{x}_n + \dot{x}_{n-1}$$
$$= h^2\begin{pmatrix} g_{1,1}\dot{y}_{h,1} - 2\phi(h\omega)\,\text{sinc}(h\omega)\,\omega\,g_{1,2}\,\text{Im}\,(e^{i\omega t}z_{h,2}) + \mathcal{O}(h) \\ \psi_1(h\omega)\,g_{2,1}\dot{y}_{h,1} - 2(\psi_1\phi)(h\omega)\,\text{sinc}(h\omega)\,\omega\,g_{2,2}\,\text{Im}\,(e^{i\omega t}z_{h,2}) + \mathcal{O}(h) \end{pmatrix}$$

where the functions $g_{i,j}$ are evaluated at $\Phi y_h(t)$ and the argument $t = nh$ is to be inserted in $\dot{y}_{h,1}$ and $z_{h,2}$. Under the condition (4.8) on $\psi_1$, we obtain as in (4.9)

$$\dot{x}_n = -W_{n-1}\dot{x}_0 + W_n\dot{x}_1 \qquad (4.10)$$
$$+ \begin{pmatrix} h\sum_{j=1}^n (n+1-j) h\, g_{1,1}(\Phi y_h(jh))\, \dot{y}_{h,1}(jh) \\ 0 \end{pmatrix} + \mathcal{O}(h).$$

Since we know from the estimates (3.12) and from the proof of Theorem 4.1 that $\Phi y_h(t) = y(t) + \mathcal{O}(h^2)$ and $\dot{y}_h(t) = \dot{y}(t) + \mathcal{O}(h^2)$, a comparison of (4.9) and (4.10) gives the result. □

## XIII.4.2 Intra-Oscillatory and Oscillatory-Smooth Exchanges

In this subsection we turn to the approximation of slow effects that take place on the time scale $\omega$. Since solutions may depart from each other exponentially, we cannot expect to obtain small point-wise error bounds on such a time scale. Instead, we take recourse to a kind of formal backward error analysis where we require that the equations determining the modulated Fourier expansion coefficients for the numerical method be small perturbations of those for the exact solution. It may be expected that methods with this property – *ceteribus paris* – show a better long-time behaviour, and this is indeed confirmed by the numerical experiments.

In the Fermi-Pasta-Ulam model, the oscillatory energy of the $j$th stiff spring is

$$I_j = \tfrac{1}{2}\dot{x}_{2,j}^2 + \tfrac{1}{2}\omega^2 x_{2,j}^2,$$

where $x_{2,j}$ is the $j$th component of the lower block $x_2$ of $x$. In terms of the modulated Fourier expansion, this is approximately, up to $\mathcal{O}(\omega^{-1})$,

$$I_j \approx \tfrac{1}{2}\left|i\omega z_{2,j}e^{i\omega t} - i\omega \bar{z}_{2,j}e^{-i\omega t}\right|^2 + \tfrac{1}{2}\omega^2\left|z_{2,j}e^{i\omega t} + \bar{z}_{2,j}e^{-i\omega t}\right|^2 = 2\omega^2 |z_{2,j}|^2.$$

The energy exchange between stiff springs as shown in Fig. 2.1 is thus caused by the slow evolution of $z_2$ determined by (3.3). This should be modeled correctly by the numerical method.

The term $g_1''(y)(z,\bar{z})$ in the differential equation for $y_1$ in (3.3) is the dominant term by which the oscillations of the stiff springs exert an influence on the smooth motion. A correct incorporation of this term in the numerical method is desirable.

Upon eliminating $y_2$ and $z_1$ in (3.3), the differential equations for $y_1$ and $z_2$ become, up to $\mathcal{O}(\omega^{-3})$ perturbations on the right-hand sides,

$$\begin{aligned} \ddot{y}_1 &= g_1(y_1, \omega^{-2}g_2(y_1,0)) + \frac{\partial^2 g_1}{\partial x_2^2}(y_1,0)(z_2,\bar{z}_2) \\ 2i\omega \dot{z}_2 &= \frac{\partial g_2}{\partial x_2}(y_1,0)\,z_2. \end{aligned} \qquad (4.11)$$

This is to be compared with the analogous equations for the modulated Fourier expansion of the numerical method, which follow from (3.11):

$$\delta_h^2 y_{h,1} = g_1(y_{h,1}, \gamma \omega^{-2} g_2(y_{h,1}, 0)) + \beta \frac{\partial^2 g_1}{\partial x_2^2}(y_{h,1}, 0)(z_{h,2}, \bar{z}_{h,2})$$

$$2i\omega \dot{z}_{h,2} = \alpha \frac{\partial g_2}{\partial x_2}(y_{h,1}, 0) z_{h,2}$$ (4.12)

with

$$\alpha = \frac{(\psi\phi)(h\omega)}{\text{sinc}(h\omega)}, \quad \beta = \phi(h\omega)^2, \quad \gamma = \frac{(\psi\phi)(h\omega)}{\text{sinc}^2(\frac{1}{2}h\omega)}.$$ (4.13)

The differential equation for $z_{h,2}$ is consistent with that for $z_2$ only if $\alpha = 1$, i.e.,

$$\psi(h\omega)\, \phi(h\omega) = \text{sinc}(h\omega).$$ (4.14)

Among all the methods (2.6) considered, only the Deuflhard/impulse method ($\psi =$ sinc, $\phi = 1$) satisfies this condition. For this method we indeed observe a qualitatively correct approximation of the energy exchange between stiff springs in Fig. 2.4, but we have also seen that the energy conservation of this method is very sensitive to near-resonances.

A correct modeling of the slow oscillatory–smooth transfer would in addition require $\beta = 1$ and possibly $\gamma = 1$. For general $h\omega$ the condition $\gamma = 1$ is, however, incompatible with (4.14).

Multi-force methods (2.13) offer a way out of these difficulties. For such methods, the coefficients of the modulated Fourier expansion satisfy (4.12) with (4.13) replaced by

$$\alpha = \frac{\sum_j \psi_j(h\omega)\, \phi_j(h\omega)}{\text{sinc}(h\omega)}, \quad \beta = \sum_j \psi_j(0)\, \phi_j(h\omega)^2,$$

$$\gamma = \sum_j \psi_j(0)\, \phi_j(h\omega)\, \frac{\sum_k \psi_k(h\omega)}{\text{sinc}^2(\frac{1}{2}h\omega)}.$$ (4.15)

The two-force method (1.23) with (1.25) has $\alpha = \beta = \gamma = 1$ as desired.

## XIII.5 Modulated Fourier Expansions

The decomposition of the exact and the numerical solution into modulated exponentials and a remainder, as derived in Sect. XIII.3, was found useful for understanding several important aspects of the numerical behaviour. Those few terms are, however, not sufficient for explaining the long-time near-conservation of the total and the oscillatory energy. The expansion can be made more accurate by adding further terms $e^{\pm 2i\omega t}$, $e^{\pm 3i\omega t}$ etc. multiplied by slowly varying functions. This leads to an asymptotic expansion which we call the *modulated Fourier expansion*. This expansion is constructed in the present section, following Hairer & Lubich (2000a). (In that paper the modulated Fourier expansion was called the frequency expansion.)

## XIII.5.1 Expansion of the Exact Solution

The following theorem extends the construction of Sect. XIII.3.1 to arbitrary order in $\omega^{-1}$.

**Theorem 5.1.** *Consider a solution $x(t)$ of (2.1) which satisfies the bounded-energy condition (2.3) and stays in a compact set $K$ for $0 \le t \le T$. Then, the solution admits an expansion*

$$x(t) = y(t) + \sum_{0<|k|<N} e^{ik\omega t} z^k(t) + R_N(t) \tag{5.1}$$

*for arbitrary $N \ge 2$, where the remainder term and its derivative are bounded by*

$$R_N(t) = \mathcal{O}(\omega^{-N-2}) \quad \text{and} \quad \dot{R}_N(t) = \mathcal{O}(\omega^{-N-1}) \quad \text{for} \quad 0 \le t \le T. \tag{5.2}$$

*The real-valued functions $y = (y_1, y_2)$ and the complex-valued functions $z^k = (z_1^k, z_2^k)$ together with all their derivatives (up to arbitrary order $M$) are bounded by*

$$\begin{aligned} y_1 &= \mathcal{O}(1), & z_1^1 &= \mathcal{O}(\omega^{-3}), & z^k &= \mathcal{O}(\omega^{-k-2}) \\ y_2 &= \mathcal{O}(\omega^{-2}), & z_2^1 &= \mathcal{O}(\omega^{-1}), & & \end{aligned} \tag{5.3}$$

*for $k = 2, \ldots, N-1$. Moreover, $z^{-k} = \overline{z^k}$ for all $k$. These functions are unique up to terms of size $\mathcal{O}(\omega^{-N-2})$. The constants symbolized by the $\mathcal{O}$-notation are independent of $\omega$ and $t$ with $0 \le t \le T$ (but they depend on $N$, $T$, on $E$ of (2.3), on bounds of the derivatives of the nonlinearity $g(x)$ on $K$, and on the maximum order $M$ of considered derivatives).*

*Proof.* We set

$$x_*(t) = y(t) + \sum_{0<|k|<N} e^{ik\omega t} z^k(t) \tag{5.4}$$

and determine the smooth functions $y(t), z(t) = z^1(t)$, and $z^2(t), \ldots, z^{N-1}(t)$ such that $x_*(t)$ inserted into the differential equation (2.1) has a small defect, of size $\mathcal{O}(\omega^{-N})$. To this end we expand $g(x_*(t))$ around $y(t)$ and compare the coefficients of $e^{ik\omega t}$. With the notation $g^{(m)}(y)z^\alpha = g^{(m)}(y)(z^{\alpha_1}, \ldots, z^{\alpha_m})$ for a multi-index $\alpha = (\alpha_1, \ldots, \alpha_m)$, there results the following system of differential equations:

$$\begin{pmatrix} \ddot{y}_1 \\ \omega^2 y_2 \end{pmatrix} + \begin{pmatrix} 0 \\ \ddot{y}_2 \end{pmatrix} = g(y) + \sum_{s(\alpha)=0} \frac{1}{m!} g^{(m)}(y) z^\alpha \tag{5.5}$$

$$\begin{pmatrix} -\omega^2 z_1 \\ 2i\omega \dot{z}_2 \end{pmatrix} + \begin{pmatrix} 2i\omega \dot{z}_1 + \ddot{z}_1 \\ \ddot{z}_2 \end{pmatrix} = \sum_{s(\alpha)=1} \frac{1}{m!} g^{(m)}(y) z^\alpha \tag{5.6}$$

$$\begin{pmatrix} -k^2 \omega^2 z_1^k \\ (1-k^2)\omega^2 z_2^k \end{pmatrix} + \begin{pmatrix} 2ki\omega \dot{z}_1^k + \ddot{z}_1^k \\ 2ki\omega \dot{z}_2^k + \ddot{z}_2^k \end{pmatrix} = \sum_{s(\alpha)=k} \frac{1}{m!} g^{(m)}(y) z^\alpha. \tag{5.7}$$

Here the sums range over all $m \geq 1$ and all multi-indices $\alpha = (\alpha_1, \ldots, \alpha_m)$ with integers $\alpha_j$ satisfying $0 < |\alpha_j| < N$, which have a given sum $s(\alpha) = \sum_{j=1}^{m} \alpha_j$.

For large $\omega$, the dominating terms in these differential equations are given by the left-most expressions. However, since the central terms involve higher derivatives, we are confronted with singular perturbation problems. We are interested in smooth functions $y, z, z^k$ that satisfy the system up to a defect of size $\mathcal{O}(\omega^{-N})$. In the spirit of Euler's derivation of the Euler-Maclaurin summation formula (see e.g. Hairer & Wanner 1997) we remove the disturbing higher derivatives by using iteratively the differentiated equations (5.5)-(5.7). This leads to a system

$$\ddot{y}_1 = \mathcal{F}_1(\dot{y}_1, y, z^1, \ldots, z^{N-1}, \omega^{-1}), \quad \dot{z}_2 = \omega^{-1}\mathcal{F}_2(\dot{y}_1, y, z^1, \ldots, z^{N-1}, \omega^{-1})$$
$$z_1 = \omega^{-2}\mathcal{G}_1(\dot{y}_1, y, z^1, \ldots, z^{N-1}, \omega^{-1}), \quad y_2 = \omega^{-2}\mathcal{G}_2(\dot{y}_1, y, z^1, \ldots, z^{N-1}, \omega^{-1})$$
$$z_1^k = \omega^{-2}\mathcal{G}_1^k(\dot{y}_1, y, z^1, \ldots, z^{N-1}, \omega^{-1}), \quad z_2^k = \omega^{-2}\mathcal{G}_2^k(\dot{y}_1, y, z^1, \ldots, z^{N-1}, \omega^{-1})$$

where $\mathcal{F}_j, \mathcal{G}_j, \mathcal{G}_j^k$ are formal series in powers of $\omega^{-1}$. Since we get formal algebraic relations for $y_2, z_1, z^k$, we can further eliminate these variables in the functions $\mathcal{F}_j, \mathcal{G}_j, \mathcal{G}_j^k$. We finally obtain for $y_2, z_2, z^k$ the algebraic relations

$$\begin{aligned}
z_1 &= \omega^{-2}\big(G_{10}(y_1, \dot{y}_1, z_2) + \omega^{-1}G_{11}(y_1, \dot{y}_1, z_2) + \ldots\big) \\
y_2 &= \omega^{-2}\big(G_{20}(y_1, \dot{y}_1, z_2) + \omega^{-1}G_{21}(y_1, \dot{y}_1, z_2) + \ldots\big) \\
z_1^k &= \omega^{-2}\big(G_{10}^k(y_1, \dot{y}_1, z_2) + \omega^{-1}G_{11}^k(y_1, \dot{y}_1, z_2) + \ldots\big) \\
z_2^k &= \omega^{-2}\big(G_{20}^k(y_1, \dot{y}_1, z_2) + \omega^{-1}G_{21}^k(y_1, \dot{y}_1, z_2) + \ldots\big)
\end{aligned} \quad (5.8)$$

and a system of real second-order differential equations for $y_1$ and complex first-order differential equations for $z_2$:

$$\begin{aligned}
\ddot{y}_1 &= F_{10}(y_1, \dot{y}_1, z_2) + \omega^{-1}F_{11}(y_1, \dot{y}_1, z_2) + \ldots \\
\dot{z}_2 &= \omega^{-1}\big(F_{20}(y_1, \dot{y}_1, z_2) + \omega^{-1}F_{21}(y_1, \dot{y}_1, z_2) + \ldots\big).
\end{aligned} \quad (5.9)$$

At this point we can forget the above derivation and take it as a motivation for the ansatz (5.8)-(5.9), which is truncated after the $\mathcal{O}(\omega^{-N})$ terms. We insert this ansatz and its first and second derivatives into (5.5)-(5.7) and compare powers of $\omega^{-1}$. This yields recurrence relations for the functions $F_{jl}^k, G_{jl}^k$, which in addition show that these functions together with their derivatives are all bounded on compact sets.

We determine initial values for (5.9) such that the function $x_*(t)$ of (5.4) satisfies $x_*(0) = x_0$ and $\dot{x}_*(0) = \dot{x}_0$. Because of the special ansatz (5.8)-(5.9), this gives a system which, by fixed-point iteration, yields (locally) unique initial values $y_1(0)$, $\dot{y}_1(0)$, $z_2(0)$ satisfying (3.5). The assumption (2.3) implies that $z_2(0) = O(\omega^{-1})$. It further follows from the boundedness of $F_{2l}$ that $z_2(t) = O(\omega^{-1})$ for $0 \leq t \leq T$. Going back to (5.7), it is seen that the functions $G_{jl}^k$ contain at least $k$ times the factor $z_2$. This implies the stated bounds for all other functions.

It remains to estimate the error $R_N(t) = x(t) - x_*(t)$. For this we consider the solution of (5.8)-(5.9) with the above initial values. By construction, these functions satisfy the system (5.5)-(5.7) up to a defect of $\mathcal{O}(\omega^{-N})$. This gives a defect of size $\mathcal{O}(\omega^{-N})$ when the function $x_*(t)$ of (5.4) is inserted into (2.1). On a finite time interval $0 \le t \le T$, this implies $R_N(t) = \mathcal{O}(\omega^{-N})$ and $\dot{R}_N(t) = \mathcal{O}(\omega^{-N})$. To obtain the slightly sharper bounds (5.2), we apply the above proof with $N$ replaced by $N+2$ and use the bounds (5.3) for $z^N$ and $z^{N+1}$. □

## XIII.5.2 Expansion of the Numerical Solution

Does the numerical solution of (2.1) have a modulated Fourier expansion similar to the analytical solution? This may of course be expected, but in Sect. XIII.3.2 we encountered difficulties in constructing the first terms of the expansion in the situation of a numerical resonance where $h\omega$ is close to an integral multiple of $\pi$. We therefore confine the discussion to the non-resonant case. We assume that $h$ and $\omega^{-1}$ lie in a subregion of the $(h, \omega^{-1})$-plane of small parameters for which there exists a positive constant $c$ such that

$$|\sin(\tfrac{1}{2}kh\omega)| \ge c\sqrt{h} \qquad \text{for } k = 1, \dots, N, \text{ with } N \ge 2. \tag{5.10}$$

This condition implies that $h\omega$ is outside an $\mathcal{O}(\sqrt{h})$ neighbourhood of integral multiples of $\pi$. For given $h$ and $\omega$, the condition imposes a restriction on $N$. In the following, $N$ is a fixed integer such that (5.10) holds. There is the following numerical analogue of Theorem 5.1.

**Theorem 5.2.** *Consider the numerical solution of the system (2.1) – (2.3) by the method (2.6) with step size $h$. Let the starting value $x_1$ be given by (2.7) with $n = 0$. Assume $h\omega \ge c_0 > 0$, the non-resonance condition (5.10), and the bounds (4.1) for $\psi(h\omega)$ and $\phi(h\omega)$. Then, the numerical solution admits an expansion*

$$x_n = y_h(t) + \sum_{0 < |k| < N} e^{ik\omega t} z_h^k(t) + R_{h,N}(t) \tag{5.11}$$

*uniformly for $0 \le t = nh \le T$. The remainder term is of the form*

$$R_{h,N}(t) = t^2 h^N \Psi r(t) \quad \text{with} \quad r(t) = \mathcal{O}\big(\phi(h\omega)^N + h^m\big), \tag{5.12}$$

*where $m \ge 0$ can be chosen arbitrarily. The coefficient functions together with all their derivatives (up to some arbitrarily fixed order) are bounded by*

$$\begin{aligned}
y_{h,1} &= \mathcal{O}(1), & z_{h,1}^1 &= \mathcal{O}(\omega^{-2}), & z_{h,1}^k &= \mathcal{O}(\omega^{-k}), \\
y_{h,2} &= \mathcal{O}(\omega^{-2}), & z_{h,2}^1 &= \mathcal{O}(\omega^{-1}), & z_{h,2}^k &= \mathcal{O}(\omega^{-k})
\end{aligned} \tag{5.13}$$

*for $k = 2, \dots, N-1$. Moreover, $z_h^{-k} = \overline{z_h^k}$ for all $k$. The constants symbolized by the $\mathcal{O}$-notation are independent of $\omega$ and $h$ with (5.10), but they depend on $E$, $N$, $m$, $c$, and $T$.*

The *proof* covers the remainder of this subsection. It constructs a function

$$x_h(t) = y_h(t) + \sum_{0 < |k| < N} e^{ik\omega t} z_h^k(t) \tag{5.14}$$

with smooth coefficient functions $y_h(t)$ and $z_h^k(t)$, which has a small defect when it is inserted into the numerical scheme (2.6). The following functional calculus is convenient for determining the coefficient functions.

**Functional Calculus.** Let $f$ be an entire complex function bounded by $|f(\zeta)| \le C\, e^{\gamma|\zeta|}$. Then,

$$f(hD)x(t) = \sum_{k=0}^{\infty} \frac{f^{(k)}(0)}{k!}\, h^k\, x^{(k)}(t)$$

converges for every function $x$ which is analytic in a disk of radius $r > \gamma h$ around $t$. If $f_1$ and $f_2$ are two such entire functions, then

$$f_1(hD)f_2(hD)x(t) = (f_1 f_2)(hD)x(t)$$

whenever both sides exist. We note $(hD)^k x(t) = h^k x^{(k)}(t)$ for $k = 0, 1, 2, \ldots$ and $\exp(hD)x(t) = x(t+h)$.

We next consider the application of such an operator to functions of the form $e^{i\omega t}z(t)$. By Leibniz' rule of calculus we have $(hD)^k e^{i\omega t} z(t) = e^{i\omega t}(hD + ih\omega)^k z(t)$. After a short calculation this yields

$$f(hD)e^{i\omega t}z(t) = e^{i\omega t} f(hD + ih\omega)z(t) \tag{5.15}$$

where $f(hD + ih\omega)z(t) = \sum_{k=0}^{\infty} f^{(k)}(ih\omega)/k! \cdot h^k z^{(k)}(t)$.

An $N$-times continuously differentiable function $x$ is replaced by its Taylor polynomial of degree $N-1$ at $t$, and $f(hD)x(t)$ is then considered up to $\mathcal{O}(h^N)$.

**Modified Equations for the Coefficient Functions.** The difference operator of the numerical method becomes in this notation

$$x(t+h) - 2\cos h\Omega\, x(t) + x(t-h) = (e^{hD} - 2\cos h\Omega + e^{-hD})x(t).$$

We factorize this operator as

$$\begin{aligned}
\mathcal{L}(hD) &:= e^{hD} - 2\cos h\Omega + e^{-hD} = 2\bigl(\cos(ihD) - \cos h\Omega\bigr) \\
&= 4\sin\bigl(\tfrac{1}{2}h\Omega + \tfrac{1}{2}ihD\bigr)\sin\bigl(\tfrac{1}{2}h\Omega - \tfrac{1}{2}ihD\bigr).
\end{aligned} \tag{5.16}$$

The function $x_h(t)$ of (5.14) should formally (up to $\mathcal{O}(h^{N+2})$) satisfy the difference scheme

$$\mathcal{L}(hD)x_h(t) = h^2 \Psi g\bigl(\Phi x_h(t)\bigr). \tag{5.17}$$

We insert the ansatz (5.14), expand the right-hand side into a Taylor series around $\Phi y_h(t)$, and compare the coefficients of $e^{ik\omega t}$. This yields the following formal equations for the functions $y_h(t)$ and $z_h^k(t)$:

$$\begin{aligned}
\mathcal{L}(hD)y_h &= h^2 \Psi \Bigl(g(\Phi y_h) + \sum_{s(\alpha)=0} \frac{1}{m!} g^{(m)}(\Phi y_h)(\Phi z_h)^\alpha \Bigr) \\
\mathcal{L}(hD + ikh\omega)z_h^k &= h^2 \Psi \sum_{s(\alpha)=k} \frac{1}{m!} g^{(m)}(\Phi y_h)(\Phi z_h)^\alpha.
\end{aligned} \tag{5.18}$$

Here, $\alpha = (\alpha_1, \ldots, \alpha_m)$ is a multi-index as in the proof of Theorem 5.1, $s(\alpha) = \sum_{j=1}^{m} \alpha_j$, and $(\Phi z)^\alpha$ is an abbreviation for the $m$-tupel $(\Phi z^{\alpha_1}, \ldots, \Phi z^{\alpha_m})$. To get smooth functions $y_h(t)$ and $z_h^k(t)$ which solve (5.18) up to a small defect, we look at the dominating terms in the Taylor expansions of $\mathcal{L}(hD)$ and $\mathcal{L}(hD + ikh\omega)$. With the abbreviations $s_k = \sin(\frac{1}{2}kh\omega)$ and $c_k = \cos(\frac{1}{2}kh\omega)$ we obtain

$$\mathcal{L}(hD) = \begin{pmatrix} 0 & 0 \\ 0 & 4s_1^2 \end{pmatrix} - \begin{pmatrix} 1 & 0 \\ 0 & 1 \end{pmatrix} (ihD)^2 + \ldots$$

$$\mathcal{L}(hD + ih\omega) = \begin{pmatrix} -4s_1^2 & 0 \\ 0 & 0 \end{pmatrix} + 2s_2 \begin{pmatrix} 1 & 0 \\ 0 & 1 \end{pmatrix} (ihD)$$

$$- c_2 \begin{pmatrix} 1 & 0 \\ 0 & 1 \end{pmatrix} (ihD)^2 + \ldots \quad (5.19)$$

$$\mathcal{L}(hD + ikh\omega) = \begin{pmatrix} -4s_k^2 & 0 \\ 0 & -4s_{k-1}s_{k+1} \end{pmatrix} + 2s_{2k} \begin{pmatrix} 1 & 0 \\ 0 & 1 \end{pmatrix} (ihD)$$

$$- c_{2k} \begin{pmatrix} 1 & 0 \\ 0 & 1 \end{pmatrix} (ihD)^2 + \ldots .$$

**Construction of the Coefficient Functions.** Under the non-resonance condition (5.10), the first non-vanishing coefficients in (5.19) are the dominant ones, and the derivation of the defining relations for $y_h$ and $z_h^k$ is the same as for the analytical solution in Theorem 5.1; see also part (b) of the proof of Theorem 4.1. We insert (5.19) into (5.18) and we eliminate recursively the higher derivatives. This motivates the following ansatz for the computation of the functions $y_h$ and $z_h^k$:

$$\ddot{y}_{h,1} = f_{10}(\cdot) + \sqrt{h}\, f_{11}(\cdot) + h\, f_{12}(\cdot) + \ldots$$

$$\dot{z}_{h,2}^1 = \frac{\psi(h\omega)h}{s_2}\Big(f_{20}(\cdot) + \sqrt{h}\, f_{21}(\cdot) + \ldots\Big)$$

$$z_{h,1}^1 = \frac{h^2}{s_1^2}\Big(g_{10}^1(\cdot) + \sqrt{h}\, g_{11}^1(\cdot) + \ldots\Big)$$

$$y_{h,2} = \frac{\psi(h\omega)h^2}{s_1^2}\Big(g_{20}^1(\cdot) + \sqrt{h}\, g_{21}^1(\cdot) + \ldots\Big) \quad (5.20)$$

$$z_{h,1}^k = \frac{h^2}{s_k^2}\Big(g_{10}^k(\cdot) + \sqrt{h}\, g_{11}^k(\cdot) + \ldots\Big)$$

$$z_{h,2}^k = \frac{\psi(h\omega)h^2}{s_{k+1}s_{k-1}}\Big(g_{20}^k(\cdot) + \sqrt{h}\, g_{21}^k(\cdot) + \ldots\Big),$$

for $k = 2, \ldots, N-1$, where the functions depend smoothly on the variables $y_{h,1}$, $\dot{y}_{h,1}$, $\phi(h\omega)z_{h,2}^1$ and on the bounded parameters $\sqrt{h}/s_k$, $s_k$, $c_k$, $\psi(h\omega)$ and $(h\omega)^{-1}$. Inserting this ansatz and its derivatives into (5.18) and comparing powers of $\sqrt{h}$ yields recurrence relations for the functions $f_{jl}^k$, $g_{jl}^k$. The functions $g_{jl}^k$ (for $k \geq 1$) contain at least $k$ times the factor $\phi(h\omega)z_{h,2}^1$, and $f_{2l}$ contains this factor at least once. Since the series in (5.20) need not converge, we truncate them after the $h^{N+m+2}$ terms.

**Initial Values.** The conditions $x_h(0) = x_0$ and $x_h(h) = x_1$ determine the initial values $y_{h,1}(0)$, $\dot{y}_{h,2}(0)$ and $z_{h,2}(0)$ in the same way as in Sect. XIII.3.2. Condition (4.1) yields again (4.6), and (2.3) then implies $z_{h,2}(0) = \mathcal{O}(\omega^{-1})$.

**Defect.** It follows from (4.1) that $h\psi(h\omega)\phi(h\omega)/s_2 = \mathcal{O}(\omega^{-1})$, so that $\dot{z}_{h,2}^1 = \mathcal{O}(\omega^{-1} z_{h,2}^1)$ by (5.20). This implies $z_{h,2}^1(t) = \mathcal{O}(\omega^{-1})$ for $t \le T$. The other estimates (5.13) are directly obtained from (5.20), which indeed yields the following more refined bounds for the coefficient functions together with their derivatives:

$$\begin{aligned} y_{h,1} &= \mathcal{O}(1), & y_{h,2} &= \mathcal{O}(\omega^{-2}) \\ z_{h,1}^1 &= \mathcal{O}(\omega^{-3}/\sqrt{h}), & z_{h,2}^1 &= \mathcal{O}(\omega^{-1}), & \dot{z}_{h,2}^1 &= \mathcal{O}(\omega^{-2}) \\ z_{h,1}^k &= \mathcal{O}(h\phi(h\omega)^k \omega^{-k}), & z_{h,2}^k &= \mathcal{O}(h\psi(h\omega)\phi(h\omega)^k \omega^{-k}). \end{aligned} \quad (5.21)$$

Consequently, the values $x_h(nh)$ inserted into the numerical scheme (2.6) yield a defect of size $\mathcal{O}(h^{N+2})$:

$$\begin{aligned} x_h(t+h) &- 2\cos(h\Omega)\, x_h(t) + x_h(t-h) = \\ &= h^2 \Psi\Big(g(\Phi x_h(t)) + \mathcal{O}\big(\phi(h\omega)^N \omega^{-N} + h^{N+m}\big)\Big). \end{aligned} \quad (5.22)$$

Standard convergence estimates then show that, on bounded time intervals, $x_n - x_h(nh)$ is of size $\mathcal{O}(t^2 h^N)$ and actually satisfies the finer estimate (5.12). This completes the proof of Theorem 5.2. □

### XIII.5.3 Expansion of the Velocity Approximation

A similar expansion holds also for the velocities. We show this for the scheme (2.11) or its equivalent one-step formulation (2.8) with (2.9).

**Theorem 5.3.** *Under the assumptions of Theorem 5.2, the velocity approximation $\dot{x}_n$ given by (2.11) has an expansion*

$$\dot{x}_n = v_h(t) + \sum_{0 < |k| < N} e^{ik\omega t} w_h^k(t) + \mathcal{O}(t^2 h^{N-1})$$

*uniformly for $0 \le t = nh \le T$, where the real-valued functions $v_h = (v_{h,1}, v_{h,2})$ and the complex-valued functions $w_h^k = (w_{h,1}^k, w_{h,2}^k)$ together with their derivatives up to arbitrary order satisfy*

$$\begin{aligned} v_{h,1} &= \dot{y}_{h,1} + \mathcal{O}(h^2), & w_{h,1}^1 &= \mathcal{O}(\omega^{-1}), & w_{h,1}^k &= \mathcal{O}(\omega^{-k}) \\ w_{h,2}^1 &= i\omega z_{h,2}^1 + \mathcal{O}(\omega^{-1}), & v_{h,2} &= \mathcal{O}(\omega^{-1}), & w_{h,2}^k &= \mathcal{O}(\omega^{-k}) \end{aligned} \quad (5.23)$$

*for $k = 2, \ldots, N-1$. Moreover, $w_h^{-k} = \overline{w_h^k}$. The constants symbolized by the $\mathcal{O}$-notation are independent of $\omega$ and $h$ with (5.10), but depend on $E$, $N$, $c$, and $T$.*

*Proof.* Let $u_h(t)$ be defined by the continuous analogue of (2.11),

$$2h \operatorname{sinc}(h\Omega) u_h(t) = x_h(t+h) - x_h(t-h). \tag{5.24}$$

Theorem 5.2 then yields that

$$\dot{x}_n = u_h(t) + \mathcal{O}(t^2 h^{N-1})$$

for $t = nh$ on bounded time intervals. Here we used that the remainder term in the lower component of (5.12) is of the form $\mathcal{O}(\psi(h\omega)(\phi(h\omega) + h)t^2 h^N)$, so that its quotient with $2h \operatorname{sinc}(h\omega)$ becomes $\mathcal{O}(t^2 h^{N-1})$ by the third of the conditions (4.1) and by (5.10). The function $u_h(t)$ can be written as

$$u_h(t) = v_h(t) + \sum_{0 < |k| < N} e^{ik\omega t} w_h^k(t). \tag{5.25}$$

We insert the relation (5.14) into $-i\sin(ihD)x_h(t) = h\operatorname{sinc}(h\Omega)u_h(t)$, which is equivalent to (5.24), and compare the coefficients of $e^{ik\omega t}$ to obtain

$$\begin{aligned}
\operatorname{sinc}(ihD) \dot{y}_{h,1} &= v_{h,1} \\
\operatorname{sinc}(ihD) \dot{y}_{h,2} &= \operatorname{sinc}(h\omega) v_{h,2} \\
(ih)^{-1} \sin(ihD - kh\omega) z_{h,1}^k &= w_{h,1}^k \\
(ih)^{-1} \sin(ihD - kh\omega) z_{h,2}^k &= \operatorname{sinc}(h\omega) w_{h,2}^k
\end{aligned} \tag{5.26}$$

for $k = 1, \ldots, N-1$. In particular, for $w_{h,2}^1$ we get

$$w_{h,2}^1 = i\omega \cos(ihD) z_{h,2}^1 - i\omega \frac{\cos(h\omega)}{\sin(h\omega)} \sin(ihD) z_{h,2}^1. \tag{5.27}$$

With the above equations, the estimates now follow with the bounds (5.21) of the coefficient functions and their derivatives, using again (4.1). □

# XIII.6 Almost-Invariants of the Modulated Fourier Expansions

The system for the coefficients of the modulated Fourier expansion of the exact solution is shown to have two formal invariants, which are related to the total and the oscillatory energy. In particular, this explains the near-conservation of the oscillatory energy over very long times. Analogous almost-invariants are shown to exist also for the modulated Fourier expansion of the numerical solution. This forms the basis for results on the long-time energy conservation of numerical methods, which will be given in Sections XIII.7 and XIII.8.

## XIII.6.1 The Hamiltonian of the Modulated Fourier Expansion

The equation (2.1) is a Hamiltonian system with the Hamiltonian

$$H(x, \dot{x}) = \tfrac{1}{2}\dot{x}^T \dot{x} + \tfrac{1}{2}x^T \Omega^2 x + U(x). \tag{6.1}$$

In the modulated Fourier expansion of the solution $x(t)$ of (2.1), denote $y^0(t) = y(t)$ and $y^k(t) = e^{ik\omega t} z^k(t)$ ($0 < |k| < N$), and let

$$\mathbf{y} = (y^{-N+1}, \ldots, y^{-1}, y^0, y^1, \ldots, y^{N-1}).$$

By (5.5)–(5.7) these functions satisfy

$$\ddot{y}^k + \Omega^2 y^k = -\sum_{s(\alpha)=k} \frac{1}{m!} U^{(m+1)}(y^0)\mathbf{y}^\alpha + \mathcal{O}(\omega^{-N}). \tag{6.2}$$

Here, the sum is over all $m \geq 0$ and all multi-indices $\alpha = (\alpha_1, \ldots, \alpha_m)$ with integers $\alpha_j$ ($0 < |\alpha_j| < N$) which have a given sum $s(\alpha) = \sum_{j=1}^m \alpha_j$, and we write $\mathbf{y}^\alpha = (y^{\alpha_1}, \ldots, y^{\alpha_m})$. We define

$$\mathcal{U}(\mathbf{y}) = U(y^0) + \sum_{s(\alpha)=0} \frac{1}{m!} U^{(m)}(y^0)\mathbf{y}^\alpha. \tag{6.3}$$

From the above it follows that $\mathbf{y}(t)$ satisfies the system

$$\ddot{y}^k + \Omega^2 y^k = -\nabla_{y^{-k}} \mathcal{U}(\mathbf{y}) + \mathcal{O}(\omega^{-N}) \tag{6.4}$$

which, neglecting the $\mathcal{O}(\omega^{-N})$ term, is the Hamiltonian system (cf. Exercise 5)

$$\dddot{y}^k = \frac{\partial \mathcal{H}}{\partial \dot{y}^{-k}}(\mathbf{y}, \dot{\mathbf{y}}), \qquad \ddot{y}^k = -\frac{\partial \mathcal{H}}{\partial y^{-k}}(\mathbf{y}, \dot{\mathbf{y}}) \tag{6.5}$$

with

$$\mathcal{H}(\mathbf{y}, \dot{\mathbf{y}}) = \frac{1}{2}\sum_{|k|<N}\left((\dot{y}^{-k})^T \dot{y}^k + (y^{-k})^T \Omega^2 y^k\right) + \mathcal{U}(\mathbf{y}). \tag{6.6}$$

**Theorem 6.1.** *Under the assumptions of Theorem 5.1, the Hamiltonian of the modulated Fourier expansion satisfies*

$$\mathcal{H}(\mathbf{y}(t), \dot{\mathbf{y}}(t)) = \mathcal{H}(\mathbf{y}(0), \dot{\mathbf{y}}(0)) + \mathcal{O}(\omega^{-N}) \tag{6.7}$$
$$\mathcal{H}(\mathbf{y}(t), \dot{\mathbf{y}}(t)) = H(x(t), \dot{x}(t)) + \mathcal{O}(\omega^{-1}). \tag{6.8}$$

*The constants symbolized by $\mathcal{O}$ are independent of $\omega$ and $t$ with $0 \leq t \leq T$, but depend on $E$, $N$ and $T$.*

*Proof.* Multiplying (6.4) with $(\dot{y}^{-k})^T$ and summing up gives

$$\sum_{|k|<N}(\dot{y}^{-k})^T(\ddot{y}^k + \Omega^2 y^k) = -\frac{d}{dt}\mathcal{U}(\mathbf{y}) + \mathcal{O}(\omega^{-N}). \tag{6.9}$$

Integrating from 0 to $t$ and using $y^{-k} = \overline{y^k}$ then yields (6.7).

By the bounds of Theorem 5.1, we have for $0 \le t \le T$

$$\mathcal{H}(\mathbf{y}, \dot{\mathbf{y}}) = \tfrac{1}{2}\|\dot{y}_1^0\|^2 + \|\dot{y}_2^1\|^2 + \omega^2\|y_2^1\|^2 + U(y^0) + \mathcal{O}(\omega^{-1}). \tag{6.10}$$

On the other hand, we have from (6.1) and (5.1)

$$H(x, \dot{x}) = \tfrac{1}{2}\|\dot{y}_1^0\|^2 + \tfrac{1}{2}\|\dot{y}_2^1 + \dot{y}_2^{-1}\|^2 + \tfrac{1}{2}\omega^2\|y_2^1 + y_2^{-1}\|^2 + U(y^0) + \mathcal{O}(\omega^{-1}). \tag{6.11}$$

Using $y_2^1 = e^{i\omega t}z_2^1$ and $\dot{y}_2^1 = e^{i\omega t}(\dot{z}_2^1 + i\omega z_2^1)$ together with $y_2^{-1} = \overline{y_2^1}$, it follows from $\dot{z}_2^1 = \mathcal{O}(\omega^{-1})$ that $\dot{y}_2^1 + \dot{y}_2^{-1} = i\omega(y_2^1 - y_2^{-1}) + \mathcal{O}(\omega^{-1})$ and $\|\dot{y}_2^1\| = \omega\|y_2^1\| + \mathcal{O}(\omega^{-1})$. Inserted into (6.10) and (6.11), this yields (6.8). $\square$

## XIII.6.2 A Formal Invariant Close to the Oscillatory Energy

In addition to the Hamiltonian $\mathcal{H}(\mathbf{y}, \dot{\mathbf{y}})$, the system for the coefficients of the modulated Fourier expansion has another formally conserved quantity. This almost-invariant depends only on the oscillating part and is given by

$$\mathcal{I}(\mathbf{y}, \dot{\mathbf{y}}) = -i\omega \sum_{0 < |k| < N} k\, (y^{-k})^T \dot{y}^k. \tag{6.12}$$

This turns out to be close to the energy of the harmonic oscillator,

$$I(x, \dot{x}) = \tfrac{1}{2}\|\dot{x}_2\|^2 + \tfrac{1}{2}\omega^2\|x_2\|^2. \tag{6.13}$$

**Theorem 6.2.** *Under the assumptions of Theorem 5.1,*

$$\mathcal{I}(\mathbf{y}(t), \dot{\mathbf{y}}(t)) = \mathcal{I}(\mathbf{y}(0), \dot{\mathbf{y}}(0)) + \mathcal{O}(\omega^{-N}) \tag{6.14}$$

$$\mathcal{I}(\mathbf{y}(t), \dot{\mathbf{y}}(t)) = I(x(t), \dot{x}(t)) + \mathcal{O}(\omega^{-1}). \tag{6.15}$$

*The constants symbolized by $\mathcal{O}$ are independent of $\omega$ and $t$ with $0 \le t \le T$, but depend on $E$, $N$ and $T$.*

*Proof.* For the vector

$$\mathbf{y}(\lambda) = \left(e^{i(-N+1)\lambda}y^{-N+1}, \ldots, e^{-i\lambda}y^{-1}, y^0, e^{i\lambda}y^1, \ldots, e^{i(N-1)\lambda}y^{N-1}\right)$$

the definition (6.3) of $\mathcal{U}$ shows that $\mathcal{U}(\mathbf{y}(\lambda))$ does not depend on $\lambda$. Its derivative with respect to $\lambda$ thus yields

$$0 = \frac{d}{d\lambda}\mathcal{U}(\mathbf{y}(\lambda)) = \sum_{0 < |k| < N} ik\, e^{ik\lambda}(y^k)^T \nabla_k \mathcal{U}(\mathbf{y}(\lambda)),$$

and putting $\lambda = 0$ we obtain

$$\sum_{0<|k|<N} ik\,(y^k)^T \nabla_k \mathcal{U}(\mathbf{y}) = 0 \tag{6.16}$$

for all vectors $\mathbf{y} = (y^{-N+1}, \ldots, y^{-1}, y^0, y^1, \ldots, y^{N-1})$.

The proof of Theorem 6.2 is now very similar to that of Theorem 6.1. We multiply the relation (6.4) with $-i\omega k(y^{-k})^T$ instead of $(\dot{y}^{-k})^T$. Summing up yields, with the use of (6.16),

$$-i\omega \sum_{0<|k|<N} k\,(y^{-k})^T (\ddot{y}^k + \Omega^2 y^k) = \mathcal{O}(\omega^{-N}). \tag{6.17}$$

The time derivative of $\mathcal{I}(\mathbf{y}, \dot{\mathbf{y}})$ of (6.12) equals

$$\frac{d}{dt}\mathcal{I}(\mathbf{y}, \dot{\mathbf{y}}) = -i\omega \sum_{0<|k|<N} k\Big((y^{-k})^T \ddot{y}^k + (\dot{y}^{-k})^T \dot{y}^k\Big). \tag{6.18}$$

In the sums $\sum_k k(y^{-k})^T \Omega^2 y^k$ and $\sum_k k(\dot{y}^{-k})^T \dot{y}^k$, the terms with $k$ and $-k$ cancel. Hence, (6.17) and (6.18) together yield

$$\frac{d}{dt}\mathcal{I}(\mathbf{y}, \dot{\mathbf{y}}) = \mathcal{O}(\omega^{-N}),$$

which implies (6.14).

With $\dot{y}^k = e^{ik\omega t}(\dot{z}^k + ik\omega z^k) = ik\omega y^k + \mathcal{O}(\omega^{-1})$, it follows from the bounds of Theorem 5.1 that
$$\mathcal{I}(\mathbf{y}, \dot{\mathbf{y}}) = 2\omega^2 \|y_2^1\|^2 + \mathcal{O}(\omega^{-1}).$$

On the other hand, using the arguments of the proof of Theorem 6.1, we have

$$I(x, \dot{x}) = \tfrac{1}{2}\|\dot{y}_2^1 + \dot{y}_2^{-1}\|^2 + \tfrac{1}{2}\omega^2\|y_2^1 + y_2^{-1}\|^2 + \mathcal{O}(\omega^{-1}) = 2\omega^2\|y_2^1\|^2 + \mathcal{O}(\omega^{-1}).$$

This proves the second statement of the theorem. □

Theorem 6.2 implies that the oscillatory energy is nearly conserved over long times:

**Theorem 6.3.** *If the solution $x(t)$ of (2.1) stays in a compact set for $0 \le t \le \omega^N$, then*
$$I(x(t), \dot{x}(t)) = I(x(0), \dot{x}(0)) + \mathcal{O}(\omega^{-1}) + \mathcal{O}(t\omega^{-N}).$$
*The constants symbolized by $\mathcal{O}$ are independent of $\omega$ and $t$ with $0 \le t \le \omega^N$, but depend on $E$ and $N$.*

*Proof.* With a fixed $T > 0$, let $\mathbf{y}_j$ denote the vector of the modulated Fourier expansion terms that correspond to starting values $(x(jT), \dot{x}(jT))$. For $t = (n + \theta)T$ with $0 \le \theta < 1$, we have by (6.15)

$$I(x(t), \dot{x}(t)) - I(x(0), \dot{x}(0))$$
$$= \mathcal{I}(\mathbf{y}_n(\theta T), \dot{\mathbf{y}}_n(\theta T)) + \mathcal{O}(\omega^{-1}) - \mathcal{I}(\mathbf{y}_0(0), \dot{\mathbf{y}}_0(0)) + \mathcal{O}(\omega^{-1})$$
$$= \mathcal{I}(\mathbf{y}_n(\theta T), \dot{\mathbf{y}}_n(\theta T)) - \mathcal{I}(\mathbf{y}_n(0), \dot{\mathbf{y}}_n(0)) +$$
$$\sum_{j=0}^{n-1} \Bigl( \mathcal{I}(\mathbf{y}_{j+1}(0), \dot{\mathbf{y}}_{j+1}(0)) - \mathcal{I}(\mathbf{y}_j(0), \dot{\mathbf{y}}_j(0)) \Bigr) + \mathcal{O}(\omega^{-1}) .$$

We note
$$\mathcal{I}(\mathbf{y}_{j+1}(0), \dot{\mathbf{y}}_{j+1}(0)) - \mathcal{I}(\mathbf{y}_j(0), \dot{\mathbf{y}}_j(0)) = \mathcal{O}(\omega^{-N}),$$
because, by the quasi-uniqueness of the coefficient functions as stated by Theorem 5.1, we have $\mathbf{y}_{j+1}(0) = \mathbf{y}_j(T) + \mathcal{O}(\omega^{-N})$ and $\dot{\mathbf{y}}_{j+1}(0) = \dot{\mathbf{y}}_j(T) + \mathcal{O}(\omega^{-N})$, and we have the bound (6.14) of Theorem 6.2. The same argument applies to $\mathcal{I}(\mathbf{y}_n(\theta T), \dot{\mathbf{y}}_n(\theta T)) - \mathcal{I}(\mathbf{y}_n(0), \dot{\mathbf{y}}_n(0))$. This yields the result. □

In a different approach, Benettin, Galgani & Giorgilli (1987) use a sequence of coordinate transformations from Hamiltonian perturbation theory to show that $I$ has only small deviations over time intervals which grow exponentially with $\omega$, in the case of an analytic potential $U$. By carefully tracing the dependence on $N$ of the constants in the $\mathcal{O}(\omega^{-N})$-terms, near-conservation of $I$ over exponentially long time intervals can be shown also within the present framework of modulated Fourier expansions; see Cohen (2000).

### XIII.6.3 Almost-Invariants of the Numerical Method

We show that the coefficients of the modulated Fourier expansion of the numerical solution have almost-invariants that are obtained similarly to the above. We denote

$$\mathbf{y}_h = (y_h^{-N+1}, \ldots, y_h^{-1}, y_h^0, y_h^1, \ldots, y_h^{N-1})$$
$$\mathbf{z}_h = (z_h^{-N+1}, \ldots, z_h^{-1}, z_h^0, z_h^1, \ldots, z_h^{N-1})$$

with $y_h^0(t) = z_h^0(t) = y_h(t)$ and $y_h^k(t) = e^{ik\omega t} z_h^k(t)$, where $y_h$ and $z_h^k$ are the coefficients of the modulated Fourier expansion of Theorem 5.2. Similar to (6.3) we consider the function

$$\mathcal{U}_h(\mathbf{y}_h) = U(\Phi y_h^0) + \sum_{s(\alpha)=0} \frac{1}{m!} U^{(m)}(\Phi y_h^0)(\Phi \mathbf{y}_h)^\alpha , \qquad (6.19)$$

where the sum is again taken over all $m \geq 1$ and all multi-indices $\alpha = (\alpha_1, \ldots, \alpha_m)$ with $0 < |\alpha_j| < N$ for which $s(\alpha) = \sum_j \alpha_j = 0$. It then follows from (5.22), multiplied with $h^{-2}\Psi^{-1}\Phi$, that the functions $y_h^k(t)$ satisfy

$$\Psi^{-1}\Phi h^{-2} \mathcal{L}(hD) y_h^k = -\nabla_{-k} \mathcal{U}_h(\mathbf{y}_h) + \mathcal{O}(h^N) , \qquad (6.20)$$

where $\mathcal{L}(hD)$ of (5.16) denotes again the difference operator of the numerical method. The similarity of these relations to (6.4) allows us to obtain almost-conserved quantities that are analogues of $\mathcal{H}$ and $\mathcal{I}$ above.

**The First Almost-Invariant.** We multiply (6.20) by $(\dot{y}_h^{-k})^T$, and as in (6.9) we obtain

$$\sum_{|k|<N} (\dot{y}_h^{-k})^T \Psi^{-1} \Phi h^{-2} \mathcal{L}(hD) y_h^k + \frac{d}{dt}\mathcal{U}_h(\mathbf{y}_h) = \mathcal{O}(h^N).$$

Since we know bounds of the coefficient functions $z_h^k$ and of their derivatives from Theorem 5.2, we switch to the quantities $z_h^k$ and we get the equivalent relation

$$\sum_{|k|<N} (\dot{z}_h^{-k} - ik\omega z_h^{-k})^T \Psi^{-1}\Phi h^{-2}\mathcal{L}(hD+ik\omega h) z_h^k + \frac{d}{dt}\mathcal{U}_h(\mathbf{z}_h) = \mathcal{O}(h^N). \quad (6.21)$$

We shall show that the left-hand side is the total derivative of an expression that depends only on $z_h^k$ and derivatives thereof. Consider first the term for $k=0$. The symmetry of the numerical method enters at this very point in the way that the expression $\mathcal{L}(hD)y = h^2\ddot{y} + c_4 h^4 y^{(4)} + c_6 h^6 y^{(6)} + \ldots$ contains only terms with derivatives of an even order. Multiplied with $\dot{y}^T$, even-order derivatives of $y$ give a total derivative:

$$\dot{y}^T y^{(2l)} = \frac{d}{dt}\left(\dot{y}^T y^{(2l-1)} - \ddot{y}^T y^{(2l-2)} + \ldots \mp (y^{(l-1)})^T y^{(l+1)} \pm \frac{1}{2}(y^{(l)})^T y^{(l)}\right).$$

Thanks to the symmetry of the difference operator $\mathcal{L}(hD)$ only expressions of this type appear in the term for $k=0$ in (6.21), with $z_h^0$ in the role of $y$. Similarly, we get for $z = z_h^k$ and $\overline{z} = z_h^{-k}$ with $0 < |k| < N$

$$\operatorname{Re} \dot{\overline{z}}^T z^{(2l)} = \operatorname{Re} \frac{d}{dt}\left(\dot{\overline{z}}^T z^{(2l-1)} - \ldots \mp (\overline{z}^{(l-1)})^T z^{(l+1)} \pm \frac{1}{2}(\overline{z}^{(l)})^T z^{(l)}\right)$$

$$\operatorname{Re} \dot{\overline{z}}^T z^{(2l+1)} = \operatorname{Re} \frac{d}{dt}\left(\dot{\overline{z}}^T z^{(2l)} - \ldots \pm (\overline{z}^{(l-1)})^T z^{(l+1)} \mp \frac{1}{2}(\overline{z}^{(l)})^T z^{(l)}\right)$$

$$\operatorname{Im} \dot{\overline{z}}^T z^{(2l+1)} = \operatorname{Im} \frac{d}{dt}\left(\dot{\overline{z}}^T z^{(2l)} - \ddot{\overline{z}}^T z^{(2l-1)} + \ldots \mp (\overline{z}^{(l)})^T z^{(l+1)}\right)$$

$$\operatorname{Im} \dot{\overline{z}}^T z^{(2l+2)} = \operatorname{Im} \frac{d}{dt}\left(\dot{\overline{z}}^T z^{(2l+1)} - \ddot{\overline{z}}^T z^{(2l)} + \ldots \pm (\overline{z}^{(l)})^T z^{(l+1)}\right).$$

Using the formulas (5.19) for $\mathcal{L}(hD+ik h\omega)$, it is seen that the term for $k$ in (6.21) has an asymptotic $h$-expansion with expressions of the above type as coefficients. The left-hand side of (6.21) can therefore be written as the time derivative of a function $\widehat{\mathcal{H}}_h[\mathbf{z}_h](t)$ which depends on the values at $t$ of the coefficient function vector $\mathbf{z}_h$ and its first $N$ time derivatives. The relation (6.21) thus becomes

$$\frac{d}{dt}\widehat{\mathcal{H}}_h[\mathbf{z}_h](t) = \mathcal{O}(h^N).$$

Together with the estimates of Theorem 5.2, this construction of $\widehat{\mathcal{H}}_h$ yields the following result.

**Lemma 6.4.** *Under the assumptions of Theorem 5.2, the coefficient functions* $\mathbf{z}_h = (z_h^{-N+1}, \ldots, z_h^{-1}, y_h, z_h^1, \ldots, z_h^{N-1})$ *of the modulated Fourier expansion of the numerical solution satisfy*

$$\widehat{\mathcal{H}}_h[\mathbf{z}_h](t) = \widehat{\mathcal{H}}_h[\mathbf{z}_h](0) + \mathcal{O}(th^N) \qquad (6.22)$$

*for* $0 \le t \le T$. *Moreover,*

$$\widehat{\mathcal{H}}_h[\mathbf{z}_h](t) = \tfrac{1}{2}\|\dot{y}_{h,1}(t)\|^2 + \mu(h\omega)\, 2\omega^2\, \|z_{h,2}^1(t)\|^2 + U(\Phi y_h(t)) + \mathcal{O}(h^2), \quad (6.23)$$

*where* $\mu(h\omega) = \mathrm{sinc}(h\omega)\phi(h\omega)/\psi(h\omega)$. □

**The Second Almost-Invariant.** By the same calculation as in the proof of Theorem 6.2 we obtain for $\mathcal{U}_h(\mathbf{y}_h(t))$ of (6.19)

$$0 = \sum_{0<|k|<N} ik\omega\, (y_h^k)^T \nabla_k \mathcal{U}_h(\mathbf{y}_h) \, .$$

It then follows from (6.20) that

$$-i\omega \sum_{0<|k|<N} k(y_h^{-k})^T \Psi^{-1} \Phi\, h^{-2} \mathcal{L}(hD) y_h^k = \mathcal{O}(h^N) \, .$$

Written in the $z$ variables, this becomes

$$-i\omega \sum_{0<|k|<N} k(z_h^{-k})^T \Psi^{-1} \Phi\, h^{-2} \mathcal{L}(hD + ik\omega h) z_h^k = \mathcal{O}(h^N) \, . \qquad (6.24)$$

As in (6.21), the left-hand expression can be written as the time derivative of a function $\widehat{\mathcal{I}}_h[\mathbf{z}_h](t)$ which depends on the values at $t$ of the function $\mathbf{z}_h$ and its first $N$ derivatives:

$$\frac{d}{dt}\widehat{\mathcal{I}}_h[\mathbf{z}_h](t) = \mathcal{O}(h^N) \, .$$

Together with the estimates of Theorem 5.2 this yields the following result.

**Lemma 6.5.** *Under the assumptions of Theorem 5.2, the coefficient functions* $\mathbf{z}_h$ *of the modulated Fourier expansion of the numerical solution satisfy*

$$\widehat{\mathcal{I}}_h[\mathbf{z}_h](t) = \widehat{\mathcal{I}}_h[\mathbf{z}_h](0) + \mathcal{O}(th^N) \qquad (6.25)$$

*for* $0 \le t \le T$. *Moreover,*

$$\widehat{\mathcal{I}}_h[\mathbf{z}_h](t) = \mu(h\omega)\, 2\omega^2\, \|z_{h,2}^1(t)\|^2 + \mathcal{O}(h^2) \, , \qquad (6.26)$$

*where again* $\mu(h\omega) = \mathrm{sinc}(h\omega)\phi(h\omega)/\psi(h\omega)$. □

Symplectic methods have $\psi(\xi) = \text{sinc}(\xi)\,\phi(\xi)$ and hence $\mu(h\omega) = 1$. To be able to also treat methods where $\mu(h\omega)$ can be small, we need to sharpen the estimates of Lemma 6.5. Close scrutiny of the equations (5.20) that determine the coefficient functions of the modulated Fourier expansion, shows that the $\mathcal{O}(h^2)$ term in (6.26) contains a factor $\phi(h\omega)^2$, and that the $\mathcal{O}(th^N)$ term in (6.25) can be put in the form $\mathcal{O}(t\phi(h\omega)^N h^N) + \mathcal{O}(th^{N+m})$ with an arbitrary integer $m \geq 0$; cf. (5.12). Assume now that

$$\phi \text{ is analytic with no real zeros other than integral multiples of } \pi. \qquad (6.27)$$

This condition ensures that $|\phi(h\omega)|^2 \geq ch^m$ for some $m$ if $h\omega$ satisfies (5.10). Under the conditions of Theorem 5.2, in particular, (4.1) and (5.10), the improved bounds of the remainder terms yield the following estimates for $\mathcal{I}_h = \widehat{\mathcal{I}}_h/\mu(h\omega)$:

$$\mathcal{I}_h[\mathbf{z}_h](t) = \mathcal{I}_h[\mathbf{z}_h](0) + \mathcal{O}(th^N) \qquad (6.28)$$

$$\mathcal{I}_h[\mathbf{z}_h](t) = 2\omega^2 \|z_{h,2}^1(t)\|^2 + \mathcal{O}(h^2). \qquad (6.29)$$

**Relationship with the Total and the Oscillatory Energy.** The almost-invariants

$$\mathcal{I}_h = \frac{1}{\mu(h\omega)}\widehat{\mathcal{I}}_h, \quad \mathcal{H}_h = \widehat{\mathcal{H}}_h - \left(1 - \frac{1}{\mu(h\omega)}\right)\widehat{\mathcal{I}}_h \qquad (6.30)$$

of the coefficient functions of the modulated Fourier expansion are then close to the total energy $H$ and the oscillatory energy $I$ along the numerical solution $(x_n, \dot{x}_n)$:

**Theorem 6.6.** *Under the conditions of Theorems 5.2 and condition (6.27),*

$$\mathcal{H}_h[\mathbf{z}_h](t) = \mathcal{H}_h[\mathbf{z}_h](0) + \mathcal{O}(th^N), \quad \mathcal{I}_h[\mathbf{z}_h](t) = \mathcal{I}_h[\mathbf{z}_h](0) + \mathcal{O}(th^N)$$
$$\mathcal{H}_h[\mathbf{z}_h](t) = H(x_n, \dot{x}_n) + \mathcal{O}(h), \quad \mathcal{I}_h[\mathbf{z}_h](t) = I(x_n, \dot{x}_n) + \mathcal{O}(h)$$

*holds for* $0 \leq t = nh \leq T$. *The constants symbolized by $\mathcal{O}$ depend on $E$, $N$ and $T$.*

*Proof.* The upper two relations follow directly from (6.22) and (6.28). Theorems 5.2 and 5.3 show

$$\omega x_{n,2} = \omega\left(e^{i\omega t}z_{h,2}^1(t) + e^{-i\omega t}z_{h,2}^{-1}(t)\right) + \mathcal{O}(h)$$
$$\dot{x}_{n,2} = i\omega\left(e^{i\omega t}z_{h,2}^1(t) - e^{-i\omega t}z_{h,2}^{-1}(t)\right) + \mathcal{O}(h).$$

With the identity $\|v + \overline{v}\|^2 + \|v - \overline{v}\|^2 = 4\|v\|^2$, this implies

$$I(x_n, \dot{x}_n) = 2\omega^2 \|z_{h,2}^1(t)\|^2 + \mathcal{O}(h).$$

A comparison with (6.29) then gives the stated relation between $I$ and $\mathcal{I}_h$. The relation between $H$ and $\mathcal{H}_h$ is proved in the same way, using in addition (6.23). $\square$

# XIII.7 Long-Time Near-Conservation of Total and Oscillatory Energy

With the results of the previous section, we can now show that the numerical method nearly preserves the total energy $H$ and the oscillatory energy $I$ over time intervals of length $C_N h^{-N+1}$, for any $N$ for which the non-resonance condition (5.10) is satisfied. Such a result is due to Hairer & Lubich (2000a).
For convenience we restate the assumptions:

- the energy bound (2.3): $\frac{1}{2}\|\dot{x}_0\|^2 + \frac{1}{2}\|\Omega x_0\|^2 \leq E$;
- the condition on the numerical solution: the values $\Phi x_n$ stay in a compact subset of a domain on which the potential $U$ is smooth;
- the conditions on the filter functions: $\psi$ and $\phi$ are even, real-analytic, and have no real zeros other than integral multiples of $\pi$; they satisfy $\psi(0) = \phi(0) = 1$ and (4.1):

$$|\psi(h\omega)| \leq C_1 \operatorname{sinc}^2(\tfrac{1}{2}h\omega), \qquad |\phi(h\omega)| \leq C_2 |\operatorname{sinc}(\tfrac{1}{2}h\omega)|, \qquad (7.1)$$
$$|\psi(h\omega)\phi(h\omega)| \leq C_3 |\operatorname{sinc}(h\omega)|;$$

- the condition $h\omega \geq c_0 > 0$;
- the non-resonance condition (5.10): for some $N \geq 2$,

$$|\sin(\tfrac{1}{2}kh\omega)| \geq c\sqrt{h} \qquad \text{for} \quad k = 1, \ldots, N.$$

**Theorem 7.1.** *Under the above conditions, the numerical solution of (2.1) obtained by the method (2.7)–(2.8) with (2.9) satisfies*

$$\begin{aligned} H(x_n, \dot{x}_n) &= H(x_0, \dot{x}_0) + \mathcal{O}(h) \\ I(x_n, \dot{x}_n) &= I(x_0, \dot{x}_0) + \mathcal{O}(h) \end{aligned} \qquad \text{for} \quad 0 \leq nh \leq h^{-N+1}.$$

*The constants symbolized by $\mathcal{O}$ are independent of $n$, $h$, $\omega$ satisfying the above conditions, but depend on $N$ and the constants in the conditions.*

*Proof.* The estimates of Theorem 6.6 hold uniformly over bounded intervals. We now apply those estimates repeatedly on intervals of length $h$, for modulated Fourier expansions corresponding to different starting values. As long as $(x_n, \dot{x}_n)$ satisfies the bounded-energy condition (2.3) (possibly with a larger constant $E$), Theorem 5.2 gives a modulated Fourier expansion that corresponds to starting values $(x_n, \dot{x}_n)$. We denote the vector of coefficient functions of this expansion by $\mathbf{z}_n(t)$:

$$\mathbf{z}_n = (z_n^{-N+1}, \ldots, z_n^{-1}, y_n, z_n^1, \ldots, z_n^{N-1})$$

(omitting the notational dependence on $h$ for simplicity). Because of the uniqueness, up to $\mathcal{O}(h^{N+1})$, of the coefficient functions of the modulated Fourier expansion constructed by (5.20), the following diagram commutes up to terms of size $\mathcal{O}(h^{N+1})$:

$$\begin{array}{ccc}
(x_n, \dot{x}_n) & \longleftrightarrow & (\mathbf{z}_n(0), \dot{\mathbf{z}}_n(0)) \\
& & \Big\downarrow \text{flow} \\
\Big\downarrow \begin{array}{c}\text{numerical}\\\text{method}\end{array} & & (\mathbf{z}_n(h), \dot{\mathbf{z}}_n(h)) \\
& & = \text{ (up to } \mathcal{O}(h^{N+1})) \\
(x_{n+1}, \dot{x}_{n+1}) & \longleftrightarrow & (\mathbf{z}_{n+1}(0), \dot{\mathbf{z}}_{n+1}(0))
\end{array}$$

The construction of the coefficient functions via (5.20) shows that also higher derivatives of $\mathbf{z}_n$ at $h$ and $\mathbf{z}_{n+1}$ at $0$ differ by only $\mathcal{O}(h^{N+1})$. From the above diagram and Theorem 6.6 we thus obtain

$$\begin{aligned}
\mathcal{H}_h[\mathbf{z}_{n+1}](0) &= \mathcal{H}_h[\mathbf{z}_n](h) + \mathcal{O}(h^{N+1}) \\
&= \mathcal{H}_h[\mathbf{z}_n](0) + \mathcal{O}(h^{N+1}).
\end{aligned}$$

Repeated use of this relation gives

$$\mathcal{H}_h[\mathbf{z}_n](0) = \mathcal{H}_h[\mathbf{z}_0](0) + \mathcal{O}(nh^{N+1}).$$

Moreover, by Theorem 6.6 the coefficient functions corresponding to the starting values $(x_n, \dot{x}_n)$ and $(x_0, \dot{x}_0)$ satisfy

$$\begin{aligned}
\mathcal{H}_h[\mathbf{z}_n](0) &= H(x_n, \dot{x}_n) + \mathcal{O}(h), \\
\mathcal{H}_h[\mathbf{z}_0](0) &= H(x_0, \dot{x}_0) + \mathcal{O}(h).
\end{aligned}$$

So we obtain

$$\begin{aligned}
H(x_n, \dot{x}_n) - H(x_0, \dot{x}_0) &= \mathcal{H}_h[\mathbf{z}_n](0) - \mathcal{H}_h[\mathbf{z}_0](0) + \mathcal{O}(h) \\
&= \mathcal{O}(nh^{N+1}) + \mathcal{O}(h),
\end{aligned}$$

which gives the desired bound for the deviation of the total energy along the numerical solution. The same argument applies to $I(x_n, \dot{x}_n)$. $\square$

The imposed bounds of $\psi$ and $\phi$ become important when $h\omega$ is close to an integral multiple of $\pi$. Are these conditions also sufficient to guarantee favourable energy behaviour uniformly in $h\omega$, arbitrarily close to multiples of $\pi$? Unfortunately the answer is negative (see Fig. 2.5 to Fig. 2.7). The analysis of the method (2.7)–(2.9) for exact resonances $h\omega = m\pi$ with integer $m$ shows that stronger conditions

$$|\psi(h\omega)| \leq C\,|\mathrm{sinc}(h\omega)|, \qquad |\psi(h\omega)\phi(h\omega)| \leq C\,\mathrm{sinc}^2(h\omega) \qquad (7.2)$$

are required. Even this is not sufficient for near-conservation of the total and the oscillatory energy for $h\omega$ near a multiple of $\pi$. For linear problems

$$\ddot{x} + \begin{pmatrix} 0 & 0 \\ 0 & \omega^2 \end{pmatrix} x = -Ax$$

with a two-dimensional symmetric matrix $A$ with $a_{11} > 0$, and with initial values satisfying the bounded-energy condition (2.3), Hairer & Lubich (2000a) show that the numerical method conserves the total energy up to $\mathcal{O}(h)$ uniformly for all times and for all values of $h\omega$, if and only if

$$\psi(\xi) = \operatorname{sinc}^2(\xi)\,\phi(\xi)\,. \tag{7.3}$$

There is *no* method (2.7)-(2.8) which approximately preserves the oscillatory energy $I$ uniformly for all $h\omega$ in a fixed open interval that contains a multiple of $2\pi$.

In summary, the bad effect of step-size resonances on the energy behaviour of the method cannot be eliminated, but it can be considerably mitigated by an appropriate choice of the filter functions $\psi$ and $\phi$.

## XIII.8 Energy Behaviour of the Störmer/Verlet Method

The results of Sections XIII.5–XIII.7 provide new insight into the energy behaviour of the classical Störmer/Verlet method. We present in this section weakened versions of results of Hairer & Lubich (2000b).

In applications, the Störmer/Verlet method is typically used with step sizes $h$ for which the product with the highest frequency $\omega$ is in the range of linear stability, but is bounded away from 0. For example, in spatially discretized wave equations, $h\omega$ is known as the CFL number, which is typically kept near 1. Values of $h\omega$ around $\frac{1}{2}$ are often used in molecular dynamics. In contrast, the backward error analysis of Chap. IX explains the long-time energy behaviour only for $h\omega \to 0$.

Consider now applying the Störmer/Verlet method to the nonlinear model problem (2.1)-(2.3),

$$x_{n+1} - 2x_n + x_{n-1} = -h^2 \Omega^2 x_n - h^2 \nabla U(x_n)\,, \tag{8.1}$$

with $h\omega < 2$ for linear stability. The method is made accessible to the analysis of Sections XIII.3–XIII.7 by rewriting it as a trigonometric method (2.6) with a *modified frequency*:

$$x_{n+1} - 2\cos(h\widetilde{\Omega})\,x_n + x_{n-1} = -h^2 \nabla U(x_n)\,, \tag{8.2}$$

where

$$\widetilde{\Omega} = \begin{pmatrix} 0 & 0 \\ 0 & \widetilde{\omega}I \end{pmatrix} \quad \text{with} \quad \sin(\tfrac{1}{2}h\widetilde{\omega}) = \tfrac{1}{2}h\omega\,. \tag{8.3}$$

The velocity approximation

$$\dot{x}_n = \frac{x_{n+1} - x_{n-1}}{2h}$$

does not correspond to the velocity approximation (2.11) of the trigonometric method, but this presents only a minor technical difficulty. We show that the following *modified energies* are well conserved by the Störmer/Verlet method:

$$H^*(x,\dot{x}) = H(x,\dot{x}) + \tfrac{1}{2}\gamma\,\|\dot{x}_2\|^2$$
$$I^*(x,\dot{x}) = I(x,\dot{x}) + \tfrac{1}{2}\gamma\,\|\dot{x}_2\|^2 \quad \text{with} \quad \gamma = \frac{1}{1-\tfrac{1}{4}(h\omega)^2} - 1. \tag{8.4}$$

Here $H$ and $I$ are again the total and the oscillatory energy of the system (2.1) (defined with the original $\omega$, not with $\widetilde{\omega}$).

**Theorem 8.1.** *Let the Störmer/Verlet method be applied to the problem (2.1)-(2.3) with a step size $h$ for which $0 < c_0 \le h\omega \le c_1 < 2$ and $|\sin(\tfrac{1}{2}kh\widetilde{\omega})| \ge c\sqrt{h}$ for $k = 1,\ldots,N$ for some $N \ge 2$ and $c > 0$. Suppose further that the numerical solution values $x_n$ stay in a region on which all derivatives of $U$ are bounded. Then, the modified energy along the numerical solution satisfy*

$$H^*(x_n,\dot{x}_n) = H^*(x_0,\dot{x}_0) + \mathcal{O}(h)$$
$$I^*(x_n,\dot{x}_n) = I^*(x_0,\dot{x}_0) + \mathcal{O}(h) \quad \text{for} \quad 0 \le nh \le h^{-N+1}. \tag{8.5}$$

*The constants symbolized by $\mathcal{O}$ are independent of $n$, $h$, $\omega$ with the above conditions.*

*Proof.* With the modified velocities $x'_n$ defined by

$$2h\,\mathrm{sinc}(h\widetilde{\Omega})\,x'_n = x_{n+1} - x_{n-1}$$

the method (8.2) becomes a method (2.6) with (2.11), or equivalently (2.7)-(2.8), with $\widetilde{\omega}$ instead of $\omega$ and with $\psi(\xi) = \phi(\xi) = 1$.

The condition $0 < c_0 \le h\omega \le c_1 < 2$ implies $|\sin(\tfrac{1}{2}kh\widetilde{\omega})| \ge c_2 > 0$ for $k = 1, 2$, and hence conditions (7.1) are trivially satisfied with $h\widetilde{\omega}$ instead of $h\omega$. We are thus in the position to apply Theorem 7.1, which yields

$$\widetilde{H}(x_n,x'_n) = \widetilde{H}(x_0,x'_0) + \mathcal{O}(h)$$
$$\widetilde{I}(x_n,x'_n) = \widetilde{I}(x_0,x'_0) + \mathcal{O}(h) \quad \text{for} \quad 0 \le nh \le h^{-N+1}, \tag{8.6}$$

where $\widetilde{H}$ and $\widetilde{I}$ are defined in the same way as $H$ and $I$, but with $\widetilde{\omega}$ in place of $\omega$. The components of the Störmer/Verlet velocities $\dot{x}_n$ and the modified velocities $x'_n$ are related by

$$\dot{x}_{n,1} = x'_{n,1}, \qquad \dot{x}_{n,2} = \mathrm{sinc}(h\widetilde{\omega})\,x'_{n,2} = \frac{\omega}{\widetilde{\omega}}\sqrt{1 - \tfrac{1}{4}h^2\omega^2}\,x'_{n,2}, \tag{8.7}$$

so that

$$\begin{aligned}
\widetilde{I}(x_n,x'_n) &= \tfrac{1}{2}\|x'_{n,2}\|^2 + \tfrac{1}{2}\widetilde{\omega}^2\|x_{n,2}\|^2 \\
&= \tfrac{1}{2}\frac{\widetilde{\omega}^2}{\omega^2}\frac{1}{1-\tfrac{1}{4}h^2\omega^2}\|\dot{x}_{n,2}\|^2 + \tfrac{1}{2}\frac{\widetilde{\omega}^2}{\omega^2}\omega^2\|x_{n,2}\|^2 \\
&= \frac{\widetilde{\omega}^2}{\omega^2}\,I^*(x_n,\dot{x}_n).
\end{aligned} \tag{8.8}$$

Similarly,

$$H^*(x_n, \dot{x}_n) = \frac{1}{2}\|\dot{x}_{n,1}\|^2 + U(x_n) + I^*(x_n, \dot{x}_n)$$
$$= \widetilde{H}(x_n, x'_n) + \left(\frac{\omega^2}{\widetilde{\omega}^2} - 1\right)\widetilde{I}(x_n, x'_n),$$
(8.9)

and hence (8.6) yields the result. □

For fixed $h\omega \geq c_0 > 0$ and $h \to 0$, the maximum deviation in the energy does not tend to 0, due to the highly oscillatory term $\frac{1}{2}\gamma\|\dot{x}_2\|$ in $H^*(x, \dot{x})$ and $I^*(x, \dot{x})$. We show, however, that *time averages* of $H$ and $I$ are nearly preserved over long time. For an arbitrary fixed $T > 0$, consider the averages over intervals of length $T$,

$$\overline{H}_n = \frac{1}{T}h \sum_{|jh|\leq T/2} H(x_{n+j}, \dot{x}_{n+j})$$

$$\overline{I}_n = \frac{1}{T}h \sum_{|jh|\leq T/2} I(x_{n+j}, \dot{x}_{n+j}).$$
(8.10)

**Theorem 8.2.** *Under the conditions of Theorem 8.1, the time averages of the total and the oscillatory energy along the numerical solution satisfy*

$$\begin{aligned}\overline{H}_n &= \overline{H}_0 + \mathcal{O}(h) \\ \overline{I}_n &= \overline{I}_0 + \mathcal{O}(h)\end{aligned} \quad \text{for} \quad 0 \leq nh \leq h^{-N+1}.$$
(8.11)

*The constants symbolized by $\mathcal{O}$ are independent of $n, h, \omega$ with the above conditions.*

*Proof.* We show

$$\overline{H}_n = H^*(x_n, \dot{x}_n) - \frac{1}{2}\frac{\gamma}{1+\gamma} I^*(x_n, \dot{x}_n) + \mathcal{O}(h)$$
$$\overline{I}_n = I^*(x_n, \dot{x}_n) - \frac{1}{2}\frac{\gamma}{1+\gamma} I^*(x_n, \dot{x}_n) + \mathcal{O}(h),$$
(8.12)

which implies the result by Theorem 8.1. Consider the modulated Fourier expansions of $x_n$ and $x'_n$ for $t = nh$ in a bounded interval. Theorem 5.3 shows that

$$x'_{n,2} = i\widetilde{\omega}\bigl(e^{i\widetilde{\omega}t}z^1_{h,2}(t) - e^{-i\widetilde{\omega}t}\overline{z^1_{h,2}(t)}\bigr) + \mathcal{O}(h), \qquad t = nh,$$

with $z^1_{h,2}(t)$ from the modulated Fourier expansion of Theorem 5.2 (with $\widetilde{\omega}$ instead of $\omega$). With (8.7) it follows that

$$\dot{x}_{n,2} = i\omega\sqrt{1 - \tfrac{1}{4}h^2\omega^2}\left(e^{i\widetilde{\omega}t}z^1_{h,2}(t) - e^{-i\widetilde{\omega}t}\overline{z^1_{h,2}(t)}\right) + \mathcal{O}(h),$$

and therefore, recalling the definition of $\gamma$,

$$\|\dot{x}_{n,2}\|^2 = \omega^2 \frac{1}{1+\gamma} \left(2\|z^1_{h,2}(t)\|^2 - 2\operatorname{Re} e^{2i\widetilde{\omega}t}z^1_{h,2}(t)^2\right) + \mathcal{O}(h).$$

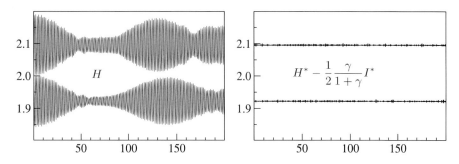

**Fig. 8.1.** Total energies (left) and their predicted averages (right) for the Störmer/Verlet method and for two different initial values, with $\omega = 50$ and $h$ such that $h\omega = 0.8$.

Theorems 5.2 and 5.3 yield
$$2\widetilde{\omega}^2 \|z^1_{h,2}(t)\|^2 = \widetilde{I}(x_n, x'_n) + \mathcal{O}(h)$$

and hence, by (8.8),
$$2\omega^2 \|z^1_{h,2}(t)\|^2 = I^*(x_n, \dot{x}_n) + \mathcal{O}(h).$$

A partial summation shows that the time average over the highly oscillatory terms $e^{2i\widetilde{\omega}t}\omega^2 z^1_{h,2}(t)^2$ is $\mathcal{O}(h)$. This finally gives

$$\frac{1}{T} h \sum_{|j| \leq T/2} \|\dot{x}_{j,2}\|^2 = \frac{1}{1+\gamma} I^*(x_n, \dot{x}_n) + \mathcal{O}(h).$$

Taking the time averages in the expressions of the definition (8.4) of $H^*$ and $I^*$ then yields (8.12). □

Figure 8.1 illustrates the above result. It shows the total energy $H$ for two different initial values on the left, and the averages as predicted by the expression on the right-hand side of (8.12) on the right picture. The initial values are as in Chap. I with the exception of $v_1(0)$ and $\dot{v}_1(0)$. We take $v_1(0) = \sqrt{2}/\omega$, $\dot{v}_1(0) = 0$ for one set of initial values and $v_1(0) = 0$, $\dot{v}_1(0) = \sqrt{2}$ for the other. The total energies at the initial values are $2.00240032$ and $2$, respectively.

## XIII.9 Exercises

1. Show that the impulse method (with exact solution of the fast system) reduces to Deuflhard's method in the case of a quadratic potential $W(q) = \frac{1}{2} q^T A q$.
2. Show that the method (2.7)–(2.8) is symplectic if and only if
$$\psi(\xi) = \mathrm{sinc}(\xi)\, \phi(\xi) \qquad \text{for} \quad \xi = h\omega.$$

3. Prove that for infinitely differentiable functions $g(t)$ the solution of $\ddot{x} + \omega^2 x = g(t)$ can be written as
$$x(t) = y(t) + \cos(\omega t)\, u(t) + \sin(\omega t)\, v(t),$$
where $y(t), u(t), v(t)$ are given by asymptotic expansions in powers of $\omega^{-1}$.
*Hint.* Use the variation-of-constants formula and apply repeated partial integration.

4. Show that the recurrence relation $e_{n+1} - 2\cos(h\Omega)\, e_n + e_{n-1} = b_n$ has the solution
$$e_{n+1} = -W_{n-1}\, e_0 + W_n\, e_1 + \sum_{j=1}^{n} W_{n-j}\, b_j$$
with $W_n = \sin(h\Omega)^{-1} \sin\bigl((n+1)h\Omega\bigr)$ (or the appropriate limit when $\sin(h\Omega)$ is not invertible).

5. Consider a Hamiltonian $H(p_R, p_I, q_R, q_I)$ and let
$$\mathcal{H}(p, q) = 2 H(p_R, p_I, q_R, q_I)$$
for $p = p_R + ip_I$, $q = q_R + iq_I$. Prove that in the new variables $p, q$ the Hamiltonian system becomes
$$\dot{p} = -\frac{\partial \mathcal{H}}{\partial \overline{q}}(p, q), \qquad \dot{q} = \frac{\partial \mathcal{H}}{\partial \overline{p}}(p, q).$$

6. Prove the following refinement of Theorem 6.3: along the solution $x(t)$ of (2.1), the modified oscillatory energy $J(x, \dot{x}) = I(x, \dot{x}) - x_2^T g_2(x)$ satisfies
$$J(x(t), \dot{x}(t)) = J(x(0), \dot{x}(0)) + \mathcal{O}(\omega^{-2}) + \mathcal{O}(t\omega^{-N}).$$

7. Define $\widehat{H}(x, \dot{x}) = H(x, \dot{x}) - \rho x_2^T g_2(x)$, $\widehat{J}(x, \dot{x}) = J(x, \dot{x}) - \rho x_2^T g_2(x)$ with $J(x, \dot{x})$ of the previous exercise and with
$$\rho = \frac{\psi(h\omega)}{\operatorname{sinc}^2(\tfrac{1}{2}h\omega)} - 1.$$
In the situation of Theorem 7.1, show that
$$\begin{aligned}\widehat{H}(x_n, \dot{x}_n) &= \widehat{H}(x_0, \dot{x}_0) + \mathcal{O}(h^2) \\ \widehat{J}(x_n, \dot{x}_n) &= \widehat{J}(x_0, \dot{x}_0) + \mathcal{O}(h^2)\end{aligned} \qquad \text{for } 0 \le nh \le h^{-N+1}.$$
Notice that the total energy $H(x_n, \dot{x}_n)$ and the modified oscillatory energy $J(x_n, \dot{x}_n)$ are conserved up to $\mathcal{O}(h^2)$ if $\rho = 0$, i.e., if $\psi(\xi) = \operatorname{sinc}^2(\tfrac{1}{2}\xi)$. This explains the excellent energy conservation of methods (A) and (D) in Figure 2.5 away from resonances.

8. Generalizing the analysis of Sect. XIII.8, study the energy behaviour of the impulse or averaged-force multiple time-stepping method of Sect. VIII.4 with a fixed number $N$ of Störmer/Verlet substeps per step, when the method is applied to the model problem with $h\omega$ bounded away from zero.

# Chapter XIV.
# Dynamics of Multistep Methods

Multistep methods are the basis of important codes for nonstiff differential equations (Adams methods) and for stiff problems (BDF methods). We study here their applicability to long-time integrations of Hamiltonian or reversible systems.

This chapter starts with numerical experiments which illustrate that the long-time behaviour of classical multistep methods is in general disappointing. They either behave as non-symplectic and non-symmetric one-step methods, or they exhibit undesired instabilities (parasitic solutions). Partitioned multistep methods, however, can have a much better long-time behaviour. They are promising methods, because in a constant step size mode they can be easily implemented, and high order can be obtained with one function evaluation per step. We shall study one-step formulations of such methods, which allow us to investigate their symplecticity and symmetry.

## XIV.1 Numerical Methods and Experiments

We present the numerical methods treated in this chapter, and we study their long-time behaviour when applied to some Hamiltonian systems.

### XIV.1.1 Linear Multistep Methods

For first order systems of differential equations $\dot{y} = f(y)$, linear multistep methods are defined by the formula

$$\sum_{j=0}^{k} \alpha_j y_{n+j} = h \sum_{j=0}^{k} \beta_j f(y_{n+j}), \tag{1.1}$$

where $\alpha_j, \beta_j$ are real parameters, $\alpha_k \neq 0$, and $|\alpha_0| + |\beta_0| > 0$. For an application of this formula we need a starting procedure which, in addition to an initial value $y(t_0) = y_0$, provides approximations $y_1, \ldots, y_{k-1}$ to $y(t_0+h), \ldots, y(t_0+(k-1)h)$. The approximations $y_n$ to $y(t_0 + nh)$ for $n \geq k$ can then be computed recursively from (1.1). In the case $\beta_k = 0$ we have an explicit method, otherwise it is implicit and the numerical solution $y_{n+k}$ has to be computed iteratively.

Since the fundamental work of Dahlquist (1956) it is common to denote the generating polynomials of the coefficients by

$$\rho(\zeta) = \sum_{j=0}^{k} \alpha_j \zeta^j, \qquad \sigma(\zeta) = \sum_{j=0}^{k} \beta_j \zeta^j. \tag{1.2}$$

For the classical theory of multistep methods we refer the reader to Chap. III of Hairer, Nørsett & Wanner (1993). We just recall some important definitions.

**Order.** A multistep method has order $p$ if, when applied with exact starting values to the problem $\dot{y} = t^q$ ($0 \leq q \leq p$), it integrates the problem without error. This is equivalent to the requirement that

$$\rho(e^h) - h\sigma(e^h) = \mathcal{O}(h^{p+1}) \qquad \text{for} \quad h \to 0. \tag{1.3}$$

**Stability.** Method (1.1) is stable if, when applied to $\dot{y} = 0$, it yields for all $y_0, \ldots, y_{k-1}$ a bounded numerical solution. This is equivalent to the requirement that the polynomial $\rho(\zeta)$ satisfies the root condition, i.e., all roots of $\rho(\zeta) = 0$ satisfy $|\zeta| \leq 1$, and those on the unit circle are simple roots. The method is called *strictly stable*, if all roots are inside the unit circle with the exception of $\zeta = 1$.

**Convergence.** If a multistep method is stable and of order $p \geq 1$, it is convergent of order $p$ for all sufficiently smooth problems. This means that, assuming starting approximations with an error bounded by $\mathcal{O}(h^p)$, the global error satisfies $y_n - y(t_0 + nh) = \mathcal{O}(h^p)$ on compact intervals $nh \leq T$.

**Symmetry.** If the coefficients of a multistep formula (1.1) satisfy

$$\alpha_{k-j} = -\alpha_j, \qquad \beta_{k-j} = \beta_j \qquad \text{for all } j, \tag{1.4}$$

then the method is called symmetric. Condition (1.4) implies that for every zero $\zeta$ of $\rho(\zeta)$ also its inverse $\zeta^{-1}$ is a zero. Hence, for stable symmetric methods all zeros of $\rho(\zeta)$ are simple and lie on the unit circle.

**Example 1.1.** We consider the pendulum equation (I.1.13), and we apply the following multistep methods: the 2-step explicit Adams method

$$y_{n+2} = y_{n+1} + h\left(\frac{3}{2}f_{n+1} - \frac{1}{2}f_n\right), \tag{1.5}$$

the 2-step backward differentiation formula (BDF)

$$\frac{3}{2}y_{n+2} - 2y_{n+1} + \frac{1}{2}y_n = hf_{n+2}, \tag{1.6}$$

and the (2-step) symmetric explicit midpoint rule

$$y_{n+2} = y_n + 2hf_{n+1}. \tag{1.7}$$

For all methods we take $y_1 = y_0 + hf_0$ as the approximation for $y(t_0 + h)$. The results of the first 108 steps are shown in Fig. 1.1. We observe that the first two methods, as expected, behave similarly as the explicit and implicit Euler method (the numerical solution spirals either outwards or inwards). This will be rigorously

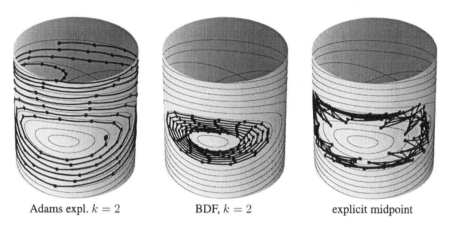

Adams expl. $k=2$     BDF, $k=2$     explicit midpoint

**Fig. 1.1.** Solutions of the pendulum problem (I.1.13); explicit Adams with step size $h = 0.5$, initial value $(p_0, q_0) = (0, 0.7)$; BDF with step size $h = 0.5$, initial value $(p_0, q_0) = (0, 0.95)$; explicit midpoint rule with $h = 0.4$ and initial value $(p_0, q_0) = (1.1, 0)$.

explained in Sect. XIV.2.1 below. However, as might not be expected, the symmetric method (1.7) does not behave like the implicit midpoint rule (cf. Fig. I.1.3), it shows undesired increasing oscillations (parasitic solutions).

After this negative experience with classical multistep methods, the obvious question is: are there multistep methods which have a long-time behaviour that is comparable to symplectic and/or symmetric one-step methods?

## XIV.1.2 Multistep Methods for Second Order Equations

Many important examples are systems of second order differential equations

$$\ddot{y} = f(y), \tag{1.8}$$

where the force $f$ is independent of the velocity $\dot{y}$. Introducing the new variable $z = \dot{y}$, we obtain the system $\dot{y} = z, \dot{z} = f(y)$ of first order equations. If we apply a multistep method (1.1) with generating polynomials $\rho^*(\zeta) = \sum_{j=0}^{k^*} \alpha_j^*$ and $\sigma^*(\zeta) = \sum_{j=0}^{k^*} \beta_j^*$ to this system, we get

$$\sum_{j=0}^{k^*} \alpha_j^* y_{n+j} = h \sum_{j=0}^{k^*} \beta_j^* z_{n+j}, \qquad \sum_{j=0}^{k^*} \alpha_j^* z_{n+j} = h \sum_{j=0}^{k^*} \beta_j^* f(y_{n+j}).$$

An elimination of the $z$-variables then yields

$$\sum_{j=0}^{k} \alpha_j y_{n+j} = h^2 \sum_{j=0}^{k} \beta_j f(y_{n+j}), \tag{1.9}$$

where $k = 2k^*$, $\rho(\zeta) = \rho^*(\zeta)^2$ and $\sigma(\zeta) = \sigma^*(\zeta)^2$. We consider here methods (1.9) which do not necessarily originate from a multistep method for first order equations, and we denote the generating polynomials of the coefficients $\alpha_j$ and $\beta_j$ again by $\rho(\zeta)$ and $\sigma(\zeta)$. From the classical theory (see Sect. III.10 of Hairer, Nørsett & Wanner 1993) we recall the following definitions and results.

**Order.** A method (1.9) has order $p$ if its generating polynomials satisfy

$$\rho(e^h) - h^2\sigma(e^h) = \mathcal{O}(h^{p+2}) \qquad \text{for} \quad h \to 0. \tag{1.10}$$

**Stability.** Method (1.9) is stable if all zeros of the polynomial $\rho(\zeta)$ satisfy $|\zeta| \leq 1$, and those on the unit circle are at most double zeros. Observe that for methods originating from (1.1) all zeros are double. The method is called *strictly stable*, if all zeros are inside the unit circle with the exception of $\zeta = 1$.

**Convergence.** If a multistep method (1.9) is stable, of order $p \geq 1$ and if the starting values are accurate enough, the global error satisfies $y_n - y(t_0 + nh) = \mathcal{O}(h^p)$ on compact intervals $nh \leq T$.

**Symmetry.** If the coefficients of (1.9) satisfy

$$\alpha_{k-j} = \alpha_j, \qquad \beta_{k-j} = \beta_j \qquad \text{for all } j, \tag{1.11}$$

then the method is symmetric. Again, for every zero $\zeta$ of $\rho(\zeta)$ the value $\zeta^{-1}$ is also a zero. Hence, stable symmetric methods have all zeros of $\rho(\zeta)$ on the unit circle and they are at most of multiplicity two.

Symmetric multistep methods for second order differential equations were first considered systematically by Lambert & Watson (1976). They analyzed in detail the application to the linear test equation $\ddot{y} = -\omega^2 y$, and they showed that for $h\omega$ in a certain interval (the so-called interval of periodicity) the numerical solution remains close to a periodic orbit. For example, the Störmer/Verlet method $y_{n+1} - 2y_n + y_{n-1} = h^2 f_n$ satisfies this property for $0 < h\omega < 2$ (see Sect. I.4.2).

Lambert & Watson (1976) observed that symmetric methods (1.9), for which $\rho(\zeta)$ has no double zero on the unit circle other than the principal zero $\zeta = 1$, have a non-vanishing interval of periodicity (Exercise 2). Quinlan & Tremaine (1990) constructed such methods of high order, and they applied them with success to the long-time integration of planetary orbits. We shall discuss this behaviour in the more general context of partitioned multistep methods.

**Stabilized Version of (1.9).** Due to the double zeros (of modulus one) of the characteristic polynomial of the difference equation $\sum_j \alpha_j y_{n+j} = 0$, we have an undesired propagation of rounding errors (especially for long-time integrations). To overcome this difficulty, we split the characteristic polynomial $\rho(\zeta)$ into

$$\rho(\zeta) = \rho_A(\zeta) \cdot \rho_B(\zeta), \tag{1.12}$$

such that each polynomial

$$\rho_A(\zeta) = \sum_{j=0}^{k_A} \alpha_j^{(A)} \zeta^j, \qquad \rho_B(\zeta) = \sum_{j=0}^{k_B} \alpha_j^{(B)} \zeta^j$$

has only simple roots of modulus one. Introducing the new variable $hz_n := \sum_j \alpha_j^{(A)} y_{n+j}$, the recurrence relation (1.9) becomes equivalent to

$$\sum_{j=0}^{k_A} \alpha_j^{(A)} y_{n+j} = hz_n, \qquad \sum_{j=0}^{k_B} \alpha_j^{(B)} z_{n+j} = h \sum_{j=0}^{k} \beta_j f_{n+j}. \qquad (1.13)$$

This formula, which for the Störmer/Verlet scheme corresponds to the one-step formulation (I.3.6), is much better suited for an implementation. If the splitting is such that $\rho'_A(1) = 1$, the discretization (1.13) is consistent with the first order partitioned system $\dot{y} = z, \dot{z} = f(y)$.

## XIV.1.3 Partitioned Multistep Methods

Motivated by the stabilized version (1.13) of multistep methods for second order equations, let us consider general partitioned systems of differential equations

$$\dot{y} = f(y, z), \qquad \dot{z} = g(y, z), \qquad (1.14)$$

where, needless to say, $y$ and $z$ may be vectors. The idea is to apply different multistep methods to different components. We thus get

$$\sum_{j=0}^{k} \alpha_j^{(A)} y_{n+j} = h \sum_{j=0}^{k} \beta_j^{(A)} f_{n+j}, \qquad \sum_{j=0}^{k} \alpha_j^{(B)} z_{n+j} = h \sum_{j=0}^{k} \beta_j^{(B)} g_{n+j}, \qquad (1.15)$$

where $f_n = f(y_n, z_n)$ and $g_n = g(y_n, z_n)$. We can take the same $k$ for both methods without loss of generality, if we abandon the assumption $|\alpha_0| + |\beta_0| > 0$.

Such a method is of order $p$, if both methods are of order $p$. It is stable (strictly stable, symmetric, ...), if both methods are stable (strictly stable, symmetric, ...).

**Example 1.2.** For our next experiment we use the symmetric methods

$$\begin{aligned}(A): \quad & y_{n+3} - y_{n+2} + y_{n+1} - y_n = h(f_{n+2} + f_{n+1}) \\ (B): \quad & z_{n+3} - z_{n+1} = 2hg_{n+2}.\end{aligned} \qquad (1.16)$$

Both methods are of order 2, and their $\rho$-polynomials $\rho_A(\zeta) = (\zeta - 1)(\zeta^2 + 1)$ and $\rho_B(\zeta) = (\zeta - 1)(\zeta + 1)$ do not have common zeros with the exception of $\zeta = 1$.

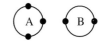

We choose the outer solar system with the data as described in Sect. I.2.3, and we apply the methods in three versions: (i) as partitioned method (AB), where the positions are treated by method (A) and the velocities by method (B); (ii) method (A) is applied to all components; (iii) method (B) is applied to all components. The numerical results are shown in Fig. 1.2. Whereas the individual methods show instabilities on rather short time intervals, the partitioned method gives a correct picture even with a large step size $h = 50$.

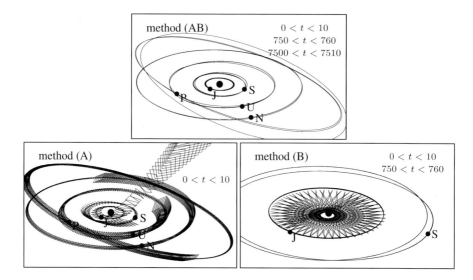

**Fig. 1.2.** Three versions of the methods (1.16) applied with step size $h = 50$ (days) to the outer solar system. For method (B) only the numerical orbits of Jupiter and Saturn are plotted. The time intervals are given in units of 10 000 days.

## XIV.1.4 Multi-Value or General Linear Methods

In contrast to one-step methods, multistep methods need the knowledge of $k > 1$ consecutive approximations $y_n, \ldots, y_{n+k-1}$ for the computation of $y_{n+k}$. With the notation $Y_n := (y_{n+k-1}, \ldots, y_n)^T$, a multistep method is thus a mapping $Y_n \mapsto Y_{n+1}$ on the space $\mathbb{R}^{kd}$ ($k$ copies of the phase space $\mathbb{R}^d$). In this way, method (1.1) can be written as

$$Y_{n+1} = DY_n + hBf(U_{n+1}) \qquad (1.17)$$

where $U_{n+1} = (y_{n+k}, \ldots, y_n)^T$, $f(U_{n+1}) = (f(y_{n+k}), \ldots, f(y_n))^T$, and

$$D = \begin{pmatrix} -\alpha_{k-1} & -\alpha_{k-2} & \cdots & -\alpha_0 \\ 1 & 0 & \cdots & 0 \\ & \ddots & \ddots & \vdots \\ & & 1 & 0 \end{pmatrix}, \quad B = \begin{pmatrix} \beta_k & \beta_{k-1} & \cdots & \beta_0 \\ 0 & 0 & \cdots & 0 \\ \vdots & \vdots & & \vdots \\ 0 & 0 & \cdots & 0 \end{pmatrix}$$

(assuming the normalization $\alpha_k = 1$). We use a sloppy notation in the sense that the matrices $D, B$ should be replaced with $D \otimes I, B \otimes I$. Observe that the eigenvalues of the matrix $D$ are exactly the zeros of the polynomial $\rho(\zeta)$. The formulation (1.17) is very convenient, and it allows us to get more insight into multistep methods.

Many results of this chapter can be extended without additional difficulties to a more general class of methods, so-called multi-value or general linear methods. They are defined by

$$Y_{n+1} = G_h(Y_n),$$

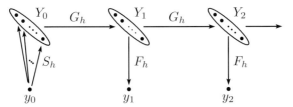

**Fig. 1.3.** Illustration of a multi-value method $Y_{n+1} = G_h(Y_n)$ with starting procedure $S_h$ and finishing procedure $F_h$.

where
$$Y_{n+1} = DY_n + hBf(U_{n+1}) \\ U_{n+1} = CY_n + hAf(U_{n+1}) \qquad (1.18)$$

with $f(U_{n+1}) = \bigl(f(u^1_{n+1}), \ldots, f(u^s_{n+1})\bigr)^T$ for $U_{n+1} = \bigl(u^1_{n+1}, \ldots, u^s_{n+1}\bigr)^T$, and $Y_n = (y^1_n, \ldots, y^k_n)$. For a computation, a starting procedure $S_h$ and a finishing procedure $F_h$, which extracts the numerical approximation $y_n$ from $Y_n$, have to be added (see Fig. 1.3). We assume the existence of a vector $e$ such that with $\mathbb{1} = (1, \ldots, 1)^T$

$$De = e, \qquad Ce = \mathbb{1} \qquad (1.19)$$

holds (preconsistency conditions). The vector $Y_n$ is then an approximation to $ey(t_n)$ (more precisely to $e \otimes y(t_n)$). In analogy to multistep methods, we call a method (1.18) *strictly stable*, if all eigenvalues of $D$ are inside the unit circle with the exception of the simple eigenvalue $\zeta = 1$. For a detailed treatment of general linear methods we refer the reader to Chap. 4 of the monograph of Butcher (1987), and to Chap. III.8 of Hairer, Nørsett & Wanner (1993).

## XIV.2 Related One-Step Methods

The main idea for studying the long-time behaviour of multistep or general linear methods is to relate them to one-step methods, and to apply the existing theory. In this section we show two ways of doing this. An extension to partitioned methods is straightforward.

### XIV.2.1 The Underlying One-Step Method

It was a surprising result when Kirchgraber (1986) proved that strictly stable multistep methods are essentially equivalent to one-step methods. Although this one-step method is "quite exotic", it is the key for a better understanding of the dynamics of strictly stable methods. An extension of Kirchgraber's result to strictly stable general linear methods is given by Stoffer (1993).

**Theorem 2.1.** *Consider a strictly stable general linear method* $Y_{n+1} = G_h(Y_n)$, *and a finishing procedure* $y_n = F_h(Y_n) = d^T Y_n + \mathcal{O}(h)$. *Assume that (1.19) and* $d^T e = 1$ *hold.*

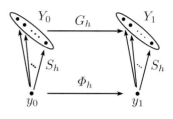

*(i) Then there exist a unique one-step method* $\Phi_h(y)$ *and a unique starting procedure* $S_h(y)$ *such that* $G_h \circ S_h = S_h \circ \Phi_h$ *and* $F_h \circ S_h = Id$ *hold.*

*(ii) The manifold* $\mathcal{M} = \{S_h(y)\,;\, y \in \mathbb{R}^d\}$ *is invariant under* $G_h$, *and it is exponentially attractive.*

*Proof.* Since the method is strictly stable, there exists a matrix $T$ such that

$$T^{-1} D T = \begin{pmatrix} 1 & 0 \\ 0 & D_0 \end{pmatrix} \quad \text{with} \quad \|D_0\| < 1,$$

and $Te_1 = e$ (where $e_1 = (1, 0, \ldots, 0)^T$). With the transformation

$$\begin{pmatrix} \xi_n \\ \eta_n \end{pmatrix} = Z_n = T^{-1} Y_n,$$

the general linear method (1.18) becomes

$$\begin{pmatrix} \xi_{n+1} \\ \eta_{n+1} \end{pmatrix} = \begin{pmatrix} \xi_n \\ D_0 \eta_n \end{pmatrix} + h T^{-1} B f(U_{n+1}). \tag{2.1}$$

with $U_{n+1} = CTZ_n + hAf(U_{n+1})$. The mapping $(\xi_n, \eta_n) \mapsto (\xi_{n+1}, \eta_{n+1})$ satisfies the assumptions of Theorem XII.3.1 with $L_{xx} = \mathcal{O}(h)$, $L_{xy} = \mathcal{O}(h)$, $L_{yx} = \mathcal{O}(h)$, and $L_{yy} = \|D_0\| < 1$. For sufficiently small $h$ this theorem yields the existence of an attractive manifold

$$\mathcal{M}_0 = \{(\xi, s(\xi))\,;\, \xi \in \mathbb{R}^d\},$$

which is invariant under the mapping (2.1). The function $s(\xi)$ is Lipschitz continuous with constant $\lambda = \mathcal{O}(h)$. The main idea is now to invert the restriction of $F_h$ onto the manifold $\mathcal{M}_0$. Indeed, due to $d^T e = 1$ and $Te_1 = e$, we have for $Z = Z(\zeta) = (\zeta, s(\zeta))^T \in \mathcal{M}_0$ that

$$y = F_h(TZ(\zeta)) = d^T TZ(\zeta) + \ldots = \zeta + g(\zeta), \tag{2.2}$$

where $g(\zeta)$ is Lipschitz continuous with constant $\mathcal{O}(h)$. By the Banach fixed-point theorem the equation (2.2) has a unique solution $\zeta = r(y)$. Putting

$$S_h(y) = TZ(r(y)) = T \begin{pmatrix} r(y) \\ s(r(y)) \end{pmatrix},$$

we have found the unique starting procedure satisfying $F_h \circ S_h = Id$ and $T^{-1} S_h(y) \in \mathcal{M}_0$. We finally define $\Phi_h = F_h \circ G_h \circ S_h$ and $\mathcal{M} = \{TZ\,;\, Z \in \mathcal{M}_0\}$, so that all statements of the theorem are verified. □

For strictly stable linear multistep methods all assumptions of the preceding theorem are satisfied (we have $e = \mathbb{1}$ and $d = e_k$). Due to the special structure of the matrix $D$, the starting procedure of Theorem 2.1 takes the form

$$S_h(y) = \left(\Phi_h^{k-1}(y), \ldots, \Phi_h(y), y\right)^T. \tag{2.3}$$

This means that taking starting values $y_j = \Phi_h^j(y_0)$ for $j = 0, \ldots, k-1$, the multistep method and the underlying one-step method yield exactly the same numerical approximations. For general starting values, the numerical solution of the multistep method tends exponentially fast to a solution of the underlying one-step method (due to the attractivity of the manifold $\mathcal{M}$).

**Example 2.2.** For a scalar linear problem $\dot{y} = \lambda y$, the application of a multistep method yields a difference equation with characteristic polynomial $\rho(\zeta) - h\lambda\sigma(\zeta)$. Denoting its zeros by $\zeta_1(h\lambda), \ldots, \zeta_k(h\lambda)$, where $\zeta_1(0) = 1$ and $|\zeta_j(0)| < 1$ for $j \geq 2$, the numerical solution can be written as

$$y_n = c_1 \zeta_1^n(h\lambda) + c_2 \zeta_2^n(h\lambda) + \ldots + c_k \zeta_k^n(h\lambda).$$

The coefficients $c_1, \ldots, c_k$ depend on $h\lambda$ and are determined by the starting approximations $y_0, \ldots, y_{k-1}$. In this situation the underlying one-step method is the mapping $y_0 \mapsto \zeta_1(h\lambda) y_0$. Observe that $\zeta_1(z)$ is in general not a rational function as we are used to with Runge-Kutta methods.

In the context of "geometric numerical integration" we are mainly interested in symplectic and/or symmetric methods which, for linear problems, are characterized by the condition $\zeta_1(-z)\zeta_1(z) \equiv 1$ (see Sect. VI.4.2). This, however, is only possible for symmetric multistep methods (Exercise 1).

## XIV.2.2 Formal Analysis for Weakly Stable Methods

Unfortunately, strictly stable multistep methods cannot be symmetric. It is therefore our aim to extend the concept of an underlying one-step method to nearly all (including weakly stable) general linear methods.

**Theorem 2.3.** *Consider a general linear method (1.18), and assume that $\zeta = 1$ is a single eigenvalue of the propagation matrix $D$. Furthermore, let $G_h(Y)$ and $F_h(Y) = d^T Y + \ldots$ have expansions in powers of $h$, and assume that (1.19) and $d^T e = 1$ hold. Then there exist a unique formal one-step method*

$$\Phi_h(y) = y + h d_1(y) + h^2 d_2(y) + \ldots$$

*and a unique formal starting procedure*

$$S_h(y) = ey + h S_1(y) + h^2 S_2(y) + \ldots,$$

*such that formally $G_h \circ S_h = S_h \circ \Phi_h$ and $F_h \circ S_h = Id$ hold.*

The formal series for $\Phi_h(y)$ is called "step-transition operator" in the Chinese literature (see e.g., Feng (1995), page 274).

*Proof.* Expanding $S_h(\Phi_h(y))$ and $G_h(S_h(y))$ into powers of $h$, a comparison of the coefficients yields

$$ed_j(y) + (I - D)S_j(y) = \ldots, \qquad (2.4)$$

where a right-hand side depends on known functions and on $d_i(y)$, $S_i(y)$ with $i < j$. Similarly, the condition $F_h(S_y(y)) = y$ leads to

$$d^T S_j(y) = \ldots . \qquad (2.5)$$

Due to the fact that $\zeta = 1$ is a single eigenvalue of $D$, and that $d^T e \neq 0$, the system (2.4)-(2.5) uniquely determines $d_j(y)$ and $S_j(y)$. □

Similar to Theorem 2.1 we have a "formal invariant manifold", which is given by $\{S_h(y) \, ; \, y \in \mathbb{R}^d\}$. However, it cannot be used for rigorous statements and it cannot be expected to be "stable" in general.

**Example 2.4.** Consider a consistent two-step method

$$\alpha_2 y_{n+2} + \alpha_1 y_{n+1} + \alpha_0 y_n = h(\beta_2 f_{n+2} + \beta_1 f_{n+1} + \beta_0 f_n),$$

and apply it to the simple system $\dot{y} = f(x)$, $\dot{x} = 1$. The $y$-component of the underlying one-step method then takes the form

$$\Phi_h(x_0, y_0) = y_0 + \sum_{j \geq 1} h^j \, a_j \, f^{(j-1)}(x_0). \qquad (2.6)$$

Putting $f(x) = e^x$ yields

$$A(\zeta) = \sum_{j \geq 1} a_j \, \zeta^{j-1} = \frac{\beta_2 e^{2\zeta} + \beta_1 e^{\zeta} + \beta_0}{\alpha_2(1 + e^{\zeta}) + \alpha_1}.$$

for the generating function of the coefficients $a_j$. Since this function has finite poles, the radius of convergence of $A(\zeta)$ is finite. Therefore, the radius of convergence of the series (2.6) has to be zero as soon as $f^{(j)}(x_0)$ behaves like $j! \, \mu \, \kappa^j$ (this is typically the case for analytic functions). Independent of the fact whether the method is strictly stable or not, the series (2.6) usually does not converge.

## XIV.2.3 Backward Error Analysis for Multistep Methods

The formal backward error analysis of Chap. IX could be directly applied to the underlying one-step method of Sect. XIV.2.2. However, due to the non-convergence of the series for $\Phi_h(y)$, difficulties arise as soon as we want to get rigorous estimates.

For ease of presentation, we consider only linear multistep methods in this subsection. We avoid the use of the underlying one-step method, and we directly search for a modified differential equation

$$\dot{\widetilde{y}} = f(\widetilde{y}) + h f_2(\widetilde{y}) + h^2 f_3(\widetilde{y}) + \ldots, \qquad \widetilde{y}(0) = y_0, \qquad (2.7)$$

such that formally the values $\widetilde{y}(jh)$ satisfy the multistep formula. Using the Lie derivative $(D_i g)(y) = g'(y) f_i(y)$ (with $f_1(y) = f(y)$) and the abbreviation $D_h = D_1 + h D_2 + h^2 D_3 + \ldots$, the solution of (2.7) with initial value $\widetilde{y}(t) = y$ satisfies

$$\widetilde{y}(t + lh) = y + \sum_{i \geq 1} \frac{(lh)^i}{i!} D_h^{i-1} F(y), \quad f\bigl(\widetilde{y}(t + lh)\bigr) = \sum_{i \geq 1} \frac{(lh)^{i-1}}{(i-1)!} D_h^{i-1} f(y),$$

where $F(y)$ denotes the right-hand side of (2.7).

**Lemma 2.5.** *For the multistep method (1.1), satisfying the normalization $\rho'(1) = 1$, the coefficient functions of the modified equation are (for $j \geq 2$) given recursively by*

$$\begin{aligned}
f_j(y) &= -\sum_{l=0}^{k} \alpha_l \sum_{i=2}^{j} \frac{l^i}{i!} \sum_{k_1 + \ldots + k_i = j} \bigl(D_{k_1} \ldots D_{k_{i-1}} f_{k_i}\bigr)(y) \\
&\quad + \sum_{l=0}^{k} \beta_l \sum_{i=1}^{j} \frac{l^{i-1}}{(i-1)!} \sum_{k_1 + \ldots + k_{i-1} + 1 = j} \bigl(D_{k_1} \ldots D_{k_{i-1}} f_1\bigr)(y).
\end{aligned}$$

*If $f(y)$ is analytic and bounded by $M$ in $B_R(y_0)$, then we have*

$$\|f_j(y)\| \leq \mu M \left(\frac{\eta M j}{R}\right)^{j-1} \quad \text{for} \quad \|y - y_0\| \leq R/2, \qquad (2.8)$$

*where $\mu$ and $\eta$ depend only on the coefficients $\alpha_j, \beta_j$ of the multistep method.*

*Proof.* Inserting the above formulas for $\widetilde{y}(t + lh)$ and $f\bigl(\widetilde{y}(t + lh)\bigr)$ into

$$\sum_{l=0}^{k} \alpha_l \widetilde{y}(t + lh) = h \sum_{l=0}^{k} \beta_l f\bigl(\widetilde{y}(t + lh)\bigr)$$

and comparing like powers of $h$ yields the recurrence relation for $f_j(y)$. The computations are closely related to those for the proof of Lemma IX.7.3.

The estimate (2.8) is obtained as in the proof of Theorem IX.7.5. We just sketch the main idea in the notation used there. With $\delta = R/(2(J-1))$ we have $\|f_j\|_j \leq \delta b_j$, where the generating function $b(\zeta) = \sum_{j \geq 1} b_j \zeta^j$ of the $b_j$ satisfies

$$b(\zeta) = \sum_{l=0}^{k} |\alpha_l| \bigl(e^{lb(\zeta)} - 1 - lb(\zeta)\bigr) + \frac{M\zeta}{\delta} \sum_{l=0}^{k} |\beta_l| \bigl(e^{lb(\zeta)} - 1\bigr).$$

By the implicit function theorem, $b(\zeta)$ is analytic and bounded in a disc of radius $c\delta/M$ centred at the origin ($c$ is a possibly small but positive constant depending only on the coefficients of the multistep method). □

If the multistep method is of order $p$, the order conditions imply that $f_j(y) = 0$ for $2 \leq j \leq p$ and $f_{p+1}(y) = C(D_1^p f)(y)$, where $C$ is the error constant of the method. As for one-step methods, this is nothing else than the local error of the method (see Sect. III.2 of Hairer, Nørsett & Wanner 1993).

We are now able to formulate the analogue of Theorem IX.7.6 which, for one-step methods, is the main ingredient for exponentially small error estimates. We truncate the modified equation, and we consider

$$\dot{\widetilde{y}} = F_N(\widetilde{y}), \qquad F_N(\widetilde{y}) = f(\widetilde{y}) + hf_2(\widetilde{y}) + \ldots + h^{N-1}f_N(\widetilde{y}) \qquad (2.9)$$

with $\widetilde{y}(0) = y_0$. Since the functions $f_j(y)$ satisfy here the same estimates as in Sect. IX.7.3, we can choose the truncation index $N$ in the same way.

**Theorem 2.6 (Hairer 1999).** *Let $f(y)$ be analytic and bounded by $M$ in $B_R(y_0)$. Consider a linear multistep method (1.1). If $h \leq h_0/4$ with $h_0 = R/(e\eta M)$, then there exists $N = N(h)$ (namely $N$ equal to the largest integer satisfying $hN \leq h_0$) such that flow $\widetilde{\varphi}_{N,t}(y_0)$ of the truncated modified equation (2.9) satisfies*

$$\left\| \sum_{j=0}^{k} \alpha_j \widetilde{\varphi}_{N,jh}(y_0) - h \sum_{j=0}^{k} \beta_j f\big(\widetilde{\varphi}_{N,jh}(y_0)\big) \right\| \leq h\gamma M e^{-h_0/h},$$

*where $\gamma$ depends only on the multistep formula.*

*Proof.* The proof closely follows that of Theorem IX.7.6, and we only point out the main differences. We consider the analytic function

$$g(h) := \sum_{j=0}^{k} \alpha_j \widetilde{\varphi}_{N,jh}(y_0) - h \sum_{j=0}^{k} \beta_j f\big(\widetilde{\varphi}_{N,jh}(y_0)\big) \qquad (2.10)$$

in a complex neighbourhood of $h = 0$. Its Taylor series coefficients vanish up to the $h^N$ term, so that the maximum principle for analytic functions can be applied to $g(h)/h^{N+1}$. To obtain an upper bound for $\|g(z)\|$ we estimate separately $\|\widetilde{\varphi}_{N,jh}(y_0) - y_0\|$ and $f\big(\widetilde{\varphi}_{N,jh}(y_0)\big)$. The rest of the proof is identical to that of Theorem IX.7.6. □

The preceding theorem shows that the exact solution of the truncated modified equation (2.9) satisfies the multistep formula up to exponentially small terms. This is a purely *local* result. In order to get information for the global error of the method, one has to study the propagation of perturbations.

For strictly stable methods, Theorem 2.1 shows the existence of an attractive invariant manifold, where the method is equivalent to a one-step method. With the techniques of this section it is possible to prove that this one-step method is exponentially close to the exact flow of the truncated modified equation. Therefore, all statements for one-step methods can be reinterpreted into statements for strictly stable multistep methods. Our conclusion is that strictly stable multistep methods behave as one-step methods (Kirchgraber 1986), and that the behaviour of the one-step method is completely described by the truncated modified equation.

For symmetric methods the situation is not so simple, since, in general, we do not have such an attractive invariant manifold. We are therefore obliged to study the propagation of errors in the higher dimensional space, and this is much more complicated. Some information will be obtained in the next subsection.

## XIV.2.4 Dynamics of Weakly Stable Methods

We consider general linear methods (1.18) that are weakly stable. This means that besides $\zeta = 1$, the matrix $D$ has further eigenvalues of modulus one. The most simple situation is where all eigenvalues of $D$ are roots of unity.

**Example 2.7.** The linear 3-step method

$$y_{n+3} - y_{n+2} + y_{n+1} - y_n = h(f_{n+2} + f_{n+1}) \qquad (2.11)$$

has characteristic polynomial $\rho(\zeta) = (\zeta - 1)(\zeta^2 + 1)$, and we apply it to the pendulum equation (I.1.13). For a better illustration of the propagation of errors we consider starting approximations $y_1, y_2$ that are rather far from the exact solution passing through $y_0$. The result is shown in Fig. 2.1.

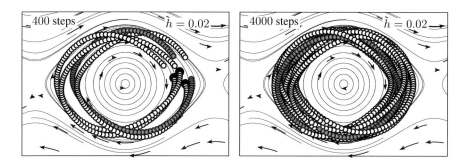

**Fig. 2.1.** Numerical solution of (2.11) applied to the pendulum equation. The initial approximations $y_0 = (1.9, 0.4)$, $y_1 = (1.7, 0.2)$, $y_2 = (2.1, 0)$ are indicated by black bullets; the solution points $y_3, y_7, y_{11}, \ldots$ in grey.

**Explanation of the Strange Behaviour in Fig. 2.1.** Considered as a multi-value method $(y_{n+2}, y_{n+1}, y_n) \mapsto (y_{n+3}, y_{n+2}, y_{n+1})$, method (2.11) has a propagation matrix $D$ with simple eigenvalues $1$, $+i$, and $-i$. They are all 4-th roots of unity, so that $D^4 = I$.

Assume that $D^m = I$ for some integer $m > 1$. The main idea is to consider the composition of $m$ consecutive steps (with step size $h/m$) as a new method. This is again a general linear method

$$\begin{aligned} Y_{n+1} &= Y_n + h\mathcal{B}f(U_{n+1}) \\ U_{n+1} &= \mathcal{C}Y_n + h\mathcal{A}f(U_{n+1}) \end{aligned} \qquad (2.12)$$

with propagation matrix $\mathcal{D} = I$, and

$$\mathcal{B} = \frac{1}{m}(D^{m-1}B, D^{m-2}B, \ldots, DB, B),$$

$$\mathcal{C} = \begin{pmatrix} C \\ CD \\ \vdots \\ CD^{m-1} \end{pmatrix}, \quad \mathcal{A} = \frac{1}{m}\begin{pmatrix} A & & & \\ CB & A & & \\ \vdots & & \ddots & \\ CD^{m-2}B & CD^{m-3}B & \cdots & A \end{pmatrix}.$$

For the method of Example 2.7 we add formula (2.11) to the same formula with $n$ augmented by one. This yields
$$y_{n+4} - y_n = h(f_{n+3} + 2f_{n+2} + f_{n+1}). \tag{2.13}$$
Replacing $h$ with $h/4$ (because we consider 4 steps of (2.11) as one step of the new method) we obtain
$$\begin{pmatrix} y_{n+4} \\ y_{n+5} \\ y_{n+6} \end{pmatrix} = \begin{pmatrix} y_n \\ y_{n+1} \\ y_{n+2} \end{pmatrix} + \frac{h}{4} \begin{pmatrix} f_{n+3} + 2f_{n+2} + f_{n+1} \\ f_{n+4} + 2f_{n+3} + f_{n+2} \\ f_{n+5} + 2f_{n+4} + f_{n+3} \end{pmatrix}, \tag{2.14}$$
where $y_{n+3}$, given by (2.11), has to be added as an internal stage. This equation is of the form (2.12).

Due to the fact that the propagation matrix is the identity, method (2.12) is a *one-step method*, and all results known for one-step methods can be applied.

**Theorem 2.8.** *Let the general linear method (1.18) satisfy (1.19) and $D^m = I$. Then, the method (2.12) is consistent with the "augmented differential equation"*
$$\dot Y = \mathcal{B} f(\mathcal{C} Y). \tag{2.15}$$
*Here $Y = (Y_1, \ldots, Y_k)^T$ with $Y_j \in \mathbb{R}^d$, and $f(\mathcal{C} Y)$ means that the function $f$ acts on the components of $\mathcal{C} Y$ individually.*

*Proof.* We insert $U_{n+1} = \mathcal{C} Y_n + \mathcal{O}(h)$ into the first formula of (2.12), and we consider the limit $h \to 0$ in the expression $(Y_{n+1} - Y_n)/h$. $\square$

Consider again the method of Example 2.7. Since $f(y_{n+j}) = f(y_{n+j-4} + \mathcal{O}(h))$ for $j \geq 4$, method (2.14) is consistent with the differential equation
$$\begin{pmatrix} \dot Y_1 \\ \dot Y_2 \\ \dot Y_3 \end{pmatrix} = \frac{1}{4} \begin{pmatrix} f(Y_4) + 2f(Y_3) + f(Y_2) \\ f(Y_1) + 2f(Y_4) + f(Y_3) \\ f(Y_2) + 2f(Y_1) + f(Y_4) \end{pmatrix}, \quad Y_4 = Y_3 - Y_2 + Y_1. \tag{2.16}$$
We plot in Fig. 2.2 the exact solution of (2.16) with initial values $Y_1(0) = y_0$, $Y_2(0) = y_1$, $Y_3(0) = y_2$ where the values $y_0, y_1, y_2$ are those from the caption of Fig. 2.1. The three components $Y_1(t), Y_2(t), Y_3(t)$ and $Y_4(t) = Y_3(t) - Y_2(t) + Y_1(t)$ are in excellent agreement with the numerical solution of Fig. 2.1. This indicates that the study of the augmented differential equation (2.15) gives much insight into the numerical solution obtained by a weakly stable method.

## XIV.2.5 Invariant Manifold of the Augmented System

We consider general linear methods satisfying
$$De = e, \quad \mathcal{C} e = 1\!\!1, \quad \mathcal{B} 1\!\!1 = e \tag{2.17}$$
for some vector $e$. The first two relations are the preconsistency conditions (1.19), and the third one implies that the underlying one-step method of Theorem 2.3 is consistent with $\dot y = f(y)$.

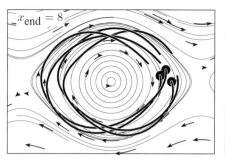 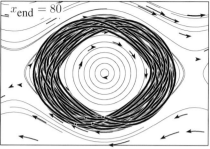

**Fig. 2.2.** Exact solution of (2.16) with $f(y)$ corresponding to the pendulum equation. The initial values are $Y_1(0) = (1.9, 0.4)$, $Y_2(0) = (1.7, 0.2)$, $Y_3(0) = (2.1, 0)$ (black bullets).

**Lemma 2.9.** *Let the general linear method (1.18) satisfy (2.17). Then, for every solution $y(t)$ of $\dot{y} = f(y)$, the function $Y(t) = ey(t)$ is a solution of (2.15). This means that $\mathcal{M} = \{ey\,;\, y \in \mathbb{R}^d\}$ is an invariant manifold of the flow of (2.15).*

*Proof.* By assumption (2.17) we have $\mathcal{C}e = \mathbb{1} \otimes \mathbb{1}$ and $\mathcal{B}(\mathbb{1} \otimes \mathbb{1}) = e$. The statement now follows from $f\big((\mathbb{1} \otimes \mathbb{1})y\big) = (\mathbb{1} \otimes \mathbb{1})f(y)$. □

In the preceding lemma, we have $e \in \mathbb{R}^k$ and $y \in \mathbb{R}^d$, and the product $ey$ is actually a tensor product if $d > 1$. We use the tensor product notation only at places where it is necessary also for $d = 1$.

If $y(t)$ is a stable solution of $\dot{y} = f(y)$, what can be said about the stability of the solution $Y(t) = ey(t)$ of (2.15)? Since we are interested in long-term integrations, and since we have seen that the numerical solution of a weakly stable method closely follows the solution of (2.15), this is an important question. For the explicit midpoint rule $y_{n+1} - y_{n-1} = 2hf_n$, Sanz-Serna (1985) presents a detailed study of properties of the augmented system.

In the rest of this chapter we mainly study the following questions:

- Can multistep and general linear methods be *symplectic*? (Sect. XIV.3).
- What is a suitable definition of *symmetric* general linear methods? (Sect. XIV.4).
- What is known about the *stability* of the (formal) invariant manifold of a general linear method; what about that of the augmented differential equation (2.15)? (Sect. XIV.5).

It is known from the study of one-step methods that, for a good long-time behaviour, the method should be either symplectic or symmetric or both. However, in the case of multi-value methods we shall see that symplecticity or symmetry alone are not sufficient. Without assuring the stability of the invariant manifold, the methods are useless.

## XIV.3 Can Multistep Methods be Symplectic?

Readers might be astonished to find a question mark in the title. The reason is that we shall present two definitions of symplecticity of multistep methods applied to a Hamiltonian system

$$\dot{p} = -H_q(p,q), \qquad \dot{q} = H_p(p,q). \tag{3.1}$$

One works in the phase space of the exact flow, the other in a higher dimensional space. But which one is suitable for symplectic multistep or general linear methods?

### XIV.3.1 Non-Symplecticity of the Underlying One-Step Method

A conjecture due to Feng Kang.     (Y.-F. Tang 1993)

A natural definition of symplecticity consists of the requirement that the underlying one-step method (Theorem 2.3) be symplectic. This means that the (truncated) modified equation (2.9) is Hamiltonian. Unfortunately, we have the following negative result.

**Theorem 3.1 (Tang 1993).** *The underlying one-step method of an irreducible linear multistep method cannot be symplectic.*

*Proof.* We show that the first perturbation term in the modified equation (2.9) is in general not Hamiltonian. From Lemma 2.5 we know that $f_{p+1}(y) = C(D_1^p f)(y)$ which (omitting the non-zero constant $C$) is given by

$$\sum_{\tau \in T, |\tau|=p+1} a(\tau) F(\tau)(y) = |\tau|! \sum_{\tau \in T, |\tau|=p+1} \frac{1}{\sigma(\tau)} b(\tau) F(\tau)(y) \tag{3.2}$$

with $b(\tau) = 1/\gamma(\tau)$ for $|\tau| = p+1$ (Theorem III.1.3 and (III.1.27)). Suppose now that (3.2) is Hamiltonian for all separable Hamiltonian vector fields $f(y) = J^{-1}\nabla H(y)$. Theorem IX.10.1 then implies

$$b(u \circ v) + b(v \circ u) = 0 \qquad \text{for all } u, v \in T \text{ with } |u| + |v| = p+1.$$

This, however, is in contradiction with

$$\frac{1}{\gamma(u \circ v)} + \frac{1}{\gamma(v \circ u)} = \frac{1}{\gamma(u)} \cdot \frac{1}{\gamma(v)},$$

which is a consequence of Corollary VI.7.2 (because the exact solution is a symplectic transformation and, as a B-series, has coefficients $a(\tau) = 1/\gamma(\tau)$). □

This negative result extends to all general linear methods (1.18). It is proved by Hairer & Leone (1998) that, among the class of one-leg methods, only the implicit mid-point rule satisfies this requirement. For general linear methods, some necessary conditions for the symplecticity of the underlying one-step method are known which are hard to satisfy. For the moment, no general linear method (not equivalent to a one-step method) is known for which the underlying one-step method is symplectic, and we conjecture that such a method does not exist, even in the class of partitioned general linear methods (treating the $p$ and $q$ variables by different methods).

## XIV.3.2 Symplecticity in the Higher-Dimensional Phase Space

We present here a second approach for the definition of symplecticity of multi-value methods. It is much inspired by the $G$-stability theory of Dahlquist (1975) for the study of stiff differential equations.

**Definition 3.2.** Let $G$ be an invertible symmetric matrix. A general linear method $Y_{n+1} = G_h(Y_n)$ given by (1.18) is called *G-symplectic* if

$$Y_{n+1}^T (G \otimes S) Y_{n+1} = Y_n^T (G \otimes S) Y_n, \qquad (3.3)$$

whenever the differential equation $\dot{y} = f(y)$ has $y^T S y$ as invariant (with symmetric $S$), i.e., the vector field satisfies $y^T S f(y) = 0$ for all $y$.

It is of course also possible to express this definition in terms of differential forms. However, as a consequence of Lemma VI.4.1 the conservation of quadratic first integrals is equivalent to symplecticity (Bochev & Scovel 1994).

In contrast to the negative results of Sect. XIV.3.1, there exist a lot of $G$-symplectic general linear methods. The most prominent examples are *one-leg methods*. They are defined by the relation

$$\sum_{j=0}^{k} \alpha_j y_{n+j} = hf\left(\sum_{j=0}^{k} \beta_j y_{n+j}\right), \qquad (3.4)$$

where the normalization $\sigma(1) = \sum_j \beta_j = 1$ is assumed. These methods have been introduced by Dahlquist (1975) in order to simplify the study of nonlinear stability of linear multistep methods. In fact, there is a close relationship between the numerical solution of (3.4) and (1.1), and their long-time behaviour is the same.

**Theorem 3.3 (Eirola & Sanz-Serna 1992).** *Every irreducible symmetric one-leg method (3.4) is G-symplectic for some matrix G.*

*Proof.* We recall that a one-leg method is irreducible if the generating polynomials $\rho(\zeta)$ and $\sigma(\zeta)$ have no common zeros.

*Construction of G.* The symmetry relation (1.4) implies $\rho(1/\zeta) = -\zeta^{-k}\rho(\zeta)$ and $\sigma(1/\zeta) = \zeta^{-k}\sigma(\zeta)$. Consequently, the polynomial $\rho(\zeta)\sigma(\omega) + \rho(\omega)\sigma(\zeta)$ vanishes for $\omega = 1/\zeta$, and contains the factor $\zeta\omega - 1$. We then define $G$ by

$$\frac{1}{2}\big(\rho(\zeta)\sigma(\omega) + \rho(\omega)\sigma(\zeta)\big) = (\zeta\omega - 1)\sum_{i,j=1}^{k} g_{ij}\zeta^{i-1}\omega^{j-1}. \qquad (3.5)$$

The matrix $G$ obtained in this way is symmetric.

*Regularity of $G$.* Applying the geometric series we get

$$\sum_{i,j=1}^{k} g_{ij}\zeta^{i-1}\omega^{j-1} = -\frac{1}{2}\big(\rho(\zeta)\sigma(\omega) + \rho(\omega)\sigma(\zeta)\big)\big(1 + \zeta\omega + \zeta^2\omega^2 + \ldots\big),$$

where the identity holds as formal power series. Suppose that the matrix $G$ is not invertible. Then there exists a vector $u = (u_0, u_1, \ldots, u_{k-1})^T$ such that $Gu = 0$. We formally replace the appearances of $\omega^{j-1}$ with $u_{j-1}$ for $j \le k$ and with zero for $j > k$. This gives an identity of the form $0 = \rho(\zeta)a(\zeta) + \sigma(\zeta)b(\zeta)$ with polynomials $a(\zeta)$ and $b(\zeta)$ of degree at most $k-1$, and we get a contradiction with the irreducibility of the method.

*G-Symplecticity.* We next replace in (3.5) $\zeta^i\omega^j$ with $y_{n+i}^T S y_{n+j}$. Together with (3.4) this yields

$$h\Big(\sum_{i=0}^{k}\beta_i y_{n+i}\Big)^T Sf\Big(\sum_{i=0}^{k}\beta_i y_{n+i}\Big) = Y_{n+1}^T(G \otimes S)Y_{n+1} - Y_n^T(G \otimes S)Y_n,$$

where $Y_n = (y_n, \ldots, y_{n+k-1})^T$. This proves (3.3) for all functions $f(y)$ satisfying $y^T Sf(y) = 0$. □

**Example 3.4.** We consider the explicit midpoint rule (1.7), which is also a one-leg method, and the 3-step method (2.11), By Theorem 3.3 the one-leg versions are $G$-symplectic. Following the constructive proof of this theorem we find

$$G = \begin{pmatrix} 0 & 1 \\ 1 & 0 \end{pmatrix} \quad \text{and} \quad G = \begin{pmatrix} 0 & 1 & 1 \\ 1 & -2 & 1 \\ 1 & 1 & 0 \end{pmatrix},$$

respectively. We apply both methods to two closely related Hamiltonian systems, namely the pendulum equation with $H(p,q) = p^2/2 - \cos q$ and a perturbed problem with $H(p,q) = p^2/2 - \cos q(1 - p/6)$, and we study the preservation of the Hamiltonian (see Fig. 3.1). The result is somewhat surprising. The midpoint rule behaves well for the perturbed problem, but shows a linear error growth in the Hamiltonian for the pendulum problem. On the other side, the weakly stable 3-step method behaves well for the pendulum equation (which is in agreement with the stable behaviour of Fig. 2.1), but has an exponential error growth for the perturbed problem. Notice that different scales are used in the four pictures.

The above example illustrates that $G$-symplecticity of a multi-value method is not sufficient for a good long-time behaviour. We shall see in Sect. XIV.5 that the stability of an invariant manifold is essential. Here, we show that for every $G$-symplectic method the augmented system (2.15) is Hamiltonian with structure matrix $G \otimes J$. This has first been discovered by Sanz-Serna & Vadillo (1987) for the explicit mid-point rule.

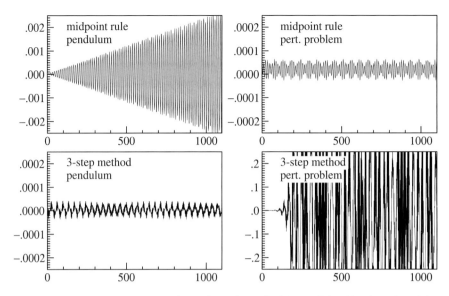

**Fig. 3.1.** Numerical Hamiltonian $H(p_n,q_n)-H(p_0,q_0)$ of the explicit mid-point rule and the 3-step method (2.11), applied with step size $h=0.01$ to the pendulum problem ($H(p,q)=p^2/2-\cos q$) and to a perturbed problem ($H(p,q)=p^2/2-\cos q(1-p/6)$) on the interval $[0,1100]$ (only every 131st step is drawn).

For the proof of such a result we need that the general linear method satisfies

$$D^T G D = G, \qquad D^T G B = C^T \Lambda \qquad (3.6)$$

for some diagonal matrix $\Lambda$. In the case of irreducible one-leg methods, where $B$ is a column vector and $C$ is a row vector, this can be seen as follows: we consider the harmonic oscillator $\dot{y}=J^{-1}y$ ($y\in\mathbb{R}^2$), we insert $Y_{n+1}=DY_n+hBf(U_{n+1})$ into the relation (3.3) with $S=I$, and we expand the expression into powers of $h$. The $h$-independent term gives $D^T G D = G$, and the $h$ term is of the form $y^T J^{-1} z = 0$ with $y=B^T G D Y_n$ and $z=CY_n$. Consequently, $y$ and $z$ are linearly dependent, so that $B^T G D = \lambda C$ for some $\lambda \in \mathbb{R}$. It seems that (3.6) is true for all irreducible $G$-symplectic methods. However, we don't have a rigorous proof for this.

**Lemma 3.5.** *Consider a general linear method (1.18) applied to a Hamiltonian system $\dot{y}=f(y)=J^{-1}\nabla H(y)$. If $D^m=I$ and (3.6) hold for an invertible symmetric matrix $G$ and a diagonal matrix $\Lambda$, then the augmented differential equation (2.15) satisfies*

$$\dot{Y}=(G\otimes J)^{-1}\nabla\mathcal{H}(Y) \quad\text{with}\quad \mathcal{H}(Y)=\frac{1}{m}\sum_{l=0}^{m-1}\sum_{i=1}^{s}\lambda_i H\Big(\sum_{j=1}^{k}c_{ij}^{(l)}Y_j\Big), \quad (3.7)$$

*where $Y=(Y_1,\ldots,Y_k)^T$, $\lambda_i$ are the diagonal entries of $\Lambda$, and $c_{ij}^{(l)}$ are the entries of $CD^l$. The system (3.7) is Hamiltonian with the structure matrix $G\otimes J$.*

*Proof.* We differentiate the function $\mathcal{H}(Y)$ of (3.7) and obtain

$$\nabla \mathcal{H}(Y) = \frac{1}{m} \sum_{l=0}^{m-1} (CD^l)^T \Lambda \nabla H(CD^l Y). \tag{3.8}$$

We next observe that (3.6) together with $D^m = I$ imply $(CD^l)^T \Lambda = GD^{m-l-1}B$. Multiplying (3.8) by $(G \otimes J)^{-1}$ proves the statement. □

We can also interpret the numerical solution $Y_n$ in the higher dimensional space in the sense of backward error analysis (Leone 2000).

**Theorem 3.6.** *Consider a G-symplectic general linear method satisfying $D^m = I$ and (3.6). Then, the numerical solution of the composed method (2.12) satisfies $Y_n = \widetilde{Y}(nh)$, where $\widetilde{Y}(t)$ is the (formal) exact solution of $\widetilde{Y}(0) = Y_0$ and*

$$\dot{\widetilde{Y}} = (G \otimes J)^{-1}\Big(\nabla \mathcal{H}(\widetilde{Y}) + h\nabla \mathcal{H}_1(\widetilde{Y}) + h^2 \nabla \mathcal{H}_2(\widetilde{Y}) + \ldots\Big).$$

*Here, $\mathcal{H}$ is the Hamiltonian of Lemma 3.5.*

*Proof.* Since (2.12) is just a one-step method for problem (2.15), all results of Chap. IX on backward error analysis and on the modified differential equation are valid here. The proof is thus the same as that for Theorem IX.3.1 with the difference that the structure matrix $J$ has to be replaced with $G \otimes J$. □

If the Hamiltonian $H(y)$ is defined globally on $\mathbb{R}^{2d}$, this result enables us to conclude that the numerical solution of a $G$-symplectic general linear method preserves the Hamiltonian $\mathcal{H}(Y)$ of (3.7) up to $\mathcal{O}(h)$ (but not necessarily $H(y)$) over exponentially long time intervals, as long as the numerical solution remains in a compact set.

# XIV.4 Symmetric Multi-Value Methods

After the disappointing non-existence result of symplectic multi-value methods (Sect. XIV.3.1), we turn our attention to symmetric methods. We know from the previous chapters that for reversible Hamiltonian systems, the long time behaviour of symmetric one-step methods is as good as that for symplectic methods.

## XIV.4.1 Definition of Symmetry

There are several definitions of symmetric general linear methods in the literature. However, they are either tailored to very special situations (e.g., Hairer, Nørsett & Wanner 1993), or they do not allow the proof of results that are expected to hold for symmetric methods. Before proposing a new definition, we establish the following Lemma.

## XIV.4 Symmetric Multi-Value Methods

**Lemma 4.1.** *For a general linear method* $Y_{n+1} = G_h(Y_n)$ *we consider two different finishing procedures* $y_n = F_h(Y_n)$ *and* $\widehat{y}_n = \widehat{F}_h(Y_n)$ :

$$
\begin{array}{ccccccc}
\widehat{y}_0 & \xrightarrow{\widehat{\Phi}_h} & \widehat{y}_1 & \xrightarrow{\widehat{\Phi}_h} & \widehat{y}_2 & \xrightarrow{\widehat{\Phi}_h} & \cdots \\
\widehat{S}_h \Big\downarrow \Big\uparrow \widehat{F}_h & & \Big\uparrow \widehat{F}_h & & \Big\uparrow \widehat{F}_h & & \\
Y_0 & \xrightarrow{G_h} & Y_1 & \xrightarrow{G_h} & Y_2 & \xrightarrow{G_h} & \cdots \\
S_h \Big\uparrow \Big\downarrow F_h & & \Big\downarrow F_h & & \Big\downarrow F_h & & \\
y_0 & \xrightarrow{\Phi_h} & y_1 & \xrightarrow{\Phi_h} & y_2 & \xrightarrow{\Phi_h} & \cdots
\end{array}
$$

*The two corresponding one-step methods* $\Phi_h(y)$ *and* $\widehat{\Phi}_h(y)$ *(given by Theorem 2.3) are then conjugate to each other, i.e.,*

$$\alpha_h^{-1} \circ \Phi_h \circ \alpha_h = \widehat{\Phi}_h \quad \text{with} \quad \alpha_h = F_h \circ \widehat{S}_h. \tag{4.1}$$

*Proof.* The equations involving the underlying one-step methods or the starting procedures have to be understood in the sense of formal series. By Theorem 2.3 we have $S_h(y) = ey + \mathcal{O}(h)$ and also $\widehat{S}_h(y) = ey + \mathcal{O}(h)$. It thus follows from $F_h \circ S_h = Id$ that $\alpha_h(y)$ is $\mathcal{O}(h)$-close to the identity and therefore invertible. □

The transformation $\alpha_h$ in the phase space is $\mathcal{O}(h)$-close to the identity. The relation $\alpha_h^{-1} \circ \Phi_h^n \circ \alpha_h = \widehat{\Phi}_h^n$, which is a consequence of (4.1), therefore implies that the numerical solutions of $\Phi_h$ and $\widehat{\Phi}_h$ remain $\mathcal{O}(h)$-close for all times. This means that the long-time behaviour of both methods is exactly the same.

Consequently, for a given general linear method $G_h$, it is sufficient to require symmetry for *one* finishing procedure only.

**Definition 4.2.** A general linear method $G_h$ is called *symmetric* if there exists a finishing procedure $F_h$ such that the underlying one-step method $\Phi_h$ of Theorem 2.3 is symmetric, i.e., $\Phi_{-h}(y) = \Phi_h^{-1}(y)$ in the sense of formal series.

**Example 4.3.** Consider the trapezoidal method in the role of $G_h$ and the explicit Euler method with step size $-\gamma h$ as finishing procedure:

$$G_h: \quad Y_{n+1} = Y_n + \frac{h}{2}\Big(f(Y_n) + f(Y_{n+1})\Big)$$
$$F_h: \quad y_{n+1} = Y_{n+1} - \gamma h f(Y_{n+1})$$

The corresponding starting procedure and underlying one-step methods are then the implicit Euler method and the following 2-stage Runge-Kutta method:

$$S_h: \quad Y_n = y_n + \gamma h f(Y_n)$$

$$\Phi_h: \quad \text{Runge-Kutta method} \quad
\begin{array}{c|cc}
\gamma & \gamma & \\
1+\gamma & 1/2+\gamma & 1/2 \\
\hline
 & 1/2+\gamma & 1/2-\gamma
\end{array}$$

The method $\Phi_h$ is symmetric only for $\gamma = 0$ or for $\gamma = 1/2$ or for $\gamma = -1/2$. This example demonstrates that the symmetry of the underlying one-step method strongly depends on the finishing procedure.

On the other hand, this example shows that the 2-stage Runge-Kutta method is symmetric in the sense of Definition 4.2 for all $\gamma$ (because it is conjugate to the trapezoidal rule). It is not symmetric according to the definition of Chap. V.

## XIV.4.2 A Useful Criterion for Symmetry

Definition 4.2 is rather impractical for verifying the symmetry of a given general linear method. We give here algebraic conditions for the coefficients $A, B, C, D$ of a general linear method (1.18), which are sufficient for the method to be symmetric. We assume that the finishing procedure $y_{n+1} = F_h(Y_{n+1})$ is given by

$$y_{n+1} = \widetilde{D} Y_{n+1} + h\widetilde{B} f(V_{n+1}), \qquad V_{n+1} = \widetilde{C} Y_{n+1} + h\widetilde{A} f(V_{n+1}), \qquad (4.2)$$

in complete analogy to method (1.18).

**Lemma 4.4 (Adjoint Method).** *Let $Y_{n+1} = G_h(Y_n)$ be the general linear method given by $A, B, C, D$ (with invertible $D$), $y_{n+1} = F_h(Y_{n+1})$ the finishing procedure given by $\widetilde{A}, \widetilde{B}, \widetilde{C}, \widetilde{D}$, and denote by $\Phi_h$ its underlying one-step method. Then, the underlying one-step method of*

$$G_h^* : \quad A^* = CD^{-1}B - A, \quad B^* = D^{-1}B, \quad C^* = CD^{-1}, \quad D^* = D^{-1}$$
$$F_h^* : \quad \widetilde{A}^* = -\widetilde{A}, \quad \widetilde{B}^* = -\widetilde{B}, \quad \widetilde{C}^* = \widetilde{C}, \quad \widetilde{D}^* = \widetilde{D}$$

*is the adjoint method $\Phi_h^* = \Phi_{-h}^{-1}$ of $\Phi_h$.*

*Proof.* Substituting $h \leftrightarrow -h$ and $Y_{n+1} \leftrightarrow Y_n$ in (1.18) yields

$$U_{n+1} = CY_{n+1} - hAf(U_{n+1}), \qquad Y_n = DY_{n+1} - hBf(U_{n+1}).$$

Extracting $Y_{n+1}$ from the second relation and inserting it into the first gives

$$U_{n+1} = CD^{-1}Y_n + h(CD^{-1}B - A)f(U_{n+1})$$
$$Y_{n+1} = D^{-1}Y_n + hD^{-1}Bf(U_{n+1}),$$

which is exactly method $G_h^*$. The same replacements in the finishing procedure

$$V_{n+1} = \widetilde{C} Y_n - h\widetilde{A} f(V_{n+1}), \qquad y_n = \widetilde{D} Y_n - h\widetilde{B} f(V_{n+1})$$

and in the diagram of Theorem 2.3 prove the statement. □

**Theorem 4.5.** *If there exist an invertible matrix $Q$ (satisfying $Qe = e$ with $e$ given by (1.19)) and a permutation matrix $P$ such that*

$$P^{-1}AP = CD^{-1}B - A, \qquad Q^{-1}BP = D^{-1}B,$$
$$P^{-1}CQ = CD^{-1}, \qquad Q^{-1}DQ = D^{-1}, \qquad (4.3)$$

*then the general linear method (1.18) is symmetric.*

*Proof.* We consider the change of variables $Y_n = Q\widehat{Y}_n$, $U_n = P\widehat{U}_n$ in the method (1.18). Since $P$ is a permutation matrix, we have $f(PU) = Pf(U)$, so that the method becomes

$$P\widehat{U}_{n+1} = CQ\widehat{Y}_n + hAPf(\widehat{U}_{n+1}), \qquad Q\widehat{Y}_{n+1} = DQ\widehat{Y}_n + hBPf(\widehat{U}_{n+1}).$$

The assumption (4.3) implies that this method is the same as the adjoint method of Lemma 4.4. Taking a finishing procedure $F_h$ in such a way that $y_{n+1} = F_h(Q\widehat{Y}_{n+1})$ is identical to the finishing procedure $y_{n+1} = F_h^*(\widehat{Y}_{n+1})$ of the adjoint method (i.e., $\widetilde{B} = 0$ and $\widetilde{D}$ such that $\widetilde{D}Q = \widetilde{D}$), we obtain $\Phi_h^* = \Phi_h$. This proves the statement. □

The sufficient condition of Theorem 4.5 reduces to the known criteria for classical methods. Let us give some examples:

- For Runge-Kutta methods we have $D = (1)$, $B = b^T$ a row vector, and $C = \mathbb{1}$. With $Q = (1)$ and $P$ the permutation matrix that inverts the elements of a vector, we get
$$b^T P = b^T, \qquad PAP = \mathbb{1}b^T - A,$$
which is the same (V.2.4).
- Multistep methods in their form as general linear methods (Sect. XIV.1.4) satisfy the condition of Theorem 4.5 if
$$\alpha_i = -\alpha_{k-i}, \qquad \beta_i = \beta_{k-i}. \tag{4.4}$$
One can take for $P$ and $Q$ the permutation matrices (inverting the elements of a vector) of dimension $k+1$ and $k$, respectively.
- One-leg methods satisfy (4.3) if the symmetry condition (4.4) holds. For $Q$ one can take the same permutation matrix as for multistep methods and $P = (1)$.

# XIV.5 Stability of the Invariant Manifold

We now introduce partitioned general linear methods. By studying the stability of the invariant manifold, we shall see why such methods have been so much better in the experiments of Sect. XIV.1.

## XIV.5.1 Partitioned General Linear Methods

For partitioned systems of differential equations

$$\dot{y} = f(y, z), \qquad \dot{z} = g(y, z), \tag{5.1}$$

we consider, in complete analogy to partitioned Runge-Kutta methods, two general linear methods and apply them to the $y$ and $z$ components, respectively. This gives

$$\begin{aligned} Y_{n+1} &= DY_n + hBf(U_{n+1}, V_{n+1}) \\ Z_{n+1} &= \widehat{D}Z_n + h\widehat{B}g(U_{n+1}, V_{n+1}) \\ U_{n+1} &= CY_n + hAf(U_{n+1}, V_{n+1}) \\ V_{n+1} &= \widehat{C}Z_n + h\widehat{A}g(U_{n+1}, V_{n+1}) \end{aligned} \qquad (5.2)$$

for a partitioned general linear method. Notice that $U_{n+1}$ and $V_{n+1}$ are super-vectors as in (1.18) and must consist of the same number of blocks, so that $f(U_{n+1}, V_{n+1})$ and $g(U_{n+1}, V_{n+1})$ make sense. However, $Y_n$ and $Z_n$ need not have the same dimension. For completing the method we have to add a starting and a finishing procedure, which can be defined by similar formulas. All results of the preceding section can be extended in a more or less straightforward fashion to partitioned methods.

A partitioned method is stable (strictly stable), if both methods are stable (strictly stable). The existence of an underlying one-step method (Theorem 2.1 and Theorem 2.3) and the backward error analysis (Theorem 2.6) can be extended without any difficulties.

Our main interest of this section are weakly stable methods. As in Sect. XIV.2.4, we restrict our discussion to methods with a propagation matrix that has only roots of unity as eigenvalues. We assume that there exist a positive integer $m$ such that $D^m = I$ and $\widehat{D}^m = I$. The composition of $m$ consecutive steps (with step size $h/m$) is a new method, given by

$$\begin{aligned} Y_{n+1} &= Y_n + h\mathcal{B}f(U_{n+1}, V_{n+1}) \\ Z_{n+1} &= Z_n + h\widehat{\mathcal{B}}g(U_{n+1}, V_{n+1}) \\ U_{n+1} &= \mathcal{C}Y_n + h\mathcal{A}f(U_{n+1}, V_{n+1}) \\ V_{n+1} &= \widehat{\mathcal{C}}Z_n + h\widehat{\mathcal{A}}g(U_{n+1}, V_{n+1}). \end{aligned} \qquad (5.3)$$

The matrices $\mathcal{A}, \mathcal{B}, \mathcal{C}$ are those of (2.12) and $\widehat{\mathcal{A}}, \widehat{\mathcal{B}}, \widehat{\mathcal{C}}$ are the corresponding matrices for the second method. If the preconsistency conditions (1.19) are satisfied for both methods, this new method is consistent with the "partitioned augmented differential equation"

$$\dot{Y} = \mathcal{B}f(\mathcal{C}Y, \widehat{\mathcal{C}}Z), \qquad \dot{Z} = \widehat{\mathcal{B}}g(\mathcal{C}Y, \widehat{\mathcal{C}}Z). \qquad (5.4)$$

**Example 5.1.** The partitioned multistep method (AB) of Example 1.2 has $\rho(\zeta) = (\zeta - 1)(\zeta^2 + 1)$ and $\widehat{\rho}(\zeta) = (\zeta - 1)(\zeta + 1)$ as generating polynomials. Hence, four consecutive steps are equivalent to a method (5.3). It is consistent with the system

$$\begin{pmatrix} \dot{Y}_1 \\ \dot{Y}_2 \\ \dot{Y}_3 \end{pmatrix} = \frac{1}{4} \begin{pmatrix} f(Y_2, Z_2) + 2f(Y_3, Z_1) + f(Y_4, Z_2) \\ f(Y_3, Z_1) + 2f(Y_4, Z_2) + f(Y_1, Z_1) \\ f(Y_4, Z_2) + 2f(Y_1, Z_1) + f(Y_2, Z_2) \end{pmatrix}$$

$$\begin{pmatrix} \dot{Z}_1 \\ \dot{Z}_2 \end{pmatrix} = \frac{1}{2} \begin{pmatrix} g(Y_2, Z_2) + g(Y_4, Z_2) \\ g(Y_3, Z_1) + g(Y_1, Z_1) \end{pmatrix}$$

where $Y_4 = Y_1 - Y_2 + Y_3$.

Also the concepts of symplecticity and symmetry can be extended to partitioned general linear methods. We refer to Exercise 6 for the definition and properties of $G$-symplectic methods, and discuss shortly the symmetry of (5.2). Definition 4.2 can be extended in a straightforward way to partitioned general linear methods. From considering uncoupled problems $\dot{y} = f(y)$, $\dot{z} = g(z)$ it can be seen that a necessary condition for the symmetry of (5.2) is that both general linear methods are symmetric. It is less clear whether this is also sufficient for the symmetry of (5.2). However, Lemma 4.4 and Theorem 4.5 are immediately extended to partitioned methods. This shows that if both methods satisfy the sufficient conditions of Theorem 4.5 with the same matrix $P$, then the partitioned method (5.2) is also symmetric (Exercise 8).

## XIV.5.2 The Linearized Augmented System

A first step for understanding the long-time behaviour of weakly stable multi-value methods is the study of the stability of the augmented differential equation (5.4).

As in Sect. XIV.2.5 we assume that

$$De = e, \qquad Ce = 1\!\!1, \qquad B1\!\!1 = e$$
$$\widehat{D}\widehat{e} = \widehat{e}, \qquad \widehat{C}\widehat{e} = 1\!\!1, \qquad \widehat{B}1\!\!1 = \widehat{e} \qquad (5.5)$$

for suitable vectors $e$ and $\widehat{e}$. This implies (see Lemma 2.9) that for every solution $y(t), z(t)$ of (5.1), the functions $Y(t) = e\, y(t)$, $Z(t) = \widehat{e}\, z(t)$ are a solution of (5.4). Since the starting value will be close to $ey_0, \widehat{e}z_0$, this is the interesting solution, whose stability has to be investigated.

We linearize (5.4) around the solution $Y(t) = e\, y(t)$, $Z(t) = \widehat{e}\, z(t)$, and consider the variational equation

$$\begin{aligned} d\dot{Y} &= \mathcal{BC}\, f_y(y,z)\, dY + \mathcal{B}\widehat{\mathcal{C}}\, f_z(y,z)\, dZ \\ d\dot{Z} &= \widehat{\mathcal{B}}\mathcal{C}\, g_y(y,z)\, dZ + \widehat{\mathcal{B}}\widehat{\mathcal{C}}\, g_z(y,z)\, dZ. \end{aligned} \qquad (5.6)$$

The special structure of this linear differential equation permits us to reduce it to problems of lower dimension.

**Theorem 5.2.** *Consider a stable partitioned general linear method (5.2). Assume (5.5), $D^m = I$, $\widehat{D}^m = I$ for a positive integer $m$, and that*

$\lambda_j, w_j, w_j^*$ $(j = 1, \ldots, k)$ *are eigenvalues and (right and left) eigenvectors of* $D$
$\widehat{\lambda}_j, \widehat{w}_j, \widehat{w}_j^*$ $(j = 1, \ldots, \widehat{k})$ *are eigenvalues and (right and left) eigenvectors of* $\widehat{D}$
*with the normalization* $w_j^* w_j = 1$ *and* $\widehat{w}_j^* \widehat{w}_j = 1$.

*Then, the solution of (5.6) can be written as*

$$(dY)(t) = \sum_{j=1}^{k} \eta_j(t)\, w_j, \qquad (dZ)(t) = \sum_{j=1}^{\widehat{k}} \zeta_j(t)\, \widehat{w}_j,$$

*where the functions $\eta_j(t)$, $\zeta_j(t)$ are solutions of the following equations:*

- if $\lambda_j = \widehat{\lambda}_j$ is an eigenvalue of $D$ and of $\widehat{D}$, then

$$\begin{aligned}\dot{\eta}_j &= \mu_j f_y(y,z)\,\eta_j + \nu_j f_z(y,z)\,\zeta_j \\ \dot{\zeta}_j &= \kappa_j g_y(y,z)\,\eta_j + \tau_j g_z(y,z)\,\zeta_j\end{aligned} \qquad (5.7)$$

(for $\lambda_1 = \widehat{\lambda}_1 = 1$ we have $\mu_1 = \nu_1 = \kappa_1 = \tau_1 = 1$).
- if $\lambda_j$ is an eigenvalue of $D$ but not of $\widehat{D}$, then

$$\dot{\eta}_j = \mu_j f_y(y,z)\,\eta_j, \qquad (5.8)$$

- if $\widehat{\lambda}_j$ is an eigenvalue of $\widehat{D}$ but not of $D$, then

$$\dot{\zeta}_j = \tau_j g_z(y,z)\,\zeta_j. \qquad (5.9)$$

*The real coefficients are given by*

$$\begin{aligned}\mu_j &= \lambda_j^{-1} w_j^* \mathcal{B}\, \mathcal{C} w_j, & \nu_j &= \lambda_j^{-1} w_j^* \mathcal{B}\, \widehat{\mathcal{C}} \widehat{w}_j, \\ \kappa_j &= \lambda_j^{-1} \widehat{w}_j^* \widehat{\mathcal{B}}\, \mathcal{C} w_j, & \tau_j &= \widehat{\lambda}_j^{-1} \widehat{w}_j^* \widehat{\mathcal{B}}\, \widehat{\mathcal{C}} \widehat{w}_j.\end{aligned} \qquad (5.10)$$

*They are called growth parameters of the method.*

*Proof.* Since the eigenvalues of $D$ and $\widehat{D}$ are simple, we have $w_j^* w_i = 0$, $\widehat{w}_j^* \widehat{w}_i = 0$ for $j \neq i$. Multiplying (5.6) from the left by $w_j^*$, we obtain

$$\dot{\eta}_j = \sum_{i=1}^{k} w_j^* \mathcal{B}\, \mathcal{C} w_i \cdot f_y(y,z)\,\eta_i + \sum_{i=1}^{\widehat{k}} w_j^* \mathcal{B}\, \widehat{\mathcal{C}} \widehat{w}_i \cdot f_z(y,z)\,\zeta_i,$$

and a similar formula for $\dot{\zeta}_j$. From the definition of $\mathcal{B}$ and $\mathcal{C}$ (see (2.12)) it follows

$$w_j^* \mathcal{B}\, \mathcal{C} w_i = \frac{1}{m}\sum_{l=0}^{m-1} \lambda_j^{m-l-1} w_j^* \mathcal{B}\, \mathcal{C} w_i\, \lambda_i^l = \frac{\lambda_j^{m-1}}{m}\, w_j^* \mathcal{B}\, \mathcal{C} w_i \sum_{l=0}^{m-1} \left(\frac{\lambda_i}{\lambda_j}\right)^l.$$

Because of $\lambda_i^m = \lambda_j^m = 1$, this expression vanishes for $i \neq j$ (observe that by stability the $\lambda_i$ are distinct). For $i = j$ we obtain $\lambda_j^{-1} w_j^* \mathcal{B}\, \mathcal{C} w_j$ which is equal to $\mu_j$. In the same way one proves the $w_j^* \mathcal{B}\, \widehat{\mathcal{C}} \widehat{w}_i$ vanishes for $\lambda_j \neq \widehat{\lambda}_i$, and it equals $\nu_j$ if $i = j$ and $\lambda_j = \widehat{\lambda}_j$, etc.

For the principal eigenvalues $\lambda_1 = \widehat{\lambda}_1 = 1$, we have $w_1 = e$, $\widehat{w}_1 = \widehat{e}$, and it follows from (5.5) that $\mu_1 = w_1^* \mathcal{B}\, \mathcal{C} w_1 = w_1^* w_1 = 1$, and similarly $\nu_1 = \kappa_1 = \tau_1 = 1$. □

**Remark 5.3.** In the case of non-partitioned general linear methods (i.e., both methods in (5.2) are the same), we have $\mu_j = \nu_j = \kappa_j = \tau_j$ and there is only one growth parameter per eigenvalue of $D$.

For linear multistep methods (1.1), growth parameters have been introduced by Dahlquist (1959) as

$$\mu_j = \frac{\sigma(\lambda_j)}{\lambda_j \rho'(\lambda_j)}. \tag{5.11}$$

Writing the multistep formula as a general linear method (see Sect. XIV.1.4), this definition coincides with the one given in Theorem 5.2 (Exercise 10).

As a consequence of the preceding theorem, only the low-dimensional systems (5.7)-(5.9) have to be investigated, if one wants to study the stability of (5.6). This allows us to better understand the behaviour of the different methods in Fig. 1.2. The outer solar system (I.2.12) has a separable Hamiltonian. Therefore it is a partitioned system of the form $\dot y = f(z)$, $\dot z = g(y)$, so that the differential equations (5.8) and (5.9) are trivially stable. For $j = 1$ the system (5.7) constitutes the variational equations of the outer solar system, which is supposed to be stable. Since the matrices $D$ and $\widehat D$ for the partitioned method (AB) of Example 1.2 do not have common eigenvalues other than $\lambda_1 = 1$, the stability of (5.6) is guaranteed.

This is not the case for the methods (A) and (B). For these methods we have $\widehat D = D$, and (5.7) has to be considered for all eigenvalues. Numerical computations of solutions of (5.7) show that for $\mu_j = \nu_j = \kappa_j = \tau_j$ and $\mu_j \notin \{0, 1\}$ the problem (5.7) is unstable. This explains the bad behaviour of these methods.

**Remark 5.4.** Although the investigation of Theorem 5.2 gives much insight into the dynamics of partitioned general linear methods, we are still far from a rigorous understanding of the long-time behaviour when the methods are applied to integrable reversible systems. In contrast to one-step methods (cf. Theorem XI.3.1), we do not know any rigorous statement concerning the long-time conservation of first integrals (depending on the action variables) or concerning the linear growth of the global error.

A partial result is given by Cano & Sanz-Serna (1998) for multistep methods (1.9) applied to equations $\ddot y = f(y)$ with periodic exact solution. There, the first terms of the asymptotic error expansion for the global error are computed, and their growth as a function of time is studied.

## XIV.5.3 Dissipatively Perturbed Hamiltonian Systems

As in the beginning of Chap. XII we consider Van der Pol's equation

$$\dot y = z, \qquad \dot z = -y + \varepsilon(1 - y^2)z \tag{5.12}$$

with a small positive parameter $\varepsilon$. The exact solution tends to a limit cycle, close to a circle of radius 2. We apply three numerical methods. The first is

$$y_{n+2} = y_n + h\bigl(\beta f_{n+2} + 2(1-\beta)f_{n+1} + \beta f_n\bigr) \tag{5.13}$$

with $\beta = 0$ (method A), the second is (5.13) with $\beta = 0.7$ (method B), and the third is the partitioned multistep method of Example 1.2 (method C). The numerical result of period 295 to period 300 is displayed in Fig. 5.1. Let us explain the different behaviour with the help of Theorem 5.2.

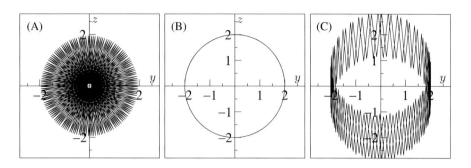

**Fig. 5.1.** The periods 295 to 300 of the numerical solution of three methods applied to Van der Pol's equation with $\varepsilon = 0.01$, initial value $y_0 = 1.4$, $z_0 = 0$, and step size $h = 0.1$.

The propagation matrix of method (5.13) has $\lambda_1 = 1$ and $\lambda_2 = -1$ as eigenvalues (the zeros of the $\rho$-polynomial) and the growth parameter corresponding to $\lambda_2 = -1$ is

$$\mu_2 = 2\beta - 1. \tag{5.14}$$

Since for multistep methods all four parameters are the same in the variational equation (5.7), it becomes for the Van der Pol equation

$$\dot{\eta} = \mu_2 \zeta, \qquad \dot{\zeta} = \mu_2 \big(-\eta - 2\varepsilon y(t)z(t)\,\eta + \varepsilon(1 - y(t)^2)\,\zeta\big),$$

where $y(t), z(t)$ is a solution of (5.12). As in Sect. XII.1 we consider the symplectic change of coordinates $y = \sqrt{2a}\sin\theta$, $y = \sqrt{2a}\cos\theta$, which transforms this equation into

$$\dot{\eta} = \mu_2 \zeta, \qquad \dot{\zeta} = \mu_2\big(-\eta - 4\varepsilon a \sin\theta \cos\theta\,\eta + \varepsilon(1 - 2a\sin^2\theta)\,\zeta\big).$$

The averaged system is then

$$\dot{\eta} = \mu_2 \zeta, \qquad \dot{\zeta} = \mu_2\big(-\eta + \varepsilon(1 - a)\,\zeta\big). \tag{5.15}$$

Close to the limit cycle (i.e., $a \approx 2$), the eigenvalues of the Jacobian are

$$-\frac{\mu_2}{2}\big(\varepsilon \pm i\sqrt{4 - \varepsilon^2}\big),$$

which implies stability of (5.15) only if $\mu_2 \geq 0$. This explains the different behaviour of methods (A) and (B) in Fig. 5.1. Indeed, the growth parameter (5.14) is negative for method (A), whereas it is positive for method (B).

In the case of the partitioned method (1.16), the eigenvalues of both propagation matrices are $\lambda_1 = 1, \lambda_{2,3} = \pm i$, and $\widehat{\lambda}_1 = 1, \widehat{\lambda}_2 = -1$, respectively. Apart from $\lambda_1 = \widehat{\lambda}_1 = 1$, the propagation matrices have no common eigenvalues. For the Van der Pol equation we have $f(y, z) = z$, so that (5.8) always gives a stable equation However, for the eigenvalue $\widehat{\lambda}_2 = -1$, we have to consider (5.9), which is

$$\dot\zeta_2 = \tau_2\varepsilon(1 - y(t)^2)\zeta_2.$$

The same symplectic change of coordinates and the use of the averaging principle leads to $\dot\zeta_2 = \tau_2\varepsilon(1-a)\zeta_2$. For $a \approx 2$ this equation is unstable, because the growth parameter of the explicit mid-point rule is $\tau_2 = -1$. Figure 5.1 nicely illustrates that the instability of method (C) occurs only in the $z$-component.

This example shows that in certain situations it is important to have all growth parameters positive. The original motivation of Dahlquist (1959) for studying the growth parameters was the following: for the stable test equation $\dot y = -y$ and for multistep methods (where $\mu_j = \nu_j = \kappa_j = \tau_j$), the variational equation (5.7) becomes $\dot\eta_j = -\mu_j\eta_j$. Hence the augmented system is stable only if all growth parameters are non-negative. The following result (Hairer 1999) gives an order barrier for such methods.

**Lemma 5.5.** *Consider a consistent, stable and symmetric multistep method (1.1). If all growth parameters are positive, then the method is A-stable. Hence its order is at most* 2.

*Proof.* For the multistep method we look at the root locus curve, which is defined by $t \mapsto \gamma(t) := \rho(e^{it})/\sigma(e^{it})$. By the symmetry of the method, it is contained in the imaginary axis. The positivity of the growth parameters implies that, close to the origin, the negative real axis belongs to the stability region. Hence the method has to be A-stable, and the second Dahlquist barrier (see Hairer & Wanner (1996), Sect. V.1) implies the order bound $p \le 2$. □

This negative result encourages us to search in the larger class of general linear methods for methods that have only positive growth parameters (Exercise 11).

## XIV.5.4 Numerical Instabilities and Resonances

> Soon after Quinlan and Tremaine's methods were published, however, Alar Toomre discovered a disturbing feature of the methods, ...
>
> (G.D. Quinlan 1999)

It is a simple task to derive multistep methods of high order. Consider, for example, methods of the form (1.9) for second order differential equations $\ddot y = f(y)$. Their order is determined by the condition (1.10). We choose arbitrarily $\rho(\zeta)$ such that $\zeta = 1$ is a double zero and the stability condition is satisfied. Condition (1.10) then gives

$$\sigma(\zeta) = \rho(\zeta)/\log^2\zeta + \mathcal{O}((\zeta-1)^p).$$

Expanding the right-hand expression into a Taylor series at $\zeta = 1$ and truncating suitably, this yields the corresponding $\sigma$ polynomial. If we take

$$\rho(\zeta) = (\zeta-1)^2(\zeta^6 + \zeta^4 + \zeta^3 + \zeta^2 + 1), \tag{5.16}$$

we get in this way Method SY8 of Table 5.1, a method proposed by Quinlan & Tremaine (1990) for computations in celestial mechanics. All methods of Table 5.1

**Table 5.1.** Symmetric multistep methods for second order problems; $k=8$ and $p=8$.

|   | SY8 | | SY8B | | SY8C | |
|---|---|---|---|---|---|---|
| $i$ | $\alpha_i$ | $12096\,\beta_i$ | $\alpha_i$ | $120960\,\beta_i$ | $\alpha_i$ | $8640\,\beta_i$ |
| 0 | 1 | 0 | 1 | 0 | 1 | 0 |
| 1 | $-2$ | 17671 | 0 | 192481 | $-1$ | 13207 |
| 2 | 2 | $-23622$ | 0 | 6582 | 0 | $-8934$ |
| 3 | $-1$ | 61449 | $-1/2$ | 816783 | 0 | 42873 |
| 4 | 0 | $-50516$ | $-1$ | $-156812$ | 0 | $-33812$ |

are 8-step methods, of order 8, and symmetric, i.e., the relations $\alpha_i = \alpha_{k-i}$ and $\beta_i = \beta_{k-i}$ are satisfied. Therefore, we present the coefficients only for $i \leq k/2$.

These methods give approximations $y_n$ to the solution of the differential equation. If also derivative approximations are needed, we get them by finite differences, e.g., for the 8th order methods of Table 5.1 we use

$$\dot{y}_n = \frac{1}{840h}\Big(672\,(y_{n+1} - y_{n-1}) - 168\,(y_{n+2} - y_{n-2}) \\ + 32\,(y_{n+3} - y_{n-3}) - 3\,(y_{n+4} - y_{n-4})\Big). \tag{5.17}$$

We apply this method to the Kepler problem (I.2.2), once with eccentricity $e = 0$ and once with $e = 0.2$, and initial values (I.2.11), such that the period of the exact solution is $2\pi$. Starting approximations are computed accurately with a high order Runge-Kutta method. We apply Method SY8 with many different step sizes ranging from $2\pi/30$ to $2\pi/95$, and we plot in Fig. 5.2 the maximum error of the total energy as a function of $2\pi/h$ (where $h$ denotes the step size). We see that in general the error decreases with the step size, but there is an extremely large error for $h \approx 2\pi/60$. For $e \neq 0$, further peaks can be observed at integral multiples of 5 and 6. It is our aim to understand this behaviour.

**Instabilities.** We put $z = q_1 + iq_2$, so that the Kepler problem becomes

$$\ddot{z} = \psi(|z|)z, \qquad \psi(r) = -r^{-3},$$

and we choose initial values such that $z(t) = e^{it}$ is a circular motion (eccentricity $e = 0$). The numerical solution of (1.9) is therefore defined by the relation

$$\sum_{j=0}^{k} \alpha_j z_{n+j} = h^2 \sum_{j=0}^{k} \beta_j \psi(|z_{n+j}|) z_{n+j}. \tag{5.18}$$

Approximating $\psi(|z_{n+j}|)$ with $\psi(1) = -\omega^2$, we get a linear recurrence relation with characteristic polynomial

$$S(\omega h, \zeta) = \rho(\zeta) + \omega^2 h^2 \sigma(\zeta).$$

The principal roots of $S(\omega h, \zeta) = 0$ satisfy $\zeta_1(\omega h) \approx e^{i\omega h}$ and $\zeta_2(\omega h) \approx e^{-i\omega h}$, and we have $|\zeta_j(\omega h)| = 1$ for all $j$ and for sufficiently small $h$, because the method

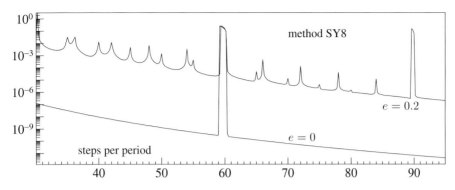

**Fig. 5.2.** Maximum error in the total energy during the integration of 2500 orbits of the Kepler problem as a function of the number of steps per period.

is symmetric (Exercise 2). As a consequence of $|\zeta_1(\omega h)| = 1$, the values $\widehat{z}_n := \zeta_1(\omega h)^n$ are not only a solution of the linear recurrence relation, but also of the nonlinear relation (5.18). Our aim is to study the stability of this numerical solution. We therefore consider a perturbed solution

$$z_n = \zeta_1(\omega h)^n \big(1 + u_n\big).$$

Using $|z_n| = 1 + \tfrac{1}{2}(u_n + \overline{u}_n) + \mathcal{O}(|u_n|^2)$ and neglecting the quadratic and higher order terms of $|u_n|$ in the relation (5.18), we get

$$\sum_{j=0}^{k}(\alpha_j + \omega^2 h^2 \beta_j)\zeta_1(\omega h)^j u_{n+j} = \frac{h^2}{2}\psi'(1) \sum_{j=0}^{k} \beta_j \zeta_1(\omega h)^j \big(u_{n+j} + \overline{u}_{n+j}\big).$$

Considering also the complex conjugate of this relation, and eliminating $\overline{u}_{n+j}$, we obtain a linear recurrence relation for $u_n$ with characteristic polynomial

$$S(\omega h, \zeta_1(\omega h)\zeta) \cdot S(\omega h, \zeta_1(\omega h)^{-1}\zeta) + \mathcal{O}(h^2). \qquad (5.19)$$

For small $h$, its zeros are close to $\zeta_1(\omega h)^{-1}\zeta_j$ and $\zeta_1(\omega h)\zeta_l$. If two of these zeros collapse, the $\mathcal{O}(h^2)$ terms in (5.19) can produce a root of modulus larger than one, so that instability occurs. This is the case, if two roots $\zeta_j, \zeta_l$ of $\rho(\zeta) = 0$ satisfy $\zeta_j \zeta_l^{-1} \approx \zeta_1^2 \approx e^{2i\omega h}$, or

$$\theta_j - \theta_l = \frac{4\pi}{N}, \qquad (5.20)$$

where $\zeta_j = e^{i\theta_j}$ and $h = 2\pi/N$.

For the Method SY8 of Table 5.1, the spurious zeros of $\rho(\zeta)$ have arguments $\pm 4\pi/5, \pm 2\pi/5$, and $\pm 2\pi/6$. With $\theta_j = 2\pi/5$ and $\theta_l = 2\pi/6$, the condition (5.20) gives $N = 60$ as a candidate for instability. This explains the experiment of Fig. 5.2 for $e = 0$. A study of the stability of orbits with eccentricity $e \neq 0$ (see Quinlan 1999) shows that instabilities can also occur when $4\pi/N$ is replaced with $2q\pi/N$ ($q = 2, 3, \ldots$) in the relation (5.20).

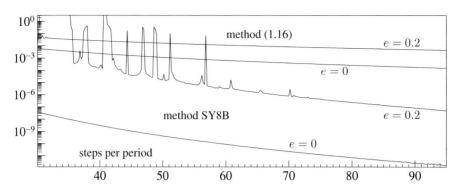

**Fig. 5.3.** Maximum error in the total energy during the integration of 2500 orbits of the Kepler problem as a function of the number of steps per period.

In order to avoid these instabilities as far as possible, Quinlan (1999) constructed symmetric multistep methods, where the spurious roots of $\rho(\zeta) = 0$ are well spread out on the unit circle and far from $\zeta = 1$. As a result he proposes Method SY8B of Table 5.1. The same experiment as above yields the results of Fig. 5.3. The $\rho$-polynomial of Method SY8B is

$$\rho(\zeta) = (\zeta - 1)^2 (\zeta^6 + 2\zeta^5 + 3\zeta^4 + 3.5\zeta^3 + 3\zeta^2 + 2\zeta + 1),$$

and the $\theta_j$ of the spurious roots are $\pm 2\pi/2.278$, $\pm 2\pi/3.353$, and $\pm 2\pi/4.678$. The condition (5.20) is satisfied only for $N \leq 23.67$, which implies that no instability occurs for $e = 0$ in the region of the experiment of Fig. 5.3.

To illustrate the importance of high order methods, we included in Fig. 5.3 the results of the second order partitioned multistep method (1.16).

## XIV.5.5 Extension to Variable Step Sizes

Variable step size multistep methods for second order differential equations $\ddot{y} = f(y)$ are of the form

$$\sum_{j=0}^{k} \alpha_j(h_n, \ldots, h_{n+k-1}) y_{n+j} = h_{n+k-1}^2 \sum_{j=0}^{k} \beta_j(h_n, \ldots, h_{n+k-1}) f(y_{n+j}),$$

where the coefficients $\alpha_j$ and $\beta_j$ are allowed to depend on the step sizes $h_n, \ldots, h_{n+k-1}$, more precisely, on the ratios $h_{n+1}/h_n, \ldots, h_{n+k-1}/h_{n+k-2}$. They yield approximations $y_n$ to $y(t_n)$ on a variable grid given by $t_{n+1} = t_n + h_n$. Such a method is of *order* $p$ (cf. formula (1.10)), if

$$\sum_{j=0}^{k} \alpha_j(h_n, \ldots, h_{n+k-1}) y(t_{n+j}) = h_{n+k-1}^2 \sum_{j=0}^{k} \beta_j(h_n, \ldots, h_{n+k-1}) \ddot{y}(t_{n+j})$$

(5.21)

for all polynomials $y(t)$ of degree $\leq p+1$. It is *stable*, if the $\rho$-polynomial with coefficients $\alpha_j(h,\ldots,h)$ (constant step size) satisfies the stability condition of Sect. XIV.1.2 (see Theorem III.5.7 of Hairer, Nørsett & Wanner (1993) and Cano & Durán (2000a)).

All methods of Sect. XIV.5.4 can be extended to symmetric, variable step size integrators. This has been discovered by Cano & Durán (2000b). For clarity of notation we let $\widetilde{\alpha}_j, \widetilde{\beta}_j$ ($j=0,\ldots,k$) be the coefficients of such a fixed step size method. Cano & Durán propose putting

$$\beta_j(h_n,\ldots,h_{n+k-1}) = \frac{h_n}{h_{n+k-1}} \widetilde{\beta}_j, \tag{5.22}$$

and to determine $\alpha_j(h_n,\ldots,h_{n+k-1})$ such that symmetry and order $k-2$ (for arbitrary step sizes) are achieved. We also suppose (5.22), but we determine the coefficients $\alpha_j(h_n,\ldots,h_{n+k-1})$ such that (5.21) holds for all polynomials $y(t)$ of degree $\leq k$. This uniquely determines these coefficients whenever $h_n > 0, \ldots, h_{n+k-1} > 0$ (Vandermonde type system) and gives the following properties.

**Lemma 5.6.** *For even $k$, let $(\widetilde{\alpha}_j, \widetilde{\beta}_j)$ define a symmetric, stable $k$-step method (1.9) of order $k$, and consider the variable step size method given by (5.22) and $\alpha_j(h_n,\ldots,h_{n+k-1})$ such that (5.21) holds for all polynomials $y$ satisfying $\deg y \leq k$. This method extends the fixed step size formula, i.e.,*

$$\alpha_j(h,\ldots,h) = \widetilde{\alpha}_j, \qquad \beta_j(h,\ldots,h) = \widetilde{\beta}_j, \tag{5.23}$$

*it satisfies the symmetry relations*

$$\begin{aligned} \alpha_j(h_n,\ldots,h_{n+k-1}) &= \alpha_{k-j}(h_{n+k-1},\ldots,h_n) \\ h_{n+k-1}^2 \beta_j(h_n,\ldots,h_{n+k-1}) &= h_n^2 \beta_{k-j}(h_{n+k-1},\ldots,h_n), \end{aligned} \tag{5.24}$$

*and it is of order $k-1$ for arbitrary step sizes. Moreover, it behaves like a method of order $k$, if $h_{n+1} = h_n\bigl(1+\mathcal{O}(h_n)\bigr)$ uniformly in $n$.*

*Proof.* The relation (5.23) for $\beta_j$ follows at once from (5.22), and for $\alpha_j$ it is a consequence of the uniqueness of the solution of the linear system for the $\alpha_j$.

The second condition of (5.24) follows directly from (5.22) and from the symmetry of the underlying fixed step size method ($\widetilde{\beta}_{k-j} = \widetilde{\beta}_j$ for all $j$). Inserting (5.22) into (5.21), replacing $y(t)$ with $y(t_{n+k}+t_n-t)$, and reversing the order of $h_n,\ldots,h_{n+k-1}$ yields

$$\sum_{j=0}^{k} \alpha_j(h_{n+k-1},\ldots,h_n)\, y(t_{n+k-j}) = h_n h_{n+k-1} \sum_{j=0}^{k} \widetilde{\beta}_j\, \ddot{y}(t_{n+k-j}).$$

Using $\widetilde{\beta}_{k-j} = \widetilde{\beta}_j$ this shows that $\alpha_{k-j}(h_{n+k-1},\ldots,h_n)$ satisfies exactly the same linear system as $\alpha_j(h_n,\ldots,h_{n+k-1})$, so that also the first relation of (5.24) is verified.

By definition, the variable step size method is at least of order $k - 1$. Under the assumption $h_{n+1} = h_n(1 + \mathcal{O}(h_n))$ the defect in (5.21) is of the form

$$h_n^{k+1} D(h_n, \ldots, h_{n+k-1}) = h_n^{k+1} D(h_n, \ldots, h_n) + \mathcal{O}(h_n^{k+2})$$

for all sufficiently smooth $y(t)$. Since the constant step size method is of order $k$, the expression $D(h_n, \ldots, h_n)$ is of size $\mathcal{O}(h_n)$, so that we observe convergence of order $k$. □

The symmetry relation (5.24) has the following interpretation: if the approximations $y_n, \ldots, y_{n+k-1}$ used with step sizes $h_n, \ldots, h_{n+k-1}$ yield $y_{n+k}$, then the values $y_{n+k}, \ldots, y_{n+1}$ applied with $h_{n+k-1}, \ldots, h_n$ yield $y_n$ as a result (since the coefficients $\alpha_j$ and $\beta_j$ only depend on step size ratios and the multistep formula only on $h_{n+k-1}^2$, the same result is obtained with $-h_{n+k-1}, \ldots, -h_n$). This is the analogue of the definition of symmetry for one-step methods.

For obtaining a good long-time behaviour, the step sizes also have to be chosen in a symmetric and reversible way (see Sect. VIII.2). One possibility is to take step sizes

$$h_{n+k-1} = \frac{\varepsilon}{2} \Big( \sigma(y_{n+k-1}) + \sigma(y_{n+k}) \Big), \tag{5.25}$$

where $\varepsilon > 0$, and $\sigma(y)$ is a given positive monitor function. This condition is an implicit equation for $h_{n+k-1}$, because $y_{n+k}$ depends on $h_{n+k-1}$. It has to be solved iteratively. Notice, however, that for an explicit multistep formula no further force evaluations are necessary during this iteration. Such a choice of the step size guarantees that whenever $h_{n+k-1}$ is chosen when stepping from $y_n, \ldots, y_{n+k-1}$ with $h_n, \ldots, h_{n+k-2}$ to $y_{n+k}$, the step size $h_n$ is chosen when stepping backwards from $y_{n+k}, \ldots, y_{n+1}$ with $h_{n+k-1}, \ldots, h_{n+1}$ to $y_n$.

**Implementation.** For given initial values $y_0, \dot{y}_0$, the starting approximations $y_1$, $\ldots, y_{k-1}$ should be computed accurately (for example, by a high-order Runge-Kutta method) with step sizes satisfying (5.25). The solution of the scalar nonlinear equation (5.25) has to be done carefully in order to reduce the overhead of the method. In our code we use $h_{n+k-1} := h_{n+k-2}^2 / h_{n+k-3}$ as predictor, and we apply modified Newton iterations with the derivative approximated by finite differences.

The coefficients $\alpha_j(h_n, \ldots, h_{n+k-1})$ have to be computed anew in every iteration. We use the basis

$$p_i(t) = \prod_{j=0}^{i-1} (t - t_{n+j}), \qquad i = 0, \ldots, k$$

for the polynomials of degree $\leq k$ in (5.21). This leads to a linear triangular system for $\alpha_0, \ldots, \alpha_k$. As noticed by Cano & Durán (2000b), the coefficients $p_i(t_j)$ and $\ddot{p}_i(t_j)$ can be obtained efficiently from the recurrence relations

$$\begin{aligned} p_0(t) &= 1, & p_{i+1}(t) &= (t - t_i) p_i(t) \\ \dot{p}_0(t) &= 0, & \dot{p}_{i+1}(t) &= (t - t_i) \dot{p}_i(t) + p_i(t) \\ \ddot{p}_0(t) &= 0, & \ddot{p}_{i+1}(t) &= (t - t_i) \ddot{p}_i(t) + 2 \dot{p}_i(t). \end{aligned}$$

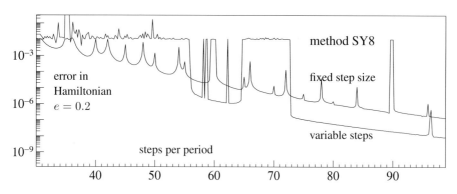

**Fig. 5.4.** Maximum error in the total energy during the integration of 2500 orbits of the Kepler problem as a function of the number of steps per period.

During the iterations for the solution of the nonlinear equation (5.25) only the values of $p_i(t_{n+k})$ have to be updated.

**Numerical Experiment.** We repeat the experiment of Fig. 5.2 with the method SY8, but this time in the variable step size version and with $\sigma(y) = \|y\|^2$ as step size monitor. We have computed 2500 periods of the Kepler problem with eccentricity $e = 0.2$, and we have plotted in Fig. 5.4 the maximal error in the Hamiltonian as a function of the number of steps per period (for a comparison we have also included the result of the fixed step size implementation). Similar to (5.17) we use approximations $\dot{y}_n$ that are the derivative of the interpolation polynomial passing through $y_n, y_{n\pm 1}, y_{n\pm 2}, \ldots$ such that the correct order is obtained. The computation is stopped when the error exceeds $10^{-2}$.

As expected, the error is smaller for the variable step size version, and it is seen that the peaks due to numerical resonances are now much less although they are not completely removed. For large step sizes, the performance deteriorates, but this is not a serious problem, because these methods are recommended only for high accuracy computations.

It should be remarked that the overhead, due to the computation of the coefficients $\alpha_j$ and the solution of the nonlinear equation (5.25), is rather high. Therefore, the use of variable step sizes is recommended only when force evaluations $f(y)$ are expensive or when constant step sizes are not appropriate. Cano & Durán (2000b) report an excellent performance of symmetric, variable step size multistep methods for computations of the outer solar system.

> Despite the resonances and instabilities, then, symmetric methods can still be a better choice than Störmer methods for long integrations of planetary orbits provided that the user is aware of the dangers.
> (G.D. Quinlan 1999)

## XIV.6 Exercises

1. Let $\zeta_1(z)$ be the principal root of the characteristic equation $\rho(\zeta) - z\sigma(\zeta) = 0$. Prove that for irreducible multistep methods the condition $\zeta_1(-z)\zeta_1(z) \equiv 1$ (in a neighbourhood of $z = 0$) is equivalent to the symmetry of the method.

2. (Lambert & Watson 1976). Prove that stable, symmetric linear multistep methods (1.9) for second order differential equations, for which the polynomial $\rho(\zeta)$ has only simple zeros (with the exception of $\zeta = 1$), has a non-vanishing interval of periodicity, i.e., the roots $\zeta_i(z)$ of $\rho(\zeta) - z^2\sigma(\zeta) = 0$ satisfy $|\zeta_i(iy)| = 1$ for sufficiently small real $y$.
   *Hint.* Simple roots cannot leave the unit circle under small perturbations of $y$.

3. Using Theorem XII.3.2, prove that the underlying one-step method of a strictly stable $p$th order linear multistep method has order $p$.

4. Consider a general linear method (1.18). If there exist an invertible symmetric matrix $G$ and a diagonal matrix $\Lambda$ such that

$$M = \begin{pmatrix} D^T G D - G & D^T G B - C^T \Lambda \\ B^T G D - \Lambda C & B^T G B - A^T \Lambda - \Lambda A \end{pmatrix} = 0, \quad (6.1)$$

then the method is $G$-symplectic.
*Hint.* Adapt the proof of Burrage & Butcher for $B$-stability (see Hairer & Wanner (1996), page 358).

5. A Runge-Kutta method can be considered as a general linear method with $D = (1)$, $C = \mathbb{1}$. Prove that the condition (6.1) is equivalent to the symplecticity condition of Chap. VI.

6. Extend the definition of $G$-symplecticity to partitioned general linear methods, and prove that the condition

$$M = \begin{pmatrix} D^T G \widehat{D} - G & D^T G \widehat{B} - C^T \Lambda \\ B^T G \widehat{D} - \Lambda \widehat{C} & B^T G \widehat{B} - A^T \Lambda - \Lambda \widehat{A} \end{pmatrix} = 0 \quad (6.2)$$

implies that the method is $G$-symplectic.

7. For the weakly stable 3-step method (2.11) compute the Hamiltonian (3.7) of the augmented differential equation. The result is

$$\mathcal{H}(Y) = H\left(\tfrac{1}{2}(Y_2 + Y_3)\right) + H\left(\tfrac{1}{2}(Y_1 - Y_2) + Y_3\right)$$
$$+ H\left(\tfrac{1}{2}(Y_3 - Y_2) + Y_1\right) + H\left(\tfrac{1}{2}(Y_1 + Y_2)\right).$$

8. Consider the partitioned general linear method (5.2) and assume that the methods $(A, B, C, D)$ and $(\widehat{A}, \widehat{B}, \widehat{C}, \widehat{D})$ satisfy both the assumptions of Theorem 4.5 with matrices $P, Q$ and $\widehat{P}, \widehat{Q}$, respectively. Prove that for $\widehat{P} = P$ the partitioned method is symmetric.

9. The combination of two symmetric multistep methods in a partitioned method yields a symmetric method.

10. Prove that for linear multistep methods the value $\mu_j$ of (5.11) is a growth parameter according to the definition given in Theorem 5.2.
11. Construct general linear methods of order $p > 2$, for which all growth parameters are positive. Find such methods, which have a smaller degree of implicitness than symmetric one-step methods of the same order.
12. Write a Maple program that checks the coefficients of Table 5.1. After defining rho:=$\rho(z)$, use the instructions
    ```
    > sigma := taylor(rho/(log(z)*log(z)),z=1,8);
    > factor(expand(convert(sigma,polynom)));
    ```
13. Construct partitioned general linear methods which are symmetric, explicit, of high order, and for which the matrices $D$ and $\widehat{D}$ have distinct eigenvalues (with the exception of 1). Compared to multistep methods, smaller dimensions of the matrices $D$ and $\widehat{D}$ are possible.

# Bibliography

L. Abia & J.M. Sanz-Serna, *Partitioned Runge-Kutta methods for separable Hamiltonian problems*, Math. Comput. 60 (1993) 617–634. *[VI.7], [IX.10]*

M.J. Ablowitz & J.F. Ladik, *A nonlinear difference scheme and inverse scattering*, Studies in Appl. Math. 55 (1976) 213–229. *[VII.2]*

M.P. Allen & D.J. Tildesley, *Computer Simulation of Liquids*, Clarendon Press, Oxford, 1987. *[I.3], [XIII.1]*

H.C. Andersen, *Rattle: a "velocity" version of the Shake algorithm for molecular dynamics calculations*, J. Comput. Phys. 52 (1983) 24–34. *[VII.1]*

V.I. Arnold, *Small denominators and problems of stability of motion in classical and celestial mechanics*, Russian Math. Surveys 18 (1963) 85–191. *[I.1], [X.1], [X.2], [X.4], [X.5], [X.7]*

V.I. Arnold, *Sur la géométrie différentielle des groupes de Lie de dimension infinie et ses applications à l'hydrodynamique des fluides parfaites*, Ann. Inst. Fourier 16 (1966) 319–361. *[VII.3]*

V.I. Arnold, *Mathematical Methods of Classical Mechanics*, Springer-Verlag, New York, 1978, second edition 1989. *[VI.1], [X.1]*

V.I. Arnold, V.V. Kozlov & A.I. Neishtadt, *Mathematical Aspects of Classical and Celestial Mechanics*, Springer, Berlin, 1997. *[X.1]*

U. Ascher & S. Reich, *On some difficulties in integrating highly oscillatory Hamiltonian systems*, in Computational Molecular Dynamics, Lect. Notes Comput. Sci. Eng. 4, Springer, Berlin, 1999, 281–296. *[V.4]*

A. Aubry & P. Chartier, *Pseudo-symplectic Runge-Kutta methods*, BIT 38 (1998) 439–461. *[X.7]*

H.F. Baker, *Alternants and continuous groups*, Proc. of London Math. Soc. 3 (1905) 24–47. *[III.4]*

G. Benettin, A.M. Cherubini & F. Fassò, *A changing-chart symplectic algorithm for rigid bodies and other Hamiltonian systems on manifolds*, SIAM J. Sci. Comput. 23 (2001) 1189–1203. *[VII.1]*

G. Benettin, L. Galgani & A. Giorgilli, *Poincaré's non-existence theorem and classical perturbation theory for nearly integrable Hamiltonian systems*, Advances in nonlinear dynamics and stochastic processes (Florence, 1985) World Sci. Publishing, Singapore, 1985, 1–22. *[X.2]*

G. Benettin, L. Galgani & A. Giorgilli, *Realization of holonomic constraints and freezing of high frequency degrees of freedom in the light of classical perturbation theory. Part I*, Comm. Math. Phys. 113 (1987) 87–103. *[XIII.6]*

G. Benettin, L. Galgani, A. Giorgilli & J.-M. Strelcyn, *A proof of Kolmogorov's theorem on invariant tori using canonical transformations defined by the Lie method*, Il Nuovo Cimento 79B (1984) 201–223. *[X.5]*

G. Benettin & A. Giorgilli, *On the Hamiltonian interpolation of near to the identity symplectic mappings with application to symplectic integration algorithms*, J. Statist. Phys. 74 (1994) 1117–1143. *[IX.3], [IX.7], [IX.8]*

B.J. Berne, *Molecular dynamics in systems with multiple time scales: reference system propagator algorithms*, in Computational Molecular Dynamics: Challenges, Methods, Ideas (P. Deuflhard et al., eds.), Springer, Berlin 1999, 297–318. *[XIII.1]*

Joh. Bernoulli, *Problème inverse des forces centrales, extrait de la réponse de Monsieur Bernoulli à Monsieur Herman*, Mém. de l'Acad. R. des Sciences de Paris (1710) p. 521, Opera Omnia I, p. 470-480. *[I.2], [X.1]*

J.J. Biesiadecki & R.D. Skeel, *Dangers of multiple time step methods*, J. Comput. Phys. 109 (1993) 318–328. *[I.3], [VIII.4], [XIII.1]*

G.D. Birkhoff, *Relativity and Modern Physics*, Harvard Univ. Press, Cambridge, Mass., 1923. *[I.5]*

G.D. Birkhoff, *Dynamical Systems*, AMS, Providence, R.I., 1927. *[X.2]*

S. Blanes, *High order numerical integrators for differential equations using composition and processing of low order methods*, Appl. Num. Math. (2001) 289–306. *[V.3]*

S. Blanes, F. Casas & J. Ros, *Symplectic integrators with processing: a general study*, SIAM J. Sci. Comput. 21 (1999) 149–161. *[V.3]*

S. Blanes, F. Casas & J. Ros, *Improved high order integrators based on the Magnus expansion*, BIT 40 (2000a) 434–450. *[IV.7]*

S. Blanes, F. Casas & J. Ros, *Processing symplectic methods for near-integrable Hamiltonian systems*, Celestial Mech. Dynam. Astronom. 77 (2000b) 17–35. *[V.3]*

S. Blanes & P.C. Moan, *Practical symplectic partitioned Runge-Kutta and Runge-Kutta-Nyström methods*, To appear in J. Comp. Appl. Math. (2002). *[V.3]*

P.B. Bochev & C. Scovel, *On quadratic invariants and symplectic structure*, BIT 34 (1994) 337–345. *[VI.4], [XIV.3]*

N. Bogolioubov & I. Mitropolski, *Les Méthodes Asymptotiques en Théorie des Oscillations Non Linéaires*, Gauthier-Villars, Paris, 1962. *[XII.2]*

N.N. Bogoliubov & Y.A. Mitropolsky, *Asymptotic Methods in the Theory of Non-Linear Oscillations*, Hindustan Publishing Corp., Delhi, 1961. *[XII.1]*

E. Bour, *L'intégration des équations différentielles de la mécanique analytique*, J. Math. Pures et Appliquées 20 (1855) 185–200. *[X.1]*

K.E. Brenan, S.L. Campbell & L.R. Petzold, *Numerical Solution of Initial-Value Problems in Differential-Algebraic Equations*, Classics in Appl. Math., SIAM, Philadelphia, 1996. *[IV.9], [VII.1]*

Ch. Brouder, *Runge-Kutta methods and renormalization*, Euro. Phys. J. C 12 (2000) 521–534. *[III.1]*

O. Buneman, *Time-reversible difference procedures*, J. Comput. Physics 1 (1967) 517–535. *[V.1]*

C. Burnton & R. Scherer, *Gauss-Runge-Kutta-Nyström methods*, BIT 38 (1998) 12–21. *[VI.8]*

K. Burrage & J.C. Butcher, *Stability criteria for implicit Runge-Kutta methods*, SIAM J. Numer. Anal. 16 (1979) 46–57. *[VI.4]*

J.C. Butcher, *Coefficients for the study of Runge-Kutta integration processes*, J. Austral. Math. Soc. 3 (1963) 185–201. *[II.1]*

J.C. Butcher, *Implicit Runge-Kutta processes*, Math. Comput. 18 (1964a) 50–64. *[II.1]*

J.C. Butcher, *Integration processes based on Radau quadrature formulas*, Math. Comput. 18 (1964b) 233–244. *[II.1]*

J.C. Butcher, *The effective order of Runge-Kutta methods*, in J.Ll. Morris, ed., Proceedings of Conference on the Numerical Solution of Differential Equations, Lecture Notes in Math. 109 (1969) 133–139. *[V.3]*

J.C. Butcher, *An algebraic theory of integration methods*, Math. Comput. 26 (1972) 79–106. *[III.1], [III.3]*

J.C. Butcher, *The Numerical Analysis of Ordinary Differential Equations. Runge-Kutta and General Linear Methods*, John Wiley & Sons, Chichester, 1987. *[III.0], [III.1], [VI.7], [XIV.1]*

J.C. Butcher, *Order and effective order*, Appl. Numer. Math. 28 (1998) 179–191. *[V.3]*

J.C. Butcher & J.M. Sanz-Serna, *The number of conditions for a Runge-Kutta method to have effective order p*, Appl. Numer. Math. 22 (1996) 103–111. *[III.1], [V.3]*

J.C. Butcher & G. Wanner, *Runge-Kutta methods: some historical notes*, Appl. Numer. Math. 22 (1996) 113–151. *[III.1]*

M.P. Calvo, *High order starting iterates for implicit Runge-Kutta methods: an improvement for variable-step symplectic integrators*, IMA J. Numer. Anal. 22 (2002) 153–166. *[VIII.6]*

M.P. Calvo & E. Hairer, *Accurate long-term integration of dynamical systems*, Appl. Numer. Math. 18 (1995a) 95–105. *[X.3]*

M.P. Calvo & E. Hairer, *Further reduction in the number of independent order conditions for symplectic, explicit Partitioned Runge-Kutta and Runge-Kutta-Nyström methods*, Appl. Numer. Math. 18 (1995b) 107–114. *[III.3]*

M.P. Calvo, A. Iserles & A. Zanna, *Numerical solution of isospectral flows*, Math. Comput. 66 (1997) 1461–1486. *[IV.3]*

M.P. Calvo, A. Iserles & A. Zanna, *Conservative methods for the Toda lattice equations*, IMA J. Numer. Anal. 19 (1999) 509–523. *[IV.3]*

M.P. Calvo, A. Murua & J.M. Sanz-Serna, *Modified equations for ODEs*, Contemporary Mathematics 172 (1994) 63–74. *[IX.9]*

M.P. Calvo & J.M. Sanz-Serna, *Variable steps for symplectic integrators*, In: Numerical Analysis 1991 (Dundee, 1991), 34–48, Pitman Res. Notes Math. Ser. 260, 1992. *[VIII.1]*

M.P. Calvo & J.M. Sanz-Serna, *The development of variable-step symplectic integrators, with application to the two-body problem*, SIAM J. Sci. Comput. 14 (1993) 936–952. *[V.3], [X.3]*

M.P. Calvo & J.M. Sanz-Serna, *Canonical B-series*, Numer. Math. 67 (1994) 161–175. *[VI.7]*

J. Candy & W. Rozmus, *A symplectic integration algorithm for separable Hamiltonian functions*, J. Comput. Phys. 92 (1991) 230–256. *[II.5]*

B. Cano & A. Durán, *Analysis of variable-stepsize linear multistep methods with special emphasis on symmetric ones*, Report 2000/9, Universidad de Valladolid, Spain (2000a). *[XIV.5]*

B. Cano & A. Durán, *An effective technique to construct symmetric variable-stepsize linear multistep methods for second-order systems*, Report 2000/10, Universidad de Valladolid, Spain (2000b). *[XIV.5]*

B. Cano & J.M. Sanz-Serna, *Error growth in the numerical integration of periodic orbits by multistep methods, with application to reversible systems*, IMA J. Numer. Anal. 18 (1998) 57–75. *[XIV.5]*

J.R. Cash, *A class of implicit Runge-Kutta methods for the numerical integration of stiff ordinary differential equations*, J. Assoc. Comput. Mach. 22 (1975) 504–511. *[II.3]*

A. Cayley, *On the theory of the analytic forms called trees*, Phil. Magazine XIII (1857) 172–176. *[III.6]*

E. Celledoni & A. Iserles, *Methods for the approximation of the matrix exponential in a Lie-algebraic setting*, IMA J. Numer. Anal. 21 (2001) 463–488. *[IV.8]*

R.P.K. Chan, *On symmetric Runge-Kutta methods of high order*, Computing 45 (1990) 301–309. *[VI.8]*

P.J. Channell & J.C. Scovel, *Integrators for Lie-Poisson dynamical systems*, Phys. D 50 (1991) 80–88. *[VII.2]*

P.J. Channell & J.C. Scovel, *Symplectic integration of Hamiltonian systems*, Nonlinearity 3 (1990) 231–259. *[VI.5]*

S. Chaplygin, *A new case of motion of a heavy rigid body supported in one point* (Russian), Moscov Phys. Sect. 10, vol. 2 (1901). *[X.1]*

M.T. Chu, *Matrix differential equations: a continuous realization process for linear algebra problems*, Nonlinear Anal. 18 (1992) 1125–1146. *[IV.3]*

S. Cirilli, E. Hairer & B. Leimkuhler, *Asymptotic error analysis of the adaptive Verlet method*, BIT 39 (1999) 25–33. *[VIII.3]*

A. Clebsch, *Ueber die simultane Integration linearer partieller Differentialgleichungen*, Crelle Journal f.d. reine u. angew. Math. 65 (1866) 257–268. *[VII.2]*

D. Cohen, *Développement asymptotique de la solution d'une équation différentielle à grandes oscillations*, Travail de diplôme, Univ. Genève, Mai 2000. *[XIII.6]*

G.J. Cooper, *Stability of Runge-Kutta methods for trajectory problems*, IMA J. Numer. Anal. 7 (1987) 1–13. *[IV.2]*

M. Creutz & A. Gocksch, *Higher-order hybrid Monte Carlo algorithms*, Phys. Rev. Lett. 63 (1989) 9–12. *[II.4]*

P.E. Crouch & R. Grossman, *Numerical integration of ordinary differential equations on manifolds*, J. Nonlinear Sci. 3 (1993) 1–33. *[IV.8]*

M. Crouzeix, *Sur la B-stabilité des méthodes de Runge-Kutta*, Numer. Math. 32 (1979) 75–82. *[VI.4]*

G. Dahlquist, *Convergence and stability in the numerical integration of ordinary differential equations*, Math. Scand. 4 (1956) 33-53. *[XIV.1]*

G. Dahlquist, *Stability and error bounds in the numerical integration of ordinary differential equations*, Trans. of the Royal Inst. of Techn. Stockholm, Sweden, Nr. 130 (1959) 87 pp. *[XIV.5]*

G. Dahlquist, *Error analysis for a class of methods for stiff nonlinear initial value problems*, Numerical Analysis, Dundee 1975, Lecture Notes in Math. 506 (1975) 60–74. *[VI.7], [XIV.3]*

G. Darboux, *Sur le problème de Pfaff*, extraît Bulletin des Sciences math. et astron. 2e série, vol. VI (1882); Gauthier-Villars, Paris, 1882. *[VII.2]*

P. Deift, *Integrable Hamiltonian systems*, in P. Deift (ed.) et al., Dynamical systems and probabilistic methods in partial differential equations. AMS Lect. Appl. Math. 31 (1996) 103–138. *[X.1]*

P. Deift, L.C. Li & C. Tomei, *Matrix factorizations and integrable systems*, Comm. Pure Appl. Math. 42 (1989) 443–521. *[IV.3]*

P. Deift, L.C. Li & C. Tomei, *Symplectic aspects of some eigenvalue algorithms*, in A.S. Fokas & V.E. Zakharov (eds.), Important Developments in Soliton Theory, Springer 1993. *[IV.3]*

P. Deift, T. Nanda & C. Tomei, *Ordinary differential equations and the symmetric eigenvalue problem*, SIAM J. Numer. Anal. 20 (1983) 1–22. *[IV.3]*

P. Deuflhard, *A study of extrapolation methods based on multistep schemes without parasitic solutions*, Z. angew. Math. Phys. 30 (1979) 177–189. *[XIII.1], [XIII.2]*

L. Dieci, R.D. Russell & E.S. van Vleck, *Unitary integrators and applications to continuous orthonormalization techniques*, SIAM J. Numer. Anal. 31 (1994) 261–281. *[IV.4]*

L. Dieci, R.D. Russell & E.S. van Vleck, *On the computation of Lyapunov exponents for continuous dynamical systems*, SIAM J. Numer. Anal. 34 (1997) 402–423. *[IV.5]*

F. Diele, L. Lopez & R. Peluso, *The Cayley transform in the numerical solution of unitary differential systems*, Adv. Comput. Math. 8 (1998) 317–334. *[IV.8]*

F. Diele, L. Lopez & T. Politi, *One step semi-explicit methods based on the Cayley transform for solving isospectral flows*, J. Comput. Appl. Math. 89 (1998) 219–223. *[IV.3]*

V. Druskin & L. Knizhnerman, *Krylov subspace approximation of eigenpairs and matrix functions in exact and computer arithmetic*, Numer. Linear Algebra Appl. 2 (1995) 205–217. *[XIII.1]*

A. Dullweber, B. Leimkuhler & R. McLachlan, *Symplectic splitting methods for rigid body molecular dynamics*, J. Chem. Phys. 107, No. 15 (1997) 5840–5851. *[VII.1]*

B.L. Ehle, *On Padé approximations to the exponential function and A-stable methods for the numerical solution of initial value problems*, Research Report CSRR 2010 (1969), Dept. AACS, Univ. of Waterloo, Ontario, Canada. *[II.1]*

E. Eich-Soellner & C. Führer, *Numerical Methods in Multibody Dynamics*, B. G. Teubner Stuttgart, 1998. *[IV.4], [VII.1]*

T. Eirola, *Aspects of backward error analysis of numerical ODE's*, J. Comp. Appl. Math. 45 (1993), 65–73. *[IX.1]*

T. Eirola & J.M. Sanz-Serna, *Conservation of integrals and symplectic structure in the integration of differential equations by multistep methods*, Numer. Math. 61 (1992) 281–290. *[XIV.3]*

L.H. Eliasson, *Absolutely convergent series expansions for quasi periodic motions*, Math. Phys. Electron. J. 2, No.4, Paper 4, 33 p. (1996). *[X.2]*

Ch. Engstler & Ch. Lubich, *Multirate extrapolation methods for differential equations with different time scales*, Computing 58 (1997) 173–185. *[VIII.4]*

L. Euler, *Du mouvement de rotation des corps solides autour d'un axe variable*, Hist. de l'Acad. Royale de Berlin, Tom.14, Année MDCCLVIII, 154–193. Opera Omnia Ser.II, Vol.8, 200–235. *[IV.1], [X.1]*

L. Euler, *Problème : un corps étant attiré en raison réciproque carrée des distances vers deux points fixes donnés, trouver les cas où la courbe décrite par ce corps sera algébrique*, Mémoires de l'Académie de Berlin pour 1760, pub. 1767, 228–249. *[X.1]*

L. Euler, *Institutionum Calculi Integralis*, Volumen Primum, *Opera Omnia*, Vol.XI. *[I.1]*

K. Feng, *On difference schemes and symplectic geometry*, Proceedings of the 5-th Intern. Symposium on differential geometry & differential equations, August 1984, Beijing (1985) 42–58. *[VI.3]*

K. Feng, *Difference schemes for Hamiltonian formalism and symplectic geometry*, J. Comp. Math. 4 (1986) 279–289. *[VI.5]*

K. Feng, *Formal power series and numerical algorithms for dynamical systems*. In Proceedings of international conference on scientific computation, Hangzhou, China, Eds. Tony Chan & Zhong-Ci Shi, Series on Appl. Math. 1 (1991) 28–35. *[IX.1]*

K. Feng, *Collected Works (II)*, National Defense Industry Press, Beijing, 1995. *[XIV.2]*

K. Feng & Z. Shang, *Volume-preserving algorithms for source-free dynamical systems*, Numer. Math. 71 (1995) 451–463. *[IV.3], [VII.3]*

K. Feng, H.M. Wu, M.-Z. Qin & D.L. Wang, *Construction of canonical difference schemes for Hamiltonian formalism via generating functions*, J. Comp. Math. 7 (1989) 71–96. *[VI.5]*

E. Fermi, J. Pasta & S. Ulam, Los Alamos Report No. LA-1940 (1955), later published in E. Fermi: Collected Papers (Chicago 1965), and Lect. Appl. Math. 15, 143 (1974). *[I.4]*

B. Fiedler & J. Scheurle, *Discretization of homoclinic orbits, rapid forcing and "invisible" chaos*, Mem. Amer. Math. Soc. 119, no. 570, 1996. *[IX.1]*

H. Flaschka, *The Toda lattice. II. Existence of integrals*, Phys. Rev. B 9 (1974) 1924–1925. *[IV.3], [X.1]*

J. Ford, *The Fermi-Pasta-Ulam problem: paradox turns discovery*, Physics Reports 213 (1992) 271–310. *[I.4]*

E. Forest, *Canonical integrators as tracking codes*, AIP Conference Proceedings 184 (1989) 1106–1136. *[II.4]*

E. Forest, *Sixth-order Lie group integrators*, J. Comput. Physics 99 (1992) 209–213. *[V.3]*

E. Forest & R.D. Ruth, *Fourth-order symplectic integration*, Phys. D 43 (1990) 105–117. *[II.5]*

L. Galgani, A. Giorgilli, A. Martinoli & S. Vanzini, *On the problem of energy equipartition for large systems of the Fermi-Pasta-Ulam type: analytical and numerical estimates*, Physica D 59 (1992), 334–348. *[I.4]*

M.J. Gander, *A non spiraling integrator for the Lotka Volterra equation*, Il Volterriano 4 (1994) 21–28. *[VII.4]*

B. García-Archilla, J.M. Sanz-Serna & R.D. Skeel, *Long-time-step methods for oscillatory differential equations*, SIAM J. Sci. Comput. 20 (1999) 930–963. *[VIII.4], [XIII.1], [XIII.2], [XIII.4]*

W. Gautschi, *Numerical integration of ordinary differential equations based on trigonometric polynomials*, Numer. Math. 3 (1961) 381–397. *[XIII.1]*

Z. Ge & J.E. Marsden, *Lie-Poisson Hamilton-Jacobi theory and Lie-Poisson integrators*, Phys. Lett. A 133 (1988) 134–139. *[VII.2], [IX.11]*

C.W. Gear & D.R. Wells, *Multirate linear multistep methods*, BIT 24 (1984) 484–502. *[VIII.4]*

W. Gentzsch & A. Schlüter, *Über ein Einschrittverfahren mit zyklischer Schrittweitenänderung zur Lösung parabolischer Differentialgleichungen*, ZAMM 58 (1978), T415–T416. *[II.4]*

S. Gill, *A process for the step-by-step integration of differential equations in an automatic digital computing machine*, Proc. Cambridge Philos. Soc. 47 (1951) 95–108. *[III.1], [VIII.5]*

A. Giorgilli & U. Locatelli, *Kolmogorov theorem and classical perturbation theory*, Z. Angew. Math. Phys. 48 (1997) 220–261. *[X.2]*

B. Gladman, M. Duncan & J. Candy, *Symplectic integrators for long-term integrations in celestial mechanics*, Celestial Mechanics and Dynamical Astronomy 52 (1991) 221–240. *[VIII.1], [IX.1]*

D. Goldman & T.J. Kaper, *Nth-order operator splitting schemes and nonreversible systems*, SIAM J. Numer. Anal. 33 (1996) 349–367. *[II.6]*

G.H. Golub & C.F. Van Loan, *Matrix Computations, 2nd edition*, John Hopkins Univ. Press, Baltimore and London, 1989. *[IV.4]*

O. Gonzalez, *Time integration and discrete Hamiltonian systems*, J. Nonlinear Sci. 6 (1996) 449–467. *[V.5]*

O. Gonzalez, D.J. Higham & A.M. Stuart, *Qualitative properties of modified equations*. IMA J. Numer. Anal. 19 (1999) 169–190. *[IX.5]*

O. Gonzalez & J.C. Simo, *On the stability of symplectic and energy-momentum algorithms for nonlinear Hamiltonian systems with symmetry*, Comput. Methods Appl. Mech. Eng. 134 (1996) 197–222. *[V.5]*

D.N. Goryachev, *On the motion of a heavy rigid body with an immobile point of support in the case $A = B = 4C$* (Russian), Moscov Math. Collect. 21 (1899) 431–438. *[X.1]*

W.B. Gragg, *Repeated extrapolation to the limit in the numerical solution of ordinary differential equations*, Thesis, Univ. of California; see also SIAM J. Numer. Anal. 2 (1965) 384–403. *[V.1]*

D.F. Griffiths & J.M. Sanz-Serna, *On the scope of the method of modified equations*, SIAM J. Sci. Stat. Comput. 7 (1986) 994–1008. *[IX.1]*

W. Gröbner, *Die Liereihen und ihre Anwendungen*, VEB Deutscher Verlag der Wiss., Berlin 1960, 2nd ed. 1967. *[III.5]*

H. Grubmüller, H. Heller, A. Windemuth & P. Tavan, *Generalized Verlet algorithm for efficient molecular dynamics simulations with long-range interactions*, Mol. Sim. 6 (1991) 121–142. *[VIII.4], [XIII.1]*

A. Guillou & J.L. Soulé, *La résolution numérique des problèmes différentiels aux conditions initiales par des méthodes de collocation*. Rev. Française Informat. Recherche Opfationnelle 3 (1969) Ser. R-3, 17–44. *[II.1]*

M. Günther & P. Rentrop, *Multirate ROW methods and latency of electric circuits*. Appl. Numer. Math. 13 (1993) 83–102. *[VIII.4]*

J. Hadamard, *Sur l'itération et les solutions asymptotiques des équations différentielles*, Bull. Soc. Math. France 29 (1901) 224–228. *[XII.3]*

E. Hairer, *Backward analysis of numerical integrators and symplectic methods*, Annals of Numerical Mathematics 1 (1994) 107–132. *[VI.7], [IX.9], [IX.10]*

E. Hairer, *Variable time step integration with symplectic methods*, Appl. Numer. Math. 25 (1997) 219–227. *[VIII.3]*

E. Hairer, *Backward error analysis for multistep methods*, Numer. Math. 84 (1999) 199–232. *[IX.9], [XIV.2]*

E. Hairer, *Symmetric projection methods for differential equations on manifolds*, BIT 40 (2000) 726–734. *[V.4]*

E. Hairer, *Geometric integration of ordinary differential equations on manifolds*, BIT 41 (2001) 996–1007. *[V.4]*

E. Hairer & P. Leone, *Order barriers for symplectic multi-value methods*. In: Numerical analysis 1997, Proc. of the 17th Dundee Biennial Conference, June 24-27, 1997, D.F. Griffiths, D.J. Higham & G.A. Watson (eds.), Pitman Research Notes in Mathematics Series 380 (1998), 133–149. *[XIV.3]*

E. Hairer & P. Leone, *Some properties of symplectic Runge-Kutta methods*, New Zealand J. of Math. 29 (2000) 169–175. *[IV.2]*

E. Hairer & Ch. Lubich, *The life-span of backward error analysis for numerical integrators*, Numer. Math. 76 (1997), pp. 441–462. Erratum: http://www.unige.ch/math/folks/hairer/ *[IX.7], [X.5]*

E. Hairer & Ch. Lubich, *Invariant tori of dissipatively perturbed Hamiltonian systems under symplectic discretization*, Appl. Numer. Math. 29 (1999) 57–71. *[XII.1], [XII.5]*

E. Hairer & Ch. Lubich, *Asymptotic expansions and backward analysis for numerical integrators*, Dynamics of Algorithms (Minneapolis, MN, 1997), IMA Vol. Math. Appl. 118, Springer, New York (2000) 91–106. *[IX.1]*

E. Hairer & Ch. Lubich, *Long-time energy conservation of numerical methods for oscillatory differential equations*, SIAM J. Numer. Anal. 38 (2000a) 414-441. *[XIII.1], [XIII.2], [XIII.5], [XIII.7]*

E. Hairer & Ch. Lubich, *Energy conservation by Störmer-type numerical integrators*, in: G.F. Griffiths, G.A. Watson (eds.), Numerical Analysis 1999, CRC Press LLC (2000b) 169–190. *[XIII.8]*

E. Hairer, Ch. Lubich & M. Roche, *The numerical solution of differential-algebraic systems by Runge-Kutta methods*, Lecture Notes in Math. 1409, Springer-Verlag, 1989. *[VII.1]*

E. Hairer, S.P. Nørsett & G. Wanner, *Solving Ordinary Differential Equations I. Nonstiff Problems, 2nd edition*, Springer Series in Computational Mathematics **8**, Springer Berlin, 1993. *[II.1], [III.0], [III.1], [III.2], [III.3], [III.6], [VII.1], [VIII.1], [VIII.7], [IX.8], [XIV.1], [XIV.2], [XIV.4], [XIV.5]*

E. Hairer & D. Stoffer, *Reversible long-term integration with variable stepsizes*, SIAM J. Sci. Comput. 18 (1997) 257–269. *[VIII.2], [IX.6]*

E. Hairer & G. Wanner, *On the Butcher group and general multi-value methods*, Computing 13 (1974) 1–15. *[III.1]*

E. Hairer & G. Wanner, *Solving Ordinary Differential Equations II. Stiff and Differential-Algebraic Problems, 2nd edition*, Springer Series in Computational Mathematics **14**, Springer-Verlag Berlin, 1996. *[II.1], [III.0], [IV.2], [VI.4], [VI.8], [VII.1], [VIII.6], [IX.5], [XIII.2], [XIV.5]*

E. Hairer & G. Wanner, *Analysis by Its History, 2nd printing*, Undergraduate Texts in Mathematics, Springer-Verlag New York, 1997. *[IX.7], [XIII.5]*

M. Hall, jr., *A basis for free Lie rings and higher commutators in free groups*, Proc. Amer. Math. Soc. 1 (1950) 575–581. *[III.3]*

Sir W.R. Hamilton, *On a general method in dynamics; by which the study of the motions of all free systems of attracting or repelling points is reduced to the search and differentiation of one central relation, or characteristic function*, Phil. Trans. Roy. Soc. Part II for 1834, 247–308; Math. Papers, Vol. II, 103–161. *[VI.1], [VI.5]*

P.C. Hammer & J.W. Hollingsworth, *Trapezoidal methods of approximating solutions of differential equations*, MTAC 9 (1955) 92–96. *[II.1], [V.1]*

F. Hausdorff, *Die symbolische Exponentialformel in der Gruppentheorie*, Berichte der Sächsischen Akad. der Wissensch. 58 (1906) 19–48. *[III.4]*

A. Hayli, *Le problème des $N$ corps dans un champ extérieur application a l'évolution dynamique des amas ouverts - I*, Bulletin Astronomique 2 (1967) 67–89. *[VIII.4]*

P. Henrici, *Discrete Variable Methods in Ordinary Differential Equations*, John Wiley & Sons, Inc., New York 1962. *[VIII.5]*

J. Hersch, *Contribution à la méthode aux différences*, Z. angew. Math. Phys. 9a (1958) 129–180. *[XIII.1]*

K. Heun, *Neue Methode zur approximativen Integration der Differentialgleichungen einer unabhängigen Veränderlichen*, Zeitschr. für Math. u. Phys. 45 (1900) 23–38. *[II.1]*

D.J. Higham, *Time-stepping and preserving orthogonality*, BIT 37 (1997) 24–36. *[IV.4]*

N.J. Higham, *The accuracy of floating point summation*, SIAM J. Sci. Comput. 14 (1993) 783–799. *[VIII.5]*

M. Hochbruck & Ch. Lubich, *On Krylov subspace approximations to the matrix exponential operator*, SIAM J. Numer. Anal. 34 (1997) 1911–1925. *[XIII.1]*

M. Hochbruck & Ch. Lubich, *A Gautschi-type method for oscillatory second-order differential equations*, Numer. Math. 83 (1999a) 403–426. *[VIII.4], [XIII.1], [XIII.2], [XIII.4]*

M. Hochbruck & Ch. Lubich, *Exponential integrators for quantum-classical molecular dynamics*, BIT 39 (1999b) 620–645. *[VIII.4]*

T. Holder, B. Leimkuhler & S. Reich, *Explicit variable step-size and time-reversible integration*, Appl. Numer. Math. 39 (2001) 367–377. *[VIII.3]*

H. Hopf, *Über die Topologie der Gruppen-Mannigfaltigkeiten und ihre Verallgemeinerungen*, Ann. of Math. 42 (1941) 22–52. *[III.1]*

W. Huang & B. Leimkuhler, *The adaptive Verlet method*, SIAM J. Sci. Comput. 18 (1997) 239–256. *[VIII.3]*

P. Hut, J. Makino & S. McMillan, *Building a better leapfrog*, Astrophys. J. 443 (1995) L93–L96. *[VIII.2]*

K.J. In't Hout, *A new interpolation procedure for adapting Runge-Kutta methods to delay differential equations*, BIT 32 (1992) 634–649. *[VIII.6]*

A. Iserles, *Solving linear ordinary differential equations by exponentials of iterated commutators*, Numer. Math. 45 (1984) 183–199. *[II.4]*

A. Iserles, H.Z. Munthe-Kaas, S.P. Nørsett & A. Zanna, *Lie-group methods*, Acta Numerica (2000) 215–365. *[IV.8]*

A. Iserles & S.P. Nørsett, *On the solution of linear differential equations in Lie groups*, R. Soc. Lond. Philos. Trans. Ser. A Math. Phys. Eng. Sci. 357 (1999) 983–1019. *[IV.7], [IV.9]*

T. Itoh & K. Abe, *Hamiltonian-conserving discrete canonical equations based on variational difference quotients*, J. Comput. Phys. 76 (1988) 85–102. *[V.5]*

J.A. Izaguirre, S. Reich & R.D. Skeel, *Longer time steps for molecular dynamics*, J. Chem. Phys. 110 (1999) 9853–9864. *[XIII.1]*

C.G.J. Jacobi, *Über diejenigen Probleme der Mechanik, in welchen eine Kräftefunction existirt, und über die Theorie der Störungen*, manuscript from 1836 or 1837, published posthumely in *Werke*, vol. 5, 217–395. *[VI.2]*

C.G.J. Jacobi, *Über die Reduktion der Integration der partiellen Differentialgleichungen erster Ordnung zwischen irgend einer Zahl Variablen auf die Integration eines einzigen Systemes gewöhnlicher Differentialgleichungen*, Crelle Journal f.d. reine u. angew. Math. 17 (1837) 97–162; K. Weierstrass, ed., C.G.J. Jacobi's Gesammelte Werke, vol. 4, pp. 57–127. *[VI.5]*

C.G.J. Jacobi, *Lettre adressée à M. le Président de l'Académie des Sciences*, Liouville J. math. pures et appl. 5 (1840) 350–355; *Werke*, vol. 5, pp. 3–189. *[IV.1], [VII.2]*

C.G.J. Jacobi, *Vorlesungen über Dynamik* (1842-43), Reimer, Berlin 1884. *[VI.1], [VI.5], [VI.6], [VI.8]*

C.G.J. Jacobi, *Nova methodus, aequationes differentiales partiales primi ordini inter numerum variabilium quemcunque propositas integrandi*, published posthumly in Crelle Journal f.d. reine u. angew. Math. 60 (1861) 1–181; *Werke*, vol. 5, pp. 3–189. *[III.5], [VII.2]*

L. Jay, *Collocation methods for differential-algebraic equations of index 3*, Numer. Math. 65 (1993) 407–421. *[VII.1]*

L. Jay, *Runge-Kutta type methods for index three differential-algebraic equations with applications to Hamiltonian systems*, Thesis No. 2658, 1994, Univ. Genève. *[VII.1]*

L. Jay, *Symplectic partitioned Runge-Kutta methods for constrained Hamiltonian systems*, SIAM J. Numer. Anal. 33 (1996) 368–387. *[II.2], [VII.1]*

R. Jost, *Winkel- und Wirkungsvariable für allgemeine mechanische Systeme*, Helv. Phys. Acta 41 (1968) 965–968. *[X.1]*

W. Kahan, *Further remarks on reducing truncation errors*, Comm. ACM 8 (1965) 40. *[VIII.5]*

W. Kahan & R.-C. Li, *Composition constants for raising the orders of unconventional schemes for ordinary differential equations*, Math. Comput. 66 (1997) 1089–1099. *[V.3], [V.6]*

B. Karasözen, *Poisson integrators*, Preprint, October 2001, METU Ankara, Turkey. *[VII.2]*

J. Kepler, *Astronomia nova αιτιολογητός seu Physica celestis, traditia commentariis de motibus stellae Martis, ex observationibus G. V. Tychonis Brahe*, Prague 1609. *[I.2]*

H. Kinoshita, H. Yoshida & H. Nakai, *Symplectic integrators and their application to dynamical astronomy*, Celest. Mech. & Dynam. Astr. 50 (1991) 59–71. *[V.3]*

U. Kirchgraber, *Multi-step methods are essentially one-step methods*, Numer. Math. 48 (1986) 85–90. *[XIV.2]*

U. Kirchgraber, F. Lasagni, K. Nipp & D. Stoffer, *On the application of invariant manifold theory, in particular to numerical analysis*, Internat. Ser. Numer. Math. 97, Birkhäuser, Basel, 1991, 189–197. *[XII.3]*

U. Kirchgraber & E. Stiefel, *Methoden der analytischen Störungsrechnung und ihre Anwendungen*, Teubner, Stuttgart, 1978. *[XII.4]*

A.N. Kolmogorov, *On conservation of conditionally periodic motions under small perturbations of the Hamiltonian*, Dokl. Akad. Nauk SSSR 98 (1954) 527–530. *[X.2], [X.5]*

A.N. Kolmogorov, *General theory of dynamical systems and classical mechanics*, Proc. Int. Congr. Math. Amsterdam 1954, Vol. 1, 315–333. *[X.2], [X.5]*

P.-V. Koseleff, *Exhaustive search of symplectic integrators using computer algebra*, Integration algorithms and classical mechanics, Fields Inst. Commun. 10 (1996) 103–120. *[V.3]*

S. Kovalevskaya (Kowalevski), *Sur le problème de la rotation d'un corps solide autour d'un point fixe*, Acta Math. 12 (1889) 177–232. *[X.1]*

V.V. Kozlov, *Integrability and non-integrability in Hamiltonian mechanics*, Uspekhi Mat. Nauk 38 (1983) 3–67. *[X.1]*

D. Kreimer, *On the Hopf algebra structure of perturbative quantum field theory*, Adv. Theor. Math. Phys. 2 (1998) 303–334. *[III.1]*

N.M. Krylov & N.N. Bogoliubov, *Application des méthodes de la mécanique non linéaire à la théorie des oscillations stationnaires*, Edition de l'Académie des Sciences de la R.S.S. d'Ukraine, 1934. *[XII.4]*

W. Kutta, *Beitrag zur näherungsweisen Integration totaler Differentialgleichungen*, Zeitschr. für Math. u. Phys. 46 (1901) 435–453. *[II.1]*

R.A. LaBudde & D. Greenspan, *Discrete mechanics – a general treatment*, J. Comput. Phys. 15 (1974) 134–167. *[V.5]*

R.A. LaBudde & D. Greenspan, *Energy and momentum conserving methods of arbitrary order for the numerical integration of equations of motion. Parts I and II*, Numer. Math. 25 (1976) 323–346 and 26 (1976) 1–26. *[V.5]*

M.P. Laburta, *Starting algorithms for IRK methods*, J. Comput. Appl. Math. 83 (1997) 269–288. *[VIII.6]*

M.P. Laburta, *Construction of starting algorithms for the RK-Gauss methods*, J. Comput. Appl. Math. 90 (1998) 239–261. *[VIII.6]*

J.-L. Lagrange, *Applications de la methode exposée dans le mémoire précédent a la solution de différents problèmes de dynamique*, 1760, Oeuvres Vol. 1, 365–468. *[VI.1], [VI.2]*

J.L. Lagrange, *Recherches sur le mouvement d'un corps qui est attiré vers deux centres fixes* (1766), Œuvres, tome II, Gauthier-Villars, Paris 1868, 67–124. *[X.1]*

J.-L. Lagrange, *Méchanique analitique*, Paris 1788. *[VI.1], [X.1]*

J.D. Lambert & I.A. Watson, *Symmetric multistep methods for periodic initial value problems*, J. Inst. Maths. Applics. 18 (1976) 189–202. *[XIV.1], [XIV.6]*

C. Lanczos, *The Variational Principles of Mechanics*, University of Toronto Press, Toronto, 1949. (Fourth edition 1970). *[VI.6]*

P.S. Laplace, *Traité de mécanique céleste II*, 1799, see Œuvres I, p. 183. *[I.5]*

F.M. Lasagni, *Canonical Runge-Kutta methods*, ZAMP 39 (1988) 952–953. *[VI.4], [VI.5], [VI.7]*

P.D. Lax, *Integrals of nonlinear equations of evolution and solitary waves*, Commun. Pure Appl. Math. 21 (1968) 467–490. *[IV.3]*

B. Leimkuhler & S. Reich, *Symplectic integration of constrained Hamiltonian systems*, Math. Comp. 63 (1994) 589–605. *[VII.1]*

B. Leimkuhler & S. Reich, *A reversible averaging integrator for multiple time-scale dynamics*, J. Comput. Phys. 171 (2001) 95–114. *[VIII.4]*

B.J. Leimkuhler & R.D. Skeel, *Symplectic numerical integrators in constrained Hamiltonian systems*, J. Comput. Phys. 112 (1994) 117–125. *[VII.1]*

P. Leone, *Symplecticity and Symmetry of General Integration Methods*, Thèse, Section de Mathématiques, Université de Genève, 2000. *[VI.7], [XIV.3]*

T. Levi-Civita, *Sur la résolution qualitative du problème restreint des trois corps*, Acta Math. 30 (1906) 305–327. *[VIII.3]*

T. Levi-Civita, *Sur la régularisation du problème des trois corps*, Acta Math. 42 (1920) 99–144. *[VIII.3]*

D. Lewis & J.C. Simo, *Conserving algorithms for the dynamics of Hamiltonian systems on Lie groups*, J. Nonlinear Sci. 4 (1994) 253–299. *[IV.8], [V.5]*

D. Lewis & J.C. Simo, *Conserving algorithms for the $N$-dimensional rigid body*, Fields Inst. Com. 10 (1996) 121–139. *[V.5]*

S. Lie, *Zur Theorie der Transformationsgruppen*, Christ. Forh. Aar. 1888, Nr. 13, 6 pages, Christiania 1888; Gesammelte Abh. vol. 5, p. 553–557. *[VII.2]*

J. Liouville, *Note à l'occasion du mémoire précédent (de M. E. Bour)*, J. Math. Pures et Appliquées 20 (1855) 201–202. *[X.1]*

L. Lopez & T. Politi, *Applications of the Cayley approach in the numerical solution of matrix differential systems on quadratic groups*, Appl. Numer. Math. 36 (2001) 35–55. *[IV.8]*

M.A. López-Marcos, J.M. Sanz-Serna & R.D. Skeel, *Cheap enhancement of symplectic integrators*, Numerical analysis 1995 (Dundee), Pitman Res. Notes Math. Ser. 344, Longman, Harlow, 1996, 107–122. *[V.3]*

A.J. Lotka, *The Elements of Physical Biology*, Williams & Wilkins, Baltimore, 1925. Reprinted 1956 under the title *Elements of mathematical biology* by Dover, New York. *[I.1]*

Ch. Lubich, *On dynamics and bifurcations of nonlinear evolution equations under numerical discretization*, in Ergodic Theory, Analysis, and Efficient Simulation of Dynamical Systems (B. Fiedler, ed.), Springer, Berlin, 2001, 469–500. *[XII.3]*

R. MacKay, *Some aspects of the dynamics of Hamiltonian systems*, in: D.S. Broomhead & A. Iserles, eds., *The Dynamics of Numerics and the Numerics of Dynamics*, Clarendon Press, Oxford, 1992, 137–193. *[VI.6]*

S. Maeda, *Canonical structure and symmetries for discrete systems*, Math. Japonica 25 (1980) 405–420. *[VI.6]*

S. Maeda, *Lagrangian formulation of discrete systems and concept of difference space*, Math. Japonica 27 (1982) 345–356. *[VI.6]*

W. Magnus, *On the exponential solution of differential equations for a linear operator*, Comm. Pure Appl. Math. VII (1954) 649–673. *[IV.7]*

G. Marchuk, *Some applications of splitting-up methods to the solution of mathematical physics problems*, Aplikace Matematiky 13 (1968) 103–132. *[II.5]*
J.E. Marsden & T.S. Ratiu, *Introduction to Mechanics and Symmetry*, Springer-Verlag, New York, 1994. *[IV.1], [VII.2]*
J.E. Marsden & M. West, *Discrete mechanics and variational integrators*, Acta Numerica 10 (2001) 1–158. *[VI.6]*
R.I. McLachlan, *Explicit Lie-Poisson integration and the Euler equations*, Phys. Rev. Lett. 71 (1993) 3043–3046. *[VII.2]*
R.I. McLachlan, *On the numerical integration of ordinary differential equations by symmetric composition methods*, SIAM J. Sci. Comput. 16 (1995) 151–168. *[II.4], [II.5], [III.3], [V.3]*
R.I. McLachlan, *Composition methods in the presence of small parameters*, BIT 35 (1995b) 258–268. *[V.3]*
R.I. McLachlan, *More on symplectic integrators*, in *Integration Algorithms and Classical Mechanics* 10, J.E. Marsden, G.W. Patrick & W.F. Shadwick, eds., Amer. Math. Soc., Providence, R.I. (1996) 141–149. *[V.3]*
R.I. McLachlan & P. Atela, *The accuracy of symplectic integrators*, Nonlinearity 5 (1992) 541–562. *[V.3]*
R.I. McLachlan & C. Scovel, *Equivariant constrained symplectic integration*, J. Nonlinear Sci. 5 (1995) 233–256. *[VII.2]*
R.I. McLachlan, G.R.W. Quispel & N. Robidoux, *Geometric integration using discrete gradients*, Philos. Trans. R. Soc. Lond., Ser. A, 357 (1999) 1021–1045. *[V.5]*
R.J.Y. McLeod & J.M. Sanz-Serna, *Geometrically derived difference formulae for the numerical integration of trajectory problems*, IMA J. Numer. Anal. 2 (1982) 357–370. *[VIII.3]*
V.L. Mehrmann, *The Autonomous Linear Quadratic Control Problem. Theory and Numerical Solution*, Lecture Notes in Control and Information Sciences, Springer-Verlag, Berlin, 1991. *[IV.5]*
R.H. Merson, *An operational method for the study of integration processes*, Proc. Symp. Data Processing, Weapons Research Establishment, Salisbury, Australia (1957) 110-1 to 110-25. *[III.1]*
S. Miesbach & H.J. Pesch, *Symplectic phase flow approximation for the numerical integration of canonical systems*, Numer. Math. 61 (1992) 501–521. *[VI.5]*
O. Møller, *Quasi double-precision in floating point addition*, BIT 5 (1965) 37–50 and 251–255. *[VIII.5]*
A. Morbidelli & A. Giorgilli, *Superexponential stability of KAM Tori*, J. Stat. Phys. 78 (1995) 1607–1617. *[X.2]*
J. Moser, *Review* MR 20-4066, Math. Rev., 1959. *[X.5]*
J. Moser, *On invariant curves of area-preserving mappings of an annulus*, Nachr. Akad. Wiss. Göttingen, II. Math.-Phys. Kl. 1962, 1–20. *[X.5]*
J. Moser, *Lectures on Hamiltonian systems*, Mem. Am. Math. Soc. 81 (1968) 1–60. *[IX.3]*
J. Moser, *Stable and Random Motions in Dynamical Systems*, Annals of Mathematics Studies. No. 77. Princeton University Press, 1973. *[XI.2]*
J. Moser, *Finitely many mass points on the line under the influence of an exponential potential — an integrable system*, Dyn. Syst., Theor. Appl., Battelle Seattle 1974 Renc., Lect. Notes Phys. 38 (1975) 467–497. *[X.1]*
J. Moser, *Is the solar system stable?*, Mathematical Intelligencer 1 (1978) 65–71. *[X.0]*
H. Munthe-Kaas, *Lie Butcher theory for Runge-Kutta methods*, BIT 35 (1995) 572–587. *[IV.8]*
H. Munthe-Kaas, *Runge-Kutta methods on Lie groups*, BIT 38 (1998) 92–111. *[IV.8]*
H. Munthe-Kaas, *High order Runge-Kutta methods on manifolds*, J. Appl. Num. Maths. 29 (1999) 115–127. *[IV.8]*
H. Munthe-Kaas & B. Owren, *Computations in a free Lie algebra*, Phil. Trans. Royal Soc. A 357 (1999) 957–981. *[IV.7]*

A. Murua, *Métodos simplécticos desarrollables en P-series*, Doctoral Thesis, Univ. Valladolid, 1994. *[IX.3]*

A. Murua & J.M. Sanz-Serna, *Order conditions for numerical integrators obtained by composing simpler integrators*, Philos. Trans. Royal Soc. London, ser. A 357 (1999) 1079–1100. *[III.1], [III.3], [V.3]*

N.N. Nekhoroshev, *An exponential estimate of the time of stability of nearly-integrable Hamiltonian systems*, Russ. Math. Surveys 32 (1977) 1–65. *[X.2], [X.4]*

N.N. Nekhoroshev, *An exponential estimate of the time of stability of nearly-integrable Hamiltonian systems. II.* (Russian), Tr. Semin. Im. I.G. Petrovskogo 5 (1979) 5–50. *[X.4]*

P. Nettesheim & S. Reich, *Symplectic multiple-time-stepping integrators for quantum-classical molecular dynamics*, in P. Deuflhard et al. (eds.), Computational Molecular Dynamics: Challenges, Methods, Ideas, Springer, Berlin 1999, 412–420. *[VIII.4]*

I. Newton, *Philosophiae Naturalis Principia Mathematica*, Londini anno MDCLXXXVII, 1687. *[I.2], [VI.1], [X.1]*

I. Newton, *Second edition of the Principia*, 1713. *[I.2], [X.1]*

K. Nipp & D. Stoffer, *Attractive invariant manifolds for maps: existence, smoothness and continuous dependence on the map*, Research Report No. 92–11, SAM, ETH Zürich, 1992. *[XII.3]*

K. Nipp & D. Stoffer, *Invariant manifolds and global error estimates of numerical integration schemes applied to stiff systems of singular perturbation type. I: RK-methods*, Numer. Math. 70 (1995) 245–257. *[XII.3]*

K. Nipp & D. Stoffer, *Invariant manifolds and global error estimates of numerical integration schemes applied to stiff systems of singular perturbation type. II: Linear multistep methods*, Numer. Math. 74 (1996) 305–323. *[XII.3]*

E. Noether, *Invariante Variationsprobleme*, Nachr. Akad. Wiss. Göttingen, Math.-Phys. Kl. (1918) 235–257. *[VI.6]*

E.J. Nyström, *Ueber die numerische Integration von Differentialgleichungen*, Acta Soc. Sci. Fenn. 50 (1925) 1–54. *[II.2]*

D. Okunbor & R.D. Skeel, *Explicit canonical methods for Hamiltonian systems*, Math. Comp. 59 (1992) 439–455. *[VI.4]*

D.I. Okunbor & R.D. Skeel, *Canonical Runge-Kutta-Nyström methods of orders five and six*, J. Comp. Appl. Math. 51 (1994) 375–382. *[V.3]*

P.J. Olver, *Applications of Lie Groups to Differential Equations*, Graduate Texts in Mathematics **107**, Springer-Verlag, New York, 1986. *[IV.6]*

B. Owren & A. Marthinsen, *Runge-Kutta methods adapted to manifolds and based on rigid frames*, BIT 39 (1999) 116–142. *[IV.8]*

B. Owren & A. Marthinsen, *Integration methods based on canonical coordinates of the second kind*, Numer. Math. 87 (2001) 763–790. *[IV.8]*

A.M. Perelomov, *Selected topics on classical integrable systems*, Troisième cycle de la physique, expanded version of lectures delivered in May 1995. *[VII.2]*

O. Perron, *Über Stabilität und asymptotisches Verhalten der Lösungen eines Systems endlicher Differenzengleichungen*, J. Reine Angew. Math. 161 (1929) 41–64. *[XII.3]*

A.D. Perry & S. Wiggins, *KAM tori are very sticky: Rigorous lower bounds on the time to move away from an invariant Lagrangian torus with linear flow*, Physica D 71 (1994) 102–121. *[X.2]*

H. Poincaré, *Les Méthodes Nouvelles de la Mécanique Céleste, Tome I*, Gauthier-Villars, Paris, 1892. *[VI.1], [X.1], [X.2]*

H. Poincaré, *Les Méthodes Nouvelles de la Mécanique Céleste, Tome II*, Gauthier-Villars, Paris, 1893. *[VI.1], [X.2]*

H. Poincaré, *Les Méthodes Nouvelles de la Mécanique Céleste. Tome III*, Gauthiers-Villars, Paris, 1899. *[VI.1], [VI.2]*

S.D. Poisson, *Sur la variation des constantes arbitraires dans les questions de mécanique*, J. de l'Ecole Polytechnique vol. 8, 15e cahier (1809) 266–344. *[VII.2]*

B. van der Pol, *Forced oscillations in a system with non-linear resistance*, Phil. Mag. 3, (1927), 65–80; *Papers* vol. I, 361–376. *[XII.4]*

J. Pöschel, *Nekhoroshev estimates for quasi-convex Hamiltonian systems*, Math. Z. 213 (1993) 187–216. *[X.2]*

F.A. Potra & W.C. Rheinboldt, *On the numerical solution of Euler-Lagrange equations*, Mech. Struct. & Mech. 19 (1991) 1–18. *[IV.5]*

M.-Z. Qin & W.-J. Zhu, *Volume-preserving schemes and numerical experiments*, Comput. Math. Appl. 26 (1993) 33–42. *[VII.3]*

G.D. Quinlan, *Resonances and instabilities in symmetric multistep methods*, Report, 1999, available on http://xxx.lanl.gov/abs/astro-ph/9901136 *[XIV.5]*

G.D. Quinlan & S. Tremaine, *Symmetric multistep methods for the numerical integration of planetary orbits*, Astron. J. 100 (1990) 1694–1700. *[XIV.1], [XIV.5]*

G.R.W. Quispel, *Volume-preserving integrators*, Phys. Lett. A 206 (1995) 26–30. *[VII.3]*

S. Reich, *Symplectic integration of constrained Hamiltonian systems by Runge-Kutta methods*, Techn. Report 93-13 (1993), Dept. Comput. Sci., Univ. of British Columbia. *[VII.1]*

S. Reich, *Numerical integration of the generalized Euler equations*, Techn. Report 93-20 (1993), Dept. Comput. Sci., Univ. of British Columbia. *[VII.2]*

S. Reich, *Momentum conserving symplectic integrators*, Phys. D 76 (1994) 375–383. *[VII.2]*

S. Reich, *Symplectic integration of constrained Hamiltonian systems by composition methods*, SIAM J. Numer. Anal. 33 (1996a) 475–491. *[VII.1], [IX.5]*

S. Reich, *Enhancing energy conserving methods*, BIT 36 (1996b) 122–134. *[V.5]*

S. Reich, *Backward error analysis for numerical integrators*, SIAM J. Numer. Anal. 36 (1999) 1549–1570. *[VIII.3], [IX.5], [IX.7]*

J.R. Rice, *Split Runge-Kutta method for simultaneous equations*, J. Res. Nat. Bur. Standards 64B (1960) 151–170. *[VIII.4]*

C. Runge, *Ueber die numerische Auflösung von Differentialgleichungen*, Math. Ann. 46 (1895) 167–178. *[II.1]*

H. Rüssmann, *On optimal estimates for the solutions of linear partial differential equations of first order with constant coefficients on the torus*, Dyn. Syst., Theor. Appl., Battelle Seattle 1974 Renc., Lect. Notes Phys. 38 (1975) 598–624. *[X.4]*

H. Rüssmann, *On optimal estimates for the solutions of linear difference equations on the circle*, Celest. Mech. 14 (1976) 33–37. *[X.4]*

R.D. Ruth, *A canonical integration technique*, IEEE Trans. Nuclear Science NS-30 (1983) 2669–2671. *[II.5], [VI.1], [VI.3], [IX.1]*

J.-P. Ryckaert, G. Ciccotti & H.J.C. Berendsen, *Numerical integration of the cartesian equations of motion of a system with constraints: molecular dynamics of n-alkanes*, J. Comput. Phys. 23 (1977) 327–341. *[VII.1], [XIII.1]*

P. Saha & S. Tremaine, *Symplectic integrators for solar system dynamics*, Astron. J. 104 (1992) 1633–1640. *[V.3]*

S. Saito, H. Sugiura & T. Mitsui, *Butcher's simplifying assumption for symplectic integrators*, BIT 32 (1992) 345–349. *[IV.9]*

J. Sand, *Methods for starting iteration schemes for implicit Runge-Kutta formulae*, Computational ordinary differential equations (London, 1989), Inst. Math. Appl. Conf. Ser. New Ser., 39, Oxford Univ. Press, New York, 1992, 115–126. *[VIII.6]*

J.M. Sanz-Serna, *Studies in numerical nonlinear instability I. Why do leapfrog schemes go unstable*, SIAM J. Sci. Stat. Comput. 6 (1985) 923–938. *[XIV.2]*

J.M. Sanz-Serna, *Runge-Kutta schemes for Hamiltonian systems*, BIT 28 (1988) 877–883. *[VI.4]*

J.M. Sanz-Serna, *Symplectic integrators for Hamiltonian problems: an overview*, Acta Numerica 1 (1992) 243–286. *[IX.1]*

J.M. Sanz-Serna, *An unconventional symplectic integrator of W. Kahan*, Appl. Numer. Math. 16 (1994) 245–250. *[VII.2]*

J.M. Sanz-Serna & L. Abia, *Order conditions for canonical Runge-Kutta schemes*, SIAM J. Numer. Anal. 28 (1991) 1081–1096. *[IV.9], [IX.10]*

J.M. Sanz-Serna & M.P. Calvo, *Numerical Hamiltonian Problems*, Chapman & Hall, London, 1994. *[VIII.6]*

J.M. Sanz-Serna & F. Vadillo, *Studies in numerical nonlinear instability III: augmented Hamiltonian systems*, SIAM J. Appl. Math. 47 (1987) 92–108. *[XIV.3]*

R. Scherer, *A note on Radau and Lobatto formulae for O.D.E:s*, BIT 17 (1977) 235–238. *[II.3]*

T. Schlick, *Some failures and successes of long-timestep approaches to biomolecular simulations*, in Computational Molecular Dynamics: Challenges, Methods, Ideas (P. Deuflhard et al., eds.), Springer, Berlin 1999, 227–262. *[XIII.1]*

C.M. Schober, *Symplectic integrators for the Ablowitz-Ladik discrete nonlinear Schrödinger equation*, Phys. Lett. A 259 (1999) 140–151. Corrigendum: Phys. Lett. A 272 (2000) 421–422. *[VII.2]*

M.B. Sevryuk, *Reversible systems*, Lecture Notes in Mathematics, 1211. Springer-Verlag, 1986. *[XI.0]*

L.F. Shampine, *Conservation laws and the numerical solution of ODEs*, Comp. Maths. Appls. 12B (1986) 1287–1296. *[IV.1]*

Z. Shang, *Generating functions for volume-preserving mappings and Hamilton-Jacobi equations for source-free dynamical systems*, Sci. China Ser. A 37 (1994a) 1172–1188. *[VII.3]*

Z. Shang, *Construction of volume-preserving difference schemes for source-free systems via generating functions*, J. Comput. Math. 12 (1994b) 265–272. *[VII.3]*

Z. Shang, *KAM theorem of symplectic algorithms for Hamiltonian systems*, Numer. Math. 83 (1999) 477–496. *[X.6]*

Z. Shang, *Resonant and Diophantine step sizes in computing invariant tori of Hamiltonian systems*, Nonlinearity 13 (2000) 299–308. *[X.6]*

Q. Sheng, *Solving linear partial differential equations by exponential splitting*, IMA J. Numer. Anal. 9 (1989) 199–212. *[II.6]*

C.L. Siegel & J.K. Moser, *Lectures on Celestial Mechanics*, Grundlehren d. math. Wiss. vol. 187, Springer-Verlag 1971; First German edition: C.L. Siegel, *Vorlesungen über Himmelsmechanik*, Grundlehren vol. 85, Springer-Verlag, 1956. *[VI.1], [VI.5], [VI.6]*

J.C. Simo & N. Tarnow, *The discrete energy-momentum method. Conserving algorithms for nonlinear elastodynamics*, Z. Angew. Math. Phys. 43 (1992) 757–792. *[V.5]*

J.C. Simo, N. Tarnow & K.K. Wong, *Exact energy-momentum conserving algorithms and symplectic schemes for nonlinear dynamics*, Comput. Methods Appl. Mech. Eng. 100 (1992) 63–116. *[V.5]*

R.D. Skeel & C.W. Gear, *Does variable step size ruin a symplectic integrator?*, Physica D60 (1992) 311–313. *[VIII.3]*

S. Sternberg, *Celestial Mechanics*, Benjamin, New York, 1969. *[X.0]*

H.J. Stetter, *Analysis of Discretization Methods for Ordinary Differential Equations*, Springer-Verlag, Berlin, 1973. *[II.3], [II.4], [V.1]*

D. Stoffer, *On reversible and canonical integration methods*, SAM-Report No. 88-05, ETH-Zürich, 1988. *[V.1], [VI.7], [VIII.2], [VIII.3]*

D. Stoffer, *Variable steps for reversible integration methods*, Computing 55 (1995) 1–22. *[VIII.2], [VIII.3]*

D. Stoffer, *General linear methods: connection to one step methods and invariant curves*, Numer. Math. 64 (1993) 395–407. *[XIV.2]*

D. Stoffer, *On the qualitative behaviour of symplectic integrators. III: Perturbed integrable systems*, J. Math. Anal. Appl. 217 (1998) 521–545. *[XII.4]*

C. Störmer, *Sur les trajectoires des corpuscules électrisés*, Arch. sci. phys. nat., Genève, vol. 24 (1907) 5–18, 113–158, 221–247. *[I.3]*

G. Strang, *On the construction and comparison of difference schemes*, SIAM J. Numer. Anal. 5 (1968) 506–517. *[II.5]*

W.B. Streett, D.J. Tildesley & G. Saville, *Multiple time step methods in molecular dynamics*, Mol. Phys. 35 (1978) 639–648. *[VIII.4]*

A.M. Stuart & A.R. Humphries, *Dynamical Systems and Numerical Analysis*, Cambridge University Press, Cambridge, 1996. *[XII.3]*

G. Sun, *Construction of high order symplectic Runge-Kutta Methods*, J. Comput. Math. 11 (1993a) 250–260. *[IV.2]*

G. Sun, *Symplectic partitioned Runge-Kutta methods*, J. Comput. Math. 11 (1993b) 365–372. *[II.2], [IV.2]*

G. Sun, *A simple way constructing symplectic Runge-Kutta methods*, J. Comput. Math. 18 (2000) 61–68. *[VI.8]*

K.F. Sundman, *Mémoire sur le problème des trois corps*, Acta Math. 36 (1912) 105–179. *[VIII.3]*

Y.B. Suris, *On the conservation of the symplectic structure in the numerical solution of Hamiltonian systems* (in Russian), In: Numerical Solution of Ordinary Differential Equations, ed. S.S. Filippov, Keldysh Institute of Applied Mathematics, USSR Academy of Sciences, Moscow, 1988, 148–160. *[VI.4]*

Y.B. Suris, *The canonicity of mappings generated by Runge-Kutta type methods when integrating the systems* $\ddot{x} = -\partial U/\partial x$, Zh. Vychisl. Mat. i Mat. Fiz. 29, 202–211 (in Russian); same as U.S.S.R. Comput. Maths. Phys. 29 (1989) 138–144. *[VI.4]*

Y.B. Suris, *Hamiltonian methods of Runge-Kutta type and their variational interpretation* (in Russian), Math. Model. 2 (1990) 78–87. *[VI.6]*

Y.B. Suris, *Partitioned Runge-Kutta methods as phase volume preserving integrators*, Phys. Lett. A 220 (1996) 63–69. *[VII.3]*

Y.B. Suris, *Integrable discretizations for lattice systems: local equations of motion and their Hamiltonian properties*, Rev. Math. Phys. 11 (1999) 727–822. *[VII.2]*

G.J. Sussman & J. Wisdom, *Chaotic evolution of the solar system*, Science 257 (1992) 56–62. *[I.2]*

M. Suzuki, *Fractal decomposition of exponential operators with applications to many-body theories and Monte Carlo simulations*, Phys. Lett. A 146 (1990) 319–323. *[II.4], [II.5]*

M. Suzuki, *General theory of fractal path integrals with applications to many-body theories and statistical physics*, J. Math. Phys. 32 (1991) 400–407. *[II.6]*

M. Suzuki, *General theory of higher-order decomposition of exponential operators and symplectic integrators*, Phys. Lett. A 165 (1992) 387–395. *[II.5], [V.6]*

M. Suzuki, *Quantum Monte Carlo methods and general decomposition theory of exponential operators and symplectic integrators*, Physica A 205 (1994) 65–79. *[V.3]*

M. Suzuki & K. Umeno, *Higher-order decomposition theory of exponential operators and its applications to QMC and nonlinear dynamics*, In: Computer Simulation Studies in Condensed-Matter Physics VI, Landau, Mon, Schüttler (eds.), Springer Proceedings in Physics 76 (1993) 74–86. *[V.3]*

W.W. Symes, *The $QR$ algorithm and scattering for the finite nonperiodic Toda lattice*, Physica D 4 (1982) 275–280. *[IV.3]*

Y.-F. Tang, *The symplecticity of multi-step methods*, Computers Math. Applic. 25 (1993) 83–90. *[XIV.3]*

Y.-F. Tang, *Formal energy of a symplectic scheme for Hamiltonian systems and its applications (I)*, Computers Math. Applic. 27 (1994) 31–39. *[IX.3]*

Y.-F. Tang, V.M. Pérez-García & L. Vázquez, *Symplectic methods for the Ablowitz-Ladik model*, Appl. Math. Comput. 82 (1997) 17–38. *[VII.2]*

W. Thirring, *Lehrbuch der Mathematischen Physik 1*, Springer-Verlag, 1977. *[X.5]*

M. Toda, *Waves in nonlinear lattice*, Progr. Theor. Phys. Suppl. 45 (1970) 174–200. *[X.1]*

J. Touma & J. Wisdom, *Lie-Poisson integrators for rigid body dynamics in the solar system*, Astron. J. 107 (1994) 1189–1202. *[VII.2]*

H.F. Trotter, *On the product of semi-groups of operators*, Proc. Am. Math. Soc.10 (1959) 545–551. *[II.5]*

M. Tuckerman, B.J. Berne & G.J. Martyna, *Reversible multiple time scale molecular dynamics*, J. Chem. Phys. 97 (1992) 1990–2001. *[VIII.4], [XIII.1]*

V.S. Varadarajan, *Lie Groups, Lie Algebras and Their Representations*, Prentice-Hall, Englewood Cliffs, New Jersey, 1974 *[III.4], [IV.6], [IV.8]*

L. Verlet, *Computer "experiments" on classical fluids. I. Thermodynamical properties of Lennard-Jones molecules*, Physical Review 159 (1967) 98–103. *[I.3], [XIII.1]*

A.P. Veselov, *Integrable systems with discrete time, and difference operators*, Funktsional. Anal. i Prilozhen. 22 (1988) 1–13, 96; transl. in Funct. Anal. Appl. 22 (1988) 83–93. *[VI.6]*

A.P. Veselov, *Integrable maps*, Russ. Math. Surv. 46 (1991) 1–51. *[VI.6]*

R. de Vogelaere, *Methods of integration which preserve the contact transformation property of the Hamiltonian equations*, Report No. 4, Dept. Math., Univ. of Notre Dame, Notre Dame, Ind. (1956) *[VI.3]*

V. Volterra, *Variazioni e fluttuazioni del numero d'individui in specie animali conviventi*, Mem. R. Comitato talassografico italiano, CXXXI, 1927; *Opere* 5, p. 1–111. *[I.1]*

J. Waldvogel & F. Spirig, *Chaotic motion in Hill's lunar problem*, In: A.E. Roy and B.A. Steves, eds., *From Newton to Chaos: Modern Techniques for Understanding and Coping with Chaos in N-Body Dynamical Systems* (NATO Adv. Sci. Inst. Ser. B Phys., 336, Plenum Press, New York, 1995). *[VIII.3]*

G. Wanner, *Runge-Kutta-methods with expansion in even powers of $h$*, Computing 11 (1973) 81–85. *[II.3], [V.2]*

R.A. Wehage & E.J. Haug, *Generalized coordinate partitioning for dimension reduction in analysis of constrained dynamic systems*, J. Mechanical Design 104 (1982) 247–255. *[IV.5]*

J.M. Wendlandt & J.E. Marsden, *Mechanical integrators derived from a discrete variational principle*, Physica D 106 (1997) 223–246. *[VI.6]*

H. Weyl, *The Classical Groups*, Princeton Univ. Press, Princeton, 1939. *[VI.2]*

H. Weyl, *The method of orthogonal projection in potential theory*, Duke Math. J. 7 (1940) 411–444. *[VII.3]*

J.H. Wilkinson, *Error analysis of floating-point computation*, Numer. Math. 2 (1960) 319–340. *[IX.0]*

J. Wisdom & M. Holman, *Symplectic maps for the $N$-body problem*, Astron. J. 102 (1991) 1528–1538. *[V.3]*

J. Wisdom, M. Holman & J. Touma, *Symplectic correctors*, in *Integration Algorithms and Classical Mechanics* 10, J.E. Marsden, G.W. Patrick & W.F. Shadwick, eds., Amer. Math. Soc., Providence, R.I. (1996) 217–244. *[V.3]*

K. Wright, *Some relationships between implicit Runge-Kutta, collocation and Lanczos $\tau$ methods, and their stability properties*, BIT 10 (1970) 217–227. *[II.1]*

H. Yoshida, *Construction of higher order symplectic integrators*, Phys. Lett. A 150 (1990) 262–268. *[II.4], [II.5], [III.4], [III.5], [V.3]*

H. Yoshida, *Recent progress in the theory and application of symplectic integrators*, Celestial Mech. Dynam. Astronom. 56 (1993) 27–43. *[IX.1], [IX.4], [IX.8]*

A. Zanna, *Collocation and relaxed collocation for the Fer and the Magnus expansions*, SIAM J. Numer. Anal. 36 (1999) 1145–1182. *[IV.7], [IV.9]*

A. Zanna, K. Engø & H.Z. Munthe-Kaas, *Adjoint and selfadjoint Lie-group methods*, BIT 41 (2001) 395–421. *[V.4], [V.6]*

K. Zare & V. Szebehely, *Time transformations in the extended phase-space*, Celestial Mechanics 11 (1975) 469–482. *[VIII.3]*

S.L. Ziglin, *The ABC-flow is not integrable for $A = B$*, Funktsional. Anal. i Prilozhen. 30 (1996) 80–81; transl. in Funct. Anal. Appl. 30 (1996) 137–138. *[VII.3]*

# Index

ABC flow, 248
Abel-Liouville-Jacobi-Ostrogradskii identity, 101, 248
Ablowitz-Ladik model, 241
action integral, 191
action-angle variables, 335
adaptive Verlet method, 265
adiabatic invariants, 18
adjoint method, 38, 133, 292, 476
– of collocation method, 134
– of Runge-Kutta method, 135
– quadratic invariants, 164
adjoint operator, 79
angular momentum, 8, 96, 97
area preservation, 6, 171, 172
Arnold-Liouville theorem, 335
attractive invariant manifold, 396, 461
attractive invariant torus, 400
– of numerical integrator, 403
averaged forces, 269
averaging
– basic scheme, 394
– perturbation series, 395
averaging principle, 392

B-series, 47, 52, 53, 205
– composition, 57
– symplectic, 199
backward error analysis, 287, 464
– formal, 287
– rigorous, 305
BCH formula, 78, 80, 298
– symmetric, 82
Bernoulli numbers, 80, 119
bi-coloured trees, 62
$B_\infty$-series, 68
Birkhoff normalization
– Hamiltonian, 350
– reversible, 384
$B(p)$, 28
Butcher group, 60

Butcher product, 71, 199

canonical, 174
– equations of motion, 169
– form, 236
– Poisson structure, 226
– transformation, 174
canonical coordinates of a Lie group
– first kind, 126
– second kind, 126
Casimir functions, 235
Cayley transform, 103, 125
central field, 330, 338, 376
characteristic lines, 231
Choleski decomposition, 142
collocation methods, 26, 119
– discontinuous, 31
– symmetric, 134
collocation polynomial, 26
commutator, 116
– matrix, 78
compensated summation, 272
complete systems, 232
completely integrable, 331
composition
– of B-series, 57
– of Runge-Kutta methods, 55
composition methods, 39, 88, 101, 177
– $\rho$-compatibility, 133
– local error, 139
– of order 2, 138
– of order 4, 140, 143
– of order 6, 141, 144
– of order 8, 145
– of order 10, 146
– order conditions, 67, 71, 76
– symmetric, 137
– symmetric-symmetric, 142
– with symmetric method, 142
conditionally periodic flow, 337
conjugate momenta, 169

conjugate symplecticity, 204
conservation
– of area, 6, 171
– of energy, 94, 160, 312, 420, 448
– of linear invariants, 95
– of mass, 94
– of momentum, 160
– of quadratic invariants, 97, 98
– of volume, 248
conserved quantity, 93
consistent initial values, 210
constant direction of projection, 153
constrained Hamiltonian systems, 211, 229, 239, 243
constrained mechanical systems, 209
continuous output, 275
coordinates
– generalized, 168
$C(q)$, 28
Crouch-Grossman methods, 121
– order conditions, 122

Darboux-Lie theorem, 230, 234, 235, 240
degrees of freedom, 4
diagonally implicit Runge-Kutta methods
– symmetric, 135
differential equations, 2
– Hamilton-Jacobi, 186
– Hamiltonian, 4, 168
– highly oscillatory, 16
– modified, 287
– on Lie groups, 115
– on manifolds, 112
– partial, linear, 231
– reversible, 131
– second order, 13, 37, 281
differential equations on manifolds, 211
– $\rho$-compatibility, 133
differential form, 174
differential-algebraic equations, 129, 209
diophantine frequencies, 344
DIRK methods
– symmetric, 135
discontinuous collocation methods, 31
discrete Euler-Lagrange equations, 192
discrete gradient methods, 159, 162
discrete Lagrangian, 192
discrete momenta, 193
dissipative systems, 391, 481
distinguished functions, 235
divergence-free vector fields, 248

eccentricity, 7

effective order
– of composition methods, 146
elementary differentials, 48, 49, 62
elementary Hamiltonian, 321
elementary weights, 51
energy
– oscillatory, 415, 420, 441, 447
– total, 170, 415, 420, 447
energy – momentum methods, 159
– for $N$-body systems, 161
energy conservation, 312, 447
energy exchange, 419, 431
equistage approximation, 278
error analysis
– backward, 287
error growth
– linear, 351, 352, 384
– of rounding errors, 273
Euler method
– -Lie, 124
– explicit, 3
– implicit, 3
– symplectic, 3, 42, 176, 213, 238, 250
Euler-Lagrange equations, 169, 191, 209
– discrete, 192
explicit symmetric methods, 136
exponential map, 117

Fermi-Pasta-Ulam problem, 17, 414
first integrals, 4, 93, 197
– long-time near-preservation, 351, 384
fixed-point iteration, 280
flow, 2
– discrete, 3
– exact, 2, 44, 186
– isospectral, 103
– numerical, 3, 44
– Poisson, 230, 234
frequencies, 337
– diophantine, 344

G-symplectic, 471
Gauss methods, 30, 97
– symmetric, 135
– symplectic, 179
Gautschi-type methods, 409, 413
general linear methods, 460
– G-symplectic, 471
– strictly stable, 461
– symmetric, 474
– weakly stable, 463, 467
generalized coordinate partitioning, 113
generating functions, 182, 184, 187, 190, 247

– for partitioned RK methods, 186
– for Runge-Kutta methods, 184
GL($n$), general linear group, 116
$\mathfrak{gl}(n)$, Lie algebra of $n \times n$ matrices, 116
gradient, 13
growth parameter, 480

Hall set, 74
Hamilton's principle, 191
Hamilton-Jacobi equation, 186, 329
Hamiltonian, 4, 169, 228
– elementary, 321
– global, 174
– local, 173, 207
– modified, 293, 319
Hamiltonian perturbation theory, 327, 342
– basic scheme, 343
– Birkhoff normalization, 350
– KAM theory, 348, 361
– perturbation series, 344
Hamiltonian systems, 4, 168
– constrained, 211, 229, 239, 243
– integrable, 328
– perturbed integrable, 342

implicit midpoint rule, 3, 30, 177, 179, 204, 239
– symmetry, 133
– symplecticity, 177
impulse method, 266, 411
– mollified, 412
index reduction, 211, 213
integrability lemma, 174
integrable systems
– Hamiltonian, 328
– reversible, 375
invariant manifold
– attractive, 396, 461
invariant torus, 335
– long-time near-preservation, 360, 386
– of numerical integrator, 371, 388, 403
– of reversible map, 387
– of symplectic map, 369
– weakly attractive, 400
invariants, 2, 4, 93
– adiabatic, 18
– linear, 95
– polynomial, 101
– quadratic, 97
– weak, 105
involution
– first integrals in, 329
irreducible

– Runge-Kutta methods, 202
isospectral flow, 103
isospectral methods, 103
iteration
– fixed-point, 280
– Newton-type, 281

Jacobi identity, 116, 227

KAM theory
– Hamiltonian, 348, 361
– reversible, 382
– reversible near-identity map, 387
– symplectic near-identity map, 369
KAM torus
– sticky, 350
Kepler problem, 7, 44, 107, 138, 208
– perturbed, 22, 256
kernel
– of processing methods, 146
kinetic energy, 168, 209
Kolmogorov's iteration, 348
Kolmogorov's theorem, 361

Lagrange equations, 169
Lagrange multipliers, 107, 209
leap-frog method, 13
Legendre transform, 169
– discrete, 193
Leibniz' rule, 227
Lie algebras, 116, 242
Lie bracket, 85, 116, 230
– differential operators, 84
Lie derivative, 83, 298, 307
– of B-series, 315
– of P-series, 318
Lie group methods, 121, 301
– symmetric, 157
Lie groups, 115, 242
– quadratic, 125
Lie midpoint rule, 124
Lie operator, 230
Lie-Euler method, 124
Lie-Poisson systems, 242, 247
Lie-Trotter splitting, 42
Lindstedt-Poincaré series, 344
linear error growth, 351, 352, 384
linear stability, 18, 477
Liouville lemma, 330
Liouville's theorem, 248
Lobatto IIIA - IIIB pair, 36, 98, 179, 197, 218, 301, 324
Lobatto IIIA methods, 30
– symmetric, 135

Lobatto IIIB methods, 33
- symmetric, 135
Lobatto IIIS, 208
Lobatto quadrature, 218
local coordinates, 110
- existence of numerical solution, 155
- symmetric methods, 154
local error, 25
- of composition methods, 139, 164
long-time behaviour
- symmetric integrators, 375, 391
- symplectic integrators, 327, 391
long-time energy conservation, 312
Lorenz problem, 164
Lotka-Volterra problem, 1, 20, 163, 229, 240, 290
Lyapunov exponents, 114

Magnus series, 118
manifolds, 105, 110, 211, 236
- symmetric methods, 149
Marchuk splitting, 42
matrix commutator, 78
matrix exponential, 117
matrix Lie groups, 115
mechanical systems
- constrained, 209
merging product, 71
methods based on local coordinates, 154
methods on manifolds, 93, 300
- symmetric, 149
midpoint rule, 120
- explicit, 457, 469
- implicit, 3, 30, 177, 179, 204, 239
- Lie, 124
- modified, 159
modified differential equation, 287
- B-series, 315
- constrained Hamiltonian system, 301
- first integrals, 300
- Lie group methods, 301
- methods on manifolds, 300
- perturbed differential equation, 402
- Poisson integrators, 297
- reversible methods, 293
- splitting methods, 298
- symmetric methods, 292
- symplectic methods, 293
- trees, 314
- variable steps, 303
modified Hamiltonian, 293, 319
modified midpoint rule, 159
modulated Fourier expansion, 422, 432

molecular dynamics, 12
mollified impulse method, 412
momenta, 169
- conjugate, 169
- discrete, 193
moments of inertia, 96
momentum
- angular, 8, 94, 96, 97, 161
- linear, 94, 161
multi-force methods, 414
multi-value methods, 460
- G-symplectic, 471
- symmetric, 474
multiple time scales, 408, 415
multiple time stepping, 266, 411
multirate methods, 266
multistep methods, 455
- backward error analysis, 464
- G-symplectic, 471
- partitioned, 459
- strictly stable, 456
- symmetric, 456, 458
- variable step sizes, 486
Munthe-Kaas methods, 123

$N$-body system, 11, 94
- energy – momentum methods, 161
Newton-type iteration, 281
Noether's theorem, 197
non-resonant frequencies, 344
non-resonant step size, 371, 435, 447
Nyström methods, 37, 65, 91, 100
- symplectic, 181

$O(n)$, orthogonal group, 116
one-leg methods, 471
one-step method, 14, 25, 175
- underlying, 461
order, 25
- of a tree, 49, 63
- of symmetric local coordinates, 155
- of symmetric projection, 150
order conditions
- composition methods, 67, 71, 76, 89, 90
- Crouch-Grossman methods, 122
- Nyström methods, 65
- partitioned RK methods, 35, 65
- processing methods, 147
- RK methods, 25, 47, 52, 54
- splitting methods, 76, 87
- symmetric composition, 143
- symmetrized, 165
ordered subtrees, 56

ordered trees, 56
oriented area, 171
orthogonal matrices, 109, 115
oscillatory differential equations, 16, 407
oscillatory energy, 18, 415, 420, 441, 447
outer solar system, 7, 10, 108

P-series, 64
– symplectic, 199, 324
parametrization
– tangent space, 113
partial differential equations
– linear, 231
partitioned Runge-Kutta methods, 34, 98, 136
– diagonally implicit, 137
– symmetric, 136
– symplectic, 180, 195, 251
partitioned systems, 3, 62
pendulum, 5, 106, 169, 173, 175, 312, 334
– double, 207
– spherical, 210, 225
perturbation series
– averaging, 395
– Hamiltonian, 344
– reversible, 380
perturbation theory
– dissipative, 391
– Hamiltonian, 327, 342
– reversible, 375
phase space, 2
Poisson
– bracket, 226, 228
– flow, 230, 234
– integrators, 237, 238, 240
– maps, 237
– systems, 226, 228
Poisson structures, 234
– canonical, 226
– general, 228
polar decomposition, 109
polynomial invariants, 101
potential energy, 169, 209
precession, 9, 22
processing
– of composition methods, 146
– order conditions, 147
projection methods, 105, 301
– standard, 106
– symmetric, 149
– symmetric non-reversible, 154
pseudo-inverse of a matrix, 112
pseudo-symplectic methods, 374

QR algorithm, 104
QR decomposition, 110
quadratic invariants, 97
quadratic Lie groups, 125
quasi-periodic flow, 337

r-RESPA method, 267, 411
Radau methods, 30
RATTLE, 216, 246, 301
resonance
– numerical, 418, 421, 483
reversibility, 211, 258
– of symmetric local coordinates, 156
– of symmetric projection, 151
reversible maps, 131, 132
reversible methods, 293
reversible perturbation theory, 375
– basic scheme, 379
– Birkhoff normalization, 384
– KAM theory, 382
– perturbation series, 380
reversible systems, 131
– integrable, 375
– perturbed integrable, 379
reversible vector fields, 132
$\rho$-compatibility condition, 133
$\rho$-reversible, 131
– maps, 132
– vector field, 131
Riccati equation, 114
right-invariant, 244
rigid body, 95, 151, 245, 247, 378, 385
Rodrigues formula, 130
rooted trees, 49
rounding error, 272
Runge-Kutta methods, 24, 25, 97, 259, 275
– $\rho$-compatibility, 133
– additive, 46
– adjoint method, 135
– irreducible, 202
– partitioned, 34, 136
– symmetric, 134
– symplectic, 178, 251
Runge-Lenz-Pauli vector, 21

Schrödinger equation
– nonlinear, 241
separable partitioned systems, 251
SHAKE, 216
simplifying assumptions, 91
sinc function, 409
singular value decomposition, 109
SL($n$), special linear group, 116, 127

$\mathfrak{sl}(n)$, special linear Lie algebra, 116
small denominators, 344
SO($n$), special orthogonal group, 116
$\mathfrak{so}(n)$, skew-symmetric matrices, 116
spherical pendulum, 210, 225
splitting
– Lie-Trotter, 42
– Marchuk, 42
– Strang, 42, 250
splitting methods, 41, 43, 86, 180, 224, 247
– $\rho$-compatibility, 133
– order conditions, 76
Sp($n$), symplectic group, 116
$\mathfrak{sp}(n)$, symplectic Lie algebra, 116
Störmer/Verlet scheme, 13, 35, 42, 177, 198, 239, 267, 299, 408
– as composition method, 136
– as Nyström method, 37
– as processing method, 147
– as splitting method, 42
– as variational integrator, 194
– energy conservation, 313, 449
– first integrals, 198, 341
– linear error growth, 352
– symmetry, 38, 133
– symplecticity, 43, 177
– variable step size, 260, 263, 265
stability
– linear, 18, 477
stability function, 181
starting approximations, 275
– order, 277
step size function, 258
step size selection
– $\rho$-reversible, 258
– standard, 255
– symmetric, 258
Stiefel manifold, 109, 114
Strang splitting, 42, 250, 298
structure constants, 242
submanifold, 105
subtrees
– ordered, 56
summation
– compensated, 272
superconvergence, 28, 33, 221
Suzuki's methods, 141
switching lemma, 72
symmetric collocation methods, 134, 164
symmetric composition, 89
– of first order methods, 138
– of symmetric methods, 138, 142
symmetric composition methods, 137

– of order 6, 144
– of order 8, 145
– of order 10, 146
symmetric Lie group methods, 157
symmetric methods, 3, 38, 131, 132, 292
– explicit, 136
– symmetric composition, 142
symmetric methods on manifolds, 149
symmetric projection, 149
– existence of numerical solution, 150
– non-reversible, 154
symmetric Runge-Kutta methods, 134, 164
symmetric splitting method, 165
symmetrized order conditions, 165
symmetry, 258
– of Gauss methods, 135
– of Lobatto, 135
– of symmetric local coordinates, 156
symmetry coefficient, 53, 63, 68
symplectic, 171, 182, 212
– B-series, 199
– P-series, 199
symplectic Euler method, 3, 42, 176, 180, 213, 238, 250, 290, 296, 299, 319
– as splitting method, 42
– energy conservation, 313
– variable step size, 262
symplectic methods, 175
– as variational integrators, 194
– based on generating functions, 189
– irreducible, 203
– Nyström methods, 181
– partitioned Runge-Kutta methods, 180, 195
– Runge-Kutta methods, 178
– variable step size, 261

tangent space, 111, 117
– parametrization, 113
$\theta$-method, 135
– adjoint, 136
three-body problem, 271, 328
time transformation, 261
time-reversible methods, 132
Toda flow, 105
Toda lattice, 340, 352, 377, 385
total differential, 174, 182
total energy, 8, 12, 17, 94, 415, 420, 447
transformations
– averaging, 394
– canonical, 174
– reversibility preserving, 376
– symplectic, 170, 171, 183, 212

trapezoidal rule, 24, 181, 204, 260
trees, 200, 314
– bi-coloured, 62
– ordered, 56
– rooted, 49
∞-trees, 68
trigonometric methods, 409
two-body problem, 7, 20
two-force methods, 414

underlying one-step method, 461

Van der Pol's equation, 391, 481
variational integrators, 191
variational problem, 191, 209
vector fields, 2

– divergence-free, 248
– reversible, 131, 132
Verlet method, 13, 35, 42, 177, 239, 267, 408, 449
– adaptive, 265
Verlet-I method, 267, 411
volume preservation, 101, 109, 206, 248, 251
volume-preserving integrators, 249

weak invariants, 105
work-precision diagrams, 138, 140, 145, 284, 285, 418, 485, 489
$W$-transformation, 208

Yoshida's methods, 141

Druck: Strauss Offsetdruck, Mörlenbach
Verarbeitung: Schäffer, Grünstadt